MW01483329

Produce Degradation

*Pathways
and Prevention*

Produce Degradation

Pathways and Prevention

Olusola Lamikanra

Syed Imam

Dike Ukuku

Taylor & Francis
Taylor & Francis Group

Boca Raton   London   New York   Singapore

A CRC title, part of the Taylor & Francis imprint, a member of the
Taylor & Francis Group, the academic division of T&F Informa plc.

Published in 2005 by
CRC Press
Taylor & Francis Group
6000 Broken Sound Parkway NW, Suite 300
Boca Raton, FL 33487-2742

International Standard Book Number-10: 0-8493-1902-1 (Hardcover)
International Standard Book Number-13: 978-0-8493-1902-0 (Hardcover)
Library of Congress Card Number 2004062867

Library of Congress Cataloging-in-Publication Data

Produce degradation : pathways and prevention / edited by Olusola
    Lamikanra, Syed Imam, and Dike Ukuku.
       p. cm.
    ISBN 0-8493-1902-1 (alk. paper)
    1. Fruit--Postharvest losses. 2. Vegetables--Postharvest losses. 3. Farm
produce--Postharvest losses. 4. Fruit--Postharvest losses--Prevention. 5.
Vegetables--Postharvest losses--Prevention. 6. Farm produce--Postharvest
losses--Prevention. I. Lamikanra, Olusola. II. Imam, Syed H. III. Ukuku, Dike.

SB360.P76 2005
630.5'6--dc22   2004062867

Taylor & Francis Group
is the Academic Division of T&F Informa plc.

Visit the Taylor & Francis Web site at
http://www.taylorandfrancis.com

and the CRC Press Web site at
http://www.crcpress.com

Preface

The produce industry is a complex and integrated network of agricultural enterprises associated with the production, transportation, processing, shipment, and storage of fruit and vegetable products. The challenges in bringing produce to supermarkets and restaurants are invisible to consumers, who expect convenience, consistency, and freshness, along with year-round abundance. This challenge is not new. Produce inherently breaks down at maturity, often before harvest. Product deterioration continues after purchase by consumers, until it is consumed. The fresh produce market remains one of the major global markets. It includes fresh fruit and vegetables; prepared salads, vegetables and dips; and fresh-cut fruits and vegetables. Annual global production of fruits and vegetables exceeds 1 trillion kg, and the economic impact is considerable.

Most books on fruits and vegetables currently on the market generally address pre- and postharvest management and handling as they relate to quality. Coverage of processes that result in the deterioration of produce quality is scattered. This book comprehensively focuses on the mechanism of produce degradation and addresses mechanisms of reactions that affect produce quality from the farm to the table. Degradative changes and conditions that favor these undesirable processes and their prevention are exhaustively reviewed from the standpoints of biochemistry, microbiology, physiology, polymer and cellular science, and genetics. The book is intended as a reference source for senior undergraduate and graduate students, and scientists, engineers, and economists in academia, government, and industry. Food scientists, plant physiologists, microbiologists, chemists, biochemists, chemical engineers, industrial engineers, nutritionists, and molecular biologists will rely on the book as an interdisciplinary resource publication.

Olusola Lamikanra
Syed H. Imam
Dike O. Ukuku

Contents

Contributors

M. L. Bari
National Food Research Institute
Food Hygiene Research Team
Kannondai, Tsukuba, Japan

Thomas H. J. Beveridge
Agriculture and Agri-Food Canada
Pacific Agri-Food Research Centre
Summerland, British Columbia, Canada

Pamela L. Brady
Institute of Food Science and
  Engineering
University of Arkansas
Fayetteville, Arkansas

Keith R. Cadwallader
Department of Food Science and Human
  Nutrition
University of Illinois
Urbana, Illinois

Jianchi Chen
USDA, ARS, PWA
San Joaquin Valley Agricultural Sciences
  Center
Exotic and Invasive Diseases and Pests
  Research
Parlier, California

Ben-Sen Chiou
USDA-ARS-WRRC
Bioproduct Chemistry and
  Engineering Research
Albany, California

Henryk Daun
Food Science Department
Cook College, Rutgers University
New Brunswick, New Jersey

Paul Everly
USDA-ARS-WRRC
Bioproduct Chemistry and
  Engineering Research
Albany, California

Shirley Gembeh
USDA-NAA-ARS-ERRC
Food Safety Intervention Technology
Wyndmoor, Pennsylvania

Gregory M. Glenn
USDA-ARS, WRRC
Bioproduct Chemistry and
  Engineering Research
Albany, California

Syed Imam
USDA-ARS, WRRC
Bioproduct Chemistry and
  Engineering Research
Albany, California

Maria Inglesby
USDA-ARS, WRRC
Bioproduct Chemistry and
  Engineering Research
Albany, California

K. Isshiki
National Food Research Institute
Food Hygiene Research Team
Kannondai, Tsukuba, Japan

S. Kawamoto
National Food Research Institute
Food Hygiene Research Team
Kannondai, Tsukuba, Japan

Arthur Klamczynski
USDA-ARS-WRRC
Bioproduct Chemistry and
  Engineering Research
Albany, California

Olusola Lamikanra
USDA-ARS-SRRC
Food Processing and
  Sensory Quality Research
New Orleans, Louisiana

Ching-Hsing Liao
USDA-NAA-ARS-ERRC
Food Safety Intervention
  Technology
Wyndmoor, Pennsylvania

Charles Ludvic
USDA-ARS-WRRC
Bioproduct Chemistry and
  Engineering Research
Albany, California

Kateřina Maštovská
MBRCCT Unit
USDA-ARS-ERRC
Wyndmoor, Pennsylvania

Aubrey Mendonca
Department of Food Science and
  Human Nutrition
Iowa State University
Ames, Iowa

Justin R. Morris
Institute of Food Science and
  Engineering
University of Arkansas
Fayetteville, Arkansas

Theobald Mosha
Department of Food Science and
  Human Nutrition
Michigan State University
East Lansing, Michigan

Brendan Niemira
USDA-NAA-ARS-ERRC
Food Safety Intervention Technology
Wyndmoor, Pennsylvania

Maria Nnyepi
Department of Food Science and
  Human Nutrition
Michigan State University
East Lansing, Michigan

William Orts
USDA-ARS, WRRC
Bioproduct Chemistry and
  Engineering Research
Albany, California

Ibrahim Sani Ozdemir
METU, Ankara
Turkey and Avignon University
Avignon, France

Glen P. Sabellano
USDA-ARS-WRRC
Bioproduct Chemistry and
  Engineering Research
Albany, California

Y. Sabina
National Food Research Institute
Food Hygiene Research Team
Kannondai, Tsukuba, Japan

Marwele Shamaila
Nestlé R&D Center, Inc.
Solon, Ohio

Justin Shey
USDA-ARS, WRRC
Bioproduct Chemistry and
  Engineering Research
Albany, California

Christopher H. Sommers
USDA-NAA-ARS-ERRC
Food Safety Intervention Technology
Wyndmoor, Pennsylvania

Peter M.A. Toivonen
Agriculture and Agri-Food Canada
Pacific Agri-Food Research Centre
Summerland, British Columbia, Canada

Dike O. Ukuku
USDA-NAA-ARS-ERRC
Food Safety Intervention Technology
Wyndmoor, Pennsylvania

Patrick Varoquaux
Ministere de L'Agriculture
INRA
Station de Technologie des Produits
  Vegetaux
Domaine St Paul
Site Agroparc
Avignon, France

Michal Voldrich
Department of Food Preservation and
  Meat Technology
FFBT, ICT
Prague, Czech Republic

Keith Warriner
Department of Food Science
University of Guelph
Guelph, Ontario, Canada

Lorraine Weatherspoon
Department of Food Science and
  Human Nutrition
Michigan State University
East Lansing, Michigan

Delilah F. Wood
USDA-ARS-WRRC
Bioproduct Chemistry and
  Engineering Research
Albany, California

Svetlana Zivanovic
Department of Food Science and
  Technology
The University of Tennessee
Knoxville, Tennesee

Editors

Olusola Lamikanra, Ph.D., is a lead scientist/research chemist at the United States Department of Agriculture-Agricultural Research Service (USDA-ARS), Southern Regional Research Center, in New Orleans, Louisiana. He is internationally recognized for his work on plant and food biochemistry and postharvest technology. His current research deals with establishing biochemical effects of fresh-cut fruit processing and storage for use in developing biochemical methods needed to assess fresh-cut fruit quality and novel technologies to improve the sensory quality and shelf life of fresh-cut fruits. He received his Ph.D. degree in food science from The University of Leeds, England. He was professor in the Division of Agricultural Sciences and director of the Center for Viticultural Science and Small Farm Development at Florida A&M University, Tallahassee (1988–1999). During this time he also served as coordinator of the annual viticultural science symposium series that catered to the southeastern viticulture industry, and edited the symposium proceedings. Dr. Lamikanra is the author of more than 100 technical publications and 60 abstracts. He is a member of the editorial board of the *Journal of Food Agriculture and Environment*. He is also editor of the multidisciplinary book *Fresh-Cut Fruits and Vegetables: Science, Technology and Market* (CRC Press).

Syed H. Imam, Ph.D., is a senior scientist/research chemist in the Bioproduct Chemistry & Engineering Research Unit at the USDA-ARS, Western Regional Research Center located in Albany, California. The Bioproduct Chemistry & Engineering Research Unit conducts interdisciplinary research on polymers derived from agricultural residues and surplus commodities to develop new and expanded markets for renewable materials. He has authored or coauthored over 85 publications and abstracts and has made over 45 scientific presentations. His research is of considerable breadth and includes fundamental studies on renewable polymers in plastics, characterization of such plastic blends, and efforts to understand biodegradable properties of the natural and blended polymers using classical and state-of-the-art biochemical methods. Dr. Imam is internationally recognized for his accomplishments in the field of polymer degradation. His research activities led him twice (1993 and 1997) to serve as a United Nations Development Program (UNDP) consultant/advisor. He is a co-editor of an ACS Book published by the Oxford University Press titled *Biopolymers: Utilizing Nature's Advanced Materials*. He is a member of the editorial advisory board of the *Journal of Environmental Polymer Degradation*.

Dike O. Ukuku, Ph.D., is a research food technologist with the USDA Food Safety Intervention Technologies Research Unit, Eastern Regional Research Center in Wyndmoor, Pennsylvania. His past research has focused on short-chain organic acids as antioxidants in lipid oxidation and functional mechanisms of antimicrobial compounds. Dr. Ukuku taught microbiology at Wayne County Community College, Detroit, Michigan, for 3 years and was employed as an assistant quality control manager for 1 year at Quaker Maid Inc., a dairy industry company in Detroit. He also worked as a microbiologist/aquatic toxicologist at a wastewater treatment plant in Detroit for 2 years. His research activities have focused on the microbiology of minimally processed cantaloupe and honeydew melons with emphasis on understanding bacterial attachment, development of novel techniques for decontamination, and minimizing bacterial transfer from melon surfaces to fresh-cut pieces.

1 Genomic Exploration of Produce Degradation

J. Chen
USDA, ARS, PWA, San Joaquin Valley Agricultural Sciences Center, 9611 South Riverbend Ave., Parlier, CA

CONTENTS

1.1   INTRODUCTION

An important aspect of learning to control microbial degradation of produce requires understanding the degradation microorganisms. Recent advancements in bacterial genome sequencing have created unprecedented amounts of genetic information. Analysis of genome data offers a new perspective on microbial research and has enhanced our understanding of produce degradation. In the genomic era, additional genes with broader functions are expected to be discovered at a faster pace. Genome study affects not only on the study of genes, but also other areas such bacterial evolution, ecology, and systematics. From a population point of view, genomics expands our understanding of model bacterial strains and helps in the exploration of bacterial strains in all ecological niches. Soft rot erwinia are a group of bacterial pathogens that cause both pre- and postharvest rot in a broad range of fresh produce. The molecular biology research on soft rot erwinia in the past two decades has led to significant discoveries of genes and mechanisms of pathogenicity (Collmer and Keen,1986; Kotoujansky, 1987; Hugouvieux-Cotte-Pattat et al., 1996; Toth et al.,

2003). In this chapter, advances in the population genomics of soft rot erwinia are used to demonstrate the importance of genome analysis in the understanding and control of produce degradation.

In the genus *Erwinia*, one noticeable development derived from genome analysis is the bacterial taxonomy. Genome sequence analyses have unveiled new phylogenetic relationships among *Erwinia* strains. Accordingly, reclassification of the bacterium has been proposed at the genus, species, and subspecies levels (Kwon et al., 1997; Hauben et al., 1998; Gardan et al., 2003). The soft rot erwinia have been regrouped into the genus *Pectobacterium*, and several subspecies have been elevated to species (Hauben et al., 1998; Gardan et al., 2003). Along this line, more accurate and sophisticated identification tools have been developed or are emerging (Louws et al., 1999).

Both *Erwinia* and *Pectobacterium* are validly published names. In this discussion, the name *Erwinia* will be used. The term *soft rot erwinia* is generically used to refer to *E. carotovora* and *E. chrysanthemi*. The former is further divided into subspecies including, *E. carotovora* subsp. *atroseptica* and *E. carotovora* subsp. *carotovora*. Detailed discussions on the taxonomy and biology of soft rot erwinia are available in a number of review publications (Starr and Chatterjee, 1972; Chatterjee and Starr, 1980; Barras et al., 1994; De Boer, 2003).

A bacterial taxon describes a population or a group of prokaryotic cells sharing a level of similarity. A defined number of strains are sampled to represent the population. Typically, one or more strains are selected as representatives for further detailed characterization and even fewer are subjected to whole genome sequencing. As such, the representative strain(s) does not contain all the genetic material in the bacterial population. During the process of environmental adaptation, a bacterial genome could be subject to base mutation, sequence insertion/deletion (indel), horizontal gene transfer, and genome rearrangement (Pym and Brosch, 2000). Such information is often related to bacterial pathogenicity and host specialization. Genetic variations are usually acquired through genomic comparison involving more strains from different sources.

Analyses of representative strains establish the main core of bacterial genomics, but gaps in information remain. Filling in these gaps relies on frequent and even extensive analyses of other bacterial strains in the population. In the past, this has been the work of identifying new genetic elements, particularly those involved in horizontal gene transfer, such as phages, plasmids, and transposons. Results from recent genomic studies show that sequence insertion/deletion or indels and genomic rearrangements are also important sources of genetic variation (Pym and Brosch, 2000; Britten et al., 2003).

The term *genome analysis* is defined as comparing and deducing information from genomic DNA. For soft rot erwinia, the wave of whole-genome sequencing has arrived. It can be certain that much new information has been generated and is being published. Reviews regarding recent research advances in the genomics of soft rot erwinia have been published (De Boer, 2003; Toth et al., 2003). Yet, it is also true that much more new information is awaiting discovery. This chapter intends to highlight some new developments in soft rot erwinia genomics and discuss bacterial genomics from the population point of view. Efforts will be made to connect the results of previous DNA-based population analyses to current genomic knowledge in an attempt to demonstrate that many "old" DNA analysis tools are still useful or even critical for soft rot erwina research in the genomic era.

1.1.1 Genotype vs. Phenotype

Characterization or identification of useful and unique features is the first step toward the study of a microorganism. Classically, the bacterial phenome is used. The phenotype of an organism can be described as the sum of its genotype and the effect of the surrounding environment:

$$\text{Phenotype} = \text{Genotype} + \text{Environment} \qquad (1)$$

However, the environmental factor is usually highly variable, particularly under field conditions, resulting in inconsistent phenotypic variations, commonly regarded as "atypical." Shifting the environmental factor to the other side of the equation, genotypic study of an organism becomes unaffected, or less affected, by the environment:

$$\text{Genotype} = \text{Phenotype} - \text{Environment} \qquad (2)$$

Although formula (1) and formula (2) may oversimplify the genome–phenome relationship, they illustrate the advantage of genomic analysis. In most cases, genotypic characterization is complementary to phenotypic characterization. In this discussion, *genome* refers to all organismal DNA. Genomic analysis can be done at both global (whole genome sequence) and partial (DNA segment) levels. In either case, the quality of analysis is directly related to the quality of the database used. A genomic database can be in the form of either a nucleotide sequence or a coding representation describing the status of DNA nucleotide variations.

DNA sequence databases are the most exchangeable among researchers. *Genome sequencing* is used to identify the linear arrangement of all bases in an organism's DNA, commonly a combination of shotgun-sequencing and a gap-closure. *Bioinformatics* is used to analyze the sequence data to predict gene function, protein and RNA structure, gene regulation, genome organization, and the phylogenetic history of genes and gene families. Because of the large volume of data, specifically developed computer software is always required to assist the analysis. One popular program for sequence comparison is BLAST (Altschul et al., 1997). There are many other programs available on the Internet or from commercial sources, depending on the goal of the analysis. Functional genomic approaches, such as gene disruption and replacement experiments, are used to confirm the biological functions predicted from sequence analysis. After all, the study of genomes, which includes DNA sequencing, gene function, and genome structure and evolution, is referred to as *genomics*.

1.2 SEQUENCE GENOMICS

1.2.1 Whole Genome Sequencing

Microbial genomics actually began in virology. The small genome size of viruses makes it feasible to determine the whole nucleotide (DNA or RNA) sequence and allows genomic comparisons that otherwise would be difficult or even impossible to perform with the early biotechnology. Viral genomics revolutionized plant virology and resulted in novel strategies for virus research and disease control. The first plant pathogenic bacterial genome sequence was completed with *Xylella fastidiosa*, a pathogen causing citrus variegated chlorosis in Brazil (Simpson et al., 2000). Because of the nutritional fastidiousness, *X. fastidiosa* was much less characterized by classical biochemical and genetic approaches than many other bacterial plant pathogens. The sequence data, however, immediately make possible the exploration of the pathogenicity, biology, epidemiology, and evolution of the organism. Since then, several other plant pathogen and plant symbiont genome sequences have been published. More genome sequencing of plant pathogens, including pathogenic fungi, has also been initiated; some of this research is in an advanced stage. The following Web sites provide some of the most up-to-date genome sequencing information:

> http://www.ncbi.nlm.nih.gov/genomes/MICROBES/Complete.html
> http://www.ncbi.nlm.nih.gov/genomes/MICROBES/InProgress.html
> http://www.ncbi.nlm.nih.gov/genomes/FUNGI/funtab.html

Genome sequencing projects have been initiated for two species of soft rot erwinia. The Sanger Institute was funded by the Scottish Office (SEERAD) to sequence the genome of *E. carotovora* subsp. *atroseptica* strain SCRI1043 (ATCC BAA-672). The sequencing is now completed (http://www.sanger.ac.uk/Projects/E_carotovora/). (Bell KS, Sebaihia M, Pritchard L, Holden MT, Hyman LJ, Holeva MC, Thomson NR, Bentley SD, Churcher LJ, Mungall K, Atkin R, Bason N, Brooks K, Chillingworth T, Clark K, Doggett J, Fraser A, Hance Z, Hauser H, Jagels K, Moule S, Norbertczak H, Ormond D, Price C, Quail MA, Sanders M, Walker D, Whitehead S, Salmond GP, Birch PR, Parkhill J, and Toth IK. 2004. Genome sequence of the enterobacterial phytopathogen *Erwinia carotovora* subsp. *atroseptica* and characterization of virulence factors. *Proc Natl Acad Sci U S A* 27:11105-10.) The genome is 5.064 Mb in length with a G+C content of 50.97%. TIGR and University of Wisconsin, funded by USDA, are collaborating in sequencing *Erwinia chrysanthemi* strain 3937 (http://www.ahabs.wisc.edu and http://www.tigr.org/tdb/mdb/mdbinprogress.html). The sequencing is anticipated to be complete in 2004. The unfinished sequence has total bases of 4,936,076 bp in 40 contigs (http://www.ncbi.nlm.nih.gov/genomes/ framik.cgi?db=genome&gi=5110).

An elaborate genome sequence comparison both globally and locally (*comparative genomics*) provides a valuable alternative to discover new genetic and phylogenetic information. The basic assumption is that many common features are shared between different bacteria. Such conserved features may even cross the border between plant and animal bacteria. For soft rot erwinia, it is assumed, and evidence indicates, that they share an enterobacterial chromosomal "backbone" derived from a common ancestor. Differences are attributed to extensive horizontal gene transfer, genome rearrangements, gene duplication, sequence insertion/deletion, and nucleotide mutation into nonfunctional pseudogenes. Bacterial "lifestyle" changes and varies from one to another in different ecological niches.

The issue of pathogenicity or virulence is always a priority for pathogen research. In the pregenomics era, many virulence genes (e.g., those involved in plant cell wall degradation, enzyme regulation, and export) were isolated by transposon mutagenesis (Hinton et al., 1989; Py et al., 1998). The role of these genes was later confirmed using simple plant assays such as stem or tuber inoculation tests, often with artificially high concentrations of bacterial cells. The question is whether such conventional approaches can identify genes expressed only under natural infection conditions. The answer to this question relies on the techniques that can comprehensively analyze every gene in the bacterial genome. Incidentally, this is one of the goals of genomic research.

Whole genome sequence comparison of different soft rot erwinia strains can identify candidate genes for pathogenicity and host specificity. The potential can be seen in the examples of *Xanthomonas* and *Xylella*. Da Silva et al. (2002) compared the complete genome sequences of *Xanthomonas axanopodis* subsp. *citri* and *Xanthomonas campestris* pv. *campestris*. The two genomes shared more than 80% of their genes. Gene order is conserved along most of their respective chromosomes. Such a high genetic similarity contrasts with their distinct disease phenotypes and host ranges. *Xanthomonas axanopodis* subsp. *citri* is the pathogen that causes citrus canker, and *Xanthomonas campestris* pv. *campestris* causes black rot in crucifers.

Sequence comparison identified a set of strain-specific genes, some of which are probably responsible for the distinct pathogenicity and host specificity profiles. Both *Xanthomonas axanpodis* pv. *citri* and *Xanthomonas campestris* pv. *campestris* have an extensive repertoire of genes for cell-wall degradation. Both genomes code for enzymes with cellulolytic, pectinolytic, and hemicellulolytic activities. *Xanthomonas campestris* pv. *campestris* has more genes involved in pectin and cellulose degradation than *Xanthomonas axanpodis* pv. *citri*. In addition, *Xanthomonas campestris* pv. *campestris* has two 1,4-b-cellobiosidases and two pectin esterases, none of which are found in *Xanthomonas axanopodis* pv. *citri*. The differences in symptoms may be correlated to the noted differences in genes for cell-wall degradation. These differences suggest that *Xanthomonas campestris* pv. *campestris* is uniquely suited to invade and colonize host tissue, a fact that may partially explain the systemic nature of its infection. In contrast, *Xanthomonas axanpodis* pv. *citri* induces a strong local response with cell proliferation and necrosis but shows little spontaneous dissemination, probably due to a smaller number of genes capable of causing a massive degeneration of host tissue (da Silva et al., 2002).

Genome comparison of two *X. fastidiosa* strains revealed a different picture with regard to population genomics. Van Sluys et al. (2003) reported the genome sequence of *X. fastidiosa* (Temecula strain), isolated from a naturally infected grapevine with Pierce's disease in California. Comparative analyses with a previously sequenced *X. fastidiosa* strain 9a5c responsible for citrus variegated chlorosis revealed that 98% of the *X. fastidiosa* Temecula genes are shared with the *X. fastidiosa* strain 9a5c genes. Furthermore, the average amino acid identity of the open reading frames in the two strains is 95.7%. Genomic differences are limited to phage-associated chromosomal rearrangements and deletions that also account for the strain-specific genes present in each genome. Genomic islands, one in each genome, were identified.

1.2.2 SAMPLE SEQUENCING

Sample sequencing is a low-cost but efficient approach to explore bacterial genomics. Large-scale sample sequencing has been applied in the genome study of several prokaryotic and eukaryotic microorganisms (Aguero et al., 2000; Sanchez et al., 2001; Mittleider et al., 2002). Using *E. carotovora* subsp. *atroseptica* strain SCRI 1039, Bell et al. (2002) constructed a BAC library and identified two DNA fragments, 2B8 and 1C22, covering almost 200 kb of the bacterial genome. BLASTx/n searches were conducted with onefold coverage sequences. Approximately 10% of the *E. carotovora* subsp. *atroseptica* sequences showed the strongest similarity to those of enterobacteria. The majority of these were found on 2B8. A number of sequences similar to rhizobacterium genes were observed. Such genes were previously unknown to *Erwinia* spp. or other enterobacteria. Conversely, strong matches in 1C22 were similar to adhesin, haemagglutanin, and haemolysin genes from mammalian pathogens such as *Neisseria meningitidis* and *Pseudomonas aeruginosa* and the plant pathogen *X. fastidiosa*. Compared to other *Erwinia* species, genes other than the targeted *hrp* and *dsp* genes, such as *pel*, *peh*, *pec* and *hec* genes, were identified. Many of them were first identified in *E. carotovora* subsp. *atroseptica*.

1.2.3 SEQUENCES FROM *rrn* OPERONS

Sequences from the *rrn* locus, particularly the gene of small subunit ribosomal RNA or 16S rDNA, is probably the most comprehensively collected DNA sequence among all the bacterial genes from most bacterial species reported. Repeated targeting of rDNA has yielded large 16S rDNA databases that provide the necessary reference material for comparative sequence analysis and hence drawing reliable conclusions on phylogeny. In fact, the reclassification and phylogenetics of soft rot erwinia depend heavily on rRNA gene sequences (Kwon et al., 1997; Hauben et al., 1998; Fessehaie et al., 2002).

Kwon et al. (1997) investigated the phylogenetic relationships of the type strains of 16 *Erwinia* species by analyzing the 16S rDNA sequences using the neighbor-joining method. Supported by moderate bootstrap values, the phylogenetic tree shows that the genus *Erwinia* forms four phylogenetic clades or clusters. Among them, cluster III consists of *E. carotovora* subspecies and *E. chrysanthemi*. The *Erwinia* phylogeny clusters are, however, intermixed with members of other genera, such as *Escherichia coli*, *Klebsiella pneumoniae*, and *Serratia marcescens*. Hauben et al (1998) also analyzed the 16S rDNA sequences of 29 plant-associated strains in Enterobacteriaceae, including *Erwina*. They proposed to include the soft rot erwinia in the genus *Pectobacterium*, a genus name proposed over half of a century ago (Waldee, 1945).

Although the validity of using a single gene for phylogenetic analysis has been questioned, the universal presence of rRNA and its sequence conservation form the basis to develop successful applications in bacterial identification (Jensen et al., 1993). When the soft rot erwinia 16S rDNA sequences are aligned, differences at the species and subspecies level are unambiguously displayed. As sequencing technology advances, sequence-base identification is becoming more practical and routine.

Alternatively, PCR methods are developed. Toth et al. (1999a) developed a one-step PCR-based method for the detection of five subspecies of *E. carotovora*, including subsp. *carotovora* and subsp. *atroseptica*, and all pathovars/biovars of *E. chrysanthemi* on plant tissue culture material.

The 16S rDNA sequence approaches are generally considered to have limited sensitivity below the species level (Vandamme et al., 1996). However, there are exceptions. Single nucleotide polymorphisms (SNPs) in this conserved region provide an opportunity to explore strain variations because the primers used are species- or subspecies-specific. Seo et al. (2002) differentiated 87 strains of *E. carotovora* subsp. *carotovora* into two groups by analysis of 16S rDNA restriction fragment length polymorphism (RFLP) generated by *Hin*fI. They found that most strains from Korea and Japan belonged to the same group. In the RFLP format, SNPs are located by restriction enzyme sites. SNPs can be included in PCR primers for strain differentiation. With real-time PCR technology, SNPs can be placed in the fluorescence-labeled probe.

In contrast, sequence polymorphisms in the 16S–23S rDNA spacer regions are more commonly recommended to differentiate closely related bacterial strains. Fessehaie et al. (2002) reported that the 16S–23S rDNA spacer regions of *Erwinia* spp. varied considerably in size and nucleotide sequence. Phylogenetic analysis showed a consistent relationship among the *Erwinia* strains tested and was roughly in agreement with the 16S rDNA data. Toth et al. (2001) studied the variation of 16S–23S rDNA spacer of soft-rot erwinia. Isolates of *E. carotovora* subspecies, *E. chrysanthemi*, and the closely related *E. cacticida* yielded unique banding patterns that clearly distinguished them from other *Erwinia* and non-*Erwinia* species tested. Within the soft rot erwinias, three PCR groups were distinguished based on differences in their banding patterns. Group I comprised *E. carotovora* subsp. *atroseptica* and subsp. *betavasculorum*. Group II comprised *E. carotovora* subsp. *carotovora*, subsp. *odorifera*, and subsp. *wasabiae* and *E. cacticida*. Groups I and II were clearly related based on the banding patterns. Group III comprised all *E. chrysanthemi* isolates.

1.2.4 OTHER GENES

Exoenzymes such as pectate lyase, pectin lyase, and polygalaturonase are the primary virulence factors of soft rot erwinia (Collmer and Keen, 1986; Hugouvieux-Cotte-Pattat et al., 1996; Py et al., 1998; Chen, 2002). Because of their economic importance, exoenzyme genes are among the earliest to be sequenced from the soft rot erwinia genome. Based on these sequences, several PCR primer sets were developed for pathogen detection (Table 1.1). Darrasse et al. (1994b) designed a primer set, Y1-Y2, based on a pectate lyase gene (*pel*) common to some strains of *E. carotovora* and *Yersinia pseudotuberculosis* (Y family). An amplified fragment of 434 bp was obtained from strains of *E. carotovora* subspecies except for *E. carotovora* subsp. *betavasculorum*, *E. chrysanthemi*, and other *Erwinia* species. Differentiation of subspecies was achieved through RFLPs of the amplicons. Another primer set, Y45-Y46 (Frechon et al., 1995), which is also from the same *Pel* gene, specifically

TABLE 1.1
Commonly Used PCR Primer Sets for Identification and Detection of Soft Rot Erwinia

Primer Name	Primer Sequence	Closest BLASTn Hit[a]	Specificity	Amplicon Size (bp)	Source
Y1	5′ ttaccggacg ccgagctgtg gcgt 3′	Erwinia carotovora pelB gene …	Erwinia carotovora	434	Darrasse et al., 1994
Y2	5′ caggaagatg tcgttatcgc gagt 3′	Erwinia carotovora pelB gene …			
y45	5′ tcaccggacg ccgaactgtg gcgt 3′	Erwinia carotovora pelB gene …	Erwinia carotovora subsp. atroseptica	439	Frechon et al., 1995
y46	5′ tcgccaacgt tcagcagaac aagt 3′	Erwinia carotovora pelB gene …			
ECA1f	5′ cggcatcata aaaacacg 3′	(Pan troglodytes BAC clone RP43-4 …)	Erwinia carotovora subsp. atroseptica	690	De Boer and Ward, 1995
ECA1r	5′ gcacacttca tccacga 3′	(Homo sapiens chromosome 8, clone …)			
ADE1	5′ gatcagaaag cccgcagcca gat 3′	Erwinia chrysanthemi pelE	Erwinia chrysanthemi	420	Nassar et al., 1996
ADE2	5′ ctgtggccga tcaggatggt tttgtcgtgc 3′	Erwinia chrysanthemi pelE			
ERWFOR	5′ acgcatgaaa tcttccaatg c 3′	(Homo sapiens BAC clone RP11-391J13)	Erwinia carotovora subsp. atroseptica	389	Smid et al., 1995
ATRREV	5′ atcgatattt gattgtc 3′	(Arabidopsis thaliana DNA)			
ERWFOR	5′ acgcatgaaa tcttccaatg c 3′	(Homo sapiens BAC clone RP11-391J13)	Erwinia chrysanthemi	450	Smid et al., 1995
CHRREV	5′ agtgctgccg tacagcacgt 3′	Erwinia chrysanthemi prt			

[a] Results from BALSTn search using GenBank nucleotide sequence database (http://www.ncbi.nlm.nih.gov). Description in parenthesis may not infer true genetic relationship.

amplified a 439-bp DNA fragment in all *E. carotovora* ssp. *atroseptica* isolates tested, but not in isolates of the other *E. carotovora* subspecies or in atypical isolates (Yahiaoui-Zaidi et al., 2003).

Building on their previous work (Ward and De Boer, 1994), De Boer and Ward (1995) developed a primer set, ECA1f-ECA1r, in which only *E. carotovora* subsp. *atroseptica* DNA was amplified. However, the genetic nature of these primers is unknown, even compared with the most currently available GenBank database (Table 1.1). Targeting the *E. chrysanthemi* group, Nassar et al. (1996) developed a primer set based on the sequence of *E. chrysanthemi* pelADE gene. A 420-bp amplified fragment was obtained for all 78 *E. chrysanthemi* strains tested. No amplified fragment was observed with the other *Erwinia* species and organisms of different genera.

Smid et al. (1995) developed three PCR primers. A combination of two, ERW-FOR-ATRREV, is specific to *E. carotovora* subsp. *atroseptica* and another combination of two, ERWFOR-CHRREV, is specific to *E. chrysanthemi*. Database searching (Table 1.1) shows that ERWFOR and ATRREV share no similarity to any known bacterial DNA sequences. In contrast, primer CHRREV falls into the locus of a protease gene from *E. chrysanthemi*. As the soft rot erwinia genome sequence database grows, the genetic nature of all PCR primers together with their amplicons can be identified. The PCR-sequencing format for pathogen detection can eliminate the nonspecific amplification problem that has been one of the major concerns in PCR application for diagnostics. Similar to sequences from the *rrn* operon, SNPs from other genomic loci, if any, could be a valuable addition to soft rot erwinia genomics.

1.3 DNA GENOMICS

1.3.1 DNA-DNA Homology

Although nucleotide sequences were not particularly identified as the target, DNA-DNA homology study is one type of genome comparison. Instead of looking for any particular locus or loci, all of the bacterial genetic contents, including coding and noncoding regions, are compared simultaneously. As a result, differences are described at the whole-genome level. The higher the DNA-DNA homology, the more closely related the bacterial strains. DNA-DNA homology is by definition a critical parameter for a bacterial species. DNA-relatedness studies of soft-rot organisms can be traced back to the early 1970s, when Brenner et al. (1973) showed that strains of *Pectobacterium carotovorum* and *Pectobacterium chrysanthemi* belonged to distinct DNA homology groups. Supported by DNA homology data, Gardan et al. (2003) recently proposed that three subspecies of *P. carotovorum* should be elevated to the species level.

Because a DNA homology test is laborious and time-consuming, it is not suitable for routine use. In a study with fluorescent pseudomonas, Cho and Tiedje (2001) developed a random genome fragment microarray in an effort to overcome the disadvantages of whole-genome DNA-DNA hybridization. Sixty to 96 genome fragments of approximately 1 kb from each of four fluorescent *Pseudomonas* species were spotted on microarrays. Genomes from 12 well-characterized fluorescent *Pseudomonas* strains were labeled with Cy dyes and hybridized to the arrays. Cluster analysis of hybridization profiles resulted in phylogenetic dendograms providing species-level to strain-level resolution. The method does not require laborious cross-hybridizations and can provide an open database of hybridization profiles.

1.3.2 RANDOM GENOME SAMPLING

Genome sequencing of a large number of bacterial strains is currently cost-prohibitive. Alternatively, whole-genome comparisons could be achieved using random, or presumptively random, sampling approaches. Some of the approaches that have been applied to study soft rot erwinia include AFLP (amplified fragment length polymorphism), RAPD (random amplified polymorphic DNA), RFLP (restriction fragment length polymorphism), and ERIC (enterobacterial repetitive intergenic consensus). Mostly, these techniques are PCR-based and they are specifically designed to differentiate closely related strains. Because of their simplicity, multiple random sampling approaches are commonly used or are used in combination with other genotypic and phenotypic techniques (Darrasse et al., 1994b; Toth et al., 1999a; Dellagi et al., 2000; Hadas et al., 2001; Seo et al., 2002; Yahiaoui-Zaidi et al., 2003).

Traditionally, RFLP analysis involves blotting restriction enzyme-digested DNA on nylon membrane. DNA polymorphisms are reviewed by hybridization with labeled probes (Chen et al., 1992). In the report of Toth et al. (1999a), RFLPs of genomic DNA of 60 strains of *E. carotovora* subsp. *atroseptica* from eight western European countries were evaluated with a DNA probe. Digestion with *Eco*RI offered no differentiation between the *E. carotovora* subsp. *atroseptica* strains. However, when *Eco*RI was replaced by *Hind*III, hybridization patterns divided the strains into two groups. Another type of RFLP study is PCR-based: PCR amplicons are digested with restriction enzymes. DNA polymorphisms are commonly resolved through agarose gel electrophoresis. When combining with specific PCR such as those using primers from the *rrn* operon, RFLP analysis is an effective tool to differentiate closely related bacterial strains (Darrasse et al., 1994b; Helias et al., 1998; Toth et al., 1999a; Hadas et al., 2001; Fessehaie et al., 2002; Seo et al., 2002; Yahiaoui-Zaidi et al., 2003).

Similar to classical RFLP anslysis, AFLP analysis is based on random genome sampling of restriction sites. The digested DNA fragments are ligated with specially designed oligomers that can be used as priming sites for PCR amplification. Based on AFLP profiles, Avrova et al. (2002) identified four clusters of soft rot erwinia. Cluster 1 contained *Erwinia carotovora* subsp. *carotovora* and *Erwinia carotovora* subsp. *odorifera*; cluster 2 contained *Erwinia carotovora* subsp. *atroseptica* and *Erwinia carotovora* subsp. *betavasculorum;* and Clusters 3 and 4 contained *Erwinia carotovora* subsp. *wasabiae* and *E. chrysanthemi* strains. *E. carotovora* subsp. *carotovora* and *E. chrysanthemi* showed a high level of molecular diversity (23 to 38% mean similarity). Others showed considerably less (56 to 76% mean similarity).

RAPD analysis compared similarities and differences among PCR-generated genomic DNA fragments flanked by defined 10-base sequences. Fragments amplified are generally ca. 0.1 to 2 kb. The electrophoretic profiles of DNA fragments based on size are scored in a binary format and variations are summarized by statistical means. Because of its technical simplicity, RAPD analysis can be used to quickly compare large numbers of bacterial strains (Chen et al., 1995; Albibi et al., 1998). DNA bands can be cloned and sequenced. Therefore, the genetic information embedded in the RAPD fragments is determined using information from the genome sequence database. This is similar to random sample sequencing, but comparison of DNA bands serves as a prescreening step to efficiently locate the strain-specific genomic loci.

Toth et al. (1999a) tested 60 strains of *Erwinia carotovora* subsp. *atroseptica* from eight western European countries and found that RAPD analysis was more sensitive than ERIC and RFLP analyses. RAPD and ERIC analyses are similar in that the former uses nonspecific primers that bind randomly to regions over the entire genome and the latter uses primers specific to a repetitive sequence. Because of its high sensitivity, RAPD analysis can be used to study pathogen host specialization. Hadas et al. (2001) performed a RAPD analysis using nine arbitrary primers. Cluster analysis from the 150 distinct DNA characters formed four clusters for the 26 *E. carotovora* subsp. *carotovora* isolates from pepper, tomato, potato, and cabbage across six countries. The clustering pattern is in agreement with the host rather than geographical origins.

Maki-Valkama and Karjalainen (1994) successfully differentiated *E. carotovora* subsp. *atroseptica* and *E. carotovora* subsp. *carotovora* and identified two separate RAPD clusters along the line of subspecies. Similarity between the 10 *E. carotovora* subsp. *atroseptica* and 10 *E. carotovora* subsp. *carotovora* strains was generally only 10 to 25%. While strain similarity within *E. carotovora* subsp. *atroseptica* was high (> 85%), isolates within the *E. carotovora* subsp. *carotovora* group showed extensive genetic diversity (< 50%).

On the other hand, Parent et al. (1996) explored the use of the RAPD technique to identify *Erwinia carotovora* from soft rot-diseased plants. *E. caratovora* and pectolytic pseudomonads such as *Pseudomonas fluorescens*, *P. marginalis*, and *P. viridiflava* are common bacteria associated with soft rot. They found that the combination of two selected primers was sufficient for adequate distinction of *E. carotovora* from pectolytic, fluorescent *Pseudomonas* species. Furthermore, *E. carotovora* subsps. *atroseptica* and *carotovora* could also be distinguished from each other.

While the reproducibility of RAPD has been controversial, its simplicity and DNA recovering capacity make the technique a good choice to isolate genome-specific DNA sequences. Maki-Valkama and Karjalainen (1994) selected three RAPD fragments from the *E. carotovora* subsps. *atroseptica* group and used them as probes for Southern hybridization. All three DNA fragments hybridized only with *E. carotovora* subsps. *atroseptica* isolates, an indication of the sequence specificity of the three RAPDs. Cloning and sequencing of the unique RAPDs together with genome sequence comparison can be an effective tool for genetic discovery. This is particularly suitable for the identification of phages, plasmids, and indels. Interestingly, Toth et al. (1999a) reported that phage typing and RAPD analysis showed a high level of diversity within *Erwinia carotovora* subsp. *atroseptica* compared to other techniques.

1.3.3 GENOMIC SUBTRACTION

In principle, genomic subtraction is identical to complete genome sequence comparison. The goal is to identify unique genomic information present in one strain but absent in the other. The most common protocol for genomic subtraction is the one from Straus and Ausubel (1990), who developed a technique for isolating the DNA that was absent in deletion mutants. The method removes from wild-type DNA the sequences that are present in both the wild-type and the deletion mutant genomes. The process is achieved by allowing a mixture of denatured wild-type and biotinylated mutant DNA to reassociate. After reassociation, the biotinylated sequences are removed by binding to avidin-coated beads.

Early application of genomic subtraction was to develop detection probes. Applying the genomic subtraction technique of Straus and Ausubel (1990), Darrasse et al. (1994) isolated six DNA sequences from a strain of *E. carotovora* subsp. *atroseptica*. One fragment was specific for typical *E. carotovora* subsp. *atroseptica* strains, two hybridized with all *E. carotovora* subsp. *atroseptica* strains and with a few *E. carotovora* subsp. *carotovora* strains, and two probes recognized only a subset of *E. carotovora* subsp. *atroseptica* strains. The last probe was absent from the genomic DNA of *E. carotovora* subsp. *carotovora* CH26 but was present in the genomes of many other strains, including those of other species and genera. Using a nonradioactive system, Ward and De Boer (1994) also developed a DNA probe that specifically detected *Erwinia carotovora* subsp. *atroseptica*.

Assuming a similar genetic background, as indicated by phenotypic similarity, genes found in one soft rot erwinia but not in others may be involved in specific pathogenesis or other plant-associated lifestyle characteristics. An example from an early study with *Listeria monocytogenes*, a food-borne human pathogen (Chen et al., 1993) can be referenced. Subtracter probe hybridization was used to screen a partial genomic library of a clinical isolate of *L. monocytogenes* against the genome DNA from *L. innocua*. *L. monocytogenes* and *L. innocua* are highly similar but the latter is not pathogenic in humans. Three clones that hybridized with genomic DNA from 174 strains of *L. monocytogenes* but not with genomic DNA from 32 strains representing other *Listeria* species were recovered. Using the limited database available then, one of the clones was identified by BLAST to be related to *in*LAB, a gene family associated with pathogenicity.

In most cases, the genetic nature of the probe DNA sequences is unknown. Darrasse et al. (1994a) cloned and sequenced their DNA probe fragments. Probably due to the limited sequence information available at the time, the genetic nature of the probe sequences was not identified, with the exception of one. This probe is homologous to the *put*P gene of *Escherichia coli*, which encodes a proline carrier.

To cope with the increase in sequence information a genomic analysis tool called microarray has recently been developed. With microarray, thousands of genes or DNA sequences can be analyzed in one hybridization experiment. In array technology, the probe or known sequence is the arrayed material, whereas the unknown or target sequence is labeled and hybridized to the array. The intensity of hybridized sequences is quantified and the resulting data are subjected to detailed statistical or

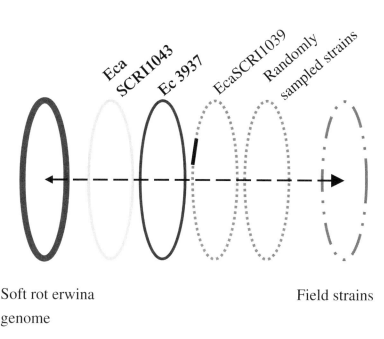

Soft rot erwina
genome

Field strains

FIGURE 1.1 A graphical illustration of the concept of population genomics. The solid circle on the left represent the population genome of soft rot erwinia, which is the compilation of whole and partial genome sequences. The two complete genome sequences (Eca SCRI1043 = *Erwinia carotovora* subsp. *atroseptica* SCRI1043 and Ec 3937 = *Erwinia chrysanthemi* 3937) serve as the system core. The leftward direction of the central line outlines the strategy to obtain the population genome. The ultimate goal of population genomics is, however, in the rightward direction (i.e., to help to resolve problems associated with strains in the fields or production).

quantitative analysis. Arrays have been fabricated on a variety of different materials, but the most common are oligonucleotide- or DNA-based arrays fabricated on glass slides or nylon membranes.

Fessehaie et al. (2003) recently applied an oligonucleotide array for the identification and differentiation of bacteria pathogenic in potatoes. In their study, oligonucleotides 16 to 24 bases long were selected from the 3′ end of the 16S gene and the 16S–23S intergenic spacer regions of bacteria pathogenic in potatoes, including *Clavibacter michiganensis* subsp. *sepedonicus*, *Ralstonia solanacearum*, and the pectolytic erwinias, including *Erwinia carotovora* subsp. *atroseptica* and *carotovora* and *E. chrysanthemi*. Genomic DNA from bacterial cultures was amplified by PCR using conserved ribosomal primers and labeled simultaneously with digoxigenin-dUTP for the hybridization experiment. Subsequent serological detection of the digoxigenin label revealed different hybridization patterns that were distinct for each species and subspecies tested. Preliminary tests also show that bacteria could be detected by hybridizing to the array amplicons from mixed cultures and inoculated potato tissue.

1.4 SUMMARY

The use of molecular biology techniques has led to significant advances in the research on soft rot erwiniae. By comparing DNA sequences and the arrangement of DNA along entire genomes, more new genetic information about the bacteria and host–pathogen interactions can be unveiled, providing a base for further function analysis. While new analytical tools such as microarray are important for genomic study, much of the technology already available can also make a significant contribution. This is particularly the case for random sampling tools that are used to test a large number of bacterial strains. For example, strain-specific DNA fragment can be deduced from RAPD analysis. The genetic nature of the specific DNA is then elucidated through sequence analysis utilizing complete genome sequence databases. As illustrated in Figure 1.1, whole genome sequences from model strains will serve as the system core to screen for new genetic information from strains of different sources. Both whole genome sequences from model strains and partial sequences from field strains form the database of the population genome of soft rot erwinia. The database will then be used to facilitate research in population genomics and to resolve problems in the pathogen control.

REFERENCES

Albibi, R., Chen, J., Lamikanra, O., Banks, D., Jarret, R.L., and Smith, B.J., RAPD finger-printing *Xylella fastidiosa* Pierce's disease strains isolated from a vineyard in North Florida, *FEMS Microbiol. Lett.*, 165, 347–352, 1998.

Altschul, S.F., Madden, T.L., Schaffer, A.A., Zhang, J., Zhang, Z., Miller, W., and Lipman, D.J., Gapped blast and psi-blast: a new generation of protein database search programs, *Nucleic Acids Res.*, 25, 3389–3402, 1997.

Aguero, F., Verdun, R.E., Frasch, A.C., and Sanchez, D.O., A random sequencing approach for the analysis of the *Trypanosoma cruzi* genome: general structure, large gene and repetitive DNA families, and gene discovery, *Genome Res.* 10, 1996–2005, 2000.

Avrova, A.O., Hyman, L.J., Toth, R.L., and Toth, I.K., Application of amplified fragment length polymorphism fingerprinting for taxonomy and identification of the soft rot bacteria *Erwinia carotovora* and *Erwinia chrysanthemi, Appl. Environ. Microbiol.*, 68,1499–1508, 2002.

Barras, F., van Gijsegem, F., and Chatterjee, A.K., Extracellular enzymes and pathogenesis of soft-rot *Erwinia, Annu. Rev. Phytopathol.*, 32, 201–234, 1994.

Bell, K.S., Avrova, A.O., Holeva, M.C., Cardle, L., Morris, W., De Jong, W., Toth, I.K., Waugh, R., Bryan, G.J., and Birch, P.R., Sample sequencing of a selected region of the genome of *Erwinia carotovora* subsp. *atroseptica* reveals candidate phytopathogenicity genes and allows comparison with *Escherichia coli, Microbiology*, 148, 1367–1378, 2002.

Brenner, D.J., Steigerwalt, A.G., Miklos, G.V., and Fanning, G.R., Deoxyribonucleic acid relatedness among erwiniae and other *Enterobacteriaceae*: the soft-rot organisms (genus *Pectobacterium* Waldee), *Int. J. Syst. Bacteriol.*, 23, 205–216, 1973.

Britten, R.J., Rowen, L., Williams, J., and Cameron, R.A., Majority of divergence between closely related DNA samples is due to indels, *Proc. Natl. Acad. Sci. USA*, 100, 4661–4665, 2003.

Chatterjee, A.K. and Starr, M.P., Genetics of *Erwinia* species, *Annu. Rev. Microbiol.*, 34, 645–676, 1980.

Chen, J., Microbial enzymes associated with fresh-cut produce, in *Fresh-Cut Fruits and Vegetables*, Lamikanra, O., Ed., CRC Press, Boca Raton, FL, pp. 249–266, 2002.

Chen, J., Chang, C.J., Jarret, R.L., and Gawel, N. Genetic variation among *Xylella fastidiosa* strains, *Phytopathology*, 82, 973–977, 1992.

Chen, J., Brosch, R., and Luchansky, J.B., Isolation and characterization of *Listeria monocytogenes*-specific nucleotide sequences, *Appl. Environ. Microbiol.*, 59, 4367–4370, 1993.

Chen, J., Lamikanra, O., Chang, C.J., and D.L. Hopkins, Randomly amplified polymorphic DNA analysis of *Xylella fastidiosa* Pierce's disease and oak leaf scorch pathotypes, *Appl. Environ. Microbiol.*, 61, 1688–1690, 1995.

Cho, J.-C. and Tiedje, J.M., Bacterial species determination from DNA-DNA hybridization by using genome fragments and DNA microarrays, *Appl. Environ. Microbiol.*, 67, 3677–3682, 2001.

Collmer, A. and Keen, N.T., The role of pectic enzymes in plant pathogenesis, *Annu. Rev. Phytopathol.*, 24, 383–409, 1986.

da Silva, A.C., Ferro, J.A., Reinach, F.C., Farah, C.S., Furlan, L.R., Quaggio, R.B., Monteiro-Vitorello, C.B., Van Sluys, M.A., Almeida, N.F., Alves, L.M., do Amaral, A.M., Bertolini, M.C., Camargo, L.E., Camarotte, G., Cannavan, F., Cardozo, J., Chambergo, F., Ciapina, L.P., Cicarelli, R.M., Coutinho, L.L., Cursino-Santos, J.R., El-Dorry, H., Faria, J.B., Ferreira, A.J., Ferreira, R.C., Ferro, M.I., Formighieri, E.F., Franco, M.C., Greggio, C.C., Gruber, A., Katsuyama, A.M., Kishi, L.T., Leite, R.P., Lemos, E.G., Lemos, M.V., Locali, E.C., Machado, M.A., Madeira, A.M., Martinez-Rossi, N.M., Martins, E.C., Meidanis, J., Menck, C.F., Miyaki, C.Y., Moon, D.H., Moreira, L.M., Novo, M.T., Okura, V.K., Oliveira, M.C., Oliveira, V.R., Pereira, H.A., Rossi, A., Sena, J.A., Silva, C., de Souza, R.F., Spinola, L.A., Takita, M.A., Tamura, R.E., Teixeira, E.C., Tezza, R.I., Trindade dos Santos, M., Truffi, D., Tsai, S.M., White, F.F., Setubal, J.C., and Kitajima, J.P., Comparison of the genomes of two *Xanthomonas* pathogens with differing host specificities, *Nature*, 417, 459–463, 2002.

De Boer, S.H., Characterization of pectolytic erwinias as highly sophisticated pathogens of plants, *Eur. J. Plant Pathol.*, 109, 893–899, 2003.

De Boer, S.H. and Ward, L.J., PCR detection of *Erwinia carotovora* subsp. *atroseptica* associated with potato tissue, *Phytopathology*, 85, 854–858, 1995.

Dellagi, A., Birch, P.R.J., Heilbronn, J., Lyon, G., and Toth, I.K., cDNA-AFLP analysis of differential gene expression in the bacterial plant pathogen *Erwinia carotovora*, *Microbiology*, 146, 165–171, 2000.

Darrasse, A., Kotoujansky, A., and Bertheau, Y., Isolation by genomic subtraction of DNA probes specific for *Erwinia carotovora* subsp. *atroseptica*, *Appl. Environ. Microbiol.*, 60, 298–306, 1994a.

Darrasse, A., Priou, S., Kotoujansky, A., and Bertheau, Y., PCR and restriction fragment length polymorphism of a *pel* gene as a tool to identify *Erwinia carotovora* in relation to potato diseases, *Appl. Environ. Microbiol.*, 60, 1437–1443, 1994b.

Fessehaie, A., De Boer, S.H., and Levesque, C.A., Molecular characterization of DNA encoding 16S-23S rRNA intergenic spacer regions and 16S rRNA of pectolytic *Erwinia* species, *Can. J. Microbiol.*, 48, 387–398, 2002.

Fessehaie, A., De Boer, S.H., and Levesque, C.A., An oligonucleotide array for the identification and differentiation of bacteria pathogenic on potato, *Phytopathology*, 93, 262–269, 2003.

Frechon, D., Exbrayat, P., Gallet, O., Guillot, E., Le Clerc, V., Payet, N., and Bertheau, Y., Sequence nuceotidiques pour la detection de *Erwinia carotovora* subsp. *atroseptica*. Institut National de la Recherche Agronomique, Institut National Agronomique Paris-Grignon, Sanofi Diagnostics Pasteur. Brevet 95, 12-803, 1995.

Gardan, L., Gouy, C., Christen, R., and Samson, R., Elevation of three subspecies of *Pectobacterium carotovorum* to species level: *Pectobacterium atrosepticum* sp. nov., *Pectobacterium betavasculorum* sp. nov. and *Pectobacterium wasabiae* sp. nov., *Int. J. Syst. Evol. Microbiol.*, 53, 381–391, 2003.

Hadas, R., Kritzman, G., Gefen, T., and Manulis, S., Detection, quantification and characterization of *Erwinia carotovora* ssp *carotovora* contaminating pepper seeds, *Plant Pathol.*, 50, 117–123, 2001.

Hauben, L., Moore, E.R., Vauterin, L., Steenackers, M., Mergaert, J., Verdonck, L., and Swings, J., Phylogenetic position of phytopathogens within the *Enterobacteriaceae*, *Syst. Appl. Microbiol.*, 21, 384–397, 1998.

Helias, V., Le Roux, A.-C., Bertheau, Y., Andrivon, D., Gauthier, J.-P., and Jouan, B., Characterisation of *Erwinia carotovora* subspecies and detection of *Erwinia carotovora* subsp. *atroseptica* in potato plants, soil and water extracts with PCR-based methods, *Eur. J. Plant Pathol.*, 104, 685–699, 1998.

Hinton, J.C.D., Sidebottom, J.M., Hyman, L.J., Perombelon, M.C.M., and Salmond, G.P.C., Isolation and characterization of transposon-induced mutants of *Erwinia carotovora* subsp. *atroseptica* exhibiting reduced virulence, *Mol. Gen. Genet.* 217, 141–148, 1989.

Hugouvieux-Cotte-Pattat, N., Condemine, G., Nasser, W., and Reverchon, S., Regulation of pectinolysis in *Erwinia chrysanthemi*, *Annu. Rev. Microbiol.*, 50, 213–257, 1996.

Jensen, M.A., Webster, J.A., and Straus, N., Rapid identification of bacteria on the basis of polymerase chain reaction-amplified ribosomal DNA spacer polymorphism, *Appl. Environ. Microbiol.*, 59, 945–952, 1993.

Kotoujansky, A., Molecular genetics of pathogenesis by soft-rot erwinias, *Annu. Rev. Phytopathol.*, 25, 405–430, 1987.

Kwon, S.-W., Go, S.-J., Kang, H.-W., Ryu, J.-C., and Jo, J.-K., Phylogenetic analysis of *Erwinia* species based on 16S rRNA gene sequences, *Int. J. Syst. Bacteriol.*, 47, 1061–1067, 1997.

Louws, F.J., Rademaker, J.L.W., and de Bruijn, F.J., The three Ds of PCR-based genomic analysis of phytobacteria: diversity, detection and disease diagnosis, *Annu. Rev. Phytopathol.*, 37, 81–125, 1999.

Maki-Valkama, T. and Karjalainen, R., Differentiation of *Erwinia carotovora* subsp. *atroseptica* and *carotovora* by RAPD-PCR, *Ann. Appl. Biol.*, 125, 301–309, 1994.

Mittleider, D., Green, L.C., Mann, V.H., Michael, S.F., Didier, E.S., and Brindley, P.J., Sequence survey of the genome of the opportunistic microsporidian pathogen, *Vittaforma corneae*, *J. Eukaryot. Microbiol.*, 49, 393–401, 2002.

Nassar, A., Darrasse, A., Lemattre, M., Kotoujansky, A., Dervin, C., Vedel, R., and Bertheau, Y., Characterisation of *Erwinia chrysanthemi* by pectolytic isozyme polymorphism and restriction fragment length polymorphism analysis of PCR-amplified fragments of *pel* genes, *Appl. Environ. Microbiol.*, 62, 2228–2235, 1996.

Parent, J.-G., Lacroix, M., Page, D., and Vezina, L., Identification of *Erwinia carotovora* from soft rot diseased plants by random amplified polymorphic DNA (RAPD) analysis, *Plant Dis.*, 80, 494–499, 1996.

Py, B., Barras, F., Harris, S., Robson, N., and Salmond, G.P.C., Extracellular enzymes and their role in *Erwinia* virulence, *Methods Microbiol.*, 27, 157–168, 1998.

Pym, A.S. and Brosch, R., Tools for the population genomics of the tubercle bacilli. *Genome Res.*, 10, 1837–1839, 2000.

Seo, S.T., Furuya, N., Lim, C.K., Takanami, Y., and Tsuchiya, K., Phenotypic and genetic diversity of *Erwinia carotovora* ssp *carotovora* strains from Asia, *J. Phytopathol.-Phytopathol. Z.*, 150, 120–127, 2002.

Sanchez, D.O., Zandomeni, R.O., Cravero, S., Verdun, R.E., Pierrou, E., Faccio, P., Diaz, G., Lanzavecchia, S., Aguero, F., Frasch, A.C.C., Andersson, S.G.E., Rossetti, O.L., Grau, O., and Ugalde, R.A., Gene discovery through genomic sequencing of *Brucella abortus. Infect. Immun.*, 69, 865–868, 2001.

Simpson, A.J., Reinach, F.C., Arruda, P., Abreu, F.A., Acencio, M., Alvarenga, R., Alves, L.M., Araya, J.E., Baia, G.S., Baptista, C.S., Barros, M.H., Bonaccorsi, E.D., Bordin, S., Bove, J.M., Briones, M.R., Bueno, M.R., Camargo, A.A., Camargo, L.E., Carraro, D.M., Carrer, H., Colauto, N.B., Colombo, C., Costa, F.F., Costa, M.C., Costa-Neto, C.M., Coutinho, L.L., Cristofani, M., Dias-Neto, E., Docena, C., El-Dorry, H., Facincani, A.P., Ferreira, A.J., Ferreira, V.C., Ferro, J.A., Fraga, J.S., Franca, S.C., Franco, M.C., Frohme, M., Furlan, L.R., Garnier, M., Goldman, G.H., Goldman, M.H., Gomes, S.L., Gruber, A., Ho, P.L., Hoheisel, J.D., Junqueira, M.L., Kemper, E.L., Kitajima, J.P., and Marino, C.L., The genome sequence of the plant pathogen *Xylella fastidiosa,* The *Xylella fastidiosa* Consortium of the Organization for Nucleotide Sequencing and Analysis, *Nature*, 406, 151–157, 2000.

Smid, E.J., Jansen, A.H.J., and Gorris, L.G.M., Detection of *Erwinia carotovora* subsp. *atroseptica* and *Erwinia chrysanthemi* in potato tubers using polymerase chain reaction, *Plant Pathol.*, 44, 1058–1069, 1995.

Starr, M.P. and Chatterjee, A.K., The genus *Erwinia*: enterobacteria pathogenic to plants and animals, *Annu. Rev. Microbiol.*, 26, 389–426, 1972.

Straus, D. and Ausubel, F.M., Genomic subtraction for cloning DNA corresponding to deletion mutations. *Proc. Natl. Acad. Sci. USA*, 87, 1889–1893, 1990.

Toth, I.K., Avrova, A.O., and Hyman, L.J., Rapid identification and differentiation of the soft rot erwinias by 16S-23S intergenic transcribed spacer-PCR and restriction fragment length polymorphism analyses, *Appl. Environ. Microbiol.*, 67, 4070–4076, 2001.

Toth, I.K., Bell, K.S., Holeva, M.C., and Birch, P.R., Soft rot erwiniae: from genes to genomes, *Mol. Plant Pathol.*, 4, 17–30, 2003.

Toth, I.K., Bertheau, Y., Hyman, L.J., Laplaze, L., Lopez, M.M., McNicol, J., Niepold, F., Persson, P., Salmond, G.P., Sletten, A., van Der Wolf, J.M., and Perombelon, M.C., 1999b. Evaluation of phenotypic and molecular typing techniques for determining diversity in *Erwinia* carotovora subspp. atroseptica, *J. Appl. Microbiol.,* 87, 770–781, 1999a.

Toth, I.K., Hyman, L.J., and Wood, J.R., A one step PCR-based method for the detection of economically important soft rot *Erwinia* species on micropropagated potato plants, *J. Appl. Microbiol.*, 87, 158–166, 1999b.

Van Sluys, M.A., de Oliveira, M.C., Monteiro-Vitorello, C.B., Miyaki, C.Y., Furlan, L.R., Camargo, L.E., da Silva, A.C., Moon, D.H., Takita, M.A., Lemos, E.G., Machado, M.A., Ferro, M.I., da Silva, F.R., Goldman, M.H., Goldman, G.H., Lemos, M.V., El-Dorry, H., Tsai, S.M., Carrer, H., Carraro, D.M., de Oliveira, R.C., Nunes, L.R., Siqueira, W.J., Coutinho, L.L., Kimura, E.T., Ferro, E.S., Harakava, R., Kuramae, E.E., Marino, C.L., Giglioti, E., Abreu, I.L., Alves, L.M., de Amaral, A.M., Baia, G.S., Blanco, S.R., Brito, M.S., Cannavan, F.S., Celestino, A.V., da Cunha, A.F., Fenille, R.C., Ferro, J.A., Formighieri, E.F., Kishi, L.T., Leoni, S.G., Oliveira, A.R., Rosa, V.E., Jr., Sassaki, F.T., Sena, J.A., de Souza, A.A., Truffi, D., Tsukumo, F., Yanai, G.M., Zaros, L.G., Civerolo, E.L., Simpson, A.J., Almeida, N.F., Jr., Setubal, J.C., and Kitajima, J.P., Comparative analyses of the complete genome sequences of Pierce's disease and citrus variegated chlorosis strains of *Xylella fastidiosa, J. Bacteriol.*, 185, 1018–1026, 2003.

Vandamme, P., Pot, B., Gillis, M., De Vos, P., Kersters, K., and Swings, J., Polyphasic taxonomy, a consensus approach to bacterial systematics, *Microbiol. Rev.*, 60, 407–438, 1996.

Waldee, E.L., Comparative studies of some peritrichous phytopathogenic bacteria, *Iowa State College J. Sci.*, 19, 435–484, 1945.

Ward, L.J. and De Boer, S.H., Specific detection of *Erwinia carotovora* subsp. *atroseptica* with a digoxigenin-labeled DNA probe, *Phytopathology*, 84, 180–186, 1994.

Yahiaoui-Zaidi, R., Jouan, B., and Andrivon, D., Biochemical and molecular diversity among *Erwinia* isolates from potato in Algeria, *Plant Pathol.*, 52, 28–40, 2003.

2 Role of Cuticles in Produce Quality and Preservation

Gregory M. Glenn, Bor-Sen Chiou, Syed H. Imam, Delilah F. Wood, and William J. Orts
United States Department of Agriculture, Agricultural Research Service, Western Regional Research Center, Albany, CA

CONTENTS

2.1 INTRODUCTION

The best ways to extend the shelf life of produce postharvest include reducing desiccation, lowering the rate of senescence and maturation, and reducing the rate of microbial infection [1]. The cuticle or the cuticular membrane plays an integral part in extending the shelf life of many different kinds of produce. The cuticle forms a continuous extracellular membrane over the epidermal cells of most aerial plant parts, including leaves and fruits [2–5]. The primary function of the cuticle is to minimize water loss [6] by mediating the wettability of the tissue surface and moisture vapor permeability [7]. Another important function of the cuticle is to prevent the loss of plant solutes through leaching [8,9]. In recent years, the cuticle has been found to possess other functions as well. It provides the first line of defense against pathogen invasion [10], acts as a shield against mechanical impact, facilitates the efficient exchange of gases, provides some protection from exposure to pesticide and fertilizer chemicals, and reduces damage from solar irradiation [11] and herbivores [6,7,12]. Some have viewed the cuticle as inert, nondynamic plant tissue because it is extracellular. In reality, the cuticle may constantly change during the lifespan of a particular fruit or vegetable.

This review chapter covers the current knowledge of plant cuticle properties and functions as well as how these might relate to the quality and preservation of produce. It is not our intent to provide a comprehensive survey of all commercial produce. For instance, produce from roots (e.g., carrots) or tubers (e.g., potato) are not discussed here because their protective skins generally have different anatomical and structural features. Rather, we primarily focus on a few representative examples that illustrate the cuticle's important role on produce quality and preservation.

2.2 GENERAL STRUCTURE AND COMPOSITION OF PLANT CUTICLES

Various microscopy techniques have been developed that provide valuable information on the structure and composition of the cuticle. These techniques include light microscopy used in conjunction with different staining methods, scanning electron microscopy (SEM), and transmission electron microscopy (TEM). More recently, confocal laser scanning microscopy techniques have been used to investigate the epicuticular wax layer [13,14] as well as to localize phenolic and flavonoid compounds in plant cuticles.

The cuticle is not simply a homogenous membrane that covers the epidermal cell layer in plants [15,16], but rather it contains a layered structure [2]. In addition, the cuticle structure and composition varies among plants and among different organs of the same plant and changes as the plant tissue grows and matures [4,15]. In one study, Bird and Gray [15] described the changes in the cuticle of developing leaf tissue. The young, juvenile tissue is covered with a highly water-repellent wax layer they referred to as the procuticle. As the leaf develops, the cuticle thickens and adds to the wax layer. The layers of cutin and cell wall polysaccharides then form to produce a lamellar structure. This lamellar structure is referred to as the cuticle proper. As the leaf tissue continues to develop, an epicuticular wax layer forms on the cuticle surface. A secondary cell wall forms beneath the primary cell wall and the primary cell wall subsequently becomes incorporated into the cuticle structure. The final stage in the development of the cuticle occurs when two layers of cutin form just outside of the secondary cell wall [4,15].

2.2.1 CUTICLE MODEL

A simplified model of the cuticle is shown in Figure 2.1. The cuticular membrane (CM) is attached to the epidermal cell wall structure (EC) by pectinaceous materials (P). The epidermal cells contain a cell membrane (ECM) and a cell wall (CW). The pectinaceous materials (P), found in the middle lamellar region, bind adjacent cell walls together as well as the cuticle to the epidermal cell wall. The cuticle or cuticular membrane (CM) consists of a lamellar structure composed primarily of cutin, but it may also contain carbohydrate polymers (LR) [12,15]. The cuticle may be isolated from the underlying tissue using pectinases. The cuticular components are produced in the epidermal cell layer and transported to the cuticular structure by an unknown mechanism [12]. Jeffree [4] reported that two cutin layers, the internal and external cuticular layers, were deposited below the cuticle proper. Bally [17] used the term *cuticle proper* to describe the inner and outer cutin layers.

The cutin component of the cuticle forms a porous, three-dimensional structure that is embedded with waxes [10,15]. Waxes that infiltrate the cutin structure also form a layer on the cuticle surface [4], called the epicuticular wax layer (EW). In some plants, additional crystalline wax structures develop on the epicuticular wax layer [4]. As the plant tissue matures, the adaxial region of the primary epidermal cell wall may become incorporated into the cuticle structure and may be replaced by a more fibrous secondary cell wall [4]. Occasionally, the cuticle and epidermal cell layers, with perhaps one or more underlying cell layers, are collectively referred to as the *skin*.

2.2.2 CUTICLE COMPOSITION

2.2.2.1 Cutin

Cutin is the major component of the cuticle and is a biopolymer that is insoluble in organic solvents [12]. Heredia [2] reported that cutin might constitute 40 to 80% of the entire mass of the cuticle. Cutin forms a three-dimensional, porous polymer structure that provides mechanical strength. There is, however, little information on chemical cross-linking or other structural information about cutin [18,19] because of the insolubility of cutin [10]. However, a portion of the cutin fraction may dissolve in strong alkali, leaving an insoluble residue refered to as *cutan* [4]. Cutan is thought to consist of cross-linked cutin and certain wax constituents. Researchers have used chemical depolymerization with $LiAlH_4$, hydrolysis with alcoholic KOH or HCl, transesterification with methanolic BF_3, and digestion with cutinases as tools to investigate the structure of cutin. These degradative approaches have established that cutin is a polyester derived primarily from esterified, hydroxylated, and epoxy hydroxylated $C_{16:1}$ and $C_{18:1}$ fatty acids [10,18].

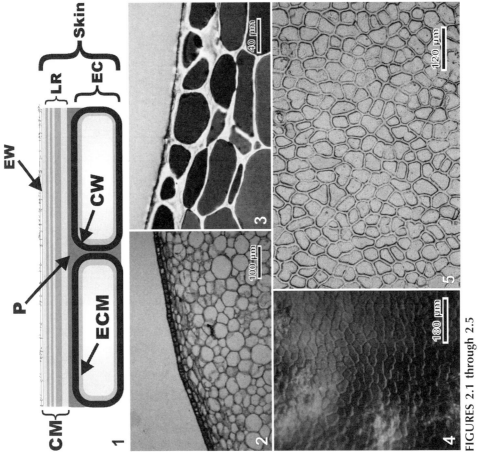

FIGURES 2.1 through 2.5

FIGURE 2.1 Model of cuticle structure in produce. The epidermal cell layer (EC) consists of epidermal cells. The cell contents are contained within the cell by the cell membrane (ECM) and cell wall (CW). The cuticular membrane (CM) consists of a lamellar region (LR) composed of a cutin-rich matrix and some cellulose and pectin material. The cutin matrix is infiltrated with waxes that may be extruded on the cuticle surface to form an epicuticular wax layer (EW). Some crystalline wax structures may also form on the cuticle surface. The cuticular membrane adheres to the epidermal cell wall by pectins (P). The cuticular membrane, pectin region, and epidermal cell layer taken together form the bulk of the skin tissue.

FIGURE 2.2 Photomicrograph of a cross-sectional view of cherry fruit stained for carbohydrates using periodic acid-Schiff's reagent. Note the carbohydrate-rich cell wall, especially in the epidermal layer.

FIGURE 2.3 Photomicrograph of a cross-sectional view of cherry fruit stained for lipids using sudan black. Note the dark staining of the cuticular membrane on the fruit surface.

FIGURE 2.4 Photomicrograph of the surface of a cherry fruit. Note the reticulate pattern of epicuticular wax on the cuticle surface. The structure was most prominent on the stylar end of the fruit and varied in quantity from one fruit to another.

FIGURE 2.5 Photomicrograph of the cuticular membrane of cherry fruit isolated using enzymes. Note the reticulate pattern of epicuticular wax similar to that shown in Figure 2.4.

2.2.2.2 Waxes

In contrast to cutin, cuticular wax is soluble in solvent and is made of a heterogeneous mixture of lipids that vary considerably among different plants. Cuticular waxes are a complex mixture of alcohols, alkanes, aldehydes, ketones, and esters made from long-chain fatty acids [4,18]. These $C_{16:1}$ and $C_{18:1}$ fatty acids are first produced in cell plastids. The fatty acids then form complex mixtures of long-chain aliphatics outside the plastids. The mixtures contain alkanes, alcohols, aldehydes, fatty acids, and esters with varying amounts of triterpenoids and hydroxycinnamic acid derivatives [18,20]. The epidermal cells then exude the waxes, which traverse their way through the cutin matrix. Upon reaching the cuticle surface, the waxes deposit as a thin, amorphous layer known as the *epicuticular wax layer*. In certain conditions, wax crystals may also form on the epicuticular layer. The structure and shape of the wax crystals depend on the wax composition [21], which is influenced by environmental and developmental factors as well as foliar applied chemicals [4,22–24].

2.3 CUTICULAR STRUCTURE AND COMPOSITION IN PRODUCE

The cuticle is only a very thin layer on the plant surface, but it represents a significant mass because it covers a large cumulative surface area of plants. In fact, Riederer [25] estimated that, on average, the total mass of cuticle produced per hectare ranged from 180 to 1,500 kg. Also, the cuticle is not simply an inert, innocuous layer on the tissue surface, but it changes considerably during the plant's development cycle. Moreover, it is affected by genetic variation, environmental factors, and chemicals.

2.3.1 Cuticular Growth and Development

The cuticle forms very early in the development of produce and plays a significant role in proper morphological development. Sieber et al. [12] noted that when the cuticle was removed or missing in developing tissue, the adjoining tissue often fused together and created deformities. The development of the cuticle often parallels the growth and development of the produce it protects. In leafy vegetables, the development of the cuticle is similar to that described previously for leaves and will not be discussed in any further detail [15]. The cuticle development in bulky, fleshy fruits is also very similar to that of leaves in the initial stages, but the cuticle continues to develop as the fruits grow and mature. A well-documented example of this is the apple [*Malus domestica* Borkh.].

Early in its development, apple fruit is protected by a thin, waxy cuticle, comparable to the procuticle structure described by Bird and Gray [15]. Eight weeks after full bloom, the cuticle grows to a thickness of approximately 6 µm. The surface remains smooth and has no distinct epicuticular wax formations. By 12 weeks, however, small epicuticular wax structures develop on the cuticle surface and subsequently increase throughout the remainder of the growing season [26]. The thickness of the cuticle increases from approximately 6 µm at 8 weeks to approximately 22 µm at 20 weeks after full bloom [26]. Even after harvest and cold storage at 4°C, the cuticle continues to grow, albeit at a reduced rate [26]. In addition, postharvest apples show a considerable increase in the amount of epicuticular wax compared to preharvest apples.

Ju and Bramlage [27] further documented changes in wax content of apple fruit cuticle during fruit ripening and storage. They observed that all wax components increased during fruit ripening, with free fatty acids and alcohols contributing the most to the increase. These waxy products formed a greasy film postclimacteric in apple as well as pear fruit [28]. These changes appear to be triggered by the climacteric rise in respiration. Experiments with Ethephon and AVG have provided evidence supporting this claim. Ethephon treatments have been shown to accelerate the climacteric rise in respiration and accelerate the changes in the wax components of the cuticle. On the other hand, AVG treatments have been shown to suppress the climacteric rise in respiration and suppress the changes in cuticular waxes [28].

2.3.2 VARIABILITY IN CUTICLE STRUCTURE AND COMPOSITION

The expression of different genes within the plant genome can cause some variability in the cuticle structure and composition on a specific plant. For instance, the cuticle on leaves of apple trees is very different from the cuticle on the fruit's surface. The mass of the fruit cuticle may be as much as $2,000 \mu g \times cm^{-2}$, whereas the mass of leaf cuticle may range from 450 to $800 \mu g \times cm^{-2}$ [2]. Natural variation in the cuticle structure can also occur in different regions of the produce itself. Farag [29] observed that in ripening cranberry fruit, the wax accumulates more in the calyx region than in other areas. Variations in the cuticle structure also occur among plants with very similar genomes. For example, Veraverbeke et al. [30] found differences in the thickness of the epicuticular wax layer among three different apple cultivars.

Variations in environmental conditions can also affect cuticle structure and development, especially the epicuticular wax layer. In some produce, crystalline wax structures are extruded to the external surface of the cuticle, giving fruits their characteristic waxy bloom [31]. For instance, the presence or absence of bloom on various regions of the apple's cuticle has been attributed to variations in the amount of incident radiation exposure. The epicuticular wax layer can partially protect the plant tissue from visible and infrared radiation by reflecting and/or refracting incident radiation [32,33]. Consequently, the shaded regions seldom have as much epicuticular wax as regions exposed to sunlight [32,33]. In addition, differences in the amount of bloom could also be influenced by variations in other microclimatic conditions [31,32].

Veraverbeke et al. [30] studied the effect of controlled atmosphere storage on cuticle development. They stored apples in controlled atmosphere storage for different time periods before removing them and determining the wax composition. They found that apples held in controlled atmosphere storage for longer periods exhibited a greater change in wax composition once they were brought out of storage and acclimated at room temperature. The alkane and ester fractions showed the largest amount of change. Using confocal laser scanning microscopy, Ververbeke et al. [30] also studied the epicuticular layer of the apple cuticle. They noted that apples stored in a controlled atmosphere environment developed cracks in the cuticle structure. The authors suggested that the cuticular cracks could have developed in response to the low oxygen level in the controlled environment. The cracks could have improved gas exchange and facilitated gas diffusion to oxygen-depleted tissue within the fruit. Once removed from storage, the fruit became less oxygen deprived but became exposed to higher temperatures and lower relative humidity, conditions that favor moisture loss. The fruit adapted by filling in the cuticular cracks to form a smooth wax layer that provided a better water vapor barrier. These results illustrate that at least in some produce, the cuticle is a dynamic structure that can adapt to the prevailing environmental conditions.

As mentioned previously, the cuticle provides a barrier against the penetration of chemicals [8]. Although chemicals applied to the produce may have beneficial effects, they may also have unintended effects. For instance, Roy et al. [34] studied the effects of calcium treatments on apple fruit. Fruit with higher calcium content were firmer and had lower respiration rates [34,35]. The authors found that the calcium content in the flesh depended on the surfactant used in the dipping solution. The most effective surfactants were those that altered the epicuticular wax [34]. However, it was not clear whether the surfactant or the calcium was directly responsible for the observed effects. The surfactant may have simply facilitated calcium penetration, which then caused the cuticular changes.

2.4 BARRIER PROPERTIES OF THE CUTICLE

2.4.1 BARRIER TO LIQUIDS AND GASES

The cuticle provides an effective barrier between the produce surface and the environment. Cuticle penetration is important for improving nutrient uptake as well as uptake of herbicides, fungicides, and other foliarly applied materials [6,9,36]. Indirectly, the cuticle affects produce quality by restricting the passage of beneficial chemicals applied on the cuticle surface while at the same time protecting the produce from chemical injury [37]. In addition, the barrier properties of the cuticle directly affect produce quality by minimizing desiccation, influencing gas exchange, and preventing the leaching of plant solutes [1,12]. Schotsmans [38] studied different methods of determining efflux of biological gases and found gases to be somewhat regulated by the fruit's skin. Banks [39] showed that the cuticle of pepper was about 10 times more permeable to CO_2 than to O_2. Amarante and Banks [40,41] found that carnauba-based wax coatings reduced the permeance of water vapor, O_2, and CO_2 in pears.

The barrier properties of the cuticle are characterized by cuticle thickness, composition, and structure. The cuticle layer on produce can be very thin (< 0.1 µm), such as those found on leafy vegetables, or can be quite thick (> 30 µm), such as those found on mature apple fruit [37,42]. Early in the development of a particular fruit or leafy vegetable, the tissue is covered with a thin cuticle containing waxes that provide an effective vapor barrier. The wax component rather than the cutin component provides the greatest moisture resistance [4]. The cutin matrix is thought to form a three-dimensional, porous structure that contributes little to the hydrophobic properties of the cuticle, even though the cutin polymer itself is rather hydrophobic [4]. These claims are supported by studies comparing moisture permeation rates of intact cuticle to cuticle with the epicuticular waxes removed by solvents. Moreover, a study using a sorghum mutant with a much thinner epicuticular wax layer than normal showed significantly higher water vapor permeability [43].

The water vapor permeation rate of the plant cuticle is important in controlling moisture loss and, indirectly, in modulating the supply of minerals, such as calcium, to the plant via the xylem [44]. High transpiration rates coincide with higher calcium levels in plant tissue [45]. In areas with high humidity, transpiration rates may be too low to provide adequate calcium to the produce, resulting in various physiological disorders [34,44,45]. Consequently, calcium supplementation may be necessary to control physiological disorders and sustain critical levels. The most effective method of achieving higher calcium levels may be to spray the surface of the produce with a dilute calcium solution [46]. Calcium sprays must come in direct contact with the deficient plant tissue since little, if any, calcium can translocate to deficient tissue [47–49]. In apples, several seasonal sprays of calcium may be needed to significantly increase the calcium content of the fruit [50–53].

One of the first requirements for cuticular penetration of spray solutions involves adequate wetting of the cuticle surface. The epicuticular wax layer is difficult to wet with aqueous solutions. Spray droplets tend to bead up and simply fall from the cuticle surface. Wetting difficulties may be compounded by the presence of crystalline wax structures that can catch spray droplets and prevent adequate contact with the plant surface. Wetting difficulties may also occur on plant surfaces with a dense distribution of trichomes or epidermal hairs [54]. To mitigate the wetting difficulties, surfactants can be added to the spray solution [55]. The surfactants lower the surface tension of spray materials and facilitate wetting of the cuticle surface. In addition, a study showed that the cuticular permeability increased even when the cuticle surface was fully saturated with a solution containing surfactants [56].

Materials can directly permeate through the cuticle itself. However, the cuticle also contains natural pathways such as stomates, lenticels, and cracks that may facilitate permeation. Glenn et al. [48] studied the cuticular pathways of calcium through enzyme-isolated apple cuticle. The cuticle from apple fruit harvested approximately 8 weeks after full bloom had a dense distribution of stoma on the fruit surface, especially at the calyx end. By 12 weeks, the stoma appeared distorted and open, and by 16 weeks the stoma had mostly developed into lenticels. Schlegel and Schonherr [50] studied the penetration of $^{45}CaCl_2$ through the cuticle surface of apple sections and found that the penetration rate depended on the stage of fruit development. During early stages of fruit development, the penetration reached its highest rate as essentially all of the calcium penetrated within 24 hours. The penetration rate dropped markedly when stomates turned to lenticels and trichomes dropped off. However, preferential sites for penetration continued to be lenticels, stomates, and trichomes.

Another study examined the different penetration pathways through the cuticle by using enzymatically isolated cuticles mounted on molten agar containing oxalic acid [26]. After the sample was cooled and became solid, a calcium solution was applied to the cuticle surface. As the calcium penetrated the cuticle, it came in contact with the agar containing oxalic acid and formed calcium oxylate crystals that were easily identified by polarized light microscopy. The results indicated that the lenticels were important pathways of calcium penetration. Further support for these conclusions involved experiments comparing calcium flux through cuticles with or without lenticels exposed to a calcium donor solution. The cuticles were mounted on a diffusion cell and the calcium flux was determined to be several times higher for the cuticles containing exposed lenticels.

Other long-season fruits also have pathways of penetration not unlike those of apples. Dietz [57] studied the structure and function of cuticles and lenticels in mango cultivars. The cuticle thickness ranged from 15.8 to 22.4 μm. In addition, epicuticular wax structures developed on the cuticle surface and it differed among the cultivars studied. Also, the number of lenticels ranged from 19 to 35 lenticels per square centimeter. Moreover, the stomates in young fruits were functional up to 21 days in some varieties and 36 days in others.

Cracks in the epicuticular layer provide another preferential pathway of cuticle penetration. Cuticular cracks have been reported in apple fruit near maturity and have been shown to increase cuticular penetration of calcium [48]. Maguire et al. [58] studied cuticular cracking in apple fruit using scanning electron, confocal, and light microscopy. The cracks observed on the cuticle surface formed a reticulate network and did not appear to completely traverse the cuticle. Also, the cracks did not appear to coincide with the blush area of the fruit. Fruit with cuticular cracks had greater water vapor permeance than fruit without any cracks. The results indicate that cuticular cracks play a major role in the water vapor permeance of apple fruit. The importance of cuticular cracks on other aspects of fruit quality is discussed in further detail elsewhere in this chapter.

Another pathway of cuticular penetration may be through small pores that traverse the cuticle. There has been some debate on whether cuticular pores really exist and whether they can facilitate cuticular penetration [25,59]. One model of the cuticle membrane describes the cuticle as a porous membrane with pores that vary in size depending on the degree of hydration [6]. Support for this model comes from data showing greater cuticular penetration of herbicides at higher relative humidity [37]. Harker and Ferguson [56] used a diffusion cell to determine cuticle permeability using enzyme-isolated cuticles. They found that cuticular permeability might be altered by using different aqueous chemical solutions. Luque et al. [60,61] studied the water permeability of isolated tomato cuticles and found that the homoionic form of the cuticle and pH affected water permeability. They also observed the cuticle swell in the presence of water and noted that permeability depended on the water content of the cuticle. Some claim that pores present in the cuticle have polar regions composed of cutin and weakly polar regions due to hydroxyl groups, unesterified carboxyl groups, acidic pectin, or cellulosic moieties. Riederer and Schreiber [25,62] reported that the bulk of the water passes through the cuticle via diffusion rather than via pathways of mass flow or polar pores present in the cutin matrix.

Cuticular penetration through stomates, lenticels, cracks, or other structures may provide sites through which mass flow of solution can occur [26,52]. Maguire et al. [63] noted that lenticular transpiration accounted for 8 to 20% of the total flow in apple fruit. However, the bulk of the cuticular penetration probably occurs by diffusion through the cuticle membrane itself, even though it is quite impermeable

[62]. Veraverbeke [30] measured water diffusion coefficients of apple flesh, cutin, and wax. Wax had the lowest values at 0.192 to 3.03×10^{-14} m^2 sec^{-1}, followed by cutin at 4.03 to 7.16×10^{-14} m^2 sec^{-1} and apple tissue at 433 to $1,120 \times 10^{-14}$ m^2 sec^{-1}. Although the diffusion of water per unit area of cuticle is very low, diffusion through the cuticle is still an important mode of penetration since the cuticle encompasses a large surface area compared to structures such as lenticels [26]. For instance, Maguire et al. [63] reported that 5- to 10-fold more water is lost by diffusion through the cuticle than through lenticels. The rate of cuticular diffusion can also be affected by environmental conditions. Chamel [37] found greater cuticular penetration of herbicides at higher relative humidity and temperature.

2.4.2 BARRIER TO PATHOGENS

Plant cells provide a substrate for the growth of number of fungi and bacteria. The cuticle can protect plants from pathogen infestation and can act as a formidable physical barrier to many microorganisms. For instance, fruit mummification in blueberries occurs when the pathogen *Monilinia vaccinii-corymbosi* infects the plant. The pathogen cannot penetrate the cuticular layer on the fruit, but it can infect the fruit by attacking the flower parts instead [64]. Conidia produced on blighted shoots are transported to the floral stigmas by wind, rain, or insects. The conidia germinate on the stigma and the hyphae grow within the stylar canal until it reaches the base of the style. The fungal mycelium then invades and begins to colonize the ovary. This colonization lasts throughout the growing season [64].

Some pathogens are able to breach the cuticle and infect the produce. The typical infection process for pathogenic fungi begins when spores attach to the cuticle surface. Fungal spores generally must first adhere to the cuticular surface of produce before they can penetrate it [65]. The pathogen adheres to the cuticle surface via intermolecular forces until a bulblike structure called the appressorium begins to develop in close contact with the cuticle surface and then provides additional bonding strength. From the appressorium, a small growing point called the penetration peg develops at the cuticle surface. As the penetration peg grows, it pierces the cuticle and epidermal cell wall and commences the infection process [65]. However, the penetration peg may not be able to pierce the cuticle if the penetration force required exceeds the adhesive strength of the pathogen to the cuticle surface. Thus, produce with a relatively thick or otherwise difficult-to-penetrate cuticle have greater resistance to infection.

Pathogens commonly secrete cutinases and other enzymes at the penetration site that may soften the cuticle and facilitate spore adhesion and penetration [66]. There are no known pathogens that can produce enzymes to degrade epicuticular waxes [65]. However, phytopathogenic fungi produce cutinases that can soften and partially degrade the cutin layer that lies beneath the epicuticular wax layer [67].

Invading pathogens apparently penetrate the epicuticular wax layer solely by mechanical force (e.g., penetration peg). Once the pathogens gain access to the cutin layers through the penetration site, they secrete cutinases. The cutinases soften the cuticle and facilitate further penetration by the penetration peg. It should be noted that the role cutinase plays in the infection process remains controversial, since some fungal mutants that lack cutinase can still infect produce [68,69]. Cruickshank et al. [70] noted that the primary mode of cuticular penetration in tomato fruit is mechanical and the cutinases may only play a minor role. In their study on fungal infection of tomatoes, they treated the epicuticular wax layer with solvents or abraded the cuticle surface before inoculating the fruit with *Colletotrichum gloeosporioides*. There appeared to be a diffusion of solutes from the underlying cell layers that stimulated the fungal growth. Infection occurred as penetration pegs eroded the cuticle, thus allowing for further infection. The authors noted that as the tomato ripened and the fruit skin turned yellow, the cuticle became resistant to cutinases. However, the penetration peg was still able to pierce the cuticle of the fruit in spite of the cutinase resistance. They concluded that cutinases might not be required for fungal infection of tomato fruit [70].

Once the penetration peg has completely penetrated the cuticle, a fine hyphal tube that spans the cuticular membrane starts to increase in diameter until it reaches its mature size and secretes enzymes into the host tissue. In diseases such as soft rot, enzymes seem to be by far the most important substances involved in spreading disease. The fungi secrete one or more sets of enzymes that collapse and break down the middle lamella and cell walls of parenchymatous tissue. Some pathogens can infect produce and remain dormant for a time before infection resumes. For instance, Prusky [71] found that Alternaria Rot in persimmon fruits stemmed from necrotic leaf spots found in the orchard. The germinating conidia penetrated the fruit cuticle directly. Subsequently, the hyphae developed intercellularly and produced dark, quiescent infections that had renewed growth during storage.

Many examples in literature illustrate that the cuticle is a formidable barrier to fungal invasion. Biles [72] found that peppers with thicker cuticle layers had greater fungal resistance. Jenks et al. [43] found that mutant sorghum with a reduced cuticle thickness had a higher susceptibility to fungal pathogens. Spotts et al. [73] studied fungal infection in pear fruit and noted that in most cases fungi typically infest pear by entering the fruit through a wound in the epidermis or stem. Pear cuticle is particularly difficult for fungi to penetrate. The infection can occur during harvest or when the fruit is exposed to contaminated water in dump tanks or in flumes used in packinghouses. Wounded fruit can get mixed with good fruit in storage and may ultimately infect other fruit.

2.5 ROLE OF CUTICLE IN HOST/PATHOGEN RESPONSE

Some produce, especially climacteric fruits, undergoes changes during ripening that make them more susceptible to fungal infection. Fungal infection may be initiated prematurely and then arrested until the host reaches the proper stage of maturity. One such illustration involves avocado fruit inoculated with *Colletotrichum gloeosporioides* at various stages of fruit ripeness [74]. In unripe fruit, the fungi produced an appressorium and a penetration peg, which only penetrated the epicuticular wax layer 4 days after inoculation. However, in ripe fruit, the penetration pegs had penetrated subcuticular tissue within 48 hours of inoculation. The fungi arrested further growth until the climacteric rise in respiration occurred. The resumption of fungi growth resulted in an enlargement of the penetration peg and a rapid deterioration of cells in the host plant. Such dormant periods of fungal growth prior to full infection, which coincided with the climacteric rise in respiration, indicated an orchestrated plant–pathogen interaction [74].

Mounting evidence indicates that the cuticle provides a store of molecules that enables the recognition of plant surfaces. As mentioned previously, cutin is a biopolymer in which C_{16} and C_{18} fatty acids are cross-linked between carboxy and ω-hydroxy groups. More recent results indicate that the ω-hydroxy fatty acids play a role in plant–pathogen interactions in addition to their role as a protective cutin layer [68]. Virulent fungi spores secrete cutinases, which results in the production of cutin monomers. The cutin monomers are known to enhance the transcription of cutinase genes in fungi as well as to stimulate the formation of the appressorium and penetration peg, ultimately leading to infection of the host plant [68,75]. In addition, the constituents of plant surface waxes can also trigger further fungal growth [75]. Controversy still exists over the importance of cutinases in fungal infection, since some fungal mutants that are unable to produce cutinases are still pathogenic. However, all the evidence taken together indicates that the plant cuticle constituents act as signals for the recognition of plant surfaces by fungal pathogens.

There is also growing evidence that cutinases elicit a defense response in the host plant [76]. Parker [76] found that fungal cutinase and other lipid esterases could protect bean leaves from disease. The lipid esterases seemed to be the most effective in inhibiting infection. However, the protective mechanism of the esterase activity was not known. Kim et al. [77] studied the interaction of pepper (*Capsicum annuum*) with the anthractnose fungus (*Colletotrichum gloeosporioides*). They found that a pepper esterase gene became highly expressed during fungal infection. The authors cloned the pepper esterase gene, expressed it in *E. coli*, and then isolated it. They inoculated the pepper with the athractose fungi amended with the pepper esterase protein. The results showed no infection from the inoculum. Furthermore, the pepper esterase was shown to inhibit the formation of the fungal appressorium by modulating the cAMP-dependent signaling pathway. The authors concluded that pepper esterase activity could provide a defense mechanism against fungal attack. In another study, Schweizer et al. [78] investigated the chemical synthesis of the major cutin monomers of barley leaves and tested the molecules to see whether they could induce defense-related genes. They showed that an mRNA induced by pathogen attack could also be induced by a particular cutin monomer. In addition, they showed that topical sprays of cutin monomers on barley and rice leaves provided resistance against *Erysiphe graminis f.* sp. *Hordei* and *M. grisea*, respectively. Since the cutin monomers exhibited no fungicidal effect, Schweizer et al. [78] concluded that the cutin monomers acted as signal molecules for the induction of disease resistance in barley.

The production of H_2O_2 is associated with the pathogen defense system in plants [79]. Fauth et al. [79] noted that fungi are able to secrete cutinases during the initial stages of fungal attack that most likely results in the formation of cutin monomers. To counter these attacks, plants have two early defense mechanisms; these include accumulation of phenolic compounds in the epidermal cell walls and apoplastic chitinase [79,80]. Fauth et al. [79] showed that fungal attack on cucumber hypocotyls treated with cutin monomers induced accumulation of cell wall phenolics that possibly required H_2O_2 production for polymerization. Accumulation of H_2O_2

just below the fungal appressoria has been demonstrated in barley [81]. Fauth et al. [79] also showed that cutin monomers made from alkaline treatments were effective in eliciting H_2O_2 in cucumber hypocotyl segments. In addition, Fauth et al. [79] found that cutin monomers enhance the activity of other H_2O_2 elicitors.

2.6 ROLE OF CUTICLE ON MECHANICAL PROPERTIES RELATED TO QUALITY

The mechanical properties of produce affect their postharvest storage life. Compression and impact stress can damage tissue, which causes cell contents to leak internally and results in discoloration and bruising. The compression and impact resistance of most produce is largely determined by the strength of the epidermal, subepidermal, and parenchyma cells that compose much of the edible tissue. Many vegetable crops can be handled, stacked several layers deep in boxes or bins, and shipped without incurring significant compression or impact damage. However, packing, storing, and shipping fruits that become soft and fleshy as they ripen becomes more difficult. Consequently, much of the produce that softens during ripening is harvested prematurely when it can still withstand some degree of impact and compressive stress. This is especially common for climacteric fruit such as tomatoes, peaches, and pears that can ripen postharvest.

The primary protection that the cuticle provides for produce other than compression and impact resistance may be resistance to damage from abrasion. Abrasion resistance reduces the chances that the produce will develop open wounds that provide sites for pathogen infestation. For produce that is soft and fleshy, the cuticle may play an even greater role in providing mechanical protection. In addition to providing abrasion resistance, the cuticle also provides support and resistance against puncture wounds.

There have been only a few limited studies on the mechanical properties of plant cuticles. Glenn and Johnston [82] studied the mechanical properties of wheat bran, which includes a thin cuticle, several layers of periderm, and an aluerone layer. The mechanical properties of wheat bran are important in obtaining good flour yield in milling operations. Typically, the grain is conditioned to a desirable moisture level to soften the endosperm and facilitate milling. A second application of moisture is added approximately 30 minutes prior to milling to hydrate the bran tissue and make it flexible and impact-resistant so that it remains relatively intact after the endosperm tissue is stripped away. Glenn and Johnston [82] studied the tensile properties of bran tissue as a function of moisture content. They showed that the elongation to break markedly increased at higher bran moisture content. Consequently, the bran tissue is more compatible with milling operations after moisture treatment.

Thompson [83] conducted a study on the rheological properties of tomato fruit skin. The study focused on understanding the cell wall properties as well as the biophysical and biochemical processes involved in fruit growth. However, the author did not examine the relationship between mechanical properties and postharvest quality or preservation. In another study on tomato fruit, Petracek and Bukovac [80] measured the mechanical properties of enzymatically isolated cuticle. Based on tensile tests, they determined that the cuticle consisted of a viscoelastic polymer network. The hydrated cuticles started to swell and became weaker and fractured more readily, but they also became more elastic, suggesting that water actually plasticized the cuticle. The moisture in tomato fruit cuticle perhaps had a plasticizing effect through the disruption of hydrogen-bonded crosslinks between chains in the polymer matrix [2]. However, it may also be due to hydration of the carbohydrate component of the cuticle, especially the pectic materials that can embrittle when they become dry. The authors also used solvents to remove the wax fraction of the cuticle to expose the cutin matrix. The cutin structure had greater elasticity but fractured more easily compared to intact cuticles [84]. Petracek and Bukovac [84] suggested that the wax acted like filler in the intact cuticle.

Round et al. [9] used atomic force microscopy (AFM) as a nonintrusive method to probe the cuticle surface of tomato fruit and characterize its nanomechanical properties. The authors showed that the surface elastic modulus decreased dramatically with increasing water content. This result was corroborated by solid-state NMR, which showed enhanced local mobility of acyl chain segments with higher water content. In another study, Luque et al. [85] examined the hydration phenomenon in isolated tomato cuticle. They studied water sorption of cuticle equilibrated at a range of relative humidity. The authors used water sorption isotherm and DSC data to conclude that water does not plasticize the tomato fruit cuticle, in contrast to the claims of Petracek and Bukovac [84]. They noted that the water sorption properties reflect the hydrophobic nature of the cuticle.

Ozgen et al. [86] noted that the stage of ripeness of cranberries influences their cuticular properties. In contrast to climacteric fruit, cranberries harvested when ripe had longer storage life than those harvested at an immature stage. The reason was that ripe fruit had a lower respiration rate and a thicker cuticle with a greater puncture resistance than unripe fruit.

2.7 ROLE OF CUTICLE IN AESTHETIC QUALITY

The structure of the epicuticular wax layer is important in determining the finish properties of produce. Produce such as apples may be coated with a film of carnauba wax or shellac to provide a glossy finish that appeals to consumers [87]. However, the films may actually worsen fruit quality. For instance, moisture from condensation or from transpiration may become entrapped in the film coating, causing the film to turn white or hazy and become unattractive. In addition, the natural wax coating from the epicuticular wax layer can either enhance or detract from the aesthetic quality [31,32]. Veraverbeke [88] mentioned that the cuticle could adversely affect quality by producing an unacceptable bloom or a sticky film. The ideal finish on the produce may vary for different commodities. For example, most consumers prefer a glossy finish on apple fruit, but generally prefer a dull finish on green beans [32]. The surface characteristics of a particular type of produce are considered such an important trait that they are often considered when developing new cultivars.

The amount of crystalline wax structure on the epicuticular wax layer has a large effect on the gloss properties of commodities. Ward [32] studied the amount of gloss in eggplants, apples, and mature green tomatoes. Of the three types of produce, the eggplant had the highest gloss readings. It also had the smoothest surface and least amount of epicuticular wax structures protruding from the surface of the cuticle. Tipton and White [89] noted that glossy leaves of the Mexican Redbud had very little epicuticular wax, whereas dull-leaf varieties had considerable amounts. Commodities such as plums have a heavy bloom of epicuticular wax on their surfaces that give the fruit a dull appearance. However, the aesthetic quality of the fruit can be improved dramatically by lightly buffing or polishing the fruit surface to remove the bloom and create a smooth surface [31].

Apart from the epicuticular wax layer, the properties of the epidermal cell layer also affect the amount of gloss on the surface of produce. An uneven epidermal cell layer has a tendency to scatter light and reduce the degree of sheen, even if the cuticle is relatively constant in thickness and has few, if any, epicuticular wax structures on its surface [87].

2.8 ROLE OF CUTICLE ON PHYSIOLOGICAL DISORDERS

2.8.1 CHILLING INJURY

Postharvest physiological disorders that affect the quality of produce are not uncommon. The many types and causes of a physiological disorder can be very specific to a particular fruit or vegetable. One common disorder is chilling injury, which reduces the storage life and quality of many kinds of produce [7]. Chilling injury is thought to stem from the inability of cell membranes to function properly at lower storage temperatures due to a phase transition in the membrane lipids [7]. A higher proportion of unsaturated fatty acids are thought to distinguish chilling resistance from susceptible produce. In one study, Aggarwal [7] enzymatically isolated cuticles from mature fruits and used DSC to examine the effect of storage temperature on phase transitions of the lipid component. The cuticular membrane underwent an endothermic transition attributed to melting of the waxes. Cuticles isolated from fruit stored in refrigeration exhibited a shift in the melting enthalpy of the waxes. Aggarwal [7] speculated that the cuticular changes could be involved in chilling injury of refrigerated fruit. However, further work is needed to support the claim.

Chilling injury in mandarins appears as small depressions or pits in the fruit surface that may cover more than 50% of the fruit's surface. In regions where pitting occurs, the crystalline wax structures normally covering the fruit surface are absent [90]. Other telltale signs of chilling injury include a collapse of the epidermal cells and other cells adjacent to the pit. A question arises as to whether the cuticular irregularities were the causal factor that preceded the chilling injury or whether the injury to the epidermal cell layer occurred first and arrested further cuticular development. Vercher et al. [90] agreed with the latter explanation. They believed the low temperature stress disrupted normal epidermal cell function. The localized injury to the epidermal and subepidermal cells ultimately caused the cells to collapse and form a pit in the fruit surface. This injury impaired the normal development of the crystalline wax structure and increased water permeability in the injured region.

2.8.2 BROWNING

A physiological disorder that occurs in Litchi (*Litchi chinensis* Sonn.) has been indirectly linked to cuticular function [91]. Litchi is a tropical fruit that commands a high price in international markets. The red color in the pericarp tissue makes it a very attractive fruit. After harvest, the pericarp tissue eventually turns brown due to polyphenol oxidase-mediated oxidation of both phenolic compounds and anthocyanins. The browning effect can be delayed by minimizing the moisture loss. Moisture loss is controlled by storing the fruit at low temperatures and high relative humidity or by applying a wrap or other moisture barrier [91]. The cuticle of the fruit itself acts as an important natural moisture barrier and contributes in preventing desiccation and the browning disorder.

2.8.3 PURPLE SPOT

An important physiological disorder that occurs in loquat fruit (*Eriobotrya japonica* Lindl.) is purple spot. The disorder affects approximately 15% of the loquat fruits in Spain and reduces the market value of the fruit by 40 to 50% [92]. Purple spot appears as a modest depression on the fruit surface. It is purple in color and has an irregular shape. Researchers have suggested that local deficiencies in calcium may lead to the development of purple spot [93]. However, Gariglio et al. [92] found no correlation between calcium concentrations in loquat fruit and the incidence of purple spot. They also investigated the water permeability of the loquat fruit cuticle to determine whether desiccation could play a role in the development of the disorder as it does in the browning disorder of litchi fruit. They used isolated fruit cuticle, but found no sign of damage or disruption of the cuticle in the damaged area compared to healthy areas. They also saw no change in the water permeability of the cuticle and concluded that other factors were responsible for the development of purple spot in loquat fruit.

2.8.4 SCALD

Superficial scald is a physiological disorder that occurs postharvest in apples and pears [94]. Superficial scald appears as a browning of the fruit skin and can affect the entire fruit surface. The accumulation of α-farnesene and perhaps other volatile compounds has been implicated in the development of scald. The cuticle can sorb lipophilic, volatile compounds such as α-farnesene and consequently can promote the development of scald [28]. Fallik et al. [95] measured the amounts of α-farnesene and other volatile compounds in apple fruit with and without the cuticle intact and noted that the cuticle provides a poor barrier to volatile compounds. However, they did not investigate the sorptive properties of the cuticle. Ju and Bramlage [28] further noted that various phenolic compounds present in fruit cuticle might decrease oxidation of α-farnesene, which would counter the development of scald [28]. Interestingly, postharvest dips of apples and pears in plant oil emulsions or coating with natural and synthetic commercial wax products have been shown to affect the physical and chemical properties of the cuticle and reduce the development of scald

[28]. The data indicate that the cuticle plays an indirect role in the development of superficial scald and that modifying the properties of the cuticle with coatings can alter scald development [28].

The cuticle has been directly or indirectly implicated in the development of various other physiological disorders. Rindstaining in oranges (*Citrus sinensis* L.) was more severe in regions where the epicuticular wax layer was disrupted [96]. In addition, brown core in pear (*Pyrus bretschneideri*) was closely correlated with changes in the composition of epicuticular waxes [97]. Moreover, chilling injury in grapefruit has been linked to the composition and amount of epicuticular wax in the fruit cuticle [98]. More research is needed to clarify the role of the cuticle in the development of each of these disorders.

2.8.5 CUTICULAR CRACKING

In their extensive review of fruit cracking, Opara et al. [99] described *cracking* as a general term for certain physical disorders of fruits. These disorders appear as fissures or fractures that penetrate the cuticle, the skin (cuticle, epidermal and subepidermal cell layers), or the flesh (skin and underlying tissue) of produce. Flesh cracking is one of the most widespread physical defects found in the production of fruit [99]. Such cracks are very visible and are easily noticed by consumers. The properties of the cuticle may be important in the initial development of flesh cracks, especially if moisture absorption through the cuticle surface initiates the cracks. The economic importance of cuticular cracking is much more difficult to assess compared to that of flesh cracks. Cuticular cracks are difficult to see with the unaided eye, and consequently the prevalence of the problem remains unknown. However, the properties of the cuticle are thought to be particularly important in cuticular cracking. The following section provides a detailed discussion of cuticular cracking in various types of produce. Particular emphasis is given to sweet cherries as an illustration of a commodity influenced considerably by cuticular cracking.

2.8.5.1 Cherry

Cherry fruit (*Prunus avium* L. cv. Bing) is covered by a thin (1 μm), smooth cuticle [100]. Cross-sectional views of the fruit show that the epidermal cell layer consists of relatively small (ca. 40 μm), rectangular cells with thick cell walls (Figure 2.2 and Figure 2.3). In addition, there are one to three layers of subepidermal cells that are oblong shaped and intermediate in size. The parenchyma cells that make up the bulk of the fruit's flesh are relatively large (ca.150 μm) and round with very thin cell walls. Meanwhile, the cuticle has a smooth surface and is nearly devoid of epicuticular wax formations or bloom. The only exception to the smooth surface is a shallow reticulate pattern of epicuticular wax that is visible near the stylar scar end of the fruit (Figure 2.4). These reticulate epicuticular wax formations can also be seen in cuticles enzymatically isolated from the fruit (Figure 2.5). The smooth, planar surface of the epidermal cell layer combined with the lack of bloom on the cuticle surface make the fruit glossy and very attractive (Figure 2.6).

The primary cause of flesh cracking in sweet cherries is due to rainfall near harvest time [101]. Borve [102] covered cherry trees with different kinds of protective coverings and were able to save about 89% of the fruit during the season. In contrast, trees with no protective covering suffered a 54% crop loss. Fruit cracking that occurs in the field during rainfall typically develops as a crescent-shaped crack near the fruit pedicel where water may accumulate. Cracks may also occur in the region of the stylar scar where the water tends to drain from other areas of the fruit and accumulate into a drop of water [101,103]. In one study, Yamaguchi [104] reported that fruit weight, flesh firmness, and skin cell size largely determined the cracking susceptibility of cherry fruit. Lane et al. [105] looked at various parameters that might explain the susceptibility and resistance of different cherry varieties to cracking. They studied mineral composition of the fruit, skin elasticity, cuticle thickness, and geometry of the mesocarp and epidermal cells. However, none of those parameters could account for the differences observed in cracking susceptibility among susceptible and resistant varieties.

Researchers disagree on whether stomates are an important penetration site for water uptake in cherry fruit. Peschel et al. [106] studied the distribution of stomates in the cherry fruit epidermal layer to determine whether stomates were important in water penetration through the cherry skin. They found a greater density of stomates near the stylar scar, where cracking often occurs, and a lower amount near the stem, another common site for cracks. They observed that the stomates did not function in mature, ripe fruit and that the critical surface tension necessary for water to enter the stomates was not achieved. Consequently, Peschel et al. [106] concluded that the stomates were unlikely to be important pathways of water absorption. In another study, Knoche et al. [107] examined the permeation rate of water through the cherry fruit surface using pericarp tissue mounted in a diffusion cell. They found a positive relationship between water conductance and stomatal density, although the stomates appeared to be only a minor pathway of water penetration [107].

Beyer et al. [103,108] studied preferential water absorption in various regions of cherry fruit. They found that the pedicel cavity and surrounding regions were important areas of water absorption. Knoche and Peschel [109] studied the transport of water through the surface of sweet cherry fruit by using hydrostatic pressure to infiltrate the fruit. They found the mature fruit had a greater water uptake, most of it through the pedicel/fruit juncture. Water or solutes have also been shown to penetrate the fruit surface through the stylar scar region [100,107,110]. As the fruit develops and increases in size, the stylar scar region also increases in size. Consequently, this region could possibly become an increasingly important site of water penetration.

Knoche et al. [107] reported that fresh cherry fruit and fruit stored for 42 days showed no difference in skin permeance. The results suggest that the cuticle underwent little or no change during storage. However, conductance increased from the cheek to the ventral suture and the stylar end. The importance of cuticular waxes in water permeance was determined by treating the cuticle surface with solvents to remove the waxes. When their surfaces were treated with cellulose acetate and

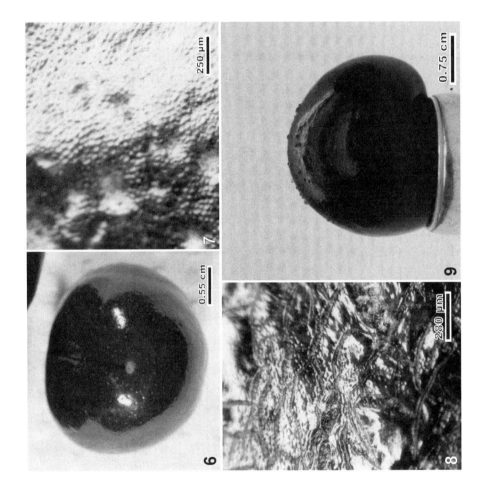

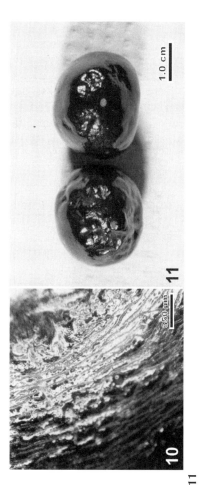

FIGURES 2.6 through 2.11

FIGURE 2.6 Photograph of a cherry fruit before exposure to water. Note the stylar scar visible as a small circular mark. The high gloss on the fruit surface is due partially to the relatively small amount of epicuticular wax and the planar surface of the epidermal cell layer.

FIGURE 2.7 Photomicrograph of the surface of a cherry fruit after being immersed in water for 30 minutes. The epidermal cells swelled and appeared stippled as water was absorbed through the fruit surface.

FIGURE 2.8 Photomicrograph of the surface of a cherry fruit after being immersed in water for several hours. The cuticle had many cracks that traversed only the cuticle. No large flesh cracks had developed even though some of the epidermal cells had ruptured and collapsed.

FIGURE 2.9 Photograph of a cherry fruit after exposure to water for several hours. The small cuticular cracks seen at higher magnication were not readily visible. No flesh cracking occurred but internal damage to the fruit was apparent by the small drops of water and cell contents exuded onto the fruit surface stemming from the high internal pressures.

FIGURE 2.10 Photomicrograph of the surface of a cherry fruit after being immersed in water for several hours. The cuticular cracks in some regions of the fruit were oriented perpendicular to the long axis of the fruit.

FIGURE 2.11 Photograph of a cherry fruit that had been exposed to 50% relatively humidity for 24 hours after exposure to water for several hours. The fruit developed a soft, flaccid texture where cuticular cracking had occurred. Eventually, the tissue collapsed in regions where the epidermal cells had collapsed or where extensive cuticular cracking occurred.

CHCl$_3$/methanol, water permeation increased 3.6- and 48.6-fold in pericarp segments, respectively. The results clearly indicate that the waxes are critical for decreasing water permeation.

Microscopy techniques have been used to document the sequence of events that lead to cuticular cracking and, subsequently, flesh cracking in cherries [100]. The first step involves moisture absorbance by the fruit. This may be accomplished by simply incubating fruit in aerated, distilled water [111]. As previously discussed, water absorption may occur either directly through the cuticle or through surface structures such as the stylar scar, pedicel, and possibly the stomates [100]. The high soluble solids content and high osmotic potential of the epidermal cells serve as a driving force to draw water into the cells, resulting in an increase in cell turgidity. The turgid cells give the once-planar surface a stippled texture. This is shown in Figure 2.7. The continued swelling of the epidermal and subepidermal cells places sufficient tensile stress on the cuticle to form small cuticular tears or cracks. The cuticular cracking can be severe and may be accompanied by a bursting or rupturing of the epidermal cells (Figure 2.8). In some cases, cuticular cracks may continue to widen and extend into the fruit's flesh, resulting in a large visible flesh crack. In other cases, the cuticular fractures sufficiently relieve the stress created by the swelling tissue so that flesh cracks do not form (see Figure 2.9). Cuticular cracking appears to be a precursor to flesh cracking [100]. As such, cuticular cracks may appear alone or in conjunction with flesh cracks, whereas flesh cracks seldom appear alone. In one study, Sekse and Ystaas [112] showed a positive correlation between the degree of flesh cracking and the number of cuticle cracks in cherry fruit.

Severe cuticular fracturing and epidermal cell damage can apparently occur in some fruit without showing any visible damage (Figure 2.9). In extreme cases, the hydraulic pressure developed in fruit exposed to water can cause small beads of water and cellular contents to exude onto the fruit surface (Figure 2.9). In addition, the cuticular cracks sometimes become oriented perpendicular to the long axis of the fruit (Figure 2.10). When fruit containing severe cuticular cracking are stored at room temperature and 50% relative humidity, they quickly lose weight and shrivel in areas where localized cuticular cracking has occurred (Figure 2.11). It is not surprising that fruit desiccation occurs as a result of cuticular cracking, because the cuticle provides the major vapor barrier to water loss.

In years when fruit cracking occurs due to heavy rains, it is a common practice to hand-sort damaged fruit. The assumption is that only severely damaged fruit contains visible flesh cracks. However, as indicated in Figures 2.8–2.10, there may be considerable damage to the cuticle and to the epidermal and subepidermal cell layers before a visible flesh crack ever develops. It is no wonder that rain-damaged fruit sorted for visible flesh cracks typically has poor shelf life and often becomes soft or infected by fungi. Borve et al. [102,113,114] studied the effect of cuticular cracking on postharvest quality of cherry fruit. They reported a high correlation between the amount of cuticular fracturing in sweet cherries and the degree of fungal infestation.

Borve et al. [102,113,114] also observed that a moderate amount of cuticular cracking might occur in cherry fruit sometime before the harvest period. Beyer et al. [115] observed cuticular cracks in cherry fruit preharvest but did not speculate on their importance or the conditions responsible for their occurrence. Sekse [101,116] described field conditions that could induce cuticular cracking, which include poor irrigation practices and preharvest rains. Cultivar and rootstock selection have also been linked to the development of cuticular cracks in sweet cherries [117–119]. Regardless of the cause, the fruit may appear to recover over time from the stress that induced the cuticular cracks in the first place. Consequently, fruit with preharvest cuticular cracks may be virtually indistinguishable from uninjured fruit at harvest time, giving the impression that it has no residual injury effects. However, the evidence shows that the presence of preharvest cuticular cracks predisposes fruit to flesh cracking, desiccation, and pathogen infection after harvest [52,100]. Borve et al. [102,113,114] observed that fruit with cuticular cracks had higher infection rates when inoculated with fungal conidial suspensions than fruit without cuticular cracks.

The pectin layer, which attaches the cuticle to the epidermal cell wall, appears to have a profound effect on cracking of cherry fruit [100]. Calcium ions and some other multivalent ions are believed to crosslink acidic pectin molecules [51,120]. In addition, calcium ions are known to delay cracking in sweet cherries [100,121,122]. Glenn et al. [100] used aerated baths of different $CaCl_2$ solutions to study the effect of calcium on cracking of cherry fruit. They found that fruit treated in solutions containing 2.5 mM $CaCl_2$ had a significantly lower cracking index than fruit treated in a water control. Fruits were also treated with a calcium chelating solution to chelate any free calcium ions. The solution containing calcium chelator (2.5 mM) increased the cracking index severalfold. Also, fruit treated in 2.5 mM solutions of both $CaCl_2$ and the calcium chelator had a cracking index similar to that of fruit treated in a water control. The study also showed that soluble pectin was leached from fruit during cracking studies. The fruit treated with the $CaCl_2$ solution had drastic reductions of leached soluble pectin. On the other hand, fruit treated in a calcium chelator solution had much higher levels of leached soluble pectin. Microscopy data indicated that swelling occurred in the region of pectin solubilization. Moreover, cuticular cracks were also associated with the swollen regions of the cuticle. The results indicate that in addition to swelling from turgid epidermal cells, the pectin-rich area of the cuticle directly accounts for much of the localized swelling and cuticular cracking. Calcium has been shown to be effective in reducing the swelling and localized stress responsible for cracking [51,100,123]. This suggests that calcium ions prevent some of the swelling and the dissolution of the pectin-rich region. In fact, field application of calcium compounds has been effective in reducing cracking of sweet cherry varieties [122].

2.8.5.2 Apple

Flesh cracking in apple fruit can incur substantial losses [124]. Cracking in the field appeared to be related to water uptake in the apple tree. Byers et al. [124] observed that flesh cracking in Stayman apples only occurs during relatively long rainy periods and after the fruit has grown to at least 5 cm in diameter. They found that applying stress to the tree, such as scoring the trunk, could reduce the amount of cracking. Applying gibberellic acid during the season also reduced cracking [124].

In addition to flesh cracking, cuticular cracking occurs in apple fruit as well. Wojcik [125] studied the epicuticular wax structure of seven apple varieties. They found that the main difference between the varieties was the occurrence of cracks in the wax layer of the cuticle. Interestingly, they found the sun-exposed side of the apple with a blush typically had fewer cracks than the shaded, unblushed side. In another study, Schirra et al. [126] used gibberellic acid sprays as a treatment to prolong postharvest life. They noted that during normal fruit development the cuticle starts out as a smooth, planar structure with very little crystalline wax structure. As the fruit develops further, the cuticle begins to form epicuticular wax platelets that are polygonal and arranged in a mosaic-type pattern. By the time the fruit has fully matured, the platelets are lifted and fine cracks of the cuticle develop. The gibberellic acid treatment delayed the formation of platelets and cracks and prolonged the fruit storage period.

A common practice after harvest involves submerging fruits for washing and conveying them to different places in the packinghouse. In some cases, fruit may be dipped in a particular bath as a postharvest treatment. Exposing fruit to prolonged dipping or washing treatments has been shown to induce cracking [124]. Byers et al. [124] observed that dipping apples in water containing surfactant resulted in higher water uptake and cracking compared to dipping them in water with no surfactant. The primary sites of water uptake included the lenticels and puncture wounds as well as the abrasions that penetrated the cuticle. In one study, Lurie et al. [127] treated Golden Delicious apples (*Malus domestica* Borkh.) with a dip in 2% $CaCl_2$ solution. One sample was heated for 4 days at 38°C before dipping. The heat treatment affected the wax surface of the apples. It appeared the heat-treated apples had their cracks filled and had a smoother surface than the apples that did not receive heat treatment. They concluded that the heat treatment softened the wax, which then filled the cracks and crevices. Consequently, less Ca penetrated into the heat-treated fruit compared to the control.

Roy et al. [128] used low-temperature scanning electron microscopy of frozen hydrated apples to view the cuticular structure. They observed cuticular cracks in older apple tissue. In a separate study, Roy et al. [129] studied infiltration of apples with $CaCl_2$ solutions. They used a control and a heat treatment at 38°C on the fruit. The results showed that the heat-treated fruit had less $CaCl_2$ infiltration. Also, the SEM micrographs showed microcracks on the control fruit, but the epicuticular wax had melted in the heat-treated sample. The results suggest that cracks on the fruit surface may be an important pathway for the penetration of $CaCl_2$ solutions. In a more in-depth study, they also showed that epicuticular wax cracks occurred in fruit stored postharvest and that the cracks became deeper and wider for fruits stored for longer periods [130]. In addition, the apples that were stored longer had a greater uptake of $CaCl_2$ solutions.

Schirra et al. [126] used a postharvest heat treatment to prolong postharvest life. Scanning electron micrographs revealed that heat treatment at 37°C actually melted the epicuticular wax so that it filled the cracks and wounds in the fruit. In addition, Veraverbeke et al. [89,131] used confocal laser scanning microscopy to perform a nondestructive analysis of apple cuticle during storage. They noted the development of cuticular cracks in fruit placed in controlled atmosphere storage. They suggested that the fruit developed cuticular cracks in response to the low oxygen and high relative humidity in the storage atmosphere. These cuticular cracks could improve air exchange between the atmosphere and oxygen-depleted tissue in the apple. Veraverbeke et al. [89,131] also observed that the cuticle structure changed when the fruits were moved from the controlled atmosphere storage to a place with normal oxygen levels. The small cuticular cracks filled in and the epicuticular wax layer on the cuticle surface became smooth. The results indicate that the cuticle is a dynamic structure that can be altered and modified by the plant to better adapt to its environment.

2.8.5.3   Tomato

Cherry tomato is a soft textured produce that can develop cuticle or skin cracks due to rainfall or poor irrigation practices [132–134]. Once skin cracks develop, the tomato can no longer be marketed because the cracked site becomes desiccated and infected with fungi [132]. Swelling and cracking susceptibility is greater in the morning when the tomato is fully hydrated [132]. In one study, Lichter et al. [132] reported that fruit thinning and deleafing also increased cracking susceptibility [135]. However, in another study, Emmons and Scott [133] found no relationship between pruning fruits and leaves and the incidence of cracking. Instead, they found that cracking potential decreased when the whole plant experienced environmental stress under hot and dry field conditions [133].

The growth rate or tomato size had no influence on cracking susceptibility [134]. On the other hand, the tomato's developmental stage had a significant effect on susceptibility to cracking. Mature-green fruits were the most susceptible, and the severity of the cracking increased the most between the mature-green and breaker stages [133]. Also, the fruit with a thicker skin, cuticle, and epidermis had a lower incidence of cracking [134].

Lichter et al. [132] used water immersion treatments to study cracking in cherry tomatoes. They found that adding calcium to the immersion solution reduced the incidence of cracking. In contrast, adding a chelating agent to the solution increased cracking [132]. The authors concluded that, similar to cherry fruit, the pectin fraction of the cuticle and epidermal cell wall plays an important role in determining the cracking potential of tomatoes.

2.8.5.4   Other Produce

Fruit cracking has also been shown to occur in other produce as well (99,134,111,136,137). Iwanami [138] reported that shallow, concentric cracks occurred during the late stage of fruit development in the cuticle of persimmons (*Diospyros kaki* Thumb.). These cracks occurred in roughly 50% of the cultivars of Japanese origin and promoted water loss. In another study, Yamada et al. [139] found that cracking susceptibility was largely genetic and constituted one of the properties considered by breeding programs when evaluating new persimmon cultivars. In most produce, moisture loss and desiccation adversely affected quality. However, in Japanese persimmon cultivars, fruit that had cuticular cracks remained high in quality and had higher soluble solids content than fruit without cuticular cracks. Apparently, the cracks effectively help increase the soluble solids content of the fruit and have few, if any, deleterious effects.

Cracking also occurs in atemoya (*Annona cherimola* Mill. × *Annona squamosa* L.) during the ripening stage or even postharvest [140]. As with cherries, apples, grapes, plums, and tomatoes, cracking in atemoya is associated with high soluble solids content and absorption of water through the fruit skin. Paull [140] observed that the neutral sugar content of the fruit increased during ripening. The high sugar concentration in the cells changed the osmotic potential and created a powerful driving force for water movement into the fruit. The increased cell turgor from water diffusion into the epidermal and subepidermal cells caused the fruit to swell and created a mechanical stress that induced cracking.

Aloni et al. [141] noted that cracking in bell peppers was preceded by cuticular cracks, similar to those found in cherry fruit. The cuticular cracks enlarged into splits or cracks that penetrated the epidermal, subepidermal, and even the paren-chyma cells. Cracking in bell peppers occurred under conditions that induced high cell turgor, such as the relatively low temperatures and high humidity at nighttime or when the crop was watered heavily after mild drought stress [141].

Storey and Price [31] used cryoscanning electron microscopy to study the epi-cuticular wax structure of d'Agen plums (*Prunus domestica* L.). They observed cuticular fractures along the long axis of the pairs of stomatal guard cells. The cuticular cracks resulted in a localized loss in moisture and led to the complete collapse of epidermal cells in the region where the cracks occurred [31].

Litchi fruit is also susceptible to desiccation and subsequent browning [4,142]. Harvested fruits can contain microcracks on their surface, and these microcracks can extend through the epidermal and subepidermal cell layers. Underhill and Simons [142] observed that microcracks on the fruit's surface were not initially responsible for the desiccation and browning damage of the fruit. In fact, desiccation initially occurred from high water vapor permeability due to cuticle damage and the presence of lenticels or a cuticle area that was simply more permeable to moisture [142]. The microcracks observed on the fruit's surface developed subsequent to the initial damage. Nevertheless, the microcracks accelerated the rate of moisture loss. In addition, the fruit with microcracks had a lower storage life due to fungal infection at the microcrack sites [142].

Mango fruit (*Mangifera indica* L.) is protected by a relatively thick cuticle that can occasionally develop cuticular cracking [17]. The cuticular cracks developed after anthesis when the cuticle was smooth and continuous, but before fruit set. By this time, a thick layer of flattened polygonal scales became evident on the epicu-ticular wax layer. As the fruit continued to develop, cracks traversed the cutin layer and formed irregular cuticular platelets. Eventually, the mature fruit developed small, erect scales. The thinnest area of the cuticle corresponded to the grooves formed by the cracks. Although the cracks did not completely traverse the cuticle, they likely provided sites for preferential transport of chemicals and potentially made the tissue more susceptible to chemical damage, blemishes, and desiccation [17].

Cracking can have a large impact on shelf life and degradation of produce. Cracking can occur when produce swells from moisture supplied either through the plant vascular tissue or directly through the cuticle surface. Cracking that results from moisture delivered through the vascular system typically occurs at times when the transpiration rates are low. The cuticle plays a role in this type of cracking because its properties affect transpiration rate and provide some mechanical strength to counter swelling and cracking. The cuticle may play an even greater role in cracking that results from absorption of moisture directly through the cuticle. In this case, the cuticle composition and its thickness affect moisture permeation rates. Also, the mechanical properties of the cuticle (strength and elasticity) determine how much stress the cuticle can withstand before it tears or cracks. Once the cuticle cracks, it allows for greater influx of water, resulting in further swelling. Eventually, deeper cracks form and penetrate the flesh.

Losses due to flesh cracking have been estimated for different kinds of produce and have considerable economic importance. However, the economic importance of cuticular cracking has not been estimated and remains difficult to assess because the cracks are so small. Nevertheless, it is clear that they often precede flesh cracking. In addition, they may also form by themselves without flesh cracking ever developing. In such instances, the produce becomes predisposed to pathogen infection and desiccation. Consequently, produce packinghouses need to be aware of produce that is susceptible to cuticular cracking. Exposure to water, even for short periods, can result in some cuticular cracking. The use of water to wash or convey fruit in packinghouses or to mist fruit on store shelves should be used with caution. In addition, shipping conditions that favor moisture condensation on the surface of the produce should also be avoided.

2.9 SUMMARY

The cuticle affects the aesthetic properties of produce and protects produce from solar radiation and chemical damage that can occur from exposure to agricultural pesticides and fertilizers. Moreover, it provides some mechanical protection from bumps, abrasions, and insect damage.

The postharvest life of produce is commonly extended by refrigeration or modified atmosphere storage, both of which reduce the rate of senescence and maturation [1]. There is some evidence that the cuticle modifies the internal gas composition of produce, which, in turn, could influence the rate of respiration and senescence. Postharvest life of produce is also affected by pathogen infestation. One way the cuticle extends the postharvest life of produce is by reducing the rate of infection by microbes [1]. The cuticle plays an important role in providing a physical barrier to pathogens. Pathogen infestation is much higher in produce with cracks that penetrate the cuticle, skin, or flesh. The cuticle also affects the events leading to infection of produce by providing a store of molecules that elicit a pathogen response and a host defense response during the initial steps of infection.

One of the best ways to extend the life of produce is to minimize water loss [1]. Because one of the primary functions of the cuticle involves controlling moisture loss, the cuticle plays a critical role in produce quality and preservation. Cracks in the cuticle layer accelerate desiccation, resulting in the loss of salable weight. Moisture loss also diminishes the quality of produce due to wilting, shriveling, softening, flaccidity, limpness, and loss of crispness and juiciness [143]. Moisture loss in citrus, for example, results in fruit that becomes soft and easily deformable [143]. In apples, a 5% loss in weight due to desiccation is all that is needed to cause the fruit to shrivel and take on a dull appearance [59,64]. The type of cracking, whether cuticular, skin, or flesh, and the susceptibility of produce to each type of cracking should be understood by those in the produce industry. Inclement weather conditions that lead to cracking are beyond control. However, cultural, handling, packing, and shipping practices can affect cracking, especially in susceptible produce, and should be controlled to optimize produce quality and preservation.

REFERENCES

1. Yaman, O. and Bayoindirli, L., Effects of an edible coating and cold storage on shelf-life and quality of cherries, *Lebensmittel-Wissenschaft und-Technologie,* 35, 146, 2002.
2. Heredia, A., Biophysical and biochemical characteristics of cutin, a plant barrier biopolymer, *Biochim. Biophys. Acta,* 1620, 1, 2003.
3. Holloway, P.J., Structure and histochemistry of plant cuticular membranes: an over-view, in *The Plant Cuticle,* Cutler, D.F., Alvin, K.L. and Price, C.E., Eds., Academic Press, London, 1982, 1–32.
4. Jeffree, C.E., Structure and ontology of plant cuticles, in *Plant Cuticles, An Integrated and Functional Approach,* Kersteins G., Ed., Bio Scientific Publishers, Oxford, U.K., 1996, 33.
5. Kolattukudy, P.E., Bio-polyester membranes of plants — cutin and suberin, *Science,* 208, 990, 1980.
6. Benavente, J., Munoz, A., Heredia, A., and Canas, A., Fixed charge and transport numbers in isolated pepper fruit cuticles from membrane potential measurements: donnan and diffusion potential contributions, *Colloids and Surfaces,* 159, 423, 1999.
7. Aggarwal, P., Phase transition of apple cuticles: a DSC study, *Thermochim. Acta,* 367, 9, 2001.
8. Martin, J.T. and Juniper, B.E., *The Cuticles of Plants,* Edward Arnold Ltd., London, 1970.
9. Round, A.N., Yan, B., Dang, S., Estephan, R., Stark, R.E., and Batteas, J.D., The influence of water on the nanomechanical behavior of the plant biopolyester cutin as studied by AFM and solid-state NMR, *Biophys. J.,* 79, 2761, 2000.
10. Fang, X., Qiu, F., Yan, B., Wang, H., Mort, A.J., and Stark, R.E., NMR studies of molecular structure in fruit cuticle polyesters, *Phytochemistry,* 57, 1035, 2001.
11. Steinmuller, D. and Tevini, M., Action of ultraviolet-radiation (uv-b) upon cuticular waxes in some crop plants, *Planta,* 164, 557, 1985.
12. Sieber, P., Schorderet, M., Ryser, U., Buchala, A., Kolattukudy, P., Metraux, J.P., and Nawrath, C., Transgenic arabidopsis plants expressing a fungal cutinase show alterations in the structure and properties of the cuticle and postgenital organ fusions, *Plant Cell,* 12, 721, 2000.
13. Fernandez, S., Osorio, S., and Heredia, A., Monitoring and visualizing plant cuticles by confocal laser scanning microscopy, *Plant Physiol. Biochem.,* 37, 789, 1999.
14. Jeffree, C.E. and Sandford, A.P., Crystalline-structure of plant epicuticular waxes demonstrated by cryostage scanning electron-microscopy, *New Phytol.,* 91, 549, 1982.
15. Bird, S.M. and Gray, J.E., Signals from the cuticle affect epidermal cell differentia-tion, *New Phytol.,* 157, 9, 2003.

16. Bukovac, M.J., Rasmussen, H.P., and Shull, V.E., The cuticle — surface-structure and function, *Scan. Electron Micro.*, Pn Part 3, 213, 1981.
17. Bally, I.S.E., Changes in the cuticular surface during the development of mango (*Mangifera indica* L.) Cv. Kensington Pride, *Sci. Hortic.*, 79, 13, 1999.
18. Kolattukudy, P.E., Biosynthetic pathways of cutin and waxes, and their sensitivity to environmental stressed, in *Plant Cuticles, an Integrated and Functional Approach,* Kersteins G., Ed., Bios Scientific Publishers, Oxford, U.K., 1996, 83.
19. Ray, A.K., Chen, Z., and Stark, R.E., Chemical depolymerization studies of the molecular architecture of lime fruit cuticle, *Phytochemistry*, 49, 65, 1998.
20. Post-Beittenmiller D., Biochemistry and molecular biology of wax production in plants, *Annu. Rev. Plant Physiol. Plant Mol. Biol.*, 47, 405, 1996.
21. Freeman, B., Albrigo, L.G., and Biggs, R.H., Ultrastructure and chemistry of cuticular waxes of developing citrus leaves and fruits, *J. Am. Soc. Hortic. Sci.*, 104, 801, 1979.
22. Prior, S.A., Pritchard, S.G., Runion G.B., Rodgers, H.H., and Mitchell, R.J., Influence of atmospheric CO_2 enrichment, soil N, and water stress on needle surface wax formation in Pinus palustris (*Pinaceae*), *Am. J. Botany,* 84, 1070, 1997.
23. Baker, E.A. and Hunt, G.M., Developmental-changes in leaf epicuticular waxes in relation to foliar penetration, *New Phytol.,* 88, 731, 1981.
24. Belding, R.D., Blankenship, S.M., Young, E., and Leidy, R.B., Composition and variability of epicuticular waxes in apple cultivars, *J. Am. Soc. Hortic. Sci.*, 123, 348, 1998.
25. Riederer, M. and Schreiber, J., Protecting against water loss: analysis of the barrier properties of plant cuticles, *J. Exp. Bot.,* 52, 2023, 2003.
26. Glenn, G.M., Poovaiah, B.W., and Rasmussen, H.P., Pathways of calcium penetration through isolated cuticles of golden delicious apple fruit, *J. Am. Soc. Hortic. Sci.,* 110, 166, 1985.
27. Ju, Z. and Bramlage, W.J., Developmental changes of cuticular constituents and their association with ethylene during fruit ripening in Delicious apples, *Postharvest Biol. Technol.*, 21, 257, 2001.
28. Ju, Z. and Bramlage, W.J., Cuticular phenolics and scald development in delicious apples, *J. Am. Soc. Hortic. Sci.,* 125, 498, 2000.
29. Farag, K. and Palta, J.P., Ultrastructure and surface morphology of cranberry plan (*Vaccinium Macrocarpon* Ait.) with reference to ethereal penetration, *Acta Hortic.*, 241, 378, 1989.
30. Veraverbeke, E.A., Lammertyn, J., Saevels, S., and Nicolai, B.M., Changes in chemical wax composition of three different apple (malus domestica borkh.) cultivars during storage, *Postharvest Biol. Technol.*, 23, 197, 2001.
31. Storey, R. and Price, W.E., Microstructure of the skin of d'agen plums, *Sci. Hort.*, 81, 279, 1999.
32. Ward, G. and Nussinovitch, A., Gloss properties and surface morphology relationships of fruits, *J. Food Sci.*, 61, 973, 1996.
33. Knuth, D. and Stosser, R., Comparison of the sun-exposed and shaded side of apple fruits.1. Cuticle, epidermal-cell size, and surface waxes, *Gartenbauwissenschaft*, 52, 49, 1987.
34. Roy, S., Conway, W.S., Buta, J.G., Watada, A.E., Sams, C.E., and Wergin, W.P., Surfactants affect calcium uptake from postharvest treatment of golden delicious apple, *J. Am. Soc. Hortic. Sci.,* 121, 1179, 1996.
35. Bangerth, F., Dilley, D.R., and Dewey, D.H., Effect of postharvest treatments on internal breakdown and respiration of apple fruits, *J. Am. Soc. Hortic. Sci.*, 97, 679, 1972.

36. Schonherr, J., Cuticular penetration of calcium salts: effects of humidity, anions, and adjuvants, *J. Plant Nutr. Soil Sci.*, 164, 225, 2001.
37. Chamel, C.E. and Chamel, A., Sorption and permeation to phenylurea herbicides of isolated cuticles of fruit and leaves. Effects of cuticular characteristics and climatic parameters, *Chemosphere*, 22, 85, 1991.
38. Schotsmana, W., Verlinden, B.E., Lammertyn, J., Peirs, A., Jancsok, P.T., Scheerlinck, N., and Nicolai, B.M., Factors affecting skin resistance measurements in pipfruit, *Postharvest Biol. Technol.*, 25, 169, 2002.
39. Banks, N.H. and Nicholson, S.E., Internal atmosphere composition and skin per-meance to gases of pepper fruit, *Postharvest Biol. Technol.*, 18, 33, 2000.
40. Amarante, C. and Banks, N.H., Ripening behaviour, postharvest quality, and physi-ological disorders of coated pears (*pyrus communis*), *N. Z. J. Crop Hortic. Sci.*, 30, 49, 2002.
41. Amarante, C., Banks, N.H., and Ganesh, S., Relationship between character of skin cover of coated pears and permeance to water vapour and gases, *Postharvest Biol. Technol.*, 21, 291, 2001.
42. Walton, T.J., Waxes, cutin and suberin, *Methods Plant Biochem.*, 4, 105, 1990.
43. Jenks, M.A., Joly, R.J., Peters, P.J., Rich, P.J., Axtell, J.D., and Ashworth, E.N., Chemically induced cuticle mutation affecting epidermal conductance to water vapor and disease susceptibility in *Sorghum bicolor* (L.) Moench, *Plant Physiol.*, 105, 1239, 1994.
44. Grusak, M.A. and Pomper, K.W., Influence of pod stomatal density and pod transpi-ration on the calcium concentration of snap bean pods, *J. Am. Soc. Hortic. Sci.*, 124, 194, 1999.
45. Qiu, Y.X., Nishina, M.S., and Paull, R.E., Papaya fruit-growth, calcium-uptake, and fruit ripening, *J. Am. Soc. Hortic. Sci.*, 120, 246, 1995.
46. Poovaiah, B.W., Glenn, G.M., and Reddy, A.S.N., Calcium and fruit softening: phys-iology and biochemistry, *Hortic. Rev.*, 10, 107, 1988.
47. Link, H., Ca-uptake and translocation by plants with special regard to apple trees, *Acta Hortic.*, 45, 53, 1974.
48. Harker, F.R. and Ferguson, I.B., Calcium-ion transport across disks of the cortical flesh of apple fruit in relation to fruit-development, *Physiol. Plant.*, 74, 695, 1988.
49. Harker, F.R. and Ferguson, I.B., Transport of calcium across cuticles isolated from apple fruit, *Sci. Hort.*, 36, 205, 1988.
50. Schlegel, Tk. and Schonherr, J., Stage of development affects penetration of calcium chloride into apple fruits, *J. Plant Nutr. Soil Sci.*, 165, 738, 2002.
51. Glenn, G.M., How cherries bruise, *Wash. St. Hortic. Assoc. Proc.*, 81, 238, 1985.
52. Glenn, G.M. and Poovaiah, B.W., Cuticular permeability to calcium compounds in golden delicious apple fruit, *J. Am. Soc. Hortic. Sci.*, 110, 192, 1985.
53. Perring, M.A., The effect of environmental and cultural practices on calcium in the apple fruits., *Comm. Soil Sci. Plant Anal.*, 10, 279, 1979.
54. Waldstein D.E. and Gut, L.J., Comparison of microcapsule density with various apple tissues and formulations of oriental fruit moth (*Lepidoptera: Tortricidae*) sprayable pheromone, *J. Econ. Entomol.*, 96, 58, 2003.
55. Fader, R.G. and Bukovac, M.J., Enhancement of transcuticular penetration of NAA with ammonium nitrate and triton X surfactants as spray additives, *Hortscience*, 32, 525, 1997.
56. Harker, F.R. and Ferguson, I.B., Effects of surfactants on calcium penetration of cuticles isolated from apple fruit, *Sci. Hort.*, 46, 225, 1991.

57. Dietz, T.J., Thimma Raju, K.R., and Joshi, S.S., Structure and development of cuticles and lenticels in fruits of certain cultivars of mango, *Acta Hortic.*, 231, 457, 1989.
58. Maguire, K.M., Lang, A., Banks, N.H., Hall, A., Hopcroft, D., and Bennett, R., Relationship between water vapour permeance of apples and micro-cracking of the cuticle, *Postharvest Biol. Technol.*, 17, 89, 1999.
59. Miller, R.H., Cuticular pores and transcuticular canals in diverse fruit varieties, *Ann. Bot.*, 51, 697, 1983.
60. Luque, P., Bruque, S., and Heredia, A., Water permeability of isolated cuticular membranes: A structural analysis, *Arch. Biochem. Biophys.*, 317, 417, 1995.
61. Luque, P., Gavara, R., and Heredia, A., A study of the hydration process of isolated cuticular membranes, *New Phytol.*, 129, 283, 1995.
62. Riederer, M. and Schonerr, J., Accumulation and transport of (2,4-dichlorophenoxy) acetic acid in plant cuticles: I, *Ecotoxicol. Environ. Saf.*, 8, 236, 1984.
63. Maguire, K.M., Banks, N.H., and Lang, A., Sources of variation in water vapour permeance of apple fruit, *Postharvest Biol. Technol.*, 17, 11, 1999.
64. Scherm, H., Ngugi, H.K., Savelle, A.T., and Edwards, J.R., Biological control of infection of blueberry flowers caused by *Monilinia vaccinii-corymbosi, Biol.l Control*, 29, 199, 2004.
65. Agrios, G.N., *Plant Pathology*, Academic Press, New York, 1988.
66. Liyanage, H.D., Koller, W., McMillan, R.T., Jr., and Kistler, H.C., Variation in cutinase from two populations of *Colletotrichum gloeosporidoides, Phytopathology*, 83, 113, 1993.
67. Kolattukudy, P.E., Enzymatic penetration of the plant cuticle by fungal pathoges, *Annu. Rev. Phytopathol.*, 23, 223, 1985.
68. Pinot, F., Benveniste, I., Salaun, J.P., Loreau, O., Noel, J.P., Schreiber, L., and Durst, F., Production *in vitro* by the cytochrome p450cyp94a1 of major c-18 cutin monomers and potential messengers in plant-pathogen interactions, enantioselectivity studies, *Biochem. J.*, 342, 27, 1999.
69. Pinot, F., Skrabs, M., Compagnon, V., Salaun, J.P., Benveniste, I., Schreiber, L., and Durst, F., Omega-hydroxylation of epoxy- and hydroxy-fatty acids by cyp94a1, possible involvement in plant defence, *Biochem. Soc. Trans.*, 28, 867, 2000.
70. Cruickshank, R.H., The influences of epicuticular wax disruption and cutinase resistance on penetration of tomatoes by *Colletotrichum gloeosporioides, J. Phytopathol.*, 143, 519, 1995.
71. Prusky, D., Etiology and histology of alternaria rot of persimmon fruits, *Phytopathology*, 71, 1124, 1981.
72. Biles, C.L., Wall, M.M., Waugh, M., and Palmer, H., Relationship of Phytophthora fruit rot to fruit maturation and cuticle thickness of New Mexican-type peppers, *Phytopathology*, 83, 607, 1993.
73. Spotts, R.A., Sanderson, P.G., Lennox, C.L., Sugar, D., and Cervantes, L.A., Wounding, wound healing and staining of mature pear fruit, *Postharvest Biol. Technol.*, 13, 27, 1998.
74. Coates, L.M., Muirhead, I.F., Irwin, J.A.G., and Gowanlock, D.H., Initial infection processes by *Colletotrichum gloeosporioides* on avocado fruit, *Mycol. Res.*, 97, 1363, 1993.
75. Kolattukudy, P.E., Rogers, L.M., Li, D., Hwang, C-S, and Flaishman, M.A., Surface signaling in pathogenesis, *Proc. Natl. Acad. Sci. USA.*, 92, 4080, 1995.
76. Parker, D.M. and Koller, W., Cutinase and other lipolytic esterases protect bean leaves from infection by rhizoctonia lani, *Mol. Plant-Micro. Inter.*, 11, 514, 1998.

77. Kim, Y.S., Lee, H.H., Ko, M.K., Ng, C.E., Bae, C.Y., Lee, Y.H., and Oh, B.J., Inhibition of fungal appresrium formation by pepper (*Capsicum annuum*) esterase, *Mol. Plant-Micro. Inter.*, 14, 80, 2001.

78. Schweizer, P., Jeanguenat, A., Whitacre, D., and Metraux, J.P., Induction of resistance in barley against *Erysiphe graminis* f.sp. *hordei* by free cutin monomers, *Physiol. Mol. Plant Path.*, 49, 103, 1996.

79. Fauth, M., Schweizer, P., Buchala, A., Markstadter, C., Riederer, M., Kato, T., and Kauss, H., Cutin monomers and surface wax constituents elicit H_2O_2 in conditioned cucumber hypocotyl segments and enhance the activity of other H_2O_2 elicitors, *Plant Physiol.*, 117, 1373, 1998.

80. Kastner, B., Tenhaken, R., and Kauss, H., Chitinase in cucumber hypocotyls is induced by germinating fungal spores and by fungal elicitor in synergism with inducers of acquired resistance, *Plant J.*, 13, 447, 1998.

81. Thordal-Christense, H., Zhank, A., Wei, Y., and Collinge, D.B., Subcellular localization of H_2O_2 in plants: H_2O_2 accumulation in papillae and hypersensitive response during the barley-powdery mildew interaction, *Plant J.*, 11, 1187, 1997.

82. Glenn, G.M. and Johnston, R.K., Mechanical properties of starch, protein and endosperm and their relationship to hardness in wheat, *Food Struct.*, 11, 187, 1992.

83. Thompson, D.S., Extensiometric determination of the rheological properties of the epidermis of growing tomato fruit, *J. Exp. Bot.*, 52, 1291, 2001.

84. Petracek, P.D. and Bukovac M.J., Rheological properties of enzymatically isolated tomato fruit cuticle, *Plant Physiol.*, 109, 675, 1995.

85. Luque, P., Gavara, R., and Heredia, A., A study of the hydration process of isolated cuticular membranes, *New Phytol.*, 129, 283, 1995.

86. Ozgen, M., Palta, J.P., and Smith, J.D., Ripeness stage at harvest influences postharvest life of cranberry fruit: physiological and anatomical explanations, *Postharvest Biol. Technol.*, 24, 291, 2002.

87. Glenn, G.M., Rom, C.R., Poovaiah, B.W., and Rasmussen, H.P., Influence of cuticular structure on the appearance of artificially waxed delicious apple fruit, *Sci. Hort.*, 42, 289, 1990.

88. Veraverbeke, E.A., Van Bruaene, N., Van Oostveldt, P., and Nicolai, B.M., Nondestructive analysis of the wax layer of apple (malus domestica borkh.) by means of confocal laser scanning microscopy, *Planta*, 213, 525, 2001.

89. Tipton, J.L. and White, M., Differences in leaf cuticle structure and efficacy among eastern redbud and Mexican redbud phenotypes, *J. Am. Soc. Hortic. Sci.*, 120, 59, 1995.

90. Vercher, R., Tadeo, F.R., Almela, V., Zaragoza, S., Primo-Millo, E., and Agusti, M., Rind structure, epicuticular wax morphology and water permeability of Fortune mandarin fruits affected by peel pitting, *Ann. Bot.*, 74, 619, 1994.

91. Jiang, Y.M. and Fu, J.R., Postharvest browning of lichti fruit by water loss and its prevention by controlled atmosphere storage at high relative humidity, *Lebensm-Wiss. U-Technol.*, 32, 278, 1999.

92. Gariglio, N., Juan, M., Castillo, A., Almela, V., and Agusti, M., Histological and physiological study of purple spot of loquat fruit, *Scientia Hortic.*, 92, 255, 2002.

93. Caballero, P., El nispero y su expansion, posibilidades y limitaciones, *Frut. Prof.*, 54, 35, 1993.

94. Whiting, M.D., Paliyath, G., and Murr, D.P., Analysis of volatile evolution from scald-developing and non-developing sides of apple fruits, *Hortscience*, 32, 457, 1997.

95. Fallik, E., Archbold, D.D., Hamilton-Kemp, T.R., Loughrin, J.H. and Collins, R.W., Heat treatment temporarily inhibits aroma volatile compound emission from Golden Delicious apples, *J. Agric. Food Chem.*, 45, 4038, 1997.

96. El-Otmani, M., Arpaia, M.L., Coggins, Jr., V.W., Pherson Jr., J.E., and O'Connell, N.V., Developmental changes in Valencia orange fruit epicuticular wax in relation to fruit position on the tree, *Sci. Hortic.*, 41, 69, 1989.

97. Huo, J.S. and Li, X.Q., Analysis of the relationship of brown core and epicuticular wax changes of Ya pear (*Pyrus bretescheideri* Rehd) by scanning electron microscope examination, *J. Hebeie Agric. Univ.*, 15, 50, 1992.

98. Nordby, H.W. and McDonald, R.E., Relationship of epicuticular wax composition of grapefruit to chilling injury, *J. Agric. Food Chem.*, 39, 957, 1991.

99. Opara, L.U., Studman, C.J., and Banks, N.H., Fruit skin splitting and cracking, *Hortic. Rev.*, 19, 217, 1997.

100. Glenn, G.M. and Poovaiah, B.W., Cuticular properties and postharvest calcium applications influence cracking of sweet cherries, *J. Am. Soc. Hortic. Sci.*, 114, 781, 1989.

101. Sekse, L., Fruit cracking in sweet cherries (*Prunus avium* L.) — some physiological-aspects — a mini review, *Sci. Hortic.*, 63, 135, 1995.

102. Borve, J., Skaar, E., Sekse, L., Meland, M., and Vangdal, E., Rain protective covering of sweet cherry trees — effects of different covering methods on fruit quality and microclimate, *Hortic. Tech.*, 13, 143, 2003.

103. Beyer, M. and Knoche, M., Studies on water transport through the sweet cherry fruit surface: v. Conductance for water uptake, *J. Am. Soc. Hortic. Sci.*, 127, 325, 2002.

104. Yamaguchi, M., Sato, I., and Ishiguro, M., Influences of epidermal cell sizes and flesh firmness on cracking susceptibility in sweet (*Prunus avium* L.) cultivars and selections, *J. Jpn. Soc. Hortic. Sci.*, 71, 738, 2002.

105. Lane, W.D., Meheriuk, M., and Mckenzie, D.L., Fruit cracking of a susceptible, an intermediate, and a resistant sweet cherry cultivar, *Hortscience*, 35, 239, 2000.

106. Peschel, S., Beyer, M., and Knoche, M., Surface characteristics of sweet cherry fruit: stomata-number, distribution, functionality and surface wetting, *Sci. Hortic.*, 97, 265, 2003.

107. Knoche, M., Peschel, S., Hinz, M., and Bukovac, M.J., Studies on water transport through the sweet cherry fruit surface: characterizing conductance of the cuticular membrane using pericarp segments, *Planta*, 212, 127, 2000.

108. Beyer, M., Peschel, S., Weichert, H., and Knoche, M., Studies on water transport through the sweet cherry fruit surface. 7. Fe3+ and al3+ reduce conductance for water uptake, *J. Agric. Food Chem.*, 50, 7600, 2002.

109. Knoche, M. and Peschel, S., Studies on water transport through the sweet cherry fruit surface. Vi. Effect of hydrostatic pressure on water uptake, *J. Hortic. Sci. Biotech.*, 77, 609, 2002.

110. Knoche, M., Peschel, S., and Hinz, M., Studies on water transport through the sweet cherry fruit surface: iii. Conductance of the cuticle in relation to fruit size, *Physiol. Plant.*, 114, 414, 2002.

111. Verner, L., A physiological study in cracking in stayman winesap apples, *J. Agric. Res.*, 51, 191, 1935.

112. Sekse, L. and Ystaas, J., Cuticular fractures in fruits of sweet cherry (*Prunus avium* L.) affect fruit quality negatively and their development is influenced by cultivar and rootstock, *Acta Hortic.*, 468, 671, 1998.

113. Borve, J., Sekse, L., and Stensvand, A., Cuticular fractures promote postharvest fruit rot in sweet cherries, *Plant Dis.*, 84, 1180, 2000.

114. Borve, J., Sekse, L., and Stensvand, A., Cuticular fractures as infection sites of *Botrytis cinerea* sweet cherry fruits, *Acta Hortic.*, 468, 737, 1998.

115. Beyer, M., Peschel, S., Knoche, M., and Knorgen, M., Studies on water transport through the sweet cherry fruit surface: IV. Regions of preferential uptake, *Hortscience*, 37, 637, 2002.

116. Sekse, L., Cuticular fracturing in fruits of sweet cherry (*Prunus avium* L.) resulting from changing soil-water contents, *J. Hortic. Sci.*, 70, 631, 1995.

117. Hovland, K.L. and Sekse, L., The development of cuticular fractures in fruits of sweet cherries (*Prunus avium* L.) can vary with cultivar and rootstock, *J. Am. Pomol. Soc.*, 57, 58, 2003.

118. Cline, J.A., Meland, M., Sekse, L., and Webster, A.D., Rain cracking of sweet cherries.2. Influence of rain covers and rootstocks on cracking and fruit-quality, *Acta Agric. Scand.*, 45, 224, 1995.

119. Cline, J.A., Sekse, L., Meland, M., and Webster, A.D., Rain-induced fruit cracking of sweet cherries. 1. Influence of cultivar and rootstock on fruit water-absorption, cracking and quality, *Acta Agric. Scand.*, 45, 213, 1995.

120. Glenn, G.M. and Poovaiah, B.W., Calcium-mediated postharvest changes in texture and cell wall structure and composition in golden delicious apples, *J. Am. Soc. Hortic. Sci.*, 115, 962, 1990.

121. Brown, G.S., Kitchener, A.E., McGlasson, W.B., and Barnes, S., The effects of copper and calcium foliar sprays on cherry and apple fruit quality, *Sci. Hortic.*, 67, 219, 1996.

122. Brown, G., Wilson, S., Boucher, W., Graham, B., and McGlasson, B., Effects of copper-calcium sprays on fruit cracking in sweet cherry (*Prunus avium*), *Sci. Hortic.*, 62, 75, 1995.

123. Yamamoto, T., Satoh, H., and Watanabe, S., The effects of calcium and naphthalene acetic-acid sprays on cracking index and natural rain cracking in sweet cherry fruits, *J. Am. Soc. Hortic. Sci.*, 61, 507. 1992.

124. Byers, R.E., Carbaugh, D.H., and Presley, C.N., Stayman fruit cracking as affected by surfactants, plant growth regulators, and other chemicals, *J. Am. Soc. Hortic. Sci.*, 115, 405, 1990.

125. Wojcik, P., Dyki, B., and Cieslinski, G., Fine structure of the fruit surface of seven apple cultivars, *J. Fruit Ornam. Plant Res.*, 5, 119, 1997.

126. Schirra, M., D'hallewin, G., Inglese, P., and La Mantia, T., Epicuticular changes and storage potential of cactus pear [*Opuntia ficus-indica* Miller (L.)] fruit following gibberellic acid preharvest sprays and postharvest heat treatment, *Postharvest Biol. Technol.*, 17, 79, 1999.

127. Lurie, S., Fallik, E., and Klein, J.D., The effect of heat treatment on apple epicuticular wax and calcium uptake, *Postharvest Biol.Technol.*, 8, 271, 1996.

128. Roy, S., Watada, A.E., Conway, W.S., Erbe, E.F., and Wergin, W.P., Low-temperature scanning electron-microscopy of frozen-hydrated apple tissues and surface organisms, *Hortscience*, 29, 305, 1994.

129. Roy, S., Conway, W.S., Watada, A.E., Sams, C.E., Erbe, E.F., and Wergin, W.P., Heat-treatment affects epicuticular wax structure and postharvest calcium-uptake in golden delicious apples, *Hortscience*, 29, 1056, 1994.

130. Roy, S., Conway, W.S., Watada, A.E., Sams, C.E., Erbe, E.F., and Wergin, W.P., Changes in the ultrastructure of the epicuticular wax and postharvest calcium uptake in apples, *Hortscience*, 34, 121, 1999.

131. Veraverbeke, E.A., Verboven, P., Scheerlinck, N., Hoang, M.L., and Nicolai, B.M., Determination of the diffusion coefficient of tissue, cuticle cutin and wax of apple, *J. Food Eng.*, 58, 285, 2003.

132. Lichter, A., Dvir, O., Fallik, E., Cohen, S., Golan, R., Shemer, Z., and Sagi, M., Cracking of cherry tomatoes in solution, *Postharvest Biol.Technol.*, 26, 305, 2002.

133. Emmons, C.L.W. and Scott, J.W., Environmental and physiological effects on cuticle cracking in tomato, *J. Am. Soc. Hortic. Sci.*, 122, 797, 1997.

134. Luque, P., Bruque, S., and Heredia, A., Water permeability of isolated cuticular membranes: a structural analysis, *Arch. Biochem. Biophys.*, 317, 417, 1995.
135. Ehret, D.L., Helmer, T., and Hall, J.W., Cuticle cracking in tomato fruit, *J. Hortic. Sci.*, 68, 195, 1993.
136. Lippert, F., Cracking symptoms of kohlrabi tubers, *J. Plant Dis. Protect.*, 106, 512, 1999.
137. Lippert, F., Method for inducing cracks in tubers of kohlrabi (*brassica-oleracea var Gongylodes* L.), *Gartenbauwissenschaft*, 60, 187, 1995.
138. Iwanami, H., Yamada, M., and Sato, A., A great increase of soluble solids concentration by shallow concentric skin cracks in Japanese persimmon, *Sci. Hortic.*, 94, 251, 2002.
139. Yamada, M., Sato, A., and Ukai, Y., Genetic differences and environmental variations in calyx-end fruit cracking among Japanese persimmon cultivars and selections, *Hortscience*, 37, 164, 2002.
140. Paull, R.E., Postharvest atemoya fruit splitting during ripening, *Postharvest Biol. Technol.*, 8, 329, 1996.
141. Aloni, B., Karni, L., Rylski, I., Cohen, Y., Lee, Y., Fuchs, M., Moreshet, S., and Yao, C., Cuticular cracking in pepper fruit. I. Effects of night temperature and humidity, *J. Hortic. Sci. Biotechnol.*, 73, 743, 1998.
142. Underhill, S.J.R., and Simons, D.H., Lychee (*Litchi-chinensis* sonn) pericarp desiccation and the importance of postharvest microcracking, *Sci. Hortic.*, 54, 287, 1993.
143. Munoz-Delgado, J.A., Problems in cold storage of citrus fruit, *Rev. Int. Froid.*, 10, 229, 1987.

3 Maturity, Ripening, and Quality Relationships

Peter M.A. Toivonen and Thomas H.J. Beveridge

Agriculture and Agri-Food Canada, Pacific Agri-Food
Research Centre, Summerland, British Columbia, Canada

CONTENTS

3.1 INTRODUCTION

The rate and extent of deterioration in fresh fruits and vegetables is very dependent on the intrinsic state of the tissues. This intrinsic state of the tissues at harvest needs to be well-defined if it is to be managed in the process of ensuring consistency in quality. Maturity, as it is referred to in this chapter, relates to what is generally called horticultural maturity (defined as the state at which the particular plant part has characteristics that are preferred for processing and/or consumption by human consumers) (Beveridge,

2003; Kader, 2003). Maturity at harvest is a very important factor in determining storage life and final quality of a fruit or vegetable product (Kader, 2003). Most fruits and mature-fruit vegetables achieve optimal eating quality if they are allowed to ripen on the plant. However, in order to withstand the rigors of shipping, some of these are harvested mature, but unripe. There is a compromise made between picking for optimal quality and picking for the best shipping condition. In most nonfruit and immature-fruit vegetables the optimum quality is reached before full maturity, but harvesting at progressively more mature stages leads to progressively more rapid deterioration after harvest (Kader, 2003). The variety of plant parts used as food adds significant complexity to the definition of optimal maturity (Figure 3.1); this unfortunately adds to the difficulty in understanding the impact of maturity on quality degradation through the postharvest continuum. The difficulty in defining maturity in unripened plant parts (leaves, immature floral parts and fruits, stems, and roots) is reflected in the lack of in-depth information on these commodities (Reid, 2002). It will be clear from the following discussion that virtually all understanding of the issue of maturity and quality is focused on ripening fruits and fruit-vegetables. It may be that handling of unripening plant parts (leaves, immature floral parts and fruits, stems, and roots) may not have required any further understanding of maturity in the past. However, because all horticultural commodities are becoming candidate components in new products such as fresh-cut salads, the need for better understanding of maturity will arise.

3.2 EFFECTS OF MATURATION AND RIPENING

3.2.1 TEXTURE

Plant growth and development is intimately related to the growth and development of the plant cell and wall, so the texture of foods derived from plant parts is ultimately determined by the plant cell, its wall, and its relationships with the surrounding cells. Jackman and Stanley (1995) view texture change in plant foods as a consequence of decreasing shear force as the fruit ripens (Figure 3.2). As the fruit progresses through the ripening phase, it eventually reaches a stage where cellular cohesiveness is failing; this stage is termed *de-bonding* (Jackman and Stanley, 1995), and fruit at this stage is normally not acceptable for use or consumption. The changes in cell wall cohesiveness with maturity have been well documented (Tu et al., 1997) and are somewhat understood at the gene expression level (Giovannoni, 2001).

Storage life is often defined by the time period over which a product can maintain a minimal acceptable firmness level (Harker et al., 2002a). The definition of the term *acceptable*, to date, has been generally more of a pragmatic one (relating to what can be measured by available instrumentation), rather than one that actually relates to human perception of quality (Harker et al., 2002a). There are exceptions to this general pattern of softening; for instance, peas, snap beans, lima beans, eggplant, broccoli, and okra actually get tougher (fibrous) or firmer with advancing maturity (Pantastico, 1975; Sams, 1999; Gross et al., 2002).

Firmness, one of the most measured textural characteristics, is quite often used as an objective criterion to define maturity level in many fruits (Pantastico et al.,

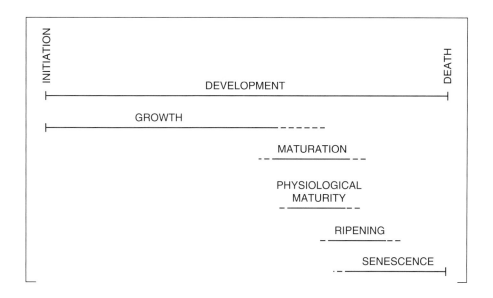

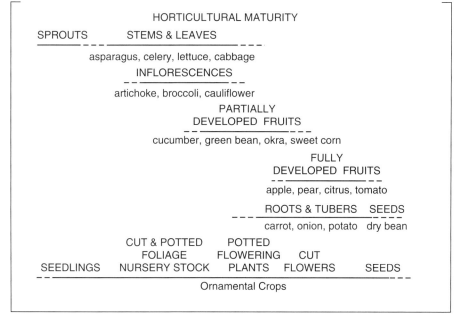

FIGURE 3.1 Stages of development and senescence based on physiological processes and usage of horticultural crops. (From Watada et al., *HortScience*, 19, 20, 1984. With permission.)

1975). Changes in firmness with advancing maturity in apples have been shown to be related to deterioration of the cell wall structure of the fruits (Tu et al., 1997). In general, a firmer fruit is assumed to be less mature and to have a more complex cell wall structure than a softer fruit. Maturity is often defined as optimal firmness in relation to a particular fruit. However, the actual threshold value for optimal

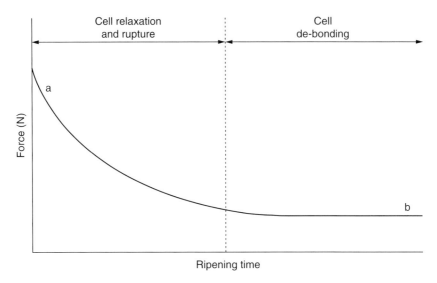

FIGURE 3.2 Texture–structure relationships in ripening tomato fruit. The graph depicts decreasing puncture force with increasing ripening time. Note that a transition occurs in the mode of tissue failure when it becomes easier for cells to separate or de-bond than to rupture, corresponding to the degradation of the middle lamella to a critical level. (From Jackman and Stanley, *Trends Food Sci. Technol.*, 6, 187, 1995. With permission.)

firmness for use is also dependent on the specific requirements for a particular use. One example where end-use parameters are important is that of pears to be used for fresh-cut slices. Chen et al. (2003) define optimal firmness as 22 to 31 N as measured by a Magness-Taylor probe. Gorny et al. (2000) define optimal firmness as 44 to 58 N using a similar instrument, about twice the previous definition. The probable reason for the difference in recommendations is that the softer fruit is intended to be stored as slices at –1.1 to –1.7°C, while the harder fruit was intended to withstand storage at 10°C as a sliced product. Fruit to be used for sliced product and stored at warmer temperatures must be firmer than fruit that will be used for product that is stored at colder temperatures since softening proceeds at a faster rate at higher temperatures (Johnston et al., 2002a). This emphasizes the importance of defining the intended use and storage conditions before attempting to define the optimal firmness or ripeness of the raw product.

One principle of postharvest handling of fruit is that product that is fully mature but not ripe will generally be firmer at harvest and will remain firmer through storage. Certainly this has been shown to be the case for apples (Knee and Smith, 1989; Silsby, 1993) and pears (Rousseas et al., 2001). The rate of softening in apples during long-term storage also increases, as does the maturity of the fruit at harvest (Tu et al., 1997). Changes in firmness in apples have been shown to have at least two components: an earlier, slow softening phase and a later, rapid softening phase (Johnston et al., 2002a). Fruits harvested later have a shorter duration of slow softening in storage than do fruits harvested earlier, while fruits of both early and late maturities have similar rapid softening phases (Johnston et al., 2002a). Also, small and large apples harvested less mature show very little difference in softening, whereas fruit harvested at more mature stages shows differentials in softening rate dependent on its size (Johnston et al., 2002b). In later-harvested apples, small fruits soften at a slower rate than do large fruits.

There are cases where fruits do not appear to soften after harvest. Strawberries are progressively softer when picked at 50%, 100%, and after full red color. No matter at which stage the fruits are harvested, strawberries do not soften after harvest for up to 5 days of cool storage plus 2 days at shelf temperature (Forney et al., 1998). A partial explanation for this observation is that postharvest softening does not occur in strawberries within a 6- to 7-day period after harvest. Similar observations have been made over short storage durations for fresh-cut apple slices. Under a short shelf-life expectation (6 days or less) softening does not appear to be significant; however, if shelf life expectations exceed 6 days, softening becomes a determinant for the end of shelf life (Kim et al., 1993). Therefore, the length of storage must be considered when evaluating whether textural changes will be important.

Another effect of harvest maturity in some fruit is the susceptibility to significant textural disorders in storage. For instance, 'Songold' plums were found to develop gel breakdown sooner and more severely if they were harvested past optimum maturity (Taylor et al., 1995). In that case, the optimal maturity was defined as fruit with an average firmness of 64 N, while fruit harvested at an average firmness of 44 N had severe incidence of the disorder.

In attempting to understand the influence of maturity at harvest on firmness and texture changes during storage, allowance must be made for understanding differences in relation to cultivar. In apples, rates of softening in response to exposure to low chilling temperatures can vary significantly for different cultivars (Johnston et al., 2001a,b). Also in apples, harvest maturity in some cultivars is extremely critical in determining the firmness and crispness in fruit after storage (Zerbini et al., 1999). In sweet cherries, softer cultivars have a shorter lag phase after pit hardening than do firmer cultivars (Choi et al., 2002). Similar findings have been reported for nectarines; firmer cultivars are described as "slow-ripening" (Brecht et al., 1984).

3.2.2 FLAVOR

As with texture, flavor can be significantly affected by the maturity of the fruit at harvest. The perception of flavor is considered to be a rather complex issue and involves the combined impact of acidity (titratable acidity and pH), soluble solids (individual sugars and sugar alcohols), and aroma volatiles (Young et al., 1996; Baldwin et al., 1998; Harker et al., 2002b). While there are general relationships among acidity, soluble solids, aroma volatiles, and flavor, the relationship of individual components to final flavor impact can be inconsistent (Harker et al., 2002b). Therefore, this discussion will focus on effects of maturity on the individual components and their change in postharvest situations.

Generation of aroma volatiles in fruit is linked with metabolism at the latter stages of maturity (Girard and Lau, 1995; Song and Bangerth, 1996; Fellman et al., 2003), and hence fruit picked at a more mature stage would be expected to have a higher production of aroma volatiles, and hence better flavor quality (Fellman et al., 2003). Flavor (aroma) generation is tightly associated with respiration and ethylene production during the climacteric stage of ripening in fruits such as apples (Song and Bangerth, 1996; Fellman et al., 2003). Knee and Hatfield (1981) found that mature but unripe apples did not synthesize acetate esters; this was due directly to inherently low rates of alcohol synthesis in mature but unripe fruit. If substrate alcohols were supplied, then acetate esters were produced by apples at all states of ripeness. Production of 2-methylbutyl acetate has been associated with "red apple aroma" and as such has been suggested as an indicator for physiological maturity in some apple cultivars (Mattheis et al., 1991; Young et al., 1996).

Generation of aroma volatiles is known to be inhibited in apples when they are placed in controlled atmosphere storage for extended periods of time, and it can take some time before volatile generation capability recovers (Girard and Lau, 1995; Fellman et al., 2003). Maturity at harvest has an influence on this process. For example, the more mature fruit of 'Redchief Delicious' apples are at harvest, the more rapid the aroma volatile regeneration after removal from controlled atmosphere storage (Fellman et al., 2003). Generation of mango aroma volatiles is similar to that of apples: greater amounts are produced at more advanced maturity (Lalel et al., 2003a,c). In addition, final eating quality is optimized if the fruit is harvested at a mature-green stage (Lalel et al., 2003a). In litchi fruit, harvesting at a more advanced state of ripeness tends to produce greater amounts of ethanol and acetaldehyde in modified atmosphere packages (Pesis et al., 2002). Increased production of ethanol and acetaldehyde was associated with greater rates of deterioration in litchis, since their accumulation was also associated with greater rates of decay in the fruit once placed under poststorage shelf conditions (Pesis et al., 2002).

The relationship of at-harvest soluble solids (sugars) to soluble solids after storage is somewhat confusing for apples. In 'Cox's Orange Pippin,' 'Jonagold,' 'Gala,' and 'Braeburn' apples, the final storage soluble solids content was similar no matter what the initial soluble solids content was when the fruit was picked, whereas 'Fuji' apples show an increase in poststorage soluble solids with later harvests (Girard and Lau, 1995; Plotto et al., 1995; Fan et al., 1997b). In addition, sensory panels indicated that late-picked 'Fuji' apples had improved overall flavor characteristics (Plotto et al., 1995). The reason for the differences found for 'Fuji' compared with 'Cox's Orange Pippin' and 'Gala' is likely related to the ripening characteristics of the cultivar. Jobling and McGlasson (1995) demonstrated that 'Fuji' ripened in a different pattern than 'Gala' and suggested that maturity indices used for 'Fuji' be based on different criteria than those for 'Gala' and other apples.

Titratable acidity declines with more advanced maturity in 'Cox's Orange Pippin' apples, and this leads to an apparent increase in sweetness of the apples, even though sugar contents do not change substantially (Knee and Smith, 1989). A similar pattern is seen with other cultivars such as 'Jonagold,' 'Gala,' 'Braeburn,' and 'Fuji,' except that soluble solids do increase somewhat with later harvests (Girard and Lau, 1995; Plotto et al., 1995). While later-harvested apples have lower titratable acidity, the relative rate of decline in titratable acidity during storage does not change with maturity (Plotto et al., 1995).

There is limited information in regard to flavor retention of vegetables in storage. Sugar content of carrots increased with harvest maturity, as did the sucrose:reducing sugar ratio, and these differences were maintained during cold storage (Fritz and Weichmann, 1979). In fresh-cut product, maturity of the raw vegetable can have an effect on the subsequent flavor retention over its shelf life. For example, fully colored, mature peppers maintained higher sugar content and better flavor than immature green peppers after they were fresh-cut into slices (López-Gálvez et al., 1997).

3.2.3 Nutritional Quality

Discussion regarding nutritional quality will focus on the vitamin and antioxidant content of fruits and vegetables. While this is a relatively new area of research investigation, results are interesting and are worthy of discussion and perhaps future research initiatives.

Ascorbic acid (vitamin C) is a water-soluble vitamin of great importance to the general nutritive quality of fruits and vegetables (Yahia et al., 2001). Ascorbic acid contents of peppers, tomatoes, apricots, peaches, and papayas have been reported to increase with more advanced maturity at harvest (Lee and Kader, 2000). In peppers and tomatoes, it has been shown that ascorbic acid increases up to full coloration; however, levels decline as the fruits are allowed to ripen beyond that stage (Soto-Zamora et al., 2001; Yahia et al., 2001). Ascorbic acid levels generally decline with harvest maturity for apples, mangoes, peas, and citrus (Lee and Kader, 2000). One hypothesis put forward regarding the decline in ascorbic acid in citrus was that the apparent content declines as the fruit increases in size, which suggests that the increase in size dilutes the total ascorbic acid content. This is supported by the observation that ascorbic acid per fruit remained constant with maturation of citrus (Lee and Kader, 2000). Since ascorbic acid generally declines in storage (Lee and Kader, 2000), the content at harvest is an important determinant of the content on the retail shelf. At all ripeness stages, the ascorbic acid content of fruit declined with increased duration in storage (Soto-Zamora et al., 2001). Similar results were obtained when using different maturities of mango in fresh-cut packages (Izumi et al., 2003).

Carotene is another important nutrient, at least partially due to its strong antioxidant power (Giovanelli et al., 1999). It is a lipid-soluble nutrient and is found in many fruits and vegetables, but it is probably highest in carrots (Salunkhe and Desai, 1984). The carotene content of carrots increased with harvest maturity, and this difference was maintained during cold storage (Fritz and Weichmann, 1979). The carotenoid (carotene and lycopene) content of tomatoes increased with the ripeness of the fruit (Giovanelli et al., 1999). However, tomatoes harvested at the mature-green stage developed a higher carotenoid (both carotene and lycopene) content than those left to ripen on the vine (Giovanelli et al., 1999). This suggests that optimal nutritional quality can be achieved in tomatoes if they are harvested at an early stage of ripening (mature-green).

Antioxidant content is influenced by the maturity of the fruit. In the case of peppers, the water-soluble antioxidant content increased with advancing ripeness stage (Jiménez et al., 2003). Levels of antioxidant capacity also increased in storage for both immature (green) and mature (red) fruit. The antioxidant constituents responsible for the water-soluble antioxidant capacity were considered to be ascorbic acid and glutathione. Ascorbic acid and glutathione (both endogenously produced antioxidants) declined with more advanced maturity in pears (Lentheric et al., 1999). This decline in antioxidant capacity was related to the increased susceptibility of the fruit to internal browning of the pears in storage. Phenolic acids and flavonoids are also components of antioxidant capacity in fruits and vegetables. Phenolic acids decline in apple peels as the fruit matures and ripens, whereas flavonoids increase (Awad et al., 2001). Tomatoes show a slight decline in phenolic acid content during on-the-vine ripening; however, if fruits are harvested at the mature-green stage, they show significant increases in phenolic acid content with ripening off-the-vine (Giovanelli et al., 1999). In carrots, phenolic acid contents increase with advancing maturity (Chubey and Nylund, 1970).

3.2.4 SHELF LIFE

The shelf life of fruits and vegetables can be determined by a number of factors, including dehydration, development of disorders, and senescence. The following discussion focuses on the influence of maturity on the development of visible changes in fruit and vegetables that signal the end of useful shelf life.

Disorders associated with fruit maturity that lead to shortened shelf life are well documented for tree fruits. Generally, the more mature an apple is when harvested, the greater susceptibility it will have to incurring injuries from controlled atmosphere storage conditions (Ferguson et al., 1999). Late-harvested 'Fuji' apples having watercore at harvest were more susceptible to internal browning than fruit harvested at a less mature state and not having visible symptoms of watercore (Fan et al., 1997a). Similar findings have been reported for internal browning disorder in 'Braeburn' apples; later harvests showed greater incidence of the disorder in storage (Elgar et al., 1997; Lau, 1997). Late-harvested 'Conference' pears stored in controlled atmosphere conditions have been shown to have a greater incidence and severity of internal flesh browning (Roelofs and deJager, 1997).

The previous paragraph indicated that more advanced maturity leads to greater flesh browning. In contrast to flesh-related disorders, skin-related disorders tend to decline as fruit matures (Toivonen, 2003). More advanced maturity in apples has been shown to be associated with significantly lower incidence of superficial scald (Toivonen, 2003). In a specific case, breaker fruit of 'Fuji' was found to be less susceptible to scald than mature-green fruit (Fan et al., 1997b).

Water loss can be a significant determinant of shelf life, and if maturity influences the water loss characteristics of a fruit, this has to be considered in the storage strategy. Late-harvested apples were found to have greater water permeance than early-harvested apples, suggesting that water loss control measures must be more important for late-harvested fruit than for early-harvested fruit (Maguire et al., 1997). The same has been found to be the case in tomatoes (Díaz-Pérez, 1997).

Senescence, expressed as visible yellowing, can be an important determinant of shelf life, and the maturity of the fruit or vegetable at harvest can often have an influence on the development of yellow coloration. Yellowing in broccoli is considered a major determinant of shelf life (Toivonen and Sweeney, 1997). As broccoli florets mature, they become more susceptible to yellowing, presumably since the florets are approaching a natural vegetative senescence stage (Watada et al., 1984; Tian et al., 1995). A similar relationship has been found with postharvest yellowing in cucumbers (Schouten et al., 1997), which are harvested as immature fruit but begin to approach a normal ripening stage if harvested too late (Watada et al., 1984).

3.3 ASSOCIATION OF HORMONAL CHANGES IN RELATION TO QUALITY

3.3.1 Endogenous Growth Hormones

Ethylene is the most studied endogenous hormone in postharvest research. Levels as low as $0.005\ \mu L\ L^{-1}$ of ethylene have been shown to have significant effects on promoting premature senescence in strawberries, oranges, lettuce, beans, Chinese cabbage, bak choi, choi sum, and gai lan (Wills, 1998). The reduction of shelf life (as determined by consumer acceptability) was proportional to the logarithmic concentration of ethylene, whether the product was stored at 0°C or 20°C (Wills, 1998). Removal of low levels of endogenously generated ethylene has been shown to significantly enhance shelf life in strawberries in overwrapped punnets (Wills and Kim, 1995). It appears that low levels of ethylene lead to premature softening and decay in fruits and vegetables that are not climacteric and show little respiratory response to ethylene (Wills and Kim, 1995; Wills, 1998).

Endogenously produced ethylene in climacteric fruits such as apples, pears, and tomatoes eventually leads to an exponential rise in ethylene synthesis, resulting in highly accentuated levels of ethylene (Figure 3.3). This dramatic rise in ethylene production is associated with rapid biochemical changes in the fruit that lead to what is generally referred to as ripening (Saltveit, 1999).

3.3.2 Exogenously Applied Hormones

3.3.2.1 Ethylene

The following discussion will highlight several points of interest in regard to the effects of exogenously applied ethylene on quality deterioration in fruits and vegetables. More detailed discussion of effects and sources of exogenous ethylene can be found in a review by Saltveit (1999).

Exogenously applied ethylene accelerates the ripening process (Abeles et al., 1992) and hence would generally be considered as detrimental to product quality retention (Saltveit, 1999). In many cases exogenous ethylene production is indeed detrimental to quality, but in some cases the application of exogenous ethylene is used to allow harvest of fruit such as bananas at a mature but unripe stage, and ethylene exposure drives ripening just before delivery of product to the consumer (Saltveit, 1999). This allows a normally damage-prone fruit to be shipped in a relatively bruise-resistant, immature stage over long distances (Kader, 2003). Pears

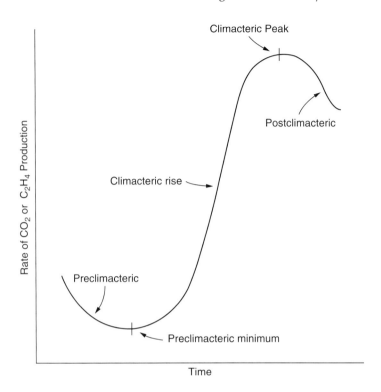

FIGURE 3.3 Phases of the climacteric period. (From Watada et al., *HortScience*, 19, 20, 1984. With permission.)

are an example of temperate fruit that requires chilling or ethylene exposure to initiate and synchronize ripening and softening so that the product will be acceptable to the consumer (Agar et al., 1999). However, exogenous ethylene exposure can also be unintentional in postharvest handling systems, the two most common examples being the production of ethylene by propane-powered fork-lifts in produce handling areas and mixed loads containing ethylene-producing fruits and ethylene-sensitive fruits and vegetables (Saltveit, 1999).

Exogenous ethylene exposure can have negative as well as positive effects on fruits and vegetables. A good example of this point is the case of tomatoes, which are often picked mature green and ripened with ethylene during transport to market (Chomchalow et al., 2002). Tomato fruits exposed to exogenous ethylene treatments to induce ripening have been shown to maintain better vitamin C content than fruit left to ripen without treatment (Chomchalow et al., 2002). This was presumably due to the fact that treated tomatoes reached the fully ripe eating stage faster than untreated fruit, and hence less time was available for vitamin C loss. However, mature green tomatoes exposed to exogenous ethylene have less vitamin C content than fruit left on the vine to ripen (Soto-Zamora et al., 2001); this was believed to be associated with higher levels of ascorbate oxidase and peroxidase activity in ethylene-treated fruit. These two enzymes are believed to reduce vitamin C content in fruit tissue (Soto-Zamora et al., 2001). These results with tomatoes show that the impact of treatments such as exogenous ethylene application can be considered positive or negative depending on the reference that is used in the comparison.

Rapid softening is a significant quality problem in apples and can be induced either by cold storage or by exogenous exposures to ethylene (Johnston et al., 2002a). However, the response to exogenous ethylene is not the same for all cultivars of apples, and three categories of response have been identified (Johnston et al., 2001b, 2002a). The first category is that of apples that exhibit rapid softening without exposure to ethylene. 'Royal Gala' and 'Cox's Orange Pippin' have been identified as belonging to this category. 'Granny Smith' softens rapidly in response to exogenous ethylene and is hence placed into a second category (Johnston et al., 2002a). 'Pacific Rose'™ belongs to a third category since it does not show rapid softening despite exposure to exogenous ethylene production. It is quite clear that the quality degradation associated with rapid softening in apples is quite variable between cultivars, and therefore it is essential to understand the specific characteristics of each cultivar that is being handled.

Exogenous ethylene can also induce production of metabolites in tissue that directly reduce the acceptability of vegetables. Exposure to ethylene generally increases phenylpropanoid pathway activities (Saltveit, 1999), and this can lead to the accumulation of phenolic compounds such as isocoumarins in carrots, which cause bitterness (Lafuente et al., 1996).

3.3.2.2 Cytokinins

Cytokinins are generally considered to be antisenescent in nature (Downs et al., 1997) and hence several studies have examined the effects of cytokinin application on senescence and shelf life in fruit and vegetables. Exogenously applied 6-benzyl-aminopurine (a cytokinin) has divergent effects on quality. Downs et al. (1997) reported that 6-benzylaminopurine (BAP) delayed yellowing and senescence-associated increases in asparagine, glutamine, and ammonium content of broccoli but had no effect on the decline in sucrose content. BAP has also been shown to slow softening and delay titratable acidity losses but had no effect on soluble solids decline in guava (Sharma and Dashora, 2001). BAP has also been shown to retard cap opening (a sign of senescence) in mushrooms (Braaksma et al., 2001). Cytokinin applications generally have a positive effect in terms of slowing deteriorative processes in both fruits and vegetables.

3.3.2.3 Gibberellic Acid

The phytoalexin columbianetin has been associated with resistance to *Botrytis* rot in celery, and exogenously applied gibbberellic acid (GA) has been shown to slow its decline in tissue during storage, resulting in up to an nine times increase in resistance to rot (Afek, 2002). This same application of GA also inhibited the accumulation of psoralens by 50%. Psoralens are associated with skin rash development in humans (Aharoni et al., 1996), so this response suggests that GA treatment prevents accumulation of compounds in celery that can lead to adverse health effects in produce handlers. GA treatment of strawberries led to increased phenylalanine ammonia lyase and tyrosine ammonia lyase activities, and these increases were, in turn, related to the total anthocyanin content in the fruit (Montero et al., 1998). Preharvest GA treatment of persimmons resulted in improved firmness retention during storage (Monterde et al., 2001); this has been found to be due to inhibition of cell wall dissolution normally experienced during ripening in that fruit (Ben-Arie et al., 1996). Postharvest application of GA resulted in improved shelf life, reduced weight loss, reduced fruit decay, and improved retention of chlorophyll and vitamin C in green chile peppers (Arora et al., 2000). Similar positive responses were found for mangoes treated with GA (Jain and Mukherjee, 2001). GA application appears to have several different effects that lead to improved quality and its retention in some fruits and vegetables.

3.3.2.4 Methyl Jasmonate

Jasmonates are classed as naturally occurring, growth-regulating oxylipins (Sembdner and Parthier, 1993) that are generally applied as gaseous methyl jasmonate. Methyl jasmonate applied postharvest enhanced the retention of total antioxidant capacity of stored blueberry fruits (Wang, 2001), accelerated ripening (ethylene production and respiration) in mangoes (Lalel et al., 2003b), and also enhanced flavor volatile production in fruit stored at 20°C. Results obtained with mangoes suggest that treated fruit should exhibit accelerated softening, but this point remains to be demonstrated. In papaya, methyl jasmonate treatment increased firmness and resistance to chilling injury at 10°C (Gonzaléz-Aguilar et al., 2003). Holding papaya at 10°C normally leads to premature softening and other disorders (Gonzaléz-Aguilar et al., 2003). Reduction in chilling injury with postharvest methyl jasmonate application has also been demonstrated in avocados, grapefruits, and bell peppers (Meir et al., 1996). Postharvest methyl jasmonate applications increased levels of β-carotene in tomatoes and apple peel (Saniewski et al., 1987; Perez et al., 1993) and inhibited sprouting in radishes, thus extending their shelf life (Wang, 1998).

On the negative side, methyl jasmonate has been shown to promote chlorophyll degradation, ethylene production, and senescence in tomatoes and apples (Saniewski et al., 1987; Perez et al., 1993; Fan et al., 1997c). However, unlike in the case of ethylene, which produces similar effects on chlorophyll, ethylene production, and senescence, methyl jasmonate does not appear to enhance softening or soluble solids loss in apples (Fan et al., 1998). In addition, methyl jasmonate application can suppress production of aroma volatiles (both alcohols and esters) in apples (Fan and Mattheis, 1999), which, again, is the opposite of the effect of exogenous ethylene application.

3.3.2.5 1-Methylcyclopropene

The compound 1-methylcyclopropene (1-MCP) is a recently discovered, man-made gaseous inhibitor of ethylene action (Blankenship and Dole, 2003) and is considered to be an anti-ethylene agent. Postharvest application of 1-MCP has been shown to produce effects that depend on the particular fruit or vegetable, the cultivar, and the specifics of the treatment and/or handling conditions. Blankenship and Dole (2003) have published an extensive review regarding the effects of 1-MCP, and hence the discussion here will be limited to raising points pertaining directly to postharvest produce degradation. Apples have been the most studied of fruits and vegetables in terms of response to 1-MCP. For example, in 'Anna,' a summer apple, 1-MCP significantly reduces the fruit's propensity to become mealy in texture (Pre-Aymard et al., 2003). However, not all apple cultivars respond to the same degree to 1-MCP treatment (Watkins et al., 2000). The level of benefit is also enhanced if the apples are held in controlled atmosphere storage conditions as opposed to cold air storage (Watkins et al., 2000). This second aspect of response is due to the fact that high carbon dioxide acts synergistically with 1-MCP in controlling softening in apples (Lu and Toivonen, 2003).

Unfortunately, not all fruit responds well to 1-MCP treatment. In preclimacteric bananas 1-MCP delays ripening-associated changes, but treated fruits do not properly degreen and their aroma volatile production is inhibited (Golding et al., 1998). Such response renders 1-MCP technology unuseful for mature, unripe bananas (Harris et al., 2000). More recently, it has been demonstrated that response to 1-MCP by partially ripened bananas is extremely variable and hence the technology would not be commercially useful for the partially ripened product (Pelayo et al., 2003).

Another aspect of 1-MCP response relates to the effect the treatment may have on storage disorder development. If disorders are increased, then reduction of quality decline may be overshadowed by an increased incidence of such a disorder. This is the case with 'Shamouti' oranges which show greater susceptibility to chilling injury after treatment with 1-MCP (Porat et al., 1999). In contrast, pineapple has a lower incidence of chilling injury-related internal browning after treatment with 1-MCP (Selvarajah et al., 2001). Thus, whether 1-MCP increases or decreases incidence of storage disorders depends on the fruit or vegetable in question.

While vegetables are not climacteric and are not expected to show a direct response to 1-MCP (Blankenship and Dole, 2003), there are situations, such as mixed loading of produce into trailers and containers, where an ethylene-sensitive vegetable such as broccoli is exposed to ethylene (Saltveit, 1999). Treatment of broccoli with 1-MCP will prevent the reduction of shelf life caused by subsequent exposure to relatively low ethylene concentrations in surrounding air (Ku and Wills, 1999; Fan and Mattheis, 2000). Therefore, 1-MCP treatment could be seen as an insurance against potential exposure to ethylene in postharvest handling situations and the resultant acceleration of quality decline in fruits and vegetables, whether or not they have climacteric ripening patterns.

3.4 RESPIRATORY METABOLISM IN RELATION TO QUALITY

3.4.1 Nonclimacteric Fruits and Vegetables

Nonclimacteric fruits and vegetables are those that do not undergo sharp rises in ethylene and respiration during the ripening process. Generally, declines in quality in such fruits and vegetables tend to be relatively gradual in nature, and hence it can be difficult to discern changes in maturity. Quality changes such as texture are

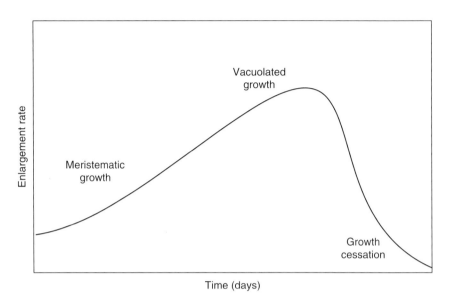

FIGURE 3.4 Time scale by which plant cell wall growth can be analyzed. Cells graduate from meristematic growth to rapid expansion accompanied by vacuolation, followed by cessation of growth. (From Cosgrove, *Plant Physiol. Biochem.*, 38, 109, 2000. With permission.)

most affected by the growth of the cells in the tissues, as illustrated in Figure 3.4. Immature fruits and vegetables are most often harvested at a point in time between meristematic growth and the end of vacuolated growth. As vacuolated growth proceeds, cell walls become thicker and more rigid (Cosgrove, 2000), and hence the tissue becomes tougher. By the time cessation of growth has been reached, the tissue is definitely fibrous and tough. Another feature of nonclimacteric fruits and vegetables is that it is more difficult to develop maturity indices for optimal harvest since they are harvested during time of growth and not after growth has ceased. Most often, size or morphology of the nonclimacteric fruit or vegetable organ is the main criterion used for determination of harvest date (Salunkhe and Desai, 1984; Gross et al., 2002). However, experience with such products has produced practical criteria for optimal harvest, but challenges still exist.

Optimal harvest of sweet corn is determined by examining two maturity characteristics. The first is ensuring that all of the kernels on the cob are fully sized. The second is that the kernels must be succulent as measured by the amount of juice that can be squeezed out of individual kernels (Weichmann, 1987). Overly mature kernels are very firm and starchy. These kernels lack succulence and are tough, and this sweet corn is not acceptable to the consumer.

Another challenge is selection of optimal harvest for winter cabbage. There have been some reports indicating that maturity that optimizes storage potential is related to the amount of nutrient reserves in the stem (core) tissue (Weichmann, 1987). However, inconsistencies between years and cultivars have rendered this approach to maturity estimation unuseful.

A third challenge is that of carrots, in which changes relating to storage ability are reached after the crop has reached a stage of maximal yield (Fritz and Weichmann, 1979). In this case, as the ratio between monosaccharides and sucrose content increases, the storage potential increases. This change has been posited as a means of monitoring "ripening" in carrots. However, it is a gradual process and difficult to monitor without expensive equipment, so carrot size is generally used as the harvest criterion for storage (Salunkhe and Desai, 1984).

3.4.2 CLIMACTERIC FRUITS

Climacteric respiratory behavior occurs only in certain fruits (Saltveit, 2002) and is characterized by a sharp, transient rise in respiration and ethylene evolution during ripening (see Figure 3.3). This dramatic transient rise implies a rapid change over a short period of time, which is another characteristic of climacteric fruit: they exhibit exponentially increasing ethylene production rates associated with rapid softening, increases in sugar, reduction of acid content, and increased aroma volatile production (Abeles et al., 1992; Saltveit, 1999). Often, development of controlled atmosphere and modified atmosphere technologies has focused on inhibiting the climacteric ripening process.

Climacteric ripening fruits are much easier to predict for harvest since a climacteric respiratory pattern is a sharp and easily discerned process, and senescence sets in quickly once the optimal ripeness has passed (Figure 3.3). In addition, there are several easily measured criteria that can be used to substantiate changes in ripeness (i.e., quality) of the fruit. The most useful criteria for estimating ripeness stage are considered to be ethylene evolution and respiration (Lau, 1985; Saltveit, 2002). These can be coupled with such criteria as firmness, starch levels, skin color, and aroma volatile production (Pantastico et al., 1975; Lau, 1985; Mattheis et al., 1991; Young et al., 1996).

3.5 QUALITY DETERIORATION MANAGEMENT THROUGH MATURITY/RIPENESS SELECTION

Maturity and stage of ripeness have been used to manage quality in all fruits and vegetables (Pantastico et al., 1975; Watada et al., 1984; Saltveit, 2002). Most of the extensive literature on the topic relates to work with apples, and despite the volume of work in that commodity, quality problems continue to persist (Toivonen, 2003). The discussion in this section will highlight considerations that may help to improve the quality prediction process.

End use and postharvest handling conditions have a significant influence on the criteria used for selecting optimal maturity. As mentioned in the earlier discussion of pears, selecting optimal maturity measured by firmness is essential to maximize shelf life of slices made from that fruit. If slices are to be handled at low temperatures (near 0°C), then fruit used to produce the slices can be relatively soft at a firmness of 23 to 31 N (Chen et al., 2003). However, if the slices are to be handled at warmer temperatures (~ 10°C), then pears that are to be sliced must have a firmness of 44 to 58 N (Gorny et al., 2000). The second temperature range is more the norm in handling of fresh-cut products, and so it is likely the pears to be used for slicing should be selected at a less ripe stage to ensure that browning and softening are adequately controlled, knowing that the flavor of the product may be sacrificed as a result. Alternately, processors may achieve a significant market advantage, in terms of flavor, if they can learn how to ensure low handling/display temperatures during distribution and marketing.

Maturity testing, using an index based on color patterns on bisected apples, in response to an iodine-potassium iodide solution, has been well developed for evaluating apples for harvest and long-term storage (Smith et al., 1979). In some cases total starch is not well estimated by a starch-iodine test, depending on the apple cultivar and year; however, this is not considered to be a serious problem (Fan et al., 1995). Lau (1985) discusses the complexity of processes that change during maturity and ripening in apples, including changes in starch content, ethylene production, firmness, soluble solids, skin and flesh color, and titratable acidity. However, in practice, it is difficult to incorporate all of these measures into an easily used tool for deciding to harvest a particular orchard at a point in time. This practical limitation might be circumvented by sampling orchards within a growing region, allowing economical determination of multiple harvest predictors (Silsby, 1993). Improvements in technologies might provide more streamlined approaches to monitoring maturity and ripening. Advances in VIS/NIR-spectroscopy offer the potential of being able to optimize harvest dates based on starch, soluble solids, titratable acidity, and firmness determined simultaneously (Peirs et al., 2000), and this may lead to an ability to improve poststorage quality over that obtained using the current starch-iodine test done in parallel with ethylene measurements (which is used as an indicator for softening potential in storage).

In stone fruit (nectarines and peaches), harvest maturity has generally been determined through subjective determination of exterior fruit color (Lavilla et al., 2002). However, this can be difficult in cultivars that show little color change during maturation and ripening. Fruit are harvested at the mature, but not ripe, stage, defined as the "well-mature stage." At this stage external ethylene application is not required to induce ripening. Ethylene is not useful as an indicator since climacteric ethylene production occurs in parallel with softening and so the fruit would already be too soft if harvested based on ethylene production (Tonutti et al., 1997). Lavilla et al. (2002) suggest that measurement of the characteristic aroma volatiles of peaches or nectarines could provide maturity indices, allowing the harvest of these fruits at a well-mature stage in a condition that ensures the least amount of injury during transport and marketing and that the fruit can ripen to produce a high-quality product for the consumer.

Another aspect that requires further study is that of mathematical modeling to predict optimal harvest dates. Hertog (2002) has pointed out that the variability of any specific population must be understood with respect to postharvest behavior since this variability will profoundly influence perceived fruit quality. It is not only important to understand the average quality change of, say, a population of fruit in a transport container, but it is also important to understand the extreme quality differences in that population. Therefore, models can be useful only if they can capture such variation. In addition, use of models may stimulate research to reduce this inherent variability, providing improved consistency in quality and improved confidence by buyers and consumers. Schouten et al. (1997) have developed a crop growth model that uses individual fruit chronological age as a predictor of the shelf life potential of greenhouse-grown cucumbers. They went further to develop a color analysis model using red and green color intensity ratios (measured with a color digital camera) as an accurate predictor of shelf life for 95% of the fruit within a single grower harvest (Schouten et al., 2002). This approach could be used for any green vegetable for which yellowing is the indicator of end of useful shelf life. Another modeling approach has been to improve the reliability of the existing starch indexing method used to estimate maturity in apples. Starch indexing often involves assessment of the relative amount of clearing (area of tissue not reactive to iodine) that is present at a point in time in an iodine-stained section of apple (Smith et al., 1979). However, that approach is subject to a myriad of errors. As an alternative, a nonlinear model can be assumed from past experience and sets of data points from the current year fitted to the model, producing a modified model that would inherently account for seasonal differences. As a result, the actual value collected at any point in time during the season becomes less important than the pattern of changes described by the model fitted with the actual value and the previously collected data points for the season. Such an approach to using starch test values has been found to greatly improve the predictive power of starch testing (Peirs et al., 2002). These examples demonstrate that modeling can improve maturity estimation in nonclimacteric (cucumber) and climacteric (apple) fruits.

3.6 SUMMARY

Maturity has a great influence on quality and perceived deterioration in storage, but it must be kept in mind that maturity is a relative measure and should be used cautiously. Factors such as crop year, specific orchard, fruit position on the tree, and temperature history during crop growth can influence the relative importance of maturity on quality retention in apples (Knee and Smith, 1989; Ferguson et al., 1999). For example, aroma may be significantly influenced by small differences in maturity in some years in some orchards, but it may not be significantly influenced until there are large maturity differences in other years (Knee and Smith, 1989). To make matters worse, not all quality characters respond similarly in relation to maturity in different years and in different orchards (Knee and Smith, 1989). It is for this reason that single-criterion approaches to measuring maturity for harvest are considered to be risky and multicharacter measures are recommended (Lau, 1985; Harker et al., 1997). This also points to the fact that the molecular and biochemical bases of maturity and ripening processes and their interactions are only superficially understood.

Proper and reliable estimation of optimal maturity for harvest or use can ensure maximal quality and quality retention for all fruits and vegetables. However, there are gaps in understanding of maturity criteria that will require further work. As in the case of apples, improvements in maturity assessment have led to significant improvements in storage quality in the past (Lau, 1985). If further improvements are to be seen over the range of horticultural commodities grown worldwide, commodities other than apples must receive increased attention.

REFERENCES

Abeles, F.B., Morgan, P.W., and Saltveit, M.E., *Ethylene in Plant Biology*, Academic Press, San Diego, CA, 1992.

Afek, U., The effect of GA_3 on psoralens and on columbianetin, a phytoalexin associated with celery resistance to pathogens during storage, *Adv. Hortic. Sci.*, 14, 189, 2002.

Aharoni, N. et al. Impact of celery age and infection by *Botrytis cinerea* on linear furanocoumarin (psoralens) content in stored celery, *Phytoparasitica*, 24, 195, 1996.

Agar, I.T., Biasi, W.V., and Mitcham, E.J., Exogenous ethylene accelerates ripening responses in 'Bartlett' pears regardless of maturity or growing season, *Postharvest Biol. Technol.*, 17, 67, 1999.

Arora, S.K. et al. Effect of gibberellic acid (GA_3) treatment on the shelf-life of chilli (*Capsicum annuum* L.) cv. Pusa Jwala, *Haryana Agric. Univ. J. Res.*, 30, 1, 2000.

Awad, M.A. et al. Flavonoid and chlorogenic acid changes in skin of 'Elstar' and 'Jonagold' apples during storage and ripening, *Sci. Hortic.*, 90, 69, 2001.

Baldwin, E.A. et al. Relationship between sensory and instrumental analysis of tomato, *J. Am. Soc. Hortic. Sci.*, 123, 906, 1998.

Ben-Arie, R. et al. Cell wall metabolism in gibberellin-treated persimmon fruits, *Plant Growth Regulat.*, 19, 25, 1996.

Beveridge, H.J.T., Maturity and quality grades for fruit and vegetables, in *Handbook of Postharvest Technology of Cereals, Fruits, Vegetables, Tea, and Spices*, Chakraverty, A.S. et al., Eds., Marcel Dekker, New York, 2003, chap. 17.

Blankenship, S.M. and Dole J.M., 1-Methylcyclopropene: A review, *Postharvest Biol. Technol.*, 28, 1, 2003.

Braaksma, A. et al., Effect of cytokinin on cap opening in *Agaricus bisporus* during storage, *Postharvest Biol. Technol.*, 23, 171, 2001.

Brecht, J.K., Kader, A.A., and Ramming, D.W., Description and postharvest physiology of some slow-ripening nectarine genotypes, *J. Am. Soc. Hortic. Sci.*, 109, 596, 1984.

Chen, P.M., Varga, D.M., and Seavert, C.F., Developing a value-added fresh-cut 'D'Anjou' pear product, *HortTechnology*, 13, 314, 2003.

Choi, C. et al. Differences in levels of pectic substances and firmness in fruit from six sweet cherry genotypes, *J. Am. Pomol. Soc.*, 56, 197, 2002.

Chomchalow, S. et al., Fruit maturity and timing of ethylene treatment affect storage performance of green tomatoes at chilling and non-chilling temperatures, *HortTechnology*, 12, 104, 2002.

Chubey, B.B. and Nylund, R.E., The effect of maturity and environment on phenolic compounds and oxidative browning in carrots, *J. Am. Soc. Hortic. Sci.*, 95, 393, 1970.

Cosgrove, D.J., Expansive growth of plant cell walls, *Plant Physiol. Biochem.*, 38, 109, 2000.

Díaz-Pérez, J.C., Changes in transpiration rates and skin permeance as affected by storage and stage of tomato fruit ripeness, in *Proc. 7th Int. CA Res. Conf., Vol. 4, Vegetables and Ornamentals*, Saltveit, M.E., Ed., University of California, Davis, 1997, 34.

Downs, C.G., Somerfield, S.D., and Davey, M.C., Cytokinin treatment delays senescence but not sucrose loss in harvested broccoli, *Postharvest Biol. Technol.*, 11, 93, 1997.

Elgar, H.J., Burmeister, D.M., and Watkins, C.B., CO_2 and O_2 effects on a browning disorder in 'Braeburn' apple, in *Proc. 7th Int. CA Res. Conf., Vol. 2, Apples and Pears*, Mitcham, E.J., Ed., University of California, Davis, 1997, 198.

Fan, X. and Mattheis, J.P., Impact of 1-methylcyclopropene and methyl jasmonate on apple volatile production, *J. Agric. Food Chem.*, 47, 2847, 1999.

Fan, X. and Mattheis, J.P., Yellowing of broccoli in storage is reduced by 1-methylcyclopropene, *HortScience*, 35, 885, 2000.

Fan, X., Mattheis, J.P., and Fellman, J.K., Responses of apples to postharvest jasmonate treatments, *J. Am. Soc. Hortic. Sci.*, 123, 421, 1998.

Fan, X. et al., Changes in amylase and total starch content in 'Fuji' apples during maturation, *HortScience*, 30, 104, 1995.

Fan, X. et al., Optimum harvest date and controlled atmosphere storage potential of 'Fuji' apples, in *Proc. 7th Int. CA Res. Conf., Vol. 2, Apples and Pears*, Mitcham, E.J., Ed., University of California, Davis, 1997a, 42.

Fan, X. et al., Evaluation of 'Fuji' apple ground color at harvest as a predictor of post-storage fruit quality, in *Proc. 7th Int. CA Res. Conf., Vol. 2, Apples and Pears*, Mitcham, E.J., Ed., University of California, Davis, 1997b, 235.

Fan, X. et al., Effect of methyl jasmonate on ethylene and volatile production by Summerred apples depends on the fruit development stage, *J. Agric. Food Chem.*, 45, 208, 1997c.

Fellman, J.K. et al., Relationship of harvest maturity to flavor regeneration after CA storage of 'Delicious' apples, *Postharvest Biol. Technol.*, 27, 39, 2003.

Ferguson, I., Volz, R., and Woolf, A., Preharvest factors affecting physiological disorders of fruit, *Postharvest Biol. Technol.*, 15, 255, 1999.

Forney, C.F. et al., Changes in strawberry fruit quality during ripening on and off the plant, *Acta Hortic.*, 464, 506, 1998.

Fritz, D. and Weichmann, J., Influence of the harvesting date of carrots on quality and quality preservation, *Acta Hortic.*, 93, 91, 1979.

Giovanelli, G. et al., Variation in antioxidant components of tomato during vine and post-harvest ripening, *J. Sci. Food Agric.*, 79, 1583, 1999.

Giovannoni, J., Molecular biology of fruit maturation and ripening, *Annu. Rev. Plant Physiol. Plant Mol. Biol.*, 52, 725, 2001.

Girard, B. and Lau, O.L., Effect of maturity and storage on quality and volatile production of 'Jonagold' apples, *Food Res. Int.*, 28, 465, 1995.

Golding, J.B. et al., Application of 1-MCP and propylene to identify ethylene-dependent ripening processes in mature banana fruit, *Postharvest Biol. Technol.*, 14, 87, 1998.

González-Aguilar, G.A., Buta, J.G., and Wang, C.Y., Methyl jasmonate and modified atmosphere packaging (MAP) reduce decay and maintain postharvest quality of papaya 'Sunrise', *Postharvest Biol. Technol.*, 28, 361, 2003.

Gorny, J.R. et al., Quality changes in fresh-cut pear slices as affected by cultivar, ripeness stage, fruit size, and storage regime, *J. Food Sci.*, 65, 541, 2000.

Gross, K.C., Wang, C.Y., and Saltveit, M., *The Commercial Storage of Fruits, Vegetables, and Florist and Nursery Crops*, U.S. Department of Agriculture, Agriculture Handbook 66, available at http://www.ba.ars.usda.gov/hb66/index.html, 2002, accessed November 8, 2002.

Harker, F.R. et al., Texture of fresh fruit, *Hortic. Rev.*, 20, 121, 1997.

Harker, F.R. et al., Sensory interpretation of instrumental measurements. 1: Texture of apple fruit, *Postharvest Biol. Technol.*, 24, 225, 2002a.

Harker, F.R. et al., Sensory interpretation of instrumental measurements. 2: Sweet and acid taste of apple fruit, *Postharvest Biol. Technol.*, 24, 241, 2002b.

Harris, D.R. et al., Effect of fruit maturity on efficacy of 1-methylcyclopropene to delay ripening in bananas, *Postharvest Biol. Technol.*, 20, 303, 2000.

Hertog, M.L.A.T.M., The impact of biological variation on postharvest population dynamics, *Postharvest Biol. Technol.*, 26, 253, 2002.

Izumi, H. et al., Physiology and quality of fresh-cut mango is affected by low O_2 controlled atmosphere storage, maturity and storage temperature, *Acta Hortic.*, 600, 833, 2003.

Jackman, R.L. and Stanley, D.W., Perspectives in the textural evaluation of plant foods, *Trends Food Sci. Technol.*, 6, 187, 1995.

Jain, S.K. and Mukherjee, S., Post-harvest application of GA_3 to delay ripening in mango (Mangifera indica L.) cv. 'Langra', *J. Eco-Physiol.*, 4, 1, 2001.

Jiang, Y., Joyce, D.C., and Terry, L.A., 1-Methylcyclopropene treatment affects strawberry fruit decay, *Postharvest Biol. Technol.*, 23, 227, 2001.

Jiménez, A. et al., Antioxidant systems and their relationship with the response of pepper fruits to storage at 20°C, *Agric. Food Chem.*, 51, 6293, 2003.

Jobling, J.J. and McGlasson, W.B., A comparison of ethylene production, maturity and controlled atmosphere storage life of Gala, Fuji and Lady Williams apples (*Malus domestica*, Borkh.), *Postharvest Biol. Technol.*, 6, 209, 1995.

Johnston, J.W. et al., Temperature induces differential softening responses in apple cultivars, *Postharvest Biol. Technol.*, 23, 185, 2001a.

Johnston, J.W. et al., Temperature and ethylene affect induction of rapid softening in 'Granny Smith' and 'Pacific Rose™' apple cultivars, *Postharvest Biol. Technol.*, 25, 257, 2001b.

Johnston, J.W., Hewett, E.W., and Hertog, M.L.A.T.M., Postharvest softening of apple (*Malus domestica*) fruit: a review, *N. Z. J. Crop Hortic. Sci.*, 30, 145, 2002a.

Johnston, J.W. et al., Harvest date and fruit size affect postharvest ripening, *J. Hortic. Sci. Biotechnol.*, 77, 355, 2002b.

Kader, A.A., Quality parameters of fresh-cut fruit and vegetable products, in *Fresh-Cut Fruits and Vegetables: Science, Technology, and Market*, Olusola Lamikanra, Ed., CRC Press, Boca Raton, FL, 2003, 11.

Kim, D.M., Smith, N.L., and Lee, C.Y., Quality of minimally processed apple slices from selected cultivars, *J. Food Sci.*, 1115, 1993.

Knee, M. and Hatfield, S.G.S., The metabolism of alcohols by apple fruit tissue, *J. Sci. Food Agric.*, 32, 593, 1981.

Knee, M. and Smith, S.M., Variation in quality of apple fruits stored after harvest on different dates, *J. Hortic. Sci.*, 64, 413, 1989.

Ku, V.V.V. and Wills, R.B.H., Effect of 1-methlcyclopropene on storage life of broccoli, *Postharvest Biol. Technol.*, 17, 127, 1999.

Lafuente, M.T. et al., Factors influencing ethylene-induced isocoumarin formation and increased respiration in carrots, *J. Am. Soc. Hortic. Sci.*, 121, 537, 1996.

Lalel, H.J.D., Singh, Z., and Tan, S.C., Glycosidically-bound aroma volatile compounds in the skin and pulp of 'Kensington Pride' mango fruit at different stages of maturity, *Postharvest Biol. Technol.*, 29, 205, 2003a.

Lalel, H.J.D., Singh, Z., and Tan, S.C., Maturity stage at harvest affects fruit ripening, quality and biosynthesis of aroma volatile compounds in 'Kensington Pride' mango, *J. Hortic. Sci. Biotechnol.*, 78, 225, 2003b.

Lalel, H.J.D., Singh, Z., and Tan, S.C., The role of methyl jasmonate in mango ripening and biosynthesis of aroma volatile compounds, *J. Hortic. Sci. Biotechnol.*, 78, 470, 2003c.

Lau, O.L., Harvest indices for B.C. apples, *B.C. Orchardist*, 7, 1a, 1985.

Lau, O.L., Influence of climate, harvest maturity, waxing, O_2 and CO_2 on browning disorders of 'Braeburn' apples, in *Proc. 7th Int. CA Res. Conf., Vol. 2, Apples and Pears*, Mitcham, E.J., Ed., University of California, Davis, 1997, 132.

Lavilla, T. et al., Multivariate analysis of maturity stages, including quality and aroma, in 'Royal Glory' peaches and 'Big Top' nectarines, *J. Sci. Food Agric.*, 82, 1842, 2002.

Lee, S.K. and Kader, A.A., Preharvest and postharvest factors influencing vitamin C content of horticultural crops, *Postharvest Biol. Technol.*, 20, 207, 2000.

Lentheric, I. et al., Harvest date affects the antioxidative systems in pear fruits, *J. Hortic. Sci. Biotechnol.*, 74, 791, 1999.

López-Gálvez, G. et al., Quality of red and green fresh-cut peppers stored in controlled atmospheres, in *Proc. 7th Int. CA Res. Conf., Vol. 5, Fresh-Cut Fruits and Vegetables and MAP*, Gorny, J.R., Ed., University of California, Davis, 1997, 152.

Lu, C. and Toivonen, P.M.A., 1-Methylcyclopropene plus high CO_2 applied after storage reduces ethylene production and enhances shelf life of 'Gala' apples, *Can. J. Plant Sci.*, 83, 817, 2003.

Maguire, K., Banks, N., and Lang, S., Harvest and cultivar effects on water vapour permeance in apples, in *Proc. 7th Int. CA Res. Conf., Vol. 2, Apples and Pears*, Mitcham, E.J., Ed., University of California, Davis, 1997, 246.

Mattheis, J.P. et al., Changes in headspace volatiles during physiological development of Bisbee Delicious apple fruit, *J. Agric. Food Chem.*, 39, 1902, 1991.

Meir, S. et al., Reduction of chilling injury in stored avocado, grapefruit, and bell pepper by methyl jasmonate, *Can. J. Bot.*, 74, 870, 1996.

Monterde, A., Del Río, M.A., and Navaro, P., Effects of giberellic acid treatment and cold storage on fruit quality of persimmon cv. 'Rojo brillante', in *Improving Postharvest Technologies of Fruits, Vegetables and Ornamentals: IIR Conference proceedings*, Artés, F., Gil, M.I., and Conesa, M.A., Eds., 2001, 315.

Montero, T. et al., Effects of gibberellic acid (GA_3) on strawberry PAL (phenylalanine ammonia-lyase) and TAL (tyrosine ammonia-lyase enzyme activities, *J. Sci. Food Agric.*, 77, 230, 1998.

Pantastico, E.B., Subramanyam, H., Bhatti, M.B., Ali, N., and Akamine, E.K., Harvest indices, in *Postharvest Physiology, Handling, and Utilization of Tropical and Subtropical Fruits and Vegetables*, Pantastico, E.B., Ed., AVI Publishing, Westport, CT, 1975, 56.

Peirs, A. et al., Prediction of the optimal picking date of different apple cultivars by means of VIS/NIR-spectroscopy, *Postharvest Biol. Technol.*, 21, 189, 2000.

Peirs, A. et al., Uncertainty analysis and modeling of the starch index during apple fruit maturation, *Postharvest Biol. Technol.*, 26, 199, 2002.

Pelayo, C. et al., Variability in responses of partially ripe bananas to 1-methylcyclopropene, *Postharvest Biol. Technol.*, 28, 75, 2003.

Perez, A.G. et al., Methyl jasmonate vapor promotes β-carotene synthesis and chlorophyll degradation in Golden Delicious apple peel, *J. Plant Growth Regulat.*, 12, 163, 1993.

Pesis, E. et al., Production of acetaldehyde and ethanol during maturation and modified atmosphere storage of litchi fruit, *Postharvest Biol. Technol.*, 26, 157, 2002.

Plotto, A. et al., 'Gala', 'Braeburn', and 'Fuji' apples: maturity indices and quality after storage, *Fruit Var. J.*, 49, 133, 1995.

Porat, R. et al., Effects of ethylene and 1-methylcyclopropene on the postharvest qualities of 'Shamouti' oranges, *Postharvest Biol. Technol.*, 15, 155, 1999.

Pre-Aymard, C., Weksler, A., and Lurie, S., Responses of 'Anna', a rapidly ripening summer apple, to 1-methylcyclopropene, *Postharvest Biol. Technol.*, 27, 163, 2003.

Reid, M.S., Maturation and maturity indices, in *Postharvest Technology of Horticultural Crops*, Kader, A.A., Ed., Publication 3311, Agriculture and Natural Resources, University of California, Oakland, 2002, 55.

Roelofs, F.P.M.M. and deJager, A., Reduction of brownheart in Conference pears, in *Proc. 7th Int. CA Res. Conf., Vol. 2, Apples and Pears*, Mitcham, E.J., Ed., University of California, Davis, 1997, 138.

Rousseas, D. et al., Effect of physiological maturity stage on the storage behaviour of 'Kantoula' pear, in *Improving Postharvest Technologies of Fruits, Vegetables and Ornamentals: IIR Conference Proceedings*, Artés, F., Gil, M.I., and Conesa, M.A., Eds., 2001, 194.

Saltveit, M.E., Effect of ethylene on quality of fresh fruits and vegetables, *Postharvest Biol. Technol.*, 15, 279, 1999.

Saltveit, M.E., Respiratory metabolism, in *The Commercial Storage of Fruits, Vegetables, and Florist and Nursery Crops,* Gross, K.C., Wang, C.Y., and Saltveit, M., Eds., U.S. Department of Agriculture, Agriculture Handbook 66, available at http://www.ba.ars.usda.gov/hb66/index.html, 2002.

Salunkhe, D.K. and Desai, B.B., *Postharvest Biotechnology of Vegetables*, Vol. I and II, CRC Press, Boca Raton, FL, 1984.

Sams, C.E., Preharvest factors affecting postharvest texture, *Postharvest Biol. Technol.*, 15, 249, 1999.

Saniewski, M., Urbanek, H., and Czapski, J., Effect of methyl jasmonate on ethylene production, chlorophyll degradation, and polygalacturonase activity in tomatoes, *J. Plant Physiol.*, 127, 177, 1987.

Schouten, R.E. et al., Keeping quality of cucumbers predicted by biological age, *Postharvest Biol. Technol.*, 12, 175, 1997.

Schouten, R.E., Tijskens, L.M.M., and van Kooten, O., Predicting keeping quality of batches of cucumber fruit based on a physiological mechanism, *Postharvest Biol. Technol.*, 26, 209, 2002.

Sembdner, G. and Parthier, B., The biochemistry and the physiological and molecular actions of jasmonates, *Annu. Rev. Plant Physiol. Plant Mol. Biol.*, 44, 569, 1993.

Selvarajah, S., Bauchot, A.D., and John, P., Internal browning in cold-stored pineapples is suppressed by a postharvest application of 1-methylcyclopropene, *Postharvest Biol. Technol.*, 23, 167, 2001.

Sharma, R.K. and Dashora, L.K., Effect of mustard oil and benzyladenine on the shelf life of guava (*Psidium guajava* L.) cv. Allahabad Safeda, *Haryana J. Hortic. Sci.*, 30, 213, 2001.

Silsby, K.J., Identifying apples harvest windows, in *Proc. 6th Internat. CA Res. Conf.*, Ithaca, NY, 1993, 554.

Smith, R.B. et al., The starch iodine test for determining stage of maturation in apples, *Can. J. Plant Sci.*, 59, 725, 1979.

Song, J. and Bangerth, F., The effect of harvest date on aroma compound production from 'Golden Delicious' apple fruit and relationship to respiration and ethylene production, *Postharvest Biol. Technol.*, 8, 259, 1996.

Soto-Zamora, G., Yahia, E.M., and Steta-Gandara, M., Ascorbic acid content in relation with ascorbate oxidase and ascorbate peroxidase in vine- or ethylene-ripened tomato fruit, in *Improving Postharvest Technologies of Fruits, Vegetables and Ornamentals: IIR Conference Proceedings*, Artés, F., Gil, M.I., and Conesa, M.A., Eds., 2001, 199.

Taylor, M.A. et al., Effect of harvest maturity on pectic substances, internal conductivity, soluble solids, and gel breakdown in cold stored 'Songold' plums, *Postharvest Biol. Technol.*, 5, 285, 1995.

Tian, M.S. et al., Effects of maturity, cytokinin, and ethylene on yellowing of broccoli after harvest, *Postharvest Biol. Technol.*, 6, 29, 1995.

Toivonen, P.M.A. and Sweeney, M., Differences in chlorophyll loss at 13°C for two broccoli (*Brassica oleracea* L.) cultivars associated with antioxidant enzyme activities, *J. Agric. Food Chem.*, 46, 20, 1997.

Toivonen, P.M.A., Effects of storage conditions and postharvest procedures on oxidative stress in fruits and vegetables, in *Postharvest Oxidative Stress in Horticultural Crops*, Hodges, D.M., Ed., Food Products Press, New York, 2003, 69.

Tonutti, P. et al., Ethylene evolution and aminocyclopropane-1-carboxylate oxidase gene expression during early development and ripening of peach fruit, *J. Am. Soc. Hortic. Sci.*, 122, 642, 1997.

Tu, R. et al., Effect of picking time and storage conditions on 'Cox's Orange Pippin' apple texture in relation to cell wall changes, *J. Hortic. Sci.*, 72, 971, 1997.

Wang, C.Y., Methyl jasmonate inhibits postharvest sprouting and improves storage quality of radishes, *Postharvest Biol. Technol.*, 14, 179, 1998.

Wang, C.Y., Improving storage quality of blueberries with methyl jasmonate, in *Improving Postharvest Technologies of Fruits, Vegetables and Ornamentals: IIR Conference Proceedings*, Artés, F., Gil, M.I., and Conesa, M.A., Eds., 2001, 206.

Watada, A.E. et al., Terminology of the description of developmental stages of horticultural crops, *HortScience*, 19, 20, 1984.

Watkins, C.B., Nock, J.F., and Whitaker, B.D., Responses of early, mid, and late season apple cultivars to postharvest application of 1-methylcyclopropene (1-MCP) under air and controlled atmosphere storage conditions, *Postharvest Biol. Technol.*, 19, 17, 2000.

Weichmann, J., Introduction, in *Postharvest Physiology of Vegetables*, Weichmann, J., Ed., Marcel Dekker, New York, 1987, 3.

Wills, R.B.H., Enhancement of senescence in non-climacteric fruit and vegetables by low ethylene levels, *Acta Hortic.*, 464, 159, 1998.

Wills, R.B.H. and Kim, G.M., Effect of ethylene on postharvest life of strawberries, *Postharvest Biol. Technol.*, 6, 249, 1995.

Yahia, E.M., Contreras-Padilla, M., and Gonzalez-Aguilar, G., Ascorbic acid content in relation to ascorbic acid oxidase activity and polyamine content in tomato and bell pepper fruit during development, maturation and senescence, *Lebensm.-Wiss. u.-Technol.*, 34, 452, 2001.

Young, H. et al., Causal effects of aroma compounds on Royal Gala flavours, *J. Sci. Food Agric.*, 71, 329, 1996.

Zerbini, P.E., Pianezzola, A., and Grassi, M., Poststorage sensory profiles of fruit of five apple cultivars harvested at different maturity stages, *J. Food Qual.*, 22, 1, 1999.

4 Mechanical Injury of Fresh Produce

Olusola Lamikanra
Southern Regional Research Center, USDA-Agricultural Research Service, New Orleans, LA

CONTENTS

4.1 INTRODUCTION

The growing demand for higher-quality fresh fruit and vegetables has revealed a worldwide problem concerning mechanical damage of products reaching the consumer through the distribution chain. Harvesting, handling, postharvest treatments, packaging, transport, and fruit distribution involve a large number of mechanical operations that subject the produce to dynamic loads. Mechanical damage is one of the most important factors that reduce produce quality. Loss due to bruising represents a substantial portion of total losses,[1] and mechanical damage has long been internationally identified as the leading cause of quality loss at wholesale and retail levels for a number of fruits and vegetables.[2,3] Minimizing injury caused by abrasion, dropping, and shearing of produce will decrease postharvest losses, improve quality, and enhance growers' returns. This chapter is a review of some factors that influence susceptibility of produce to mechanical damage and the biochemical effects of such injury on produce quality.

During the preharvest stage of development, fruits and vegetables may be subjected to mechanical stresses from a variety of sources (e.g., wind, rain, hail, and compression of developing fruit due to adjacent fruit or limbs), and these stresses can have a pronounced effect on the eventual appearance of the product. Mechanical stress can be separated into two general types: mechanical perturbment and physical wounding.[4] The former, while not resulting in a direct wound to the product, can result in distinct alterations in its growth and development, altering the eventual size and shape of the product. Physical wounding is perhaps the more readily recognized form of mechanical damage in that it causes a physical injury (a mechanical failure) to the tissue. Types of mechanical stress include friction, impact, and compression, and these lead to one or more types of tissue failure (cleavage, slip, bruising, and buckling), which result in subtle to pronounced alterations in the appearance of the product.[4]

Injury may also be related to harvesting, hauling, and packing, resulting in cuts and bruises of differing severity. Minor wounds provide entry points for postharvest pathogens, which invariably leads to decay. Postharvest, mechanically induced damage may be caused by loading compression, impact, and vibration as in normal packing lines where damage results from forces such as pressure between fruit and machinery, surface abrasion, and handling.[5] Other factors that determine the differences in bruising susceptibility to a given force include packaging, transportation and harvesting methods, storage conditions, and produce characteristics.[6]

Fresh-cut processing involves the deliberate wounding of fruit and vegetable tissues to provide a product that requires less handling by the consumer. Fresh-cut products are usually washed, packaged, and maintained with refrigeration. The demand for convenience and freshlike produce has resulted in the rapid growth of this industry. Fresh-cut produce is the fastest growing food category in the supermarket, with an estimated retail market of $14 billion in the year 2003. Cutting removes the natural protection of the epidermis and destroys the internal compartmentation that keeps cells separate. The disruption of tissue and cell integrity that results from fresh-cut processing decreases produce's shelf life through physiological and biochemical changes similar to those seen in tissues of fruits and vegetables that are not fresh-cut.

4.2 WOUND RESPONSE

Responses to mechanical damage involve the generation, translocation, perception, and transduction of wound signals to activate the expression of wound-inducible genes.

There is clear evidence that plant tissues have evolved a highly sensitive and efficient system for monitoring changes in their environment. Surviving cells in a wounded plant must perceive a "death message" from killed cells in order to start the signal processing that results in defense responses. Many structurally different molecules play regulatory roles in "wound signaling," including the oligopeptide system,[7] oligosaccharides released from the damaged cell wall,[8] and molecules with hormonal activity such as jasmonates,[9] ethylene,[10] and abscisic acid.[11] A molecular response is evoked in the local vicinity of a stimulus and in other parts of the plant, often at distances far removed from the initial site of stimulation. Thus, a stimulus can cause local and long-range effects. The long-range or systematic effects can be rapid and involve a massive amplification of response, as local changes occurring within one cell or a small group of cells producing systemic changes throughout the entire plant.[12] Wounding of plant cells and tissues always causes bioelectric responses. The vicinity of a wound, for example, becomes transiently polarized.[13] Wounded cells induce at least one of three kinds of responses in the receptor cell: (1) slow depolarization lasting more than 10 min, (2) action potentials, and (3) small spikes. The first of these response types is ubiquitous. This transient depolarization is usually the first step in a chain of reactions that could be propagated over large distances in the plant. Cell turgor pressure is needed for the depolarization that is generated only at the nodal end of the receptor cell, not at the flank. The "death message" from killed cells contains information that turgor pressure has been lost.[14] Wound signals may also be transmitted rapidly over considerable distances through propagated electrical signals at speeds that could range from 0.5 cm/h in cut lettuce to 0.13 cm/sec in wounded *Pueraria lobata* leaves.[15]

Inward current at the wound site and outward current through the unwounded parts, resulting in a change of polarity in cells, involve ions in the plant cell. Calcium ions, for example, are involved in many signal transduction pathways.[16] The control of numerous cellular reactions by calcium occurs with high spatial and temporal precision in the cytoplasm. Many of the effects of calcium are exerted through proteins containing EF-hand domains for Ca^{2+}-binding as a common structural motif.[17] Calmodulin, one of the calcium-binding proteins, is found in all eukaryotes and is known to be a primary transducer of the intracellular calcium signal. It mediates a number of calcium-regulated events in eukaryotic cells. It has been observed that when the plant cell is activated by external stimuli, the intracellular levels of free calcium are increased.[18] Calcium activates calmodulin by binding to its four calcium-binding domains, thus causing a conformational change. The calcium–calmodulin complex can then regulate the activity of many plant enzymes, including NAD kinase, Ca^{2+}-ATPase, H^+-ATPase, and protein kinase.[18]

Calcium levels within the cells are also important because they profoundly affect stability of cellular membranes. The essential role of calcium in delaying plant senescence is largely associated with its stabilizing influence on cell membranes[19] and the induction of membrane lipid catabolism released as a result of wounding as in fresh-cut processing.[20] Abscisic acid (ABA), a plant hormone involved in the response of plants to reduced guard cell turgor, diminishes the aperture of the stomatal pore and thereby contributes to the ability of the plant to conserve water. Cytosolic Ca^{2+} is involved in the signal transduction pathway that mediates the reduction in guard cell turgor elicited by ABA, and ABA uses a Ca^{2+}-mobilization pathway that involves cyclic adenosine 5′-diphosphoribose (cADPR).[21] A number of compounds that participate in the transmission of stress signals such as nicotinamide, an inhibitor of cADPR action, may offer defensive actions in wounded plant tissues. Nicotinamide and its metabolites appear to play a role in DNA hypomethylation, as a link between various types of stressors and the induction of plant defensive metabolism.[22]

Wounding induces expression of genes encoding defense-related proteins involved in wound healing. Plants react to mechanical injury produced by abiotic or biotic agents by activating a set of responses that include, in most cases, the transcriptional activation of wound responsive (WR) genes. Some of these genes are expressed in the vicinity of the wound site, while others are also systematically activated in the undamaged parts of the plant tissue. Some of them may have a defensive role against organisms that attack the plant tissue and prevent subsequent pathogen invasion, while others may work to facilitate wound healing. Some of the induced proteins are enzymes of phenolic metabolism, such as phenylalanine ammonia-lyase (PAL), whose increased activity leads to the accumulation of phenolic compounds (e.g., chlorogenic acid, dicaffeoyl tartaric acid, and isochlorogenic acid) and tissue browning. Wounding of iceberg lettuce leaves increases PAL activity 6- to 12-fold over 24 h at 10°C and leads to a threefold increase in the total phenolic content within 3 days. There may be a hierarchical order to the plant's response to different abiotic stresses.[23] Plants respond differentially to wounding and pathogens using distinct signaling pathways; wound signals are transmitted to jasmonic acid (JA), which induces basic pathogenesis-related (PR) proteins, whereas pathogenic signals cause, in addition to JA, accumulation of salicylic acid (SA), which stimulates production of acidic PR proteins.[24]

Jasmonic acid has been considered to be a key signal molecule in this pathway. Allene oxide synthase (AOS; hydroperoxide dehydratase; EC 4.2.1.92) catalyzes the first step in the biosynthesis of jasmonic acid from lipoxygenase-derived hydroperoxides of free fatty acids. The addition of silver thiosulfate, an ethylene action inhibitor, prevented the wound-induced expression of both AOS and proteinase inhibitor II (PIN II). Products of hydroperoxide lyase affected neither AOS nor PIN II but induced expression of prosystemin.[25] Systemin, ABA, ethylene, and electrical current have been suggested to function by transmitting the wound signal to JA.[26] Ethylene treatment may, however, inhibit the methyl-JA-induced expression. The defensive gene expression induced by signal transduction through the octadecanoid pathway is suppressed by stress-induced ethylene that is produced as a consequence of wounding.[27] Ethylene also inhibits its own biosynthesis by decreasing 1-aminocyclopropane-1-carboxylate (ACC) synthase transcript levels via a negative feedback loop.[28] ACC is the product of ACC synthase and the immediate precursor to ethylene. Different jasmonic acid-dependent and -independent wound signal transduction pathways have been identified recently and partially characterized. Components of these signalling pathways are mostly similar to those implicated in other signalling cascades in eukaryotes and include reversible protein phosphorylation steps, calcium/calmodulin-regulated events, and production of active oxygen species. Indeed, some of these components involved in transducing wound signals also function in signalling other plant defence responses, suggesting that cross-talk events may regulate temporal and spatial activation of different defences.[29] Other compounds, including the oligopeptide systemin, oligosaccharides, and other phytohormones such as ABA have also been proposed to play a role in wound signaling.[29]

4.3 TISSUE DAMAGE

4.3.1 WOUND SEVERITY

The extent of product damage varies according to a number of conditions. Variety, temperature, water content, specific gravity, and degree of maturity are some of these. Mechanical parameters of fruits and vegetables depend on cell dimension and cell wall structure. The initial effect of an impact sufficient to cause damage is cell rupture and/or loss of tissue integrity extending from the organ surface at the impact site to several cell layers below. External bruises are caused by failure of the outer pericarp due to cell rupture or fracture between cells when the tissue receives an impact above the bio-yield point. Normally, the tissue affected by the impact is softer than surrounding tissues. Internal bruising is, however, caused by impacts lower than the bio-yield point if the impact energy is transmitted to the underlying locular tissue with sufficient force so as to impair ripening. In tomatoes, this is indicated by subsequent abnormal color development and increased gel viscosity.[30] Sargent et al.,[31,32] in a study of four tomato cultivars, determined that internal injury could be cumulative. Tomatoes handled at the breaker stage were significantly more susceptible to internal bruising than those handled mature-green. Internal bruising can be easily detected by the unusual increase in ethylene (C_2H_4) production (wound C_2H_4) several hours later. The CO_2 respiration rate may also be used to evaluate internal fruit injury.[33]

Internal bruising significantly affects chemical composition and physical properties of pericarp and locule tissues, but not placental tissue. In bruised tomato locule tissue, carotenoids, vitamin C, and titratable acidity were 37, 15, and 15% lower, respectively, than in unbruised fruit. For bruised pericarp tissue, vitamin C content was 16% lower than for unbruised tissue, whereas bruising increased electrolyte leakage and extractable polygalacturonase activity by 25 and 33%, respectively. Evidence of abnormal ripening following impact bruising was confined to locule and pericarp tissues and may be related to the disruption of cell structure and altered enzyme activity.[34] Firmness plays a critical role in resistance to mechanical damage resistance of apples. Apples tested using destructive (static and impact penetrometers), nondestructive (sphere probe penetrometer), and subjective techniques indicated that during harvesting the highest coefficient of variation was 16.6% for bruising resistance.[35] During storage the highest coefficients of variation were 19% for rupture force and 18% for penetration depth. Puchalski and Bieluga[36] also demonstrated that cultivars with the firmest flesh consisting of smaller cells and intercellular spaces are more resistant to mechanical damage. During storage, the resistance factor fell as a result of ripening processes that weakened the intercellular bonds and decreased the firmness of the fruit. The damage resistance of apples, however, cannot be solely assessed from firmness measurements. Absorbed energy at impact often correlates linearly to bruise volume of apple fruit. The elasticity theory postulated by Roudot et al.[37] permits calculation of the limit strength of apple fruit flesh on the basis of the diameter of the impact surface and the depth of tissue damage. Pressure marks on apples appear to have no effect on parasitic spoilage, although some physiological defects that may be variety-dependent could occur.

Damage inflicted on the fruit 100 to 150 days postharvest led to greater spoilage rate and flesh-browning than damage occurring at harvest time.[38] The pH of bruised apple tissue was significantly higher and °Brix significantly lower than the pH and °Brix of undamaged tissue regardless of the source.[39]

Apricots as well other stone fruit such as peaches and nectarines are dense, with a low volume of intercellular air space, and are susceptible to deep bruising. In peaches the bruise symptoms inside the flesh are cone-shaped with radial fractures.[5,40,41] Fruit impact, under the low to moderate impact energies used, often has negligible effects on fruit respiration and ethylene production in peaches. Bruise incidence increases with volume and drop height, especially with advancing stage of ripeness. Maness et al.[5] suggested that the severity of peach fruit bruise could be determined by either bruise incidence or bruise volume of mesocarp tissue. An assessment of the effects of mechanical handling on peaches using cultivars of Big Top, Caldesi 2000, Centry, and Rich Lady peaches subjected to impacts representative of the conditions observed at the critical points in packing lines by means of a simple drop-test device revealed that at the highest impact level (180 g, 2.20 m/sec) damaged fruit did not exceed 18% and the average dimension of the flesh did not exceed 10 mm in diameter (Big Top) and 6 mm (Centry) in depth. In general, repeated drops did not seem to cause substantial additional damage.[42] Degree of ripeness has a significant effect on quality and shelf life of strawberries. Picking berries when they are partially ripe can reduce mechanical damage during postharvest transport and extend shelf life compared with picking fully ripened berries. However, this needs to be balanced with the reduced sensory and nutritional quality obtained with partially ripened berries.[43]

Tissue damage appears to increase with increased freezing stress. The severity of damage to each of the tissues varied seasonally. Chlorophyll fluorescence emissions were lower with higher freezing stress (except during November and December, when test temperatures were not low enough to significantly damage the seedlings) and showed a strong relationship with morphological assessments of freezing stress. The bulky nature and high moisture content of yam tubers make them vulnerable to mechanical damage during production, handling, and storage operations. Mechanical damage susceptibility is also influenced by the modulus of deformability, bio-yield strength, rupture strength, and density of yam tubers.[44] Harvested cassava roots undergo a rapid postharvest deterioration that is associated with mechanical damage and incipient wound healing. Physiological and biochemical changes that take place following harvesting mainly include activation of preexisting enzymes, degradation of membrane lipids, gene expression, and signal transmission from the sites of wounding.[45] Wound responses in cassava do not remain localized at wound surfaces in roots when they are held at low storage relative humidity but spread through the roots, causing discoloration of the vascular tissue and storage parenchyma.[46] Roots stored at high relative humidity show a more typical wound response with localized production of phenols and periderm formation. Mechanical damage during harvesting and transport produces injuries leading to internal discoloration and rotting decay associated with microbial infections, especially at a high temperature and relative humidity of storage. Deterioration can be delayed by root "curing" (wound healing) by storage at 25 to 35°C and 80 to 50% relative humidity. Chemical treatments such as dipping in calcium hypochlorite, ethyl alcohol, benomyl, or Dichloran do not yield consistent results.[46] It is postulated that the wound healing response, effective in many other storage tissues, is insufficient in cassava to prevent further damage.[45] It is suggested that genetic manipulation of cassava might be beneficial in regulating wound response genes. Potato tuber cells with high turgor pressure (nearly 0.4 MPa) when hit by a pendulum striker created shatter bruises. With a decline in turgor to < 0.2 MPa, little or no shatter bruising occurred. Effects of cultivar on severity of tissue damage was demonstrated by Laerke et al.,[47] who found that tubers of the potato cultivar Maris Piper (MP) were 10 times more blackspot-susceptible than those of the cultivar Colmo (CM). The membranes of MP leaked more K^+, Mg^{2+}, and Ca^{2+} than those of CM. Both cultivars showed a very fast increase in K^+ leakage immediately after impact. Thereafter, CM seemed to reconstitute the membranes, while a high efflux of K^+ from MP continued for up to 32 h of incubation. It has also been reported that potato tubers from 0 and 10% potassium treatments had a significantly lower Young's modulus (a measure of stiffness) than those grown at full-strength Hoagland's solution. Significantly lower failure stress and shock waves of impact were also propagated through both of them.[48]

4.3.2 MECHANICAL HANDLING

Several reports have shown that bruising is linearly related to impact energy,[49] and mechanically stressed fruit exhibit visible degeneration of mesocarp and endocarp during storage. Other factors that determine differences in bruising susceptibility of fruit include methods of packaging, transportation, and harvesting; storage conditions; and fruit characteristics. Mechanical stress itself can also modify fruit metabolism, with important consequences on quality of the product. Mechanical handling on grading lines is one of the most hazardous operations as far as mechanical damage is concerned. Fruit bruising occurs as an overall effect of two combined factors: machine roughness and intrinsic fruit susceptibility.[50] While electronic fruits similar to the fruit in size and shape may be used in grading lines to assess machine roughness,[51] relationships between bruise susceptibility and physical and rheological properties of fruit are used to assess intrinsic fruit susceptibility.[52] Thus, bruise susceptibility increases for lower curvature radius of surfaces in contact, higher tissue turgidity, higher viscoelastic behavior, and lower fruit firmness.[53] Beilza et al.[53] recently described a method to estimate bruise probability by means of logistic regression that can be used for prediction and simulation models to describe the flow of fruits along the grading line and to evaluate impact at each transfer point. The model was also used to estimate the percentage of damaged fruit at the end of the line as a function of its intrinsic susceptibility.

The steps at which onions are most likely to suffer damage are generally the drops between conveyor belts at various stages of harvest and postharvest handling. Holik[54] determined the degree of damage for onions in which 35- to 50-g onions, dried in the field, were dropped from heights of 0 to 1,000 mm. Whereas onions seemed to tolerate falls of up to 400 mm, falls of around 1,000 mm caused considerable mechanical damage. Machinery and storage containers that minimize the height of drops during onion harvest and handling appear to be best suited for minimizing mechanical damage in onions. Mechanical damage also stimulates sprouting of onions during storage.[55] An impact rate of 1.5 m/sec onto a firm surface and a static load of 400 N are considered critical from the viewpoint of subsequent keeping quality of watermelons.[56] A study of the damage caused to Bykovskii 22 watermelons by mechanical harvesting and transport during storage at 12°C (80% relative humidity) for 1 month, followed by 1 month at –6°C (90% relative humidity), indicated that the greatest losses occurred in abraded fruit (none acceptable after 1 month) and in bruised fruit (67% acceptable after 1 month, 17% after 2 months), while controls showed 100% acceptability after 30 days, 97% after 40 days, and 87% after 60 days. Solo papaya (*Carica papaya* L.) fruit removed at different points from a commercial packing house showed that skin injury due to mechanical damage increased as fruit moved through the handling system.[57] The occurrence of green islands — areas of skin that remain green and sunken when the fruit is fully ripe — was apparently induced by mechanical injury. Skin injury was seen in fruit samples in contact with the sides of field bins, but not in fruit taken from the center of the bins. When bruise-free fruit at different stages of ripeness (5 to 50% yellow) were dropped from heights of 0 to 100 cm onto a smooth steel plate to simulate drops and injury incurred during commercial handling, no skin injury occurred, although riper fruit showed internal injury when dropped from > 75 cm. Fruit (10 to 15% yellow) dropped onto sandpaper from a height of 10 cm had skin injury symptoms similar to those seen on fruit from the commercial handling system. These results suggest that abrasion and puncture injury were more important than impact injury for papaya fruit.

4.3.3 Fresh-Cut Produce

Wound severity in cut produce may also be affected by cultivar, preharvest crop management, physiological maturity, and degree of cutting-induced injury.[58] The extent of physical damage that results from cutting fruit tissue affects the intensity of physiological stress, microbial growth, and product shelf life. Inner tissue exposure facilitates contamination by the epithelial microflora. It also increases respiration rates, decompatmentalization of enzymes and substrates, total moisture loss, and overall sensory quality.[59,60] Very sharp cutting tools limit the number of injured cells, unlike blunt cutting instruments, which induce injury to cells many layers below the actual cut.[15] Fruit pieces prepared with sharp cutting tools retained marketable visual qualities longer than the same fruit processed with dull cutters.[61] Pre- and postharvest conditions, tissue anatomy, cell-to-cell adhesion, cell turgor, and cell wall strength are factors contributing to the bruising response,[62] and the magnitude of these factors seems to be affected by ripening and variety.[63]

4.5 HARVESTING

Harvest methods and conditions influence mechanical stress. Kolodyaznaya and Zakatova[64] indicated that mechanically harvesting carrots could considerably increase the percentage of mechanically damaged tubers and that susceptibility may vary with cultivars. Storage of carrots for 180 days, even in a cold environment, resulted in 55% absolute wastage for mechanical harvesting, with 8% additional wastage after finishing off. Wastage was 7% for manual harvesting. Mechanical damage during harvest and preparation affects the internal as well as external quality of carrots. Damage of mechanically harvested carrots may be reduced by reducing the fall distance of carrots between parts of the preparation equipment and reducing the number of right angles through which carrots must pass.[65] Carrots harvested by hand or machine and given additional mechanical stress by shaking in a transport simulator, however, showed no significant difference from machine-harvested carrots with regard to chemical or sensory variables. Principal component analysis showed only slightly different placing of these samples in the score plot.[66] Experiments examining causes of damage also showed that the number of damaged carrots was increased by the frequency of impacts; harvested carrots packed in 3-kg plastic bags broke more easily than those packed in 1-kg bags. The damage caused by falling depended on the surface; the effects of a 30-cm drop onto steel were greater than those of a 90-cm drop onto a conveyor belt. The degree of damage also varied with orientation of the carrot; falls onto the point caused considerably more damage than those onto the side or head.[65] Hand-lifted boxes containing carrot samples also stored better than machine-lifted ones, although storage life did not correlate with obvious mechanical damage.[67]

The need to harvest potato tubers at maturity to decrease mechanical damage, and for favorable storage conditions, was indicated by Muresan et al.[68] They reported that mechanical damage resulted in wet and dry infections that decreased the total value of potato tubers harvested for bulk sale. Temporary storage in sacks in the field raised temperature, increased necrotic diseases, and gave time for the appearance of disease symptoms and injury suberization. Kundzicz,[69] however, found that the influence of harvest date on the storage loss was considerable. In a study that used three cultivars, Sokol, Sowa, and Pola, the highest losses during storage of Sokol and Sowa occurred for potatoes harvested at later dates. For the Pola variety, the largest losses occurred for harvesting at the earliest date. Based on the observation that early-maturing varieties (harvested late August to mid-September) suffer less damage than varieties harvested a month later, Schuhmann and Oertel[70] concluded that the resistance of tubers to mechanical damage is genetically controlled with strong environmental effects. Haulm destruction seems to be effective in reducing peeler-type damage. A comparison of the effect of a number of reciprocating riddle diggers with elevator diggers indicated that the elevator digger with a main web (elevator chain) and without an agitator had the lowest mechanical damage index. The differences among the elevator diggers tested were not significant.[71] Harvesters with rotary working elements tested in clay soil to determine whether mechanical damage of potatoes could be reduced showed that they improved soil separation by 1.5 to 2 times compared to a conventional harvester, and mechanical damage to tubers did not exceed 3%.[72] High soil NPK levels could also result in an increased mechanical damage index and proportion of damaged tubers.[73]

A comparison of fruit quality of sour cherry (*Prunus*) cv. Nefris after mechanical and manual harvesting indicated that mechanical damage of fruits was below 2% under both harvesting methods. Fruit elasticity and fruit firmness assessment indicated that harvesting methods had no effects on these parameters.[74] Flexible curved fingers of predetermined curvature and stiffness used for harvesting Valencia oranges removed about 90% of mature oranges; 64.1 to 77.6% showed no mechanical damage, 2.0 to 3.9% had severe mechanical damage (damage evaluation 3 weeks after harvest), and 0.3 to 2.7% showed decay 7 weeks after harvest (storage at 4.4°C).[75] Cushion materials installed on mechanical harvesters in the 1960s through the 1970s to prevent splitting and bruising of fruit have been replaced by smooth, hard surfaces. In a study to determine the effect of cushion and hard surfaces on mechanical damage, drop tests were used to evaluate cherries' firmness sensitivity and identified cushion materials that can reduce firmness loss during mechanical harvesting. Firmness loss averaged 28% for a 0.9-m (3-foot) drop onto a hard surface compared to 6 to 10% for four cushion materials.[76] Cherries dropped 4 m (13 feet) onto a tightly stretched tarp without cushion material resulted in a loss of firmness of 35%, but only 14 to 28% showed loss of firmness with cushion material. To minimize mechanical damage and maintain firmness, hard harvester surfaces that cherries impact should be covered with adequate cushioning.

Mechanical stress received during harvest of cucumbers (*Cucumis sativus* L.) was demonstrated to affect physiological degeneration of the placental tissues and their suitability for use in some pickled products. In experiments to compare rolling and dropping of cucumbers, rolling had a greater effect on refreshed delayed light emission (RDLE) from chlorophyll than did dropping. After 48 hours, RDLE suppression persisted and starch granules were no longer evident in chloroplasts from mechanically stressed fruit, but very electron-dense inclusions had developed in the chloroplasts.[76] The quality of concord grapes is affected by mechanical harvesting and handling. Grapes with brush (vascular bundle) intact had less damage for a given number of drops than did grapes with brush removed. Weight loss occurred in Concord grapes and increased with severity of mechanical damage, increased storage temperature, and time. Bulk density increased with severity of mechanical damage and appears to have the best potential as a means of rapidly determining the degree of damage.[77] Translucence or water soaking of the flesh is a problem in pineapples (*Ananas comosus* L.) marketed fresh. Affected fruit is more prone to postharvest mechanical damage and the flavor is poor. Fruit translucency and preharvest conditions that predispose fruit to translucency were studied by Paull[78] who developed a model to show the best relationship between preharvest weather and translucency. A period of 2 to 3 months before harvest was found to be crucial in the development of fruit translucency at harvest and crown growth. Translucency was more severe and had a higher incidence when maximum and minimum temperatures 3 months before harvest were low (15 to 23°C), relative to that at high temperatures (20 to 29°C). Fruit with larger crowns had a lower incidence and severity of translucency.

Effect of harvesting and subsequent handling on mechanical damage of pairs was tested using cylindrical tissue samples of two pear cultivars (D'Anjou and Bosc) in two sizes.[79] Using dynamic axial compression at four strain rates, it was determined that strain rate significantly affected the failure stress, failure strain, and the secant elastic modulus, while it had no effect on the shock wave speed or tissue toughness. Larger pears had higher failure stresses and lower failure strains, and cultivar significantly affected all of the tissue failure properties except shock wave speed.

4.6 STORAGE

Mechanically stressed fruit exhibits visible degeneration of mesocarp and endocarp during storage. Storage polyamines in fruit have been reported to be involved in alleviating a number of stress conditions associated with fruit storage. Putrescine, for example, tends to accumulate in plant organs exposed to chilling stress.[79] Polyamine accumulation also occurs in response to external mechanical bruising of mandarin cultivars.[80] The role of polyamines in extending shelf life and the reduction of mechanical damage in plums (*Prunus salicina* Lindl.) was demonstrated by Pérez-Vicente et al.[81] Activation of the polyamine biosynthesis pathway during storage indicated accumulation of cell wall putrescine and spermidine, which are possibly responsible for the greater firmness of putrescine-treated plums. It was suggested that the increase in free spermidine levels could act as a physiological marker of mechanical damage.

Apricots (*Prunus armeniaca* cv. Mauricio) infiltrated with exogenous putrescine (at 1 mM) before storage (5 days at 10°C) and then mechanically damaged showed higher firmness and a delayed color change than the controls.[82,83] The treatment was effective in reducing the damaged area and volume after compression. Endogenous spermidine was at similar levels in treated fruits, both damaged and undamaged. The most significant effect was observed in damaged apricots, in which an increase in spermidine levels was observed in response to mechanical damage, whereas undamaged apricots had the lowest concentrations of this polyamine during storage. Lemon fruit under postharvest putrescine and calcium treatments also maintained higher firmness values and more resistance to peel rupture than control fruit during storage.[84] Treated lemons showed less deformation when the compression force (50 N) to induce mechanical damage was applied. Treated and damaged fruits showed a decline in polyamine content while ABA and color changed in parallel with maturation during storage, unlike in control fruits in which increases in spermine and ABA levels occurred as a consequence of mechanical damage.

Mechanical damage considerably increases alpha-solanine glycoalkaloids in potato tubers. Storage of mechanically damaged tubers further increased glycoalkaloids in the first 5 months of storage but decreased to initial values in the subsequent 3 months.[85] Steroid glycoalkaloid content decreased during cold storage and increased significantly under shop conditions in damaged tubers. Peeling and removal of bruises reduced the content to an acceptable level of < 40 mg/kg.[86] In a study by Rataj and Dzupin[87] to determine the relationship between the effect of storage of potatoes on the density of tubers, the shape coefficient, and the air force necessary for penetrating part of a tuber at quasistasis, it was concluded that the density and axial compression tests showed no significant changes throughout the duration of storage. Moderate changes observed in the dimensions and in the shape coefficient were slight increases.

Storage properties of fruit may vary depending on cultivar and packaging material used. At the time of harvest, Tsu Li and Ya Li pears resisted mechanical damage nearly as well as Chojuro pears, but they became more susceptible to bruising in cold storage. Twentieth Century pears were most sensitive to impact and compression bruising. Increased time in the ripening room produced more softening and increased the bruise resistance of Chojuro and Twentieth Century pears.[88] Bruising of packaged tomatoes is affected by the nature and volume of the packaging material used. Tomatoes on the bottom of boxes exhibit worse bruises. The separated clapboards and the metal corners of wooden boxes cause the most severe injuries. Tomatoes packaged in smaller cardboard containers are usually less susceptible to bruising than those in plastic and wooden boxes.[89] The detrimental effect of Brazilian "K-wooden boxes" for storage and transportation of tomatoes on shelf-life and quality (weight, shelf life, color, mechanical damage, firmness, and relative water content) due to their rough surface was demonstrated by Luengo et al.[90] These boxes also allow pathogen colonization because of the excessive number of fruit layers and lateral cut openings. The damage was significantly different and lower using Embrapa boxes, reducing postharvest losses in tomato fruits. Polypropylene plastic containers, however, when compared with other plastic and wooden containers, resulted in less mechanical damage to fruits and did not affect their chemical properties.[91]

Tests conducted by Timm et al.[92] using bulk bin designs (two hardwood, one plywood, and two plastic) and two trailer suspension systems (steel-spring and air-cushion) indicated that apple fruit positioned in the middle of each bin during transportation had similar bruise and abrasion damage levels regardless of the bin design, suspension system, or trip distance. Abrasion damage on the sidewall apples varied among bin design, suspension system, and trip distance. Damage to sidewall fruit was significantly less using the air-cushion system. Their results indicate that higher apple quality can be maintained if bulk apples are transported in plastic bins of these designs and on semi-trailers having air-cushion suspension.

Mechanical abrasion adversely affects the physiology, green life, and commercial quality of bananas. This includes the physiological development of the fruits (respiration and ethylene production) and final quality. Air humidity level may be negatively correlated with the severity of the symptoms.[93] Akkaravessapong et al.,[94] however, found that the relative humidity (RH) at which bananas are stored did not affect the bruise resistance coefficient, CO_2 or ethylene production, or starch or sugar content, but low RH significantly increased the rate of water loss by three- to fourfold. The susceptibility of the fruits to mechanical damage increased rapidly on the second day after ripening had been initiated. The magnitude of the change was four- to eightfold. Although RH did not affect susceptibility to mechanical injury, the tissues damaged at low RH dried to a black color, while those damaged at high RH remained light brown. Under low RH conditions, desiccation is the dominant causative factor in postharvest browning of rambutan (*Nephelium lappaceum* L.). Landrigan et al.[95] infiltrated mature fruits with known enzyme inhibitors, then either mechanically damaged them or left them undamaged before storage at 20°C at 95 or 65% RH. All fruits at low RH browned severely. At high RH, infiltration with water, but not with the enzyme inhibitors salicylhydroxamic acid and catalase, led to a large increase in browning. It is inferred that enzymes were involved in browning in damaged tissue under high RH.

Mechanically damaged vegetables should be separated in the field, particularly from batches intended for longer storage. Studies on damaged vegetables revealed that losses due to molds were 5 to 11 times higher in carrots, 1.5 to 1.8 times higher in white cabbage, and 2 to 4 times higher in onions than in corresponding undamaged vegetables.[96] Carrot slices obtained from freshly harvested roots are also more sensitive to mechanical damage and short-term storage than those prepared from roots previously stored. Carrot root slices stored for 4 days at 20°C reacted with a strong accumulation of total phenols, especially chlorogenic acid. Synthesis of phenols was accompanied by an increase in phenylalanine ammonia lyase activity, wound-induced respiration, and ethylene production.[97]

4.7 DETECTION

The variability in factors that affect tissue damage within a population of produce has led to efforts to determine damage susceptibility. A number of sampling techniques have been described to quantify the scope of the problem for specific industries.[98,99] Bollen et al.[62] used a logistic function to describe the relationship between impact energy and the likelihood of a certain bruise size occurring. The approach takes into account the natural variability of bruising within any given population. The slope of the logit was shown to be a characteristic of variety, with the logit offset describing season and maturity variation for apples. The method also provided a good prediction technique for the susceptibility of varieties of nectarines and potatoes.[100]

Using measurements of visible infrared light passing through the longitudinal midsection of whole cucumber fruit, Miller et al.[101] quantified on a unitless sigmoid scale from 1 to 10 the effect of mechanical injury on fruit quality. Unbruised cucumbers exhibited transmission values between 2 and 3, regardless of cultivar. Mechanical-stress treatment that simulated bruising incurred during harvesting and handling of cucumbers caused a decline in the internal quality of the fruit and was associated with an increase to a value of 6 in light transmission compared to that of unstressed fruit. Light transmission increased as the severity of stress applied to the fruit increased. Fruits exhibiting high transmission values were judged by human sensory evaluations to be of lower quality than those exhibiting low transmission values.

Bruises may take place beneath the peel and may be difficult to detect by visual or automatic color sorting, especially for fruits such as dark-colored apples. X-ray imaging,[102,103] magnetic resonance imaging (MRI)[104], and near-infrared (NIR) reflectance[105,106] are among the successful techniques used for defect grading. Near-infrared imaging was used by Upchurch et al.[107] to characterize the influence of time, bruise type, and severity on the NIR reflectance from bruised and unbruised regions on Delicious and Golden Delicious apples. Within 24 h after inducing damage, a maximum contrast in NIR reflectance occurred for both impact and compression types of bruises. The contrast decreased until it equaled the contrast for an unbruised region, and changes in contrast after 1 day were more gradual for impact-type bruises than for bruises created by compression. Surfaces of mechanically injured apples have lower reflectance than those of undamaged apples in both the near-infrared and visible spectra, but the injuries are detected more easily in the near-infrared spectrum. Based on this, Damerow[108] concluded that the reflectance index at 800 nm could be used to distinguish the grade of injury. Recently, Varith et al.[109] detected bruising in apples using thermal imaging of dropped apples stored at 26°C for 48 h. The temperature differences between bruised and sound tissues were possibly due to the differences in thermal diffusivity. Under steady-state temperature, thermal imaging did not detect bruises, indicating that the temperature differences were not due to emissivity differences.[109]

Electrical conductivity has been investigated in relation to percentage of pared fruit or damage indexes prepared by manually paring strawberries and stimulating bruising with a three-dimensional vibrator. The conductivity was associated with the percentage of pared fruit (0 to 40%) or damage indexes (1 to 5) of strawberries, with correlation coefficients of 0.938 and 0.917 at $P \leq 0.05$, respectively. Furthermore, the bruising caused by the three-dimensional vibrator resulted in increases in electrical conductivity in response to vibrating time and vibrating force. The decrease in electrical conductivity of strawberries treated with the vibrator during 2 days of storage at 25°C and 80% RH may be due to wound healing of the fruit. Electrical conductivity exhibited potential for quantitatively evaluating damage of strawberries during transportation and marketing.[110]

Analysis based on determining scar tissue, suberin, and resistance to fungi may be used for evaluating the storage quality of potato tuber lots. In a study of tuber lots of the variety Nevskii harvested manually and mechanically, tubers without mechanical damage and visible disease symptoms formed scar tissue more quickly and had thicker scar tissue and a thicker suberin layer than the rest.[111] Both a muriatic acid dip test and X-ray computerized tomography have been found to provide some indication of damage of sweet onions under certain conditions. Muriatic acid has been found to cause the flesh of a freshly harvested sweet onion to appear yellow and become a sticky gel in the vicinity of damage, thus enhancing the visibility of damage. X-rays also aid viewing of the internal structure of a sweet onion and thus damage that may have occurred.[112]

4.8 METABOLIC CHANGES

4.8.1 RESPIRATION

An increase in respiration rate is one of the first responses to mechanical injury. In general, respiration rates are inversely related to the shelf life of produce.[113,114] Fresh-cut fruits and vegetables best illustrate the effect of wounding on respiration rate. The increase in respiration rate following fresh cutting normally results in decreased product shelf life relative to that of the whole fruit.[114,115] Fresh-cut products from a number of fruits show increased respiration rates relative to the whole fruit.[115-118] The respiratory rate response to wounding, however, may depend on the commodity. The rise in respiration after cutting or wounding may be related to α-oxidation of long-chain fatty acids that results in CO_2 release.[114,116,120] Cutting does not appear to influence the respiration rate of oranges during refrigerated storage,[121] while cut

Golden Delicious apple slices continued to respire intensely until at least day 4, when there was a peak in CO_2.[122] Longitudinally cut green-tip bananas have higher respiration rates during storage than those sliced into transverse or obtuse sections.[123] Maturity may also influence respiration rate of damaged plant tissues. Allong et al.[124] reported that in mature-green Julie and Graham mangoes, respiration rates are highest in cut fruit immediately after slicing, decrease significantly within the first 12 h of storage at both 5 and 10°C, and remain at levels above that of intact fruit throughout the storage period. The effect of slicing on half-ripe and firm-ripe fruit was an initial increase in respiration rate followed by a decline to levels equal to that of intact fruit. 1-Aminocyclopropane-1-carboxylic acid (ACC) increases in cut mangoes relative to the whole fruit.[125]

Respiration rates of cut produce generally increase with an increase in storage temperature and ripeness.[61,117,126] The Q10 in several fresh-cut commodities stored at 0 to 10°C are usually higher; only a few are lower than in the whole product.[115] The Q10 is greater in the 10 to 20°C temperature range for most cut fruits and vegetables, apparently because of the rapid deterioration at these temperatures. Respiration in mechanically damaged apricots varies depending on storage temperature and fluctuations in temperature.[127] Forced-air cooling may be used to reduce the respiration rate of mechanically damaged fruit. Damaged plums before precooling showed a respiration rate double that of damaged fruit after precooling during storage.[83] Respiration rates are reported to increase with storage time in fresh-cut tomatoes[128] and melons.[129] Fresh-cut pears maintained respiration activity over a period of 15 days at 3°C.[130] Reduction of storage temperature of damaged lettuce from 37 to 10°C caused a decrease in respiration rate, but reducing the temperature further to 4°C did not affect the respiration rate.[131] Lowering atmospheric oxygen concentration reduced respiration rate. Storage in a low-oxygen atmosphere will generally decrease respiration rate.[125,126,132]

The increase in respiration rate in freshly wounded plant tissue appears to be accompanied by rapid cell expansion. Asparagus cells respond to wounding and to the culture medium by rapid cell expansion on day 3 postisolation, which precedes cell division by 24 h.[133] Cell expansion was accompanied by a large rise in respiration rate and a massive increase in RNA synthesis through the generation of wound-enriched mRNA populations.

4.8.2 ETHYLENE

Ethylene is synthesized naturally during plant development, fruit ripening, leaf senescence, and responses to stress and pathogens. Synthesis is by reactions involving a cycle in which the amino acid methionine participates in the formation of ACC, ACC synthase being a key enzyme in this pathway. Ethylene production is stimulated when plant tissues are injured. Wounding increases activity of ACC synthase and results in the accumulation of ACC that is subsequently oxidized to ethylene. Wound ethylene is believed to be involved in the increased respiration of tissues,[134] which differs among fruit products.[58]

Ethylene exerts its action through a complex regulation of its own biosynthesis, perception, and signal transduction[135,136] that leads to dramatic changes in gene expression.[137] Ethylene-responsive genes appear to be differentially regulated during abscission, organ senescence, and wounding. Ruperti et al.[138] isolated four genes from peaches (*Prunus persica*, Batsch) that were up-regulated during propylene-induced abscission of young fruit. DNA and deduced protein sequences of four selected clones, termed *Prunus persica* Abscission zone (*PpAz*), revealed homology to thaumatin-like proteins (*PpAz8* and *PpAz44*), to proteins belonging to the PR4 class of pathogenesis-related (PR) proteins (*PpAz89*), and to fungal and plant β-D-xylosidases (*PpAz152*).

The potential effects of wound ethylene are dependent on the type and physiology of the tissue in question. Fruit maturity influences wound-induced ethylene response. Wound-generated ethylene, if sufficient, may start the climacteric respiratory response that could cause fruit to immediately start to ripen. Wound ethylene, which is higher in preclimacteric and climacteric than in postclimacteric tissues,[139] may accelerate deterioration and senescence in vegetative tissues and promote ripening of climacteric fruit.[20] In melons, ethylene production was higher in the cut fruit at the preclimacteric phase,[140] whereas in the postclimacteric phase cutting resulted in a reduction of ethylene production. The accelerated climacteric phase caused by ethylene production results in a difference in physiological age between intact and sliced tissue and contributes to softening and color during storage.[114,141,142] Compression levels and number of days in storage were found to have no significant effect on ethylene levels in lettuce.[142] Quality deterioration may result from increased ethylene production, which may induce higher cellular metabolism and higher enzymatic activity.[144] Changes in ethylene production with storage time and temperature are also product-specific.[61,115,129] Air humidity level may be negatively correlated with the severity of the symptoms of mechanical damage in bananas, including respiration and ethylene production.[91] In apricots, ethylene started to rise after 12 hours under the impact area of dropped fruit (cv. San Castrese), and the sound area on the opposite side produced more ethylene 6 h later.[127] Putrescine-treated apricot fruits (damaged and undamaged) had a repressed ethylene emission, but damaged fruits produced higher levels of ethylene.[83]

ACC synthase transcript levels could be decreased by ethylene application that leads to decreased ethylene synthesis.[28] It is also likely that ethylene reduces bruise severity through another mechanism such as the induction of cell division. Product pretreatment with ethylene may thus reduce bruise severity. In potatoes, pretreatment of tubers with ethylene reduced bruising severity that could not be linked to a drop in polyphenol oxidase (catechol oxidase) activity, tyrosine concentration, or the biochemical potential for melanin formation. Postimpact treatment of tubers with ethylene does not appear to reduce bruising damage.[145]

4.8.3 PHENOLIC COMPOUNDS

Phenolic compounds appear to be important in a wide range of reactions that result from injury and a loss of organizational resistance between substrates and enzymes within the cell. They contribute to resistance of plants to mechanical stress by participating in lignification of cell walls surrounding the injured zone and inhibition of microbial growth and germination of spores. Phenolics present in plant tissues may be used for synthesis of other phenolic compounds (phytoalexins) that aid the healing process. Immediately after injury, oxidation of preexisting phenolics occurs, leading to an initial decrease in phenolic content. O-quinones produced usually have antimicrobial properties and they readily undergo oxidative polymerization. Following this, there is a significant increase in phenolic content.[146]

When browning occurs, constituent phenols are oxidized to produce a quinone or quinonelike compound that polymerizes, forming brown pigments. These unsaturated brown polymers are generally referred to as melanins or melaninodins. Among the compounds believed to be important as substrates are chlorogenic acid, neochlorogenic acid, catechol, tyrosine, caffeic acid, phenylalanine, protocatechin, and dopamine.[147] Some induced proteins in wounded plant tissues are enzymes of phenolic metabolism, such as phenylalanine ammonia-lyase (PAL), whose increased activity leads to the accumulation of phenolic compounds (e.g., chlorogenic acid, dicaffeoyl tartaric acid, and isochlorogenic acid) and tissue browning. Wounding of iceberg lettuce leaves increases PAL activity 6- to 12-fold over 24 h at 10°C and leads to a threefold increase in the total phenolic content within 3 days.[23] Soaking cut lettuce in hypertonic solutions appeared to render the tissue insensitive to wound induction of PAL and caused a reduction in the wound-induced accumulation of phenolic compounds. The hypertonic solution did not cause the loss of a portion of the wound signal through efflux of water from the tissue, but rather induced a general stress-related resistance to further abiotic stresses.[148] The synthesis and accumulation of phenolic compounds in lettuce as a consequence of wounding can be suppressed by a heat-shock treatment,[149] although heat shock-treated cut lettuce produced more phenolic compounds than unwounded lettuce.[150]

Bruising pear fruit after 120 days of storage caused a 30% increase in chlorogenic acid and a 50% increase in catechin, but no increase in p-coumaric acid derivatives.[151] After 3 days of storage of whole heads and excised midrib sections of iceberg, butter leaf, and romaine lettuce (*Lactuca sativa* L.) at 5 and 10°C, only 5-caffeoylquinic acid (chlorogenic acid), 3,5-dicaffeoylquinic acid (isochlorogenic acid), caffeoyltartaric acid, and dicaffeoyltartaric acid were detected in wounded lettuce midribs. Of these four compounds, chlorogenic acid accumulated to the highest level in all three lettuce types.[152] Storage of root slices of carrot cv. Flakoro at 5 or 20°C resulted in considerable accumulation of soluble phenols, particularly chlorogenic acid, and of ethylene. Isocoumarin also accumulated in the peel, particularly at 20°C.[153] The wound responses observed in cassava include increased activity of PAL, peroxidase (POD), and polyphenoloxidase (PPO); formation of phenols/polyphenols including leucoanthocyanidins, catechins, scopoletin, and condensed tannins; and often the formation of a wound periderm.

4.9 ENZYMATIC EFFECTS

A number of enzymes are known to contribute to flavor biogenesis that could affect sensory quality and shelf life of fresh fruits and vegetables. Flavor production by the lipoxygenase (LOX) pathway, for example, is generally quiescent unless triggered by maceration or cell damage.[154] The resulting production of volatile aldehydes of chain lengths C6 and C9 that is widespread in fruits and vegetables[155–157] proceeds rapidly when plant cells are disrupted in the presence of oxygen. The effects of LOX on the quality of fresh-cut fruits and vegetables and the effects of the enzyme on flavor and aroma compounds have been reviewed.[158,159]

Mechanically stressed fruits and vegetables exhibit visible degeneration of mesocarp and endocarp during storage.[84] In addition to increases in metabolic rate, respiration, and gene expression, severalfold increased activity of enzymes such as polygalacturonase (PG) and polyphenol oxidase (PPO) occurs.[160] Mechanical injury increases metabolic and pectin esterase activity in cantaloupe melon.[161] Polygalacturonase extracts from *Phomopsis cucurbitae*, a latent infection pathogen, produced little maceration in netted muskmelon tissue until fruits were 50 days postanthesis (10 days postharvest). In contrast, PGs from *Rhizopus stolonifer*, a wound pathogen, produced high levels of maceration at all stages of fruit development from 20 to 50 days postanthesis.[162] Pectic enzymes have received considerable attention regarding their involvement in the softening of cell wall components. Firmness retention is an important quality parameter in fresh-cut produce.[163–166] One of the most obvious changes that occur during the softening of fruits is the progressive solubilization and depolymerization of pectic substances.[167–169] The effect of pectic enzymes on freshness and shelf life of fresh-cut fruits and vegetables was also recently reviewed.[159]

Most changes in color during the storage of fresh-cut products are enzyme-mediated. The most common are related to PPO activity located in the chloroplast thylakoid membranes.[170,171] PPO action often results in the formation of highly reactive quinones that can then react with amino and sulfhydryl groups of proteins and enzymes as well as with other substrates such as chlorogenic acid derivatives and flavonoids. These secondary reactions may bring about changes in physical, chemical, and nutritional characteristics and may also affect the sensory properties of fruits and vegetables. Quinones contribute to the formation of brown pigments by participating in polymerization and condensation reactions with proteins.[146,172–174] It has been suggested that expression of closely related heterologous genes can be used to prevent enzymatic browning in a wide variety of food crops without the application of various food additives. Coetzer et al.[175] demonstrated that PPO activity of Russet Burbank potatoes could be inhibited by sense and antisense PPO RNAs expressed from a tomato PPO cDNA under the control of the 35S promoter from the cauliflower mosaic virus. Their results indicate that expression of tomato PPO RNA in sense or antisense orientation inhibits PPO activity and enzymatic browning in the potato cultivar. Tissue printing indicates that the PPO enzyme is distributed throughout the potato tuber. Following impact injury, both tissue printing and quantitative electron microscopy revealed that there was no increase in the level of the enzyme, although there was subcellular redistribution of PPO. This redistribution was first apparent at 12 h after impact, as determined by the use of confocal immunolocation, and coincided with loss of membrane integrity.[176]

Impact-induced blackspots in potato (*Solanum tuberosum*) tubers are still unexplained at the cellular level. Blackspot bruise is a physiological disorder of potato tubers resulting from mechanical damage to tissues during handling. It is well known that blackspots develop through enzymatic oxidation of phenols by PPO, leading to spontaneous polymerization and subsequent formation of the dark pigment melanin. Tubers produced under 10 and 0% potassium regimens had almost twice as much free tyrosine as those grown at full-strength Hoagland's solution. The 0% potassium tubers had significantly higher PPO activity than the control.[48] Many studies have reported good correlation between free tyrosine and enzymatic browning elicited by cutting, maceration, or homogenization.[177,178]

Mechanical damage of peaches increased activity of PAL, PPO, and peroxidase (POD) and lignin synthesis in cell walls. Application of $CaCl_2$ at the site of injury increased the concentration of bound calcium in cell walls, delayed the peaks in enzyme activity, stimulated synthesis of neutral sugars, and reduced the degree of esterification of cell wall pectins.[179] In bananas, the peel shows high levels of activity early in development but activity declines until ripening starts and then remains constant. PPO activity in fruit does not appear to be substantially induced after wounding or treatment with 5-methyl jasmonate, suggesting that browning of banana fruit during ripening results from the release of preexisting PPO enzyme synthesized very early in fruit development.[180]

Red discoloration in chicory leaves may result from mechanical damage. Gillis et al.[181] noted that red discoloration was induced by applying mechanical loads on leaves under constant pressure at ambient conditions. Compression and plate contact induced severe dark pink and brown discolorations as a result of mechanical damage as well as the development of russet spotting in lettuce.[143] Browning of individual lettuce heads was observed, although the treatments had no significant effects on the development of internal tissue browning. Browning was most prominent on the white midrib tissue. An increase in heat-shock temperature from 20 to 70°C in excised midrib segments of iceberg lettuce (cv. Salinas) caused a reduced increase in PAL activity and the accumulation of phenolic compounds in the excised midrib segments, as well as browning. Synthesis of chlorogenic acid, dicaffeoyl tartaric acid, and isochlorogenic acid was significantly reduced by these heat-shock treatments. These treatments also decreased polyphenol oxidase and peroxidase activities.[182]

Peroxidase enzymes are also able to contribute to enzymatic browning based on their ability to accept a wide range of hydrogen donors such as polyphenols.[183] They are able to oxidize catechins,[184] hydoxycinnamic acid derivatives and flavans,[185,186] and flavonoids.[183] The presence of PPO enzyme, however, enhances POD-mediated browning reactions.[183] The production of large amounts of reactive oxygen species is one of the earliest defence responses against mechanical damage or pathogen attack, but it also exposes the plant cells to serious oxidative stress. Morimoto et al.[187] in a model study to determine H_2O_2 metabolism using cells of *Seutellaria baicalensis* found that in response to an elicitor (such as yeast extract) the cells immediately initiate the hydrolysis of baicalein 7-*O*-beta-D-glucuronide by beta-glucuronidase, and the released baicalein is then quickly oxidized to 6,7-dehydroba-icalein by peroxidases. Hydrogen peroxide is effectively consumed during the per-oxidase reaction. The beta-glucuronidase inhibitor saccharic acid 1,4-lactone significantly reduced the H_2O_2-metabolizing ability of the cells, indicating that beta-glucuronidase, which does not catalyze the H_2O_2 degradation, plays an important role in the H_2O_2 metabolism. Baicalein, however, predominantly contributed to H_2O_2 metabolism. Because beta-glucuronidase, cell wall peroxidases, and baicalein pre-exist in *S. baicalensis* cells, their constitutive presence enables the cells to rapidly induce the H_2O_2-metabolizing system.

Changes that occur in POD activity in wounded and fresh-cut fruits significantly contribute to their product quality.[182,188–190] In addition to the role played by POD in the flavor of fresh fruits and vegetables, POD isozymes are involved in many cell alterations.[191] They appear to influence flesh firmness through catalysis of the cross-linking between tyrosine residues of the cell wall extensins and ferulic acid substit-uents of pectins[192] and the synthesis of lignin and suberin polymers.[192,193] The anionic peroxidase associated with the suberization response in potato (*Solanum tuberosum* L.) tubers during wound healing is a 45-kDa, class III (plant secretory) peroxidase that is localized in suberizing tissues and shows a preference for feruloyls such as those that accumulate in tubers during wound healing. In contrast, the cationic peroxidase(s) induced in response to wound healing in potato tubers is present in both suberizing and nonsuberizing tissues and does not discriminate between hydroxy-cinnamates and hydroxycinnamyl alcohols.[194]

Lipoxygenase (linoleate: oxygen oxireductase; EC.1.13.11.12) (LOX) is present in most plant tissues and, in the presence of oxygen, catalyzes oxidation of polyun-saturated fatty acids (PUFA) containing a *cis,cis*-1,4-pentadiene structure. Allene oxide synthase (AOS; hydroperoxide dehydratase; EC 4.2.1.92) catalyzes the first step in the biosynthesis of jasmonic acid from lipoxygenase-derived hydroperoxides of free fatty acids. Rangel et al.[195] demonstrated that wounding causes local and systemic induction of LOX activity in passion fruit (*Passiflora edulis f. flavicarpa*) leaves, while exposing intact plants to methyl jasmonate (MJ) vapor provoked a much stronger response. Based on immunocytochemical localization studies using leaf tissue from MJ-treated plants that showed that the inducible LOX was compart-mented in large quantities in the chloroplasts of mesophyll cells associated with the stroma, it was suggested that the wound response in passion fruit may be mediated by a chloroplast 13-LOX, a key enzyme of the octadecanoid defense-signaling pathway.

Conjugated hydroperoxy acids (HPO) produced by LOX catalysis undergo metabolism by hydroperoxide lyase (HPO lyase). HPO lyase catalyzes the cleavage of HPO to aldehydes, such as *cis*-3-noneal and hexanal from linoleic acid HPO and *cis*-3,*cis*-6-nonadienal and *cis*-3-hexenal from linolenic acid HPO.[196,197] Besides hydroperoxidation, formation of oxoacids and ketodienes is catalyzed by LOX.[198,199] Hydroperoxide isomerase could also catalyze the isomerization of hydroperoxide as an intermediate reaction pathway.[198] Internal bruising may alter aroma volatile pro-files in fruit tissues. Individual volatile profiles of the three tissues in bruised tomato fruit were significantly different from those of corresponding tissues in undropped, control fruit, notably, *trans*-2-hexenal from pericarp tissue; 1-penten-3-one, *cis*-3-hexenal, 6-methyl-5-hepten-2-one, *cis*-3-hexenol and 2-isobutylthiazole from locule tissue; and 1-penten-3-one and beta-ionone from placental tissue. Alteration of volatile profiles was most pronounced in the locule tissue, which was more sensitive to internal bruising than the other tissues. Changes observed in the volatile profiles appear to be related to disruption of cellular structures.[200]

Plants degrade cellular materials during senescence and under various stresses. Hayashi et al.[201] reported the precursors of two stress-inducible cysteine proteinases, RD21 (product of responsive-to-desiccation gene 21) and a vacuolar processing enzyme, that are specifically accumulated in transgenic *Arabidopsis thaliana* (ecotype Columbia) plants. They are surrounded with ribosomes and thus are assumed to be directly derived from the endoplasmic reticulum (ER). These ER bodies are considered to be proteinase-sorting systems that assist the plant cell under various stress conditions.[202] Heat shock in mechanically wounded plants appears to lead to the disruption of ER lamellae that has been hypothesized to cause destabilization of otherwise stable mRNA associated with ER-bound polyribosomes.

Putrescine N-methyltransferase (PMT) catalyzes the first committed step in the biosynthesis of pyrrolinium ring-containing alkaloids. Using mechanically damaged plums, Pérez-Vicente et al.[81] demonstrated the activation of the polyamine biosynthesis pathway, showing an accumulation of cell wall putrescine and spermidine as a result of wounding. Infiltration of exogenous polyamines can ameliorate mechanical damage, increase fruit's firmness, and reduce ethylene production and respiration rate. Fruits under putrescine and calcium treatments maintain higher firmness values and are more resistance to peel rupture than control fruits during storage.[84] They also show less deformation when the compression force to induce mechanical damage is applied. Exogenous polyamines can inhibit ethylene production in several climacteric fruit and prolong the postharvest shelf life of whole and mechanically damaged fruit. This has been reported for plums,[81] tomatoes,[203] peaches,[204] avocados, and pears.[205] Inhibition of ethylene production after polyamine treatment is due to the inhibition of ACC synthase.[205] The contents of the diterpene alkaloid lappaconitin in various parts of *Aconitum septentrionale* plants (leaves, stems, whole aboveground part, roots) when studied in relation to the extent of mechanical damage indicate that lappaconitin appears to be synthesized mainly in roots and that the change of its content in plant parts after damage to the above-ground part is connected with an activation of growth processes.[206]

4.10 TEXTURE

Lignification of injuries is an important component of a plant's defense against postharvest diseases. Wounding triggers a variety of biochemical and developmental pathways in plants that collectively harden injury sites against infection. A response commonly observed in plants is the accumulation of aldehyde-selective reagents in cells adjacent to injuries. This material has been called lignin, lignin-like, phenolic polymers, and wound gum.[207]

Mechanical injury to avocado (*Persea americana* Mill.) pericarp will initiate a meristem and the production of periderm. Injury to tissues deep within the pericarp results in cellular differentiation of parenchyma with various degrees of cell wall thickening. Sclereid-like cells with thick, lignified walls and prominent pits can be formed in tissue normally occupied by thin-walled, oil-filled parenchyma.[208] Wound healing in developing apple fruit has been largely associated with wound periderm formation, which is lacking in fruit wounded after harvest.[209] Wounds in mature

Golden Delicious and Granny Smith apples exhibited formation of wall thickenings extending four to six cell layers from the wound. Cell walls near healed wounds stained positive for phenolic substances, tannins, lignins, and callose after 38 days at 5°C or 14 days at 20°C. These compounds have been associated with wound healing in many plants and have been implicated in resistance to infection and colonization in many host/pathogen systems.[210] Spotts et al.[211] reported that wounded pear tissue rapidly accumulated callose and tannins as well as gums, but tests for lignin were negative. It was suggested that this could possibly result from the test's not allowing sufficient time for lignin to accumulate in the wound tissue or that lignin was lost during fixation of the tissue in formalin-acetic acid-ethanol. Lignification of cells in the injured area increases resistance to infection of injured peel in oranges and lemons.[212] Oleocellosis is a physiological rind disorder of citrus fruit commonly caused by mechanical damage and the consequent rupture of the epidermis above oil glands. It is an unattractive surface blemish caused by phytotoxic effects of released rind oils. The blemish is characterized by rind collapse and darkening and is attributed to cellular damage. Released surface oil appears to infiltrate the rind via the ruptured epidermis, resulting in rapid degeneration of cortical, but not epidermal, cell contents.[213] Wounding also increases lignin formation in the cell walls of peaches.[179]

Bland et al.[100] compared the healing of potato cv. Russet Burbank tuber cores with that of cuts and bruises on whole tubers. Water loss from cut surfaces on whole tubers decreased more rapidly than that from cores of tissue. Bruises made on whole tubers also healed more slowly than cut surfaces on tubers.

4.11   TEMPERATURE EFFECT

Good management of temperature can reduce the physiological response of the tissue to bruising and control the appearance of bruising symptoms. Temperature greatly affects tissue response to mechanical damage. Baritelle and Hyde[214] have shown that an increase in handling temperature of both potatoes and pears increased the failure strain of the tissue, while in apples no effect was observed. Moreover, temperature can influence tissue resistance to bruising by affecting cell turgor. An increase in relative turgor reduces strain failure of tissue.[214] Contradictory results have, however, been obtained in relation to fruit temperature and mechanical damage. For example, the resistance of sweet cherries to compression damage decreased linearly with increasing temperature, but the incidence of impact-induced surface pitting increased linearly as fruit temperature decreased.[215] Impact bruising damage was greatest in several varieties of sweet cherries at low temperatures, while vibration damage was not influenced by fruit temperature.[216] In blueberries, the loss of major quality attributes increased with increasing damage levels and increasing storage temperature.[217] The mechanical strength of frozen potato tissue decreases with an increase in the number of temperature fluctuations and in most cases is lower in packed samples. Moisture loss was greatest in the –18 to –6°C range for prepacked samples. Changes in the maximum compression force, a measure of mechanical damage, occur with storage temperature.[218]

A primary consequence of mechanical damage is increased loss of water and water-soluble nutrients. This results in shriveling of produce and increased susceptibility to decay-causing pathogens. The use of different precooling methods, such as hydrocooling and vacuum cooling in broccoli,[219, 220] showed beneficial effects in terms of increased firmness and reduced weight loss. Weight loss has also been studied in several fruits, such as lemons,[84] oranges,[221] and blueberries.[217] Weight loss in mechanically damaged fruit is a consequence of modification of tissue permeability and the occurrence of small cracks connecting internal and external atmospheres, allowing the interchange of atmospheric gases, particularly water vapor.[222] Delaying the start of precooling also resulted in greater weight loss and lower tissue firmness in strawberry fruit.[223] Forced-air cooling led to a reduction in the respiration rate of mechanically damaged plums. Damaged fruit before precooling showed a respiration rate double that of damaged fruit after precooling during storage. Damaged plums before precooling showed higher weight losses and lower firmness than damaged fruit after precooling. Precooling Santa Rosa plums after harvesting and before manipulation (transportation to the packing house, handling in the packing-house and during storage or transportation) can help to maintain fruit quality and prolong shelf life.[83] With respect to color, mechanical damage before precooling significantly reduced chroma values; plums were darker and less bright than those damaged after precooling. These color changes resulted in less attractive fruit. Polyphenol leakage and tissue browning usually accompany mechanical damage.[83] In kiwifruit,[118] reducing the temperature soon after impact decreased ethylene production and the appearance of injury symptoms.

Temperatures of 15 to 25°C affected development of pink rot caused by *Phytophthora erythroseptica* in wounded potatoes. Infections in unwounded tubers started at 15°C, whereas in wounded tubers infection started at 10°C. Incidence of pink rot was high at high temperature and high inoculum levels.[224] The susceptibility of potato tubers to bruising is also greatest when they are harvested at a low temperature. Tubers harvested at a low temperatures also display a greater severity of bruising after impact. This detrimental effect of low harvest temperature appears to be unrelated to polyphenol oxidase activity, catalase activity, or the content of ascorbic acid but seems to occur through an effect on the permeability of cell membranes.[225] Prestorage at low temperatures also reduces storage losses in mechanically damaged potatoes. Weight loss was the major component of total loss.[226] Prestorage tended to increase tuber firmness and reduced the incidence of blue spots. Total loss, weight loss, and dry rot increased with increasing severity of wounding; total losses were lower in fruit stored at 5°C than in fruit stored at 7°C.

Heating dropped papaya fruit at 48°C for approximately 6 hours or until the fruit core temperature (FCT) reached 47.5°C aggravated the severity of mechanical skin injury. Delays in the application of heat treatment from dropping did not reduce the severity of skin injury significantly, except for fruit heated 24 hours after dropping. Waxing fruit alleviated the severity of skin injury, whether applied before or after the heat treatment.[57] Impact bruising thresholds could be affected by delay after harvest and temperature. In bananas, compression and impact bruising increased from 93 to 120 µJ following a 2-day delay after harvest and from 74 to 104 µJ occurred as a result of elevating temperature from 19 to 30°C.[227] Delay after harvest and temperature had the opposite effect on compression damage threshold. Incidence of tip rot increases with increasing severity of nonvisible impact damage to asparagus spear tips. Lallu et al.[228] reported that impact on apical tissues after drops from 0, 50, 100, and 150 mm resulted in 0, 34, 36, and 64% tip rot, respectively, after 5 days at 20°C and 93 to 95% RH. Washing spears after impact increased the incidence of tip rot. Although adverse physiological stress may be a factor involved in the expression of tip rot, results indicated that physical damage may be a major contributing factor, exploited by microorganisms present on the asparagus spears and in packhouse wash water, leading to spoilage. Drop experiments using apricots showed that ethylene production could be greatly affected by low temperature.[127] Ethylene was found to increase more in fruits that were impacted at 4°C and then stored at 18°C than in fruits that were kept continuously at 18°C. Respiration was also affected by temperature, but not as greatly as ethylene production. L (lightness) and b (yellowness) values decreased significantly in the injured flesh compared to the sound flesh, especially in fruit impacted at 4°C and then stored at 18°C.

The combined use of perforated polypropylene (PPP) and intermittent warming (IW) appears to be a practical method to protect peaches from mechanical damage due to transfer from cold to warm rooms and vice versa. In a study to determine the effect of IW on quality of peaches cold-stored in PPP, IW slightly increased senescence but extended shelf life, and the peaches were preferred for color and flavor.[229] IW also allowed normal ripening and prevented wooliness. Controlled atmosphere (CA) storage at 5 to 8% CO_2 by volume prevented the spread of *Botrytis cinerea* in headed cabbages grown under extreme weather conditions, harvested by hand or machine, and subjected to mechanical damage before cold storage (-0.5 to 0°C). Total storage losses were lower under CA storage than under normal cold storage.[230] Temperature is the most critical factor influencing postharvest quality of strawberries. Precooling as soon as possible after harvest and storage at low temperature are considered essential for reducing the respiration rate of strawberries and susceptibility to mechanical damage during transport. It is also recommended that trucks with air-spring suspension be used to transport strawberries to minimize vibrational damage.[43] The postharvest response of wild lowbush blueberries (*Vaccinium angustifolium* Ait. and *V. myrtilloides* Michx.) to mechanical damage and storage temperature is mainly an increased number of shrivelled or split berries.[217] In general, the major quality attributes (firmness, microbial growth, hue, bloom, split, and unblemished berries) deteriorate with increasing damage levels and increasing storage temperature without significant interaction.

In a study of the effect of mechanical damage of greenhouse lettuce on some biochemical processes occurring during storage, Leja and Mareczek[153] found that total phenols and amino acids and enzyme activity increased during storage at either 20°C for 4 days or 5°C for 7 days. In autumn-grown lettuces, PPO and POD activities were slightly higher in wounded lettuces than in intact ones. Free amino acid concentrations increased in response to wounding, particularly in lettuces stored at the higher temperature. Membrane permeability increased after wounding in lettuces grown in autumn, and spring-grown lettuces were less sensitive to wounding than autumn-grown ones. Activity of wound-induced PAL activity in lettuce browning increases with an increase in storage temperature from 0 to 25°C. PAL stimulates phenolic metabolism and consequent tissue browning. A heat shock at 50°C for 90 sec protected fresh-cut lettuce tissue against browning, helped retain greenness, and decreased subsequent production of phenolics when applied either after or before wounding.[149]

4.12 SUMMARY

The most important key to quality maintenance of fresh fruits and vegetables is careful handling before and after they are harvested. Careless handling may cause damage to produce that could be either internal or external. The effects of the injury are immediate in that wound signals are rapidly propagated to adjacent and distant tissues. Subsequent effects involve abnormal physiological breakdown during handling and storage. Bruises and other mechanical damage not only detract from the appearance of the product but are good avenues of entrance for decay organisms. Variety, preharvest crop management, temperature, water content, and physiological maturity are some of the factors that influence susceptibility to mechanical damage. Appropriate harvesting implements and conditions are also essential for maintaining product quality.

Postharvest mechanically induced damage may be caused by overfilling of crates, excess movement of harvested product during in-field transport, loading compression, impact and vibration from forces such as pressure between fruit and machinery, and surface abrasion and handling. Symptoms of injuries, particularly minor cuts and bruises incurred during harvesting, handling, grading, and packaging, may be difficult to detect at the early stages. Subsequent physical and physiological changes, the severity of which is influenced by factors such as the commodity, severity of injury, maturity, storage temperature, and relative humidity during storage, handling, and transportation, could lead to considerable product deterioration that is quite visible by the time the products reach retail or consumer levels. Protective measures throughout pre- and postharvest handling can reduce damage. Knowledge of the types of injuries and their response mechanisms will help us to devise methods to limit the undesirable physical and physiological effects of mechanical damage. Advances in detecting mechanical injury-related disorders and future research in this area should improve screening and sorting methods and reduce the percentage of product lost due to mechanical damage.

REFERENCES

1. Baugher, T.A., Hogmire, H.W., Jr., and Lightner, G.W., Determining apple packout losses and impact on profitability, *Appl. Agric. Res.*, 5, 343–349, 1990.
2. FAO, Prevention of postharvest food losses: Fruits, vegetables and root crops, Food and Agriculture Organisation of the United Nations, Rome, Italy, 1989.
3. Wright, W.R. and Billeter, B.A., Marketing loss of selected fruits and vegetables at wholesale, retail and consumer levels in the Chicago area, *U.S. Dep. Agric. Res. Rep.*, 1017, 1975.

4. Kays, S.J., Preharvest factors affecting appearance, *Postharv. Biol. Technol.*, 15, 233–247, 1999.
5. Maness, N.O., Brusewitz, G.H., and McCollum, T.G., Impact bruise resistance comparison among peach cultivars, *HortScience,* 27, 1008–1011, 1992.
6. Ericsson, N.A. and Tahir, I.I., Studies on apple bruising. 2. The effects of fruit characteristics, harvest date, *Acta Agric. Scand. B*, SP46, 214–217, 1996.
7. Pearce, G., Strydom, D., Johnson, S., and Ryan, C.A., A polypeptide from tomato leaves induces wound-inducible proteinase inhibitor proteins, *Science*, 253, 895–898, 1991.
8. Bishop, P.D., Makus, D.J., Pearce, G., and Ryan, C.A., Proteinase inhibitor-inducing factor activity in tomato leaves resides in oligosaccharides enzymically released from cell walls Wound hormone pest attack, *Proc. Natl. Acad. Sci. U.S.A.*, 78, 3536–3540, 1981.
9. Farmer, E.E. and Ryan, C.A., Interplant communication: Airborne methyl jasmonate induces synthesis of proteinase inhibitors in plant leaves, *Proc. Natl. Acad. Sci. U.S.A.*, 87, 7713–7716, 1990.
10. O'Donnell, P.J., Calvert, C., Atzorn, R., Wasternack, C., Leyser, H.M.O., and Bowles, D.J., Ethylene as a signal mediating the wound response of tomato plants, *Science*, 274, 1914–1917, 1996.
11. Peña-Cortés, H., Sánchez-Serrrano, J.J., Mertens, R., and Willmitzer, L., Abscisic acid is involved in the wound-induced expression of the proteinase inhibitor II gene in potato and tomato, *Proc. Natl. Acad. Sci.* U.S.A., 86, 9851–9855, 1989.
12. Bowles, D., The wound response of plants, *Curr. Biol.*, 3, 165–167, 1991.
13. Julien, J.L., Desbiez, M.O., Jaegher, G., and de Frachisse, J.M., Characteristics of the wave of depolarization induced by wounding in *Bidens pilosa* L., *J. Exp. Bot.*, 42, 131–137, 1991.
14. Shimmen, T., Electrical perception of "death message" in Chara: Involvement of turgor pressure, *Plant Cell Physiol.* 42, 366–373, 2002.
15. Saltveit, M.E., Physical and physiological changes in minimally processed fruits and vegetables, in *Phytochemistry of Fruits and Vegetables*, Tomas-Barberan, F.A. and Robins, R.J., Eds., Clarendon Press, Oxford, U.K., 1997, pp. 205–220.
16. Poovaiah, B.W., Biochemical and molecular aspects of calcium action, *Acta Hortic.*, 326, 139–147, 1993.
17. Weinstein, H. and Mehler, E.L., Ca^{2+}-binding and structural dynamics in the function of calmodulin, *Annu. Rev. Physiol.*, 56, 213–236, 1994.
18. Allan, E. and Hepler, P.K., Calmodulin and calcium-binding proteins, in *The Biochemistry of Plants, A Comprehensive Treatise*, Vol. 15, Academic Press, New York, 1989, pp. 455–484.
19. Ferguson, I.B., Calcium in plant senescence and fruit ripening, *Plant Cell Environ.*, 7, 477–489, 1984.
20. Brecht, J.K., A physiology of lightly processed fruits and vegetables, *HortScience*, 30, 18–22, 1995.
21. Leckie, C.P., McAinsh, M.R., Allen, G.J., Sanders, D., Alistair, M., and Hetherington, A.M., Abscisic acid-induced stomatal closure mediated by cyclic ADP-ribose*, Plant Biol.*, 26, 15837–15842, 1998.
22. Berglund, T. and Ohlsson, A.B., Defensive and secondary metabolism in plant tissue cultures, with special reference to nicotinamide, glutathione and oxidative stress, *Plant Cell Tissue Organ Cult.*, 43, 137–145, 1995.
23. Saltveit, M.E., Wound induced changes in phenolic metabolism and tissue browning are altered by heat shock, *Postharv. Biol. Technol.*, 21, 61–69, 2000.

24. Sano, H., Seo, S., Koizumi, N., Niki, T., Iwamura, H., and Ohashi, Y., Regulation by cytokinins of endogenous levels of jasmonic and salicylic acids in mechanically wounded tobacco plants, *Plant Cell Physiol.*, 37, 762–769, 1996.
25. Sivasankar, S., Sheldrick, B., and Rothstein, S.J., Expression of allene oxide synthase determines defense gene activation in tomato, *Plant Physiol.*, 22, 1335–1342, 2000.
26. Shigemi, S., Hiroshi, S., and Yuko, O., Jasmonic acid in wound signal transduction pathways, *Physiologia-Plantarum*, 101, 740–745, 1997.
27. Zhu, S.K., Salzman, R.A., Koiwa, H., Murdock, L.L., Bressan, R.A., and Hasegawa, P.M., Ethylene negatively regulates local expression of plant defense lectin genes, *Physiologia-Plantarum*, 104, 365–372, 1998.
28. Peck, S.C. and Kende, H., Differential regulation of genes encoding 1-aminocyclo-propane-1-carboxylate (ACC) synthase in etiolated pea seedlings: Effects of indole-3-acetic acid, wounding, and ethylene, *Plant Mol. Biol.*, 38, 977–982, 1998.
29. Leon, J., Rojo, E., and Sanchez-Serrano, J.J., Wound signalling in plants, *J. Exp. Bot.*, 52, 1–9, 2001.
30. Sargent, S.A., Brecht, J.K., and Zoellner, J.J., Sensitivity of tomatoes at mature-green and breaker ripeness stages to internal bruising, *J. Am. Soc. Hortic. Sci.*, 117, 119–123, 1992.
31. Sargent, S.A., Brecht, J.K., Zoellner, J.J., and Chau, K.V., Reducing mechanical damage to tomatoes during handling and shipment, *Am. Soc. Agric. Eng.*, 89, 6616, 1989.
32. Sargent, S.A., Brecht, J.K., and Zoellner, J.J., Assessment of mechanical damage in tomato packinglines, *Am. Soc. Agric. Eng.*, 89, 6060, 1989.
33. Burkner, P.F., Chesson, J.H., and Brown, G.K., Padded collecting surfaces for reducing citrus fruit injury, *Trans. ASAE*, 15, 627–629, 1972.
34. Moretti, C.L., Sargent, S.A., Huber, D.J., Calbo, A.G., and Puschmann, R., Chemical composition and physical properties of pericarp, locule, and placental tissues of tomatoes with internal bruising, *J. Am. Soc. Hortic. Sci.*, 123, 656–660, 1998.
35. Bieluga, B. and Puchalski, C., Methods of determining resistance to mechanical damage, in *Proc. 3rd Int. Conf. Phys. Prop. Agric. Mater.*, Vol. 44, Prague, 1985, pp. 37–44.
36. Puchalski, C., and Bieluga, B., Testing apple resistance to dynamic loading. *Zesz. Problem. Postepow Nauk. Rolniczych*, 399, 187–191, 1993.
37. Roudot, A.C., Grotte-Nicolas M., Duprat, F., and Arakelian, J., Comparison of firmness and resistance to mechanical damage in two apple cultivars during cold storage, *Sci. Aliments*, 9, 319–333, 1989.
38. Stolle, G. and Wahl, H., The effect of mechanical damage on the keeping quality of apples, *Gartenbau B*, 19, 209, 1972.
39. Dingman, D.W., Growth of *Escherichia coli* O157:H7 in bruised apple (*Malus domestica*) tissue as influenced by cultivar, date of harvest, and source, *Appl. Environ. Microbiol.*, 66, 1077–1083, 2000.
40. Brusewitz, G.H., McCollum, T.G., and Zhang, H., Impact bruise resistance of peaches, *Trans. Am. Soc. Agric. Eng.*, 34, 962–965, 1991.
41. Menesatti, P. and Paglia, G., Development of a drop damage index of fruit resistance to damage, *J. Agric. Eng. Res.* 80, 1–12, 2001.
42. Berardinelli, A., Guarnieri, A., Phuntsho, J., and Ragni, L., Fruit damage assessment in peach packing lines, *Appl. Eng. Agric.* 17, 57–62, 2001.
43. Mokkila, M., Sariola, J., and Hagg, M., The key factors in the harvest and postharvest treatments of strawberries, *VTT Res. Notes*, 1955, 55, 1999.
44. Nwandikom, G.I., Yam tuber resistance to mechanical damage, *Agric. Mech. Asia Africa Latin Am.*, 21, 33–36, 1990.

45. Beeching, J.R., Dodge, A.D., Moore, K.G., Phillips, H.M., and Wenham, J.E., Physiological deterioration in cassava: Possibilities for control, *Trop. Sci.*, 34, 335–343, 1994.

46. Booth, R.H., Storage of fresh cassava (*Manihot esculenta*). I. Post-harvest deterioration and its control, *Exp. Agric.*, 12, 103–111, 1976.

47. Laerke, P.E., Brierley, E.R., and Cobb, A.H., Impact-induced blackspots and membrane deterioration in potato (*Solanum tuberosum* L.) tubers, *J. Sci. Food. Agric.*, 80, 1332–1338, 2000.

48. McNabnay, M., Dean, B.B., Bajema, R.W., and Hyde, G.M., The effect of potassium deficiency on chemical, biochemical and physical factors commonly associated with blackspot development in potato tubers, *Am. J. Potato Res.*, 76, 53–60, 1999.

49. Barreiro, P., Steinmetz, V., and Ruiz-Altisent, M., Neural bruise prediction models for fruit handling and machinery evaluation, *Comput. Electron. Agric.*, 18, 91–103, 1997.

50. García, J.L., Riquielme, F., Ruiz-Altisent, M., and Barreiro, P., Study of grading lines for stone fruits and citrus using two instrumented spheres, in *Some Cooperatives in the Region of Murcia (Spain)*, Proc. of AgEng'96 Conf. Agric. Eng., Paper 96F-038, 1996.

51. Garcia, J.L., Ruiz-Altisent, M., and Barreiro, P., Factors influencing mechanical properties and bruise susceptibility of apples and pears, *J. Agric. Eng. Res.*, 61, 11–17, 1995.

52. García, F. and Ruiz-Altisent, M., Effects of precooling and degreening treatments on the susceptibility of peach and citrus to handling damage, in *Proc. 5th Int. Symp. Fruit, Nut and Vegetable Prod. Eng.*, California, 1997.

53. Bielza, C., Barreirob, P., Rodríguez-Galianoa, M.I., and Martínc, J., Logistic regression for simulating damage occurrence on a fruit grading line, *Comput. Electron. Agric.*, 39, 95–113, 2003.

54. Holik, K., Problems of mechanical damage to onions, *Mechanizace-Zemedelstvi*, 37, 304–305, 1987.

55. Finger, F.L. and Casali, V.W.D., Harvest, curing and storage of onion, *Informe Agropecuario*, 23, 93–98, 2002.

56. Rakov, E. and Bykovskii, Y., Mechanical damage and keeping quality of watermelons, *Kartofel' Ovoshchi*, 10, 34–35, 1983.

57. Quintana, M.E.G. and Paull, R.E., Mechanical injury during postharvest handling of Solo papaya fruit, *J. Am. Soc. Hortic. Sci.*, 118, 618–622, 1993.

58. Toivonen, P.M.A. and DeEll, R., Physiology of fresh-cut fruits and vegetables, in *Fresh-Cut Fruits and Vegetables: Science, Technology and Market*, Lamikanra, O., Ed., CRC Press, Boca Raton, FL, 2002.

59. Watada, A.E. and Qi, L., Quality of fresh-cut produce, *Postharv. Biol. Technol.*, 15, 201–205, 1999.

60. Alzamora, S.M., Tapia, M.S, and López-Malo, M.A., *Minimally Processed Fruits and Vegetables: Fundamental Aspects and Applications*, Aspen Publishers, Gaithersburg, MD, 2002.

61. Portela, S.I. and Cantwell, M.I., Cutting blade sharpness affects appearance and other quality attributes of fresh-cut cantaloupe melon, *J. Food. Sci.*, 66, 1265–1270, 2001.

62. Bollen, A.F., Cox, N.R., Dela Rue, B.T., and Painter, D.J., A descriptor for damage susceptibility of a population of produce, *J. Agric. Eng. Res.*, 78, 391–395, 2001.

63. Thiagu, R., Chand, N., and Ramana, K.V.R., Evolution of mechanical characteristics of tomatoes of two varieties during ripening, *J. Sci. Food Agric.*, 62, 175–183, 1993.

64. Kolodyaznaya, V.S. and Zakatova, G.N., Storage ability of tubers after mechanical harvesting, *Kartofel' Ovoshchi*, 9, 21–22, 1983.

65. Mempel, H., Washed carrots—quality from harvest…to consumer?, *Gemuse Munchen*, 34, 402–404, 1998.

66. Siljasen, R., Bengtsson, G.B., Hoftun, H., and Vogt, G., Sensory and chemical changes in five varieties of carrot (*Daucus carota* L.) in response to mechanical stress at harvest and post-harvest, *J. Sci. Food Agric.*, 81, 436–447, 2001.

67. Derbyshire, D., Problems of wastage in cool-stored carrots, *Commercial Grower*, 3928, 703&710, 1971.

68. Muresan, S., Olariu V., and Donescu, V., Preparing potatoes for storage — an important stage in reducing quantitative and qualitative losses, *Prod. Veg. Hortic.*, 32, 19–22, 1983.

69. Kundzicz, K., Storability and mechanical tuber damage of several potato varieties harvested at various dates with Z-644 harvester, *Biul. Inst. Ziemniaka*, 33, 137–147, 1985.

70. Schuhmann, P. and Oertel, H., Damage to potatoes during production procedures, *Tagungsber. Akad. Landwirtsch. Swissensch. Deutsch. Demokratisch. Repub.*, 250, 5–13, 1986.

71. Hemmat, A. and Taki, O., Potato losses and mechanical damage by potato diggers in the Fereidan region of Isfahan, *J. Sci. Tech. Agric.*, 5, 195–209, 2001.

72. Skwarski, B., Effect of potato harvester working sets on the extent of mechanical damage to tubers, *Zesz. Problem. Postepow Nauk Rolniczych*, 399, 227–230, 1993.

73. Grzeskiewicz, H., Gruczek, T., and Gojski, B., Influence of mineral fertilization level on the mechanical tuber damage incurred during harvest under commercial field conditions, *Biul. Inst. Ziemniaka*, 33, 67–72, 1985.

74. Szymczak, J.A. and Wawrzynczak, P., Manual picking and mechanical harvesting of sour cherry fruit 'Nefris' and fruit quality, *Folia-Horticulturae*, 13, 111–119, 2001.

75. Chen, P., Mehlschau, J., and Ortiz-Canavate, J., Harvesting Valencia oranges with flexible curved fingers, *Trans. ASAE*, 25, 534–537, 1982.

76. Timm, E.J. and Guyer, D.E., Tart cherry firmness and quality changes during mechanical harvesting and handling, *Appl. Eng. Agric.*, 14, 153–158, 1998.

77. Marshall, D.E., Levin, J.H., and Cargill, B.F., Properties of Concord grapes related to mechanical harvesting and handling, *Trans. Am. Soc. Agric. Eng.,* 14, 373–376, 1971.

78. Paull, R.E. and Reyes, M.E.Q., Preharvest weather conditions and pineapple fruit translucency, *Scientia-Horticulturae*, 66, 59–67, 1996.

79. Baritelle, A. and Hyde, G.M., Strain rate and size effects on pear tissue failure, *Trans. ASAE*, 43, 95–98, 2000.

80. Valero, D., Martinez, D., Riquelme, F., and Serrano, M., Polyamine response to external mechanical bruising in two mandarin cultivars, *HortScience*, 33, 1220–1223, 1998.

81. Pérez-Vicente, A., Martínez-Romero, D., Carbonell, A., Serrano, M., Riquelme, F., Guillén, F., and Valero, D., Role of polyamines in extending shelf life and the reduction of mechanical damage during plum (*Prunus salicina* Lindl.) storage, *Postharv. Biol. Technol,* 25, 25–32, 2002.

82. Martinez-Romero, D., Serrano, M., Carbonell, A., Burgos, L., Riquelme, F., and Valero, D., Effects of postharvest putrescine treatment on extending shelf life and reducing mechanical damage in apricot, *J. Food Sci.*, 67, 1706–1712, 2002.

83. Martínez-Romero, D., Castillo, S., and Valero, D., Forced-air cooling applied before fruit handling to prevent mechanical damage of plums (*Prunus salicina* Lindl.), *Postharv. Biol. Technol.*, 28, 135–142, 2003.

84. Martínez-Romero, D., Valero, D., Serrano, M., Martinez-Sanchez, F., and Riquelme, F., Effects of post-harvest putrescine and calcium treatments on reducing mechanical damage and polyamines and abscisic acid levels during lemon storage, *J. Sci. Food Agric.*, 79, 1589–1595, 1999.

85. Zrust, J., The glycoalkaloid content in potato tubers (*Solanum tuberosum* L.) as affected by cultivation technology and mechanical damage, *Rostlinna Vyroba.*, 43, 509–515, 1997.

86. Petersen, H.W., Christiansen, J., and Nielsen, S., Effects of light and mechanical damage on the steroid glycoalkaloid and chlorophyll contents of ware potatoes during lifting and storage under shop conditions, *SP Rapport*, 3, 15, 1994.

87. Rataj, V. and Dzupin, R., The influence of storage time on the change of technological properties of potato tubers, *Acta Technol. Agric.*, 1, 17–19, 1998.

88. Chen, P., Ruiz, M., Lu, F., and Kader, A.A., Study of impact and compression damage on Asian pears, *Trans. ASAE*, 30, 1193–1197, 1987.

89. de Castro, L.R., Cortez, L.A.B., and Jorge, J.T., Packaging influence on the development of mechanical injuries in tomatoes, *Cienc. Technol. Aliment.*, 21, 26–33, 2001.

90. Luengo, R.F.A., Moita, A.W., Nascimento, E.F., and Melo, M.F., Reduction of tomato post-harvest losses stored in three different types of boxes, *Hortic. Brasil.*, 19, 151–154, 2001.

91. Nantes, J.F.D. and Durigan, J.F., Evaluation of plastic containers used for packaging, transport and storage of tomatoes, *Rev. Brasil. Armazen.*, 25, 23–30, 2000.

92. Timm, E.J., Brown, G.K., and Armstrong, P.R., Apple damage in bulk bins during semi-trailer transport, *Appl. Eng. Agric.*, 12, 369–377, 1996.

93. Santana Llado, J.D. and Marrero-Dominguez, A., The effects of peel abrasion on the postharvest physiology and commercial life of banana fruits, *Acta Hortic.*, 490, 547–553, 1998.

94. Akkaravessapong, P., Joyce, D.C., and Turner, D.W., The relative humidity at which bananas are stored or ripened does not influence their susceptibility to mechanical damage, *Scientia-Horticultural*, 52, 265–268, 1992.

95. Landrigan, M., Morris, S.C., and Gibb, K.S., Relative humidity influences postharvest browning in rambutan (*Nephelium lappaceum* L.), *HortScience*, 31, 417–418, 1996.

96. Nikolaeva, M.A., Losses during storage of mechanically damaged vegetables, *Konservn. Ovoshchesush. Prom-st'*, 8, 14–15, 1983.

97. Maria, L., Anna, M., Renata, W., and Stanislaw, R., Phenolic metabolism in root slices of selected carrot cultivars, *Plant Physiol. Acta Physiol.-Plant.*, 19, 319–325, 1997.

98. Schulte, N., Timm, E., Ladd, J., and Brown, G., Peach and pear impact damage thresholds: A progress report, Paper 916620, ASAE Winter Meeting, Chicago, 1991.

99. Matthew, R. and Hyde, G.M., Potato impact damage thresholds, *Trans. ASAE*, 40, 705–709, 1997.

100. Bland, W.L., Tanner, C.B., and Maher, E.A., Vapor conductance of wounded potato tuber tissue, *Am. Potato J.*, 64, 197–204, 1987.

101. Miller, A.R., Kelley, T.J., and White, B.D., Nondestructive evaluation of pickling cucumbers using visible-infrared light transmission, *J. Am. Soc. Hortic. Sci.*, 120, 1063–1068, 1995.

102. Schatzki, T.F., Haff, R.P., Young, R., Can, I., Le, L.C., and Toyofuku, N., Defect detection in apples by means of x-ray imaging, *Trans. ASAE*, 40, 1407–1415, 1997.

103. Thomas, P., Kannan, A., Degwekar, V.H., and Ramamurthy, M.S., Non-destructive detection of seed weevil-infested mango fruits by x-ray imaging, *Postharv. Biol. Technol.*, 5, 161–165, 1995.

104. Zion, B., Chen, P., and McCarthy, M.J., Imaging analysis technique for detection of bruises in magnetic resonance images of apples, Paper 93–3084, ASAE Annual International Summer Meeting, Spokane, WA, 1993.

105. Wen, Z. and Tao, Y., Dual-camera NIR/MIR imaging for stem-end/calyx identification in apple defect sorting, *Trans. ASAE*, 43, 449–452, 2000.

106. Greensill, C.V. and Newman, D.S., An experimental comparison of simple NIR spectrometers for fruit grading applications, *Appl. Eng. Agric.*, 17, 69–76, 2001.

107. Upchurch, B.L., Throop, J.A., and Aneshansley, D.J., Influence of time, bruise-type, and severity on near-infared reflectance from apple surfaces for automatic bruise detection, *Trans. ASAE*, 37, 1571–1575, 1994.

108. Damerow, L., Characteristics of reflection on apples after mechanical damage, Paper 993084, ASAE-CSAE-SCGR Annual International Meeting, 1999.

109. Varith, J., Hyde, G.M., Baritelle, A.L., Fellman, J.K., and Sattabongkot, T., Non-contact bruise detection in apples by thermal imaging, *Innovat. Food Sci. Emerg. Technol.*, 4, 211–218, 2003.

110. Jiang, Y., Shiina, T., Nakamura, N., and Nakahara, A., Electrical conductivity evaluation of postharvest strawberry damage, *J. Food Sci.*, 2066, 1392–1395, 2001.

111. Eliseeva, L.G., Neverov, A.N., Moiseev, Yu-V., and Latushkin, V.V., Rapid method for evaluating the storage quality of tubers, *Kartofel' Ovoshchi.*, 1, 29, 1996.

112. Maw, B.W., Hung, Y.C., Tollner, E.W., Smittle, D.A., and Mullinix, B.G., Detecting impact damage of sweet onions using muriatic acid and x-rays, *Appl. Eng. Agric.*, 11, 823–826, 1995.

113. Kader, A.A., Respiration and gas exchange of vegetables, in *Postharvest Physiology of Vegetables*, Weichmann, J., Ed., Marcel Dekker, New York, 1987, pp. 25–43.

114. Rolle, R.S. and Chism, G.W., III, Physiological consequences of minimally processed fruits and vegetables, *J. Food Qual.*, 10, 157–177, 1987.

115. Watada, A.E., Ko, N.P., and Minott, D.A., Factors affecting quality of fresh-cut horticultural products, *Postharv. Biol. Technol.*, 9, 115–125, 1996.

116. Rosen, J.C. and Kader, A.A., Postharvest physiology and quality maintenance of sliced pear and strawberry fruits, *J. Food Sci.*, 54, 656–659, 1989.

117. Gorny, J.R., Cifuentes, R.A., Hess-Pierce, B., and Kader, A.A., Quality changes in fresh-cut pear slices as affected by cultivar, ripeness stage, fruit size, and storage regime, *J. Food Sci.*, 65, 541–544, 2000.

118. Mencarelli, F., Massantini, R., and Botondi, R., Influence of impact surface and temperature on the ripening response of kiwifruit, *Postharv. Biol. Technol.*, 8, 165–177, 1996.

119. Laties, G.G., The onset of tricarboxylic acid cycle activity with aging in potato slices, *Plant Physiol.*, 39, 654–663, 1964.

120. Laties, G.G. and Hoelle, C., The α-oxidation of long chain fatty acids as a possible component of basal respiration of potato slices, *Phytochemistry*, 6, 49–57, 1967.

121. Rocha, A.M.C.N, Brochado, C.M., Kirby, R., and Morais, A.M.M.B., Shelf-life of chilled cut orange determined by sensory quality, *Food Control*, 6, 317–322, 1995.

122. Senesi, E. and Pastine, R., Firmness, colour and atmosphere inside the packages of minimally processed apple slices prepared at different times after harvest and packed under different conditions, *Riv. Frutticolt. Ortofloricolt.*, 59, 57–61, 1997.

123. Abe, K., Tanase, M., and Chachin, K.J., *Jpn. Soc. Hortic. Sci.*, 67, 123–129, 1998.

124. Allong, R., Wickham, L.D., and Mohammed, M., Effect of slicing on the rate of respiration, ethylene production and ripening of mango fruit, *J. Food Qual.*, 24, 405–419, 2001.

125. Tovar, B., Garcia, H.S., and Mata, M., Physiology of pre-cut mango. I. ACC and ACC oxidase activity of slices subjected to osmotic dehydration, *Food Res. Int.*, 34, 207–215, 2001.

126. DeMartino, G., Massantini, R., Botondi, R., and Mencarelli, F., Temperature affects impact injury on apricot fruit, *Postharv. Biol. Technol.*, 25, 145–149, 2002.

127. Gorny, J.R., Hess-Pierce, B., and Kader, A.A., Effects of fruit ripeness and storage temperature on the deterioration rate of fresh-cut peach and nectarine slices, *Hort-Science*, 33, 110–113, 1998.

128. Hakim, A., Batal, K.M., Austin, M.E., Gullo, S., and Khatoon, M., Quality of packaged fresh-cut tomatoes, *Adv. Hortic. Sci.*, 14, 59–64, 2000.

129. Luna-Guzmán, I., Cantwell, M., and Barrett, D.M., Fresh-cut cantaloupe: Effects of CaCl$_2$ dips and heat treatments on firmness and metabolic activity, *Postharv. Biol. Technol.*, 17, 201–213, 1999.

130. Senesi, E., Galvis, A., and Fumagalli, G., Quality indexes and internal atmosphere of packaged fresh-cut pears (Abate Fetel and Kaiser varieties), *It. J. Food Sci.*, 11, 111–120, 1999.

131. Takeuchi, K., Hassan, A.N., and Frank, J.F., Penetration of *Escherichia coli* O157:H7 into lettuce as influenced by modified atmosphere and temperature, *Food Environ. Sanit.*, 64, 1820–1823, 2001.

132. Bai, J.H., Saftner, R.A., Watada, A.E., and Lee, Y.S., Modified atmosphere maintains quality of fresh-cut cantaloupe (*Cucumis melo* L.), *J. Food Sci.*, 66, 1207–1211, 2001.

133. Harikrishna, K., Paul, E., Darby, R., and Draper, J., Wound response in mechanically isolated asparagus mesophyll cells: A model monocotyledon system, *J. Exp. Bot.*, 42, 791–799, 1991.

134. Yang, S.F. and Pratt, H.K., The physiology of ethylene in wounded plant tissue, in *Biochemistry of Wounded Plant Tissues*, Kahl, G., Ed., De Gruyter, Berlin, 1978, pp. 595–622.

135. Nakatsuka, A., Murachi, S., Okunishi, H., Shiomi, S., Nakano, R., Kubo,Y., and Inaba, A., Differential expression and internal feedback regulation of 1-aminocyclopropane-1-carboxylate synthase, 1-aminocyclopropane-1-carboxylate oxidase, and ethylene receptor genes in tomato fruit during development and ripening, *Plant Physiol.*, 118, 1295–1305, 1998.

136. Barry, C.S., Llop-Tous, M.I., and Grierson, D., The regulation of 1-aminocarboxylic acid synthase gene expression during the transition from system-1 to system-2 ethylene synthesis in tomato, *Plant Physiol.*, 123, 979–986, 2000.

137. Chang C. and Shockey, J.A., The ethylene-response pathway: Signal perception to gene regulation, *Curr. Opin. Plant Biol.*, 2, 352–358, 1999.

138. Ruperti, B., Cattivelli, L., Pagni, S., and Ramina, A., Ethylene-responsive genes are differentially regulated during abscission, organ senescence and wounding in peach (*Prunus persica*), *J. Exp. Bot.*, 53, 429–437, 2002.

139. Abeles, F.B., Morgan, P.W., and Saltveit, M.E., *Ethylene in Plant Biology*, 2nd ed., Academic Press, San Diego, CA, 1992.

140. Hoffman, N.E. and Yang, S.F., Enhancement of wound-induced ethylene synthesis by ethylene in preclimacteric cantaloupe, *Plant Physiol.*, 69, 317–322, 1982.

141. Abe, K. and Watada, A.E., Ethylene absorbent to maintain quality of lightly processed fruits and vegetables, *J. Food Sci.*, 56, 1589–1592, 1991.

142. Watada, A.E., Abe, K., and Yamauchi, N., Physiological activities of partially processed fruits and vegetables, *Food Technol.*, 20, 116–122, 1990.

143. Turner, N.J. and Irving, D.E., The effects of mechanical damage on the physiological responses of cool-stored lettuce, 2nd Austral. Lettuce Industry Conf., Paddock to Plate, Gatton, Queensland, Australia, 2002.

144. Reyes, V.G., Improved preservation systems for minimally processed vegetables, *Food Austr.*, 48, 87–90, 1996.

145. Brierley, E.R., Edgell, T., Wiltshire, J.J.J. and Cobb, A.H., The influence of temperature on factors affecting susceptibility to bruising at harvest, *Aspects Appl. Biol.*, 52, 309–314, 1998.

146. Shaidi, F. and Naczk, M., Phenolic compounds in fruits and vegetables, in *Food Phenolics*, Shaidi, F. and Naczk, M., Eds., Technomic Publishing Co., Lancaster, PA, 1995, pp. 75–107.

147. Amiot, M.J., Fleuriet, A., Cheynier, V., and Nicolas, J., Phenolic compounds and oxidative mechanisms in fruit and vegetables, in *Phytochemistry of Fruit and Vegetables,* Tomás-Barberán, F.A. and Robins, R.J., Eds., Clarendon Press, Oxford, U.K., 1997, pp. 51–85.

148. Ho, M.K. and Saltveit, M.E., Wound-induced increases in phenolic content of fresh-cut lettuce is reduced by a short immersion in aqueous hypertonic solutions, *Postharv. Biol. Technol.*, 29, 271–277, 2003.

149. Loaiza-Velarde, J.G. and Saltveit, M.E., Heat shocks applied either before or after wounding reduce browning of lettuce leaf tissue, *J. Am. Soc. Hortic. Sci.*, 126, 227–234, 2001.

150. Ho, M.K. and Saltveit, M.E., Antioxidant capacity of lettuce leaf tissue increases after wounding, *J. Agric. Food Chem.*, 50, 7536–7541, 2002.

151. Blankenship, S.M. and Richardson, D.G., Changes in phenolic acids and internal ethylene during long-term cold storage of pears, *J. Am. Soc. Hortic. Sci.*, 110, 336–339, 1985.

152. Tomas-Barberan, F.A., Loaiza-Velarde, J., Bonfanti, A., and Saltveit, M.E., Early wound- and ethylene-induced changes in phenylpropanoid metabolism in harvested lettuce, *J. Am. Soc. Hortic. Sci.*, 122, 399–404, 1997.

153. Leja, M. and Mareczek, A., Effect of mechanical damage of greenhouse lettuce on some biochemical processes occurring in it during storage, *Folia Hortic.*, 7, 75–89, 1995.

154. Gardner, H.W., How the lipoxygenase pathway affects the organoleptic properties of fresh fruit and vegetables, in *Flavor Chemistry of Lipid Foods*, Min, D.B. and Smouse, T.H., Eds., American Oil Chemists Society, Champaign, IL, 1989, pp. 98–112.

155. Kim, I. and Grosch, W., Partial purification of hydroperoxide lyase from fruits of pear, *J. Agric. Food Chem.* 29, 1220–1225, 1981.

156. Gray, D.A., Prestage, S., Linforth, R.S.T., and Taylor, A.J., Fresh tomato specific fluctuations in the composition of lipoxygenase-generated C6 aldehydes, *Food Chem.* 64, 149–155, 1999.

157. Perez, A.G., Sanz, C., Olias, R., and Olias, J.M., Lipoxygenase and hydroperoxide lyase activities in ripening strawberry fruits, *J. Agric. Food Chem.*, 47, 249–253, 1999.

158. Beaulieu, J.C. and Baldwin, E.A., Flavor and aroma of fresh cut fruits and vegetables, in *Fresh-Cut Fruits and Vegetables: Science, Technology and Market*, Lamikanra, O., Ed., CRC Press, Boca Raton, FL, 2002, pp. 391–425.

159. Lamikanra, O., Enzymatic effects on flavor and texture of fresh-cut fruits and vegetables, in *Fresh-Cut Fruits and Vegetables: Science, Technology and Market*, Lamikanra, O., Ed., CRC Press, Boca Raton, FL, 2002, pp. 125–185.

160. Miller, A.R., Physiology, biochemistry and detection of bruising (mechanical stress) in fruits and vegetables, *Postharv. News Inform.*, 3, 53–58, 1992.

161. Halloran, N., Kasim, M.U., Cagiran, R., Abak, K., and Buyukalaca, S., Determination of mechanical injury and effects of bruising on the postharvest quality of cantaloupes, *Acta Hortic.*, 492, 105–111, 1999.

162. Bruton, B.D., Conway, W.S., Gross, K.C., Zhang, J.X., Biles, C.L., and Sams, C.E., Polygalacturonases of a latent and wound postharvest fungal pathogen of muskmelon fruit, *Postharv. Biol. Technol.*, 13, 205–214, 1998.

163. Agar, I.T., Massantini, R., Hess-Pierce, B., and Kader, A.A., Postharvest CO_2 and ethylene production and quality maintenance of fresh-cut kiwifruit slices, *J. Food Sci.*, 64, 433–440, 1999.

164. Gorny, J.R., Hess-Pierce, B., and Kader, A.A., Quality changes in fresh-cut peach and nectarine slices as affected by cultivar, storage atmosphere and chemical treatments, *J. Food Sci.*, 64, 429–432, 1999.

165. Ji, H.H. and Gross, K.C., Surface sterilization of whole tomato fruit with sodium hypochlorite influences subsequent postharvest behavior of fresh-cut slices, *Postharv. Biol. Technol.*, 13, 51–58, 1998.

166. Senesi, E. and Pastine, R., Pre-treatments of ready-to-use fresh cut fruits, *Indust. Aliment.*, 35, 1161–1166, 1996.

167. Aspinall, G.O., Chemistry of cell wall polysaccharides, in *The Biochemistry of Plants — A Comprehensive Treatise*, Preiss, J., Ed., Academic Press, New York, 1980, pp. 473–500.

168. McNeil, M., Darvill, A.G., Fry, S., and Albersheim, P., Structure and function of the primary cell wall of plants, *Annu. Rev. Biochem.*, 53, 625–663, 1984.

169. Bacic, A., Harris, P.J., and Stone, B.A., Structure and function of plant cell walls, in *The Biochemistry of Plants — A Comprehensive Treatise*, Preiss, J., Ed., Academic Press, New York, 1980, pp. 297–232.

170. Sherman, T.D., Vaughn, K.C., and Duke, S.O., A limited survey of the phylogenetic distribution of polyphenol oxidase, *Phytochemistry*, 30, 2499–2506, 1991.

171. Hind, G., Marshak, D.R., and Coughlan, S.J., Spinach thylakoid polyphenol oxidase: Cloning, characterization, and relation to a putative protein kinas, *Biochemistry*, 34, 64, 1995.

172. Mayer, A.M. and Harel, E., Polyphenol oxidases in plants, *Phytochemistry*, 18, 193–215, 1979.

173. Mathew, A.G. and Parpia, H.A.B., Food browning as a polyphenol reaction, *Adv. Food Res.*, 19, 75–145, 1971.

174. Vamos-Vigyazo, L., Polyphenol oxidase and peroxidase in fruits and vegetables. *CRC Crit. Rev. Food Sci. Nutr.*, 15, 49–127, 1981.

175. Coetzer, C., Corsini, D., Love, S., Pavek, J., and Tumer, N., Control of enzymatic browning in potato (*Solanum tuberosum* L.) by sense and antisense RNA from tomato polyphenol oxidase, *J. Agric Food Chem.*, 49, 652–657, 2001.

176. Partington, J.C., Smith, C., and Bolwell, G.P., Changes in the location of polyphenol oxidase in potato (*Solanum tuberosum* L.) tuber during cell death in response to impact injury: Comparison with wound tissue, *Planta*, 207, 449–460, 1999.

177. McGarry, A., Hole, C. C., Drew, R.L.K., and Parsons, N., Internal damage in potato tubers: A critical review, *Postharv. Biol. Technol.*, 8, 239–258, 1996.

178. Stevens, L.H. and Davelaar, E., Biochemical potential of potato tubers to synthesize blackspot pigments in relation to their actual blackspot susceptibility, *J. Agric. Food Chem.*, 45, 4221–4226, 1997.

179. de Souza, A.L.B., Chitarra, M.I.F., Chitarra, A.B., Machado, J.-da-C., and de Souza, A.L.B., Biochemical reactions in mechanically damaged peaches treated with CaCl2 at the site of the injury, *Cienc. Agrotecnol.*, 23, 658–666, 1999.
180. Gooding, P.S., Bird, C., and Robinson, S.P., Molecular cloning and characterisation of banana fruit polyphenol oxidase, *Planta*, 213, 748–757, 2001.
181. Gillis, N., Verlinden, B.E., Hecke, P., Baerdemaeker, J., Nicolai, B.M., de Baerdemaeker, J., Ben-Arie, R., and Philosoph-Hadas, S., Finite element analysis of mechanical damage in chicory (*Cichorium intybus* L.), *Acta Hortic.*, 553, 527–529, 2001.
182. Loaiza-Velarde, J.G., Tomas-Barbera, F.A., and Saltveit, M.E., Effect of intensity and duration of heat-shock treatments on wound-induced phenolic metabolism in iceberg lettuce, *J. Am. Soc. Hortic. Sci*, 122, 873–877, 1997.
183. Richard-Forget, F.C., and Gauillard, F.A., Oxidation of chlorogenic acid, catechins, and 4-methylcatechol in model solutions by combinations of pear (*Pyrus communis* cv. Williams) polyphenol oxidase and peroxidase: A possible involvement of peroxidase in enzymatic browning, *J. Agric. Food Chem.* 45, 2472–2476, 1997.
184. Lopez-Serrano, M. and Ros-Barcelo, A., Kinetic properties of (dextro)-catechin oxidation by a basic peroxidase isoenzyme from strawberries, *J. Food Sci.*, 62, 676–679, 1997.
185. Robinson, D.S., Peroxidases and catalases in foods, in *Oxidative Enzymes in Foods*, Robinson, D.S. and Eskin, N.A.M., Eds., Elsevier, London, pp 1-47, 1991.
186. Nicholas, J.J., Richard, F.F., Goupy, P., Amiot, M.J., and Aubert, S., Enzymatic browning reactions in apple and apple products, *CRC Crit. Rev. Food Sci. Nutr.* 34, 109–157, 1994.
187. Morimoto, S., Tateishi, N., Matsuda, T., Tanaka, H., Taura, F., Furuya, N., Matsuyama, N., and Shoyama, Y., Novel hydrogen peroxide metabolism in suspension cells of *Scutellaria baicalensis* Georgi, *J. Biol. Chem.*, 273, 12606–12611, 1998.
188. Svalheim, C. and Robertsen, B., Induction of peroxidases in cucumber hypocotyls by wounding and fungal infection, *Physiol. Plant*, 78, 261–267, 1990.
189. Lamikanra O. and Watson, M.A., Effects of ascorbic acid on peroxidase and polyphenoloxidase activities in fresh-cut cantaloupe melon, *J. Food Sci.*, 66, 1283–1286, 2001.
190. Lamikanra, O. and Watson, M.A., Cantaloupe melon peroxidase: Characterization and effects of additives on activity, *Nahrung*, 44, 168–172, 2000.
191. Yamamoto, E., Bokelman, G.H., and Lewis, N.G., Phenylpropanoid metabolism in cell walls: An overview, *ACS Symp. Ser.*, 399, 68–88, 1989.
192. Oosterveld, A., Beldman, G., and Voragen, A.G.J., Oxidative cross-linking of pectic polysaccharides from sugar beet pulp, *Carbohydr. Res.*, 328, 199–207, 2000.
193. Quiroga, M., Guerrero, C., Botella, M.A., Barcelo, A., Amaya, I., and Medina, M.I., A tomato peroxidase involved in the synthesis of lignin and suberin, *Plant Physiol.*, 122, 1119–1127, 2000.
194. Bernards, M.A., Fleming, W.D., Llewellyn, D.B., Priefer, R., Yang, X., Sabatino, A., and Plourde, G.L., Biochemical characterization of the suberization-associated anionic peroxidase of potato, *Plant Physiol.* 121, 135–145, 1999.
195. Rangel, M., Machado, O.L.T., Cunha, M., and Jacinto, T., Accumulation of chloroplast-targeted lipoxygenase in passion fruit leaves in response to methyl jasmonate, *Phytochemistry*, 60, 619–625, 2002.
196. Grun, I.U., Barbeau, W.E., and Crowther, J.B., Changes in headspace volatiles and peroxide values of undeodorized menhaden oil over 20 weeks of storage, *J. Agric. Food Chem.*, 44, 1190–1194, 1996.
197. Wardale, D.A. and Galliard, T., Subcellular localization of lipoxygenase and lipolytic acyl hydrolase enzymes in plants, *Phytochemistry*, 14, 2323–2329, 1975.

198. Vick, B.A. and Zimmerman, D.C., Oxidative systems for modification of fatty acids: The lipoxygenase pathway, in *The Biochemistry of Plants*, Stumpf, P.K. and Conn, E.E., Eds., Academic Press, New York, 1980, pp. 53–90.

199. Sanz, L.C., Perez, A.G., Rios, J.J., and Olias, J.M., Positional specificity of ketones from linoleic acid aerobically formed by lipoxygenase isozymes from kidney bean pea, *J. Agric. Food Chem.*, 41, 696–699, 1993.

200. Moretti, C.L., Baldwin, E.A., Sargent, S.A., and Huber, D.J., Internal bruising alters aroma volatile profiles in tomato fruit tissues, *HortScience*, 37, 378–382, 2002.

201. Hayashi, Y., Yamada, K., Shimada, T., Matsushima, R., Nishizawa, N.K., Nishimura, M., and Hara-Nishimura, I., The ER body, a novel endoplasmic reticulum-derived structure in Arabidopsis, *Plant Cell Physiol.*, 42, 894–899, 2001.

202. Matsushima, R., Hayashi, Y., Kondo, M., Shimada, T., Nishimura, M., and Hara-Nishimura, I., An endoplasmic reticulum-derived structure that is induced under stress conditions in Arabidopsis, *Plant Physiol.*, 130, 1807–1814, 2002.

203. Law, D.M., Davies, P.J., and Mutschler, M.A., Polyamine-induced prolongation of storage in tomato fruits, *Plant Growth Regul.*, 10, 283–290, 1991.

204. Martínez-Romero, D., Valero, D., Serrano, M., Burló, F., Carbonell, A., Burgos, L., and Riquelme, F., Exogenous polyamines and gibberellic acid effects on peach (*Prunus persica* L.) storability improvement, *J. Food Sci.* 65, 288–294, 2000.

205. Kakkar, R.K. and Rai, V.K., Plant polyamines in flowering and fruit ripening, *Phytochemistry*, 33, 1281–1288, 1933.

206. Fedorov, N.I., Starukhin, F.N., Migranova, I.G., Isangulova, A.A., and Nikitina, V.S., Influence of mechanical damage of plants of *Aconitum septentrionale* Koelle on their contents of lappaconitin, *Rastit. Resur.*, 33, 62–67, 1997.

207. Stange, R.R, Jr. and McDonald, R.E., A simple and rapid method for determination of lignin in plant tissues — Its usefulness in elicitor screening and comparison to the thioglycolic acid method, *Postharv. Biol. Technol.*, 15, 185–193, 1999.

208. Schroeder, C.A., *Proc. 2nd World Avocado Congr.*, 1992, pp. 485–488.

209. Skene, D.S., Wound healing in apple fruits: The anatomical response of 'Cox's Orange Pippin' at different stages of development, *J. Hortic. Sci.*, 56, 145–153, 1981.

210. Bostock, R.M. and Stermer, B.A., Perspectives on wound healing in resistance to pathogens, *Annu. Rev. Phytopathol.*, 27, 343–371, 1989.

211. Spotts, R.A., Sanderson, P.G., Lennox, C.L., Sugar, D., and Cervantes, L.A., Wounding, wound healing and staining of mature pear fruit, *Postharv. Biol. Tech.*, 13, 27–36, 1998.

212. Baudoin, A.B.A.M. and Eckert, J.W., Development of resistance against *Geotrichum candidum* in lemon peel injuries, *Phytopathology*, 75, 174–179, 1985.

213. Knight, T.G., Klieber, A., and Sedgley, M., Structural basis of the rind disorder oleocellosis in Washington navel orange (*Citrus sinensis* L. Osbeck), *Ann. Bot.*, 90, 765–773, 2002.

214. Baritelle, A.L. and Hyde, G.M., Commodity conditioning to reduce impact bruising, *Postharv. Biol. Technol.*, 21, 331–339, 2001.

215. Patten, K.D. and Patterson, M.E., Fruit temperature effects on mechanical damage of sweet cherries, *J. Am. Soc. Hortic. Sci.*, 110, 215–219, 1985.

216. Crisoto, C.H., Garner, D., Doyle, J., and Day, K.R., Relationship between fruit respiration, bruising susceptibility, and temperature in sweet cherries, *HortScience*, 28, 132–135, 1993.

217. Sanford, K.A., Lidster, P.D., McRae, K.B., Jackson, E.D., Lawrence, R., Stark, R., and Pragne, R.K., Lowbush blueberry quality in response to mechanical damage and storage temperature, *J. Am. Soc. Hortic. Sci.*, 116, 47–51, 1991.

218. Alvarez, M.D. and Canet, W., Effect of temperature fluctuations during frozen storage on the quality of potato tissue (cv. Monalisa), *Z. Lebensmit. Untersuch. Forsch. A.*, 206, 52–57, 1998.

219. Klieber, A., Jewell, L., and Simbeya, N., Ice or an ice-replacement agent does not improve refrigerated broccoli storage at 1C, *HortTechnology*, 3, 317–318, 1993.

220. Gillies, S.L, Cliff, M.-A., Toivonen, P.M.A, and King, M.C., Effect of atmosphere on broccoli sensory attributes in commercial MAP and microperforated packages, *J. Food Qual.*, 20, 105–115, 1997.

221. Miller, W.M. and Burns, J.K., Interrelationship of impact regimes with citrus fruit quality, *Pap. Am. Soc. Agric. Eng.*, Winter 1991.

222. Woods, J.L., Moisture loss from fruits and vegetables, *Postharv. News Info.* 1, 195–199, 1990.

223. Nunes, M.C.N., Morais, A.M.M.B., Brecht, J.K., and Sargent, S.A., Quality of strawberries after storage is reduced by a short delay to cooling, in *Harvest and Postharvest Technologies for Fresh Fruits and Vegetables*, Kushwaha, L., Serwatowski, R., and Brook, R., Eds., American Society of Agricultural Engineers, St. Joseph, MI, 1995, pp. 15–22.

224. Salas, B., Stack, R.W., Secor, G.A., and Gudmestad, N.C., The effect of wounding, temperature, and inoculum on the development of pink rot of potatoes caused by *Phytophthora erythroseptica, Plant Dis.*, 84, 1327–1333, 2000.

225. Brierley, E.R, Edgell, T., Cobb, A.H., Dale, M.F.B., Dewar, A.M., Fisherr, S.J., Haydock, P.P.J., Jaggard, K.W., May, M.J., Smith, H.G., Storey, R.M.J., and Wiltshire, J.J.J., The influence of ethylene treatments on the blackspot bruising susceptibility of stored potato tubers, *Asp. Appl. Biol.*, 52, 215–222, 1998.

226. Ronsen, K., Tuber wound damage in relation to storage losses of potatoes. *Forskning og Forsk; Landbruket,* 32, 85–96, 1981.

227. Banks, N.H. and Joseph, M., Factors affecting resistance of banana fruit to compression and impact bruising, *J. Sci. Food. Agric.*, 56, 315–323,1991.

228. Lallu, N., Yearsley, C.W., and Elgar, H.J., Effects of cooling treatments and physical damage on tip rot and postharvest quality of asparagus spears, *N. Z. J. Crop Hortic. Sci.*, 28, 27–36, 2000.

229. Fernendez-Trujillo, J.P. and Artes, F., Intermittent warming during cold storage of peaches packed in perforated polypropylene, *Lebensmit. Wissensch. Technol.*, 31, 38–43, 1998.

230. Nuske, D. and Muller, H., Preliminary results on the industrial-type storage of headed cabbage under CA storage conditions, *Nachrichten. Pflanzensch. DDR*, 38, 185–187, 1984.

5 Packaging and Produce Degradation

Patrick Varoquaux
INRA, Avignon, France and Ecole Nationale Supérieure
Agronomique de Montpellier, Montpellier, France

Ibrahim Sani Ozdemir
METU, Ankara, Turkey and Avignon University, Avignon, France

CONTENTS

0-8493-1902-1/05/$0.00+$1.50
© 2005 by CRC Press

5.1 INTRODUCTION

Consumer demand for fresh fruit and vegetables has triggered the need for procedures, including packaging, that maintain the freshness, safety, and quality attributes
of these commodities. Research aimed at improving the quality and extending the
shelf-life of these commodities has been conducted in the fields of physiology,
microbiology, and nutrition.

The first way of reducing postharvest spoilage is to lower the temperature. Low
temperature exponentially reduces the physiological activities of both plant tissues
and microorganisms. The effects of refrigeration are limited by the sensitivity of
some fruits and vegetables to chilling injury. The other technique, most often associated with the first one, is modification of the storage atmosphere.

Controlled atmosphere (CA) is most often used for long-term storage of fruits
(a few months) or for transportation in mobile containers (a few weeks). Modified
atmosphere packaging (MAP) permits atmosphere modification in smaller containers. The internal atmosphere may be changed actively, flushing the packaging with
a gas mixture at sealing, and/or passively by natural equilibration between the
respiration of the plant tissue and diffusion through the permeable membrane of the
pack or tray. This technique is more difficult to handle than CA because the gas
transmission rate (permeability divided by the thickness of the film) of the films or
exchange membranes must be properly optimized; otherwise, MAP may be ineffi-
cient or even detrimental [1]. For example, a film with a permeability that is too
low for O_2 and for CO_2 leads to anoxia and fermentation of plant tissues [2]. The
same phenomenon may occur if the package is stored at a temperature higher than
that of the recommended cold storage [3].

In order to define the optimal permeances (or gas transmission rates) of the
packaging film for O_2, CO_2, and water vapor, the respiration rate of plant tissue must
first be measured as a function of temperature and atmosphere, and then the O_2 and
CO_2 concentrations that maximize shelf life must be determined. The best combination of films, product, and temperature is computed and verified experimentally.
The standardized unit for gas partial pressure in the atmosphere is the Pascal (Pa).
Since 1 kPa = 0.00987 atm, it is possible to approximate 1 kPa to 1% of the
considered gas in the storage atmosphere under atmospheric pressure.

In this chapter we review the effects of gases on respiration (aerobic and anaerobic catabolisms), on growth of spoilage microorganisms, and on quality attributes
of fruits and vegetables. A general procedure to optimize modified atmosphere
packaging is proposed.

5.2 MAP OPTIMIZATION

5.2.1 RESPIRATION RATES

Fresh fruits and vegetables are living tissues and must stay alive until their final
consumption by consumers. Respiration is a catabolic process that supplies the
energy (ATP and ADP) for the plant to thrive and, after harvest, to survive. This
energy is mostly provided by the oxidation of organic metabolites: sugars, lipids,

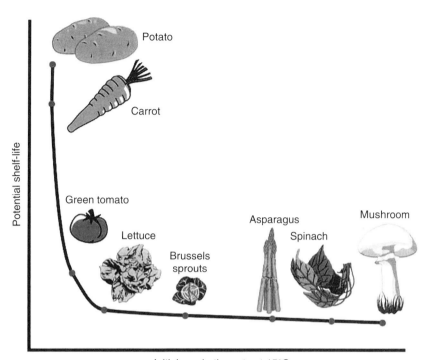

FIGURE 5.1 Relationship between the respiration rate of some vegetables and their potential shelf life. (From Varoquaux, P., Les films à perméabilité aux gaz ajustable: Application aux fruits et légumes, in *Les Emballages Actifs*, Gontard, N., Ed., Lavoisier, Tec & Doc, Paris, 2000, Chap. 5.7.)

and amino acids. The metabolic pathways involved in aerobic respiration are glycolysis, the tricarboxylic acid cycle (or Krebs cycle), and the electron transport system [4]. Within the cells, the energy potential (ATP + 1/2 ADP)/(ATP + ADP + AMP) remains constant even in the case of energy stress. When cells consume their constitutive proteins and membrane lipids [5]. The damaged membranes will gradually lose their active transfer properties and the cell will die. King and Morris [6] reported that the aerobic catabolism of broccoli kept at 20°C results in a large loss of sugars (up to 30%), organic acids, and even proteins during the first 6 h of storage. After 12 h, the warning signs of energy stress appear (i.e., increase in the pool of free amino acids such as glutamine and asparagine).

The potential shelf life of plant tissue after harvest is closely related to its respiration rate (Figure 5.1). The simplified relationship shown in this figure takes into account only spoilage of physiological origin. The shelf life of plant tissue may be limited by other mechanisms such as dehydration or microbial proliferation.

Moreover, accurate assessment of the respiration rate and other respiratory parameters of fruits and vegetables to be packed under modified atmosphere is essential for the optimization of MAP systems.

5.2.1.1 How to Measure Respiration Rates

Respiration of plant tissue is measured in millimoles of O_2 consumed or CO_2 produced per kilogram of tissue per hour. Other units may be used for respiration rate, such as milliliters or milligrams of gas per kilogram per hour. Because the volume of gases is dependent on temperature, the unit measure of millimoles per kilogram per hour should be preferred, especially when the effect of temperature on respiration rate is considered.

The respiratory quotient (RQ) is the ratio RR_{CO_2}/RR_{O_2} when respiration rates are measured in millimoles or milliliters per kilogram per hour. Whereas respiration rates of different plant tissues, assessed under air, vary a lot, RQ ranges from 0.8 to 1.3 [4] depending on the nature of the catabolite. RQ is equal to 1 if the catabolic substrates are carbohydrates. Catabolism of lipids results in a lower RQ and, conversely, in catabolism of organic acid the RQ is higher than unity. It should be pointed out that RQ measured in a closed system method yields an abnormally low RQ, especially at low temperature, because of the dissolution of CO_2 produced in the plant tissue during the equilibration of the system. High RQ is also the sign of a switch to anaerobic catabolism. For example, a fermentative respiratory pathway displays an RQ higher than 2.

Techniques for measurement of respiration rates were reviewed recently by Fonseca et al. [8], who detailed three main methods: (1) the closed or static system, (2) the flowing or flushed system, and (3) the permeable system.

In the closed system, the plant tissue is placed in a gas-tight container fitted with a sampling port (a silicone septum, for example). The weight of the plant tissue and volume of the container are known. The initial atmosphere is usually normal air but can be actively modified at closing of the container using an atmosphere generator. Using gas chromatography or any sensitive gas analyzer, changes in O_2 and CO_2 within the containers are monitored for several hours at regular time intervals. The respiration rate of the commodity is proportional to the absolute value of the slope of the O_2 depletion and CO_2 production curves. The closed system is simple and fast, but it does not allow measurements under stable atmospheric conditions and the internal pressure within the container changes if RQ deviates from unity.

The flushing technique overcomes this drawback. A continuous flow of air or any gas or mixture of gases (flow rate must be constant and accurately measured) ensures the permanent renewal of the atmosphere in the containers. Comparison of gas composition at the entry and exhaust points permits calculation of the respiration rate over a long duration. However, this technique is time-consuming and requires a considerable amount of gas.

The permeable system consists of placing the plant tissue in a flexible pack of known volume, permeability, and diffusing area. The pack is sealed and stored at a constant temperature. The O_2 and CO_2 respiration rates of the plant tissue are proportional to the steady-state concentrations, respectively. As described by Fonseca et al. [8], all of these systems were submitted to modifications to increase their reliability and accuracy.

The last method to determine respiration rate is based on a controlled atmosphere system in which O_2 input and CO_2 scrubbed are accurately measured. The respirometer described by Varoquaux et al. [9] and Benkeblia et al. [10] consists of two airtight stainless steel, cylindrical vessels and a temperature-controlled bath with an accuracy of $\pm\,0.1°C$. Plant tissues are placed in one of the vessels, the lid of which is tightly secured. The head space of the whole instrument is flushed for 0.5 h (repeated after CO_2 equilibration, if necessary) with a ternary gas mixture of preset composition. Once the target atmosphere has been achieved, the vents are shut and the program is run. Internal gas is pumped through a gas analyzer to measure CO_2 and O_2 partial pressures. If this partial pressure is 0.1 kPa higher than the preset value of CO_2 partial pressure, the computer activates an electronic valve that directs the head space atmosphere through a CO_2 trap filled with a 0.1 N sodium hydroxide solution until the initial CO_2 concentration in the sample vessel is restored. The trapping of excess CO_2 results in a proportional decrease in pressure, which is detected by a highly sensitive differential pressure probe. The pressures in each vessel are then balanced by the injection of pure O_2 through a mass flow meter into the vessel containing the plant tissue samples. The CO_2 trapped in the sodium hydroxide is continuously measured by conductivity. The computer logs the O_2 injection times and the changes in conductance of the carbonated sodium hydroxide. At the end of a run, when the O_2 consumption has been constant for at least 2 h, the computer calculates O_2 and CO_2 respiration rates in millimoles per kilogram per hour. The instrument (like all other respirometers) is not able to achieve headspace levels of 0% CO_2, since the minimum CO_2 partial pressure at steady-state is reached when the rate of CO_2 production by the plant tissue equilibrates with the CO_2 trapping rate. The minimum CO_2 partial pressure that can be obtained is about 0.3 kPa in the normal configuration of the instrument.

5.2.1.2 Influence of Temperature

The respiration rate of plant tissues increases exponentially with temperature. Gore's law (log RR is proportional to temperature), an approximation of Arrhenius' law, allows the calculation of Q_{10} (Figure 5.2). Q_{10} is the multiplication factor for the RR

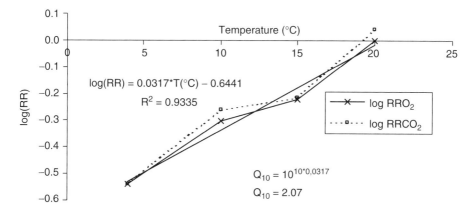

FIGURE 5.2 Changes in the natural log of RR (mmoles per kilogram per hour) as a function of temperature in degrees Celsius in litchis. (From Varoquaux, P. et al., *Fruits*, 57, 313, 2002.)

(RR_{O_2} or RR_{CO_2}) when temperature increases by 10°C. The example in Figure 5.2 shows RR (RR_{O_2} or RR_{CO_2}) of litchis measured at four temperatures. The Q_{10} of oxygen consumption is equal to 2.07, which is consistent with previous results for litchis ranging from 2.04 to 2.8. The Q_{10} of most plant tissues ranges from 2 to 3 [11] but may exceed 3, as with blueberries [12]. Benkeblia et al. [10] reported that the Q_{10} of onion respiration rates were 1.67 and 1.84 for RR_{O2} and RR_{CO2}, respectively, and as a consequence a rise in temperature will increase the RQ.

5.2.1.3 Influence of Gas Composition on Respiration Rate and Physiological Disorders

Respiration rates also depend on the O_2 and CO_2 composition of the atmosphere. The respiration rate of plant tissues is also reduced with decreasing O_2 and increasing CO_2 concentrations. In fermentative catabolism (anaerobic pathway), ethanol production requires decarboxylation of pyruvate to CO_2 without O_2 uptake [8]. A high respiratory quotient (above 2) is a good indicator of anaerobic catabolism [12]. It is established that high CO_2 concentrations inhibit several enzymes of the Krebs cycle, including succinate dehydrogenase [14]. This would inhibit the aerobic pathway and result in the accumulation of succinic acid, which is toxic to plant tissue [15]. CO_2 levels as low as 5% may induce physiological disorders in the common mushroom [16], and asparagus exhibits surface pitting when stored in atmospheres above 10% CO_2 [17]. The average CO_2 toxicity threshold ranges from 7 to 30% depending on plant and storage factors. Crisp head lettuce in storage with elevated CO_2 is strongly affected by O_2 concentration [18]; however, this is not the case for romaine lettuce [19]. Cultivation conditions such as irrigation, climate, and fertilization can modify plant tissue susceptibility to CO_2 injury. Krahn [20] found that the outer leaves of crisp head lettuce are not injured by 2% CO_2 but the inner leaves and midribs show damage. The effect of CO_2 on cell ultrastructures [21] and membranes [22] could account for its toxicity. It can be postulated that CO_2 dissolution, which enhances acidity in the cell medium, may participate in the physiological disorder. Optimum concentrations of CO_2 should minimize the respiration rate without danger of anaerobic metabolism.

Commodities vary widely in their tolerance of different atmospheres [23]. A classification of fresh fruits and vegetables according to their tolerance to reduced O_2 and elevated CO_2 has been presented by Kader et al. [24]. According to Zagory and Kader [25], numerous vegetables such as tomatoes, bell peppers, artichokes, and cabbages do not tolerate over 2% of CO_2 in their storage atmosphere. Apples, peaches, peas, cauliflower, eggplants, carrots, and radishes can be stored with up to 5% CO_2. Leeks, asparagus, beans, onions, cucumbers, garlic, potatoes, strawberries, and raspberries still survive under 10% CO_2. Lastly, some plant tissues such as cherry and broad-leafed endive are very resistant to this gas (up to 20%). It should be noted that most fresh-cut plants are less sensitive to CO_2 than their intact counterparts [26].

Figure 5.3 shows some results for broccoli florets (cv. Emperor). The increase in CO_2 concentration (0, 5, and 12%) inhibits broccoli's respiration rate. Without any carbon dioxide (0%) the increase in respiration rate (RR_{O_2}) as a function of O_2

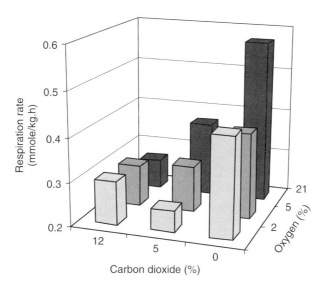

FIGURE 5.3 Changes in respiration rate (RR_{O_2}) in millimoles per kilogram per hour of broccoli florets cv. Emperor as a function of percentage of O_2 and CO_2 in the atmosphere after 7 d of storage at 0°C. (From Zagory, D. and Kader, A.A., *Food Technol.*, 42, 70, 1988. With permission.)

content (2, 5, and 21%) is very important, but under 12% CO_2 the effect of oxygen is no longer noticeable. Respiration is considered to be a succession of enzymatic reactions of Michaelian type and is characterized by an apparent Km (appKm), which is the substrate concentration that gives half the maximal rate of the reaction [27]. The use of the Michaelis equation to model the respiration rate of plant tissue is not rigorous because plant respiration is a multienzyme, multisubstrate system; moreover, the determination of appKm also takes into account the diffusion of oxygen through the epidermis and inner tissues, which does not obey the same equation. In addition, the substrate concentration to consider should be the one in the reaction medium and not its concentration in the surrounding atmosphere. The results reported in Figure 5.3 do not permit even a rough estimate of appKm. Values of appKm are usually measured by fitting the Michaelis-Menten model to sets of data describing gas exchange of plant tissues as a function of O_2 concentration. Considering all possible cumulative errors due to intra- and intervariabilities of batches of fruits and vegetables, appKm cannot be accurately assessed using this approach [10]. For this reason the respirometer described previously was slightly modified in order to permit the direct assessment of O_2 consumption rate as a function of the residual O_2 partial pressure in the head space, which allows the determination of the apparent Km. First, the CO_2 was continuously trapped; second, the O_2 injected to balance the pressure between the two vessels was replaced with pure nitrogen. Under these conditions, the respiration of the plant tissues causes a progressive decrease in O_2 content in the vessel under a constant partial pressure of CO_2. The instrument permits the recording of the O_2 depletion as a function of time (Figure 5.4). This curve is fitted to a polynomial equation, the derivative of which is the

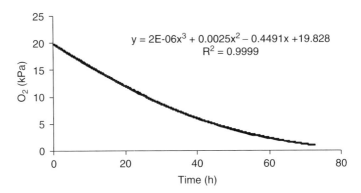

FIGURE 5.4 Changes in oxygen concentration in the respirometer storing litchis at 20°C as a function of time (h). The experimental data and the polynomial fitting equation are superimposed. (From Varoquaux, P. et al., *Fruits*, 57, 313, 2002.)

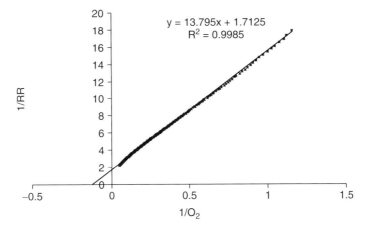

FIGURE 5.5 Apparent Michaelis constant (appKm in percentage of O_2) of litchi at 20°C and 0% CO_2. Lineweaver-Burk double reciprocal coordinates. (From Varoquaux, P. et al., *Fruits*, 57, 313, 2002.)

respiration rate. It is therefore possible to measure RR as a function of O_2 partial pressure and to calculate the appKm. The respiration rate is represented in Figure 5.5 as a function of oxygen concentration in double reciprocal coordinates ($1/RR_{O_2} = f(1/O_2)$). A linear regression, fitted to the experimental points, allows the calculation of the appKm. The appKm values of litchis at 20°C and tomatoes at 10°C are very high (37 and 23.2% O_2, respectively). Hertog et al. [28] reported the intermediary appKm values of 3.76% for apples and 2.7% for chicory chicons. Peppelenbos et al. [29] found higher appKm for apples (6.26%) and similar values for tomatoes and chicory chicons. However, modified atmospheres will not affect the respiration rate of some plant tissues, such as mushroom and lettuce, because their appKm is much lower than 0.1% [9]. Benkeblia et al. [10] claimed that the respiration of onions did not follow the Michaelis-Menten equation because the appKm varied from 14% for high O_2 concentration (from 10 to 20%) to 3% for oxygen concentration lower than 7%. The temperature dependence of appKm was reported by Cameron et al. [12] and Ratti et al. [30]. Song et al. [31] measured the respiration rate of six blueberry cultivars and reported appKm values ranging from 0.1 to 0.78% at 15°C and from 0.1 to 5.2% at 25°C, confirming the very large variability in the determination of appKm with traditional techniques. Based on a dynamic model of gas exchange of respiring produce, Hertog et al. [28] stated that appKm was independent of temperature. They postulated that the observed increase in appKm with temperature was probably due to interfering phenomena such as the onset of fermentation at high temperature or was within the margin of error of the experiment. Benkeblia et al. [10], using the respirometer described above, found appKm values of onion respiration that were reproducible and highly dependent on temperature (1.6% at 4°C and 6.3% at 20°C).

The effect of CO_2 on respiration rate is quantified by its inhibition constant (Ki as a percentage or in kilopascals). Ki is the CO_2 concentration or partial pressure in the surrounding atmosphere that reduces the respiration rate twofold compared with a control sample without inhibitor. Few Ki values are available in the literature; nevertheless Peppelenbos and Van't Leven [32] found 7.68% for apples but much higher concentrations for tomatoes (268% CO_2) and endive (chicon) (181%). It is obvious that CO_2, at concentrations usable for fruit and vegetable MAP, will not have any visible effect on the respiration of the two latter commodities. The calculation of Ki depends on the type of inhibition that CO_2 displays with the enzymes involved in the respiration chain. This issue was discussed by Peppelenbos and Van't Leven [32], but the determination of competitive, noncompetitive, and uncompetitive Ki is too complex for practical use.

5.2.1.4 Influence of Other Parameters

The most marked effect of bruising or wounding (fresh-cut processing) plant tissues is the activation of their catabolism, which includes respiration rate and ethylene production. The response depends on the magnitude of the stress. The O_2 respiration rate of shredded endive is only 1.2 times that of the intact organ [33]. This ratio increases to 1.4 for broccoli florets [34] and to 2 for shredded iceberg lettuce [35]. For more drastic stresses such as grating, respiration reaches 3 to 7 times that of unaltered tissue in carrot [36].

Other interfering phenomena can change the apparent respiration rate of plants, the most important of which is duration of exposure [37]. As shown in Figure 5.6a, the apparent respiration rates of bean sprouts measured in air at various temperatures from 1 to 20°C immediately after harvesting are consistent with the Arrhenius law. After a lag period depending on storage temperature (from 7 d at 1°C to immediate increase at 20°C), the apparent respiration rates of bean sprouts increased from 0.1 to 1.9 mmole·kg^{-1}·h^{-1} at 1°C in 4 d and from 2 to 8 mmole·kg^{-1}·h^{-1} at 20°C after 1 d. Thereafter, their apparent respiration rate stabilized and decreased. This particular behavior strongly suggested that the increase in apparent RR was mainly due to a substantial and synchronous (Figure 5.6b) growth of microorganisms and that the subsequent decrease corresponded to a progressive death of plant tissues and a

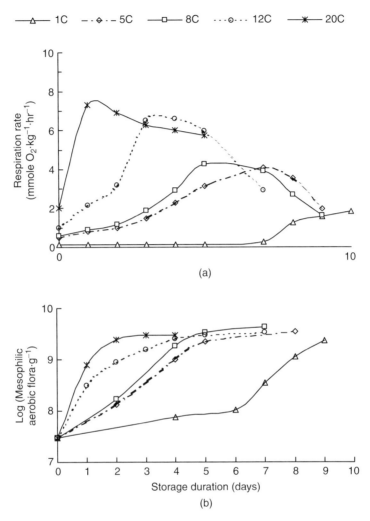

FIGURE 5.6 a. Respiration rate of bean sprouts (mmol·kg^{-1}·h^{-1}) as a function of storage duration at different temperatures. b. Growth of mesophilic aerobic flora (colony forming units per gram) as a function of storage duration at different temperatures. (From Varoquaux, P. et al., *J. Sci. Food Agric.*, 70, 224, 1996.)

stabilization of microbial growth, visible in Figure 5.6b. This increase in apparent RR during storage was previously reported by Chambroy [33], who stated that the apparent RR of shredded endive leaves in air was stable for over 9 d at 5 and even 10°C but increased with time at higher temperatures. The gas composition in 35 µm polypropylene packs with the same endive leaves did not reach a steady-state atmosphere at temperatures higher than 10°C. Chambroy [33] stated that the temperature for distribution of fresh-cut endive leaves should not exceed 10°C. According to French regulations [39], the maximum distribution temperature of these commodities is 4°C. After storage of plant tissues under various atmospheres, the measurement

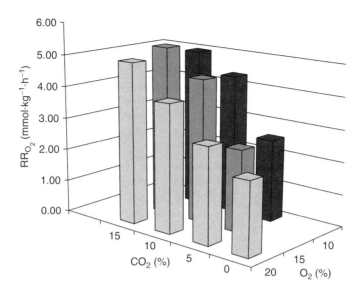

FIGURE 5.7 Respiration rate at 10°C of mushroom (millimoles per kilogram per hour) as a function of storage atmosphere composition. (From Lopez-Briones, G., Augmentation de la durée de vie du produit frais: Application au champignon *Agaricus bisporus*, Ph.D. thesis, Reims University, France, 1991.)

of respiration rate in air is a good index of the physiological condition of the commodity. Figure 5.7 shows the effect of storage of mushrooms at 10°C for 7 d in controlled atmosphere containing from 0 to 15% CO_2 on the respiration rate measured at 10°C under normal atmosphere [40]. There is a good linear relationship between CO_2 concentration in the CA and the increased respiration rate (regression coefficient = 0.965), but the respiration after exposure to CA is not significantly affected by the O_2 concentrations of the CA from 5 to 20%. The temporary inhibition of mushroom catabolism by CO_2 results in an enhanced respiration when the product is again placed under normal atmosphere. This increase in respiration may be considered as a stress response, suggesting that CO_2 concentrations above 5% have a marked phytotoxic effect on mushrooms.

5.2.2 OPTIMAL GAS COMPOSITION

5.2.2.1 Determination Procedure (Controlled Atmosphere)

The optimal gas composition for MAP is determined using a large range of atmosphere compositions. The CA facilities are based on three different principles:

1. Diffusive membrane
   CA facilities fitted with a diffusive membrane permit the passive equilibration of the storage atmosphere to target O_2 and CO_2 composition, depending on the diffusive properties of the membrane (area and O_2 and CO_2 permeabilities). Under commercial conditions only silicone membranes were used,

limiting the range of a steady-state atmosphere. The system requires accurate knowledge of the respiratory parameters of the commodity to be stored. The area of the diffusion membrane must be recalculated as a function of temperature, respiration rate, and weight of the product [41]. The facility is cheap and does not consume any gas or chemicals but is difficult to master.

2. Flushing method

The flushing method is usable only for laboratory facilities. The CA chamber is flushed with the storage atmosphere from an atmosphere generator connected to N_2, O_2, and CO_2 gas bottles. The flow rate to achieve a determined atmosphere is dependent on the CO_2 production and O_2 depletion rates in the CA room. The system is easy to run but is gas-consuming. The flushing method can be improved using a system described by Varoquaux et al. [13]. In this CA facility designed for research purposes, plant tissues are placed in airtight cabinets in which initial atmospheres are created with a system allowing the injection of gases. The atmospheres in the cabinets are regularly analyzed and then adjusted to preset values. The atmosphere in each cell is successively sampled. If the concentrations of the different gases are ± 0.1 kPa from the preset values, the computer orders the opening of the valves of mass flow controllers to input the exact amount of N_2, O_2, and CO_2 necessary to restore the preset composition. The algorithms are based on the dilution law since the system operates at about 100 kPa. Since the three mass flow controllers can be opened at the same time, the system saves time and gas. This equipment allows the storage of fruits and vegetables in a controlled atmosphere, which can then be analyzed at intervals for their sensory, microbiological, and physiological qualities.

3. Scrubbing method

This is the principle of most current commercial CA facilities. The atmosphere is regulated by a CO_2 scrubber of different types (diethanolamine, potassium carbonate, adsorption on activated charcoal or zeolites). Oxygen concentration is equilibrated by injection of air. The universal respirometer previously detailed is based on this method, but O_2 depletion is compensated by injection of pure O_2.

5.2.2.2 Effect of Gas on Ethylene Biosynthesis and Action

Maturation mechanisms of climacteric and nonclimacteric [42] fruits and vegetables have been reported in Chapter 4 of this book. Ethylene (C_2H_4) is produced by all plant tissues, either intact or stressed [43]. Harvesting may also be considered as an injury stress [44]. Exposure of climacteric plant organs to exogenic ethylene triggers maturation and senescence. A reduction in ethylene production and sensitivity associated with the composition of storage atmosphere can extend the shelf life of climacteric commodities. Ethylene production is reduced by either low O_2 or high CO_2, or both, and the effects are additive (Figure 5.8). Oxygen is a substrate of 1-aminocyclopropane-1-carboxylate (ACC) oxidase. Marcellin [45] claimed that a

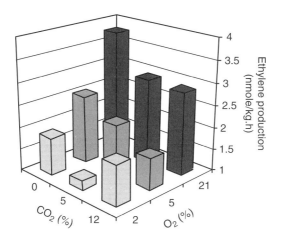

FIGURE 5.8 Ethylene production rate in broccoli as a function of atmosphere composition. (From Zagory, D. and Kader, A.A., *Food Technol.*, 42, 70, 1988. With permission.)

50% reduction in C_2H_4 production was obtained at 1% O_2. This condition is attained in ultra-low oxygen (ULO) storage of apples. Low oxygen partial pressure, also generated in hypobaric storage, inhibits the activity of ACC oxidase [46] and, more surprisingly, stimulates that of ACC synthase [47,48]. Hypoxia also reduces the expression of genes involved in the maturation process and regulated by ethylene [49]. Carbon dioxide, at an optimal concentration of 2%, sustains the *in vivo* activity of ACC oxidase [50,51]. High CO_2 concentrations competitively inhibit ethylene's effects, preventing the autocatalytic induction of ACC synthase [52].

5.2.2.3 Effect of Other Gases

Some gases, including nitric oxide (NO) or nitrous oxide (N_2O), were investigated for reduction of ethylene production and expression.

Short exposure (2 to 24 h) to high NO concentration in anoxia can extend the shelf life of climacteric and nonclimacteric plant tissues [53]. Nitric oxide, a natural plant growth regulator, markedly delays senescence and maturation [54]. These authors postulated a possible role of NO as a natural senescence-delaying plant growth regulator agent regulating ethylene emission. Nitrous oxide at 80% with 20% oxygen delayed maturation of tomatoes and avocados [55]. In both cases an inhibition of ethylene synthesis and action was noted, but the effects of N_2O were not reproducible.

Ethanol vapor also inhibits tomato maturation [56]. Ethanol reduces both ethylene production and action [57]. This beneficial effect of ethanol vapor has not been confirmed on numerous other fruits, including bananas, honeydew and cantaloupe melons, peaches, pears, and plums [58]. These authors found that exposing fruit to ethanol vapor resulted in either an accelerated or a reduced maturation rate.

Atmospheric relative humidity may also interfere with ethylene production and maturation. Postharvest partial dehydration of avocados [59] and bananas [60] favors ethylene production (for more information see Chapter 10).

5.2.2.4 Effect of Gas on Microorganisms

The literature on the effect of MAs on the microbial spoilage of produce is well documented. There exists a large variety of research papers, reviews, and book chapters devoted to the effect of atmospheric compositions, packaging films, and temperature on the microbial population of produce. The reader can also refer to Chapters 13 and 16 of this book for detailed information about microbial spoilage of produce. In the light of scientific research and "commercial conventional wisdom" [61], it can be concluded that the initial microbial load of produce is a major factor that affects the quality and shelf life of fresh-cut fruits and vegetables. Bolin et al. [62] have shown that produce that has a high initial microbial load is more prone to spoilage than produce with a low microbial load. The relation between microbial load and spoilage has led scientists to develop predictive shelf-life models that are based on the growth of microbial populations and their relationship to temperature, atmospheric composition, etc. [63–66]. Therefore, controlling microbial spoilage of produce in MAP by MAs can be considered as a supplementary application to the temperature control and disinfection processes aimed at decreasing initial microbial loads, such as washing with chlorinated water [67–69], gamma irradiation [70], electrolyzed water [71], frozen acidic electrolyzed water [72], UV-C [73], and ozone [74,75]. On the other hand, the studies carried out by Babic et al. [76–78] show that the correlation between high microbial loads and spoilage is not always valid. In another words, produce might have good quality ratings at the end of the shelf life but have high microbial loads. Intactness of the produce, lack of injury, and strict temperature control can be counted as the main factors that prevent spoilage by opportunistic pectinolytic microorganisms even at high populations [61]. The frequency of contradictory results in the literature indicates the difficulty of microbial management in fresh-cut and MAP technologies, which need to be well "tailored" [79]. The complex interaction and dynamic structure of microorganisms, bacteria, yeasts, and molds that make up the microbial flora of the produce and their close relations with produce physiology and morphology force scientists and producers to think in a multidimensional manner. Microflora are often produce-specific and have a dominant group due to their competitive nature [80]. Disinfection processes, selective action of MAs on microorganisms, and temperature play important roles in the final structure of microflora. Bennik et al. [79] have shown that the psychotropic pathogen *Listeria monocytogenes* grew better on disinfected than on undisinfected chicory endive due to the loss of competitive bacteria.

The gases used in MAP, namely O_2 and CO_2, have different modes of action on microorganisms. In commercial applications, atmospheres low in O_2 and high in CO_2 concentrations are preferred, not only due to their beneficial effects on the produce's sensorial and physiological quality, but also to their potential in controlling microbial growth [24]. When it is used at or below 21%, the effect of O_2 on microorganisms in MAs is limited to the creation of aerobic conditions that inhibit the growth of strict anaerobic microorganisms. Nevertheless, most bacteria in fruits are Gram-negative (see Chapter 13), and microorganisms such as pseudomonads and *Leuconostoc* spp. will not be affected by the reduction of the O_2 gas concentration [81]. Bennik et al. [79] showed that *Pseudomonas* counts on chicory endive at 8°C were lower under 0% O_2 than under 21% O_2, regardless of the CO_2 concentration, contrary to a 1.5% O_2 atmosphere, which gave results similar to those with 21% O_2. In another study Bennik et al. [82] found that 1.5 or 21% O_2 at 8°C did not significantly affect maximum specific growth rate or maximum population densities of the food-borne pathogens *Aeromonas hydrophila, Yersinia enterocolitica, Listeria monocytogenes,* and *Bacillus cereus.* Abdul-Raouf et al. [83] have reported that an atmosphere of 3% O_2 and 97% N_2 had no apparent effect on populations of *E. coli* O157:H7 inoculated onto shredded lettuce, sliced cucumbers, and shredded carrots. Oxygen depletion does not prevent yeast growth, except in the case of strict anaerobiosis, which is not to be used with plant tissues [84].

As demonstrated by Vial [85], lowering O_2 from 10 to 1.5% does not significantly reduce the total yeast and bacteria count on slices of kiwi fruit stored at 10°C for 10 d [85]. However, a strict anaerobiosis will trigger anaerobic catabolism of plant tissue and, if the pH is neutral or slightly alkaline, can lead to *Clostridium botulinum* growth [86]. But accumulation of *C. botulinum* toxin in the absence of visual decay has not been reported even on mushrooms [87] and tomatoes [88]. The results of recent studies have shown that O_2 can be used as a "hurdle" in MAP, when it is used at concentrations above 21%. Day [89] and Barry-Ryan et al. [90] found that high O_2 could extend the shelf life of fresh fruits and vegetables, but Kader and Ben-Yehosua [91] failed to show such an enhancement of storability. Creation of oxidative stress by high oxygen concentrations induces the generation of intracellular reactive oxygen species that affect vital cell components and reduce cell viability [92]. However, the effectiveness of these atmospheres is highly related to the antioxidant capacity of the microorganisms to challenge the stress conditions, which can even be different among the strains of the same species [93]. The study carried out by Amanatidou et al. [92] shows that high oxygen concentrations (80% or 90%, balanced with N_2) have no inhibitory effect on the *in vitro* growth rate of some selected pure cultures of bacteria and yeasts (*Lactococcus lactis, Leuconostoc mesenteroides, Lactobacillus plantarum, Aureobacterium strain 27, Pseudomonas fluorescens, Enterobacter agglomerans, Listeria monocytogenes, Salmonella enteritidis, Escherichia coli,* and *Candida sake*) except *Salmonella typhimurium* and *Candida guilliermondii,* whose growth rate and/or maximum yield were reduced significantly under these atmospheres. Gonzalez-Roncero and Day [94] reported similar effects of 99% O_2 on *Pseudomonas fragi, Aeromonas hydropylia, Yersinia enterocolitica,* and *Listeria monocytogenes.* However, Jacxsens et al. [95] observed retarded *in vitro* growth rates of *Pseudomonas fluorescens, Candida lambica, Botrytis cinerea, Aspergillus flavus,* and *Aeromonas caviae* (HG4) under 70, 80, and 95% O_2 at 4°C, whereas *Erwinia carotovora* was stimulated under increasing O_2 concentrations. Wszelaki and Mitcham [96] found a 100% O_2 atmosphere to be the most effective in reducing the mycelia growth of *Botrytis cinerea* at 5°C. However, the same oxygen concentration was reported to have a stimulatory effect on *Penicillium* decay on grapefruit [91]. When combined with high CO_2 concentrations, hyperoxygenation becomes more effective in reducing the growth of most of the microorganisms that were counted above [91,92,94,96,97].

Many researchers have reported a bacteriostatic effect of elevated carbon dioxide atmospheres. Carbon dioxide (20 kPa) was reported to be effective in reducing the

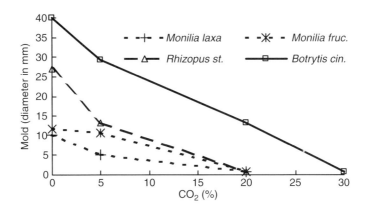

FIGURE 5.9 Mold growth under varying CO_2 concentrations. (From Chambroy, Y. and Souty, M., Rôle du CO_2 sur le comportement des fruits à noyau. Utilisation des atmosphères modifiées, in *Proc. Symp. Qualité Post-Récolte et Produits Dérivés chez les Fruits à Noyau*, Vendrell, M. and Audergon, J.M. Eds., Lerida, Spain, 1994, p. 139.)

microbial counts of bacteria of the genera *Leuconostoc*, *Pseudomonas*, *Micrococcus*, and *Bacillus* on prickly pear cactus. [98] Gunes and Hotchkiss [99] found that at abusive temperatures (15 and 20°C), elevated CO_2 (greater than or equal to 15%), and reduced O_2 (less than 1%), MAs inhibited the growth of *E. coli* O157:H7 on apple slices. Bennik et al. [82] observed decreases in the specific growth rates of *Aeromonas hydrophila*, *Yersinia enterocolitica*, *Listeria monocytogenes*, and a cold-tolerant strain of *Bacillus cereus* with increasing CO_2 (0, 5, 20, and 50%) concentrations regardless of the O_2 concentrations (1.5 or 20%).

An increase in CO_2 concentration inhibits yeast growth and significantly reduces mesophilic flora compared to air storage. The CO_2 protective effect against molds occurs for concentrations above 5% on a synthetic culture medium [100]. El Halouat et al. [101] found that 40 and 80% CO_2 did not support the growth of *Aspergillus niger*. The growth of most molds is completely inhibited above 20% CO_2, but some, such as *Botrytis cinerea*, resist up to 30% (Figure 5.9). Yeasts on inoculated fruits are more resistant to CO_2. *Monilia laxa*, which is totally inhibited by 20% when cultivated *in vitro*, manages to survive and grows in atmospheres with up to 30% CO_2 when inoculated on peaches (Figure 5.10). Vial [85] also found that the protective effect of CO_2 against yeasts and bacteria occurs for concentrations of 20%. If normal atmosphere is restored after exposure to high CO_2 concentrations, the molds resume their growth, proving that CO_2 has no lasting effect (Figure 5.10).

Carbon dioxide is a soluble gas both in water and fat, and its impact on microorganisms depends on its degree of solubility in produce. In MAP, the solubility of carbon dioxide depends on the ratio of headspace to medium volume, buffering capacity of the medium, pH, water activity (aw), and the temperature [102]. As carbon dioxide dissolves in cells, it hydrates to H_2CO_3 and so reduces the pH and damages cell activity. Lowenadler [102] and Devlieghere et al. [103] have shown that decrease in the growth rate of some bacteria, *Pseudomonas fragi*, *Yersinia enterocolitica*, and *Lactobacillus sake*, is well correlated with increasing CO_2 solubility in the medium.

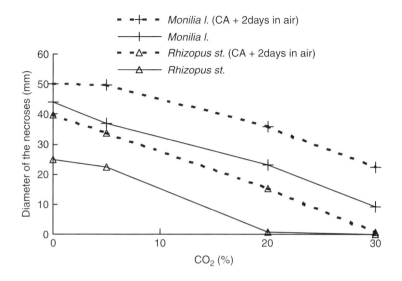

FIGURE 5.10 Diameter of necroses (in millimeters) caused by *Monilia* and *Rhizopus* inoculated on peaches maintained under various controlled atmospheres at 10°C (full line). Dotted lines present data obtained from peaches kept for 2 d in normal air after controlled atmosphere. (From Nguyen The, personal communication. With permission.)

However, this mode of action of carbon dioxide makes it phytotoxic for much produce and causes physiological disorders. Therefore, produce tolerance to elevated CO_2 should be taken into account before applying this kind of atmosphere.

Some other atmospheric gases such as noble gases and nitrous oxide (N_2O) have also been tested for their antimicrobial effects [89]. Qadir and Hashinaga [104] showed that N_2O delayed and reduced lesion growth rate on some fruits inoculated with *Alternaria alternata, Penicillium expansum, Botrytis cinerea, Fusarium oxysporum* f. sp. *Fragariae, Rhizopus stolonifer, Geotrichum candidum, Fusarium oxysporum* f. sp. *lycopersici*, and *Colletotrichum acutatum*, regardless of the physiological nature of the fruits or group of fungi.

5.2.2.5   Effect of Gas on Quality Attributes

5.2.2.5.1   Firmness

The firmness of plant tissue is a combination of the internal pressure of the cells (or turgor) and the resistance of the cell walls, including middle lamella integrity. Therefore, either plasmolysis or disruption of the middle lamella results in firmness breakdown. The mechanisms involved in these phenomena are detailed in Chapters 2 and 5 of this book.

The effect of atmosphere composition on climacteric fruit is closely associated with ethylene production, but CA and MAP show specific effects on the firmness retention of some plant tissue; mushroom, strawberry, and kiwi fruit slices will be taken as examples.

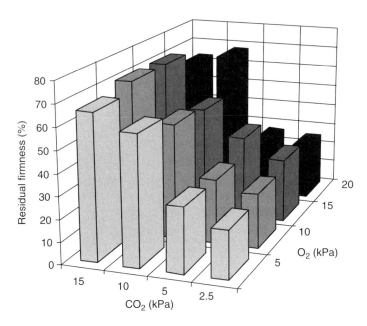

FIGURE 5.11 Residual firmness of the mushroom *Agaricus bisporus* L. as a function of atmosphere composition after 7 d of storage at 10°C. Loss in firmness was measured as the percentage of final firmness/initial firmness, both in newtons. This percentage is highly correlated with both the opening of the cap and change in the length of the pileus. (From Lopez-Briones, G. et al., *Int. J. Food Sci. Technol.,* 27, 493, 1992.)

Aging of the common mushroom *Agaricus bisporus* is characterized by a soft and spongy texture partly due to cell growth and water migration. The senescence mechanism of mushrooms after harvest is still unclear but CO_2 has a most important effect on their texture (Figure 5.11). Oxygen apparently does not play a part in the process. As a consequence, CO_2 also exhibits a marked effect on mushroom development, as previously pointed out by Burton et al. [105]. Mushrooms stored for 7 d at 10°C under 15% CO_2 did not break their veil whatever the O_2 content. An unbroken veil is a major quality criterion for French consumers. Therefore, CO_2 in the marketing containers of mushrooms should be maintained at the highest level compatible with the preservation of their whiteness.

It has been demonstrated that controlled atmosphere containing 2% O_2 and 5% CO_2 was beneficial for intact kiwi fruit storage [106]. Fresh-cut processing of kiwi fruit is responsible for a rapid decrease in firmness of the slices, but the mechanism of the texture breakdown due to injury stress is not fully understood [107]. Vial [85] demonstrated that MAP with CO_2 concentrations up to 60% did not significantly affect their softening.

Maybe the most pronounced effect of CO_2 on firmness has been reported for strawberries. Harker et al. [108] found that strawberries stored under air at 0°C were firmer than at harvest and firmness was further enhanced irreversibly by CO_2 (5 to 40%) treatments, which confirms the findings of Larsen and Watkins [109] and Smith and Skog [110]. They owed this enhancement to increased cell-to-cell adhesion, which increased by 60% after CO_2 treatment. The changes in the pH of the apoplast and consequent precipitation of soluble pectins were thought to be responsible for the gain in cell-to-cell adhesion. However, whether the pH of the apoplast was increased or decreased due to CO_2 treatment was not clearly stated. A decrease in the resistance of the apoplast was observed and was supposed to be due to the accumulation of H^+ and HCO_3^- ions. Holcroft and Kader [111] postulated that changes in the intracellular pH of produce treated with elevated CO_2 atmospheres depend on the size and pH of the vacuole and concentration of weak organic acids that affect the buffering capacity of the cells. They observed an increase in intracellular pH of strawberries, which have large and acidic vacuoles, with unchanged firmness. Yong and Soon [112] reported a similar increase in pH for fig fruit accompanied by enhancement in firmness. On the other hand, the pH of cut lettuce tissues, which do not accumulate high concentrations of organic weak acids, decreased after treatments of elevated CO_2 [113]. Wszelaki and Mitcham [96] and Perez and Sanz [114] have reported similar beneficial effects of high-oxygen atmospheres on firmness of strawberries. However, the mode of action of high-oxygen atmospheres on firmness retention of strawberries is not well understood.

Gas composition in the storage atmosphere may have an indirect effect on firmness. For example, the fungistatic effect of CO_2 on models prevents or delays the rot of plant tissue. In some cases the indirect effect of gas composition on decay is more complex.

Pectinolytic *Pseudomonas fluorescens* and *Pseudomonas viridiflava* are well known as soft rot bacteria on stored vegetables [115]. They may also induce spoilage of shredded endive leaves [116]. Strains of pectinolytic *Pseudomonas marginalis* isolated from fresh-cut broad-leafed endives show a strong spoilage capacity on the commodity though these bacteria are present in both spoiled and apparently sound packs [80]. CA enriched in CO_2 up to 50% reduces the *in vitro* growth of pseudomonads and *Erwinia* spp. Surprisingly, the same atmosphere does not modify epiphytic proliferation of *Pseudomonas marginalis* in green salad leaves, but CA or MA containing over 15% CO_2 reduces or eliminates soft rot on endive leaves that were previously inoculated with heavily concentrated *Pseudomonas marginalis* suspension. The same phenomenon is observed when leaves are inoculated with a sterile (ultrafiltered) growth medium of this bacterium. The presence, in the ultrafiltrate, of active pectinolytic enzyme can account for the spoilage. Increasing CO_2 to 20% reduces the necrosis, and at 40% it prevents any damage. It is remarkable that MA, with high CO_2 concentrations up to 40%, has no effect on the soft rot induced by *Aspergillus niger* growth medium. Because *P. marginalis* produces pectate lyases [117] with an optimum pH of 6 to 8, and *Aspergillus niger* a polygalacturonase active at pH 4 to 5, it is postulated that the beneficial effects of CO_2 on soft rot induced by *P. marginalis* are due to the acidification of cell medium by dissolved CO_2.

5.2.2.5.2 Flavor

At least three mechanisms are responsible for off-flavors in MAP fruits and vegetables: (1) enzymatic action, (2) a switch to anaerobic catabolism, and (3) microbial spoilage.

5.2.2.5.2.1   Enzymatic Action

Enzymatic peroxidation of unsaturated fatty acids (namely, linoleic and linolenic acids) is the most dramatic example of the biochemical alteration of the natural flavor of either fresh or fresh-cut vegetables. This peroxidation by oxygen is catalyzed by lipoxidase and leads to the formation of volatile aldehydes and ketones [118]. It has been demonstrated that the concentrations of *n*-hexanal, a byproduct of hydroperoxide degradation, is well correlated with postharvest development of off-flavor in peas [119]. Bruising of peas and snap beans has been shown to be an important factor in the development of delayed off-flavor. Hand-shelled peas do not deteriorate in flavor as rapidly as vine-shelled peas. The hydroperoxides are very unstable and may be cytotoxic, affecting particularly proteins and membranes [120]. The positive and negative effects of lipoxygenase activity on aroma and flavor compounds of fruits and vegetables were recently reviewed by Lamikanra [121].

5.2.2.5.2.2   Switch to Anaerobic Catabolism

Anaerobic catabolism results not only in the production of ethanol and CO_2 but also in a drastic modification of the aromatic profile of the stored commodity. Ke et al. [122] demonstrated that exposure of strawberries to hypoxic conditions ($O_2 < 0.25\%$) resulted in the production of ethanol, ethyl acetate, and acetaldehyde during storage. Guichard et al. [123] studied the effect of controlled atmosphere on the volatile compounds of strawberries stored for 7 d at 10°C. They found that some aromatic molecules such as methyl-ethyl acetate, ethyl propionate, and ethyl hydroxy-3-butanoate were not present originally but appeared during storage (i.e., ethyl hexanoate, ethyl octanoate, butyl acetate, and most volatile compounds responsible for off-flavor) proportionally to CO_2 concentration. Conversely, some compounds, including volatile acids such as butanoic, hexanoic, and octanoic acids decreased with increasing CO_2. Furaneol or dimethyl-2,5-hydroxy-4-(2H)-furanose is considered to be a key compound in strawberry aroma [124] (Figure 5.12), but its concentration in fruit, even at the highest CO_2 concentration, remained constant during storage. It seems that CO_2 significantly alters strawberry flavor only at concentrations higher than 20% [122,123].

Hyperoxygenation is also detrimental to strawberry flavor. Perez and Sanz [114] stored strawberries at 8°C in four different atmospheres containing from 5 to 90% O_2 and 10 to 20% CO_2. After 1 week of storage they found high contents of off-flavor–related compounds in the samples stored in high CO_2–high O_2 atmospheres. They concluded that hyperoxygenation was responsible for the alteration of ester biosynthesis and suggested that stress induced by high CO_2 and stress induced by high O_2 had an additive effect on strawberry flavor alteration.

The switch to anaerobic catabolism changes the taste of all plant tissues. As shown in Figure 5.13, both high CO_2 and low O_2 can promote ethanol production in litchis, which can be stored for 4 weeks under 15% CO_2 (with 5% O_2) at 1°C without developing a noticeable off-flavor. Keeping stone fruits under carboxic anaerobiosis for 2 to 5 d modifies their aroma with a relative increase in their ester and terpeneol contents and a decrease in their C6 aldehydes and ketones. A moderate exposure to CO_2 would improve the flavor of apricots [124a]. As a general rule it is advisable to prevent any catabolic deviation of intact and fresh-cut fruit because this physiological disorder provokes tissue necrosis and favors yeast growth [125].

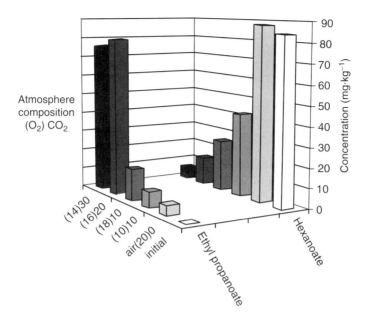

FIGURE 5.12 Hexanoate and ethyl propanoate as a function of atmosphere composition. (From Guichard, E. et al., *Sci. Aliments*, 12, 83, 1992. With permission.)

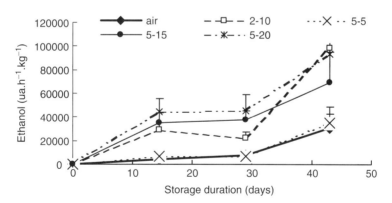

FIGURE 5.13 Changes in ethanol production of litchis as a function of storage duration at 1°C under different atmosphere compositions. (From Varoquaux, P. et al., *Sci. Aliments*, 10, 127, 1990.)

5.2.2.5.2.3 Off-Flavor of Microbial Origin
Off-flavor occurring on spoiled shredded carrots is associated with the following characteristics: MA with excessive CO_2 concentrations (over 30%) and low O_2 content (below 1%) and high counts of acid lactic bacteria (LAB) and yeasts along with the accumulation of ethanol and acetic and lactic acids. The main microbial contaminant was identified as *Leuconostoc mesenteroides* [126]. However, the initial

LAB counts did not markedly differ from one pack to another, and Carlin et al. [3] demonstrated that the growth in both LAB and yeast was faster when the CO_2 content within the packs increased above 20% and O_2 decreased below 1%. These authors [3] claimed that microorganisms were not the primary cause of spoilage because the growth of *L. mesenteroides* on a sterile medium was not affected by either O_2 or CO_2 concentration in the range from 1 to 10% O_2 and from 1 to 40% CO_2.

Storage of shredded carrots in CO_2-enriched and O_2-deprived atmosphere resulted in high K^+ leakage [3], and Barry-Ryan et al. [90] demonstrated that the main factor in the switch to anaerobic catabolism of MAP shredded carrots is O_2 depletion. It may be postulated that both CO_2 accumulation and O_2 depletion participate synergistically in physiological disorder. Other electrolytes and nutrients, including sugars, were also expelled from the carrot tissues. This exudate provided microorganisms with a good growth substrate. Modified atmosphere containing less than 20% CO_2 and more than 2% O_2 prevented physiological disorder and, as a consequence, microbial spoilage. Conversely, very permeable films favored high respiration and induced a fast consumption of sugars, which caused a noticeable loss in palatability of the shredded carrots. It should be noted that the MA passively generated in packs is highly dependent on the storage temperature because the Q_{10} of plant tissue respiration rate is higher than that of the gas diffusion rate through a polymeric film. At low temperature (from 1 to 3°C) physiological activity and bacterial growth are sufficiently reduced to delay the spoilage for over a week even with the least permeable film (OPP, oriented polypropylene, from 30 to 40 μm in thickness) used to pack fresh-cut carrots in Europe.

5.2.2.5.3   Color

Color changes during the storage and distribution of fresh fruits and vegetables result from many mechanisms: yellowing, or conversely greening, of green tissues, enzymatic browning, oxidation of carotenoids and flavonoids, etc.

5.2.2.5.3.1   Chlorophylls

The chlorophyll molecule consists of a magnesium-chelated tetrapyrrole with a fat-soluble "tail" (phytol). The change from bright green to a yellowish color is due to the replacement of the magnesium atom by hydrogen to form pheophytin [127]. Chlorophyll in plant tissue is protected from acidic cytoplasm by its linkage with the protein. This reaction (pheophytinization) is responsible for the degreening of broccoli during storage [34]. Piagentini et al. [128] found that the type of packaging film (OPP and LDPE) did not affect chlorophyll retention in fresh-cut spinach. Destruction of chlorophyll by exposure to ethylene has been reported and was correlated with an increase in chlorophyllase activity [129]. The chlorophyll changes may result from the loss of membrane integrity that occurs with senescence hastened by ethylene [130]. Other degradative pathways may participate in chlorophyll breakdown, such as chlorophyll oxidase, lipase, and lipoxidase. The results reported by Watada et al. [120] suggest that the chlorophyll degradation mechanism probably differs among plant species (more information is given in Chapter 8). Degreening of plant tissues can be reduced by lowering ethylene production and keeping the cell membranes in good physiological condition. Wang et al. [131] found that the storage of green asparagus at 1°C in CA with 5% CO_2 and high relative humidity resulted in increased pH and better retention of chlorophyll than control samples stored in air. These effects became more pronounced as the concentration of CO_2 in the atmosphere was increased. The degradation products of chlorophyll in CA-stored asparagus were exclusively pheophytins. It may be postulated that in spite of lowering the pH, which expectedly favors pheophytinization, CO_2 has a more protective effect against other chlorophyll degradation pathways.

Conversely, chlorophyll synthesis in vegetables exposed to light after harvest, such as white asparagus, endive (chicon), and potatoes is associated with decreasing quality. Greening of some plant tissues is accompanied by the production of the poisonous glycoalkaloids solanine and chaconine [132]. It has been demonstrated that green discoloration can be prevented by maintaining these sensitive commodities in modified atmosphere containing 15% CO_2 [133,134]. Experiments on endive (chicon) showed that the oxygen depletion likely plays a major role in greening. This alteration is prevented by maintaining plant tissues in atmosphere with less than 1% oxygen. The effects of high CO_2 and low oxygen on the greening and degreening mechanisms are still unclear.

5.2.2.5.3.2 Enzymatic Browning

Enzymatic browning is the main color deterioration in bruised or injured plant tissues [135]. The mechanisms of enzymatic browning are detailed in Chapters 2 and 11 of this book. The reactions result in the formation of brown, red, or yellow molecules that may also degrade the appearance of intact tissues such as litchi and mushroom, but overall it is one of the main causes of spoilage in fresh-cut processing and distribution.

One of the most important problems with litchis is the loss of the red coloration of the peel during storage. Commercialization becomes impossible after browning, so it is very important to maintain this red color. As an example, Figure 5.14 shows

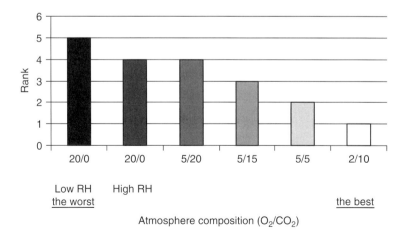

FIGURE 5.14 Effect of gas composition on litchi fruit discoloration assessed by sensory analysis (the lower the rank, the higher the quality) after 6 weeks at 1°C. (From Varoquaux, P. et al., *Fruits*, 57, 313, 2002.)

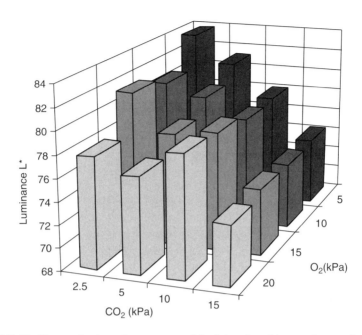

FIGURE 5.15 Changes in the color parameter L* of *Agaricus bisporus* L. as a function of O_2 and CO_2 concentrations. Color was measured after 7 d of storage at 10°C. Discoloration was evaluated by measurement of L* in the L*a*b* CIE color system. (From Lopez-Briones, G. et al., *Int. J. Food Sci. Technol.*, 27, 493, 1992.)

the brown discoloration of litchi fruit estimated by sensory analysis as a function of atmospheric composition. Results reported in Figure 5.14 show the detrimental effect of both dehydration and high CO_2 concentrations. A low oxygen concentration is also beneficial to the external color of litchi fruit. The color of mushroom caps measured as the L* parameter of the CIE system is well correlated with the visual appearance of mushroom [136]. White strain cultivars of mushrooms with L values > 80 would be acceptable to wholesalers. Those with L < 70 would be rejected by French consumers. As shown in Figure 5.15, there is a good relationship between the luminance of the pileus surface after 7 d of storage at 10°C and the CO_2 content of CA [40]. It is possible that very low percentages of CO_2 (less than 2.5%) improve color compared to controls kept under normal atmosphere. Oxygen concentrations from 5 to 20% have no marked effect on external color. This browning phenomenon not only affects the surface color but also the color of inner tissues. Internal browning as a function of CA composition follows the same pattern as external browning, but to a lesser extent. However, CO_2 concentration is not the only factor involved in the brown discoloration of mushrooms. The availability of O_2 at the intracellular level and O_2 affinity for other oxidases may play a role in color deterioration. The influence of O_2 is also significant but much less than that of CO_2. These conclusions confirm previous results of Smith [137] and Tomkin [138] on external color. Nichols and Hammond [139] found a linear regression between the internal color of mushrooms and their respiratory rate in prepacked, overwrapped punnets. These authors also concluded that there was an inverse relationship between internal color and CO_2 concentration for a given film type. This relationship was confirmed by Lopez-Briones et al. [140].

Green salads, especially fresh-cut ones, are highly sensitive to enzymatic browning. Lowering O_2 concentration to less than 2% will reduce discoloration. Polyphenol oxidase is not active at O_2 concentrations below 1%, but this hypoxic condition may trigger anaerobic catabolism. It is also possible to inject 5 to 10% CO_2 into the bag at sealing (actively modified atmosphere). CO_2 inhibits polyphenoloxidase activity and prevents discoloration during the equilibration phase of the passively modified atmosphere. The optimal gas composition for the storage of lettuce depends on many parameters, including the cultivar, season of growing, pedoclimatic conditions, etc. A 2% O_2 and 5% CO_2 atmosphere may be recommended. Higher CO_2 concentrations may be used for lettuce insensitive to russet spotting [141].

Increasing CO_2 and/or decreasing O_2 concentration in the storage atmosphere may have opposite effects on different quality attributes of the commodities. The optimum gas composition will therefore depend on the market requirements.

5.3 PACKAGING CONDITIONS

5.3.1 MODIFIED ATMOSPHERE PACKAGING

Knowing the respiratory parameters of plant tissues (respiration rates RR_{O_2} and RR_{CO_2}, respiratory quotient RQ, Q_{10} of respiration, and apparent Km), the conditions of packaging, and the optimal gas concentration for controlled atmosphere storage, it is possible to determine the optimal gas transmission rate of packaging film (in milliliters of O_2 or CO_2 per square meter per 24 h per atmosphere) with several mathematical models that were extensively reviewed by Fonseca et al. [8]. Among all available semiempirical models, those based on a Michaelis-Menten type equation are the most commonly used [142]. CO_2 inhibition is also considered in some of them, but for practical reasons the interference of CO_2 is most often skipped due to the lack of available literature data and the limited effect of CO_2 variation that can be tolerated by most plant tissues. As stated by Fonseca et al. [8], this simple Michaelis-Menten model is a simplification that tends to fit the experimental data well.

In all models gas diffusion through a film follows Fick's laws, and oxygen concentration within a pack is governed by the following equation:

$$x = x_0 - a \int_0^t \frac{RRm \cdot x}{Km_{app} + x} * \frac{m}{V} dt + b \int_0^t \frac{Po_2 \cdot A}{V}(x_e - x)dt \qquad (5.1)$$

where
x = concentration of oxygen within the pack at any given time t (%)
x_0 = initial concentration of oxygen (%)
x_e = concentration of oxygen outside the pack (generally 21%)

[RRm] = maximum respiration rate (mmol·kg^{-1}·h^{-1})
Km$_{app}$ = apparent Michaelis constant (% O$_2$)
P$_{O_2}$ = effective oxygen transmission coefficient or permeance (mL·m^{-2}·24 h^{-1}·atm^{-1})
t   = time (h)
a   = a coefficient 2.24
b   = a coefficient 0.4 E-4
V   = volume of the container (L)
m   = mass of commodity (kg)
A   = surface area of packaging over which diffusion can occur (m^2)

The quantity of carbon dioxide produced by plant tissues (RR$_{CO_2}$) is equal to that of consumed oxygen multiplied by the respiratory quotient (RQ).

After a certain storage time (equilibration phase), the respiration of plant tissue will balance diffusive exchanges through the film and Equation (1) can be simplified:

$$a \cdot \frac{RRm \cdot x_s}{Km_{app} + x_s} \cdot \frac{m}{V} = b \frac{P_{O_2} \cdot A}{V}(x_e - x_s) \tag{5.2}$$

where x$_s$ = O$_2$ concentration at steady-state (%).

$$RRm = RR_0 \cdot Q_{10R}^{\frac{T}{10}} \tag{5.3}$$

where
RR$_0$ = maximum respiration rate of the plant tissue at 0°C (mmol·kg^{-1}·h^{-1})
Q$_{10R}$ = Q$_{10}$ of respiration
T   = temperature (°C)

Combining Equation 5.2 and Equation 5.3 gives Equation 5.4:

$$P_{O_2} = \frac{a}{b} \cdot \frac{m \cdot RR_0 \cdot Q_{10}^{\frac{T}{10}}}{A(x_e - x_s)} \cdot \frac{x_s}{x_s + Km_{app}} \tag{5.4}$$

This equation shows that, at equilibrium, oxygen concentration does not depend on the headspace or on the initial gas concentration. The injection of gas when closing the package or the punnet (actively modified atmosphere) only permits the steady-state to be attained more rapidly, thereby shortening or avoiding the detrimental equilibration phase when the produce is sensitive to enzymatic browning or lipid peroxidation.

In Equation 5.4, x$_s$ is the optimal O$_2$ concentration previously determined by controlled atmosphere studies and P$_{O_2}$ is the permeance to O$_2$ of the film that will give x$_s$ at equilibrium and at temperature T [142]. It is therefore possible to calculate the permeance of the film at 23°C (usual temperature for measurements of permeability and permeance of packaging films):

$$P_{O_2,(23°C)} = P_{O_2}T \cdot Q_{10F}^{\frac{23-T}{10}} \tag{5.5}$$

where $Q_{10F} = Q_{10}$ of permeance of the film for oxygen.

Unfortunately, the constants Q_{10F} for O_2 and CO_2 transmission rates are difficult to obtain and generally are not given by the film suppliers. Q_{10F} of microperforated films ranges between 1.05 and 1.15, while that of polymeric nonperforated films is higher, between 1.4 and 1.8 [11]. If Q_{10p} is known, Equation 5.5 may be used to determine the theoretical permeance of the film at 23°C. Using very simple mathematical software it is possible to find the optimal permeability for O_2 and CO_2 [143].

5.3.2 PRINCIPAL PROPERTIES OF FILMS

Microperforated films with a large range of permeability to oxygen, from 1 to 200,000 mL $O_2 \cdot m^2 \cdot 24$ $h^{-1} \cdot atm^{-1}$, are now available. However, this kind of film is devoid of selectivity to gases ($S = P_{CO_2}/P_{O_2} = 1$ for all microperforated films) and the sum of the O_2 and CO_2 partial pressures is always equal to 21% (or kPa) provided RQ = 1 [144]. Numerous commodities need storage atmospheres simultaneously poor in O_2 and CO_2 and, therefore, films with a selectivity much greater than 1 [145]. As an example, the selectivity of oriented polypropylene (OPP) film is 3 and that of low-density polyethylene up to 7 [146]. It is possible to modify the permeance of these nonperforated polymeric films to O_2 (and to CO_2 at the same rate) by manufacturing films with different thicknesses, since the gas transmission rate of a film is proportional to the reciprocal of its thickness (not valid for microperforated films). However, for commercial and mechanical reasons, the thickness of a film must be in the range 15 to 100 μm.

The selectivity of new hydrophilic films such as Pebax®, Sympatex®, and Hytrel® ranges from 5 to 30 depending on the density of the hydrophilic groups in the polymer, the relative humidity, and the temperature. The selectivity of biofilms, such as wheat gluten, maize zein, and methyl cellulose, can reach 30 [147], which allows almost all combinations of gas composition at steady-state.

It is therefore possible with all of the acquired data to predict the theoretical changes in the atmospheric composition of an optimized modified atmosphere package. Many phenomena may interfere, such as microbial growth, switch to partially anaerobic catabolism, normal senescence, and climacteric crisis. Moreover, the respiration rate of the common mushroom may vary from 1 to 2 as reported by Beit-Halachmy and Mannheim [148]. This large between-batch difference in respiration rate of the same produce could be due to its stage of maturity [149]. This variability makes the prediction of the mathematical model unreliable, and the respiration rate of plant tissues should be determined on each batch and optimum packaging conditions recalculated.

5.3.3 SIMULATION OF MODIFIED ATMOSPHERES

Equation 5.1 above allows the prediction of changes in O_2 concentration in a package. This equation is valid only if the storage temperature is stable. In the case of temperature abuse, the effect of temperature on respiration rate and on the O_2 and CO_2 transmission rates of films must be introduced in Equation 5.1 with Equation 5.4. It should be noted that when distribution occurs at ambient temperature, the optimal permeability of the film must be calculated for the highest temperature likely to be encountered. The modified atmosphere reached at lower temperatures, however, will not be optimal for preserving the sensory qualities of the commodity. Software developed in the Netherlands predicts changes in modified atmospheres for the case of variable temperature (linear interpolation) [28]. After this last step, the packaging parameters should be tested, since numerous variables are not taken into account in the theoretical calculations, which are based on an extreme simplification of fruit and vegetable physiology. For example, models do not take into account microbial growth which may be responsible for a fourfold increase in the apparent respiration rate of soybean sprouts within 14 h at 20°C [38]. Some models [29] can also simulate catabolic deviation (commonly called anaerobic or fermentative catabolism), but many of the parameters required in the model are difficult to determine experimentally. Numerous other parameters can also interfere; for example, some plant tissues are sensitive to carbon dioxide, maturation (stage of maturity and ripeness), chilling injury, etc.

5.4 CONCLUSION

The effects of gases on the quality attributes of fruits and vegetables have been investigated and commercially used for more than 70 years. Yet the question posed recently by Salveit [150] about the possibility of finding an optimal controlled atmosphere is still controversial. Salveit [150] defined this optimal CA as "the storage conditions that produce the best quality product," but as stated by Harker et al. [151], the quality of fruits and vegetables is a relative concept depending on consumer beliefs, attitudes, perceptions, and preferences. Therefore, the quality keeping targets of CA will have to be adapted to the consumer's expectations. Moreover, as discussed in this chapter, modifying gas composition may have opposite effects on quality attributes. The natural variability in the physiological responses of plant tissues to gases and their changes during storage make the choice of a stable optimal atmosphere empirical in real conditions. Nevertheless, new developments are designed to overcome this problem. Veltman et al. [152] proposed a dynamic control system that monitors ethanol concentration in the headspace of the CA room and regulates the O_2 concentration during apple storage. This system takes into account batch variability and physiological changes during storage but ignores most other causes of spoilage.

In any case, CA storage of fruit is continuously improving and, when conducted reasonably in terms of duration and temperature, this technique gives satisfactory results [153]. Modified atmosphere packaging is still a more difficult issue. Because the atmosphere is regulated passively, the variability of respiratory parameters is of major importance. Optimization of the MAP system would require the determination of the respiration rate of each batch to be packed. It should also be recalled that some phenomena can interfere with the product respiration rate. Fruits and vegetables are contaminated by a wide range of microorganisms, the growth of which is more or less affected by the storage atmosphere. Their growth will depend mostly on temperature, relative humidity, and host-parasite resistance, which is extremely complex to model. No model to date has integrated the effect of microorganisms on respiration rate. Therefore, the prediction of atmosphere changes in the MAP system and the optimization of packaging conditions is only possible when microorganisms do not interfere with plant catabolism. To overcome this limitation it is recommended that the shelf-life duration (sell-by date) be matched to the lag period of microorganism growth at the distribution temperature. This difficulty underlines the uttermost importance of either efficient disinfecting of packed commodities ([154] and in this book Chapter 16) or active films with antimicrobial activity [155]. The dynamic response of stored fruits and vegetables to their environmental conditions (e.g., exposure to endogenous ethylene or changes in CO_2 and hypoxia sensitivities due to natural senescence) is not taken into account in any proposed mathematical simulations of MAP [150]. These considerations emphasize the role of new active packaging material with special functions, such as ethylene scavenging and new pretreatment of climacteric plant organs with 1-methylcyclopropene (1-MCP) [156] or any other process reducing ethylene production or effects.

Moreover, some atmosphere compositions that are easy to generate and maintain in CA are almost out of reach in MAP. For example, low O_2–low CO_2 atmospheres require diffusion membranes with very high permselectivity such as hydrophilic films, either biofilm or polymeric membranes, which are not available on the market or do not meet the minimal conditions for practical use.

In the case of temperature abuse, the Q_{10} of the respiration rate (from 2 to 3) does not match the Q_{10} of the gas diffusion rate through the diffusion membrane (from 1 to 2), thus resulting in a detrimental deviation of the atmosphere. Hydrophilic film may match this difference in Q_{10} for CO_2 but not for O_2 diffusion rate.

The choice of a packaging film depends on many other parameters not reviewed in this chapter, including the following.

5.4.1 WATER TRANSMISSION RATE

This characteristic of the film will govern the dehydration rate of the commodity. A moderate dehydration may be beneficial to some vegetables such as onion bulbs [157] and mushrooms [158,159] but detrimental to most other fruits and vegetables [160,161]. Conversely, excessive relative humidity favors fungal and bacterial decay [162].

5.4.2 CONSUMER ACCEPTANCE OF APPEARANCE AND HANDLING

The choice of a film to pack fresh-cut commodities depends more often on the physical appearance of the film, which must be glossy, transparent, and crispy, than on the film's permeability to gases. To match the commercial quality of polypropylene to high-respiring commodities the membrane is microperforated.

5.4.3 Industrial Acceptability of the Film's Mechanical Resistance, Compatibility with Packaging Machines, Thermo-Sealability, and Printing Qualities

More research, especially on MAP produce, is needed (1) to make packaging of fruits and vegetables more predictable and reliable, and (2) to select fruit and vegetable cultivars with physiological characteristics adapted to MAP and CA techniques (i.e., large sugar reserves, low respiration rate, and high resistance to hypoxia and elevated CO_2 concentrations).

REFERENCES

1. Myers, R.A., Packaging considerations for minimally processed fruits and vegetables, *Food Technol.*, 43, 129, 1989.
2. Kader, A.A., Biochemical and physiological basis for effects of controlled and modified atmospheres on fruits and vegetables, *Food Technol.*, 40, 99, 1986.
3. Carlin, F. et al., Modified atmosphere packaging of fresh "ready-to-use" grated carrots in polymeric films, *J. Food Sci.*, 4, 1033, 1990.
4. Kader, A.A., Respiration and gas exchange of vegetables, in *Postharvest Physiology of Vegetables*, Weichman, J., Ed., Marcel Dekker, New York, 1987, Chap. 3.
5. Brouquisse, R. et al., Study of glucose starvation in excised maize roots tips, *Plant Physiol.*, 96, 619, 1992.
6. King, G.A. and Morris, S.C., Early compositional changes during postharvest senescence of broccoli, *J. Am. Soc. Hort. Sci.*, 119, 1000, 1994.
7. Varoquaux, P., Les films à perméabilité aux gaz ajustable: Application aux fruits et légumes, in *Les Emballages Actifs*, Gontard, N., Ed., Lavoisier, Tec & Doc, Paris, 2000, Chap. 5.
8. Fonseca, S.C., Oliveira, F.A.R., and Brecht, J.K., Modeling respiration rate of fresh fruits and vegetables for modified atmosphere packages: A review. *J. Food Eng.*, 52, 99, 2002.
9. Varoquaux, P. et al., Respiratory parameters and sugar catabolism of mushroom (*Agaricus bisporus* Lange), *Postharv. Biol. Technol.*, 16, 51, 1999.
10. Benkeblia, N. et al., Respiratory parameters of onion bulbs (*Allium cepa*) during storage: Effects of ionising radiation and temperature, *J. Sci. Food Agric.*, 80, 1772, 2000.
11. Exama, A., Abul, J., and Lencki, R.W., Suitability of plastic films for modified atmosphere packaging of fruits and vegetables, *J. Food Sci.*, 6, 1365, 1993.
12. Cameron, A.C. et al., Modified atmosphere packaging of blueberry fruit: Modeling respiration and package oxygen partial pressures as a function of temperature, *J. Am. Soc. Hortic. Sci.*, 119, 534, 1994.
13. Varoquaux, P. et al., Procedure to optimize modified atmosphere packaging for fruits, *Fruits*, 57, 313, 2002.
14. Ranson, L., Walker, D.A., and Clark, I.D., The inhibition of succinic oxidase by high CO_2 concentration, *Biochem. J.*, 66, 57, 1957.
15. Bendall, D.S., Ranson, S.L., and Walker, D.A., Effects of CO_2 on the oxidation of succinate and reduced diphosphopyridine nucleotide by Ricinus mitochondria, *Biochem. J.*, 76, 221, 1960.
16. Lopez-Briones, G., Augmentation de la durée de vie du produit frais: Application au champignon *Agaricus bisporus*, Ph.D. thesis, Reims University, France, 1991.

17. Lipton, W.J., Carbon dioxide induced injury of Romaine lettuce stored in controlled atmosphere, *Hortscience*, 22, 461, 1987.
18. Steward, J.K. and Uota, M., Carbon dioxide injury to lettuce as influenced by carbon monoxide and oxygen levels, *Hortscience*, 7, 327, 1972.
19. Lipton, W.J., Recommendations for CA storage of broccoli, Brussels sprouts, cabbage, cauliflower, asparagus and potatoes, *Michigan State Univ. Hortic. Rep.,* 28, 277, 1977.
20. Krahn, T.R., Improving the keeping quality of cut head lettuce, *Acta Hortic.*, 62, 79, 1977.
21. Frenkel, C. and Patterson, M.E., Effect of carbon dioxide on ultra-structure of Bartlett pears, *Hortscience*, 9, 338, 1974.
22. Sears, D.F. and Eisenberg, R.M., A model representing a physiological role of CO_2 at the cell membrane, *J. Gen. Physiol.*, 44, 869, 1961.
23. Lougheed, E.C., Interaction of oxygen, carbon dioxide, temperature and ethylene that may induce injuries in vegetable, *Hortscience*, 22, 791, 1987.
24. Kader, A.A., Zagory, D., and Kerbel, E.L., Modified atmosphere packaging of fruits and vegetables, *Crit. Rev. Food Sci. Nutr.*, 28, 1, 1989.
25. Zagory, D. and Kader, A.A., Modified atmosphere packaging of fresh produce, *Food Technol.*, 42, 70, 1988.
26. Varoquaux, P. and Wiley, R.C., Biological and biochemical changes in minimally processed refrigerated fruits and vegetables, in *Minimally Processed Refrigerated Fruits & Vegetables*, Wiley, R.C., Ed., Chapman & Hall, New York, 1994, Chap. 6.
27. Talasila, P.C., Cameron, A.C., and Joles, D.W., 1994, Frequency distribution of steady state oxygen partial pressures in Modified Atmosphere packages of cut broccoli, *J. Am. Soc. Hortic. Sci.,* 119, 556, 1994.
28. Hertog, M.L.A.T.M. et al., A dynamic and generic model of gas exchange of respiring produce: the effects of oxygen, carbon dioxide and temperature, *Postharv. Biol. Technol.*, 14, 335, 1998.
29. Peppelenbos, H.W. et al., Modeling oxidative and fermentative carbon dioxide production of fruits and vegetables, *Postharv. Biol. Technol.*, 9, 283, 1996.
30. Ratti, C., Raghavan, G.S.V., and Gariépy, Y., Respiration model and modified atmosphere packaging of fresh cauliflower, *J. Food Eng.*, 28, 297, 1996.
31. Song, Y., Kim, H.K., and Yam, K.L., Respiration rate of blueberry in modified atmosphere at various temperatures, *J. Am. Soc. Hortic. Sci.*, 117, 925, 1992.
32. Peppelenbos, H.W. and Van't Leven, J., Evaluation of four types of inhibition for modelling the influence of carbon dioxide on oxygen consumption of fruits and vegetables, *Postharv. Biol. Technol.*, 7, 17, 1996.
33. Chambroy, Y., Physiologie et température des produits frais découpés, *Rev. Gén. Froid*, 3, 78, 1989.
34. Ballantyne, A., Stark, R., and Selman, J.D., Modified atmosphere packaging of broccoli floret, *Int. J. Food Sci. Technol.*, 23, 353, 1988.
35. Ballantyne, A., Stark, R., and Selman, J.D., Modified atmosphere packaging of shredded lettuce, *Int. J. Food Sci. Technol.*, 23, 267, 1988.
36. Carlin, F. et al., Effect of controlled atmosphere on microbial spoilage, electrolyte leakage and sugar content on fresh, "ready-to-use" grated carrots, *Int. J. Food Sci. Technol.*, 25, 110, 1990.
37. Brecht, J.K. et al., Maintaining optimal atmosphere conditions for fruits and vegetables throughout the postharvest handling chain, *Postharv. Biol. Technol.*, 27, 87, 2003.
38. Varoquaux, P. et al., Modified atmosphere packaging of fresh beansprouts, *J. Sci. Food Agric.*, 70, 224, 1996.

39. Anon., Guide de bonnes pratiques hygiéniques concernant les produits végétaux prêts à l'emploi dits de "IV' gamme," *Bull. Off. Rép. Franç.*, 4, 221, 1988.
40. Lopez-Briones, G. et al., Storage of common mushroom under controlled atmospheres, *Int. J. Food Sci. Technol.*, 27, 493, 1992.
41. Kok, R., and Raghavan, G.S.V., A mathematical model of a Marcellin type storage system, *Acta Hortic.*, 157, 31, 1985.
42. Alonso, J.M., Chamarro, J., and Granell, A., Evidence for the involvement of ethylene in the expression of specific RNAs during maturation of the orange, a non-climacteric fruit, *Plant Mol. Biol.*, 29, 385, 1995.
43. Salveit, M.E., Yang S.F., and Kim, W.T., History of the discovery of ethylene as a plant growth substance, in *Discovery in Plant Biology*, Vol. 1, Kang, S.D. and Yang, S.F., Eds., World Scientific Publ., Singapore, 1998, 47.
44. Kato, M. et al., Wound-injured ethylene synthesis in stem of harvested broccoli and its effect in senescence and ethylene synthesis in broccoli florets, *Postharv. Biol Technol.*, 24, 69, 2002.
45. Marcellin, P., Conservation des fruits en atmosphère contrôlée: situation actuelle, *Rev. Gén. Froid*, 76,155, 1986.
46. Yip, W.K., Jiao, X.Z., and Yang, S.F., Dependence of *in vivo* ethylene production rate on 1-aminocyclopropane-1-carboxylic acid content and oxygen concentrations, *Plant Physiol.*, 88, 553, 1988.
47. Buffer, G. and Bangeth, F., Effects of propylene and oxygen on the ethylene producing system of apples, *Physiol. Plant.*, 58, 486, 1983.
48. Gorny, J.R. and Kader, A.A., Controlled-atmosphere suppression of ACC synthase and ACC oxidase in 'Golden delicious' apples during long-term cold storage, *J. Am. Soc. Hortic. Sci.*, 121, 751, 1996.
49. Kanellis, A.K., Solomon, T., and Roubelakis-Angelakis, K., Suppression of cellulase and polygalacturonase and induction of alcohol dehydrogenase isoenzymes in avocado fruit mesocarp subjected to low oxygen stress, *Plant Physiol.*, 96, 269, 1990.
50. Smith, J.J., Ververidis, P., and John, P., Characterization of the ethylene-forming enzyme partially purified from melon, *Phytochemistry*, 31, 1485, 1992.
51. John, P., Ethylene biosynthesis: The role of 1-amino-1-carboxylate (ACC) oxidase during development and the effects of various tree factors, *Aust. J. Agric. Res.*, 17, 465, 1997.
52. Buffer, G., Ethylene enhanced 1-amino-cyclopropane 1-carboxylic acid synthase activity in ripening apples, *Plant Physiol.* 75, 192, 1984.
53. Leshem, Y.Y. and Wills, R.B.H., Harnessing senescence delaying gases nitric oxide and nitrous oxide: A novel approach to postharvest control of fresh horticultural produce, *Biol. Plant.*, 41, 1, 1998.
54. Leshem, Y.Y., Wills, R.B.H., and Veng-Va Ku, V., Evidence for the function of the free radical gas – nitric oxide (NO) – as an endogenous maturation and senescence regulating factor in higher plants, *Plant Physiol. Biochem.*, 36, 825, 1998.
55. Gouble, B., Fath, D.H., and Soudain, O., Nitrous oxide inhibition of ethylene production in ripening and senescing fruits, *Postharv. Biol. Technol.*, 5, 311, 1995.
56. Salveit, M.E. and Mancarelli, F., Inhibition of ethylene synthesis and action in ripening tomato fruit by ethanol vapors, *J. Am. Soc. Hortic. Sci.*, 113, 572, 1988.
57. Salveit, M.E., Effect of alcohols and their interaction with ethylene on the ripening of epidermal pericarp discs of tomato fruit, *Plant Physiol.*, 90, 167, 1989.
58. Ritenour, M.A. et al., Ethanol effects on the ripening of climacteric fruits, *Postharv. Biol. Technol.*, 12, 35, 1997.
59. Adato, I. and Gazit, S., Water deficit stress, ethylene production and ripening in avocado fruits, *Plant Physiol.* 53, 45, 1974.

60. Littmann, M..D., Effect of water loss on the ripening of climacteric fruits, *Queensl. J. Agr. Anim. Sci.*, 29, 131, 1971.

61. Zagory, D., Effects of post-processing handling and packaging on microbial populations, *Postharv. Biol. Technol.*, 15, 313, 1999.

62. Bolin, H.R. et al., Factors affecting the storage stability of shredded lettuce, *J. Food Sci.*, 42, 1319, 1977.

63. Van Impe, J.F. et al., Predictive microbiology in a dynamic environment: A system theory approach, *Int. J. Food Microbiol.*, 25, 227, 1995.

64. Garcia-Gimeno, R. M. et al., Determination of packed green asparagus shelf-life, *Food Microbiol.*, 15, 191, 1998.

65. Guerzoni, M.E. et al., Shelf-life modeling of fresh-cut vegetables, *Postharv. Biol. Technol.*, 9, 195, 1996.

66. Jacxens, L., Devlieghere, F., and Debevere, J., Temperature dependence of shelf-life as affected by microbial proliferation and sensory quality of equilibrium modified atmosphere packaged fresh produce, *Postharv. Biol. Technol.*, 26, 59, 2002.

67. Beuchat, L.R. and Brackett, R.E., Survival and growth of *Listeria monocytogenes* on lettuce as influenced by shredding, chlorine treatment, modified atmosphere packaging and temperature, *J. Food Sci.*, 55, 755, 1990.

68. Delaquis, P. et al., Survival and growth of *Listeria monocytogenes* and *Eschericihia coli* O157:H7 in ready-to-eat iceberg lettuce washed in warm chlorinated water, *J. Food Prot.*, 65, 459, 2002.

69. Li, Y. et al., Changes in appearance and natural microflora on iceberg lettuce treated in warm, chlorinated water and then stored at refrigeration temperature, *Food Microbiol.*, 18, 299, 2001.

70. Hagenmaier, R.D. and Baker, R.A., Low-dose irradiation of cut iceberg lettuce in modified atmosphere packaging, *J. Agric. Food Chem.*, 45, 2864, 1997.

71. Park, C.M. et al., Pathogen reduction and quality of lettuce treated with electrolyzed oxidizing and acidified chlorinated water, *J. Food Sci.*, 66, 1368, 2001.

72. Koseki, S., Fujiwara, K. and Itoh, K., Decontaminative effect of frozen acidic electrolyzed water on lettuce, *J. Food Prot.*, 65, 411, 2002.

73. Allende, A. and Artès, F., UV-C radiation as novel technique for keeping quality of fresh processed "Lollo Rosso" lettuce, *Food Res. Int.*, 36, 739, 2003.

74. Palou L. et al., Effects of continuous 0.3 ppm ozone exposure on decay development and physiological responses of peaches and table grapes in cold storage, *Postharv. Biol. Technol.*, 24, 39, 2002.

75. Singh, N., Bhunia, A.K., and Stroshine, R.L., Efficacy of chlorine dioxide, ozone, and thyme essential oil or a sequential washing in killing *Escherichia coli* O157:H7 on lettuce and baby carrots, *Lebensm.–Wiss. Technol.*, 35, 720, 2003.

76. Babic, I. et al., The yeast flora of stored ready-to-use carrots and their role in spoilage, *Int. J. Food Sci. Technol.*, 27, 473, 1992.

77. Barriga, M.I., Trachy, G., and Simard, R.E., Microbial changes in shredded iceberg lettuce stored under controlled atmospheres, *J. Food Sci.*, 56, 1586, 1991.

78. O'Connor-Shaw, R.E. et al., Shelf-life of minimally processed honeydew, kiwifruit, papaya, pineapple and cantaloupe, *J. Food Sci.*, 59, 1202, 1994.

79. Bennik, M.H.J. et al., Microbiology of minimally processed, modified-atmosphere packaged chicory endive, *Postharv. Biol. Technol.*, 9, 209, 1996.

80. Nguyen The, C. and Carlin, F., Alterations microbiologiques des légumes prêts à l'emploi, Proc. 2nd Int. Conf. Plant Dis., Bordeaux, France, 1, 743, 1988.

81. Potter, L. et al., Survival of bacteria during oxygen limitation, *Int. J. Food Microbiol.*, 55, 11, 2000.

82. Bennik, M.H.J. et al., Growth of psychrotrophic foodborne pathogens in a solid surface model system under the influence of carbon dioxide and oxygen, *Food Microbiol.*, 12, 509, 1995.

83. Abdul-Raouf, U.M., Beuchat, L.R., and Ammar, M.S., Survival and growth of *Echerichia coli* O157:H7 on salad vegetables, *Appl. Environ. Microbiol.*, 59, 1999, 1993.

84. Miller, M.W., Yeast in food spoilage: an update, *Food Technol.*, 1, 76, 1979.

85. Vial, C., Contribution à la mise au point de tranches de kiwi de 4ème gamme: etudes biochimiques et microbiologiques, Ph.D. thesis, Montpellier II University, France, 1992.

86. Braconnier, A. et al., Growth and germination of proteolytic *Clostridium botulinum* in vegetable-based media, *J. Food Prot.*, 66, 833, 2003.

87. Malizio, C.J. and Johnson E.A., Evaluation of the botulism hazard from vacuum-packaged Enoki mushrooms (*Flammulina velutipes*), *J. Food Prot.*, 54, 20, 1991.

88. Hotchkiss, J.H. et al., The relationship between botulinal toxin production and spoilage of fresh tomatoes held at 13 and 23 °C under passively modified and controlled atmospheres and air, *J. Food Prot.*, 55, 522, 1992.

89. Day, B.P.F., High oxygen modified atmosphere packaging for fresh prepared produce, *Postharv. News Info.*, 7, 31, 1996.

90. Barry-Ryan, C., Pascussi, J.M., and O'Beirne, D., Quality of shredded carrots as affected by packaging film and temperature storage, *J. Food Sci.*, 65, 726, 2000.

91. Kader, A.A. and Ben Yehoshua, S., Effects of superatmospheric oxygen levels on postharvest physiology and quality of fresh fruits and vegetables, *Postharv. Biol. Technol.*, 20, 1, 2000.

92. Amanatidou, A., Smid, E.J., and Gorris, L.G.M., Effect of elevated oxygen and carbon dioxide on the surface growth of vegetable associated micro-organisms, *J. Appl. Microbiol.*, 86, 429, 1999.

93. Amanatidou, A. et al., Antioxidative properties of *Lactobacillus sake* upon exposure to elevated oxygen concentrations, *FEMS Microbiol. Lett.,* 203, 1, 87, 2001.

94. Gonzalez-Roncero M.I. and Day B.P.F., The effect of elevated oxygen and carbon dioxide modified atmosphere on psychotropic pathogens and spoilage microorganisms associated with fresh prepared produce, Campden and Chorleywood Food Research Association, Chipping Campden, UK Research Summary Sheet 98, 1998.

95. Jacxens L. et al., Effect of modified atmosphere packaging on microbial growth and sensorial qualities of fresh-cut produce, *Int. J. Food Microbiol.* 71, 197, 2001.

96. Wszelaki, A.L. and Mitcham, E.J., Effect of superatmospheric oxygen on strawberry fruit quality and decay, *Postharv. Biol. Technol.*, 20,125, 2000.

97. Amanatidou, A. et al., High oxygen and high carbon dioxide modified atmospheres for shelf-life extension of minimally processed carrots, *Food Chem. Toxicol.*, 65, 61, 2000.

98. Guevera, J.C. et al., Effect of elevated concentrations of CO_2 in modified atmosphere packaging on the quality of prickly pear cactus stems (*Opuntia* spp.), *Postharv. Biol. Technol.*, 29,167, 2003.

99. Gunes G.G. and Hotchkiss J.H., Growth and survival of *Escherichia coli* O157:H7 on fresh-cut apples in modified atmospheres at abusive temperatures, *J. Food Prot.*, 65, 1641, 2002.

100. Chambroy, Y. and Souty, M., Rôle du CO2 sur le comportement des fruits à noyau: utilisation des atmosphères modifiées, in *Proc. Symp. Qualité Post-Récolte et Produits Dérivés chez les Fruits à Noyau,* Vendrell, M. and Audergon, J.M. Eds., Lerida, Spain, 1994, 139.

101. El Halouat, A. et al., Effect of modified atmosphere packaging and perspectives on the shelf-life of high moisture prunes and raisins, *Int. J. Food Microbiol.*, 41, 177, 1998.

102. Lowenadler, J., Dissolved CO_2 and its growth inhibiting effect on microorganisms in MAP model systems, *SIK Rapp.*, 604, 69, 1994.
103. Devlieghere, F., Debevere, J., and Van Impe, J., Effect of dissolved carbon dioxide and temperature on the growth of *Lactobacillus sake* in modified atmospheres, *Int. J. Food Microbiol.*, 41, 231, 1998.
104. Qadir, A. and Hashinaga, F., Inhibition of postharvest decay fruits by nitrous oxide, *Postharv. Biol. Technol.*, 22, 279, 2001.
105. Burton, K.S., Frost, C.E., and Nichols, R., A combination plastic permeable film system for controlling post-harvest mushroom quality, *Biotechnol. Lett.*, 9, 529, 1987.
106. Arpaia, M.L., Mitchell, F.G., and Kader, A.A., Effects of 2% O2 and varying concentrations of CO_2 with or without ethylene on the storage performances of kiwifruit, *J. Am. Soc. Hortic. Sci.*, 110, 200, 1985.
107. Varoquaux, P. et al., Change in firmness of kiwifruit after slicing, *Sci. Aliments*, 10, 127, 1990.
108. Harker, F.R. et al., Physical and mechanical changes in strawberry fruit after high carbon dioxide treatments, *Postharv. Biol. Technol.*, 19, 139, 2000.
109. Larsen, M. and Watkins, C.B., Firmness and concentrations of acetaldehyde, ethyl acetate and ethanol in strawberries stored in controlled and modified atmospheres, *Postharv. Biol. Technol.*, 5, 39, 1995.
110. Smith, R.B. and Skoog, L.J., Postharvest carbon dioxide treatment enhances firmness of several cultivars of strawberry, *Hortscience*, 27, 420, 1992.
111. Holcroft, D.M. and Kader, A.A., Controlled atmosphere-induced changes in pH and organic acid metabolism may affect color of stored strawberry fruit, *Postharv. Biol. Technol.*, 17, 19, 1999.
112. Yong, S.P. and Soon, T.J., Effects of CO_2 treatments within polyethylene film bags on fruit quality of fig fruits during storage, *J. Kor. Soc. Hortic. Sci.*, 41, 618, 2000.
113. Siriphanich, J. and Kader, A.A., Effects of CO_2 on total phenolics, phenylalanine ammonia lyase and polyphenol oxidase in lettuce tissue, *J. Am. Soc. Hortic. Sci.*, 110, 249, 1985.
114. Perez, A.G. and Sanz, C., Effect of high-oxygen and high-carbon dioxide atmospheres on strawberry flavor and other quality traits, *J. Agric. Food Chem.*, 49, 2370, 2001.
115. Lund, B.M., Bacterial contamination of food crops, *Asp. Appl. Biol.*, 17, 71, 1988.
116. Nguyen The, C. and Prunier, J.P., Involvement of Pseudomonads in "ready-to-use" salads deterioration, *Int. J. Food Sci. Technol.*, 24, 47, 1989.
117. Lund, B.M., Bacteria spoilage, in *Postharvest Pathology of Fruits and Vegetables*, Dennis, C., Ed., Academic Press, London, 1983, p. 218.
118. Hildebrand, D.F., Lipoxygenases, *Physiol. Plant.*, 76, 249, 1989.
119. Bengtsson, B.L., Bosund, I., and Rasmussen, I., Hexanal and ethanol formation in peas in relation to off-flavor development, *Food Technol.*, 21, 478, 1967.
120. Watada, A.E., Abe, K., and Yamauchi, N., Physiological activities of partially processed fruits and vegetables, *Food Technol.*, 44, 116, 1990.
121. Lamikanra, O., Enzymatic effects on flavor and texture of fresh-cut fruits and vegetables, in *Fresh-Cut Fruits and Vegetables: Science, Technology and Market*, Lamikanra, O., Ed., CRC Press, Boca Raton, FL, 2002, Chap. 6.
122. Ke, D. et al. Effect of short-term exposure to low O2 and high CO_2 atmospheres on quality attributes of strawberries, *J. Food Sci.*, 1, 50, 1991.
123. Guichard, E. et al., Effect of carbon dioxide concentration on aroma of strawberries after storage, *Sci. Aliments*, 12, 83, 1992.

124. Pickenhagen, W. et al., Estimation of 2,5-dimethyl-4-hydroxy-3(2H)-furanone (fura-neol) in cultivated and wild strawberries, pineapples and mangoes, *J. Sci. Food Agric.*, 32, 1132, 1981.

124a. Bitter, S. et al., Influence d'une anaérobiose carbonique sur l'arôme de deux variétés d'abricots, *Sci. Aliments*, 10, 846, 1990.

125. Varoquaux, P., Fruits frais prêts à l'emploi dits de quatrième gamme, in *Technologies de Transformation des Fruits*, Albagnac, G., Varoquaux, P., and Montigaud, J.C., Eds., Lavoisier, Paris, 2002, Chap. 5.

126. Carlin, F. et al., Microbial spoilage of fresh "ready-to-use" grated carrots, *Sci. Aliments*, 9, 371, 1989.

127. Tan, C.T. and Francis, F.J., Effect of processing temperature on pigments and color of spinach, *J. Food Sci.*, 27, 232, 1962.

128. Piagentini, A.M., Güemes, D.R., and Pirovani, M.E., Sensory characteristics of fresh-cut spinach preserved by combined factors methodology, *J. Food Sci.*, 67, 1544, 2002.

129. Amir-Shapira, D., Goldshmidt, E.E., and Altman, A., Chlorophyll catabolism in senescing plant tissues *in vivo* breakdown intermediates suggest different degradative pathways for citrus fruit and parsley leaves, *Proc. Natl. Acad. Sci. USA*, 84, 1901, 1987.

130. Rolle, R.S., and Chism, G.W., Physiological consequences of minimally processed fruits and vegetables, *J. Food Quality*, 10, 157, 1987.

131. Wang, S.S., Haard, N.F., and DiMarco, G.R., Chlorophyll degradation during con-trolled-atmosphere storage of asparagus, *J. Food Sci.*, 34, 657, 1971.

132. Hutchings, J.B., Chemistry of food colour, in *Food Colour and Appearance*, Hutchings, J.B., Ed., Chapman & Hall, London, 1994, p. 367.

133. Ziegler, R., Schanderl, S.H., and Markakis, P., Gamma irradiation and enriched CO_2 atmosphere storage effects on the light-induced greening of potato, *Food Sci.*, 33, 533, 1968.

134. Patil, B.C., Singh, B., and Salunkhe, D.K., Formation of chlorophyll and solanine in Irish potato (*Solanum tuberosum* L.) tubers and their control by gamma radiation and CO_2 enriched packaging. *Lebensm.-Wiss. Technol.*, 4, 123, 1971.

135. Mayer, J.J., Polyphenoloxidase in plants. Recent progress, *Phytochemistry*, 26, 11, 1987.

136. Gormley, T.R., Chill storage of mushrooms, *J. Sci. Food Agric.*, 26, 401, 1975.

137. Smith, W.H., Storage in reduced oxygen atmospheres, *Rep. Dilton Covent Garden Lab.*, London, 1967.

138. Tomkin, R.G., The storage of mushrooms, *MGA Bull.*, 202, 534, 1966.

139. Nichols R. and Hammond, J.B.W., Storage of mushrooms in prepacks: The effect of changes in CO_2 and O_2 on quality, *J. Sci. Food Agric.*, 25, 1371, 1973.

140. Lopez-Briones, G. et al., Modified atmosphere packaging of common mushroom, *Int. J. Food Sci. Agric.*, 28, 57, 1993.

141. Varoquaux, P., Mazollier, J., and Albagnac, G., The influence of raw material char-acteristics on the storage life of fresh cut butterhead lettuce, *Postharv. Biol. Technol.*, 9, 127, 1996.

142. Lee, D.S., Model for fresh produce respiration in modified atmospheres based on principles of enzymatic kinetics, *J. Food Sci.*, 56, 1580, 1991.

143. Doyon, G. et al., Estimation rapide des perméabilités au gaz carbonique et à l'oxygène des matériaux plastiques utilisés pour les produits horticoles frais: Calculatrice HP et PC compatible, *Ind. Alim. Agric.*, 4, 229, 1996.

144. Varoquaux, P., L'emballage des végétaux vivants emballages physiologiques atmo-sphère auto-régulée, in *L'emballage des Denreés Alimentaires de Grande Consom-mation*, Multon, J.L. and Bureau, G., Eds., Lavoisier, Paris, 1998, Chap. 41.

145. Fishman, S. et al., Model for gas exchange dynamics in modified atmosphere package of fruits and vegetables, *J. Food Sci.*, 66, 1083, 1995.
146. Rogers, C.E., Permeability and chemical resistance, in *Engineering Design for Plastics*, Krieger, R.E., Ed., Baur, New York, 1975, 682.
147. Gontard, N. et al., Influence of relative humidity and film composition on oxygen and carbon dioxide permeabilities of edible films, *J. Agric. Food Chem.*, 44, 1064, 1996.
148. Beit-Halachmy, I. and Mannheim, C.H., Is modified atmosphere packaging beneficial for fresh mushrooms?, *Lebensm. Wiss. Technol.*, 25, 426, 1992.
149. Hammonds, J.B.W. and Nichols, R., Changes in respiration and soluble carbohydrates during post-harvest storage of mushroom (*Agaricus bisporus*), *J. Sci. Food Agric.*, 26, 835, 1975
150. Saltveit, M.E., Is it possible to find an optimal controlled atmosphere?, *Postharv. Biol. Technol.*, 27, 3, 2003.
151. Harker, F.R., Gunson, F.A., and Jeager, S.R., The case for fruit quality: An interpretive review of consumer attitudes and preferences for apples, *Postharv. Biol. Technol.*, 28, 333, 2003.
152. Veltman, R.H. et al. Dynamic control system (DCS) for apples (*Malus domestica* Berkh. Cv "Elstar": Optimal quality through storage based on product response, *Postharv. Biol. Technol.*, 27, 79, 2003.
153. Peppelenbos, H., How to control the atmosphere, *Postharv. Biol. Technol.*, 27, 1, 2003.
154. Heard, G.M., Microbiology of fresh-cut produce, in *Fresh-Cut Fruits and Vegetables: Science, Technology and Market*, Lamikanra, O., Ed., CRC Press, Boca Raton, FL, 2002, Chap. 7.
155. Cuq, B. and Redl, A., Activités antimicrobiennes des films d'emballage, in *Les Emballages Actifs*, Gontard, N., Ed., Lavoisier Tec & Doc, Paris, 2000, Chap. 3.
156. Dong, L. et al., Effect of 1-methylcyclopropene on ripening of 'Canino' apricots and 'Royal Zee' plum, *Postharv. Biol. Technol.*, 24, 135, 2002.
157. Benkeblia, N. and Varoquaux, P., Modified atmosphere packaging of onion bulbs (*Allium cepa* L.), *Sci. Aliments*, 21, 297, 2001.
158. Roy, S., Ananthewaran, R.C., and Beelman, R.B., Modified atmosphere and modified humidity packaging of fresh mushroom, *J. Food Sci.*, 61, 391, 1996.
159. Barron, C. et al., Modified atmosphere packaging of cultivated mushroom (*Agaricus bisporus* L.) with hydrophilic films, *J. Food Sci.*, 67, 251, 2002.
160. Ben-Yeshoshua, S. et al., Mode of action of plastic film in extending life of lemon and bell pepper fruits by alleviation of water stress, *Plant. Physiol.*, 73, 83, 1983.
161. Ben-Yeshoshua, S., Transpiration, water stress and gas exchange, in *Postharvest Physiology of Vegetables*, Weichman, J., Ed., Marcel Dekker, New York, 1987, 113.
162. Geeson, J., Packaging to keep produce fresh, *Nutr. Food Sci.*, 123, 2, 1990.

6 Flavor and Volatile Metabolism in Produce

Keith R. Cadwallader
University of Illinois at Urbana-Champaign, Department of Food
Science and Human Nutrition, Urbana, IL 61801

CONTENTS

0-8493-1902-1/05/$0.00+$1.50
© 2005 by CRC Press

6.1 INTRODUCTION

The flavor of a food is an important quality attribute and is often the sole reason why a product is accepted or rejected by the consumer. Food flavor is linked to culture and plays a predominant role in the lives of all human beings. The term *flavor* is actually the combined or integrated perception of aroma (odor) and taste, and to a lesser extent pain or nerve response (e.g., heat of capsaicin, astringency of tea, cooling of menthol, etc.), texture and mouth-feel, and overall appearance. It is sometimes difficult to differentiate between aroma and taste as being two separate physiological senses since they are often (i.e., during eating) perceived simultaneously. Of these two senses, the perception of aroma is undoubtedly the most complicated. Most people can distinguish thousands of individual aroma impressions, while there are only four (or five, i.e., umami) basic taste qualities.

Fruit and vegetable flavors are composed a wide range of chemical compounds, from nonvolatile *taste-active* (including both inorganic and organic compounds) to volatile *aroma-active* organic molecules. Most often it is the aroma components that are the predominant contributors to the distinctive flavor of fruits and vegetables. It is this reason that most flavor research has focused on the study of the occurrence of volatile aroma components, including their origin through biosynthetic or other chemical reactions.

Before delving into the chemistry and biochemistry of the various types of fruit and vegetable flavor systems, it is worthwhile to spend a moment on some basic aroma terminology. First, it must be made clear that not all volatile compounds have perceivable odors when present in specific food products. While it is possible to identify hundreds of volatile compounds in a single food product, only a few of these may actually be "aroma-active" and, therefore, contribute substantially to the overall flavor. Often a volatile compound has a detectable odor at very high concentration but cannot be detected at the low level at which it is found in a food product. Therefore, when we speak of aroma substances, we generally refer to only those volatile compounds that have perceivable odors at relatively low concentrations. Sometimes a food contains only one or just a few volatile compounds, so called *character-impact components*, which are responsible for its characteristic aroma; however, in many cases a large number of compounds make a seemingly equal contribution to the aroma.

In the following discussions, the flavor chemistry of fruits and vegetables will be presented, along with current knowledge in the area of metabolism and postharvest effects on aroma composition.

6.2 FLAVOR COMPONENTS OF FRUITS AND VEGETABLES

6.2.1 NONVOLATILE TASTE COMPOUNDS

Sweetness, sourness, saltiness, bitterness, and astringency are some key flavor attributes of fruits and vegetables [1]. Sweetness in fruits and vegetables is usually due to the presence of various sugars, such as glucose, fructose, and sucrose. Total sugars are usually expressed as percentage of soluble solids, which may be directly related to sweetness (e.g., citrus). Sourness (or acidity) is usually measured as percentage of titratable acidity, which is most often expressed in terms of the major organic acid(s) found in a particular fruit or vegetable (e.g., citric acid in orange). Compounds responsible for bitter and astringent flavors are often more difficult to determine since many compound classes may be involved (e.g., phenolic compounds, tannins, flavonoids, and alkaloids). During ripening, the sweetness of some fruits may increase due to starch to sugar conversions (e.g., banana and apple), while at the same time acidity, bitterness, and/or astringency generally decreases [2]. In fact, percentage of soluble solids (°Brix) and °Brix-to-acid ratio are often used as indices of ripeness and flavor quality.

6.2.2 VOLATILE AROMA COMPOUNDS

Numerous compound classes occur naturally (endogenously) in the volatile profiles of fruits and vegetables and, depending on species or variety, one or more classes may predominate. Among the thousands of volatile compounds identified to date only a relative few actually contribute to the distinctive aromas of fruits and vegetables. The great diversity of chemical structures indicates the involvement of numerous chemical reactions in their formation. In fruits and vegetables there are numerous common reactions that produce similar or overlapping aroma profiles. The following section is devoted to the discussion of the biogenesis of important classes of volatile constituents of produce.

6.2.3 AROMA BIOGENESIS

The biosynthesis of volatile compounds in plants is a dynamic process. The activities of the various enzymes and pathways change at different stages of maturity and ripeness. Many volatile compounds are generated in plants through biosynthetic processes or during processing of these materials. During fruit and vegetable ripening the plant undergoes many catabolic reactions to form volatiles as secondary products.

Acyl pathways play a central role in the metabolism of fatty acids, terpenes, amino acids, and carbohydrates [3,4]. β-Oxidation of linolenyl-CoA results in the formation of volatiles important in the aroma of pears and other fruit. Hydroxylated fatty acids (C_8–C_{12}) are also formed via β-oxidation and may cyclize (spontaneously) to form γ- and δ-lactones. Branched-chain esters are formed via amino acid metabolism to give typical fruity aromas. In strawberries, the character-impact compound 2,5-dimethyl-4-hydroxy-3(2H)-furanone is formed from metabolism of carbohydrates. Metabolism of carotenoids leads to the formation of aroma-significant C_{10} and C_{13} norisoprenoids. Lipoxygenase-derived compounds play an important role in the generation of aromas in green plants. Aromas of some vegetables such as onions are formed by the action of specific enzymes released after tissue disruption.

It is not possible to cover all of the possible compounds and pathways involved in the generation of fruit and vegetable aromas. The biogenesis of aroma in plants has been the subject of extensive review [3–6]. This discussion will present only a few examples of mechanisms involved in the biogenesis of important volatile classes involved in the aromas of fruits and vegetables.

6.2.3.1 Esters

Esters are one of the largest and most important classes of aroma compounds, especially in fruit flavors. Most of the intense (low threshold) esters are synthesized during ripening from free amino acids in the fruit (or from alcohols). For the most part, biosynthesis of esters in plants is mediated via biochemical processes involving acetyl CoA transferase [7]. Acetyl CoA transferase is a coenzyme-A-dependent enzyme that catalyzes the transfer of an acyl moiety residing on an acyl CoA onto a corresponding alcohol [8]. Ester production is, therefore, dependent on alcohols derived from lipids, free amino acids, lipoxygenase pathways, or fermentation (e.g., ethanol). Unsaturated esters, such as methyl and ethyl decadienoates of Bartlett pears, are derived via (E,Z)-2,4-decadienoyl-CoA by β-oxidation of unsaturated fatty acids [3]. In apples, β-oxidation of fatty acids is responsible for generation of some straight-chain esters from acetic, butanoic, and hexanoic acids, which also may undergo reduction to alcohols prior to esterification [9,10]. Amino acid catabolism accounts for methyl branched-chain esters as well as alcohols, acids, ketones, and sulfur-containing and aromatic compounds [3,9]. For example, transamination of L-leucine leads to the formation of α-ketoisocaproate, which undergoes decarboxylation to form 3-methylbutanoyl-CoA. Further catabolism of 3-methylbutanoyl-CoA gives rise to 3-methylbutanoate esters, 3-methyl-1-butanol, and 3-methylbutyl esters [3,5]. Likewise, methionine may be transformed into thio-esters important in some fruit aromas, such as pineapples [11]. Fatty acid degradation via the lipoxygenase pathway is responsible for formation of hexanal, (E)-2-hexenal, and (Z)-3-hexenal, which after further isomerization, oxidation, and/or reduction can participate in ester synthesis or other metabolic reactions [10].

6.2.3.2 Lactones

The γ- and δ-lactones impart intense fruity, peachy, or coconut-like aromas and are important aroma constituents of several fruits. Apricot aroma can be largely attributed to γ-octa- and γ-decalactones [12,13]. β-Oxidation, reduction, chain elongation, and hydroxylation are involved in the biosynthesis of lactones [14]. The biosynthesis of lactones is sometimes chiroselective, leading to formation of predominantly the *R* enantiomer in some fruit, such as apricots [15], while in other fruit such as pineapples both the *R* and *S* enantiomers may be present [14], indicating that various biosynthetic pathways may be involved in the formation of lactones.

6.2.3.3 Terpenoids

Terpenes are ubiquitous components of fruits and vegetables, although they may not necessarily contribute to aroma. For example, in some fruits such as strawberries and tomatoes terpenes make essentially no aroma impact. On the other hand, terpenes make a substantial contribution to citrus, carrot, and blueberry aromas.

The arrangement of the groups of five carbon atoms found within terpenes is related to the structure of the molecule isoprene [16]. The key step in the biosynthesis of most monoterpenes is the cyclization of an allylic pyrophosphate catalyzed by enzymes collectively known as cyclases [17,18]. Most naturally occurring monoterpenes are oxygenated (e.g., camphor, menthol, citral, and carvone). Oxygenated monoterpenes are of particular importance because they impart characteristic aroma notes to essential oils, such as citrus. Oxygenated monoterpenes may be derived from the monoterpene hydrocarbons via oxygenation involving multifunction oxidases and molecular oxygen [19].

6.2.3.4 Norisoprenoids

Norisoprenoids comprise of a large group of C_{13}-carotenoid-derived volatile compounds that make a substantial contribution to the aromas of many fruits and vegetables [20,21]. For the most part, the biochemical mechanisms involved in carotenoid catabolism are still not well understood. The formation of norisoprenoids is believed to initiate via enzymatic cleavage (dioxygenase system) of a C_{40}-carotenoid (e.g., β-carotene) chain at the 9,10 double bond to form C_{13} fragments, which then undergo enzymatic and acid-catalyzed conversions to yield both bound forms (glycosides) of C_{13}-norisoprenoids and volatile aglycones, such as the well-known β-damascenone and β-ionone [20]. β-Damascenone has an extremely low threshold of 2 pg/g in water [22]. Therefore, this compound probably makes a significant contribution to the aroma of any fruit or vegetable in which it has been found. Many C_{13}-norisoprenoids, including β-damascenone and β-ionone, dominate the volatile profile of some fruit, such as starfruit [23].

Although norisoprenoids represent the major carotene-derived aroma compounds, cleavage of the carotenoid occurs not only at the 9,10 position; there are a number of other products formed with a different number of carbon atoms, such as C_{10}-norterpenoids, that arise from cleavage of the 7,8 double bond [21,21,24].

Norterpenoids, such as the marmelo lactones and marmelo oxides, are important aroma components of quince [25].

6.2.3.5 Furanones

The predominant furanones include 2,5-dimethyl-4-hydroxy-3(2*H*)-furanone (DMHF), commonly referred to as strawberry ketone or Furaneol® (registered trademark of Firmenich S.A, Geneva, Switzerland), and its derivatives, which are widely distributed in both fruits and vegetables [26]. DMHF is especially important in strawberries [27,28], pineapples [29], muscadine grapes [30], and tomatoes [31]. At low concentrations DMHF has a fruity or strawberry-like note and at higher concentrations its aroma becomes caramel- or burnt sugar-like [26]. DMHF occurs naturally in its free form (aglycone), as 2,5-dimethyl-4-methoxy-3(2*H*)-furanone (mesifuran), and in two glycosidically bound forms (DMHF glucoside and DMHF malonyl-glucoside) [26]. Two other DMHF-like furanones include 5-methyl-4-hydroxy-3(2*H*)-furanone (norfuraneol) and 4,5-dimethyl-3-hydroxy-2(5H)-furanone (sotolon).

The DMHF glycosides are themselves odorless and are considered the probable precursors to DMHF. Biosynthesis of DMHF, via its glucoside, from deoxyketoses by strawberry callus cultures has been demonstrated [26,32,33]. Isotopic precursor studies [34] and radiotracer studies [35] with detached ripening strawberries were used to demonstrate the biosynthesis of DMHF and its glycosides from D-glucose or D-fructose, via D-fructose-6-phosphate as an intermediate. Mesifuran is formed in strawberry by enzymatic *O*-methylation of DMHF by *O*-methyltransferase, with *S*-adenosyl-L-methionine providing the methoxy group [36–38].

6.2.3.6 Sulfur-Containing Compounds

Sinigrin and other glucosinolates (thioglucosides) are major precursors of sulfurous aroma components of cruciferous vegetables, such as broccoli, cabbage, brussels sprouts, and others. Upon disruption of the plant tissue, the thioglucoside sinigrin is hydrolyzed by the enzyme myrosinase to form allyl isothiocyanate, allyl thiocyantate, allyl cyanide, and 1-cyano-2,3-epithiopropane [39,40]. In both wasabi and horseradish several thioglucosides are hydrolyzed by myrosinase to yield numerous volatile compounds, such as isothiocyanates, thiocyanates, nitriles, and others [41]. Isothiocyanates are characteristic aroma compounds in wasabi and horseradish [41]. The main component, allyl isothiocyanate, is believed to be primarily responsible for the pungency of both plants, but other isothiocyanates are important as well. Phenethyl isothiocyanate is a characteristic aroma component of horseradish, while unique and differentiating *green* notes of wasabi are due to a number of ω-methylthioalkyl and ω-alkenyl isothiocyanates [41].

S-Alkyl- and *S*-alkenylcysteine sulphoxides are important nonvolatile aroma precursors of the genus *Allium*, which includes onion, garlic, leek, and shallot [4,42]. Alliinases (C-S-lyases), released upon tissue disruption, catalyze the hydrolysis of the nonvolatile precursors to yield pyruvate, ammonia, and a range of volatile sulfur-containing constituents with pungent and characteristic aromas [43].

Ethyl 3-methylthiopropanoate, methyl 3-methylthiopropanoate, and 3-methylthio-propyl acetate are important thio-esters found in pineapple [11] and passion fruit [44]. Thio-esters are formed from sulfur-containing amino acids as described above for the branched-chain esters. Various other types of sulfur aroma compounds, such as thiols, mercapto-alcohols, and mercapto-ketones, are aroma constituents of several fruits, such as durian [45], passion fruit [46,47], black current [48], and orange [49].

6.2.3.7   Aromatic Compounds

L-Phenylalanine, derived via the shikimic acid pathway, serves as an important precursor for the biosynthesis of aromatic compounds. This amino acid can give rise to 2-phenylacetyl-CoA in a reaction analogous to that described above for leucine metabolism to form esters. 2-Phenylacetyl-CoA is further converted to 2-phenylacetate esters or is reduced to 2-phenylethanol and 2-phenylethyl esters [3]. Other compounds derived from the shikimic acid pathway include cinnamic aldehyde, cinamyl alcohol, and esters. Reduction of cinnamyl alcohol leads to formation of propenyl- and allylphenols including eugenol, methyleugenol, and elemicin found in bananas [3].

6.2.3.8   Lipoxygenase-Derived Alcohols and Aldehydes

Tissue disruption in green plants releases a group of enzymes that act upon the available polyunsaturated fatty acids (PUFAs) to yield C_6, C_8, and C_9 carbonyls and alcohols with distinctive green, cut-leaf-like, and melon-like aromas [50]. These compounds are derived from PUFAs, such as linoleic and linolenic acids, through the action of endogenous lipoxygenase, hydroperoxide lyases, Z-3/E-2-enal isomerases, and alcohol dehydrogenases [51,52]. The 9-lipoxygenase acting on linolenic acid produces a n-9 hydroperoxide. Hydrolysis of this hydroperoxide by a specific hydroperoxide lyase leads to formation of mainly (Z,Z)-3,6-nonadienal (watermelon-like aroma), which can undergo spontaneous or enzyme-catalyzed isomeration to (E,Z)-2,6-nonadienal (cucumber-like aroma). These aldehydes may undergo reduction to their corresponding alcohols. The conversion of the aldehydes to their corresponding alcohols is a significant step since it leads to a general decline in aroma intensity due to alcohols having somewhat higher odor detection thresholds than the aldehydes. In tomato, for example, linoleic acid and α-linolenic acid are converted to the n-13 hydroperoxides, which undergo hydrolysis via hydroperoxide lyase to yield the C_6 aldehydes hexanal and (Z)-3-hexenal, respectively [53,54]. (Z)-3-Hexenal can then undergo isomerization to form (E)-2-hexenal. These aldehydes, and their corresponding alcohols formed by action of alcohol dehydrogenase, are often found among the volatiles of tomato homogenates [54].

6.3   FRUIT AND VEGETABLE FLAVOR SYSTEMS

Fruit and vegetable flavors are composed of complex mixtures of volatile chemicals derived from numerous biochemical pathways. In addition, some nonvolatile compounds, such as sugars and organic acids, impart sweet and sour tastes, respectively.

Individual aroma-significant volatile components are generally present at low parts-per-million (micrograms per gram), parts-per-billion (nanograms per gram), and sometimes parts-per-trillion (picograms per gram) levels, making the detection and identification of these compounds difficult. Nonetheless, due to modern advances in analytical chemistry, especially application of instrumental-sensory techniques such as gas chromatography-olfactometry (GCO) combined with GC-mass spectrometry (MS), significant progress has been made toward our understanding of the flavor chemistry of fruits and vegetables. This section provides a concise review of the flavor chemistry of some selected fruits and vegetables, with emphasis placed on discussion of character-impact aroma components. More extensive reviews and compilations on fruit and vegetable flavor systems can be found elsewhere [55–60].

6.3.1 FRUIT FLAVORS

6.3.1.1 Pome Fruit

Among the various pome fruits, apples and pears are of greatest economic importance. Berger [55] covered the published literature on apple and pear flavor up to about 1991. The following discussion will focus on apple flavor since it has undergone extensive study in recent years. Pear (*Pyrus* spp.) flavor has been the subject of some recent studies [61–63].

6.3.1.1.1 Apple

It is nearly impossible to accurately describe the exact chemistry of apple flavor, because each cultivar can have a unique flavor character. Furthermore, there also is the factor of ripeness to consider. That is, some varieties, such as Granny Smith, are intentionally eaten in their "green" or unripe state, while many other varieties are preferably consumed in their ripe stage. A large number of volatile compounds have been reported as aroma-impact components of apples. β-Damascenone has been established by analysis of over 40 cultivars to be a predominant aroma component of apple aroma [64,65]. Aldehydes, alcohols, and esters are other predominant components. These compounds are formed in a manner that parallels respiration (climacteric) to its maximum [66]. Secondary aroma compounds, such as hexanal and (*E*)-2-hexenal, which provide *green* or unripe notes, are formed by lipoxygenase action after disruption of the cells during cutting or chewing. 2-Methylbutanoic acid and corresponding methyl to hexyl esters, 2-methylbutanol, and 2-methylbutyl acetate have been reported as aroma-impact components of apples [67–69].

6.3.1.2 Stone Fruit

Stone fruits of commercial importance belong to the genus *Prunus*, subfamily Prunoideae, and include apricots, cherries, peaches, and plums. A detailed discussion of the recent literature on peach and nectarine flavor will be presented here. Studies on apricot [70], cherry [71], and plum [70,72] flavors have been recently published.

6.3.1.2.1 Peach and Nectarine

Extensive investigation on peach and nectarine flavor has led to the identification of over 100 volatile constituents [73]. Lactones, in particular γ- and δ-decalactones, have been established as aroma-impact components, but numerous other aroma constituents, such as C_6 aldehydes, aliphatic alcohols, and terpenes, may contribute to the aromas of peaches and nectarines [73–75]. Recently, β-damascenone and 6-dodecenyl-γ-lactone were reported as additional predominant aroma components of peach aroma [75].

6.3.1.3 Melon, Cucumber, and Squash

The key aroma components of cantaloupe (a.k.a. muskmelon or rockmelon) (*Cucumis melo*) [76–81] and honeydew (*Cucumis inodorus*) [82] cultivars have been reported [76–78]. The fruity-smelling branched esters ethyl 2-methylbutanoate and methyl 2-methylpropanoate were reported as predominant aroma compounds. Additional important esters include ethyl 2-methylpropanoate, 2-methylpropyl acetate, and several thio-esters that vary among varieties.

In contrast to cantaloupe and honeydew melons, watermelon (*Citrullis vulgaris*) [83,84], cucumber (*Cucumis sativus*) [76,85], and pumpkin and squash (*Cucurbita* spp.) [86] do not contain esters as part of their characteristic aromas. Instead, their typical aroma constituents are primarily generated upon tissue disruption via lipoxygenase pathways, leading to compounds having characteristic cucumber- and watermelon-like aromas, such as (*E,Z*)-2,6-nonadienal and (*Z,Z*)-3,6-nonadienal, respectively, among numerous other C_6 and C_9 volatile aldehydes and alcohols that affect the aromas.

6.3.1.4 Tomato

The flavor of tomato (*Lycopersicon esculentem* Mill.) is mainly attributed to its aroma volatiles, sugars, and acids. DMHF is an essential component of fresh tomato aroma [31], but many additional compounds have been reported as major contributors, including (*Z*)-3-hexenal, hexanal, β-ionone, β-damascenone, 1-penten-3-one, 3-methylbutanal, (*E*)-2-hexenal, phenylacetaldehyde, 2-phenylethanol, 2-isobutyl-thiazole, 6-methyl-5-hepten-2-ol, 1-nitrophenylethane, 3-(methylthio)propanal (methional), (*E*)-2-heptenal, (*E,Z*)- and (*E,E*)-2,4-decadienal, *trans*- and *cis*-epoxy-(*E*)-2-decenal, 1-octen-3-one, 3-methylbutanoic acid, and (*Z*)-1,5-octadien-3-one [87–92]. In regard to tomato flavor quality (preference), Mayer et al. [92] reported that higher levels of the 2,4-decadienal isomers, 1-penten-3-one, and DMHF had a positive influence, whereas phenylacetaldehyde, 2-phenylethanol, and 2-isobutyl-thiazole negatively affected fresh tomato aroma.

6.3.1.5 Small Fruit

The small fruits are composed of over 10 genera and include strawberries, raspberries, blackberries, blueberries, and others. Latrasse [56] reviewed the flavor chemistry of 20 species of wild and cultivated berries. The flavor of berries was also the subject of a review by Honkanen and Hirvi [93]. This section will be restricted to the discussion of strawberry flavor, since this fruit has undergone extensive study in recent years. The reader is encouraged to review the recent literature on raspberry (*Rubus idaeus* L.) [94–97], blackberry (*Rubus fructicosus*) [98], and blueberry (*Viccinium* spp.) [99,100] flavors.

TABLE 6.1
Key Aroma Components of Strawberries on the Basis
of Gas Chromatography-Olfactometry and Sensory Studies

Schieberle [106][a]	Schieberle and Hofmann [27][a]	Ulrich et al. [107][b]	Ulrich et al. [108][c]
DMHF[d]	DMHF[d]	Acetone	Methyl butanoate
(Z)-3-hexenal	(Z)-3-hexenal	Methyl butanoate	Ethyl butanoate
Methyl butanoate	Methyl butanoate	Ethyl butanoate	Methyl hexanoate
Ethyl butanoate	Ethyl butanoate	Butyl acetate	Ethyl hexanoate
Ethyl 2-methylbutanoate	Ethyl 2-methylpropanoate	2-Heptanone	Linalool
Ethyl 3-methylbutanoate	2,3-Butanedione	Methyl hexanoate	Mesifuran
Methyl 2-methylbutanoate	Methyl 2-methylbutanoate	(E)-2-hexenal	Butanoic acid
Methyl 3-Methylbutanoate	Methyl 3-Methylbutanoate	Ethyl hexanoate	2-Methylbutanoic acid
Acetic acid		(Z)-3-hexen-1-ol	Hexanoic acid
2,3-Butanedione		Linalool	2-Phenyl ethanol
Butanoic acid		Mesifuran	DMHF
3-Methylbutanoic acid		Butanoic acid	γ-Decalactone
2-Methylbutanoic acid		2-Methylbutanoic acid	
Ethyl 2-methylpropanoate		hexanoic acid	
Mesifuran		2-Phenyl ethanol	
		DMHF	
		γ-Decalactone	

Note: Components are not necessarily listed in order of importance.

[a] Fresh strawberry juice.
[b] Thirteen strawberry varieties.
[c] Cultivated and wild strawberries.
[d] 2,5-Dimethyl-4-hydroxy-3(2H)-furanone.

6.3.1.5.1 Strawberry

Among the various small fruit, strawberry (*Fragaria ananassa*) flavor has by far received the greatest attention. More than 360 volatile compounds have been identified in strawberries [32,56]. Determination of the relative importance of various volatile compounds to the aroma of fresh strawberries has been the subject of numerous investigations [27,101–111]. Aldehydes, esters, ketones, alcohols, acids, terpenes, and lactones have long been recognized as important contributors to strawberry aroma. Recent studies that employed GCO and sensory studies of model aroma systems have been able to more exactly define the important strawberry aroma components (Table 6.1). As mentioned previously, DMHF is a key aroma component, and fruity esters and green-smelling aldehydes such as (Z)-3-hexenal are necessary for typical strawberry flavor.

6.3.1.6 Citrus

Much of what is known about citrus flavors is the result of studies performed on peel essential oils, essence oils, aqueous essences, and juices, with considerably less

TABLE 6.2
Predominant Aroma Components of Major Citrus Varieties

Orange[a]	Grapefruit[b]	Mandarin[c]	Lemon[d]	Lime[e]
Acetaldehyde	Ethyl butanoate	Acetaldehyde	Citral (mixture of neral and geranial)	Citral (mixture of neral and geranial)
Decanal	p-1-Methene-8-thiol	Octanal	Geranial	Germacrene B
(Z)-3-Hexenal	1-Heptane-3-one	Decanal	β-Pinene	Linalool
Ethyl butanoate	Ethyl butanoate	Thymol	γ-Terpinene	β-Pinene
Ethyl 2-methyl-propanoate	4-Mercapto-4-methyl-pentane-2-one	Methyl N-methyl-anthranilate	Geranyl acetate	β-Myrcene
(S)-Ethyl 2-methyl-butanoate	trans-4,5-Epoxy-(E)-2-decenal	β-Pinene	Neryl acetate	
Wine lactone[f]	Wine lactone	γ-Terpinene	Bergamotene	
(R)-Limonine		α-Sinensal	Caryphyllene	
(R)-α-Pinene		Ethyl-2-methyl-butanoate	α-Bisabolol	
trans-4,5-Epoxy-(E)-2-decenal		3-sec-Butyl-2-methoxypyrazine	Carvyl ethyl ether	
		3-Isopropyl-2-methoxypyrazine	8-p-Cymenyl ethyl ether	
		(Z)-3-Hexenal	Fenchyl ethyl ether	
			Linalyl ethyl ether	
			Myrcenyl ethyl ether	
			α-Terpinyl ethyl ether	
			Methyl epijasmonate	

Note: Components are not necessarily listed in order of importance.

[a] Data from Hinterholzer and Schieberle [113] and Buettner and Schieberle [114].
[b] Data from Buettner and Schieberle [49].
[c] Data from Shaw [57], Wilson and Shaw [115], and Schieberle et al. [116].
[d] Data from Mussinan et al. [119].
[e] Data from Clark et al. [117] and Chisholm et al. [118].
[f] 3α,4,5,7α-Tetrahydro-3,6-dimethyl-2(3H)-benzofuranone.

effort spent on fresh citrus fruit [57,112]. The flavor of fresh citrus is a result of the combination of peel essential oil, which is expressed during rupture of the ductless oil glands located in the outer portion (flavedo) of the fruit, and volatile constituents present within the fruit in juice sacs [57]. A large number of compound classes contribute to citrus flavors, including terpenes, alcohols, aldehydes, esters, ketones, and others. Results of extensive compositional studies on the various essential oil fractions and juices have indicated that only a relatively few volatile compounds are important in characterizing the flavors of individual citrus varieties. Predominant aroma constituents of five major citrus varieties are presented in Table 6.2.

A more extensive review of the flavor chemistry of the various citrus varieties was presented by Shaw [57]. Orange and grapefruit flavors are delicate and complex, with no clear set of character-impact compounds yet identified [49,113,114]. Both flavors are the result of a specific balance of several aroma compounds, and grapefruit is distinguished from orange primarily on the basis of the sulfur-containing compounds 1-*p*-menthene-8-ol and 4-mercapto-4-methylpentane-2-one. Nootkatone, a peel oil constituent, is reported to possess a grapefruit-like character [57] but was found to be of only moderate importance in grapefruit juice [49]. Mandarin (tangerine) flavor is not as clearly defined as orange and grapefruit; methyl *N*-methylanthranilate, β-pinene, γ-terpinene, and α-sinensal have been reported as potential contributing compounds to its unique character [57,115]. A recent study indicated the importance of 3-*sec*-butyl-2-methoxypyrazine, ethyl 2-methylbutanoate, 3-isopropyl-2-methoxypyrazine, and (Z)-3-hexenal in the aroma of clementine segments [116]. Lemon flavor is dominated by citral, and other volatile components have somewhat less impact. Citral also predominates in lime flavor, with possibly germacrene B, linalool, β-pinene, and β-myrcene providing enough complexity to differentiate lime from lemon flavor [117,118].

6.3.1.7 Tropical Fruit

Among the various tropical fruits, bananas and pineapples are of greatest economic importance. The aroma of these fruits have, therefore, undergone extensive investigation. In recent years, only relatively few studies have been conducted on the flavor of some other important tropical fruits, including avocado (*Persea americana* L.) [120,121], guava (*Psidium* spp.) [122–128], mango (*Mangifera indica*, Anacardiaceae) [129–131], passion fruit (*Passiflora* spp.) [44,47,132,133], and kiwifruit (*Actinidia* spp.) [134–136]. The following discussion will present the current understanding of the key aroma constituents of banana and pineapple fruit.

6.3.1.7.1 Banana

The flavor chemistry of banana (*Musa sapientum* L.) has been extensively studied for over 30 years. Over 350 volatile compounds have been identified, but only a few components are considered to greatly affect banana aroma [55]. Esters, especially 3-methylbutyl acetate, and to a lesser extent 3-methylbutyl butanoate and various 3-methylbutanoate esters, are important aroma constituents [137,138]. Ester biogenesis in banana has received recent attention [138,139]. Some alcohol and aldehyde constituents (e.g., 3-methyl-1-butanol and 3-methylbutanal) may also be important in banana aroma [133,137,140].

6.3.1.7.2 Pineapple

Pineapple (*Ananas comosus* L.) flavor has been the subject of extensive study. Esters were reported as the major volatile constituents of green and ripened pineapples [11]. Takeoka et al. [11] used results of quantitative studies and calculated aroma activity values to establish DMHF, methyl 2-methylbutanoate, ethyl 2-methylbutanoate, ethyl acetate, ethyl hexanoate, ethyl butanoate, ethyl 2-methylpropanoate, methyl hexanoate, and methyl butanoate as important contributors of pineapple aroma. Additional thiol-esters and phenolic and furanoid compounds have recently been identified in pineapple [142,143].

6.3.2   Vegetable Flavors

Whitfield and Last [59] reviewed the subject of volatile constituents of vegetables and covered the published literature up to about 1991. Since that time much progress has been made in our understanding of vegetable flavors. Most studies have dealt with the flavor and aroma composition of cooked or prepared (e.g., fermented) vegetables. This seems reasonable since most vegetables are consumed after cooking or further processing. A discussion of the flavor and volatile components of all major vegetable crops is beyond the scope of this chapter. Instead, this section will present recent findings on some selected vegetables that are commonly consumed in their raw (uncooked) state, such as lettuce, celery, carrot, bell pepper, broccoli, and onion.

6.3.2.1   Lettuce

The green leafy portion of lettuce (*Lactuca sativa* L.) is mostly consumed in its raw state as a salad component. To date, very few studies have been conducted on the volatile components of lettuce. Murray and Whitfield [144] reported the occurrence of earthy/bell pepper-like smelling 2-isopropyl-, 2-*sec*-butyl-, and 2-methylpropyl-3-methoxypyrazines in the headspace of raw lettuce. Volatile emissions of (*Z*)-3-hexenal, (*Z*)-3-hexen-1-ol, and (*Z*)-3-hexenyl acetate from lettuce have been reported [145].

6.3.2.2   Celery

The main volatile constituents of raw celery (*Apium gaveolens* L.) stalk are monoterpene and sesquiterpene hydrocarbons, accounting for over 80% of the volatile oil. Additional major components include monoterpene alcohols, phthalides, and terpene carbonyls [146]. Among these, the phthalides are considered the most important with respect to aroma [147]. 3-Butylphthalide and some hydrogenated derivatives, sedanolide (3-butyl-3α,4,5,6-tetrahydrophthalide), 3-isobutylidene-3α,4,5,6-tetrahydrophthalide, and 3-isobutylidene-3α,4-dihydrophthalide, were reported as character-impact aroma components of celery [148,149].

6.3.2.3   Carrot

More than 90 volatile compounds have been identified in carrot (*Daucus carota* L.) [150]. Mono- and sesquiterpenes are the predominant volatile compounds, comprising over 98% of the total volatile composition [150,151]. Characteristic aroma components of carrot include α- and β-pinene, β-myrcene, γ-terpinene, terpenolene, 6-methyl-5-hepten-2-one, β-bisabolene, β-ionone, and myristicin [150,152–154].

6.3.2.4   Bell Pepper

Bell or sweet pepper (*Capsicum annuum*, var. grossum Sendt) is often consumed in the fresh state in salads or as cut vegetables in vegetable platters. Fresh bell pepper aroma is best characterized by 2-methylpropyl-3-methoxypyrazine, which has an earthy aroma reminiscent of bell peppers [155,156]. Other compounds reported to contribute to raw bell pepper aroma include (*E,Z*)-2,6-nonadienal, (*E,E*)-2,4-deca-dienal, and 2-*sec*-butyl-3-methoxypyrazine [144,155].

6.3.2.5 Broccoli

Broccoli (*Brassica oleracea* var. *itallica*) is one of the cruciferous vegetables, which include cabbage, cauliflower, mustard, and horseradish. These products often have characteristically irritating, pungent, and lachrymatory odors. As explained earlier, the flavor constituents of crucifers are primarily formed through enzymatic processes in disrupted tissues or may form during cooking. In raw broccoli, tissue disruptions lead to formation of numerous isothiocyanates from glucosinolate precursors. The isothiocyanates are unstable and are destroyed during cooking to allow nitriles and other sulfur-containing degradation products and rearrangement products to predom-inate. In a recent study, the major volatiles in blanched broccoli were identified, including 3-methyl-2-pentanone, hexanal, heptanal, cyclopentanecarboxaldehyde, ethyl acetate, 3-methylbutanal, 3-butenenitrile, 2-methylbutanal, dimethyl trisulfide, and dimethyl disulfide [157]. In raw broccoli, cysteine lyase action on sulfur-con-taining amino acid, *S*-methylcysteine *S*-oxide, can generate methanethiol, dimethyl sulfide, dimethyldisulfide, and dimethyltrisulfide [158]. These same volatiles may be formed by thermal degradation of *S*-methylcysteine and its *S*-oxide derivative [159].

6.3.2.6 Onion

Members of the genus *Allium*, such as onions, garlic, leeks, chives, and shallots, have characteristically strong, pungent, and penetrating odors that are formed fol-lowing tissue damage as discussed above. Onion (*A. sepa*) flavor is dominated by volatile sulfur-containing compounds that arise from the enzymatic hydrolysis of *S*-alk(en)yl-L-cysteine *S*-oxide [42]. For example, breakdown of the major flavor precursor, *S*-1-propenyl-L-cysteine *S*-oxide, leads to formation of lachrymatory com-pounds, such as propanethial *S*-oxide [160–162]. A series of sulfinates are then formed from various sulfenic acid intermediates to give the characteristic aroma of onion [161].

6.4 FACTORS INFLUENCING FLAVOR AND FLAVOR STABILITY

The aroma profiles can change rapidly during postharvest storage of fresh fruits and vegetables, particularly for climacteric fruits, in which the dominant volatiles may be quite different in the unripe fruit, ripe fruit, or overripe fruit [163]. Some produce, especially those designed for vegetative reproduction (carrot, potato, and onion) and generative reproduction (seeds and fruits) are well-equipped to live detached from the host plant. These products contain relatively large amounts of carbohydrate reserves that enable the maintenance of respiration and energy production. Other harvested produce, such as leaves (spinach) or whole plants (lettuce or endive) do not contain sufficient storage reserves and are susceptible to rapid senescence and wilting [164].

The fact that fruits and vegetables are living, respiring tissues has major consequences related to shelf life and storage. Product respiration, transpiration, and ethylene production are major factors responsible for the deterioration of fresh produce. Temperature may have the greatest effect on respiration rate and thus deterioration rate of produce. Other factors that can be controlled to maximize storage period include relative humidity and gas composition. All of the above parameters are commodity-specific; each fruit or vegetable is unique and is affected by both genetic factors (e.g., species, cultivar, etc.) and stage of maturation or ripeness. Postharvest practices prior to storage may also affect the behavior of produce during storage. Understanding the physiology of fresh produce is fundamental to understanding its stability and likely shelf life. Factors that affect stability, shelf life, and quality attributes such as appearance, texture, and flavor of fruits and vegetables have been reviewed [1,165]. The following discussion addresses some important intrinsic and external factors that affect the flavor quality of fruits and vegetables.

6.4.1 TEMPERATURE

Temperature affects respiration rate, and likewise volatile production is known to increase with increasing temperature within a moderate range (e.g., between 0 and 30°C in the case of apples) [166]. However, high-temperature storage is not conducive to prolonged shelf life for most produce. In general, respiration rate is controlled by lowering the temperature during storage or by controlled or modified atmospheric storage with or without active packaging technology. Humidification also is necessary for some produce especially susceptible to moisture loss (e.g., leafy vegetables), but the impact on flavor is not well documented. Another advantage of low-temperature storage is the reduced growth rate of postharvest pathogens and spoilage microorganisms. Ideal storage temperatures varies from product to product and recommended conditions for over 50 fruits and vegetables have been tabulated [167].

Most fruits and vegetables cannot survive cold temperature storage (chilling injury) at near-freezing temperatures, and many are intolerant to low temperatures well above freezing.

Refrigeration may also slow, limit, or halt (e.g., in the case of chilling injury) the development of aroma volatiles in ripening fruits. Temperature conditioning can affect the response of some produce. For example, tomatoes are susceptible to chilling injury when stored below about 12 to 13°C [168], resulting in several metabolic disorders leading to decline in soluble solids [169] and flavor loss [170]. However, chilling injury can be prevented in mature green tomatoes by high-temperature conditioning (36 to 40°C for 3 d) before cold storage (2 to 3°C) for up to 3 weeks [171].

6.4.2 ETHYLENE

The best-quality fruits are produced when the concentration of ethylene, CO_2, and O_2 in the atmosphere and the duration of exposure, temperature, and humidity are carefully controlled and maintained at optimal levels [172]. The detrimental effects of ethylene on quality center on its ability to alter or accelerate the same natural processes of development, ripening, and senescence that are viewed as beneficial in a different context [172].

Ethylene may be controlled by various strategies such as isolating ethylene-sensitive produce away from commodities that produce high levels of ethylene (or by storage of unripe fruit away from ripe ones), providing adequate ventilation, use of ethylene scrubbers (catalytic or adsorption mechanisms), use of ethylene antagonists such as CO_2, or by destruction of ethylene by reaction with ozone [167].

The inhibition of ethylene biosynthesis or action will inhibit not only ripening but also the production of characteristic aroma compounds. Intentional exposure of ethylene is primarily used to ripen harvested climacteric fruit [172]. In general, the effect of ethylene on taste and aroma is caused by stimulating the ripening of fruit [173]. Fruits ripened in this way most often have inferior flavor profiles since production of aroma compounds is diminished [174,175], although some exceptions have been reported [73]. In the case where fruit have been stored for a prolonged period in an atmosphere in which ethylene activity has been suppressed, it may take some time after reintroduction to air in order for volatile production to return to normal and reestablish the characteristic aroma profile [172].

6.4.3 Controlled and Modified Atmospheric Storage

Fruits and vegetables are living materials, and rate of respiration is a critical factor in determining shelf life. Thus, the greater the respiration, the shorter the shelf life. Immature produce such as peas and beans have greater respiration rates and, therefore, shorter shelf lives, whereas the opposite is true for mature storage organs (rhisomes) such as potatoes and onions. Excessive respiration in nonstorage tissues, such as leafy vegetables or immature flower produce such as broccoli, leads to rapid degradation due to rapid depletion of energy reserves. Climacteric fruit, such as apples, bananas, and tomatoes, can be harvested in the unripe stage and then artificially ripened at a later stage. The respiration rate of these fruits must be carefully controlled and monitored (e.g., temperature and ethylene levels), because the rate is known to increase dramatically during a short time period, leading to overripening, senescence, and generation of uncharacteristic volatile compounds or off-odors.

The basic principle behind the use of controlled (CA) and modified atmospheres (MA) during storage is to reduce the rate of respiration, reduce microbial growth, and retard enzymatic deterioration by manipulating the gaseous environment surrounding the produce. Generally, this is achieved by reducing the concentration of O_2, which is required for respiration, or by adding an inhibitory gas such as CO_2. CA storage generally refers to use of decreased O_2 and increased CO_2 with continuous monitoring and adjustment of the gas composition. CA storage of produce has been recently reviewed and optimal conditions have been determined for various fruits and vegetables [167,176].

In contrast to CA, the gas composition used in MA packaging (MAP) is neither monitored nor adjusted. Eventually an equilibrium gas composition will be reached in MAP, which is a result of a balance between metabolic rates of the produce and diffusion characteristics of the packaging materials [164]. In MAP, polymeric films are used to control movement of respiratory gases. Depending on the film, lowered O_2 and raised CO_2 levels within the package can lead to a reduction in respiration of the produce and potentially longer shelf life. One potential drawback to MAP is the potential for O_2 levels to fall too low and cause production of off-odors due to fermentation.

Although respiration rate is important, its reduction is not the only beneficial effect of altered atmospheres. Increased CO_2 levels also influence (suppress) ethylene production and thus affect ripening and flavor development [172,177]. Volatile production tends to peak within a few days during the beginning of storage, in association with ripening, and then tends to drop rapidly as substrates are depleted and senescence sets in. CA-stored produce generally exhibit a slow rate of aroma production over a prolonged period, except under low O_2 levels, in which case anaerobiosis occurs, leading to formation of ethanol and other volatiles. Brackman et al. [178] reported that apple cultivars whose aroma precursors are derived principally from fatty acid metabolism (ester cultivars) were more susceptible to aroma suppression under CA than apples that biosynthesize aroma primarily from amino acids (alcohol-type cultivars). Yahia et al. [179] reported that loss of volatiles from apples stored under CA can be considerable and the severity is dependent on the atmosphere composition and duration of storage. For example, low-ethylene CA storage suppressed production of butanoate, 2-methylbutanoate, pentanoate, and hexanoate esters, while acetate esters were unaffected [179]. CA storage can reduce aroma production when apples are ripened after storage with varying effects on the different aroma constituents [66,180]. CA storage under high CO_2 and low O_2 suppressed synthesis of branched-chain aroma components [181]. In apples, the generation of esters, which takes place mainly in the peel of intact fruit, is oxygen-dependent [182].

Alternatives to the usual high-CO_2, low-O_2 CA and MAP practices have been developed in recent years. Use of novel noble gas mixtures in MAP (e.g., argon, 70 to 90%; CO_2, 0 to 20%; O_2, 0 to 15%) was shown to improve quality and lengthen shelf life of prepared salads [183]. On the other hand, strawberries stored in MAP with argon containing high CO_2 and high O_2 levels experienced altered ester synthesis [184].

CA and MA are not necessary for some vegetables such as potatoes, carrots, garlic, and onions. On the other hand, use of CA and MA has been found to benefit broccoli, cabbage, lettuce, asparagus, and brussels sprouts [185].

6.4.3.1 MAP of Fresh-Cut Fruits and Vegetables

One area where MAP has been utilized is for fresh-cut or minimally processed produce, such as salads and fruit pieces, to prevent rapid deterioration that occurs once fresh produce has been cut [186,187]. Prepared or fresh-cut fruits and vegetables represent a group of short-shelf-life produce that has experienced considerable growth in the past decade [188,189]. Examples range from ready-to-eat washed, sliced, chopped, or shredded prepared fruits or vegetables, such as salads and coleslaw mixes, to diced, cubed, or segmented fruit and stir-fry products. The biochemistry, physiology, microbiology, and quality factors in fresh-cut fruit were the subject of a recent book [190].

The characteristics of these products differ considerably from their intact counterparts. The act of cutting or shredding the tissues results in a large cut surface area
of damaged cells, leading to greater instability and much shortened shelf life of only
a few days, since cut or damaged produce has a respiration rate higher than that of
undamaged tissue [191]. In addition, physical injury such as wounding or bruising
can stimulate ethylene production [172].

Flavor and appearance have been observed to deteriorate faster in various cut
vegetables (lettuce, carrot, celery, endive, and radish) than in undamaged tissues
[191]. Discoloration, off-flavors, and drying of product are the main determinants
of shelf life [192]. In terms of flavor, both aroma and taste components may be
affected. Decreases in sucrose, glucose, fructose, and malic acid were observed for
grated carrot during storage at 2°C [193]. Lamikanra and Richard [194] observed a
decline in esters and in β-ionone and geranylacetone during storage of fresh-cut
cantaloupe. Factors affecting flavor and shelf life of fresh-cut fruits and vegetables
have been reviewed recently [195].

6.4.4 AROMA MODIFICATION BY EXOGENOUS VOLATILES

In addition to the well-known ethylene modulating compounds 1-methylcycloproene
and methyl jasmonate [196], various other volatile compounds have been shown to
affect the production of volatiles of whole and fresh-cut fruits during storage.
Treatment or CA exposure of strawberries with acetaldehyde vapor can suppress
mold development and improve color [197]. However, acetaldehyde treatment can
alter the volatile profiles of strawberries and other fruits. Acetaldehyde treatment of
feijoa fruit was shown to increase ethyl butanoate concentrations during storage [198].
The nature and extent of the volatile profile change depends on the acetaldehyde level
used, the duration of treatment, and the biochemical pathways involved [199].

6.4.5 COATINGS AND WRAPS

Coating the surface of fruits and vegetables with waxes or other suitable material can
have beneficial effects in terms of improved moisture retention and reduction of CO_2
and O_2 exchange. The effect may be positive or negative in terms of flavor, since reduced
respiration is often desirable, but a coating that is too thick may reduce respiration to
the point where fermentation is favored, leading to off-flavors. Initial studies indicated
that citrus fruits did not respond well to waxing [200], but more recent investigations
with shellac coatings indicated some advantages of coating in regard to moisture retention and appearance, with minimal effects on volatile composition [201]. Application
of high-density polyethylene film (wrap) is a good alternative to waxes since it provides
good water vapor barrier properties while not affecting the movement of respiratory
gases and contributing to the development of off-odors [202].

6.4.6 CURING

Under proper storage conditions (curing) some root and tuber crops, such as sweet
potatoes and Irish potatoes, have the ability to heal minor wounds incurred during
harvest and handling.

Curing with respect to bulb crops, such as onions and garlic, is much different from what is used for root and tuber crops. In this case curing is a dehydration process that readies the produce for prolonged storage. The effect of curing on flavor of bulb crops is uncertain; however, curing is known to affect the flavor potential of some tuber crops, such as sweet potatoes [203]. In this case, curing of sweet potatoes increases the levels of amylotyic enzymes that are responsible for starch degradation to maltose during baking, leading to a sweeter flavor as well as increased levels of thermally derived (via Maillard reaction) aroma components [204].

6.4.7 IRRADIATION

For more than 30 years there have been many studies on the application of irradiation for improving the keeping quality of produce. Low-dose gamma irradiation delays ripening by reducing respiration in some fruits and vegetables, and irradiated produce remains edible for longer periods because senescence is delayed [205]. In addition, low-dose irradiation functions as an effective quarantine method against fruit flies and other insects, and fungal development and microbial spoilage may also be delayed. In tuber crops, such as potatoes, onions, carrots, etc., low-dose irradiation is an effective means of suppressing sprouting and controlling fungal and bacterial pathogens, which prolongs shelf life. In general, when doses are kept within recommended allowances, application of low-dose irradiation is an effective means for shelf-life extension in fruits and vegetables without significant deleterious changes in flavor [206,207].

6.4.8 OFF-FLAVOR AND TAINTS

Off-odors can be formed in fresh produce by microbial and chemical transformations during harvesting, processing, packaging, and storage. They also may be introduced (taints) by absorption of alien chemicals from water used in processing, from packaging materials, and from the environment in which the produce has been grown, processed, and stored [208].

Understanding the physiology of fresh produce is fundamental to understanding its stability and likely shelf life. Factors that affect stability and shelf life and quality attributes such as appearance, texture, and flavor of fruits and vegetables have been reviewed.

6.4.8.1 Anaerobiosis and Off-flavor Development

Off-odors can result if fruit or vegetable tissues undergo anaerobic respiration (fermentation or anaerobiosis) [209], leading to formation of lactic acid, acetic acid, ethanol, and other volatiles [210,211]. Furthermore, CO_2 in aqueous solution is a strong oxidant that causes bleaching of color and generation of off-odors.

Storage of fruits under an anaerobic environment (anoxia) has been shown to result in off-flavor development [212]. Anaerobic respiration leads to accumulation of acetaldehyde and ethanol, which increases with time of exposure in static N_2 (or decreased O_2). In commodities such as apples the anaerobic profile included high concentrations of ethanol [213]. In addition, anoxic conditions that favor ethanol production were reported to inhibit the formation of ethanol-derived esters in Red-chief Delicious apples [214]. The effects may be at least partially reversible due to possible conversion of alcohols to esters or other modifications via oxidation upon return of the fruit to air [166,212,214]. Anaerobic pretreatment may have a positive influence on flavor in some instances. It has been shown to increase aroma-relevant volatiles in some fruit, such as citrus [215], and was shown to increase levels of ethyl butanoate and ethyl acetate in feijoa [198].

Some fresh produce is more susceptible to the development of off-odors than other commodities. For example, strong off-odors have been reported for broccoli stored in O_2 levels less than 0.5% [216–219]. Forney et al. [217] monitored the formation of volatiles by broccoli stored under CA conditions (0.5% O_2) and found ethanol, acetaldehyde, methanethiol, hydrogen sulfide, octane, methyl acetate, ethyl acetate, and dimethyl sulfide. Methanethiol was considered to be primarily responsible for the off-odor. Likewise, scallions during storage at 0 to 5°C developed propanethiol S-oxide [220].

6.4.8.2 Ethylene-Induced Off-Flavors

Ethylene is produced in copious amounts to mediate the defense response of stressed, diseased, and injured tissues. Parsnip roots exposed to ethylene developed higher levels of total phenolics and bitterness (after cooking) compared with controls [221]. Similarly, exposure of mature carrots to ethylene resulted in higher content of bitter isocoumerin in the peel [222]. Injured carrots (sliced, cut, or bruised) exposed to ethylene showed greater rates of isocoumarin accumulation than intact (uninjured) specimens. This enhanced ethylene sensitivity has been observed in mechanically wounded lettuce [223] and has major implications in relation to prepared salads and other fresh-cut produce.

6.4.8.3 Sanitizers and Chemical Treatments

Prior to grading, processing, and packaging, produce is often washed with chlorinated (50 to 100 ppm) or ozonated water [224] to help control fungal and bacterial pathogens, but this practice may lead to off-odors [225]. For example, chlorine disinfection is known to cause formation of compounds with musty and medicinal off-odors, such as trichloroanisoles [208]. Fungicides and sprouting inhibitors are approved for use on some produce [226] and provide another potential source of off-odors. For example, potatoes treated with isopropyl-N-(3-chlorophenyl)carbamate and HiSol carrier (mixture of aromatic hydrocarbons) were found to impart an off-odor caused by high levels of benzene derivatives, some of which were components of the HiSol carrier [227].

It should be noted, however, that musty/earthy aromas are not always negative, but can be associated with the characteristic and desirable flavor of some produce, such as beets, in which geosmin is responsible for the characteristic aroma in both raw and cooked beets [228]. Geosmin also has been reported as a normal volatile component of sweet corn [229]. Earthy smelling 2-methoxy-3-isopropylpyrazine is important in the aromas of raw potatoes [230], raw peas [231], and other raw vegetables [144].

6.4.8.4   Microbial Sources

Produce inhabited by pathogenic microorganisms may exhibit pathogen-specific volatile profiles or volatiles associated with characteristic change in the principal substrates attacked [232]. When a disease organism is present, the volatile profile will include those of the produce (normal metabolites and background) plus metabolites of active pathogens as a result of their metabolism and interaction with the host material. Volatile monitoring has been demonstrated for the detection and differentiation of pathogens in stored potatoes [232–234]. Infection-specific volatiles associated with diseased potatoes (soft rot, *Erwinia carotovara* var. *atroseptica*) included methanol, ethanol, acetone, 1-butanol, 2-butanol, 2-butanone, and 3-hydroxy-2-butanone (acetoin) [233,235]. Soft rot (*Clostridium* infection) in stored potatoes was reported to cause pigsty-like off-odors due to formation of *p*-cresol, indole, and skatole [236].

6.5   CONCLUSIONS

Development and application of advanced instrumental-sensory techniques in the last two decades has led to tremendous progress toward our understanding of the flavor chemistry of fruits and vegetables. This is exemplified by recent studies on strawberry flavor that have led to identification of key aroma-impact components and elucidatation of pathways involved in their formation. Such knowledge, coupled with a greater understanding of important intrinsic and external factors that affect the flavor quality of fruits and vegetables, will lead to development of improved postharvest storage practices, with the ultimate goal of providing higher-quality produce with long shelf lives.

REFERENCES

1. Aked, J., Fruits and vegetables, in *The Stability and Shelf-Life of Food*, Kilcast, D. and Subramanian, P., Eds., Woodhead Publishing Limited, Cambridge, UK, 2000, p. 249.
2. Tucker, G.A., Introduction, in *Biochemistry of Fruit Ripening*, Taylor, J.E. and Tucker, G.A., Eds., Chapman and Hall, London, 1993, p. 53.
3. Tressl, R. and Albrecht, W., Biogenesis of aroma compounds through acyl pathways, in *Biogeneration of Aromas*, Parliment, T.H. and Croteau, R., Eds., ACS Symposium Series 317, American Chemical Society, Washington, DC, 1986, p. 114.
4. Schreier, P., Biogeneration of plant aromas, in *Development in Food Flavors*, Birch, G.G. and Lindley, M.G., Eds., Elsevier Applied Science, New York, 1986, p. 89.
5. Wyllie, S.G., Leach, D.N., Nonhebel, H.N., and Lusunzi, I., Biochemical pathways for the formation of esters in ripening fruit, in *Flavour Science: Recent Developments*, Taylor, A.J. and Mottram, D.S., Eds., The Royal Society of Chemistry, Cambridge, UK, 1996, p. 52.
6. Leahy, M.M. and Roderick, R.G., Fruit flavor biogenesis, in *Flavor Chemistry: Thirty Years of Progress*, Teranishi, R., Wick, E.L., and Hornstein, I., Eds., Klewer Academic/Plenum Publishers, New York, 1999, p. 275.

7. Shalit, M., Katzir, N., Tadmor, Y., Larkov, O., Burger, Y., Shalekhet, F., Lastochkin, E., Ravid, U., Amar, O., Edelstein, M., Karchi, Z., and Lewinsohn, E., Acetyl-CoA: Alcohol acetyltransferase activity and aroma formation in ripening melon fruits, *J. Agric. Food Chem.*, 49, 794, 2001.
8. Olias, J.M., Sanz, C., Rios, J.J., and Pérez, A.G., Substrate specificity of alcohol acyltransferase from strawberry and banana fruit, in *Fruit Flavors: Biogenesis, Characterization, and Authentication*, Rouseff, R.L. and Leahy, M.M., Eds., ACS Symposium Series 596, American Chemical Society, Washington, DC, 1995, p. 134.
9. Rowan, D.D., Lane, H.P., Allen, J.M., Fielder, S., and Hunt, M.B., Biosynthesis of 2-methylbutyl, 2-methyl-2-butenyl and 2-methylbutanoate esters in Red Delicious and Granny Smith apples using deuterium labeled substrates, *J. Agric. Food Chem.*, 44, 3276, 1996.
10. Rowan, D.D., Allen, J.M., Fielder, S., and Hunt, M.B., Biosynthesis of straight-chain ester volatiles in Red Delicious and Granny Smith apples using deuterium labeled precursors, *J. Agric. Food Chem.*, 47, 2553, 1999.
11. Takeoka, G., Buttery, R.G., Flath, R.A., Teranishi, R., Wheeler, E.L., Wieczorek, R.L., and Guentert, M., Volatile constituents of pineapple (*Ananas comosus* [L.] Merr.), in *Flavor Chemistry: Trends and Developments*, Teranishi, R., Buttery, R.G., and Shahidi, F., Eds., ACS Symposium Series 388, American Chemical Society, Washington, DC, 1989, p. 223.
12. Tang, C.S. and Jennings, W.C., Lactonic compounds of apricot, *J. Agric. Food Chem.*, 16, 252, 1968.
13. Chairotte, G., Rodriguez, F., and Crozet, J., Characterization of additional volatile flavor components of apricot, *J. Food Sci.*, 46, 1898, 1981.
14. Engel, K.-H., Heidlas, J., Albrecht, W., and Tressl, R., Biosynthesis of chiral flavor and aroma compounds in plants and microorganisms, in *Flavor Chemistry: Trends and Developments*, Teranishi, R., Buttery, R.G., and Shahidi, F., Eds., ACS Symposium Series 388, American Chemical Society, Washington, DC, 1989, p. 8.
15. Guichard, E., Chiral γ-lactones, key compounds to apricot flavor, in *Fruit Flavors: Biogenesis, Characterization, and Authentication*, Rouseff, R.L. and Leahy, M.M., Eds., ACS Symposium Series 596, American Chemical Society, Washington, DC, 1995, p. 258.
16. Ramosvaldivia, A.C., Vanderheijden, R., and Verpoorte, R., Isopentyl diphosphate isomerase — a core enzyme in isoprenoid biosynthesis — a review of its biochemistry and function, *Nat. Prod. Rep.*, 14, 591, 1997.
17. Croteau, R., Biosynthesis of cyclic monoterpenes, in *Biogeneration of Aromas*, Parliment, T.H. and Croteau, R., Eds., ACS Symposium Series 317, American Chemical Society, Washington, DC, 1986, p. 134.
18. Davis, E.M. and Croteau, R., Cyclization enzymes in the biosynthesis of monoterpenes, sesquiterpenes, and diterpenes, in *Biosynthesis: Aromatic Polyketides, Isoprenoids, Alkaloids (Top. Curr. Chem.)*, Leeper, F.J. and Vederas, J.C., Eds., Verlag/Hersteller: Springer, Berlin, 209, 53-85, 2000.
19. Karp, F., Mihaliak, C.A., Harris, J.L., and Croteau, R., Monoterpene biosynthesis: Specificity of the hydroxylations of (–)-limonene by enzyme preparations from peppermint (*Mentha piperita*), spearmint (*Mentha spicata*), and perilla (*Perilla frutescens*) leaves. *Arch. Biochem. Biophys.*, 276, 219, 1990.
20. Winterhalter, P., Carotenoid-derived aroma compounds: Biogenentic and biotechnological aspects, in *Biotechnology for Improved Foods and Flavors*, Takeoka, G.R., Teranishi, R., Williams, P.J., and Kobayashi, A., Eds., ACS Symposium Series 637, American Chemical Society, Washington, DC, 1996, p. 295.

21. Winterhalter, P. and Rouseff, R.L., Carotenoid-derived aroma compounds: An intro-
    duction, in *Carotenoid-Derived Aroma Compounds*, Winterhalter, P. and Rouseff,
    R.L., Eds., ACS Symposium Series 802, American Chemical Society, Washington,
    DC, 2002, p. 1.
22. Buttery, R.G., Teranishi, R., and Ling, L.C., Identification of damascenone in tomato
    volatiles, *Chem. Industr. (Lond.)*, 238, 1988.
23. Winterhalter, P. and Schreier, P., The generation of norisoprenoid volatiles in starfruit
    (*Averrhoa carambola* L.): A review, *Food Rev. Int.*, 11, 237, 1995.
24. Knapp, H., Straubinger, M., Sting, C., and Winterhalter, P., Analysis of norisoprenoid
    aroma precursors, in *Carotenoid-Derived Aroma Compounds*, Winterhalter, P. and
    Rouseff, R.L., Eds., ACS Symposium Series 802, American Chemical Society, Wash-
    ington, DC, 2002, p. 20.
25. Tsuneya, T., Ishihara, M., Shiota, H., and Shiga, M., Volatile components of quince
    fruit (*Cydonia oblonga* Mill.), *Agric. Biol. Chem.*, 47, 2495, 1983.
26. Zabetakis, I., Gramshaw, J.W., and Robinson, D.S., 2,5-Dimethyl-4-hydroxy-2H-
    furan-3-one and its derivatives: Analysis, synthesis and biosynthesis — A review,
    Food Chem., 65, 139, 1999.
27. Schieberle, P. and Hofmann, T., Evaluation of the character impact odorants in fresh
    strawberry juice by quantitative measurements and sensory studies on model mix-
    tures, *J. Agric. Food Chem.*, 45, 227, 1997.
28. Sanz, C., Richardson, D.G., and Pérez, A.G., 2,5-Dimethyl-4-hydroxy-3(2H)-fura-
    none and derivatives in strawberries during ripening, in *Fruit Flavors: Biogenesis,
    Characterization, and Authentication*, Rouseff, R.L. and Leahy, M.M., Eds., ACS
    Symposium Series 596, American Chemical Society, Washington, DC, 1995, p. 268.
29. Pickenhagen, W., Velluz, A., Passert, J.-P., and Ohloff, G., Estimation of 2,5-dimethyl-
    4-hydroxy-3(2H)-furanone (furaneol) in cultivated and wild strawberries, pineapples
    and mangoes, *J. Sci. Food Agric.*, 32, 1132, 1981.
30. Baek, H.H., Cadwallader, K.R., Marroquin, E., and Silva, J.L., Identification of
    predominant aroma compounds in muscadine grape juice, *J. Food Sci.*, 62, 249, 1997.
31. Buttery, R.G., Takeoka, G.R., Krammer, G., and Ling, L.C., Identification of 2,5-
    dimethyl-4-hydroxy-3(2H)-furanone (furaneol) and 5-methyl-4-hydroxy-3(2H)-fura-
    none in fresh and processed tomato, *Lebensm.-Wiss. Technol.*, 27, 592, 1994.
32. Zabetakis, I. and Holden, M.A., Strawberry flavor: Analysis and biosynthesis, *J. Sci.
    Food Agric.*, 74, 421, 1997.
33. Zabetakis, I., Gramshaw, J.W., and Robinson, D.S., The biosynthesis of 2,5-dimethyl-
    4-hydroxy-2H-furan-3-one and its derivatives in strawberry, in *Flavour Science:
    Recent Developments*, Mottram, D.S. and Taylor, A.J., Eds., The Royal Society of
    Chemistry, Cambridge, UK, 1996, p. 90.
34. Wein, M., Lewinsohn, E., and Schwab, W., Metabolic fate of isotopes during the
    biological transformation of carbohydrates to 2,5-dimethyl-4-hydroxy-3(2H)-fura-
    none in strawberry fruits, *J. Agric. Food Chem.*, 49, 2427, 2001.
35. Roscher, R., Bringmann, G., Schreier, P., and Schwab, W., Radiotracer studies on the
    formation of 2,5-dimethyl-4-hydroxy-3(2H)-furanone in detached ripening straw-
    berry fruits, *J. Agric. Food Chem.*, 46, 1488, 1998.
36. Roscher, R., Schreier, P., and Schwab, W., Metabolism of 2,5-dimethyl-4-hydroxy-
    3(2H)-furanone in detached ripening strawberry fruits, *J. Agric. Food Chem.*, 45,
    3202, 1997.
37. Lavid, N., Schwab, W., Kafkas, E., Koch-Dean, M., Bar, E., Larkov, O., Ravid, U., and
    Lewinsohn, E., Aroma biosynthesis in strawberry: S-Adenosylmethionine: Furaneol
    O-methyltransferase activity in ripening fruits, *J. Agric. Food Chem.*, 50, 4025, 2002.

38. Wein, M., Lavid, N., Lunkenbein, S., Lewinsohn, E., Schwab, W., and Kaldenhoff, R., Isolation, cloning and expression of a multifunctional O-methyltransferase capable of forming 2,5-dimethyl-3(2H)-furanone, one of the key aroma compounds in strawberry fruits, *Plant J.*, 31, 755, 2002.

39. Bones, A.M. and Rossiter, J.T., The myrosinase-glucosinolate system, its organization and biochemistry, *Physiol. Plant.*, 97, 194, 1996.

40. Shofran, B.G., Purrington, S.T., Breidt, F., and Fleming, H.P., Antimicrobial properties of sinigrin and its hydrolysis products, *J. Food Sci.*, 63, 621, 1998.

41. Masuda, H., Harada, Y., Tanaka, K., and Nakajima, M., Characterization odorants of wasabi (*Wasabia japonica* matum), Japanese horseradish, in comparison with those of horseradish (*Armoracia rusticana*), in *Biotechnology for Improved Foods and Flavors*, Takeoka, G.R., Teranishi, R., Williams, P.J., and Kobayashi, A., Eds., ACS Symposium Series 637, American Chemical Society, Washington, DC, 1996, p. 67.

42. Hanum, T., Sinha, N.K., Guyer, D.E., and Cash, J.N., Pyruvate and flavor development in macerated onions (*Allium cepa* L.) by γ-glutamyl transpeptidase and exogenous C-S lyase, *Food Chem.*, 54, 183, 1995.

43. Bacon, J.R., Moates, G.K., Ng, A., Rhodes, M.J.C., Smith, A.C., and Waldron, K.W., Quantitative analysis of flavour precursors and pyruvate levels in different tissues and cultivars of onion (*Allium cepa*), *Food Chem.*, 64, 257, 1999.

44. Weber, B., Maas, B., and Mosandl, A., Stereoisomeric flavor compounds. 72. Stereoisomeric distribution of some chiral sulfur-containing trace components of yellow passion fruit, *J. Agric. Food Chem.*, 43, 2438, 1995.

45. Jiang, J., Choo, S.Y., Omar, N., and Ahamad, N., GC-MS analysis of volatile compounds in durian (*Durio zibethinus* Murr.), in *Food Flavors: Formation, Analysis and Packaging Influences*, Contis, E.T., Ho, C.-T., Mussinan, C.J., Parliment, T.H., Shahidi, F., and Spanier, A.M., Eds., Elsevier, New York, 1998, p. 345.

46. Engel, K.-H. and Tressl, R., Identification of new sulfur-containing volatiles in yellow passion fruits (*Passiflora edulis* f. *flavicarpa*), *J. Agric. Food Chem.*, 39, 2249, 1991.

47. Weber, B., Dietrich, A., Maas, B., Marx, A., Olk, J., and Mosandl, A., Stereoisomeric flavour compounds. LXVI. Enantiomeric distribution of the chiral sulfur-containing alcohols in yellow passion fruits, *Z. Lebensm. Unters. Forsch.*, 199, 48, 1994.

48. Riguad, J., Etievant, P., Henry, R., and Latrasse, A., 4-Methoxy-2-methyl-2-mercaptobutane, a major constituent of the aroma of the blackcurrent bud (*Ribes nigrum* L.), *Sci. Aliments*, 6, 213, 1986.

49. Buettner, A. and Schieberle, P., Characterization of the most odor-active volatiles in fresh hand-squeezed juice of grapefruit (*Citrus paradise* Macfayden), *J. Agric. Food Chem.*, 47, 5189, 1999.

50. Hatanaka, A., The fresh green odor emitted by plants, *Food Rev. Int.*, 12, 303, 1996.

51. Fuessner, I. and Wasternack, C., Lipoxygenase catalyzed oxygenation of lipids, *Fett/Lipid*, 100, 146, 1998.

52. Grechkin, A., Recent developments in biochemistry of the plant lipoxygenase pathway, *Prog. Lipid Res.*, 37, 317, 1998.

53. Riley, J.C.M., Willemot, C., and Thompson, J.E., Lipoxygenase and hydroperoxide lyase activities in ripening tomato fruit, *Postharv. Biol. Technol.*, 7, 97, 1996.

54. Gray, D.A., Prestage, S., Linforth, R.S., and Taylor, A.J., Fresh tomato specific fluctuations in the composition of lipoxygenase-generated C6 aldehydes, *Food Chem.*, 64, 149, 1999.

55. Berger, R.G., Fruits I, in *Volatile Compounds in Foods and Beverages*, Maarse, H., Ed., Marcel Dekker, New York, 1991, p. 283.

56. Latrasse, A., Fruits III, in *Volatile Compounds in Foods and Beverages*, Maarse, H., Ed., Marcel Dekker, New York, 1991, p. 329.
57. Shaw, P.E., Fruits II, in *Volatile Compounds in Foods and Beverages*, Maarse, H., Ed., Marcel Dekker, New York, 1991, p. 305.
58. Winterhalter, P., Fruits IV, in *Volatile Compounds in Foods and Beverages*, Maarse, H., Ed., Marcel Dekker, New York, 1991, p. 389.
59. Whitfield, F.B. and Last, J.H., Vegetables, in *Volatile Compounds in Foods and Beverages*, Maarse, H., Ed., Marcel Dekker, New York, 1991, p. 203.
60. Rouseff, R.L. and Leahy, M.M., Eds., *Fruit Flavors: Biogenesis, Characterization, and Authentication*, ACS Symposium Series 596, American Chemical Society, Washington, DC, 1995.
61. Takeoka, G.R., Buttery, R.G., and Flath, R.A., Volatile constituents of Asian pear (*Pyrus serotine*), *J. Agric. Food Chem.*, 40, 1925, 1992.
62. Guntert, M., Krammer, G., Sommer, H., and Werkhoff, P., The importance of the vacuum headspace method for the analysis of fruit flavors, in *Flavor Analysis Developments in Isolation and Characterization*, Mussinan, C.J. and Morello, M.J., Eds., ACS Symposium Series 705, American Chemical Society, Washington, DC, 1998, p. 38.
63. Rapparini, F. and Predieri, S., Pear fruit volatiles, *Hortic. Rev.*, 28, 237, 2003.
64. Cunningham, D.G., Acree, T.E., Barnard, J., Butts, R.M., and Braell, P.A. Charm analysis of apple volatiles, *Food Chem.*, 19, 137, 1986.
65. Roberts, D.D. and Acree, T.E., Developments in the isolation and characterization of β-damascenone precursors from apples, in *Fruit Flavors: Biogenesis, Characterization, and Authentication*, Rouseff, R.L. and Leahy, M.M., Eds., ACS Symposium Series 596, American Chemical Society, Washington, DC, 1995, p. 190.
66. Dirinck, P., De Pooter, H., and Schamp, N., Aroma development in ripening fruits, in *Flavor Chemistry: Trends and Developments*, Teranishi, R., Buttery, R.G., and Shahidi, F., Eds., ACS Symposium Series 388, American Chemical Society, Washington, DC, 1989, p. 23.
67. Young, H., Gilbert, J.M., Murray, S.H., and Ball, R.D., Causal effects of aroma compounds on Royal Gala apple flavours, *J. Sci. Food Agric.*, 71, 329, 1996.
68. Schumacher, K., Asche, S., Heil, M., Mittelstädt, F., Dietrich, H., and Mosandl, A., Methyl-branched flavor compounds in fresh and processed apples, *J. Agric. Food Chem.*, 46, 4496, 1998.
69. Bult, J.H.F., Schifferstein, H.N.J., Roozen, J.P., Dalmau-Boronat, E., Voragen, A.G.J., and Kroeze, J.H.A., Sensory evaluation of character-impact components in an apple model mixture, *Chem. Senses*, 27, 485, 2002.
70. Gómez, E. and Ledbetter, C.A., Development of volatile compounds during fruit maturation: Characterization of apricot and plum × apricot hybrids, *J. Sci. Food Agric.*, 74, 541, 1997.
71. Poll, L., Peterson, M.B., and Nielson, G.S., Influences of harvest year and harvest time on soluble solids, titratable acidity, anthocyanin content and aroma components of cherry (*Prunus cerasus* L. cv. "Stevnsbaer"), *Eur. Food Res. Technol.*, 216, 212, 2003.
72. Gómez, E. and Ledbetter, C.A., Comparative study of the aromatic profiles of two different plum species: *Prunus salicina* Lindl and *Prunus simonii* L., *J. Sci. Food Agric.*, 65, 111, 1994.
73. Aubert, C., Günata, Z., Ambid, C., and Baumes, R., Changes in physicochemical characteristics and volatile constituents of yellow- and white-fleshed nectarines during maturation and artificial ripening, *J. Agric. Food Chem.*, 51, 3083, 2003.

74. Rizzolo, A., Lombardi, P., Vanoli, M., and Polesello, S., Use of capillary gas chromatography sensory analysis as an additional tool for sampling technique comparison in peach aroma analysis, *J. High Resolut. Chromatogr.*, 18, 309,1995.

75. Derail, C., Hofmann, T., and Schieberle, P., Difference in key odorants of handmade juice of yellow-fleshed peaches (*Punus persica* L.) induced by the workup procedure, *J. Agric. Food Chem.*, 47, 4742, 1999.

76. Schieberle, P., Ofner, S., and Grosch, W., Evaluation of potent odorants in cucumbers (*Cucumis sativus*) and muskmelons (*Cucumis melo*) by aroma extract dilution analysis, *J. Food Sci.*, 55, 193, 1990.

77. Wyllie, S.G., Leach, D.N., Wang, Y., and Shewfelt, R.L., Key aroma compound of melons: Their development and cultivar dependence, in *Fruit Flavors: Biogenesis, Characterization, and Authentication*, Rouseff, R.L. and Leahy, M.M., Eds., ACS Symposium Series 596, American Chemical Society, Washington, DC, 1995, p. 248.

78. Wyllie, S.G., Leach, D.N., Wang, Y., and Shewfelt, R.L., Sulfur volatiles in *Cucumis melo* cv. Makdimon (Muskmelon) aroma. Sensory evaluation by gas chromatography-olfactometry, in *Sulfur Compounds in Foods*, Mussinan, C.J. and Keelan, M.E., Eds., ACS Symposium Series 564, American Chemical Society, Washington, DC, 1994, p. 36.

79. Wang, Y., Wyllie, S.G., and Leach, D., Chemical changes during the development and ripening of the fruit of *Cucumis melo* (cv. Makdimon), *J. Agric. Food Chem.*, 44, 210, 1996.

80. Jordán, M.J., Shaw, P.E., and Goodner, K.L., Volatile components in aqueous essence and fresh fruit of *Cucumis melo* cv. Athena (muskmelon) by GC-MS and GC-O, *J. Agric. Food Chem.*, 49, 5929, 2001.

81. Fallik, E., Alkali-Tuvia, S., Horev, B., Copel, A., Rodov, V., Aharoni, Y., Ulrich, D., and Schultz, H., Characterization of 'Galia' melon aroma by GC and mass spectrometric sensor measurements after prolonged storage, *Postharv. Biol. Technol.*, 22, 85, 2001.

82. Buttery, R.G., Seifert, R.M., Ling, L.C., Soderstrom, E.L., Ogawa, J.M., and Turnbaugh, J.G., Additional aroma constituents of honeydew melon, *J. Agric. Food Chem.*, 30, 1208, 1982.

83. Yajima, I., Sakakibara, H., Ide, J., Yanai, T., and Hayashi, K., Volatile flavor components of watermelon (*Citrullus vulgaris*), *Agric. Biol. Chem.*, 49, 3145, 1985.

84. Cai, M., Biogeneration of watermelon aroma compounds, M.S. thesis, Mississippi State University, Mississippi State, MS, 1997.

85. Palma-Harris, C., McFeeters, R.F., and Fleming, H.P., Solid-phase microextraction (SPME) technique for measurement of degeration of fresh cucumber flavor compounds, *J. Agric. Food Chem.*, 49, 4203, 2001.

86. Anderson, J.F., Composition of the floral odor of *Cucurbita maxima* Dechesne (Cucurbitaceae), *J. Agric. Food Chem.*, 35, 60, 1987.

87. Buttery, R.G., Teranishi, R., and Ling, L.C., Fresh tomato aroma volatiles: A quantitative study, *J. Agric. Food Chem.*, 35, 540, 1987.

88. Buttery, R.G., Teranishi, R., Flath, R.A., and Ling, L.C., Fresh tomato volatiles, in *Flavor Chemistry: Trends and Developments*, Teranishi, R., Buttery, R.G., and Shahidi, F., Eds., ACS Symposium Series 388, American Chemical Society, Washington, DC, 1989, p. 213.

89. Stern, D.J., Buttery, R.G., Teranishi, R., Ling, L., Scott, K., and Cantwell, M., Effect of tomato storage and ripening on fresh tomato quality, part I, *Food Chem.*, 49, 225, 1994.

90. Guth, H. and Grosch, W., Evaluation of important odorants in foods by dilution techniques, in *Flavor Chemistry: 30 Years of Progress,* Teranishi, R., Wick, E.L., and Hornstein, I., Eds., Klewer Academic/Plenum Publishers, New York, 1999, p. 377.

91. Mayer, F., Takeoka, G., Buttery, R., Nam, Y., Naim, M., Bezman, Y., and Rabinowitch, H., Aroma of fresh field tomatoes, in *Freshness and Shelf Life of Foods*, Cadwallader, K.R. and Weenen, H., Eds., ACS Symposium Series 836, American Chemical Society, Washington, DC, 2003, p. 144.

92. Mayer, F., Takeoka, G., Buttery, R., Whitehand, L., Bezman, Y., Naim, M., and Rabinowitch, H., Differences in the aroma of selected fresh tomato cultivars, in *Handbook of Flavor Characterization: Sensory Analysis, Chemistry, and Physiology,* Deibler, K.D. and Delwiche, J., Eds., Marcel-Dekker, New York, 2004, p. 189.

93. Honkanen, E. and Hirvi, T., The flavour of berries, in *Food Flavours, Part C: The Flavor of Fruits*, Morton, I. and MacLeod, A., Eds., Elsevier, Amsterdam, 1990, p. 125.

94. Pabst, A., Barron, D., Etiévant, P., and Schreier, P., Studies on the enzymatic hydrolysis of bound aroma constituents from raspberry fruit pulp, *J. Agric. Food Chem.,* 39, 173, 1991.

95. Borejska-Wysocki, W. and Hrazdina, G., Biosynthesis of p-hydroxyphenylbutan-3-one in raspberry fruits and tissue cultures, *Phytochemistry,* 35, 623, 1994.

96. de Ancos, B., Ibañez, E., Reglero, G., and Cano, M.P., Frozen storage effects on anthocyanins and volatile compounds of raspberry fruit, *J. Agric. Food Chem.,* 48, 873, 2000.

97. Gerasimov, A.V., Application of planar chromatography with the computer processing of chromatograms to the analysis of flavoring materials using the determination of 4-(4-hydroxyphenyl)-2-butanone (raspberry ketone) as an example, *J. Anal. Chem.,* 56, 370, 2001.

98. Klesk, K. and Qian, M., Aroma extract dilution analysis of cv. Marion (*Rubus* spp. hyb) and cv. Evergreen (*R. laciniatus* L.) blackberries, *J. Agric. Food Chem.,* 51, 3436, 2003.

99. Baloga, D.W., Vorsa, N., and Lawter, L., Dynamic headspace gas chromatography-mass spectrometry analysis of volatile flavor compounds from wild diploid blueberry species, in *Fruit Flavors: Biogenesis, Characterization, and Authentication*, Rouseff, R.L. and Leahy, M.M., Eds., ACS Symposium Series 596, American Chemical Society, Washington, DC, 1995, p. 235.

100. Di Cesare, L.F., Nani, R., Proietti, M., and Griombelli, R., Volatile composition of the fruit and juice of some blueberry cultivars grown in Italy (in Italian), *Indust. Alimet.* 38, 277, 1999.

101. Schreier, P., Quantitative composition of volatile constituents in cultivated strawberries, *Fragaria ananassa* cv. Senga Sengana, Senga Litessa and Senga Gourmella, *J. Sci. Food Agric.,* 31, 487, 1980.

102. Dirinck, P., De Pooter, H.L., Willaert, G.A., and Schamp, N.M., Flavor quality of cultivated strawberries: The role of the sulfur compounds, *J. Agric. Food Chem.,* 29, 316, 1981.

103. Hirvi, T., Mass fragmentographic and sensory analyses in the evaluation of the aroma of some strawberries varieties, *Lebesm.-Wiss. Technol.,* 16, 157, 1983.

104. Larsen, A.C. and Poll, L., Odour thresholds of some important aroma compounds in strawberries, *Z. Lebemsm. Unters. Forsch.,* 195, 120, 1992.

105. Pérez, A.G., Rios, J.J., Sanz, C., and Olías, J.M., Aroma components and free amino acids in the strawberry variety Chandler during ripening, *J. Agric. Food Chem.,* 40, 2232, 1992.

106. Schieberle, P., Heat-induced changes in the most odour-active volatiles of strawberries, in *Trends in Flavour Research*, Maarse, H. and Van der Heij, D.J., Eds., Elsevier, Amsterdam, 1994, p. 345.

107. Ulrich, D., Rapp, A., and Hoberg, E., Analyse des erdbeeraromas — Quantifizierung der flüchtigen komponenten in kulturerdbeervarietäten und der walderdbeere, *Z. Lebesm. Unters. Forsch.*, 200, 217, 1995.

108. Ulrich, D., Hoberg, E., Rapp, A., and Kecke, S. Analysis of strawberry flavour – discrimination of aroma types by quantification of volatile compounds, *Z. Lebesm. Unters. Forsch. A*, 205, 218, 1997.

109. Gomes da Silva, M.D.R. and Chaves das Neves, H.J., Complementary use of hyphenated purge-and-trap gas chromatography techniques and sensory analysis in the aroma profiling of strawberry (*Fragaria ananassa*), *J. Agric. Food Chem.*, 47, 4568, 1999.

110. Hakala, M.A., Lapveteläinen, A.T., and Kallio, H., Volatile compounds of selected strawberry varieties analyzed by purge-and-trap headspace GC-MS, *J. Agric. Food Chem.*, 50, 1133, 2002.

111. Urruty, L., Giraudel, J.-L., Lek, S., Roudeillac, P., and Montury, M., Assessment of strawberry aroma through SPME/GC and ANN methods: Classification and discrimination of varieties, *J. Agric. Food Chem.*, 50, 3129, 2002.

112. Nagy, S. and Shaw, P.E., Factors affecting flavor of citrus fruits, in *Food Flavors, Part C: The Flavor of Fruits,* Morton, I.D. and MacLeod, A.J., Eds., Elsevier, Amsterdam, 1990, p. 93.

113. Hinterholzer, A. and Schieberle, P., Identification of the most odor-active volatiles in fresh, hand-extracted juice of Valencia late oranges by odor dilution techniques, *Flavour Fragrance J.*, 13, 49, 1998.

114. Buettner, A. and Schieberle, P., Evaluation of aroma differences between hand-squeezed juices from Valencia late and naval oranges by quantitation of key odorants and flavor reconstitution experiments, *J. Agric. Food Chem.*, 49, 2387, 2001.

115. Wilson, C.W., III and Shaw, P.E., Importance of thymol, methyl N-methylanthranilate, and monoterpene hydrocarbons to the aroma and flavor of mandarin cold-pressed oils, *J. Agric. Food Chem.*, 29, 494, 1981.

116. Schieberle, P., Mestres, M., and Buettner, A., Characterization of aroma compounds in fresh and processed mandarin oranges, in *Freshness and Shelf Life of Foods*, Cadwallader, K.R. and Weenen, H., Eds., ACS Symposium Series 836, American Chemical Society, Washington, DC, 2003, p. 162.

117. Clark, B.C., Jr., Chamblee, T.S., and Iacobucci, G.A., HPLC isolation of the sesquiterpene hydrocarbon germacrene B from lime peel oil and its characterization as an important flavor impact constituent, *J. Agric. Food Chem.*, 35, 514, 1987.

118. Chisholm, M.G., Wilson, M.A., Gaskey, G.M., Jell, J.A., and Cass, D.M., Jr., The identification of aroma compounds in key lime oil using solid-phase microextraction and gas chromatography-olfactometry, in *Gas Chromatography-Olfactometry: The State of the Art, Leland*, J.V., Schieberle, P., Buettner, A., and Acree, T.E., Eds., ACS Symposium Series 782, American Chemical Society, Washington, DC, 2001, p. 100.

119. Mussinan, C.J., Mookherjee, B.D., and Malcolm, G.I., Isolation and identification of the volatile constituents of fresh lemon juice, in *Essential Oils*, Mookherjee, B.D. and Mussinan, C.J., Eds., Allured, Wheaton, IL, 1981, p. 199.

120. Sinyinda, S. and Gramshaw, J.W., Volatiles of avocado fruit, *Food Chem.*, 62, 483, 1998.

121. Pino, J.A., Rosado, A., and Aguero, J., Volatile components of avocado (*Persea americana* Mill.) fruits, *J. Essent. Oil Res.*, 12, 377, 2000.

122. Vernin, G., Vernin, C., Pieribattesti, J.C., and Roque, C., Analysis of the volatile compounds of *Psidium cattleianum* Sabine fruit from Reunion Island, *J. Essent. Oil Res.*, 10, 353, 1998.

123. Pino, J.A., Ortega, A., and Rosado, A., Volatile constituents of guava (*Psidium guajava* L.) fruits from Cuba, *J. Essent. Oil Res.*, 11, 623, 1999.

124. Yen, G.-C. and Lin, H.-T., Changes in volatile flavor components of guava juice with high-pressure treatment and heat processing and during storage, *J. Agric. Food Chem.*, 47, 2082, 1999.

125. Paniandy, J.C., Chane-Ming, J., and Pieribattesti, J.C., Chemical composition of the essential oil and headspace solid-phase microextraction of the guava fruit (*Psidium guajava* L.), *J. Essent. Oil Res.*, 12, 153, 2000.

126. Pino, J.A., Marbot, R., and Vázquez, C., Characterization of volatiles in strawberry guava (*Psidium cattleianum* Sabine) fruit, *J. Agric. Food Chem.*, 49, 5883, 2001.

127. Pino, J.A., Marbot, R., and Vázquez, C., Characterization of volatiles in Costa Rican guava [*Psidium friedrichsthalianum* (Berg) Niedenzu] fruit, *J. Agric. Food Chem.*, 50, 6023, 2002.

128. Jordán, M.J., Margaría, C.A., Shaw, P.E., and Goodner, K.L., Volatile components and aroma active compounds in aqueous essence and fresh pink guava fruit puree (*Psidium guajava* L.) by GC-MS and multidimensional GC/GC-O, *J. Agric. Food Chem.*, 51, 1421, 2003.

129. Koulibaly, A., Sakho, M., and Crouzet, J., Variability of free and bound volatile terpenic compounds in mango, *Lebesm.-Wiss. Technol.*, 25, 374, 1992.

130. Ollé, D., Baumes, R.L., Bayonove, C.L., Lozano, Y.F., Sznaper, C., and Brillouet, J.-M., Comparison of free and glycosidically linked volatile components from poly-embryonic and monoembryonic mango (*Mangifera indica* L.) cultivars, *J. Agric. Food Chem.*, 46, 1094, 1998.

131. Lalel, H.J.D., Singh, Z., and Tan, S.C., Aroma volatiles production during fruit ripening of 'Kensington Pride' mango, *Postharv. Biol. Technol.*, 27, 323, 2003.

132. Werkhoff, P., Güntert, M., Krammer, G., Sommer, H., and Kaulen, J., Vacuum head-space method in aroma research: Flavor chemistry of yellow passion fruits, *J. Agric. Food Chem.*, 46, 1076, 1998.

133. Jordán, M.J., Goodner, K.L., and Shaw, P.E., Volatile components in tropical fruit essences: Yellow passion fruit (*Passiflora adulis* Sims f. *flavicarpa* Degner) and banana (*Musa sapientum* L.), *Proc. Fla. State Hortic. Soc.*, 113, 284, 2000.

134. Young, H. and Paterson, V.J., Characterisation of bound flavour components in kiwi-fruit, *J. Sci. Food Agric.*, 68, 257, 1995.

135. Jordán, M.J., Margaria, C.A., Shaw, P.E., and Goodner, K.L., Aroma active compo-nents in aqueous kiwi fruit essence and kiwi fruit puree by GC-MS and multidimen-sional GC/GC-O, *J. Agric. Food Chem.*, 50, 5386, 2002.

136. Talens, P., Escriche, I., Martinez-Navarrete, N., and Chiralt, A., Influence of osmotic dehydration and freezing on the volatile profile of kiwi fruit, *Food Res. Int.*, 36, 635, 2003.

137. Cosio, R. and Rene, F., Volatile compounds from bananas: Comparative study of two extraction methods [French], *Sci. Aliments*, 16, 383, 1996.

138. Wyllie, S.G. and Fellman, J.K., Formation of volatile branched esters in banana (*Musa sapientum* L.), *J. Agric. Food Chem.*, 48, 3493, 2000.

139. Liu, T.-T. and Yang, T.-S., Optimization of solid-phase microextraction analysis for studying change of headspace flavor compounds in banana during ripening, *J. Agric. Food Chem.*, 50, 653, 2002.

140. Jordán, M.J., Tandon, K., Shaw, P.E., and Goodner, K.L., Aromatic profile of aqueous banana essence and banana fruit by gas chromatography-mass spectrometry (GC-MS) and gas chromatography-olfactometry (GC-O), *J. Agric. Food Chem.*, 49, 4813, 2001.

141. Umano, K., Hagi, Y., Nakahara, K., Shoji, A., and Shibamoto, T., Volatile constituents of green and ripened pineapple (*Ananas comosus* [L] Merr.), *J. Agric. Food Chem.*, 40, 599, 1992.

142. Takeoka, G.R., Buttery, R.G., Teranishi, R., Flath, R.A., and Guntert, M., Identification of additional pineapple volatiles, *J. Agric. Food Chem.*, 39, 1848, 1991.

143. Teai, T., Claude-Lafontaine, A., Schippa, C., and Cozzolino, F., Volatile compounds in fresh pulp of pineapple (*Ananas comosus* [L] Merr.) from French Polynesia, *J. Essent. Oil Res.*, 13, 314, 2001.

144. Murray, K.E. and Whitfield, F.B., The occurrence of 3-alkyl-2-methoxypyrazines in raw vegetables, *J. Sci. Food Agric.*, 26, 973, 1975.

145. Charron, C., Cantliffe, D.J., Wheeler, R.M., Manukian, A., and Heath, R.R., Photosynthetic photon flux, photoperiod, and temperature effects on emissions of (Z)-3-hexenal, (Z)-3-hexenol, and (Z)-3-hexenyl acetate from lettuce, *J. Am. Soc. Hortic. Sci.*, 121, 488, 1996.

146. Ludwiczuk, A., Najda, A., Wolski, T., and Baj, T., Chromatographic determination of the content and composition of extracts and essential oils from the fruits of three varieties of stalk celery (*Apium gaveolens* L. var. dulce Mill. Pers.), *J. Planar Chromatogr.-Modern Tlc.*, 14, 400, 2001.

147. MacLeod, A.J., MacLeod, G., and Subramanian, G., Volatile aroma constituents of celery, *Phytochemistry*, 27, 373, 1988.

148. Uhlig, J.W., Chang, A., and Jen, J.J., Effect of phthalides on celery flavor, *J. Food Sci.*, 52, 658, 1987.

149. Bartschat, D., Beck, T., and Mosandl, A., Stereoisomeric flavor compounds. 79. Simultaneous enantiomeric selective analysis of 3-butylphthalide and 3-butylhexahydrophthalide stereoisomers in celery, celeriac, and fennel, *J. Agric. Food Chem.*, 45, 4554, 1997.

150. Kjeldsen, F., Christensen, L.P., and Edelenbos, M., Changes in volatile compounds of carrots (*Daucus carota* L.) during refrigerated and frozen storage, *J. Agric. Food Chem.*, 51, 5400, 2003.

151. Kjeldsen, F., Christensen, L.P., and Edelenbos, M., Quantitative analysis of aroma compounds in carrot (*Daucus carota* L.) cultivars by capillary gas chromatography using large volume injection technique, *J. Agric. Food Chem.*, 49, 4342, 2001.

152. Buttery, R.G., Seifert, R.M., Guadagni, D.G., Black, D.R., and Ling, L.C., Characterization of some volatile constituents of carrots, *J. Agric. Food Chem.*, 16, 1009, 1968.

153. Seifert, R.M. and Buttery, R.G., Characterization of some previously unidentified sesquiterpenes in carrot roots, *J. Agric. Food Chem.*, 26, 181, 1979.

154. Schnitzler, W.H., Broda, S., and Schaller, R.G., Screening for important monoterpenes in carrots (*Daucus carota* L. ssp. sativa) by headspace gas-chromatography-olfactometry, *Angew. Bot.*, 77, 53, 2003.

155. Buttery, R.G., Seifert, R.M., Guadagni, D.G., and Ling, L.C., Characterization of some volatile constituents of bell peppers, *J. Agric. Food Chem.*, 17, 1322, 1969.

156. Luning, P.A., Yuksel, D., and Roosen, J.P., Sensory attributes of bell peppers (*Capsicum annum*) correlated with the composition of volatile compounds, in *Trends in Flavour Research*, Maarse, H. and Van der Heij, Eds., Elsevier, Amsterdam, 1994, p. 241.

157. Kallio, H., Raimoaho, P., and Virtalaine, T., Emission of blanched broccoli volatiles in headspace during cooking, in *Flavor Chemistry of Ethnic Foods*, Shahidi, F. and Ho, C.-T., Eds., Kluwer, New York, 1999, p. 111.

158. Obenland, D.M. and Aung, L.H., Cysteine lyase activity and anaerobically-induced sulfur gas emission from broccoli florets, *Phyton*, 58, 147, 1996.

159. Kubec, R., Drhova, V., and Velisek, J., Thermal degradation of S-methylcystein and its sulfoxide – important flavor precursors of *Brassica* and *Allium* vegetables, *J. Agric. Food Chem.*, 46, 4334, 1998.

160. Ueda, Y., Tsubuku, T., and Miyajima, R., Composition of sulfur-containing components in onion and their flavor characters, *Biosci. Biotechnol. Biochem.*, 58, 108, 1994.

161. Randle, W.M., Onion flavor chemistry and factors influencing flavor intensity, in *Spices: Flavor Chemistry and Antioxidant Properties*, Risch, S.J. and Ho, C.-T., Eds., ACS Symposium Series 660, American Chemical Society, Washington, DC, 1996, p. 41.

162. Yoo, K.S. and Pike, L.M., Determination of flavor precursor compound S-alk(en)yl-L-cysteine sulfoxides by an HPLC method and their distribution in *Allium* species, *Sci. Hortic.*, 75, 1, 1998.

163. Morton, I.D. and MacLeod, A.J., Eds., *Food Flavours, Part C. The Flavors of Fruits*, Elsevier, Amsterdam, 1994, p. 19.

164. DeEll, J.R., Prange, R.K., and Peppelenbos, H.W., Postharvest physiology of fresh fruits and vegetables, in *Handbook of Postharvest Technology: Cereals, Fruits, Vegetables, Tea, and Spices*, Chakraverty, A., Mujumdar, A.S., Raghavan, G.S.V., and Ramaswamy, H.S., Eds., Marcel Dekker, New York, 2003, p. 455.

165. Jongen, W., Ed., *Fruit and Vegetable Processing: Improving Quality*, CRC Press, Boca Raton, FL, 2002.

166. Dixon, J. and Hewett, E.W., Factors affecting apple aroma/flavor volatile concentration: A review, *N. Z. J. Crop Hortic. Sci.*, 28, 155, 2000.

167. Rennie, T.J., Vigneault, C., DeEll, J., and Raghavan, G.S.V., Cooling and storage, in *Handbook of Postharvest Technology: Cereals, Fruits, Vegetables, Tea, and Spices*, Chakraverty, A., Mujumdar, A.S., Raghavan, G.S.V., and Ramaswamy, H.S., Eds., Marcel Dekker, New York, 2003, p. 505.

168. Hobson, G.E. and Davies, J.N., The tomato, in *The Biochemistry of Fruit and Their Products*, Vol. 2, Hulme, A.C., Ed., Academic Press, New York, 1971, Chap. 13.

169. Craft, C.C. and Heintze, P.H., Physiological studies of mature green tomatoes in storage, *Proc. Am. Soc. Hortic. Sci.*, 64, 343, 1954.

170. Kader, A.A., Morris, L.L., Stevens, M.A., and Albright-Holten, M., Composition and flavor quality of fresh market tomatoes as influenced by some postharvest handling procedures, *J. Am. Soc. Hortic. Sci.*, 103, 6, 1978.

171. Lurie, S. and Klein, J.D., Acquisition of low-temperature tolerance in tomatoes by exposure to high-termperature stress, *J. Am. Soc. Hortic. Sci.*, 116, 1007, 1991.

172. Saltveit, M.E., Effect of ethylene on quality of fresh fruits and vegetables, *Postharv. Biol. Technol.*, 15, 279, 1999.

173. Watada, A.E., Aulenbach, B.B., and Worthington, J.T., Vitamins A and C in ripe tomatoes as affected by stage of ripeness at harvest and by supplementary ethylene, *J. Food Sci.*, 41, 856, 1976.

174. McDonald, R.E., McCollum, T.G., and Baldwin, E.A., Prestorage heat treatments influence free sterols and flavor volatiles in tomatoes stored at chilling temperature, *J. Am. Soc. Hortic. Sci.*, 121, 531, 1996.

175. Scriven, F.M., Gek, C.O., and Wills, R.B.H., Sensory differences between bananas ripened without and with ethylene, *HortScience*, 24, 983, 1989.

176. Thompson, A.K., *Controlled Atmosphere Storage of Fruits and Vegetables*, Blackwell Science, Berlin, 1998.
177. Kader, A.A., Zagory, D., and Kerbel, E.L., Modified atmosphere packaging of fruits and vegetables, *Crit. Rev. Food Sci. Nutr.*, 28, 1, 1989.
178. Brackman, A., Streif, J., and Bangerth, F., Relationship between a reduced aroma production and lipid metabolism of apples after long-term controlled-atmosphere storage, *J. Am. Soc. Hortic. Sci.*, 118, 243, 1993.
179. Yahia, E.M., Liu, F.W., and Acree, T.E., Changes of some odour-active volatiles in low-ethylene controlled atmosphere stored apples, *Lebensm.-Wiss. Technol.*, 24, 145, 1991.
180. Willaert, G.A., Dirinck, P.J., De Pooter, H.L., and Schamp, N.N., Objective measurement of aroma quality of Golden Delicious apples as a function of controlled-atmosphere storage time, *J. Agric. Food Chem.*, 31, 809, 1983.
181. Knee, M. and Hatfield, S.G.S., The metabolism of alcohols by apple fruit tissue, *J. Sci. Food Agric.*, 32, 593, 1981.
182. Guadagni, D.G., Bomben, J.L., and Hudson, J.S., Factors influencing the development of aroma in apple peels, *J. Sci. Food Agric.*, 22, 110, 1971.
183. Spencer, K.C. and Humphreys, D.J., Argon packaging and processing preserves and enhances flavor, freshness, and shelf-life of foods, in *Freshness and Shelf Life of Foods*, Cadwallader, K.R. and Weenen, H., Eds., ACS Symposium Series 836, American Chemical Society, Washington, DC, p. 270.
184. Pérez, A.G. and Sanz, C., Effect of high-oxygen and high-carbon-dioxide atmospheres on strawberry flavor and other traits, *J. Agric. Food Chem.*, 49, 2370, 2001.
185. Mazza, G. and Jayas, D.S., Controlled and modified atmosphere storage, in *Food Shelf Life Stability: Chemical, Biochemical, and Microbiological Changes*, Eskin, N.A.M. and Robinson, D.S., Eds., CRC Press, New York, 2001, p. 149.
186. Day, B., Novel MAP for fresh prepared produce, *Eur. Food Drink Rev.*, Spring, 73, 1996.
187. Portela, S.I. and Cantwell, M.J., Quality changes of minimally processed honeydew melons stored in air or controlled atmosphere, *Postharv. Biol. Technol.*, 14, 351, 1998.
188. Garrett, E., Fresh-cut produce: Tracks and trends, in *Fresh-Cut Fruits and Vegetables: Science, Technology, and Market*, Lamikanra, O., Ed., CRC Press, New York, 2002, p. 1.
189. Brocklehurst, T.F., Delicatessen salads and chilled prepared fruit and vegetable products, in *Shelf Life Evaluation of Foods*, Man, C.M.D. and Jones, A.A., Eds., Aspen Publishers, Gaithersburg, MD, 1999, p. 87.
190. Lamikanra, O., Ed., *Fresh-Cut Fruits and Vegetables: Science, Technology, and Market*, CRC Press, New York, 2002.
191. Priepke, P.E., Wei, L.S., and Nelson, A.I., Refrigerated storage of prepackaged salad vegetables, *J. Food Sci.*, 41, 379, 1976.
192. Williams, A.P., Chilled combined foods and chilled meals, in *Chilled Foods. The State of the Art*, Gormley, T.R., Ed., Elsevier Applied Science, New York, 1990, p. 225.
193. Carlin, F., Nguyen-The, C., Chambroy, Y., and Reich, M., Effects of controlled atmospheres on microbial spoilage, electrolyte leakage and sugar content of fresh 'ready-to-use' grated carrots, *Int. J. Food Sci. Technol.*, 25, 110, 1990.
194. Lamikanra, O. and Richard, O.A., Effect of storage on some volatile aroma compounds in fresh-cut cantaloupe melon, *J. Agric. Food Chem.*, 50, 4043, 2002.
195. Beaulieu, J.C. and Baldwin, E.A., Flavor and aroma of fresh-cut fruits and vegetables, in *Fresh-Cut Fruits and Vegetables: Science, Technology, and Market*, Lamikanra, O., Ed., CRC Press, New York, 2002, p. 391.

196. Fan, X. and Mattheis, J.P., Impact of 1-methylcyclopropene and methyl jasmonate on apple volatile production, *J. Agric. Food Chem.*, 47, 2847, 1999.
197. Morris, J.R., Cawthon, D.L., and Buescher, R.W., Effect of acetaldehyde on postharvest quality of mechanically harvested strawberries for processing, *J. Am. Hortic. Soc.*, 104, 262, 1979.
198. Pesis, E., Zauberman, G., and Avissar, I., Induction of certain volatiles in feijoa fruit by postharvest application of acetaldehyde or anaerobic conditions, *J. Sci. Food Agric.*, 54, 329, 1991.
199. Pesis E. and Avissar, I., Effect of postharvest application of acetaldehyde vapour on strawberry decay, taste and certain volatiles, *J. Sci. Food Agric.*, 52, 377, 1990.
200. Davis, P.L., Chace, W.G., Jr., and Cubbedgem R.H., Factors affecting internal O_2 and CO_2 concentratons of citrus fruits, *HortScience*, 2, 168, 1967.
201. Baldwin, E.A., Nisperos-Carriedo, M., Shaw, P.E., and Burns, J.K., Effects of coatings and prolonged storage conditions on fresh orange flavor volatiles, degree brix, and ascorbic acid levels, *J. Agric. Food Chem.*, 43, 1321, 1995.
202. Ben Yehoshua, S., Transpiration, water stress and gas exchange, in *Postharvest Physiology of Vegetables*, Weichman, J., Ed., Marcel Dekker, New York, p. 113.
203. Hamann, D.D., Miller, N.C., and Purcell, A.E., Effects of curing on the flavor and texture of baked sweet potatoes, *J. Food Sci.*, 45, 992, 1980.
204. Sun, J.-B., Severson, R.F., Scholtzhauer, W.S., and Kays, S.J., Identifying critical volatiles in the flavor of baked 'Jewel' sweetpotatoes [*Ipomoea batatas* (L.) Lam.], *J. Am. Soc. Hortic. Sci.*, 120, 468, 1995.
205. Blank, G. and Cumming, R., Irradiation, in *Food Shelf Life Stability: Chemical, Biochemical, and Microbiological Changes*, Eskin, N.A.M. and Robinson, D.S., Eds., CRC Press, New York, 2001, p. 87.
206. Kobayashi, A., Itagaki, R., Tokitomo, Y, and Kubota, K., Changes of aroma character of irradiated onion during storage, *Nippon Shokuhin Kogyo Gakkaishi*, 41, 682, 1994.
207. Lacroix, M., Marcotte, M., and Ramaswamy, H., Irradiation of fruits, vegetables, nuts and spices, in *Handbook of Postharvest Technology: Cereals, Fruits, Vegetables, Tea, and Spices*, Chakraverty, A., Mujumdar, A.S., Raghavan, G.S.V., and Ramaswamy, H.S., Eds., Marcel Dekker, New York, 2003, p. 623.
208. Whitfield, F.B., Food off-flavours: Cause and effect, in *Developments in Food Flavours*, Birch, G.G. and Lindley, M.G., Eds., Elsevier Applied Science, New York, 1986, p. 249.
209. Bolin, H.R. and Huxsoll, C.C., Effect of preparation procedures and storage parameters on quality retention of salad-cut lettuce, *J. Food Sci.*, 56, 60, 1991.
210. Carlin, F., Nguyen-The, C., Chambroy, Y., and Reich, M., Microbiological spoilage of 'ready-to-use' grated carrot, *Sci. Aliments*, 9, 339, 1989.
211. Perata, P. and Alpi, A., Plant responses to anaerobiosis, *Plant Sci.*, 93, 1, 1993.
212. Richardson, D.G. and Kosittrakun, M., Off-flavor development of apples, pears, berries, and plums under anaerobiosis and partial reversal in air, in *Fruit Flavors: Biogenesis, Characterization, and Authentication*, Rouseff, R.L. and Leahy, M.M., Eds., ACS Symposium Series 596, American Chemical Society, Washington, DC, 1995, p. 211.
213. Mattheis, J.P., Buchanan, D.A., and Fellman, J.K., Change in apple fruit volatiles after storage at atmospheres inducing anaerobic metabolism, *J. Agric. Food Chem.*, 39, 1602, 1991.
214. Rudell, D.R., Mattinson, D.S., Mattheis, J.P., Wyllie, S.G., and Fellman, J.K., Investigations of aroma volatile biosynthesis under anoxic conditions and in different tissues of "Redchief Delicious" apple fruit (*Malus domestica* Borkh.), *J. Agric. Food Chem.*, 50, 2627, 2002.

215. Shaw, P.E., Moshonas, M.G., and Pesis, E., Changes during storage of oranges pretreated with nitrogen, carbon dioxide and acetaldehyde in air., *J. Food Sci.*, 56, 469, 1991.

216. Lipton, W.J. and Harris, C.M., Controlled atmosphere effects on market quality of stored broccoli (*Brassica aleracea* L. Itilica group), *J. Am. Soc. Hortic. Sci.*, 99, 200, 1974.

217. Forney, C.F., Mattheis, J.P., and Austin, R.K., Volatile compounds produced by broccoli under anaerobic conditions, *J. Agric. Food Chem.*, 39, 2257, 1991.

218. Hansen, M., Buttery, R.G., Stern, D.J., Cantwell, M.I., and Ling, L., Broccoli stored under low-oxygen atmosphere: Identification of higher boiling volatiles, *J. Agric. Food Chem.*, 40, 850, 1992.

219. Derbali, E., Makhlouf, J., and Vezina, L.-P., Biosynthsis of sulfur volatile compounds in broccoli seedlings stored under anaerobic conditions, *Postharv. Biol. Technol.*, 13, 191, 1998.

220. Yamane, A., Yamane, A., and Shibamoto, T., Propanethiol S-oxide content in scallions (*Allium fistulosum* L. variety Caespitosum) as a possible marker for freshness during cold storage, *J. Agric. Food Chem.*, 40, 1010, 1994.

221. Shattuck, V.I., Yada, R., and Lougheed, E.C., Ethylene-induced bitterness in stored parsnips, *HortScience*, 23, 912, 1988.

222. Lafuente, M.T., Lopez-Galvez, G., Cantwell, M., and Yang, S.F., Factors influencing ethylene-induced isocoumarin formation and increased respiration in carrots, *J. Am. Soc. Hortic. Sci.*, 121, 537, 1996.

223. Ke, D. and Saltveit, M.E., Wound-induced ethylene production, phenolic metabolism and susceptibility to russet spotting in iceberg lettuce, *Phys. Plant.*, 76, 412, 1989.

224. Buechat, L., Surface disinfection of raw produce, *Diary Food Environ. Sanitation*, 12, 6, 1992.

225. Delaquis, P.J., Stewart, S., Cliff, M., Toivonen, P.M., and Moyls, A.L., Sensory quality of ready-to-eat lettuce washed in warm, chlorinated water, *J. Food Qual.*, 23, 553, 2000.

226. Eckert, J.W. and Ogawa, J.M., Recent developments in the chemical control of postharvest disease, *Acta Hortic.*, 269, 477, 1990.

227. Mazza, G. and Pietrzak, E.M., Headspace volatiles and sensory characteristics of earthy, musty flavoured potatoes, *Food Chem.*, 36, 97, 1990.

228. Acree, T.E., Lee, C.Y., Butts, R.M., and Barnard, J., Geosmin, the earthy component of table beet odor, *J. Agric. Food Chem.*, 24, 430, 1976.

229. Buttery, R.G., Ling, L.C., and Chan, B.G. J., Volatiles of corn kernels and husks: Possible corn ear worm attactants, *Agric. Food Chem.*, 26, 866, 1978.

230. Buttery, R.G. and Ling, L.C., Earthy aroma of potatoes, *J. Agric. Food Chem.*, 21, 745, 1973.

231. Murray, K.E., Shipton, J., and Whitfield, F.B., 2-Methoxypyrazines and the flavor of green peas (*Pisum sativum*), *Chem. Ind. (Lond.)*, 897, 1970.

232. Waterer, D.R. and Pritchard, M.K., Monitoring of volatiles: A technique for detection of soft rot (*Erwinia carotovora*) in potato tubers, *Can. J. Plant Pathol.*, 6, 165, 1984.

233. Waterer, D.R. and Pritchard, M.K., Volatile monitoring as a technique for differentiating between *E. carotovora* and *C. sepedonium* infections in stored potatoes, *Am. Potato J.*, 61, 345, 1984.

234. Alvo, P., Dodds, G., Raghavan, G.S.V., and Kushhalappa, A.C., Potential applications of volatile monitoring in storage, in *Handbook of Postharvest Technology: Cereals, Fruits, Vegetables, Tea, and Spices*, Chakraverty, A., Mujumdar, A.S., Raghavan, G.S.V., and Ramaswamy, H.S., Eds., Marcel Dekker, New York, 2003, p. 585.

235. Varns, J.L. and Glynn, M.T., Detection of disease in stored potatoes by volatile monitoring, *Am. Potato J.*, 56, 185, 1979.
236. Whitfield, F.B., Last, J.H., and Tindale, C.R., Skatole, indole, and p-cresol: Components in off-flavored frozen french fries, *Chem. Industr. (Lond.)*, 4662, 1982.

7 Produce Color and Appearance

Henryk Daun
Food Science Department, Cook College, Rutgers University,
New Brunswick, NJ

CONTENTS

7.1   IMPORTANCE OF PRODUCE APPEARANCE

The term *produce* is used to describe all plant farm products. Appearance as applied to produce may be defined as the look or outward aspect of these products. The importance of produce appearance and its relationship to the perception of quality cannot be overemphasized. It may be used to make judgments related to the degree of plant development, deviation from normal growth, ripening, and deterioration.

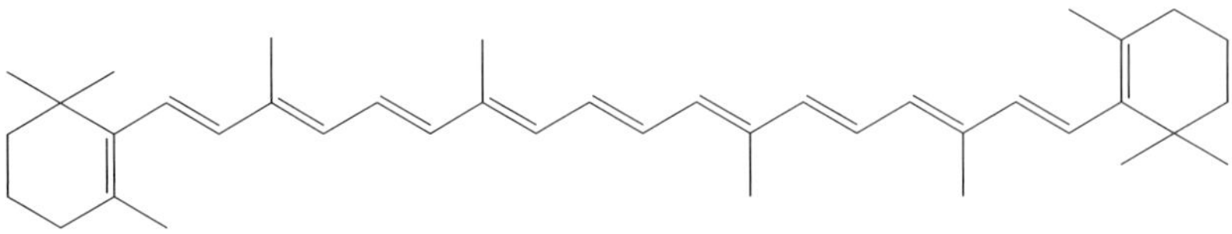

FIGURE 7.1 β-Carotene.

Generally, appearance includes geometric and color characteristics. The determination of appearance may be made by visual observation and instrumental measurements. Detailed general principles and discussion of appearance and its measurement may be found in the classic monograph by Hunter (1975). Additional concepts related to appearance were advanced by Hutchings (1999). Due to the large number of attributes and their interdependence in the case of plant products it would be not practical to attempt a comprehensive appearance description. Only attributes that are critical to quality should be selected and determined. However, it should be kept in mind that consideration of produce color alone without appropriate attention to other attributes of appearance such as size, shape, surface characteristics, and others might lead to superficial, incomplete, and erroneous judgments. In this chapter the emphasis will be on chemical aspects of produce color. The major classes of compounds contributing to the color of fruits and vegetables include carotenoids, chlorophylls, benzopyran derivatives, and betalains.

7.2 CAROTENOIDS

The carotenoids are yellow, orange, and red pigments widely distributed in nature. Most of the more than 100 million tons of annual production of these compounds (Isler, 1971) takes place in the oceans via synthesis by seaweeds and algae. On land, carotenoids are formed by plants and almost always occur together with chlorophylls. The importance of these compounds comes from their multifarious functions, including light energy absorption, protection against photosensitized oxidation, and oxygen transport. Carotenoids are also singlet oxygen quenchers and regulators of plant growth. A separate area of interest is the role of these compounds in nutrition, promoting health and disease prevention. Some carotenoids are vitamin A precursors. The provitamin A activity remains the most important function of a selected group of carotenoids, especially beta-carotene (see Figure 7.1). Vitamin A deficiency is still a cause of blindness, premature death, and other health problems in numerous underdeveloped countries. In recent years attention of many researchers has been focused on carotenoids as compounds preventing development of cancer and heart diseases. An important aspect of carotenoids is their contribution to the aesthetic value of food. In produce, carotenoids make up a class of pigments contributing to the beautiful yellow to red spectrum of colors stepwise synthesized during the development stages of fruits and vegetables, culminating in ripening, and degrading during senescence. Genetic nature, environmental conditions, and agricultural cultivation practices significantly influence the qualitative and quantitative composition of carotenoids in produce. Postharvest treatments, packaging, industrial processing, and storage affect retention and changes of these compounds in food products.

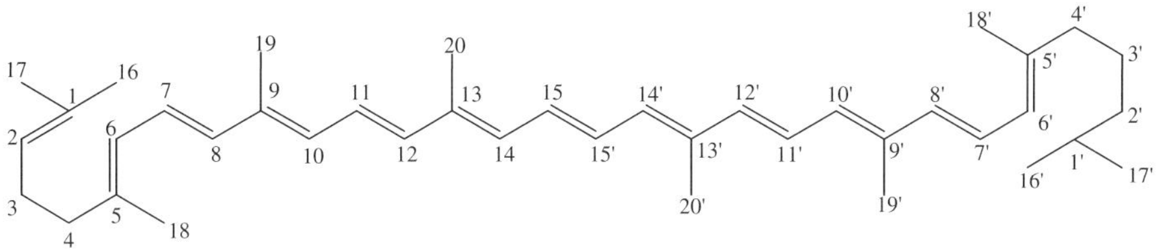

FIGURE 7.2 Lycopene.

Numerous carotenoid preparations derived from both natural sources (annatto, saffron, tomato, and paprika) and synthetic compounds (beta-carotene, canthaxanthin, astaxanthin, beta-apo-8-carotenal, and beta-apo-8-carotenoic acid ethyl ester) are available commercially as food colorants. Animals do not have the ability to synthesize carotenoids and depend on ingestion of these compounds from feed. Subsequent transformation of carotenoids in animals may lead to the formation of characteristic body pigments.

7.2.1 CHEMICAL DEFINITION AND STRUCTURE

According to the tentative rules of the International Union of Pure and Applied Chemistry (IUPAC) Commission on Nomenclature of Organic Chemistry,

> carotenoids are a class of hydrocarbons (carotenes) and their oxygenated derivatives (xanthophylls) consisting of eight isoprenoid units. These units are joint in such a manner that the arrangement is reversed at the center of the molecule, such that the two central methyl groups are in a 1,6 positional relationship and the remaining non-terminal methyl groups are in a 1,5 positional relationship (IUPAC, 1971).

This definition was affirmed in the rules for the nomenclature of carotenoids published in 1975 (IUPAC, 1975). These 13 rules include the definition of the class of compounds, stem name "carotene," specific names and end group designations, numbering of carotenoid hydrocarbons, definitions of norcarotenoids and secocarotenoids, changes in hydrogenation level, oxygenated derivatives, numbering of oxygenated derivatives, *retro* nomenclature, Apo nomenclature, higher carotenoids, stereochemistry, and trivial names.

Numerous definitions of carotenoids found in literature are derived from the classic definition published by Karrer and Tucker (1950). Traditional names of carotenoids are frequently used to avoid the relative complexity of IUPAC's systematic nomenclature. It is desirable that in scientific publications traditional names be accompanied by systematic names. The structure of lycopene, shown on Figure 7.2, is commonly used to formally derive structures of other carotenoids. Numbering of carbons in the lycopene formula is used for all carotenoids. Almost all of over 600 naturally occurring carotenoids may be shown as compounds deduced from lycopene by one or more of the following: hydrogenation, dehydrogenation, cyclization, and oxidation. They are soluble in lipids and not soluble in water, with the exception of some that have strong polar groups such as norbixin with two carboxylic groups in its molecule.

7.2.2 OCCURRENCE AND BIOSYNTHESIS

Detailed descriptions of the content of carotenoids in various species of fruits and vegetables may be found in several monographs (Karrer and Tucker, 1950; Goodwin, 1976; Gross, 1991). Goodwin and Goad (1970) proposed the following seven patterns of carotenoid distribution, which frequently merge into one another: (1) insignificant amounts of carotenoids; (2) small amounts found in chloroplasts, mainly beta-carotene, lutein, violaxanthin, and neoxanthin; (3) comparatively large amounts of acyclic lycopene and partly saturated related compounds; (4) relatively large amounts of beta-carotene and its derivatives, cryptoxanthin and zeaxanthin; (5) unusually large amounts of epoxides; (6) unique carotenoids (e.g., capsanthin); and (7) poly-*cis*-carotenoids.

From the point of view of aesthetic quality of the produce, carotenoid synthesis is especially important during ripening. At this stage the formation of carotenoids accelerates by chloroplasts changing to chromoplasts with concurrent degradation of chlorophyll. The synthesis of carotenoids may be summarized in the following manner. Initially, mevalonic acid is formed. The next steps involve formation of geranylgeranyl and phytoene and desaturation of phytoene to lycopene. Cyclization of the end groups of lycopene leads to the formation of cyclic carotenes. Further steps may include incorporation of oxygen with generation of xanthophylls (Gross, 1991). The final stages of biosynthesis leading to structural modifications involve complex enzymatic transformations that are not fully understood. The end result of these reactions is the high carotenoid content of the ripened produce and optimal color, which varies from commodity to commodity. In many fruits and vegetables the color at the ripened stage is critical for consumer acceptability and perception of quality.

Red pepper is an important example of this phenomenon. Several recent studies concentrated on the changes in carotenoid composition of different varieties of red pepper during ripening (Minguez-Mosquera and Hornero-Mendez,1994a,b; Hornero-Mendez and Minguez-Mosquera, 2000a,b). Chloroplast carotenoid pigments, including lutein and neoxanthin, vanish, together with chlorophylls. At the same time, new carotenoids characteristic of chromoplasts are synthesized. Marin et al. (2004), as part of their broader study of antioxidant constituents of sweet pepper, identified and quantified carotenoids at four maturity stages (immature green, green, immature red, and red). Neoxanthin, lutein (predominating at this stage), *cis*-lutein, violaxanthin, beta-cryptoxanthin, and beta-carotene were found in immature green peppers. In the green stage the same compounds were present, with beta-carotene predominating. In the immature red stage lutein and *cis*-lutein were not detected but new compounds occurred, including capsanthin-5-6,-epoxide capsanthin, antheraxanthin, cucurbitaxanthin A, and zeaxanthin. In the red stage, capsorubin, *cis*-capsantin, and *cis*-zeaxanthin were also detected. The content of capsanthin increased 19 times from the immature stage. Also, the total amount of pigments increased approximately five times. This is an example of a biosynthesis pattern in a produce wherein unique carotenoids are synthesized at the maturity stage.

The qualitative and quantitative carotenoid composition and color of carrots also has an impact on consumer preferences. The content of carotenes in carrots is influenced by genotype and numerous other factors (Heinonen, 1990). In a recent study by Surles et al. (2004), carotenoid profiles of specialty carrots were determined and compared with consumer acceptance. The amounts of alpha-carotene, beta-carotene, lycopene, and lutein were determined in the following six carrot varieties: high-beta-carotene orange, orange, purple, red, yellow, and white. Beta-carotene content ranged from 18.5 to 0.006 mg/100 g of fresh carrots. Sensory evaluation showed that high-beta-carotene orange and white varieties were favored.

While anthocyanins and related compounds contribute to the color of wines, carotenoids are considered to be important precursors of compounds influencing aroma. The qualitative and quantitative composition of carotenoids in mature grapes is influenced by variety, soil, and climate (Razungles et al., 1988; Marais et al., 1989; Guedes de Pinho et al., 2001). Mendez-Pinto et al. (2004) analyzed carotenoids in grapes of three port winemaking varieties of *Vitis vinifera* L., cv. Tinta Barroca, Touriga Francesa, and Tinta Roriz. Twenty-eight compounds were found in extracts of all three cultivars utilizing reversed phase and normal phase HPLC-DAD. Seven previously reported carotenoids (neochrome, neoxanthin, violaxathin, flavoxanthin, zeaxanthin, lutein, and beta-carotene) were identified. A new type of neochrome and two geometrical isomers of lutein and beta-carotene were tentatively identified. The remaining 17 compounds need to be analyzed further for identification.

7.2.3 POSTHARVEST DEGRADATION

Degradation of carotenoids in produce may be a part of natural biological senescence or may occur as a result of postharvest commercial handling, packaging, processing, storage, and food preparation. Due to the large number of conjugated double bonds in carotenoids these compounds are susceptible to various biochemical and chemical changes.

The enzymatic oxidation of carotenoids is an important aspect of carotenoid degradation in nature and in commercial postharvest handling, storage, and processing of produce. As early as 1932, Andre and Hou gave the name *lipoxidase* to the enzyme from soybeans responsible for bleaching carotenoids in bread dough (Gross, 1991). Subsequently it was demonstrated that oxidation of carotenoids occurs when lipoxygenase simultaneously oxidizes unsaturated fatty acids. The presence of lipoxygenase in produce was reported by Pinsky et al. (1971). Eskin et al. (1977) isolated lioxygenase from various food plants. According to Robinson et al. (1995), co-oxidation of carotenoids occurs as a result of the action of an activated form of lipoxygenase. Several volatile compounds, including beta-ionone and beta-ionone epoxide, were identified as a result of bleaching with soybean lipoxygenase (Grosch et al., 1977). There is only limited information available on the nonvolatile products formed by co-oxidation of carotenoids. Wu and Robinson (1999) reported tentative identification of several nonvolatile compounds generated by lipoygenase-catalyzed co-oxidation of beta-carotene, including apocarotenal, epoxycarotenal, apocaroten-one, and epoxycarotenone. GC/MS and HPLC combined with photodiode array detection allowed determination of a large number of high-molecular-weight compounds. The authors proposed a mechanism that involves random attack along the alkene chain of the carotenoid by a lipoxygenase-generated linoleolylperoxyl radical followed by further chemical oxidation and rearrangements.

Other than enzymatic chemical changes of carotenoids during thermal processing, irradiation, canning, drying, and storage are also very important. There is a significant number of early publications concerning thermal degradation of carotenoids. Most frequently, losses of beta-carotene and lycopene were reported earlier without identification of the degradation products. Typically these losses were proportional to the temperature and time of processing. Numerous authors reported volatile thermal degradation products. Toluene and xylene were reported as thermal degradation products of beta-carotene in more than 10 publications between 1963 and 1980. In addition, many researchers found 2,6-dimethylnaphthalene in thermally treated carotenoids. Ionene and alpha- and beta-ionones, beta-cyclocitral, 5,6-epoxy-beta-ionone, and dihydroactinidiolide have also been reported (Schreier et al., 1979). A mechanism for the formation of toluene, xylene, and dimethylcyclodecapentaene from beta-carotene was advanced by Schwieter et al. (1969). In this mechanism a rearrangement of double bonds in the polyene chain is followed by the formation of a four-ring intermediate, which undergoes cleavage to form toluene. Xylene and dimethylcyclodecapentaene are formed in a similar way. The authors did not provide information on the structure of the remaining part of the beta-carotene molecule after formation of toluene, xylene, and related compounds. A mechanism for the formation of ionene was postulated by Ohloff (1971). Only a limited number of studies are available on the formation of nonvolatile thermal degradation products of carotenoids. Halaby and Fagerson (1971) reported small amounts of polycyclic aromatic hydrocarbons formed from beta-carotene at temperatures between 400 and 700°C, which is not relevant to food processing. During simulated commercial deodorization of palm oil, Ouyang et al. (1980) identified beta-13-apo-carotenone, beta-15-apo-carotenal, and beta-14-apo-carotenal. Onyewu et al. (1982) identified two nonpolar compounds, dimethyl-(trimethylcyclohex-1-enyl) octatetraene and tri-methyl-(trimethylcyclohex-1-enyl) dodecahexaene, in a model system simulating time and temperature conditions of palm oil deodorization. In a subsequent publication Onyewu et al. (1986) described thermal degradation products of beta-carotene formed under time and temperature conditions used in various methods of food processing. Temperatures of 210, 155, and 100°C and times of 4 and 1 h and 15 and 5 min were investigated. At 210°C, 97.8, 97.2, 93.9, and 91.1% degradation of beta-carotene occurred, respectively. At 155 and 100°C the losses were 30.6, 16.7, 8.3, and 8.3%, and 8.3, 7.7, 0.9, and 0.9%, respectively. The presence of over 70 non-volatile thermal decomposition products of beta-carotene was reported and several compounds of relatively complicated structure were identified in samples heated for 4 h at 210°C.

Drying is frequently used in the processing of some produce. Perez-Galvez et al. (2004) investigated changes of carotenoids during simulated traditional drying of red pepper fruits. The temperature was increased from 25°C to approximately 45°C during the first 24 h. After 144 h the temperature was decreased slowly over 48 h. The content of eight carotenoids was analyzed by HPLC. Three different stages were observed in relation to the determined amount of carotenoids. During the first 24 h there was a decrease of approximately 20%. After 5 d, recovery of the biosynthetic capability of the fruits was observed. During the following 48 h a sharp increase in carotenoids took place. Since the final high content of carotenoids is an important quality requirement, the authors concluded that a careful control of temperature and humidity may improve carotenoid retention.

Among several groups of compounds possessing antioxidant activity (Kaur and Kapoor, 2001) carotenoids are an important category. In a recent study Matsufuji et al. (2004) examined the antioxidant effects of paprika pigments on oxidation of linoleic acid at 37°C in darkness or exposed to fluorescent light. The addition of paprika pigments significantly suppressed the oxidation of linoleic acid in darkness and only slightly under light. The antioxidant effect was higher with increased concentration of pigments from 0.02 to 2%. The authors suggested that discoloration of paprika pigments may be used as a visual indicator of the oxidative deterioration of oils. It is worth mentioning that the author of this chapter used, many years ago, discoloration of paprika pigments to estimate the progress of oxidation of rapeseed oil with and without antioxidants from wood smoke.

7.2.4 Progress in Analytical Methodology

An excellent and comprehensive description of analytical aspects of carotenoids may be found in *Advances in HPLC and HPLC-MS of Carotenoids* by Taylor et al. (1990), *Isolation and Analysis* by Schiedt and Liaaen-Jensen (1995), *Mass Spectrometry* by Enzell and Back (1995), and chapters of *Carotenoids*, Volumes 1A (Isolation and Analysis) and 1B (Spectroscopy) edited by Britton et al., and in other related monographs. Only selected publications after 1995 will be mentioned here.

Bhosale et al. (2004) used resonance Raman spectroscopy for a rapid, nondestructive estimation of carotenoids in selected fruits and vegetables. The results were compared with data obtained by extraction and HPLC. A strong correlation was observed in the case of fruit juices and tomatoes at the same stage of ripening. Spinach leaves showed very high Raman counts despite lower carotenoid content. The method demonstrates potential value in some instances but also presents important limitations.

Cortes et al. (2004) described a method for identification and quantification of carotenoids, including isomers, in fruit and vegetable juices. The authors used HPLC with an ultraviolet-diode array detector. Seventeen different *cis*- and *trans*-carotenoids were identified in orange juice and a mixture of orange and carrot juice. For identification, spectral and retention time data from standards and literature values were utilized. The authors claimed that their proposed method is fast, sensitive, reliable, accurate, and reproducible for determination of *cis*- and *trans*-carotenoids in produce juices.

Felicissimo et al. (2004) applied ToF-SIMS (time-of-flight secondary ion mass spectrometry) and XPS (x-ray photoelectron spectroscopy) to analyze annatto in seeds and seed extracts. The analysis of seeds without any sample treatment showed the presence of bixin and its characteristic fragments. When an extract of seeds was analyzed, the results revealed the same components. Internal parts of seeds did not show the presence of the colorant. A prolonged exposure to light led to significant degradation of bixin. Several minor constituents and products of bixin degradation were observed, but their precise assignment requires further analysis.

7.3 CHLOROPHYLLS

Chlorophylls are green pigments of ubiquitous occurrence in nature. It is estimated that plants, algae, and some microorganisms produce over 1 billion tons of chlorophylls per year (Schwartz and Lorenzo, 1990). Chlorophylls in photosynthesis allow the sun's energy to be utilized to synthesize carbohydrates and in this way change light energy into chemical energy that is indispensable for the survival of all living organisms. In higher plants chlorophylls a and b are imbedded in the highly structured lamellae of chloroplasts. The chlorophyll molecules interact with each other, as well as with proteins and lipids. Most of the research on chlorophylls is concentrated on the biochemistry of photosynthesis and is beyond the scope of this review. The chlorophylls are important constituents of produce. In fruits they usually degrade during ripening and allow other colors to be unveiled or formed. In green vegetables retention of color is of paramount importance for consumer acceptance.

7.3.1 Chemical Definition and Structure

Porphin can be considered a basic structure from which chlorophylls may be derived. Porphin consists of four pyrolle rings linked by carbon atoms. According to the IUPAC nomenclature there are four pyrolle rings, A, B, C, and D, in the porphin molecule. The carbon atoms are numbered from 1 to 20 starting from ring A. Fisher proposed the most commonly used nomenclature and numbering of carbon atoms (see Figure 7.3). The pyrolle rings are numbered I through IV. Peripheral carbon atoms on the porphin rings are numbered from 1 to 8. The bridging carbons are designated as Alpha, Beta, Gamma, and Delta. Alternatively, the pyrolle rings may be assigned letters A, B, C, and D. When a fifth isocyclic ring is added to porphin the resulting structure is called *phorbin*. The name *porphyrin* is used for substituted porphins. Chlorophyll is considered a porphyrin with a centrally located Mg^{2+} ion. In position 7 of the fourth ring there is a propionic acid substituent esterified with phytol, a C-20 monounsaturated isoprenoid alcohol, that is responsible for the hydrophobic nature of the chlorophyll molecule. In positions 1, 5, and 8, methyl groups are present. A vinyl group is located in position 2. In position 9 there is a carbonyl group and in position 10 an acetyl group. The isomer chlorophyll a contains a methyl group in position 3, whereas an aldehyde group occurs in this position in chlorophyll b. Isomers of chlorophyll a and b are found in products that undergo heat treatment. These isomers are formed by inversion of the carbometoxy group at C-10. Some trivial names are widely accepted in chlorophyll chemistry and were adopted here from Jackson (1976):

Phyllins: chlorophyll derivatives containing magnesium
Pheophytins: the magnesium-free derivatives of chlorophylls
Chlorophyllide: the acid derivative resulting from enzymic or chemical hydrolysis of the C-7 propionate ester
Chlorophyllase: the enzyme present in leaves that catalyzes hydrolysis of the C-7 propionate ester
Pheophorbides: the products containing a C-7 propionic acid resulting from removal of magnesium and hydrolysis of phytyl ester

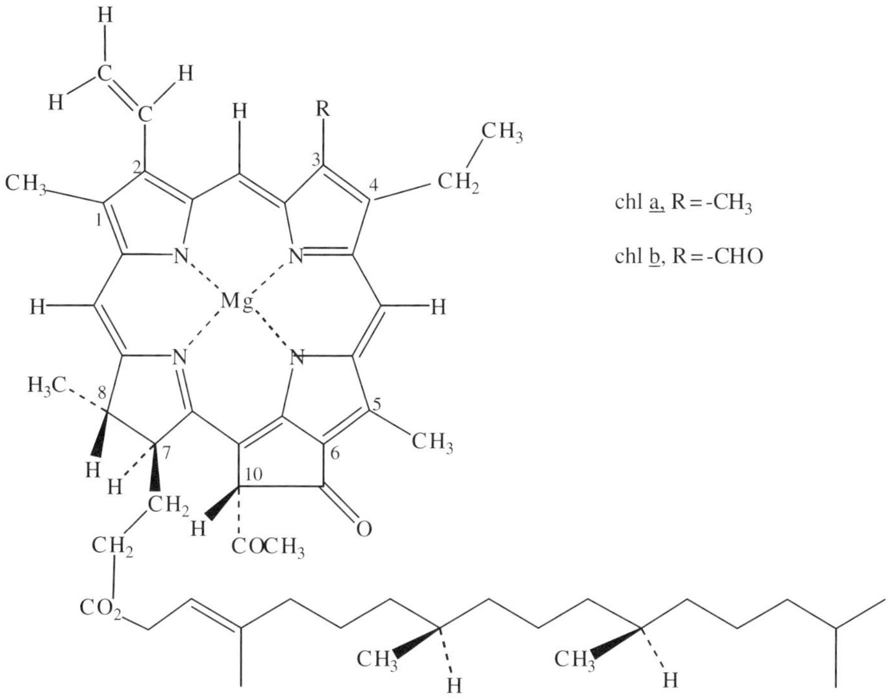

FIGURE 7.3 Chlorophyll.

chl _a_, R =-CH$_3$

chl _b_, R =-CHO

"Meso" compounds: derivatives in which the C-2 vinyl group has been reduced to to ethyl

"Pyro" compounds: derivatives in which the C-10 carbomethoxy group has been replaced by hydrogen

Chlorins e: derivatives of pheophorbide resulting from cleavage of the isocyclic ring

Rhodins g: the corresponding derivatives from pheophorbide b

7.3.2 DEGRADATION OF CHLOROPHYLLS DURING SENESCENCE, STORAGE, AND PROCESSING

The past research on chlorophylls related to fruits and vegetables is available in reviews and monographs (Simpson, 1985; Schwartz and Lorenzo, 1990; Gross, 1991).

The senescence process of chlorophyll is not yet fully understood. Its maximum concentration, which occurs in the summer, decreases stepwise until it totally disappears in the autumn. The degradation of chlorophyll a proceeds faster than that of chlorophyll b. Pheophytin and pheophorbide are the initial degradation products. Oxidation of the ring structure leads to formation of chlorines, followed by formation of colorless products (Gross, 1987). Amir-Shapira et al. (1987) investigated chlorophyll changes during senescence in citrus peel and parsley leaves. Chlorophyllide a was detected in citrus peels with a concurrent increase in chlorophyllase activity. A different path of chlorophyll degradation, with formation of pheophytin, was observed in parsley leaves. According to the older publications reviewed by Gross (1991) the most common degradation of chlorophyll during thermal processing, freezing, and storage of green vegetables is the conversion of chlorophyll to pheophytins with a color change to olive-brown. The duration of treatment, temperature, and concurrent release of acids accelerates degradation. In frozen-pack peas and string beans 60 to 85% of chlorophyll changed to pheophytin. In canned green beans all chlorophyll was converted to pheophytin. Several studies demonstrated that blanching, which involves various forms of heat treatment to inactivate enzymes, has a beneficial effect on retention of chlorophyll during processing and storage. Dehydration causes chlorophyll degradation. During dehydration of spinach 26% of chlorophyll was converted to pheophytin. Blanching before dehydration increased the loss to 44%. Sterilization of canned green produce has a detrimental effect on chlorophyll. After sterilization of green peas for 25 min at 120°C all chlorophyll was degraded to pheophytin.

The initial change in the chlorophyll molecule subjected to heat is isomerization by inversion of the C-10 carbomethoxy group (vonElbe and Schwartz, 1996). The chlorophyll a and b isomers can be separated on a C-18 reverse-phase HPLC column. Heating spinach leaves for 10 min at 100°C leads to conversion of 5 to 10% of chlorophyll a and b to pheophytin a and b (Schwartz et al., 1981). The olive-brown pheophytin is formed by magnesium ions' being displaced by two hydrogen ions. Formation of pheophytin from chlorophyll a occurs more rapidly than that from chlorophyll b. This effect is attributed to the electron-withdrawing effect of the C-3 formyl group in chlorophyll b. Pheophytin is subject to further change by the replacement of the C-10 carbomethoxy group with a hydrogen atom, leading to formation of pyropheophytin. Reverse-phase HPLC allows separation of pyropheophytins a and b from corresponding pheophytins. It was demonstrated that pyropheophytins a and b are responsible for the olive-green color of many commercially canned vegetables (Schwartz and Lorenzo, 1991). The hydrogen ions of pheophytins and pyropheophytins are easily displaced by zinc or copper ions. The green-colored complexes are stable in acidic solutions. Copper complexes of pheophytin and pheophorbide are available commercially but are not allowed for food use in the U.S.

7.4 BENZOPYRAN DERIVATIVES

7.4.1 CHEMICAL DEFINITION AND STRUCTURE

Benzopyran derivatives in produce include compounds containing a central pyran ring and two adjacent aromatic rings. Selected basic structures [see Figure 7.4 (a)-(g)] include flavan, flavan-3-ol, flavanone, flavanonol, flavone, flavone-3-ol, and chalcone. By adding functional groups to these structures at various positions, numerous compounds may be derived. Flavonoids derived from these structures play an important role in the physiology and postharvest natural and man-made changes in produce. Many of these polyphenolic compounds have strong antioxidative properties and are postulated to be beneficial for human health. Although anthocyanins are

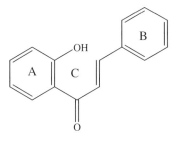

a) FLAVAN

b) FLAVAN-3-OL

c) FLAVANONE

d) FLAVANONOL

e) FLAVONE

f) FLAVONE-3-OL

g) CHALCONE

FIGURE 7.4 (a) Flavan; (b) flavan-3-ol; (c) flavanone; (d) flavanonol; (e) flavone; (f) flavone-3-ol; (g) chalcone.

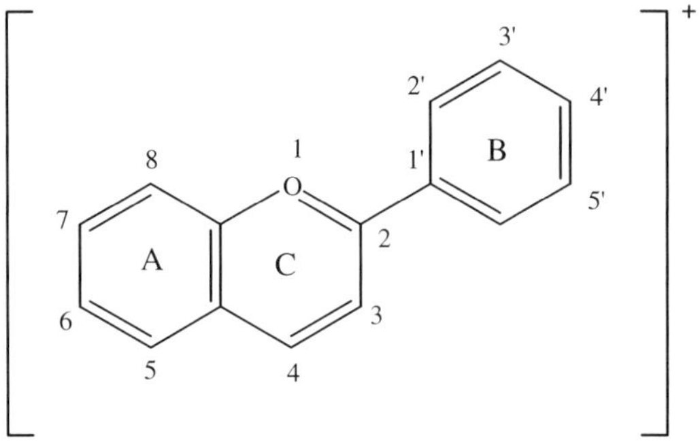

FIGURE 7.5 Flavylium cation.

most important in terms of the color of produce, other polyphenols may contribute to yellow and brown colors directly or their molecules may condense with anthocyanins and cause so-called "co-pigmentation." Numerous publications cover research related to many kinds of polyphenols, making it very difficult and sometimes impossible to divide the reviews into separate parts related to individual groups such as anthocyanins or other flavonoids.

7.4.2 CHEMICAL PROPERTIES AND OCCURRENCE OF ANTHOCYANINS

Many authors consider anthocyanins as a special subgroup of flavonoids with a basic structure of a falvylium cation (see Figure 7.5) that contains a pyran ring with a positive charge. Anthocyanins may be defined as "glycosylated polyhydroxy and polymethoxy derivatives of 2-phenylbenzopyrylium (flavylium) salts" (Brouillard, 1982). An addition of hydroxy and methoxy functional groups in specified positions of the flavylium cation leads to compounds known as anthocyanidins. Approximately 20 anthocyanidins are known, but only six are considered important in produce. These are pelargonidin, cyanidin, delfinidin, peonidin, petunidin, and malvinidin (Francis, 1985a). These anthocyanidins occurring in produce are shown in Figure 7.6 (a)-(f). Since anthocyanidins do not contain sugar residues they may be considered aglycones of anthocyanins. Formation of esters with one or more sugars leads to formation of anthocyanins. Sugars occuring most commonly in anthocyanins are glucose, rhamnose, galactose, xylose, and arabinose. These sugars may further react to form esters. *P*-coumaric, caffeic, ferulic, sinapic, *p*-hydroxybenzoic, malonic, and acetic acids participate most frequently in this reaction (Brouillard, 1982).

Anthocyanins are water-soluble components of cell sap that give red and blue colors to many fruits and vegetables. The color of anthocyanins is considered the most important quality attribute in many processed food products. Numerous factors, including the structure of the anthocyanins involved, concentration, solvent, pH, the presence of other substances, and temperature, influence the nature and stability of this color. The number and location of the hydroxy and methoxy groups in the anthocyanin molecule affect its color properties.

In water solutions the molecules of anthocyanins exist in several forms in a dynamic equilibrium that may be changed by a change in the parameters of the system. In a slightly acidic pH range at room temperature the following four species are present: the blue quinonoidal base, the red flavylium cation, the colorless pseudo-base or carbinol, and the colorless chalcone (Brouillard and Delaporte, 1997).

Acidity of the system as expressed by pH has the dominant influence on the color of produce containing anthocyanins. In the range of pH 2 to 4 the flavylium cation is mostly responsible for the color, along with some contribution from the quinonoidal base. With an increase of pH a large bathochromic shift occurs and the quinonoidal base increasingly influences the color. With molecules containing more than one OH group the existing equilibrium becomes more complex. An elegant discussion of this topic may be found in a chapter written by Brouillard (1982).

Temperature significantly influences the degradation of anthocyanins, especially in the presence of oxygen. This effect varies depending on the molecular structure of the specific compounds. Anthocyanins containing more methoxy groups or sugars are more stable than those containing more hydroxy groups (Mazza and Miniati, 1993). Three possible mechanisms have been suggested for the thermal degradation of anthocyanins (Jackman and Smith, 1992). In the first, the quinonoidal base that is in equilibrium with the flavylium cation leads to a coumarin derivative and B-ring derivative. In the second, the flavylium cation transforms to a carbinol base, chalcone, and, finally, brown polymeric compounds. In the third, the first two steps are similar to those in the second mechanism but they result in low molecular degradation products of chalcone instead of polymers.

Light significantly influences the stability of anthocyanins.

Numerous chemicals that are used in processing of foods containing fruits may interact with anthocyanins. Ascorbic acid is frequently added to fruit juices to increase its nutritional value. However, anthocyanins and ascorbic acid cause mutual degradation.

Sulfites and sulfur dioxide added sometimes for preservation purposes are known to cause loss of color of anthocyanins. Under some circumstances the color may be restored (Francis, 1989). Anthocyanins are known to form complexes with metals. These complexes may change the color of anthocyanins. The formation of complexes may occur during processing in metal equipment or by adding some salts. Complexes of metals with anthocyanins contribute to the wide variety of colors of plants in nature. Molecules of anthocyanins may form complexes with molecules of other organic compounds. This phenomenon is known as co-pigmentation and may involve flavonoids, amino acids, proteins, pectin, carbohydrates, and polyphenols (Francis, 1989).

Vaccinium berries constitute an important group of small fruits containing biologically active compounds (Prior et al., 1998; Kalt et al., 1999a,b; Moyer et al., 2002). These compounds include anthocyanins and flavonols, most commonly as glycosides, and acetylated glycosides. Taruscio et al. (2004) investigated *Vaccinium* fruit samples from cultivated and undomesticated colonies within the northwestern U.S. Total phenolics (TPH), total anthocyanins (ACY), and their antioxidant capacity were determined. Total phenolics ranged from 0.81 to 2.84 mg/g of fresh weight. Total anthocyanins were lowest in red huckleberry and wild cranberry (0.11 and 031 mg/g fresh weight, respectively), and the evergreen huckleberry and oval leaf blueberry that contained the highest TPH also showed the greatest ACY (3.64 and 3.07 mg/g of fresh weight, respectively). The selected bioactive compounds analyzed by

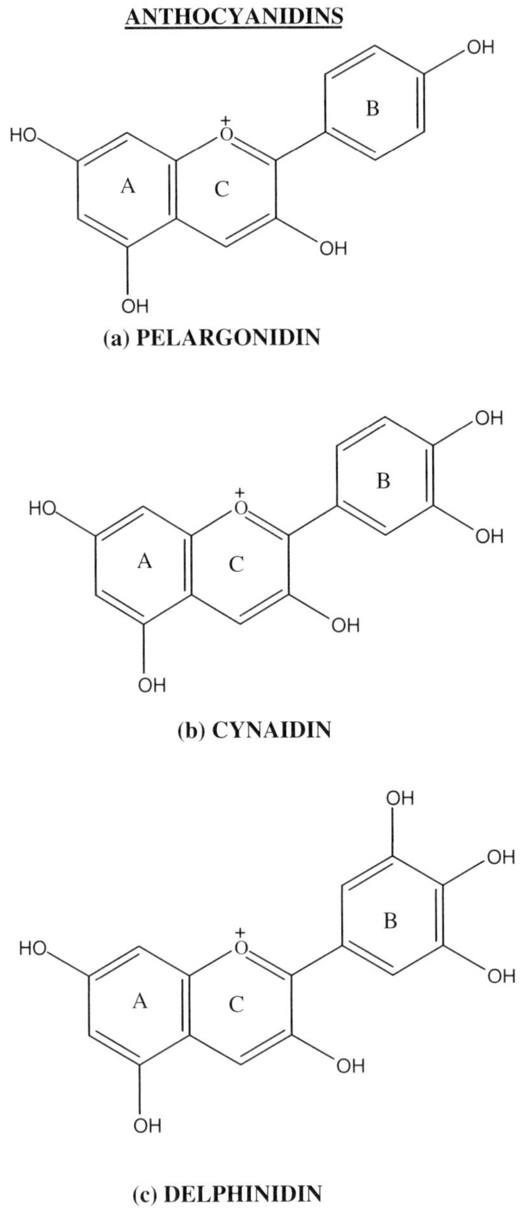

ANTHOCYANIDINS

(a) PELARGONIDIN

(b) CYNAIDIN

(c) DELPHINIDIN

FIGURE 7.6 (a) Pelargonidin; (b) cynaidin; (c) delphinidin; (d) peonidin; (e) petunidin; (f) malvidin.

HPLC equipped with photodiode array and mass spectrometric detectors included flavan-3-ols and flavonols (catechin, epicatechin, myricetin, and quercetin), phenolic acids (caffeic, chlorogenic, p-coumaric, ferulic, and p-hydroxybenzoic), and anthocyanidins (cyaniding, delphinidin, malvinidin, peonidin, and petunidin).

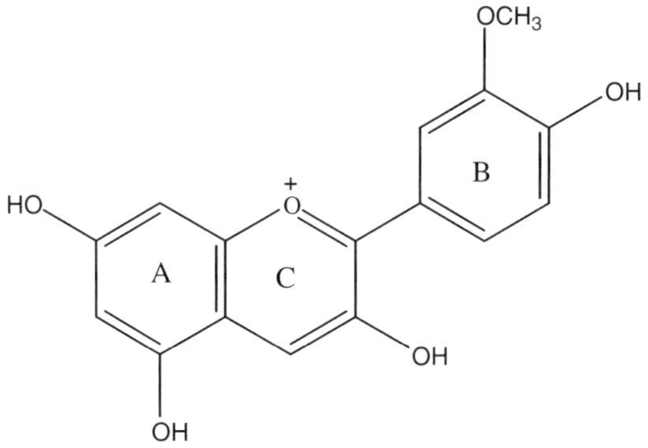

(d) PEONIDIN

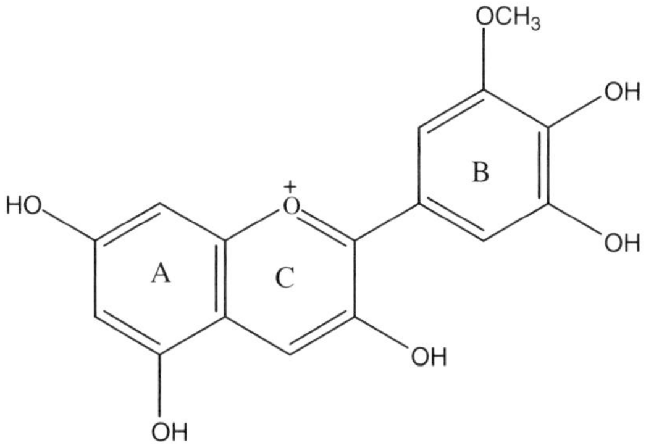

(e) PETUNIDIN

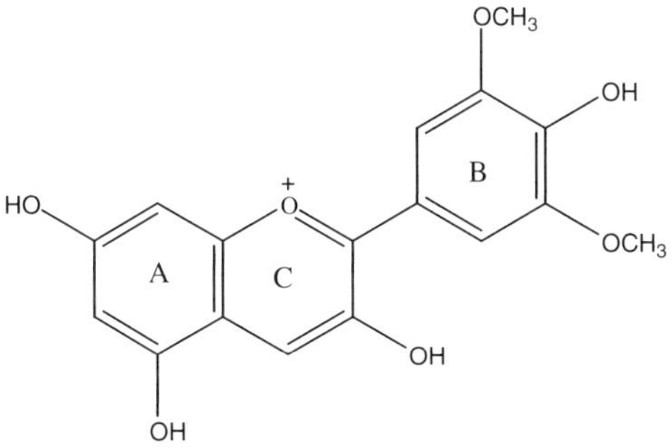

(f) MALVIDIN

FIGURE 7.6 (continued)

Several varieties of anthocyanin-pigmented rice have been developed through genetic engineering. Hyun and Chung (2004) isolated two anthocyanidins from dark purple grains of rice (*Oryza sativa* cv. Heugjinjubyeo). They identified these compounds as cyanidin and malvidin and demonstrated that they inhibit growth of human monocytic leukemia cells.

Black chokeberry (*Aronia melanocarpa* Elliot) fruit extract showed antioxidant activity both *in vitro* and *in vivo*. Matsumoto et al. (2004) measured radical scavenging activity of the extract using DPPH stable radicals. The red pigment fraction did scavenge over 44% of DPPH radicals at a concentration of 25 µg/mL compared with a control solution. The HPLC profile of the red pigment fraction gave three major peaks, the largest one identified as cyanidin-3-glucoside. The black chokeberry extract administered at 2 g/kg of body weight of rats showed significant antiulcerative properties.

Biosynthesis of radiolabeled anthocyanins, proanthocyanidins, and other flavonoids using cell suspensions of two plant species, ohelo (*Vaccinum pahalae*) and grape (*Vitis* hybrid, Bailey Alicant A), was reported by Yousef et al. (2004). Uniformly labeled (^{14}C) sucrose was used in the medium. In the flavonoid-rich fractions of ohelo and grape cultures, an average of 14 or 15% of the of the radioactivity was recovered, respectively. HPLC-MS and NMR were used to characterize obtained fractions. Proanthocyanidins ranging from monomers to dodecamers were determined from ohelo cell cultures. Small amounts of several anthocyanins were also detected in these samples. Grape cell cultures contained more anthocyanins and fewer proanthocyanidins. Peonidin 3-*O*-glucoside and cyaniding *p*-coumarylglucoside were dominant in grape fractions. Several other anthocyanins occurred in small amounts. The reported biosynthesis allowed the researchers to obtain in a reproducible manner and in sufficient amounts uniquely labeled phytochemicals suitable for *in vivo* experiments.

7.4.3 POSTHARVEST AND PROCESSING CHANGES OF ANTHOCYANINS

Retaining color in products containing anthocyanins is a paramount task for the food industry. The factors affecting color stability of anthocyanins include nature of compounds used, their concentration, pH, light, storage conditions, presence of phenolic compounds, and other product constituents or additives. These topics are covered in classic monographs, such as *Stability of Anthocyanins in Foods* by Markakis (1982). The occurrence, properties, stability, and changes of anthocyanins in various foods remain subjects of numerous current publications (Eiro and Heinonen, 2002; Garzon and Wrolstad, 2002; Lopez-Da-Silva et al., 2002; Kahkonen et al., 2003; Kirca and Bekir, 2003; Schwartz et al., 2003; Talcott et al., 2003).

Rein and Heinonen (2004) studied color enhancement and stability of four berry juices: lingonberries (*Vaccinium vitis-idaea* L.), cranberries (*Vaccinium oxycoccus* L.), strawberries (*Fragaria ananasa*), and raspberries (*Rubus ideaeus* L.). For copigmentation, ferulic, sinapic, and rosmarinic acids were added in an amount 10 times greater than that of anthocyanins in juices. Grape skin and black carrot extracts containing 3% anthocyanins and 97% carbohydrate carriers as well as non-anthocyanin extract of rosemary were also added at 2 g/L of juice. The juices were stored in sealed tubes in daylight at room temperature for 103 d. The absorption spectra in the visible range from 450 to 600 nm were recorded during storage of the samples. The profile of anthocyanins was also monitored using HPLC. All phenolic acids improved color and its retention in the four investigated juices during the 103 d of storage compared to untreated samples. Sinapic acid was most effective in strawberry juice, leading to color intensity after storage higher than the initial value. The effectiveness of ferulic and synaptic acids was equal in the case of raspberry juice. Rosmarinic acid showed the best effect in lingonberry and cranberry juices. Unenhanced strawberry and raspberry juice color was not detectable after 36 and 51 d, respectively, under experimental conditions. The color of lingonberry and cranberry juices was 33 and 19%, respectively, of its original intensity under the same conditions. In general, the addition of grape skin and black carrot extracts to the four juices significantly increased initial color intensity but led to some problems later. The presence of grape skin extract caused brown discoloration of strawberry juice, leading to total color deterioration after 51 d. The addition of commercial non-anthocyanin rosemary extract caused similar problems. HPLC results showed a decrease of total peak area of anthocyanins. Untreated raspberry and strawberry juices retained after storage only about 1% of their initial anthocyanins. Lingonberry and cranberry juices contained 13 and 10%, respectively. The addition of phenolic acids to the juices resulted in new HPLC peaks. Identification of these peaks was not conducted in this study.

It is believed that the decrease of the content of anthocyanins during aging of red wine is due to an interaction with pyruvic acid and formation of visitins A (Vivar-Quintana et al., 2002). Antioxidant and biological activities of three most abundant anthocyanins in red wines and relevant visitins A were investigated by Garcia-Alonso et al. (2004). Visitins A were synthesized according to previously described methods of Romero and Bakker (1999). Delphinidin-3-glucoside, petunidin-3-glucoside, and malvinidin-3-glucoside exhibited higher antioxidant activity than the corresponding visitins A.

Characterization of 20-, 3-, and 1-year-old Port wines was conducted by high-resolution NMR and high-resolution diffusion-ordered spectroscopy (DOSY) in order to identify changes during storage (Nilsson et al., 2004). High-resolution DOSY allowed identification of an increased number of compounds. The wine samples of various ages differed in their content of organic acids, some amino acids, and aromatic species. A significant decrease of high molecular aromatics in the oldest wine is consistent with the formation and precipitation of anthocyanin-based polymers during aging of red wines.

7.4.4 PROGRESS IN ANALYTICAL METHODOLOGY

The rapid progress in analytical methodology allows improved, faster, and more precise qualitative determination of anthocyanins. The application of HPLC to analyze anthocyanins was reported and reviewed previously (Giusti et al., 1999; da Costa et al., 2000; Merken and Beecher, 2000; Chandra et al., 2001; Mullen et al., 2002). There are still major challenges related to preparation of the samples for analysis, especially changes of anthocyanins during hydrolysis and lack of reference compounds.

Zhang et al. (2004) reported a comparison of an application of HPLC of anthocyanins before and after sample hydrolysis for determination of anthocyanins and anthocyanidins in bilberry extracts. In the direct analysis approach bilberry extract powder was dispersed in a 1:1 water/methanol solution containing 2% HCl. This solution was found to give the best extraction yields. After passing through a 0.45-μm PTFE filter, samples were injected into HPLC. The identification was conducted using HPLC coupled with DAD and a mass spectrometer. A UV-vis detector found over 15 peaks and MS identified 9. In the hydrolysis approach, samples were dispersed and filtered as in direct analysis, placed in a dry bath, and hydrolyzed at 100°C for 30 min, 1 h, and 2 h. After hydrolysis, samples were cooled immediately to room temperature and analyzed by HPLC. After 30 min of hydrolysis most of anthocyanins were converted to anthocyanidins, with approximately 10% remaining. After 1 and 2 h all anthocyanin peaks almost completely disappeared. Five major peaks were identified as delphinidin, cyanidin, petunidin, peonidin, and malvinidin. The authors concluded that the analysis of unhydrolyzed samples provides an anthocyanin profile useful for characterization of raw materials, whereas the acid hydrolysis approach allows quantification of individual anthocyanidins in botanical extracts.

Parejo et al. (2004) used HPLC equipped with DAD coupled to negative electrospray ionization (ESI) tandem mass spectrometry (MS/MS) to analyze an extract from dried fennel residue after steam distillation of essential oils. Forty-two phenolic compounds were identified, including 27 not previously reported in fennel. Among newly identified compounds were hydoxycinnamic acid derivatives, flavonoid glycosides, and flavonoid aglycons.

Anthocyanins are valuable compounds of pigmented orange juices. An economically feasible method of removing these compounds from citrus by-products may lead to their recovery and utilization in food and pharmaceutical industries. Scordino et al. (2004) studied equilibrium and kinetic data of adsorption of cyanidin-3-glucoside chloride on 13 commercial resins. The presence of hesperidin, which is a dominant flavanone in orange fruit, had to be considered. The experimental equilibrium data fitted well into Langmuir and Freundlich isotherms. It was demonstrated that the resin EXA-118 was most effective to obtain a highly concentrated extract of anthocyanins from pigmented orange juice.

7.4.5 OCCURRENCE AND ANTIOXIDANT PROPERTIES OF FLAVONOIDS

The increased interest in the flavonoids and phenolic acids other than anthocyanins in fruits and vegetables comes from convincing epidemiological and other studies that show health-promoting and disease-preventing effects of these compounds (Rice-Evans and Parker, 1998).

Marin et al. (2004) investigated antioxidant constituents of sweet pepper (*Capsicum annum* L.) at four maturity stages (immature green, green, immature red, and red). The study included phenolics (hydroxycinnamic acids and flavonoids), vitamin C (ascorbic acid and dehydroascorbic acid), and carotenoids. Utilization of modern analytical techniques allowed the separation and identification of a significant number of compounds. The HPLC-DAD analysis of the methanol extract of the pepper pericarp showed peaks of numerous polyphenols and UV spectra typical for hydroxycinnamic acids, flavonols, flavons, and C-glycosylflavones. The combination of DAD and ESI-MS detectors provided information necessary for structural identification. Five hydroxycinnamic derivatives and 23 flavonoids were characterized and quantified. Some of these compounds were not previously reported. Significant differences in the qualitative and quantitative content of phenolic compounds were found among different maturity stages. The number and amount of phenolics was very high in the immature green pepper while green, whereas immature red and red peppers showed four- to fivefold reductions in these compounds.

The chemical composition of different apple cultivars determines their usefulness for various products. The classification of cider apple varieties is based on total content of polyphenols. These compounds are considered responsible for color and for balance of bitterness and astringency in ciders. Five technological groups are recognized: sweet, bittersweet, semiacid, semiacid-bitter, and acid-bitter. Alonso-Salces et al. (2004) determined the polyphenolic composition of 31 Basque cider apple cultivars in pulp, peel, and juice. The concentration of these compounds differed widely depending on the cultivars. A higher concentration of total polyphenols was found in peel than in pulp in all varieties. Depending on the cultivar, the peel:pulp ratio ranged from 1.5 to 5.4. The authors recommend that ciders be made using mixtures of different apple cultivars in order to obtain balanced composition for proper fermentation, flavor, color, and stability.

Characterization of flavonols in cranberry (*Vaccinium macrocarpon*) powder was reported by Vvedenskaya et al. (2004). Freeze-dried cranberry powder was extracted twice with 80% acetone/water, partially evaporated, defatted with hexane, and further extracted with ethyl acetate. The ethyl acetate extract was evaporated and separated further using Sephadex LH-20 column chromatography. The fraction containing flavonols was further separated by a developed HPLC procedure. Twenty-two distinct peaks were observed that were determined by UV-vis and mass spectra to correspond to flavonol glycoside conjugates. The following six compounds previously not reported in cranberry and cranberry products were identified: myricetin-3-beta-xylopyranoside, quercetin-3-beta-glucoside, quercetin-3-alpha-arabinopyranoside, 3-methoxyquercetin-3-alpha-xylopyranoside, quercetin-3-*O*-6-*p*-coumaroyl)-beta-galactoside, and quercetin-3-*O*-(6-benzoyl)-beta-galactoside. This work represents an important contribution to the identification of new flavonoid compounds in cranberries.

Olsson et al. (2004) studied selected antioxidants, low-molecular-weight carbohydrates, and total antioxidant capacity in four cultivars of strawberries at three ripening stages, unripe (white green), ripe (red), and fully ripe (dark red), over two seasons. For the measured content of antioxidants and total antioxidant capacity there was a two- to fivefold variation among cultivars. Lower concentrations of chlorogenic acid, *p*-coumaric acid, quercetin, and kaempferol were found in unripe berries than in riper berries.

Kosar et al. (2004) investigated phenolic components of strawberry cultivars Camarosa, Dorit, Chandler, and Osmanli and their hybrids. The effects of maturation were analyzed at three stages (green, pink, and ripe). The authors presented quantitative data on acids (*p*-hydroxybenzoic, *p*-coumaric, and ellagic), anthocyanins (cyanidin-3-glucoside and pelargonidin-3-glucoside), and flavonoids (kaempferol, quercetin, and myricetin). Ellagic acid in the green stage, pelargonidin-3-glucoside and *p*-coumaric acid in the pink stage, and pelargonidin-3-glucoside and *p*-coumaric acid in the ripe stage were determined to be the major constituents. Significant changes in phenolic compounds were observed during maturation, including a large increase in *p*-coumaric acid.

7.4.6 POSTHARVEST AND PROCESSING CHANGES OF FLAVONOIDS

Significant amounts of fruits and vegetables are subjected to various postharvest operations before consumption. Different stages of handling include storage, industrial processing, distribution, and preparation for consumption. All of these steps affect physical and chemical characteristics of produce. One of the concerns is the possible loss of nutritionally important and health-protecting ingredients. Scientists and the food industry are continuously making attempts to reduce processing losses of beneficial bioactive compounds present in fruits and vegetables.

Spanos et al. (1990) reported that diffusion extraction at various temperatures leads to a significant increase of phenolic compounds in apple juice compared to traditional straught pressing.

Van der Sluis et al. (2002) reported that using conventional methods of processing apples into juice causes a significant degradation of flavonoids and 90 to 97% loss of the fruit's antioxidant activity. The same authors (Van der Sluis et al., 2004) proposed a novel production method for apple juice that includes alcohol extraction of pulp or pomace and evaporating the alcohol and enriching the juice with the extracted polyphenols. The enriched juice showed five times higher antioxidant activity compared to conventionally obtained juice. The triangle sensory test showed a significant taste difference between enriched and conventional juice samples at 0.1% level.

Traditional methods applied to decide the optimal time for harvesting grapes for wine production include determination of weight, sugar content, and acidity (Gonzalez-Sanjose et al., 1991). Recently, more attention is being paid to the content of polyphenols due to their influence on color, bitterness, and astringency. Perez-Magarino and Gonzalez-San Jose (2004) investigated the influence of degree of grape ripening on wine composition, quality, and changes during aging. Two varieties of grapes were used (Tinto Fino and Cabernet Sauvignon). The grapes were harvested at three stages of ripening (the first stage was based mainly on sugar content, and the second and third stages were 1 and 2 weeks after the first stage). The aging was conducted for 1 year in American oak barrels and followed for 6 months in bottles. The results show that the delay in harvesting grapes between 1 and 2 weeks under the conditions of these particular experiments led to production of wines that contained higher levels of anthocyanin derivatives and to better color quality and stability. The results also showed that if grapes are allowed to ripen too much the wine's quality benefits are lost.

Numerous studies in the past showed significant losses of color and nutrients during thermal processing of fruits and vegetables. Recently, it was reported that in some instances thermal processing might have beneficial effects. Total antioxidant activity of tomatoes (Dewanto et al., 2002) was increased as a result of thermal processing. Jiratanan and Liu (2004) studied the influence of typical commercial canning conditions on antioxidant activity of table beets and green beans. When heated at 115°C, initially phenolics decreased by 12%, but during further processing the content raised to the level of unprocessed beets and after 45 min increased further by 14%. Thermal processing had no effect on total antioxidant activity of beets. In the case of green beans, after 10 min at 115°C there was a 40% decrease in total free phenolic compounds. After 20 min no further decrease was noticed, and after 40 min there was some increase in phenolics. The total antioxidant activity of green beans decreased by 9% after 10 min of processing and remained unchanged after further heating for 40 min. According to other authors (Price et al., 1998) the flavonoids in green beans are not degraded but leach into the water during heat processing. Due to the complex nature of changes in different kinds of produce, no general conclusions concerning the fate of phenolic compounds during heat processing can be made at this time.

The content of procyanidins in frozen and canned Ross clingstone peaches and in the syrup used for canning over a 3-month period was investigated by Hong et al. (2004). Optimized analytical methods allowed quantification of oligomers through octamers. The following amounts of procyanidins were found in frozen peeled peaches: monomers 19.59 mg/kg, dimers 39.59 mg/kg, trimers 38.82 mg/kg, and tetramers through octamers 17.81, 12.43, 10.62, 3.94, and 1.75 mg/kg, respectively. Thermal treatment resulted in losses of all oligomers; the highest loss (30%) was in hexamers and hectamers, and octamers were not detectable. A significant amount of the oligomers up to the hexamers lost during heating was detected in the syrup. Previously reported studies did not quantify oligomers larger than tetramers. During 3 months of storage the losses of procyanidins were time-related, and oligomers larger than tetramers were not detected.

Goncalves et al. (2004) studied the effect of ripeness and storage of cherries on the content of phenolic compounds. Samples of sweet cherry cultivars Burlat, Saco, Summit, and Van were randomly harvested by hand in 2001 and 2002 at two stages of ripeness (partially ripe and ripe). The color of skin was used as the main indicator of maturity. After harvest, weight, skin color (measured by colorimeter), soluble solids, titratable acidity, and pH were recorded. The samples were prepared for analyses after 0, 5, 10, 15, 20, 25, and 30 d of storage at 1 to 2°C and 0, 3, and 6 d at room temperature (15°C). Total content of phenolics was determined and at the same time samples were prepared for HPLC analysis. Pitted and freeze-dried cherry samples were extracted by 60% methanol, filtered, and injected into the HPLC. The system was equipped with DAD and identification was conducted using spectral and retention time characteristics. The cultivar Saco contained the highest amount of phenolics (227 mg/100 g of fruit), whereas the cultivar Van contained the lowest amount (124 mg/100 g fruit). During storage at a low temperature of 1 to 2°C the content of phenolic acids decreased, but it increased at 15°C. Anthocyanins in the cultivar Van increased from 47 to 230 mg/100 g of fresh weight during storage at 15°C. Phenolic acids were higher in all samples from 2001 and anthocyanins were higher in 2002, indicating an important influence of seasonal conditions.

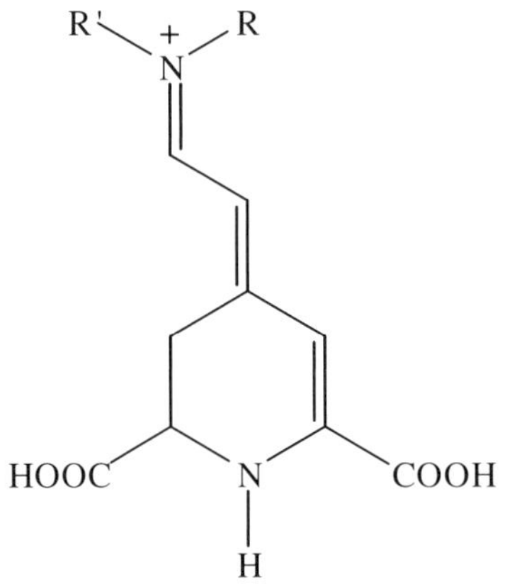

FIGURE 7.7 Betalains.

7.5 BETALAINS

Betalains are purple, red, pink, orange, and yellow water-soluble pigments. They occur exclusively in 13 families of Caryophyllales (Mabry et al., 1963), an order of dicotyledonous plants in the relatively small subclass Caryophyllidae characterized by free-central or basal placentation. Betalains occur also in some higher fungi. Betalains may be found in various organs of higher plants (Rosendal-Jensen et al., 1989), including roots, stems, leaves, bracts, flowers, fruits, and seeds. More than 50 naturally occurring betalains are known (Piatelli and Minale, 1964).

7.5.1 CHEMICAL DEFINITION AND STRUCTURE

The term *betalains* was introduced by Mabry and Dreiding (1968). They are immonium derivatives of betalamic acid (Piatelli, 1981). The chromophore of betalains is a protonated 1,2,4,7,7-pentasubstituted 1,7-diazaheptamethin system (Piatelli, 1976). Betalain's general formula is shown on Figure 7.7. There are two subgroups of betalains: red-violet betacyanins (see Figure 7.8) and yellow betaxanthins (see Figure 7.9). The structures of betacyanins may be derived from an aglycone betanidin with a formula in which the R group is –OH. For example, if the R group is –glucose, the formula represents betanin, or if the R group is –glucose 2-glucoronic acid-glucose, the formula represents amaranthin. The common names of betalains are assigned according to their botanical genus. For example, betacyanin obtained from *Beta vulgaris* is named "betanin" and betacyanin from *Amaranthus tricolor* is named "amaranthin." In all betacyanins two chiral carbons (C-2 and C-5) are present. Depending on the configuration on carbon 15, in the case of betanine and amaranthin the two epimers are betanine and isobetanin or amaranthin and isoamaranthin, respectively. In betacyanins an aromatic substituent contributes additional conjugation,

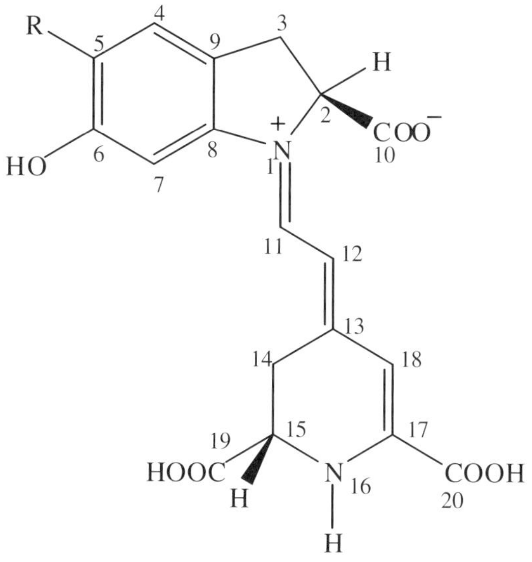

FIGURE 7.8 Betacyanins.

resulting in a bathochromic shift of maximum light absorption to a higher wave length of 540 nm, causing color to be red. Betaxanthins constitute the second subgroup of betalains. Due to the missing extended conjugation in betaxanthins the maximum light absorption is approximately 480 nm and the resulting color is yellow.

7.5.2 Chemical Changes During Processing and Storage

The stability of betalains is affected by pH, temperature, light, water activity, oxygen, and interaction with other compounds. Various factors affecting betalains' stability were reported in a comprehensive review of carotenoids, anthocyanins, and betalains published by Delgado-Vargas et al. (2000). The betalains are relatively stable in a wide range of pHs from 3.5 to 7.0. The optimal pH range is 5.5 to 5.8 (Huang and vonElbe, 1987). Mild alkaline conditions lead to the degradation of betanin and formation of betalamic acid (BA) and cyclodopa-5-O-glucoside (CDG). These two compounds are also formed during heating of betanine under acidic conditions. The formation of BA and CDG is reversible, with optimal regeneration at a pH range from 4.0 to 5.0 (Huang and vonElbe, 1985). In water solutions, heating causes a graduated reduction of red color and brown discoloration of betanin. The degradation rate increases after exposure to light (vonElbe et al., 1974). Ascorbic acid and isoascorbic acid decrease the light-induced oxidative degradation of betalains, and metal chelators such as EDTA or citric acid increase the stabilizing effect of ascorbic acid (Bilyk et al., 1981; Attoe and vonElbe, 1984). Water activity significantly influences the stability of betalains (vonElbe, 1987). The degradation of betanin in beet powder follows first-order kinetics, with the lowest rate at low water activity

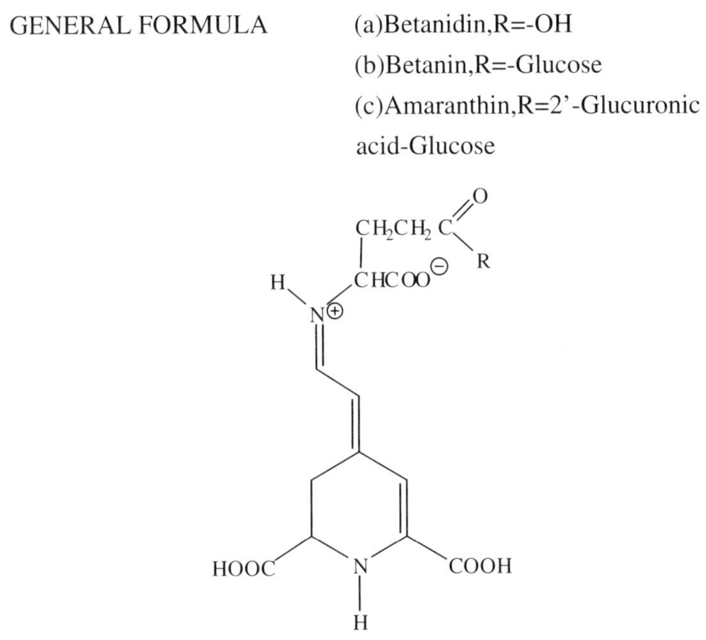

GENERAL FORMULA   (a)Betanidin,R=-OH

(b)Betanin,R=-Glucose

(c)Amaranthin,R=2'-Glucuronic

acid-Glucose

FIGURE 7.9 Vulgaxanthin-I, R = NH2; vulgaxanthin-II, R = OH.

(Cohen and Saguy, 1983). Degradation of betanin is influenced by oxygen in the air. Betanin solutions exposed to air for 6 d at 15°C degraded 15% more than samples under nitrogen (vonElbe et al., 1974). All the above-mentioned factors related to betalains' stability as well as potential off-flavor formation and economic feasibility must be taken into account when red beet and other preparations are used commercially as food colorants.

Jiratanan and Liu (2004) investigated antioxidant activity of processed table beets (*Beta vulgaris* var. *conditiva*) and green beans (*Phaseolus vulgaris* L.). Sliced beets were mixed to obtain a homogeneous sample, packaged in 8.8- × 6.2-cm cans, sealed, and heat-treated for 15, 30, and 45 min at 105, 115, and 125°C. The commercial condition for canning beets is 30 min at 115°C. The antioxidant activity of beets processed under commercial conditions remained constant despite an 8% loss of vitamin C and 60% loss of color. Changes of betacyanin and betaxanthin after thermal processing were measured by spectrophotometry. After 115°C thermal treatment for 15 min there was an approximate 50% loss of betacyanin. During the 115°C heating, the half-lives of betacyanin and betaxanthin were 22 and 26 min, respectively. The half-lives of pure pigments reported in the literature are approximately 50% of the half-lives reported by the authors for beet puree. The authors concluded that the antioxidants in beets may help to preserve the pigments.

Swiss chard is a potential edible, betalain-containing source of food colorant preparations. Kugler et al. (2004) investigated the betacyanin and betaxanthin composition of Swiss chard (*Beta vulgaris* l. ssp.*cicla* (L.) lef. cv. Bright Lights). Bright Lights is a mixture cultivar that produces differently colored plants. RP-HPLC and positive ion electrospray mass spectrometry were used for analyses. Samples of purple, red-purple, yellow-orange, and orange petioles were freeze-dried, extracted with 80% aqueous methanol, filtrated, concentrated in vacuum, and subjected to further separation and identification. Colorimetric measurements of the extracts were also recorded. Nineteen betaxanthins and nine betacyanins were identified in the extracts. Histamine-betaxanthin and alanine-betaxanthin were novel compounds not reported previously. Tyramine-betaxanthin and 3-methoxytyramine-betaxanthin were identified for the first time in Swiss chard. The compounds identified in purple petioles included betanin, isobetanin, betanidin, and isobetanidin.

7.6 CONCLUDING REMARKS

In the past few current years significant progress has been made in the chemistry of pigments and related compounds occurring in fruits and vegetables. Most of the research has been done in the area of carotenoids and polyphenolic compounds due to the presence of epidemiological and other data related to promotion of health and disease prevention, not necessarily due to the color function of these substances. A significant number of publications in the biochemistry of chlorophylls has been devoted to photosynthesis. Relatively less work related to betalains has been reported. To refresh the reader's memory, at the beginning of each section of this review short introductory information is included on chemical definitions and structures of selected pigments. Only a few figures with basic structures considered to be essential are presented in the text. Numerous and sometimes complex chemical structures and tables are available in many referenced publications. The introductory information in each section is followed by summaries of the most current reports. These reports are not equally distributed among all subjects of the review and are more or less numerous in some areas depending on the available publications. An effort was made to provide enough information to reduce the need to read the full articles. However, readers who are interested in more details will find them in referenced original publications, which in most instances also contain many citations to the past work in the specific areas. They may be especially important for those who have an interest in health aspects of produce pigments and related compounds that were not covered in this review due to the space limitation. Several monographs that contain comprehensive coverage of the past work on chemistry of pigments of fruits and vegetables are also cited here. The purpose of this review will be fulfilled if it contributes at least in part to the dissemination of current progress in the field of chemistry of produce pigments and motivates some young researchers to work in this area.

REFERENCES

Alonso-Salces R, Barranco A, Abad B, Berrueta L, Gallo B, and Vicente R (2004). Polyphenolic profiles of basque cider apple cultivars and their technological properties. *J. Agric. Food Chem.* 52, 2938–2952.

Amir-Shapira K, Goldschmidt E, and Altman A (1987). Chlorophyll catabolism in senescing plant tissues: *in vivo* breakdown intermediates suggest different degradative pathways for citrus fruit and parsley leaves. *Proc. Natl. Acad. Sci.* 84, 1091–1905.

Attoe EL and vonElbe JH (1984). Oxygen involvement in betanine degradation: Effect of antioxidants. *J. Food Sci.* 50, 106–110.

Bhosale P, Ermakou IV, Ermakova MR, Gellermann W, and Bernstein PS (2004). Resonance raman quantification of nutritionally important carotenoids in fruits, vegetables, and their juices in comparison to high-pressure liquid chromatography analysis. *J. Agric. Food Chem.* 52, 3281–3285.

Bilyk A, Kolodijand MA, and Sapers GM (1981). Stabilization of red beet pigments with isoascorbic acid. *J. Food Sci.* 46, 1616–1617.

Brouillard R (1982). Chemical structures of anthocyanins, in *Anthocyanins as Food Colors*, Markakis P, Ed., Academic Press, New York.

Chandra A, Rana J, and Li Y (2001). Separation, identification, quantification, and method validation of anthocyanins in botanical supplement raw materials by HPLC and HPLC-MS. *J. Agric. Food Chem.* 49, 3515–3521.

Cohen E and Saguy I (1983). Effect of water activity and moisture content on the stability of beet powder pigments. *J. Food Sci.* 48, 703–707.

Cortes C, Esteve M, Frigola A, and Torregrosa F (2004). Identification and quantification of carotenoids including geometrical isomers in fruit and vegetables juices by liquid chromatography with ultraviolet-diode array detection. *J. Agric. Food Chem.* 52, 2103–2212.

da Costa C, Horton D, and Margolis S (2000). Analysis of anthocyanins in foods by liquid chromatography, liquid chromatography-mass spectrometry, and capillary electrophoresis. *J. Chromatogr. A* 881, 403–410.

Delgado-Vargas F, Jimenez AR, and Paredes-Lopez O (2000). Natural pigments: Carotenoids, anthocyanins, and betalains-characteristics, biosynthesis, processing, and stability. *Crit. Rev. Food Sci. Nutr.* 40, 173–289.

Dewanto V, Wu XZ, Adom KK, and Liu RH (2002). Thermal processing enhances the nutritional value of tomatoes by increasing total antioxidant activity. *J. Agric. Food Chem.* 50, 3010–3014.

Eiro M and Heinonen M (2002). Anthocyanin color behavior and stability during storage: Effect of intermolecular copigmentation. *J. Agric. Food Chem.* 50, 7461–7466.

Enzell C and Back S (1995). Mass spectrometry, in *Carotenoids*, Britton G, Liaaen-Jensen S, and Pfander H, Eds., Birkhauser Verlag, Basel, pp. 261–320.

Eskin MAN, Grossman S, and Pinsky A (1977). Biochemistry of hpoxygenase in relation to food quality. *Crit. Rev. Food Sci. Nutr.* 9, 1–40.

Felicissimo M, Bittencourt C, Houssian L, and Pireaux J-J (2004). Time-of-flight secondary ion mass spectrometry and x-ray photoelectron spectroscopy analyses of *bixa orellana* seeds. *J. Agric. Food Chem.* 52, 1810–1814.

Francis F (1989). Food colorants: Anthocyanins. *Crit. Rev. Food Sci. Nutr.* 28, 273–314.

Francis F (1985). Pigments and other colorants in *Food Chemistry,* Fennema, O Ed. 2nd ed. Marcel Dekker, Inc. 545, 563–565.

Garcia-Alonso M, Rimbach G, Rivas-Gonzalo J, and de Pascual-Teresa S (2004). Antioxidant and cellular activities of anthocyanins and their corresponding vitisins A: Studies in platelets, monocytes, and human endothelial cells. *J. Agric. Food Chem.* 52, 3378–3384.

Garzon G and Wrolstad R (2002). Comparison of the stability of pelargonidin-based anthocyanins in strawberry juice and concentrate. *J. Agric. Food Chem.* 67, 1288–1299.

Giusti M, Rodriguez-Saona L, Griffin D, and Wrolstad R (1999). Electrospray and tanden mass spectroscopy as tools for anthocyanin characterization. *J. Agric. Food Chem.* 47, 4657–4664.

Goncalves B, Landbo A-K, Knudsen D, Silva A, Moutinho-Pereira J, Rosa E, and Meyer A (2004). Effect of ripeness and postharvest storage onphenolic profiles of cherries (*Prunus avium* L.). *J. Agric. Food Chem.* 52, 523–530.

Gonzalez-San Jose, ML, Barron, LJR, Junquera, B and Roberedo, M (1991). Application of principal component analysis to ripening indices for wine grapes. *J. Food Comp. Anal.* 4, 245–255.

Goodwin T (1976). Distribution of carotenoids, in *Chemistry and Biochemistry of Plant Pigments*, Goodwin T, Ed., Academic Press, New York.

Goodwin T and Goad L (1970). Carotenoids and triterpenoids, in *The Biochemistsry of Fruits and Their Products*, Hugme A., Ed., Academic Press, New York, pp. 305–368.

Grosch W, Weber F, and Fischer K (1977). Bleaching of carotenoid by the enzyme lipoxygenase. *Ann. Technol. Agric.* 26, 133–137.

Gross J (1987). *Pigments in Fruits*, Alden Press, Oxford.

Gross J (1991b). *Pigments in Vegetables, Chlorophylls and Carotenoids*, Van Nostrand Reinhold, New York.

Gudes de Pinho P, Ferreira S, Mendez Pinto M, Benitez J, and Hoge T (2001). Determination of carotenoid profiles in grapes, musts, and fortified wines from four varieties of vitis vinifera. *J. Agric. Food Chem.* 49, 5484–5488.

Halaby G and Fagerson I (1971). Polycyclic aromatic hydrocarbons in heat-treated foods: Pyrolysis of some lipids, beta-carotene and cholesterol, in *Proceedings of the 3rd International Congress of Food Science Technologists*, Chicago, IL, p. 820.

Heinonen M (1990). Carotenoids and provitamin A activity of carrot (*Daucus carota* L.). *J. Agric. Food Chem.* 38, 609–612.

Hong Y-H, Barrett D, and Mitchell A (2004). Liquid chromatography/mass spectrometry investigation of the impact of thermal processing and storage on peach procyanidins. *J. Agric. Food Chem.* 52, 2366–2371.

Hornero-Mendez D and Minguez-Mosquera MI (2000). Carotenoid pigments in *Rosa mosqueta* hips, and alternative carotenoid source for foods. *J. Agric. Food Chem.* 48, 825–828.

Hornero-Mendez D and Minguez-Mosquera MI (2000). Xanthophyll esterification accompanying carotenoid overaccumulation in chromoplast of *Capsicum annuum* ripening fruits is a constitutive process and useful for ripeness index. *J. Agric. Food Chem.* 48, 1617–1622.

Huang A and vonElbe JH (1985). Kinetics of the degradation and regeneration of betanine. *J. Food Sci.* 50, 1115–1120, 1129.

Huang A and vonElbe JH (1987). Effect of pH on the degradation and regeneration of betanine. *J. Food Sci.* 52, 1689–1693.

Hunter R (1975). *The Measurement of Appearance*, John Wiley & Sons, New York.

Hutchings J (1999). *Food Color and Appearance*, Aspen Publications, Gaithersburg, MD.

Hyun J and Chung H (2004). Cyanidin and malvidin from *Oryza sativa* cv. heugninjubyeo mediate cytotoxicity against human monocytic leukemia cells by arrest of G_2/M phase and induction of apoptosis. *J. Agric. Food Chem.* 52, 2213–2217.

Isler O (1971). Introduction, in *Carotenoids*, Isler O, Ed., Birkhauser Verlag, Basel, pp. 13–16.

IUPAC (1971). Tentative rules for the nomenclature of carotenoids, in *Carotenoids*, Isler O, Ed., Birkhauser Verlag, Basel, pp. 851–864.

IUPAC (1975). IUPAC commission on the nomenclature of organic chemistry and IUPAC-IUB commission on biochemical nomenclature. *Pure Appl. Chem.* 41, 407.

Jackman R and Smith J (1992). Anthocyanins and betalains, in *Natural Food Colorants*, Hendry GAF and Houghton JD, Eds., Kluwer, New York.

Jackson A (1976). Structure, properties and distribution of chlorophyll, in *Chemistry and Biochemistry of Plant Pigments*, Goodwin T, Ed., Academic Press, New York, pp. 1–63.

Jiratanan T and Liu RH (2004). Antioxidant activity of processed table beets (*Beta vulgaris var, conditiva*). and green beens (*Phaseolus vulgaris* L.). *J. Agric. Food Chem.* 52, 2659–2670.

Kahkonen M, Heinimak J, Ollilainen V, and Heinonen M (2003). Berry anthocyanins: Isolation, identification and antioxidant activities. *J. Sci.. Food Agric.* 83, 1403–1411.

Kalt W, Forney C, Martin A, and Prior R (1999a). Antioxidant capacity, vitamin C, phenolics and anthocyanins after fresh storage of small fruits. *J. Agric. Food Chem.* 47, 4368–4644.

Kalt W, McDonald J, Ricker R, and Lu X (1999b). Anthocyanin content and profile within and among bluberries species. *Can. J. Plant Sci..* 79, 617–623.

Karrer P and Tucker E (1950). *Carotenoids*, Elsevier, New York.

Kaur C and Kapoor H (2001). Antioxidants in fruits and vegetables: The millennium's health. *Int. J. Food Sci.. Technol.* 36, 703–725.

Kirca A and Bekir C (2003). Degradation kinetics of anthocyannis in blood orange juice and concentrates. *Food Chem.* 81, 583–587.

Kosar M, Kafkas E, Paydas S, and Baser K (2004). Phenolic composition of strawberry genotypes at different maturation stages. *J. Agric. Food Chem.* 52, 1586–1589.

Kugler F, Stintzing F, and Carle R (2004). Identification of betalains from petioles of differently colored swiss chard (*Beta velgaris* L. ssp. *cicla* [L.] Alef. Cv. bright lights) by high-performance liquid chromatography-electrospray ionization mass spectrometry. *J. Agric. Food Chem.* 52, 2975–2981.

Lopez-Da-Silva F, de Pascual-Teresa S, Rivas-Gonzalo J, and Santos-Buelga C (2002). Identification of anthocyanin pigments in strawberry (*cv Camarosa*) by LC using DAD and ESI-MS detection. *Eur. Food Res. Technol.* 214, 248–253.

Mabry T and Dreiding A (1968). The betalaines, in *Recent Advances in Phytochemistry*, Mabry T, Alston E, and Runelkles VC, Eds., Appleton Century Crofts, New York.

Mabry T, Taylor A, and Turner B (1963). The betacyanins and their distribution. *Phytochemistry* 2, 61–64.

Marais J, van Wyk C, and Rapp A (1989). Carotenoids in grapes, in *Proceedings of the 6th International Flavor Conference*, Rethymnon, Crete, Greece, Charalambous G, Ed., Elsevier, Amsterdam, pp. 71–85.

Marin A, Ferreres F, Tomas-Barberan F, and Gil M (2004). Characterization and quantitation of antioxidant constituents of sweet pepper (*Capsicum annuun* L.). *J. Agric. Food Chem.* 52, 3861–3869.

Markakis P (1982). Stability of anthocyanin in food. In: *Anthocyanins as food colors*, Markakis P Ed. Academic Press, London, pp. 163–181.

Matsufuji H, Chino M, and Takeda M (2004). Effects of paprika pigments on oxidation of linoleic acid stored in the dark or exposed to light. *J. Agric. Food Chem.* 52, 3601–3605.

Matsumoto M, Hara H, Chiji H, and Kasai T (2004). Gastroprotective effect of red pigments in black chokeberry fruit (*Aronia melanocarpa* Elliot) on acute gastric hemorrhagic lesions in rats. *J. Agric. Food Chem.* 52, 2226–2229.

Mazza G and Miniati E (1993). *Anthocyanins in Fruits, Vegetables, and Grains*, CRC Press, Boca Raton, FL.

Mendes-Pinto M, Silva-Ferreira A, Oliveira M, Beatriz P, and Guedes de Pinho P (2004). Evaluation of some carotenoids in grapes by reversed- and normal-phase liquid chromatography: A qualitative analysis. *J. Agric. Food Chem.* 52, 3182–3188.

Merken H and Beecher G (2000). Measurement of food flavonoids by high-performance liquid chromatography: A review. *J. Agric. Food Chem.* 48, 577–599.

Minguez-Mosquera MI and Hornero-Mendez D (1994a). Formation and transformation of pigments during the fruit ripening of *Capsicum annuum* cv. *Bola* and *Agridulce, J. Agric. Food Chem.,* 42, 38–44.

Minguez-Mosquera MI and Hornero-Mendez D (1994b). Changes in carotenoid esterification during the fruit ripening of *Capsicum annuum* cv. *Bola. J. Agric. Food Chem.,* 42, 640–644.

Moyer R, Hummer K, Finn C, Frei B, and Wrolstad R (2002). Anthocyanins, phenolics, and antioxidant capacity in diverse small fruits: *Vacciniu, Rubus* and *Ribes. J. Agric. Food Chem.* 50, 519–525.

Mullen W, Lean M, and Crozier A (2002). Rapid characterization of anthocyanins in red raspberry fruit by high-performance liquid chromatography coupled to single quadruple mass spectrometry. *J. Chromatogr. A* 2002, 63–70.

Nilsson M, Duarte I, Almeida C, Delgadillo I, Goodfellow B, Gil M, and Morris G (2004). High-resolution NMR and diffusion-ordered spectroscopy of port wine. *J. Agric. Food Chem.* 52, 3736–3743.

Ohloff G (1971). Classification and genesis of food flavors, in *Proceedings of the 3rd International Conference of Food Science Technologists,* Chicaco, p. 368.

Olsson M, Ekvall J, Gustavsson K, Nilsson J, Pillai D, Sjoholm I, Svensson U, Akesson B, and Nyman M (2004). Antioxidants, low molecular weight carbohydrates, and total antioxidant capacity in strawberries (*Fragaria* × *ananassa*): Effects of cultivar, ripening, and storage. *J. Agric. Food Chem.* 52, 2490–2498.

Onyewu P, Daun H, and Ho C (1982). Formation of two nonpolar thermal degradation products of beta-carotene. *J. Agric. Food Chem.* 30, 1147.

Onyewu P, Ho C, Daun H (1986). Characterization of beta-carotene thermal degradation products in model food system. *JAOCS* 63, 1437–1441.

Ouyang J, Daun H, Chang S, andHo C (1980). Formation of carbonyl compounds from beta-carotene during palm oil deodorization. *J. Food Sci.* 45, 1214–1217.

Parejo I, Jauregui O, Sanchez-Rabaneda F, Viladomat F, Bastida J, and Codina C (2004). Separation and characterization of phenolic compounds in fennel (*Foeniculum vulgare*) using liquid chromatography-negative electrospray ionization tandem mass spectrometry. *J. Agric. Food Chem.* 52, 2679–3687.

Perez-Galvez A, Hornero-Mendez D, and Minquez-Mosquera M (2004). Changes in the carotenoid metabolism of capsicum fruits during application of modelized slow drying process or paprika production. *J. Agric. Food Chem.* 52, 518–522.

Perez-Magarino S and Gonzales-San Jose M (2004). Evolution of flavanols, anthocyanins, and their derivatives during the aging of red wines elaborated from grapes harvested at different stages of ripening. *J. Agric. Food Chem.* 52, 1181–1189.

Piatelli M (1976). Betalains, in *Chemistry and Biochemistry of Plant Pigments*, Goodwin T, Ed., Academic Press, New York, pp. 560–596.

Piatelli M (1981). The betalains: Structure, biosynthesis and chemical taxonomy, in *The Biochemistsry of Plants: A Comprehensive Treatise. Secondary Plant Products*, Academic Press, New York.

Piatelli M and Minale L (1964). Pigments of centrospermae. III. Distribution of betacyanins. *Phytochemistry* 3, 547–551.

Pinsky A, Grossman S, and Trop M (1971). Lipoxygenase content and antioxidant activity of soem fruits and vegetables. *J. Food Sci.* 36, 571–572.

Price K, Colquhoun I, Barnes K, and Rhodes M (1998). Composition and content of flavonol glycosides in green beans and their fate during processing. *J. Agric. Food Chem.* 46, 4898–4903.

Prior R, Cao G, et al. (1998). Antioxidant capacity as influenced by total phenolic and anthocyanin content, maturity, and variety of *Vaccinum* species. *J. Agric. Food Chem.* 46, 2686–2693.

Razungles A, Bayonove C, and Cordonnier R (1988). Grape carotenoids: Changes during the maturation period and localization in mature grapes. *Am. J. Enol. Viticult.* 1, 44–48.

Rein M and Heinonen M (2004). Stability and enhancement of berry juice color. *J. Agric. Food Chem.* 52, 3106–3114.

Rice-Evans CA, and Parker, L, Eds., (1998). *Flavonoids in Health and Disease,* Marcel Dekker Inc. NY.

Robinson D, Wu Z, Casey R, and Domoney C (1995). Lipoxygenases and the quality of foods. *Food Chem.* 54, 33–43.

Romero C and Bakker J (1999). Interactions between grape anthocyanins and pyruvic acid, with effect of pH and acid concentration on anthocyanin composition and color in model solutions. *J. Agric. Food Chem.* 47, 3130–3139.

Rosendal-Jensen S, Johl-Nielsen B, and Dahlgren R (1989). Use of chemistry in plant classification. *SJPPL* 1, 66–89.

Schiedt K and Liaaen-Jensen S (1995). Isolation and analysis, in *Carotenoids*, Britton G, Liaaen-Jensen S, and Pfander H, Eds., Birkhauser Verlag, Basel, pp. 81–108.

Schreier P, Drawert F, and Bhiwapurkar S (1979). Volatile compounds formed by thermal degradation of beta-carotene. *Chem. Microbiol. Technol. Lebens.,* 6, 90–91.

Schwartz M, Wabnitz T, and Winterhalter P (2003). Pathway leading to the information of anthocyanin-vinylphenol adducts and related pigments in red wines. *J. Agric. Food Chem.* 51, 3682–3687.

Schwartz S and Lorenzo T (1991). Chlorophyll stability during aseptic processing and storage. *J. Food Sci.* 56, 1056–1062.

Schwartz S, Woo S, and vonElbe JH (1981). High performance liquid chromatography of chlorophylls and their derivatives in fresh and processed spinach. *J. Agric. Food Chem.* 29, 533–535.

Schwieter U, Englert G, Rigassi N, and Vetter W (1969). Physical organic methods in carotenoid research. *Pure Appl. Chem.* 20, 365.

Scordino M, Di Mauro A, Passerini A, and Maccarone E (2004). Adsorption of flavonoids on resins: Cyanidin 3-glucoside. *J. Agric. Food Chem.* 52, 1965–1972.

Schwartz S and Lorenzo T (1990). Chlorophylls in foods. *Crit. Rev. Food Sci. Nutr.* 29, 1–17.

Simpson K (1985). Chemical changes in natural food pigments, in *Chemical Changes in Food During Processing*, Richardson T and Finley J. Eds., AVI Publishing, Westport, CT, p. 409.

Spanos G, Wrolsted R, and Healtherbell D (1990). Influence of processing and storage on the phenolic composition of apple juice. *J. Agric. Food Chem.* 38, 1572–1579.

Surles R, Weng N, Simon P, and Tanumihardjo S (2004). Carotenoids profile and sensory evaluation of specialty carrots (*Daucus Carota,* L.) of various colors. *J. Agric. Food Chem.* 52, 3417–3421.

Talcott S, Brenes C, Pires D, and Del Pozo-Insfran D (2003). Phytochemical stability and color retention of copigmented and processed muscadine grape juice. *J. Agric. Food Chem.* 51, 957–963.

Taruscio T, Barney D, and Exon J (2004). Content and profile of flavanoid and phenolic acid compounds in conjunction with the antioxidant capacity for a variety of northwest *Vaccinium* berries. *J. Agric. Food Chem.* 52, 3169–3176.

Taylor R, Yelle L, Harris J, and Marenchic I (1990). Advances in HPLC and HPLC-MS on carotenoids and retinoids, in *Carotenoids: Chemistry and Biology*, Krinsky N, Mathews-Roth M, and Taylor R, Eds., Plenum Press, New York, pp. 105–123.

van der Sluis A, Dekker M, Skrede G, and Jongen W (2002). Activity and concentration of polyphenolic antioxidants in apple juice. 1. Effect of exisiting production methods. *J. Agric. Food Chem.* 50, 7211–7219.

van der Sluis A, Dekker M, Skrede G, and Jongen W (2004). Activity and concentration of polyphenolic antioxidants in apple juice. 2. Effect of novel production methods. *J. Agric. Food Chem.* 52, 2840–2848.

Vivar-Quintana A, Santos-Buelga C, and Rivas-Gonzalo J (2002). Anthocyanin-derived pigments and colour of red wines. *Anal. Chim. Acta* 458, 147–155.

vonElbe JH (1987). Influence of water activity on pigment stability in food products, in *Water Activity: Theory and Applications to Food*, Rockland L and Belichat L, Eds., Marcel Dekker, New York, pp. 55–83.

vonElbe JH, Maine I, and Amundson C (1974). Color stability of betanin. *J. Food Sci.* 39, 334–337.

vonElbe JH and Schwartz S (1996). Colorants, in *Food Chemistry*, Fennema O, Ed., Marcel Dekker, New York, pp. 651–722.

Vvedenskaya I, Rosen R, Guido J, Russell D, Mills K, and Vorsa N (2004). Characterization of flavonols in cranberry (*Vaccinium macrocarpon)* powder. *J. Agric. Food Chem.* 52, 188–195.

Wu Z and Robinson D (1999). Co-oxidation of β-carotene catalyzed by soy bean and recombinant pea lipoxygluases. *J. Agric. Food Chem.* 47, 4899–4906.

Yousef G, Seigler D, Grusak M, Rogers R, Knight C, Kraft T, Erdman J, Jr, and Lila M (2004). Biosynthesis and characterization of ^{14}C-enriched flavonoid fractions from plant cell suspension cultures. *J. Agric. Food Chem.* 52, 1138–1145.

Zhang Z, Kou X, Fugal K, and McLaughlin J (2004). Comparison of HPLC methods for determination of anthocyanins and anthocyanidins in bilberry extract. *J. Agric. Food Chem.* 52, 688–691.

8 Nutrient Loss

Lorraine Weatherspoon Theobald Mosha and Maria Nnyepi

Michigan State University, East Lansing, MI

CONTENTS

0-8493-1902-1/05/$0.00+$1.50
© 2005 by CRC Press

8.1   INTRODUCTION

Fruits and vegetables contain small to significant amounts of several key nutrient (e.g., carbohydrates, vitamins, and minerals) and nonnutrient substances (e.g., phytochemicals) that are critical to human health. In addition, fruits and vegetables are composed of living tissues that are metabolically active and for this reason their composition changes constantly. The rate and extent of such changes depend on the physiological role of the substances involved and the stage of maturity of the fruits and vegetables (Salunkhe et al., 1991). Although fruits and vegetables can make a significant contribution to human nutrition, they are not recommended as the sole source of nourishment. Rather, they are used advantageously to correct nutrition deficiencies (e.g., the role of citrus fruits in the correction of scurvy) or to complement other foods.

     Several of the compounds in fruits and vegetables are very sensitive to produce handling and the treatment processes that occur between the time produce leaves the farm to the time of purchase/consumption. Therefore, when the nutritional content of fruits and vegetables is considered, pre- and postharvest handling and treatment factors that influence nutrient composition should be taken into account. A few of the factors of importance in this respect include stage of maturity and ripening of fruits and vegetables during harvesting, disinfecting treatments, storage, and processing treatments.

     The maturity or ripening of fruits and vegetables as described by Salunkhe et al. (1991) is accompanied by a series of changes in color, texture, flavor, and chemical composition. These changes may or may not be associated with degradative or synthetic processes. In this chapter, nutrients common in fruits and vegetables and the pathways for nutrient losses will be discussed. In addition, factors that influence the nutrient content of fresh fruits and vegetables and nutritional quality, as well implications of produce processing, will be addressed.

8.2 OVERVIEW OF NUTRITIONAL VALUE OF PRODUCE

Fruits and vegetables are the major sources of dietary fiber, minerals, vitamins, and other potentially beneficial compounds. In addition, they contribute water, carbohydrates (more in fruits than in vegetables, for the most part), and, to a smaller extent, some protein and fats in selected products such as avocados and olives (Salunkhe et al., 1991). Seeds and legumes are not the focus of this book, but these foods would make more substantial protein and fat contributions than fresh fruits and vegetables. Many fruits and vegetables are excellent sources of provitamin A, vitamin C, vitamin E, magnesium, potassium, and folate as well as numerous biologically active phytochemicals and dietary fiber (Broekmans et al., 2000). Phytochemicals are defined as substances found in edible fruits and vegetables ingested by humans daily in gram quantities that exhibit a potential for modulating human metabolism (Jenkins, 2002). Table 8.1 lists the richest fruit and vegetable sources of specific

TABLE 8.1
Richest Fruit and Vegetable Sources of Selected Nutrient and Nonnutrient Compounds

Substance	Richest source
Vitamin C	Citrus and other fruits, green vegetables, potatoes
Vitamin E	Vegetable oils, avocados
Folates	Green leafy vegetables, potatoes, oranges
Vitamin K	Green leafy vegetables
Calcium, iron, magnesium	Green leafy vegetables
Potassium	Bananas, vegetables and fruits generally
Fiber, pectin, polysaccharides	Fruits and vegetables generally
Monounsaturated fatty acids	Olive oil
Alpha, beta-carotene	Carrots, green leafy vegetables, yellow/orange fleshed fruits
Beta-cryptoxanthin	Oranges and related fruits
Lutein	Yellow/green vegetables
Lycopene	Tomatoes
Flavonoids	Onions, apples, green beans
Flavanoids	Peaches, strawberries
Anthocyanins	Red/purple berries
Glucosinolates	Brassicas
Alkenyl cysteine sulfoxides	Alliums
Glycoalkaloids	Potatoes, aubergines
Fouranocoumarines	Parsnips, celery
Cyanogenic glycosides	Cassava, *Prunus* spp., butter beans

compounds. However, apart from one or two exceptions, these compounds are also present (in varying amounts) in most other fruits and vegetables. Most of the nutrients are essential for the human body. The amount of each of the nutrients required by the body depends on factors such as age, body mass, gender, physiological state, health status, and level of physical activity of individuals (National Academy of Sciences, 1999). Some nutrients are required only in small quantities, while others are required in relatively large amounts. Some of the key metabolic roles of these nutrients or nutrition-related components include healthy vision, healthy red blood cells, skin and bone strength and maintenance, enhanced immune system (some components in fruits and vegetables have the capacity to act as antioxidant, antiviral, and antibacterial agents), healthy neuromuscular and gastrointestinal systems, anti-inflammatory and anticancer agents, and cardiovascular health (Attaway, 1994; Middleton and Kandaswami, 1994; Meydani et al., 1997; Groff and Gropper, 2000; Volpe, 2000; Jongen, 2002; Fragakis, 2003).

8.2.1 MACRONUTRIENTS

8.2.1.1 Water

Water is an essential component of all body tissues and critical to the physiologic processes of digestion, absorption, and excretion. It is key to the structure and function of the circulatory system, maintains physical and chemical constancy of intracellular and extracellular fluids, and has a direct role in the maintenance of body temperature. Survival without water is limited to 10 days in moderate weather (Mahan and Escott-Stump, 2000). It is the most abundant component of fruits and vegetables. On average, fruits and vegetables contain 80 to 85% water. Grapes, strawberries, and tomatoes, for example, have a water content as high as 82, 90, and 93%, respectively (Holland et al., 1992). The maximum water content in fruits and vegetables may vary between fruits and vegetables of the same kind due to structural differences. These structural differences arise from the influence of agronomic conditions on structural differentiation of fruit and vegetable tissues (Salunkhe et al., 1991).

8.2.1.2 Proteins

The structure of humans is based on proteins, which are also a source of energy (Mahan and Escott-Stump, 2000). The quality of a protein in food is largely determined by the number and amount of essential amino acids, which cannot be synthesized by the body. A protein that contains all 10 essential amino acids (valine, threonine, tryptophan, isoleucine, leucine, methionine, lysine, phenylalanine, histidine, and arginine) is referred to as a *complete protein* (Potter, 1986). Apart from being a complete protein, a high-quality protein must have all the amino acids fully bioavailable to the body in the correct proportions. Incomplete proteins may be supplemented with the essential amino acids either in the form of synthetic compounds or as protein concentrates from natural sources.

Proteins usually constitute less than 1% of the fresh weight of fruits and less than 5% of the weight of fresh vegetables. The protein content of fresh fruits and vegetables is generally of low biological value compared to that of animal products.

However, some vegetables such as potatoes have highly digestible proteins, although the total protein content is low (2%) (Salunkhe and Kadam, 1991). Compared to other fresh vegetables, mushrooms have a high protein content. The protein content of mushrooms approaches that of beans, which ranges from 16 to 25% on a dry weight basis (Bano and Rajarathnam, 1988).

Protein in fruits and vegetables is also present in the form of simple nitrogenous substances that may be present in the uncombined form. For example, asparagine and glutamine and their acids aspartic and glutamic acids are abundant in fruits and vegetables such as strawberries, tomatoes, and citrus fruits. Asparagine is also very high in apples and pears, and oranges are rich in proline (Salunkhe et al., 1991). The protein content in these fruits and vegetables can be estimated by multiplying the total nitrogen content by a factor of 6.25. This figure is based on the fact that protein contains about 16% nitrogen.

8.2.1.3 Carbohydrates

Carbohydrates play a pivotal role in human energy modulation (especially blood glucose, can affect lipid and vitamin and mineral absorption, have an impact on fecal bulk and colonic motility, and can have protective influences (e.g., against cancer) (Mahan and Escott-Stump, 2000). Vegetables are generally lower sources of carbohydrates than are fruits. Carbohydrates consist of polysaccharides such as starch and fiber (cellulose, hemicellulose, and pectic material), and also of disaccharides and monosaccharides such as the sugars sucrose, fructose, and glucose (Wardlaw and Kessel, 2002). The amount of each of these constituents can change significantly, especially during ripening of fruits. Sugars are usually abundant when a fruit reaches full maturity. In fruits containing starch, all the starch is usually hydrolyzed to glucose and fructose, but in some fruits such as peaches, apricots, and nectarines sucrose is the main sugar (Wills et al., 1983). Traces of other mono- and disaccharide sugars such as xylose, arabinose, mannose, galactose, and maltose may also be present in fruits in small quantities. Sorbitol, a polyol related to sugars in structure, known for its laxative effect, is also present in relatively high concentrations in pears and plums (Wrolstad and Sallenberger, 1981). However, there are fruits, such as strawberries, that do not have sorbitol at all (Wrolstad and Shallenberger, 1981).

Fruits and vegetables are excellent sources of dietary fiber. Dietary fiber is the structural material of plant cells that is resistant to the digestive enzymes in the human stomach or intestines. It includes the structural polysaccharides of the cell wall such as cellulose, hemicelluloses, lignin, gums, pectins, and mucins with different chemical, physiochemical, and physiological properties (Ross et al., 1985; Dreher, 1987). Cellulose and hemicellulose, which have numerous functional uses in foods, are derived from the cell walls of fruits and vegetables. Pectin, which is also found in cell walls of fruits, is used commercially for the manufacture of jams and jellies because of its gelatinous or thickening properties. Pectin is also commonly extracted from the white spongy layers of citrus fruit skins, especially grapefruits and lemons.

The dietary and health benefits of fiber vary widely depending on the characteristics of polysaccharides. Total dietary fiber can be further classified as soluble or insoluble. Soluble fiber includes noncellulosic polysaccharides, pectins, gums, and mucilages (Dreher, 1987). They possess the capacity to imbibe water and swell, and thus they contribute to dietary bulkiness and viscosity in foods and may improve fecal incontinence (Prakash, 1995; Bliss et al., 2001). Soluble dietary fiber, particularly pectins and gums, also causes distension of the stomach, which results in feelings of satiety. Soluble fiber increases the activity of gastric enzymes by protecting them from degradation. It also delays absorption of glucose and fat after a meal, and it increases fecal loss of bile acids by binding and promoting their loss through fecal matter. In this respect, soluble fiber has been shown to have a potentially important role in glucose and fat metabolism and hence cardiovascular disease and diabetes (Florholmen et al., 1982; Anderson et al., 2000; Burke et al., 2001; Clark et al., 2002; Jenkins et al., 2002). Among the different fractions of fiber, lignin has the strongest binding capacity. Insoluble dietary fiber provides a carbohydrate substrate, which stimulates growth of bacteria in the large bowel. It increases fecal bulk and reduces transit time, both of which are important in cancer risk reduction (Chavan and Kadam, 1989; Dickerson, 1993).

In summary, vegetables and fruits provide the dietary fiber essential for bowel movement and possibly for decreasing the risk of health problems such as constipation and fecal incontinence, coronary heart disease, appendicitis, colon cancer, diabetes, diverticulosis, gallstones, and obesity as well as assisting in improving blood glucose control and insulin utilization in individuals with diabetes and decreasing blood pressure (Florholmen et al., 1982; Roy and Chakrabarti, 1993; Anderson et al., 2000; Bliss et al., 2001; Burke et al., 2001; Fleischauer and Arab, 2001; Clark et al., 2002; Hsing et al., 2002; Jenkins et al., 2002; Southon and Faulks, 2002).

8.2.1.4 Lipids

The lipid/fat content of fruits and vegetables is generally low and varies with the type of fruits or vegetables (olives and avocados are especially high in fat content). Lipids/fats and carbohydrates are the major sources of energy for humans. However, since fruits and vegetables are not good sources of fats, they contribute to total energy mostly through their carbohydrate content. To a limited extent, the proteins and organic acids found in fruits and vegetables can also serve as a source of energy (Holland et al., 1992).

8.2.2 Micronutrients

Fruits and vegetables contain significantly more micronutrients than proteins and fat (as depicted in Table 8.1). Generally, vegetables have a much higher content of minerals than fruits, while some fruits have a higher content of some vitamins relative to vegetables (e.g., vitamin C or ascorbic acid in citrus fruit). When viewed within the context of the total food supply, fruits and vegetables contribute a significant amount of the micronutrients compared to macronutrients (Figure 8.1): 90% of the dietary vitamin C, more than 50% of the vitamin A, and more than 35% of vitamin B_6. Hence, the importance of fruit and vegetables in human nutrition is clearly evident.

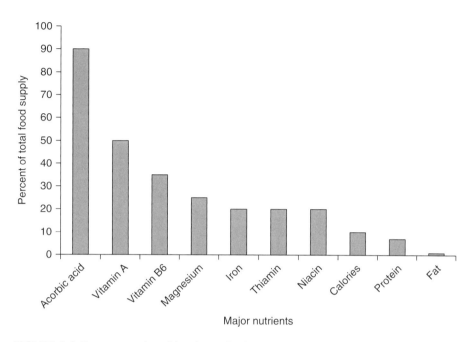

FIGURE 8.1 Percentage of nutritional contribution of fruits and vegetables of the total food supply. (Modified from Salunkhe et al. (1991), *Storage, Processing, and Nutritional Quality of Fruits and Vegetables*, Vol. 1, CRC Press, Boca Raton, FL.)

8.2.2.1 Minerals

About 14 mineral elements are considered to be essential for good nutrition, primarily as body process regulators (Mahan and Escott-Stump, 2002): calcium, sodium, zinc, iodine, copper, phosphorus, potassium, sulfur, fluoride, manganese, iron, magnesium, cobalt, and chloride. The concentration of minerals in fruits and vegetables varies depending on agronomical practices, types of soils, and the stage of maturity (Salunkhe et al., 1991). Overall, vegetables contribute moderate amounts of potassium, phosphorus, and calcium, while fruits contribute moderate amounts of potassium to the diet. Potassium in fruits occurs in combination with various organic acids (Hugo and Du, 1969). In fruits, the potassium/organic acid balance controls the pH of the tissue.

8.2.2.2 Vitamins

Vitamins are organic nutrients that are usually required in small quantities. They play various roles in biochemical reactions in cells within the body. Fruits and vegetables contribute 20 to 90% of various vitamins in the food supply (see Figure 8.1). For example, they are good sources of provitamin A, thiamin, riboflavin, niacin, pantothenic acid, and folic acid (Salunkhe et al., 1991). In general, deep yellow, red, orange, and dark green fruits and vegetables, such as mangoes, carrots, sweet potatoes, spinach, and tomatoes, are rich in provitamin A, while most green leafy vegetables are good sources of folate. On average, green leafy vegetables provide about 40 to 80 μg of folic acid per 100 g of edible portions (Salunkhe et al., 1991). Folate is also present in significant amounts in fruits such as strawberries, oranges, and grapefruits. Other fruits such as strawberries, black currants, and citrus fruits are moderate to good sources of ascorbic acid (Salunkhe et al., 1991). Plums and tomatoes are good sources of niacin and pantothenic acid.

Some vitamins, especially vitamins B_{12}, D, and E, are not present in significant amounts in fruits. Vitamin B_{12} is found in moderate amounts in vegetables, but it is not present in fruits. Vitamin D is also absent in fruits, while vitamin E occurs only in small quantities in fruits (Holland et al., 1992). Considerable differences in vitamin content are reported between fruit and vegetable species and varieties as well as within the same variety grown under different environmental conditions. Climate, soil, and agronomical practices affect the level of vitamins in fruits and vegetables. The application of fertilizer to plants, for example, affects the concentration of provitamin A in the leaves and fruit.

8.3 HEALTHY NONNUTRIENT COMPOUNDS

8.3.1 FLAVOR COMPOUNDS

Apart from the major nutrients, there are compounds in fruits and vegetables that play significant roles in food consumption and health. These compounds influence the appearance, color, taste, and aroma of fruits and vegetables, thus affecting consumption and subsequent nutrient benefit from these foods significantly. Flavor compounds in vegetables include sugars, amino acids, organic acids, aromatic compounds, hydrocarbons, aldehydes, acetals, ketones, alcohols, esters, and sulfur compounds (volatiles). The aromatic compounds in onions and garlic, for example, explain the common use of these vegetables as flavoring agents in a variety of soups, sausage, and curries. With respect to garlic, the volatile sulfur compound diallyl disulfide is responsible for the garlicky aroma. Diallyl disulfide is produced by the action of allinase on the amino acid allin present in garlic (Roy and Chakrabarti, 1993). In addition to flavor, garlic has been investigated as a potential health supplement for cancer prevention (Fleischauer and Arab, 2001; Hsing et al., 2002).

In fruits, organic acids play a role in taste through the sugar/acid ratio. Sugars provide the sweetness and organic acids the sourness in fruits. The main organic acids present in fruits are citric and malic acids. Citrus fruits, strawberries, pears, and tomatoes predominantly contain citric acid, while apples, plums, cherries, and apricots primarily contain malic acid. In peaches, these two acids are present in equal amounts. Grapes differ from the other fruits because they contain high levels of tartaric acid. In general, there is a decrease of acidity during ripening in all fruits. Bitterness in fruits, though not common, is a characteristics of flavonoids, such as naringin in grapefruits and limonene in citrus fruits. The aroma of fruits is a key factor in assessing quality as well as identity (Salunkhe et al., 1991). The major chemical compounds associated with aroma are esters of aliphatic alcohols and short-chain fatty acids.

8.3.2 PIGMENTS

In addition to flavor compounds, fruits and vegetables also contain a variety of pigments, some of which will be discussed in greater detail in section 8.3.3 (Phytonutrients). Garden beets owe their characteristic red color to betacyanin, which strongly overshadows the yellow pigment betaxanthin also found in beets (Roy and Chakrabarti, 1993). Other pigments such as chlorophyll, carotenoids, and anthocyanins are responsible for the color of other fruits and vegetables. Chlorophyll provides the green color of most green vegetables and fruits (e.g., spinach and apples), while carotenoids provide the orange color of citrus fruit, apricots, and peaches, and lycopene the red color of tomatoes. Anthocyanins provide the naturally red, blue, or purple colors of sweet potatoes, apples, plums, grapes, and strawberries (Roy and Chakrabarti, 1993). These pigments have been shown to have a potential role in disease prevention; lycopene from tomatoes has been suggested to help prevent prostate cancer (Bowen et al., 2002; Hadley et al., 2002; Kucuk et al., 2002; Giovannuci et al., 1999) and anthocyanins from tart cherries have been posited as antioxidants and anti-inflammatory agents (Wang et al., 1999).

8.3.3 PHYTONUTRIENTS

For the past 20 years, we have heard that fruits and vegetables are the cornerstone of health, supplying us with a wealth of vitamins, minerals, fiber, and complex carbohydrates. More recently, scientists have found another group of compounds within fruits and vegetables, phytonutrients or phytochemicals, which occur naturally in plants. Phytonutrients are biologically active organic substances in plants that give them their color, flavor, odor, and protection against plant diseases. During the past few years, scientists have discovered that many of these plant chemicals may also promote human health and protect the body against diseases (Van Duyen and Pivonka, 2000). Studies have consistently found that eating greater amounts of fruits and vegetables may reduce the risk of heart attacks, macular degeneration (the chief cause of blindness in adults), and most cancers (Gillman et al., 1995; Voorrips et al., 2000; Riboli and Norat, 2003). Unlike the traditional nutrients (protein, fat, vitamins, and minerals), phytonutrients are not "essential" for life; thus, some scientists prefer to call them *phytochemicals*. Phytonutrients are classified into 11 classes, some of which contain hundreds of different phytochemicals: carotenoids (e.g., carotene, α-, β-, γ-carotene, lutein, lycopene, zeaxanthin, β-cryptoxanthin, and capsanthin); dietary fiber; glucosinolates, indols, and isothiocyanates; inositol phophates; polyphenols (e.g., flavonoids, isoflavones, anthocyanidins, and catechins); phenol and cyclic compounds (e.g., terpenes); phytoestrogens; plant sterols; protease inhibitors; saponins; and sulfide- and thiol-containing compounds (e.g., sulphoraphane and allylic sulfides). Phytochemicals are found in a variety of foods such as onions, garlic, leeks, celery, apples, cranberries, raspberries, blueberries, some nuts such as almonds, red wine, tea, peppers, parsley, spinach, chives, broccoli, cabbage, kale, cauliflower, soybeans, tomatoes, citrus fruits, carrots, brussels sprouts, turnips, papaya, pineapples, whole grains, grapes, beans, legumes, cherries, strawberries, sweet potatoes, peaches, apricots, watermelons, guavas, and pumpkin (Salunkhe et al., 1991; Van Duyen and Pivonka, 2000).

TABLE 8.2
Common Phytonutrients, Potential Food Sources and Associated Postulated Health Benefits

Phytonutrient	Health benefit	Potential food sources
Flavonoids	Fights oxidation and blood clots	Apples, citrus fruits, cranberries, grapes, broccoli, celery, onions, tea, red wine
Carotenoids	Fights oxidation	Yellow/red fruits and vegetables: papaya, carrots, peppers, tomatoes, dark green leafy vegetables (e.g., spinach)
Allyl sulfides	May reduce blood cholesterol, helps liver detoxify carcinogens	Chives, garlic, leeks, onions
Isothiocyanates	May block carcinogens from damaging DNA	Cruciferous vegetables: broccoli, cabbage, sunflower
Indoles	May convert estrogen into less cancer-promoting form of the hormone	Cruciferous vegetables: broccoli, cabbage, sunflower
Terpenes	May help the liver to detoxify carcinogens	Citrus fruits: oranges, tangerines, limes, lemons
Isoflavones	May block entry of estrogen into cells, reducing the risk of breast, colon, or ovarian cancers; may alleviate menopausal symptoms	Soy

The proposed mechanisms by which phytonutrients act to protect human health include serving as antioxidants, enhancing the immune response, enhancing cell-to-cell communication, altering estrogen metabolism, converting to vitamin A (α-, β-, and γ-carotene are metabolized to retinal), causing cancer cells to die, repairing DNA damage caused by smoking and other environmental toxicants, and detoxifying carcinogens through the activation of the cytochrome P450 and Phase II enzyme systems (Gillman et al., 1995; Voorrips et al., 2000; Riboli and Norat, 2003). Table 8.2 shows some of the common phytonutrients, their food sources, and some associated health benefits.

Evidence that consuming phytonutrients in fruits and vegetables protects human health is accumulating from large epidemiological studies, human feeding studies, and cell culture studies (Gillman et al., 1995; Van-Duyen and Pivonka, 2000; Voorrips et al., 2000; Riboli and Norat, 2003). For example, in a study by Gillman et al. (1995), fruits and vegetables were linked to decreased risk of stroke — both hemorrhagic and ischemic. They observed that each increment of three daily servings of fruits and vegetables equated to a 22% decrease in the risk of stroke, including transient ischemic attack. In other studies by Gaziano et al. (1995) and Colditz et al. (1985), elderly men whose intake of dark green and deep yellow vegetables put them in the highest quartile for consumption of these vegetables had 46% decrease in the risk of heart disease relative to men who ranked in the lowest quartile. Men in the highest quintile had about 70% lower risk of cancer than did their counterparts in the lowest quintile. Men in the highest quartile or quintile consumed more than two servings (> 2.05 and > 2.2) of dark green or deep yellow vegetables a day; those in the lowest quartile or quintile consumed less than one serving daily (< 0.8 and < 0.7). This suggests that small but consistent changes in the type of vegetable consumption can make important changes in health outcomes.

Consumption of tomatoes and tomato products has also been linked to decreased risk of prostate cancer. A study by Giovannucci et al. (1995) showed that men in the highest quintile for consumption of tomato products (10 or more servings per week) had about a 35% decrease in risk of prostate cancer compared to counterparts whose consumption put them in the lowest quintile (1.5 or fewer servings of tomato products per week). It was also shown in a study involving 5,500 Italian people that eating tomatoes, which are rich in lycopene, was more effective in preventing digestive tract cancers than eating green vegetables. Individuals who ate tomatoes at least seven times a week had half the risk of developing these cancers compared to those who ate tomatoes only once a week. For a similar study involving the consumption of lutein-rich vegetables, spinach or collard greens, people in the highest quintile for consumption of these vegetables had a 46% reduction in the risk of age-related macular degeneration compared to those in the lowest quintile (consumed these vegetables less than once per month) (Seddon et al., 1994).

Flavonoid consumption has been linked to lower risk of heart disease in some studies. In a study of elderly Dutch men, those consuming the highest tertile of flavonoids had a 58% lower risk of heart disease than that of their counterparts who consumed the lowest tertile of flavonoids (Hertog et al., 1993). Those in the lowest tertile consumed 19 mg or less of flavonoid per day, whereas those in the highest tertile consumed approximately 30 mg per day or more. Similarly, Finnish subjects with the highest quartile of flavonoid intake had a risk of mortality from heart disease that was about 27% (for women) and 33% (for men) lower than that of subjects in the lowest quartile (Knekt et al., 1996). In the Iowa Women's Health Study involving more than 41,000 women, it was found that a diet consisting of garlic, fruits, and vegetables reduced the risk of colon cancer by 35% (Steinmetz et al.,1994). A similar study in the Shandong Province in China showed that people who ate garlic and onions regularly experienced almost 40% less stomach cancer compared to those who ate them occasionally (Giovannucci, 1999). However, in other studies the protective effect of flavonoids could not be confirmed. For example, in a study of Welsh men, flavonoid intake did not predict a lower rate of ischemic heart disease and was weakly positively associated with ischemic heart disease mortality (Hertog et al., 1997). In a separate study involving U.S male health professionals, the data also did not support a strong link between the intake of flavonoids and protection from coronary heart disease (Rimm et al., 1996).

In these studies and many others (Terry et al., 2001; Hsing et al., 2002; Riboli and Norat, 2003), fruits and vegetables consumption has been linked with a lower risk for chronic diseases including specific cancers and heart disease. However, media and consumer interest in phytonutrients and functional foods is far ahead of the established proof that documents the health benefits of these foods or food components for humans. Phytonutrient research is experiencing remarkable growth. It is hoped that more scientific information on phytonutrient consumption and human health will be forthcoming in the near future. For now, it appears that an effective strategy for reducing the risk of cancer, heart disease, and other chronic diseases is to increase consumption of phytonutrient-rich foods including fruits, vegetables, and whole grains.

8.5 ENZYMES AND HORMONES IN FRUITS AND VEGETABLES

Enzymes are important biological compounds that catalyze chemical reactions in both fruits and vegetables. Common enzymes in fruits and vegetables include phenolase, lipases, oxidases, and phenoloxidases found in apples, pears, grapes, strawberries, avocados, and figs (Salunkhe, 1991). They cause quality/structural deterioration of produce, which in turn affects acceptability and consumption. These enzymes are responsible for the discoloration of cut surfaces of fruits and vegetables when exposed to air. Other enzymes include polygalacturonase in tomatoes and peptic esterase in citrus fruits and tomatoes. These pectolytic enzymes are responsible for softening of fruits during ripening.

The ethylene hormone produced or used in the fruit ripening process has shown some effects on the nutritive value of fruits and vegetables. Ethylene has been reported to increase concentration of vitamin C/ascorbic acid in fruits when used to aid ripening. For example, the average content of ascorbic acid was different in tomatoes harvested mature-green compared to those harvested red-ripe (Saltveit, 1999). Variations in ascorbic acid content were large. The differences among cultivars were greater between ripe fruits harvested as mature-green or red-ripe. The ascorbic acid content of fruits harvested mature-green and ripened with the aid of ethylene was higher than for untreated fruit. Similarly, the content of ascorbic acid was significantly higher in papaya fruit ripened with the aid of ethylene than in controls left to ripen on their own (Bal et al., 1992). However, this effect was not consistent among fruits from different growers or fruits harvested at different times of the year. In both cases, the effect of ethylene was not directly on ascorbic acid only, but also through the stimulation of the other ripening parameters. As a result, the fruit ripened quicker, and there was less time for the loss of ascorbic acid.

Therefore, enzymes and hormones in fruits and vegetables might have an impact on nutrient content or losses. It is important to consider the extent to which these are present or used when determining nutrient differences or content in fruits and vegetables and processed products.

8.6 PATHWAYS OF NUTRIENT LOSS

The major pathways for potential nutrient loss in fruits and vegetables are summarized in Figure 8.2.

8.6.1 PREHARVEST

The postharvest storage life of fresh produce or processed products depends on preharvest agronomical and environmental conditions. These are important determinants of quality and nutritional value of fresh produce. Procedures such as early or late harvesting of fruits and vegetables have significant effects on the nutritional quality of produce (Salunkhe and Kadam, 1998).

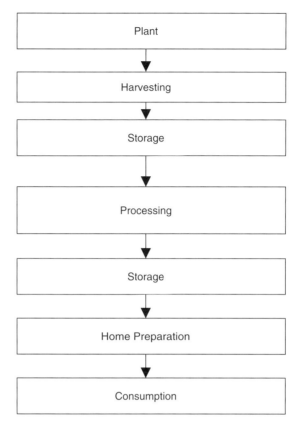

FIGURE 8.2 Possible route from the farm to consumption where nutrients can be lost from fruits and vegetables. (Modified from Salunkhe and Kadam (1998), *Handbook of Vegetables Science and Technology: Production, Composition, Storage and Processing*, Marcel Dekker, New York.)

8.6.2 HARVESTING

Harvesting, especially when done mechanically, can cause significant physical damage to the fruit and vegetable tissues, which may enhance nutrient losses. The bruising of apples during harvesting, for example, causes enzymatic browning, which in turn leads to loss of both physical and nutritional quality (Salunkhe and Kadam, 1998).

8.6.3 TRANSPORTATION

Transportation of produce from farms to the storage rooms may cause significant loss of nutrients if the produce is not properly packaged. For example, the transportation of fruits and vegetables in crates without bedding can cause crushing or bruising, which degrades both their nutritional and market value. Similarly, the mode of transportation (e.g., refrigerated or unrefrigerated) may also have an impact on fruit and vegetable quality and nutrient content, especially for temperature-sensitive products such as lettuce (Salunkhe et al., 1991).

8.6.4 STORAGE

Storage conditions (temperature, atmosphere, relative humidity) are critical for maintaining the nutritional quality of fruits and vegetables. Handling produce at high temperatures enhances the loss of heat-sensitive nutrients, particularly vitamin C, but also other components such as carotene, chlorophyll, and lycopene. The storage of produce at too low temperatures may also cause nutrient loss as a result of chilling injury. Thus, storage at too high or too low temperatures can decrease the nutritional and market quality of produce. The delay in cooling of some produce has also been reported to impair the physical appearance and biochemical processes (Nunes et al., 1995). Storage conditions may also enhance degradation processes such as transpiration, respiration, and degreening. Therefore, it is imperative that adequate storage conditions be maintained from harvest until consumption.

8.6.5 PROCESSING

Processing conditions may affect the nutrient quality of fruits and vegetables by enhancing losses of vitamins and minerals. Thermoprocessing of fruits and vegetables, for example, results in significant losses of vitamin C, thiamin, and vitamin A activity (Salunkhe et al., 1991). Light or minimal processing of produce such as fresh cutting makes the produce highly perishable because a large proportion of the surface area is exposed. If storage conditions such as temperature, atmosphere, relative humidity, and sanitation are not maintained during processing of produce, significant losses of physical, chemical, and market qualities occur (Watada et al., 1996). In some studies, significant losses of vitamin C were observed in stored fresh-cut potatoes, butterhead lettuce, and spinach (Varoquaux et al., 1996; Tudela et al., 2002; Evelyn et al., 2003).

8.6.6 PACKAGING

Packaging of fresh produce is important because it controls the atmosphere surrounding the produce (Watada, 1987; Watada et al., 1996; Watada and Qi, 1999). Proper packaging slows down the degradation processes of transpiration and respiration. Packaging also provides sanitary conditions for produce and minimizes the need for use of disinfectants (Zhang and Farber, 1996). When fruits and vegetables are packaged in permeable polyethylene bags, the relative humidity in the surrounding atmosphere increases, leading to a decrease in transpiration. Water loss is rapid at low relative humidity (RH) and slower at higher RH because the air in the room contains less water vapor than it can hold at the prevailing temperature. Thus, water vapor is readily transferred from the humid interior of the fruits and leafy vegetables to the relatively dry air. Excessive transpiration causes loss of water-soluble nutrients such as some vitamins and minerals and weight of the produce, which are important for market and nutritional value. Packaging also protects the produce from rapid evaporation caused by a high velocity of air circulating in the storage rooms (Zagory, 1999).

8.7 FACTORS INFLUENCING NUTRIENT CONTENT
OF FRUITS AND VEGETABLES

Nutrient content of fruit and vegetables is influenced by many factors such as stage of maturity, degree of ripeness, storage and handling conditions, and packaging materials and methods.

8.7.1 MATURITY AND RIPENESS

During maturation, the last stage of development, fruits and vegetables attain full size and optimum eating quality. Ripening is the terminal period of maturation when fruits develop the flavor, texture, and aroma that contribute to optimum eating quality. Thus, a fruit can be mature but unripe, and indeed many fruits are harvested while mature but unripe. This process is accompanied by softening, loss of astringency, biosynthesis of aroma volatiles, and conversion of starch to sugar. The ripening process in climacteric fruits is usually initiated by a growth regulator, ethylene. Ethylene, which is physiologically synthesized in the fruit, must reach a certain concentration/level before the ripening process can be triggered (Klein, 1989; Saltveit, 1999). Some growth regulators in fruits have been shown to promote ethylene synthesis. During the course of ripening, enzymes are synthesized that are used in the ripening process, but the enzymes cannot act until enough ethylene is present (Ryall and Pentzer, 1974). The stage of maturity of a fruit has a big influence on the nutritional quality. A substantial amount of research has been done on the composition of fruits and vegetables at different stages of development, mostly on apples and pears, but also on various other fruits and vegetables.

The important changes as the fruit goes through the last stages of maturity and ripening, including an increase in sugars; a decrease in acidity; a decrease in starch (apples, pears, and bananas); an increase in oil content in the case of avocados and nuts; changes in pectic constituents that result in softening, loss of tannins, and astringency; development of volatiles that contribute to flavor and aroma; formation of pigments in the skin and flesh; loss of the green pigment chlorophyll in the skin and flesh; an increase in ascorbic acid followed by a decrease; and an increase in β-carotene (provitamin A) (Watada et al., 1976). For example, the β-carotene (pro-vitamin A) content of tomatoes has been shown to vary directly with the ripeness of the fruit at harvest (Watada et al., 1976). However, fruits harvested mature-green, the stage at which most fresh market tomatoes are harvested, did not differ in vitamin A activity among various cultivars. The skin of many fruits also develops wax as they mature, giving the fruit protection against moisture loss. The percentage of soluble solids increases in fruits as sugar content increases. Some of these properties of fruits have been used as indices for harvest maturity and standardization for marketing. Notably, the sugar:acid ratio of citrus fruits, oil content of avocados, sugar content of grapes, and the firmness of the flesh and color of pears, peaches, and plums have been used as indices for nutritional and market quality (Aked, 2002).

Vegetables for fresh use and for processing should be of a satisfactory size and texture. Optimum maturity differs depending on the type of vegetable and the method

TABLE 8.3
Percentage of Fresh Fruits and Vegetables in Supermarket Display Cases Held at Their Optimum Temperature Range during the Summer

Sample Commodities	Percentage		
	Below Range	Within Range	Above Range
Apples, lettuce, cabbages	0	7	93
Mandarins, snap beans, summer squash	11	37	52
Winter squash, peppers	41	48	11
Sweet potatoes, bananas, green tomatoes	67	33	0

Source: Adapted from Paull, R.E. (1999), *Postharv. Biol. Technol.*, 15:263277.

of processing for the fresh market. For example, peas, corn, and green beans should have attained full size but still be tender, whereas mature potatoes are more desirable than those less mature (Salunkhe et al., 1991). Small cabbage heads are higher in ascorbic acid than large ones (Salunkhe et al., 1991). Small turnip leaves have relatively more thiamin and riboflavin content than ascorbic acid but less carotene than large leaves. In general, broccoli, cabbage, and cauliflower have the maximum content of nutritive value at the harvest time. It is exceedingly important that vegetables in general, and crucifers in particular, be harvested at the proper stage of maturity. Early-maturing fruits and vegetables usually have a shorter storage life (Salunkhe and Kadam, 1998). The exact degree of maturity at which a given cultivar of a species should be harvested also depends on the purpose for which it is used.

Fruits and vegetables harvested for the local fresh market must be harvested at prime maturity. When harvested for shipping or processing, they should be harvested a little earlier, for example, a week. However, if harvested too early and then stored for a long time, they may shrivel and the characteristic aroma will disappear, and if they are harvested too late and then stored for a longer period, they might have higher-than-average concentrations of sulfur compounds and low ascorbic acid and other water-soluble vitamins (Salunkhe and Kadam, 1998).

8.7.2 STORAGE CONDITIONS

8.7.2.1 Temperature and Relative Humidity

Temperature and humidity are the most important storage parameters that influence the nutritional quality of both fruits and vegetables. Most fruits and vegetables are exposed to unfavorable temperatures during the handling chain from harvest to the wholesalers and retailers. Fresh fruits and vegetables probably receive the greatest temperature abuse by retailers and consumers (Table 8.3). Temperature abuse is a function of time and temperature during holding and the relative perishability of a particular commodity. Some produce, for example, apples and cabbages, are often displayed at improper temperatures in retail stores, but they do not lose quality as rapidly as others such as strawberries and broccoli (Paull, 1999).

Temperature has a strong influence on the nutritional composition of fruits and vegetables, especially the vitamins. Loss of vitamin C is generally more rapid at higher storage temperatures and slower in acid fruit than in more neutral commodities (Watada, 1987). About 40% of vitamin C is lost from tangerines at higher storage temperatures (7 to 13°C) over 8 weeks, while there is a negligible loss in lemons at the same storage period. The loss is, however, significant at 24°C. In grapefruit stored at 8 and 12°C, the loss is not significant for up to 2 months (Paull, 1999). Immature potatoes show a rapid loss of vitamin C during the first weeks of storage, while mature tubers show minimal loss.

Leafy vegetables lose vitamin C postharvest, but it is frequently unclear if this loss is due to temperature or water loss. Kale, collards, turnip greens, spinach, grape, cabbage, and snap beans exposed to conditions favorable for water loss have more rapid water loss. However, wilting is much less important than temperature. The loss of vitamin C in kale is 0.32% per hour at 10°C and 0.05% per hour at 0°C with slow wilting, but when exposed to rapid wilting conditions the rate is 0.69 and 0.11% per hour, respectively (Paull, 1999). Optimal temperature needs to be maintained to minimize loss of vitamin C, particularly because 97% of the vitamin C in the U.S. diet is derived from fresh fruits and vegetables (Goddard and Matthews, 1979).

There are few reports on vitamin B_1 (thiamin) and niacin losses during storage. Potato tubers have an insignificant loss of vitamin B_1 and niacin after 30 weeks of storage at 5°C and a slight loss at 10°C. Small losses are found during storage of green beans, peaches, and sweet potatoes (Watada, 1987; Elkin, 1979). Folic acid losses of up to 40% can occur in potatoes stored at 7.5°C for 8 months (Augustin et al., 1978).

A key temperature-related factor is relative humidity (RH). Many studies in storage facilities indicate that temperature is strictly controlled, but the relative humidity is not. Relative humidity (RH) is influenced by air exchange rate, temperature distribution in the storage room, the type of commodity, and the packing materials used. There are some practical difficulties in maintaining RH in large storage rooms within a narrow range, particularly when the RH is high (Paull, 1999). These difficulties occur more when the RH is high because a small fluctuation in temperature (< 0.5°C) can result in condensation on cool surfaces. RH has a significant impact on the nutritional quality of fruits and vegetables during storage. Wilting of leafy vegetables leads to loss of vitamin C (Paull, 1999). The loss of vitamin C in kale increases under slow wilting conditions from 0.05 to 0.11% per hour under high wilting (low RH) conditions. Reducing water loss not only reduces leaf yellowing, but it also increases sweetness and retards protein degradation and the loss of vitamin C in *Brassica juncea* (Lazan et al., 1987). Carotene (provitamin A) loss is enhanced by storage at 0°C and a rapid wilting rate vs. a slow wilting of from 5 to 0% for kale and 13 to 2% for collards (Ketsa and Pangkool, 1994).

8.7.2.2 Packaging

Packaging is important in reducing nutrient losses in fruits and vegetables. Modified atmosphere packaging is particularly important because it reduces the rate of respiration in fruits and vegetables (Salunkhe and Kadam, 1998). Respiration is a complex process in which sugars and starch are converted into CO_2 and water. Respiration causes loss of weight (as some of the food components are oxidized), loss of sweetness, and change in texture. Generally, during storage of fruits and vegetables, there is an initial increase in sugar concentration, followed by a decrease (Potter, 1986). This is typical of climacteric fruits; the increase is due to the breakdown of polysaccharides during respiration. For example, starch hydrolysis in bananas results in equal concentrations of glucose and fructose and a little sucrose. When bananas are stored for a long time, the three sugars decrease in amount. Mangoes also have shown a large increase in sucrose, which is later converted to reducing sugars (Salunkhe and Kadam, 1998). In nonclimacteric fruits, changes in sugar content are slight and slow.

8.8 STABILITY OF NUTRIENTS AND OTHER HEALTHY COMPOUNDS IN FRUIT AND VEGETABLES DURING STORAGE

Both nutrients and nonnutrient compounds may be adversely affected during storage, as follows.

8.8.1 PROTEINS

The storage of fruit and vegetables influences the stability of proteins, amino acids, and other nitrogenous compounds. Generally, proteolysis occurs early in storage with a corresponding increase in free amino acids in fruits and vegetables (Salunkhe, 1974; Salunkhe et al., 1991). The free amino acids are then metabolized into nitrogen-storage compounds. These nitrogen-storage compounds are degraded at high temperature to liberate ammonia, which is accompanied by a rapid decrease in amino acid concentrations. Monitoring of this sequence of reactions has been used to characterize the degree of freshness of fruits and vegetables when they are stored at low temperatures. The actual storage conditions also affect the stability of amino acids in fruits and vegetables. For example, storage of potatoes in a high-nitrogen or in a low-oxygen and high-CO_2 atmosphere increases the level of free amino acids (Salunkhe and Kadam, 1998). The quantity of proteins, nitrates, and nitrites and free amino acids in fruits and vegetables are mostly governed by preharvest conditions such as climate, agronomic practices, varieties, soil fertility, maturity, and photosynthesis. Nitrogenous compounds, nitrates and nitrites, have received much attention in recent years because they have been associated with the methemoglobinemia condition in humans (Salunkhe et al., 1991). The conversion of nitrates to nitrites is the hazardous principle of nitrate accumulation in fruits and vegetables (Salunkhe et al., 1991). Nitrate formation is usually a postharvest process. During storage, nitrate may be reduced to nitrite, but the conversion rate depends mostly on the storage conditions.

The higher the storage temperatures, the higher the quantity of nitrites in fruits and vegetables. Nitrate reductase activity declines quickly or remains steady during storage of produce. Treatment with high CO_2 does not affect the nitrate concentration; however, packaging in nitrogen has been shown to reduce the nitrate content and keep the enzyme nitrate reductase activity high (Salunkhe, 1974; Salunkhe et al., 1991). Frozen and canned fruits and vegetables do not form nitrites (Salunkhe et al., 1991; Salunkhe and Kadam, 1998).

8.8.2 CARBOHYDRATES

During storage, reduction in firmness of fruits and vegetables may be caused by the degradation of insoluble protopectins to the more soluble pectic acid and pectin. If no ripening occurs in storage, changes in the pectic substances are minimal. For example, in grapefruits, no detectable changes are observed in soluble pectin or protopectin substances during storage for up to 6 weeks because no ripening occurs during storage (Salunkhe et al., 1991).

In produce, storage temperature and treatments also influence the production of volatile materials. Controlled-atmosphere storage lowers the production of volatile compounds, and fruits usually evolve volatiles fairly late after removal from a controlled atmosphere. Under very low oxygen conditions, off-flavors may develop in fruits due to accumulation of acetaldehydes. During storage, some enzymes also increase their activity. For example, catalase, pectinesterase, cellulase, and amylase increase their activity during storage. Oxidases usually exhibit a reduction in activity. The activity of most enzymes is dependent on storage temperature and maturity of the stored fruits. Fully mature fruits usually show higher activities of catalase and pectinesterase and a lower oxidase activity during storage compared to fruits picked at a relatively less mature stage. Conversely, starch hydrolytic enzyme activity is greater in immature fruits than in fully mature ones (Salunkhe et al., 1991).

8.8.3 LIPIDS

Storage also affects the lipid fraction in fruits and vegetables. Fruits usually become greasy in storage. The increase in the amount of waxy constituents of the cuticle is most noticeable in apples. Wax analysis shows that the oily fraction increases at a faster rate than the ursolic acid (Salunkhe, 1974). The nonvolatile esters are produced most rapidly during the early stages of storage, while the volatile esters appear much later. With prolonged storage, the soft wax fraction of the cuticle accumulates, increasing amounts of unsaturated compounds. The increase in wax, however, depends on the variety of fruit. For example, the surface wax of *Cox's Orange Pippin* apple variety remains constant in quantity and composition, while in the *Bramley* apple variety there is an increase in the amount of surface wax and esters (Salunkhe, 1974). As a whole, saturated fatty acids increase as the storage period increases and higher unsaturated fatty acids such as linolenic, linoleic, and oleic acids are metabolized rapidly during the early stages of the storage period (Salunkhe, 1974; Salunkhe et al., 1991). This may have implications for some products such as dehydrated potato products, because a high percentage of low-molecular-weight polyunsaturated fatty acids may cause oxidative off-flavors. Controlled-atmosphere storage increases the content of palmitic and palmitoleic acids and decreases oleic acid (Salunkhe et al., 1991). The percentage of polyunsaturated fatty acids in avocados is therefore higher in a more oxygenated atmosphere (Salunkhe et al., 1991).

8.8.4 Vitamins and Minerals

Storage of fruits and vegetables also affects the stability of vitamins. The type and amount of vitamins lost depend on the prior treatment of the produce before storage and the storage conditions. For example, trimming, dicing, or slicing of fruits and vegetables before storage enhances losses of water-soluble vitamins (vitamins C, B-group, E) through leaching. Vitamin C also becomes susceptible to degradation by oxidation (Salunkhe et al., 1991). Thermal processing treatment prior to storage significantly reduces the amount of thermolabile vitamins such as vitamin C and thiamine. Canning of produce prior to storage also causes degradation of the thermolabile vitamins. Storage temperatures and time are also important for the retention of vitamins. For example, retention of thiamine in canned spinach stored for 12 months at 50°F was 96%, but retention decreased to 71% when the fresh vegetable was stored at 50°F for the same length of time (Kadam and Salunkhe, 1998). In general, the decrease in vitamin content is more rapid at higher than at low storage temperatures. Retention of other vitamins, including vitamin C, niacin, carotene, and riboflavin, is also adversely affected by the storage time and conditions. Retention of minerals is not significantly affected during storage of fruits and vegetables. However, prestorage treatments such as thermoprocessing, canning, and chopping may reduce the quantity of minerals in fruits and vegetables. High-temperature storage over an extended period of time may also enhance mineral redox reactions and interactions, forming complexes that are not bioavailable (Salunkhe et al., 1991).

8.8.5 Pigmentation

Storage affects the content of pigments in fruits and vegetables, including chlorophyll and carotenoids. Generally, due to ongoing respiration, chlorophyll content decreases while concentration of other pigments may either increase or decrease depending on the storage temperatures, maturity, and variety. For example, in sweet potatoes, carotene content and total carotenoid pigments decreased during storage at low temperatures. Very small, immature tomatoes stored at 10°C maintained chlorophyll longer than larger, mature ones. Carotene content (provitamin A) shows little loss in sweet potatoes during 4 months of storage at 24°C, while significant losses of β-carotene occurs in kale (17%), collards (30%), turnip greens, and grapes held at 10°C instead of 0°C. Carrots show an increase in carotene during the first months of storage, even allowing for water loss and storage at different temperatures (Paull, 1999). There is a steady increase in lycopene and other pigments during tomato ripening at 15 and 30°C, while at less than 1°C and above 30°C no lycopene synthesis takes place (Paull, 1999).

Therefore, the stability of nutrients and other healthy components is dependent on several storage conditions, including duration, temperature, and atmospheric conditions. The primary storage objective should be directed at fostering ideal conditions to enhance nutrient retention and bioavailability (e.g., β-carotene), decrease losses (e.g., vitamin C), and increase consumer acceptability via adequate but not excessive ripening.

8.9 PRODUCE TREATMENT AND NUTRIENT QUALITY

8.9.1 PREHARVEST TREATMENT

Environmental conditions and agronomical treatments during the growth of the plant influence the quality of produce. The influence of soil moisture and temperature, fertilization, insecticides, and fungicides on the nutritional composition of fruits and vegetables has been widely recognized (Salunkhe et al., 1991). The climate of the crop areas, soil type, irrigation, pruning, thinning, and rootstock also exert an influence on the nutritional composition of the produce. Fertilizers, growth regulators, and pesticides have all been known to influence the quality of produce to varying degrees. Marked effects on yields and quality have been found in fruits and vegetables to which different types of fertilizers have been applied. Growth regulators have been known to modify size, shape, duration of maturation, and other growth characteristics. Pesticide residues give rise to flavor taints in fresh and processed products. If excessive amounts are used, fruits and vegetables may even produce toxic metabolites that may not necessarily be destroyed during processing. The same variety of plant grown in different locations will sometimes vary in composition, taste, and desirability. *Valencia* orange juice from California is, in general, more deeply colored and more acidic than that from Florida. Grapefruits from Florida have higher naringin content and a more pungent tang than those grown in California and Arizona. The latter possess a stronger acidic taste due to higher citric acid content. Likewise, Texas-grown grapefruits have higher soluble solids:acid ratio and lower naringin content than fruits from other states.

The ascorbic acid content of turnips and rutabagas is influenced by variations in soil fertility and moisture of the locations where they are grown (Salunkhe and Kadam, 1998). Growth in general is also influenced by the soil temperature because of the increased uptake of minerals. The nitrogen, phosphorus, and potassium are higher in the leaves when large quantities of minerals are available in the soil. The literature relating to the effect of fertilization upon the nutrient content is generally contradictory, presumably because of the variations in soil chemistry, water availability, and other environmental factors. It can thus be stated that adequate fertilization with nitrogen, phosphorus, potassium, calcium, magnesium, manganese, boron, and iron is essential for normal growth and adequate yields of high-quality nutritious crops (Winsor and Adams, 1975).

Seasons also influence the nutritional quality of produce. In general, the effect of season and light on the synthesis of any nutrient in a plant depends on the plant species, the amount and duration of light exposure, and temperature and soil fertility. In pineapples, for instance, summer fruit is generally richer in volatile flavor constituents than winter fruits. In tomatoes, the quantity of sugar increases as the growing season progresses (Winsor and Adams, 1975). The carotene content of carrots tends to decrease during fall and winter and increase in the early spring. Likewise, the ascorbic acid content of turnip greens is highest in the spring and lowest in the fall, but in other greens, such as kale, broccoli, and collards, the ascorbic acid content is not affected by the season. It has also been found that carotene content decreases with an increase in light intensity; thiamin content increases when trace elements are included in the fertilizer, while riboflavin content also increases with high light intensity and trace elements in the soil. Thiamin synthesis in plants is stimulated by light and generally occurs in the leaves and increases in concentration until the plant matures (Winsor and Adams, 1975; Salunkhe and Kadam, 1998). Therefore, the nutrients that seem to be most affected by season or light exposure are carbohydrates, carotene/provitamin A, ascorbic acid, thiamin, and niacin.

8.9.2 POSTHARVEST TREATMENT

8.9.2.1 Heat Treatments

During the last few years, there has been increasing interest in the use of postharvest heat treatments to control insect pests, prevent fungal rot, and affect ripening. Part of this interest is due to consumer demand for chemical-free produce. The demand for restricted use of chemicals has increased the interest in postharvest heat treatments that can be used as nondamaging physical treatment substitutes for chemicals. There are three methods commonly used to heat commodities: hot water, vapor heat, and hot air.

8.9.2.1.1 Hot Water Blanching

Hot water dips and sprays were originally used for fungal control, but their use has now been extended to include disinfection of insects. Hot water dips are effective for fungal pathogen control because fungal spores and latent infections are either on the surface or in the first few cell layers under the peel of the fruit or vegetable. Postharvest dips to control decay are often applied for only a few minutes at temperatures higher than heat treatments designed to kill insect pests located at the interior of a commodity. Only the surface of the commodity requires heating. Many fruits and vegetables tolerate exposure to water temperatures of 50 to 60°C for up to 10 min, but shorter exposure at these temperatures can control many postharvest plant pathogens (Lurie, 1999). In contrast, hot water dips for fruits require 90 min of exposure to 46°C. A combination of fungicide and heating may be applied to the produce simultaneously to increase pathogen destruction. For example, the use of the fungicides thiabendazole and imazalil with hot water has been very effective in citrus fruits (Lurie, 1999).

Hot water dips have been used for disinfecting fruits from insects as well. Since hot water is a more efficient heat transfer medium than hot air, when properly circulated through a load of fruits a uniform temperature profile is established in the bath. For disinfection, a longer treatment is necessary than for fungal control because the total fruit, not just the surface, has to be brought to the proper temperature. Procedures have been developed to disinfect a number of subtropical and tropical fruits from various species of fruit fly (Paull, 1994). The time of immersion can be as long as 1 h or more at temperatures below 50°C, in contrast to many antifungal treatments, which are for less time at temperatures above 50°C (Lurie, 1998).

A recent extension of the hot water treatment has been the development of a hot water spray machine (Fallik et al., 1996). This is a technique designed to be part of a sorting line, whereby the commodity is moved by means of brush rollers through a pressurized spray of hot water. By varying the speed of the brushes and the number of nozzles spraying the water, the commodity can be exposed to high temperatures for 10 to 60 sec. The water is recycled, but because of the temperatures used (50 to 70°C) organisms that are washed off the product into the water do not survive.

8.9.2.1.2 Vapor Heat

Vapor heat was developed specifically for insect control. It is a method of heating fruit with air saturated with water vapor at temperatures of 40 to 50°C to kill insect eggs and larvae as a sanitary and quarantine treatment before fresh market shipment (Animal and Plant Health Inspection Service, 1985). Heat transfer is by condensation of water vapor on the cooler fruit surface. This procedure was first used to kill the Mediterranean (*Ceratitis capitata* Wiedemann) and the Mexican (*Anastrepha ludens* Loew) fruit fly in a chamber without forced air (Lurie, 1999). However, once ethylene dibromide and methyl bromide came into use as inexpensive chemical fumigants, vapor heat was abandoned. With the ban on use of ethylene bromide in 1984, and the imminent removal of methyl bromide from use in 2010, the use of vapor heat has resurfaced (Gaffney et al., 1990). In modern facilities, the vapor heat includes forced air, which circulates through the pallets and heats the commodity faster than vapor heat without forced air. The treatment consists of a period of warming (approach time) that can be faster or slower depending on the commodity's sensitivity to high temperatures. This is followed by a holding period when the interior temperature of produce reaches the desired temperature for the length of time required to kill the insect. The last part is the cooling down period, which can be air cooling (slow) or hydrocooling (fast). There are thus a number of components of the treatment that can be manipulated to obtain the best combination for elimination of the insect pest with little to no damage of the commodity.

8.9.2.1.3 Hot Air

Hot air has been used for both fungal and insect control and to study the response of commodities to high temperatures. Hot air can be applied by placing fruit or vegetables in a heated chamber with a ventilating fan or by applying forced hot air where the speed of air circulation is precisely controlled. Hot air, whether forced or not, heats more slowly than hot water immersion or forced vapor heat, although forced hot air will heat the produce faster than a regular heating chamber. The hot air chamber has been utilized to study physiological changes in fruits and vegetables in response to heat and to develop quarantine procedures (Gaffney and Armstrong, 1990; Klein and Lurie, 1992). The high humidity in vapor heat can sometimes damage the fruit or vegetable being treated, while the slower heating time and lower humidity of forced hot air causes less damage. A high temperature, forced air quarantine treatment to kill the Mediterranean fruit fly, melon fly, and oriental fruit fly on papayas has been developed (Lurie, 1999). This procedure may, however, require rapid cooling after the heat treatment to prevent fruit injury, especially in forced hot air treatment for citrus.

Exposure to high-temperature forced or static air can also decrease fungal infections. This requires long-term heating, from 12 to 96 h at temperatures ranging from 38 to 46°C, and is therefore unlikely to become commercially attractive. However, the potential for hot air treatment to benefit commodity physiology, as well as prevent both insect and fungal invasion, justifies further development of these treatments.

Ripening of most climacteric fruits is characterized by softening of flesh, an increase in the sugar:acid ratio, enhanced color development, and increased respiratory activity and ethylene production. Exposing fruits and vegetables to high temperatures attenuates some of these processes while enhancing others. This anomalous situation results in heated fruit's being more advanced in some ripening characteristics than unheated fruits while maintaining quality longer during shelf life at room temperature.

The inhibition of ripening by heat may be mediated by its effect on the ripening hormone, ethylene. Hot air treatment of 35 to 40°C inhibits ethylene synthesis within hours in both apples and tomatoes (Biggs et al., 1988; Klein, 1989). The inhibition of ethylene formation is reversed when the fruits are removed from heat (Biggs et al., 1988) and often the level of ethylene rises to higher levels than in unheated fruits (Lurie and Klein, 1992). During the heating period, not only is endogenous ethylene production inhibited, but fruits do not respond to exogenous ethylene as well (Yang et al., 1990). This indicates either loss or inactivation of ethylene receptors or the inability to transfer the signal to the subsequent series of events leading to ripening. No information is available on the response of ethylene receptors to heat, but it has been shown that the expression of tomato ripening genes is inhibited by high temperature (Picton and Grierson, 1988).

Specific mRNAs associated with ripening process enzymes were found to disappear during a 38°C hot air treatment of tomatoes and to reappear during recovery from heat (Lurie et al., 1996). These included ACC oxidase, polygalacturonase, and lycopene synthase. Fruits subjected to extended hot air treatments often soften more slowly than unheated fruits. For example, plums, pears, avocados, and tomatoes soften more slowly when held continuously at temperatures of 30 and 40°C than at 20°C. The rate of softening increased when heated fruits were returned to 20°C, but it was still less than that of unheated fruits (Lurie, 1998). In cell wall studies of apples, less soluble pectin and more insoluble pectin was found after exposure to 38°C air for 4 d than in fruit that had not been heated (Lurie, 1998). This indicated that degradation of uronic acid was inhibited (Lurie, 1998).

In addition, in the heated apples, less calcium was present in the water-soluble pectin and more was bound to the cell wall. It was thought that this was the result of the activity of pectin esterase creating more sites for calcium binding. However, a study of heated and unheated fruits showed a similar degree of esterification (Klein and Lurie, 1992). During the heating period, arabinose and galactose content decreased with no accompanying decrease in uronic acid. It was thus possible that loss of neutral sugar side chains during the heat treatment may lead to closer packing of the pectin strands and in turn hinder enzymic cleavage during and after storage, which resulted in firmer fruit (Lurie, 1998). The decrease in the rate of softening may be due to inhibition of the synthesis of cell wall hydrolytic enzymes such as polygalacturonase and α- and β-galactosidase (Lurie, 1998). For example, in tomatoes, mRNA for polygalacturonase was absent in fruits during heat treatment of 1 to 3 d at 38°C and reappeared after the fruits were removed from heat (Lurie et al., 1996). Depending on the length of heat treatment, heated tomato fruits may recover and soften to the same extent as unheated fruits or remain firmer than unheated fruits.

Heat treatment also affects flavor characteristics of fruits and vegetables. For example, titratable acidity declines in apples held for 3 or 4 d at 38°C, while the concentration of soluble solids is not affected by heating. The same effect has been shown in other fruits such as nectarines and strawberries, while in other fruits, such as tomatoes and grapefruits, the effect was not seen. Heat treatment also increases sugar content of some fruits and vegetables. For example, when squash fruits were subjected to 3-h heat treatment at 30°C, their sucrose content increased significantly. The heat-treated squash were perceived as sweeter by a test panel (Bycroft et al., 1997). Likewise, heat-treated Golden Delicious apples treated for 4 d at 38°C were perceived as crisper, sweeter, and overall more acceptable than unheated fruit. In the apples, the sweetness was due more to a decrease in acidity than to an increase in sugar content.

8.9.3 CHILLING TREATMENTS

Overexposure of produce before and during harvest to high environmental temperatures might be harmful to the quality of fruits and vegetables, such as when produce is harvested during the summer. Precooling is one of the methods used to remove heat from fruits and vegetables when they are taken from the field. The aim of precooling is to quickly slow down the respiration rate, minimize microbial growth, and reduce transpiration rate. Various methods are used for chilling produce. These include air cooling, hydrocooling, and vacuum cooling. Air cooling involves forced cold air (~ 1.5°C) flow through the produce in cold rooms. Hydrocooling involves submerging the fruits or vegetables in ice-cold water to immediately lower their temperature. This is done for only a limited time and is followed by another cooling method such as air cooling. Vacuum cooling is a rapid method of cooling vegetables. The vacuum cooler consists of a large autoclave with steam injectors that works by evaporative cooling. In this case, since water can be removed from the produce, it must be properly packaged in nonporous material to minimize wilting (Salunkhe et al., 1991).

Improper chilling treatments can result in chilling injuries in both fruits and vegetables. A chilling injury is a disorder induced by low, but not freezing, temperatures that occurs in certain susceptible fruits and vegetables. Usually this damage can occur in tropical fruits and vegetables when they are stored at low refrigerated temperatures. Chilling injury may occur most often in sweet potatoes, bananas, apples, olives, oranges, tomatoes, avocados, cucumbers, lemons, limes, mangos, melons, papayas, eggplants, and other tropical and subtropical fruits and vegetables. Chilling injury induces decay by enhancing tissue breakdown, which would lead to nutrient losses. Some of the chilling injury effects observed in various fruits and vegetables are internal browning (apples, avocados, olives, pineapples), pitting (oranges, avocados, lemons, limes, melons, papayas, tomatoes), brown stains (oranges, bananas, cucumbers, lemons, mangos), surface scalds (apples, eggplants) and soft rots (apples, tomatoes, melons) (Salunkhe et al., 1991). Chilling, similar to heat treatment, can potentially have negative physical and nutritional effects that can be minimized by strict temperature control.

8.9.4 Waxing

Waxing of certain fruits and vegetables reduces their rate of respiration and enhances product gloss, improving merchandising and marketing (Salunkhe et al., 1991). For example, waxing of cassava tubers extends the market life of the roots for up to 1 month by reducing the rate of gas transfer between the tissues and the atmosphere (Salunkhe et al., 1991; Kadam and Salunkhe, 1998). Waxing is applied by foaming, dipping, or brushing. Waxing emulsions may contain some fungicide to protect the produce from fungal attack. They also reduce losses due to transpiration and respiration. In some fruits, for example citrus, dyes are added to the waxes to enhance the merchandising quality and prolong shelf life by partially covering lenticels and stomata, thus reducing water and water-soluble nutrient losses due to transpiration and respiration (Salunkhe et al., 1991).

8.9.5 Coating with Edible Films

Preservation of fruits and vegetables by using edible films is becoming more popular nowadays, although information on the nutrient retention of preserved fruits and vegetables preserved by edible films is not widely available. Edible films can act as barriers to prevent losses of flavor, texture, and nutrients (McCarthy and Matthews, 1994). For example, a casein-based film with a small amount of ascorbic acid added for increased protection against browning kept a sample of small pieces of sliced and peeled apples fresh for six d, while unfilmed apple pieces shriveled and turned brown within 2 h. Likewise, avocados treated by edible films remained fresh for 6 d without blackening (McCarthy and Matthews, 1994). Another edible film made of vegetable oil, cellulose, and an emulsifier is used in fruits and vegetables as an antioxidant atmospheric barrier. In a laboratory experiment with mature green tomatoes, only 40% of the tomatoes treated with the film ripened after 14 d of storage, compared with 100% ripening for the untreated tomatoes. Similar experiments with oranges and carambolas showed excellent results. This emulsified edible film has proved to be an invaluable aid to the fresh and minimally processed fruits and vegetable industry in preventing spoilage and retaining nutrients (Sanchez, 1990). With the increasing trend toward buying foods that are ready to use and easily stored, edible films have a potential benefit to consumers, restaurants, and others in the food industry who want to purchase fresh and minimally processed fruits and vegetables (USDA, 1989).

8.10 NUTRITIONAL IMPLICATIONS OF PRODUCE PROCESSING

The aim of produce processing is to improve the storage life and extend the availability of perishable foods over a longer period of time. The food processing techniques used, however, may limit the availability of some essential nutrients. Maximizing nutrient retention during thermal processing of produce has been a considerable challenge for the food industry. The losses of nutrients as a result of processing have been divided into three categories: intentional, accidental, and inevitable. Intentional losses occur due to intentional removal of parts of fruits and vegetables, for example, during peeling. Accidental or avoidable losses occur as a result of inadequate control and handling of food materials. The major concern from a food-processing point of view is the inevitable losses that represent the loss of heat-labile nutrients destroyed to some degree by heat. The extent of these losses depends on the nature of the thermal process (blanching, pasteurization, or sterilization), the raw materials, and preprocessing preparation, because operations such as size reduction (slicing and dicing) result in increasing losses through increasing the surface-to-volume ratio. All water-soluble vitamins and minerals as well as some parts of soluble proteins and carbohydrates are susceptible to losses during produce processing. In any processing operation, therefore, the emphasis is on reducing the inevitable nutrient losses through adoption of appropriate time–temperature processing conditions as well as environmental factors (pH and concentrations) that will ensure maximum retention of the nutrients.

Because *processing* in fresh fruits and vegetables is a broad term that is difficult to define, for the scope of this chapter those treatments to fruits and vegetables that begin after harvest before consumption will be discussed. These treatments include, but are not limited to, heating, refrigeration, dehydration, chemical treatment, and radiation. Changes in the composition of raw fruits and vegetables, which may decrease their nutritive value, can occur after harvest, during transportation, holding, handling, processing, and subsequent storage and distribution. After they are harvested, fruits and vegetables are still physiologically active. Enzymatic and respiratory processes may bring about profound changes unless they are controlled. The demand for fresh, healthier, convenience-type fresh-cut fruits and vegetables that are minimally processed has increased tremendously in the past decade (Zhang and Farber, 1996; Watada et al., 1996). In the U.S., for instance, the sales of fresh-cut fruits and vegetables exceeded U.S.$19 billion in the year 1999. The minimally processed fruits and vegetables, however, are generally more perishable than the original raw materials.

Injury stress due to procedures such as peeling, cutting, shredding, trimming, slicing, coring, and grating greatly increase tissue respiration and lead to various biochemical deteriorations such as browning, off-flavor development, and texture breakdown. Moreover, minimal processing may increase microbial spoilage of the product through transfer of skin microflora to the fruit and vegetable tissues (Bolin and Huxsoll, 1989; Pittia et al., 1999). For example, fresh-cut vegetables support the survival and growth of foodborne pathogens such as *Listeria monocytogens*, and these foods have been implicated in outbreaks of foodborne illness (Nguyen-The and Carlin, 1994).

Improvement of the shelf life of fresh and minimally processed fruits and vegetables may be achieved by applying one or a combination of many of the classic preservation procedures such as heat treatment, refrigeration, dehydration, fermentation, and additives. The effectiveness of one or several combinations of these preservation methods on the shelf-life extension of minimally processed products has been evaluated for many different categories of foods (Bolin and Huxsoll, 1989; Pittia et al., 1999). The choice of an optimum multiple preservation method is very difficult due to the various and complex biochemical and microbiological changes that should be inhibited or eliminated. It is thus reasonable to state that the type of equipment used, blanching method, time, temperature, maturity, and cultivar of the produce would influence the degree to which the nutritive composition changes.

8.10.1 BLANCHING

One of the ancient and most often used heat treatments of fresh fruits and vegetable is blanching. This is a short heat treatment of fruits and vegetables to kill microorganisms on the surface of the produce and inactivate some of the thermolabile enzymes that cause quality deterioration. Blanching also helps to expel gases from the produce, in preparation for canning. Blanching can be done by immersing the produce in hot water (water blanching), passing a stream of steam through the produce (steam blanching), or microwaving the produce for a short time (microwave blanching). The influence of blanching on the degradation of the nutritive composition varies from product to product. Water blanching has been shown to be associated with enormous loss of the water-soluble vitamins such as ascorbic acid, thiamin, riboflavin, and niacin. Steam and microwave blanching has been shown to be more effective in the retention of the water-soluble vitamins (Mosha et al., 1995). For example, green vegetables blanched by the microwave method showed significantly higher retention (40 to 60%) of carotene and water-soluble vitamins relative to conventional hot water blanching. It was also more effective in inactivating antinutritional factors such as trypsin and chymotrypsin inhibitors than was hot water blanching (Mosha et al., 1995; Mosha and Gaga, 1999). Broccoli packaged in plastic bags and microwave-blanched for 20 to 30 sec indicated 100% inactivation of catalase and peroxidase enzymes. There was also 100% retention of ascorbic acid relative to 37 to 100% retention in steam blanching and 24 to 93% retention in water blanching. In these cases, the losses of vitamin C were mostly due to leaching rather than to destruction (Salunkhe, 1974).

Although heat treatments are generally effective in inhibiting enzymatic reactions and in reducing microbial loads, they are rarely used for stabilizing minimally processed fruits and vegetables due to their negative effects on flavor, texture, and fresh-like quality. Conversely, the development of suitable heat treatments associated with a low negative impact could be of great interest in the production of minimally processed fruits and vegetables with a longer shelf-life that could be used as semi-manufactured and final products. Low temperature blanching treatments that are associated with activating pectin methylesterase and cross-linking with cations have shown a good retention of color and texture in fresh fruits and vegetables (Bourne, 1987). Therefore, the type of heat treatment needs to be considered on multiple levels for optimum product quality (physical and nutritional), as well as safety, and would obviously be product- and process-specific.

8.10.2 REFRIGERATION AND FREEZING

Cold storage is one of the oldest methods for preserving shelf life and nutrients in fresh fruits and vegetables. It involves keeping the produce at low temperatures (at least between 0 and 10°C) (Salunkhe, 1974; Paull et al., 1997). Refrigeration helps to minimize water loss from the produce and slows down the rate of respiration, enzyme activity, and growth and proliferation of microorganisms. Fresh fruits and vegetables are usually packaged in permeable polyethylene bags to minimize evaporation and control relative humidity. Refrigerated produce also may have received other treatments such as blanching, irradiation, waxing, and disinfection by chemicals and controlled-atmosphere packaging. Low temperature treatments have been shown to be effective in retaining the thermolabile nutrients in fruits and vegetables. For example, ascorbic acid, folate, and carotene in broccoli, cabbage, cauliflower, kale, collard and turnip greens, radishes, and kohlrabi were destroyed rapidly at room temperature but significantly retained under refrigeration (Salunkhe, 1974). A sample of brussels sprouts stored at 0°C for 72 d showed lower respiration rate and maintained a satisfactory organoleptic quality compared to that stored at 14 to 20°C, which decolorized and decayed (Salunkhe, 1974).

Minimally processed fruits and vegetables might be more perishable than unprocessed produce during refrigeration because of the potential for subjection to some physical stress. They also have a higher rate of respiration than unprocessed, intact produce (Watada et al., 1996). Consequently, minimally processed fresh fruits and vegetables should be held at lower temperatures than those recommended for intact commodities. Although 0°C generally is the desirable temperature for most minimally processed fruits and vegetables, many are prepared, shipped, and stored at 5°C, and sometimes as high as 10°C. This is mainly due to concerns about chilling injury, which occurs around 0°C. The deterioration of quality caused by enzyme activity and infection by pathogens that occurs when a produce is stored outside the chilling temperature is higher than it would have been if the produce had been stored at chilling temperatures. It is strongly recommended, therefore, that chilling-sensitive, minimally processed products be held at chilling temperatures at which injury from chilling will be of less consequence than the deterioration that occurs at nonchilling temperatures (Watada et al., 1996).

Freezing is a cooling process that involves storing the produce at temperatures below 0°C. Fruits and vegetables are usually frozen after blanching. When frozen without blanching, significant losses of ascorbic acid occur during storage (Salunkhe, 1974). Freezing generally does not cause significant nutrient losses if proper packaging and processing conditions are observed. However, during thawing, losses may occur. In strawberries and other berries, freezing without pretreatment practically causes no change in the nutritive value (Beattle and Wade, 1996). A change in the form of ascorbic acid has been reported during freeze storage of some produce. Freeze storage of broccoli and cauliflower resulted in a slight oxidation of ascorbic acid to dehydro-ascorbic and 2, 3-diketogulonic acids (Salunkhe, 1974).

8.10.3 Dehydration

Dehydration is also an old processing method for fruits and vegetables that dates back to ancient times. It involves exposing the produce to elevated temperatures to reduce its water content, and hence its water activity. This helps to slow down the rate of proliferation of lethal microorganisms and degradation by biochemical processes (e.g., enzymic browning). Dehydration can be achieved by conventional or unconventional methods. Conventional methods include sun drying, oven drying, and controlled solar drying. Sun drying involves exposing produce to direct sunlight and heat, while with controlled solar drying the produce is exposed to elevated temperatures (e.g., 40 to 60°C) but is protected from direct sunlight. Light can cause degradation of ultraviolet light-sensitive vitamins, including riboflavin, carotene, and thiamin in foods that are dried in the open sun. Controlled solar drying has been associated with a greater retention of vitamin C, riboflavin, thiamin, α-carotene, β-carotene, and total carotenoids relative to sun drying. It also results in a product of superior sensory quality (Mosha et al., 1995, 1997). The most common unconventional drying method is freeze-drying. This process involves freezing the blanched product and removing the moisture rapidly under vacuum by sublimation. Dehydration is achieved at a low temperature because of the vacuum condition, and this reduces degradation of the thermolabile nutrients. Since the product is dried without thawing, the losses of nutrients and flavor associated with thawing and heating are minimized.

Oxidation is the primary cause of nutrient losses during dehydration and drying. This can be reduced by drying fruits and vegetables in a vacuum, or by flushing the product with nitrogen gas and shortening the dehydration time without increasing the temperature. Freeze-drying of fruits and vegetables, which is carried out in the absence of oxygen, does not result in appreciable loss of vitamin C (IFT, 1986). Dehydration of broccoli resulted in a 40% loss of thiamin but insignificant losses of riboflavin, niacin, and pantothenic acid (Salunkhe, 1974). Dehydration of fruits and vegetables with high fat content yields more unstable products. The oil in the produce, if present (e.g., in olives), may undergo oxidative rancidity, leading to off-flavors during storage. Blanching of fruits and vegetables prior to dehydration is important because it helps to inactivate the enzymes. During dehydration of cabbage, ascorbic acid is oxidized to dihydroascorbic acid and 2,3-diketogulonic acid, which in turn reacts nonenzymatically with free amino acids to form a red-brown compound that causes discoloration in the dried product. During this process, carbonyl groups are also released that impart an objectionable flavor. Treating the produce with some chemicals prior to drying may enhance retention of some thermolabile nutrients. For example, soaking cabbage in 1% sodium sulfite for 30 sec prior to drying significantly improved retention of ascorbic acid. Similarly, soaking broccoli and brussels sprouts in sodium chloride solution prior to dehydration enhanced the retention of ascorbic acid and carotene considerably (Salunkhe, 1974).

8.10.4 FERMENTATION

From the Roman times, people knew how to preserve vegetables for long periods without the use of freezers or canning machines. This was done through the process of fermentation. There are three forms of fermentation: lactic acid, acetic acid, and alkaline fermentation.

8.10.4.1 Lactic Acid Fermentation

Lactic acid fermentation is carried out by lactic acid bacteria, which are Gram-positive, nonrespiring, non-spore-forming, and coccoid in structure (Teitelbaum and

TABLE 8.4
Major Lactic Acid Bacteria in Fermented Plant Products

Homofermenter	Facultative Homofermenter	Obligate Homofermenter
Enterococcus faecum	*Lactobacillus bavaricus*	*Lactobacillus brevis*
Enterococcus faecalis	*Lactobacillus casei*	*Lactobacillus buchneri*
Lactobacillus acidophilus	*Lactobacillus coryniformis*	*Lactobacillus cellobiosus*
Lactobacillus lactis	*Lactobacillus curvatus*	*Lactobacillus confusus*
Lactobacillus delbrueckii	*Lactobacillus plantarum*	*Lactobacillus coprophilus*
Lactobacillus leichmannii	*Lactobacillus sake*	*Lactobacillus fermentatum*
Lactobacillus salivarius		*Lactobacillus sanfrancisco*
Streptococcus bovis		*Lactobacillus dextranicum*
Streptococcus thermophilus		*Lactobacillus mesenteroides*
Pediococcus acidilactici		*Lactobacillus paramesenteroides*
Pediococcus damnosus		
Pediococcus pentocacus		

Source: Battcock, M. and Azam-Ali (1998), Food and Agriculture Organization, Rome.

Walker, 2002). They produce lactic acid as the major end product of the fermentation of carbohydrates. They are the most important bacteria in desirable food fermentation and are responsible for the fermentation of pickled (fermented) vegetables such as sauerkraut. Historically, bacteria from the genera *Lactobacillus, Leuconostoc, Pediococcus*, and *Streptococcus* have been the main species involved in food fermentation (Hartley et al., 2001; Teitelbaum and Walker, 2002). Several more strains of bacteria have been identified but play a minor role in lactic acid fermentation. Lactic acid bacteria carry out the conversion of carbohydrate to lactic acid plus carbon dioxide and other organic acids using only a small amount of oxygen (i.e., they are microaerophilic). Thus, changes that they effect do not cause drastic changes in the composition of the food. Some of the *Lactobacillus* family produce only lactic acid (homofermentative), while others produce lactic acid and other volatile compounds and small amounts of alcohol (heterofermentative) (Battcock and Azam-Ali, 1998; Hartley et al., 2001). Homofermenters produce lactic acid through the glycolytic (Embden-Meyerhof) pathway, while heterofermenters produce lactic acid and appreciable amounts of ethanol, acetate, and carbon dioxide via the 6-phosphogluconate/phosphoketolase pathway. Table 8.4 shows the major lactic acid bacteria in fermented plant products. Although all species of lactic acid bacteria are capable of producing lactic acid, *L. plantarum*, a homofermenter, produces the highest acidity in all vegetable fermentations and plays the major role (Battcock and Azam-Ali, 1998).

8.10.4.2 Acetic Acid Fermentation

Acetic acid fermentation involves *Acetobacter* bacterial species. *Acetobacter* converts alcohol formed in food by action of yeast on sugars to acetic acid. This process occurs in excess oxygen. The most desirable product of acetic acid fermentation is vinegar (Battcock and Azam-Ali, 1998). The most commonly known fermented produce is sauerkraut. Sauerkraut, which is prepared from cabbage, was consumed from the Roman times and was highly prized for its delicious taste as well as medicinal properties. Apart from cabbage, other fruits and vegetables used in acid fermentation include cucumbers, beets, green tomatoes, peppers, lettuce, eggplant, onion, squash, carrots, and turnips. For example, pickled green tomatoes, peppers, and lettuce are widely eaten in Russia and Poland, while in Asia, pickled preparations of cabbage, turnips, eggplant, cucumber, onion, squash, and carrot are popular. The popular Korean *kimchi* is a fermented condiment of cabbage with other vegetables and seasonings. In America, popular acid-fermented products include corn and cucumber relishes, fruit chutneys, and watermelon rind (Battcock and Azam-Ali, 1998; Hartley et al., 2001).

8.10.4.3 Alkaline Fermentation

Alkaline fermentation is carried out by the *Bacillus* species. Many bacterial species belong to this family, such as *Bacillus subtilis*, *Bacillus lichifornis*, and *Bacillus pumilius,* but *B. subtilis* is the dominant species (Battcock and Azam-Ali, 1998; Hartley et al., 2001). Alkaline fermentation bacteria act by hydrolyzing protein in food to form amino acids and peptides and releasing ammonia in the process, which increases the alkalinity of the food. Like acid fermentation, alkaline fermentation alters the pH of food, making it unsuitable for growth of other microorganisms that may decompose or spoil the food. Alkaline fermentations are more common with protein-rich foods such as soybeans and other legumes, although there are a few examples utilizing fruits and vegetables. For example, watermelon seeds (*Ogiri* in Nigeria) and green vegetables (*Semayi* in Indonesia and *Kawal* in Sudan) are used as substrates for alkaline fermentation (Battcock and Azam-Ali, 1998). Alkaline-fermented products are important in providing protein-rich, low-cost condiments from green vegetables, seeds, and beans that contribute to the diet of millions of people, especially in developing countries.

Preservation of fruits and vegetables by the process of fermentation has numerous advantages beyond those of simple preservation. Fermented products retain most of their ascorbic acid contents; however, the retention depends on the fermentation temperature and the fermenting time. When properly packaged and stored in cool, dark place, most fermented products would retain their good texture and flavor. When fermented foods are exposed to light and high temperature there is discoloration, degradation of vitamin C, and off-flavor development (Salunkhe, 1974).

8.10.5 PROBIOTICS

The role of probiotics in foods, and indeed in some fermented produce products, has also been recognized as beneficial to health, especially in the past decade, and will be discussed in this section.

Probiotics are beneficial bacteria, such as *Lactobacillus acidophilus* and *Bifidobacterium bifidum* (Teitelbaum and Walker, 2002). Beneficial bacteria generally include three lactic acid genera, namely, *Bifidobacterium (bifidum, breve, infantis, longum,* and *adolescentis*); *Enterococcu*s (*facecum* and *faecalis*)*;* and *Lactobacillus* (*acidophilus, paracasei, rhamnosus*, and *reuteri*) (Kawese, 1982; Rasic, 1983; Teitelbaum and Walker, 2002). Probiotic bacteria avoidably alter the intestinal microflora balance, inhibit growth of harmful bacteria, promote good digestion, boost immune function, and increase resistance to infection. People with flourishing intestinal colonies of beneficial bacteria are better equipped to fight the growth of disease-causing bacteria (De Semone et al., 1993). Lactobacilli and bifidobacteria maintain a healthy balance of intestinal flora by producing organic compounds such as lactic acid, hydrogen peroxide, and acetic acid that increase the acidity of the intestine and inhibit the reproduction of many harmful bacteria (Kawese, 1982; Rasic, 1983). Probiotic bacteria also produce bacteriocins, which act as natural antibiotics to kill undesirable microorganisms (Barefoot and Klaenhammer, 1983).

Immune function tends to decline with age. A double-blind clinical study involving daily supplementation of elderly people with *Bifidobacterium lactis* (a particular strain of bifidobacteria) resulted in a significant enhancement of the immune function (Arunachalam et al., 2000). Benefits were apparent after only 6 weeks of supplementation. Research has also shown that both topical and oral use of acidophillus can prevent yeast infection caused by *candida* overgrowth (Ekmer et al., 1996). For example, regular ingestion of probiotic bacteria may help to prevent vaginal yeast infection (Hilton et al., 1992; Reid et al., 1996).

Diarrhea flushes intestinal microorganisms out of the gastrointestinal tract, leaving the body vulnerable to opportunistic diseases. Replenishing the beneficial bacteria with probiotic supplements can help prevent new infections. The incidence of "traveler's diarrhea" caused by pathogenic bacteria in drinking water or undercooked foods can be reduced by the preventive use of probiotics (Scarpignato and Rampal, 1995). Most people associate lactobacilli with *L. acidophilus*, the most popular species in this group of probiotic bacteria. However, research shows that other *Lactobacillus* species may be beneficial as well. For example, *L. rhamnosus* and *L. plantarum* appear to be protective bacteria. They are involved in the production of several "gut nutrients" such as short-chain fatty acids and the amino acids arginine, cysteine, and glutamine (Bengmark, 1996). These bacteria have been reported to remove toxins from the gut and exert a beneficial effect on cholesterol levels (Bengmark, 2000). Likewise, a probiotic *Saccharomyces boulardii* has been shown to prevent diarrhea in several human trials (Golledge and Riley, 1996). A double-blind research study with critically ill patients found that this strain of yeast prevented diarrhea when 500 mg were taken per day (Bleichner et al., 1997). Probiotics are also important in recolonizing the intestine during and after antibiotic use. Probiotic supplements replenish the beneficial bacteria, preventing up to 50% of infections occurring after antibiotic use (Louzeau, 1993).

Probiotics also promote healthy digestion. Enzymes secreted by probiotic bacteria aid digestion. Acidophilus is a source of lactase, the enzyme needed to digest the sugar lactose, which is lacking in lactose-intolerant people (McDonough et al., 1987). Fructo-oligosaccharides are naturally occurring carbohydrates that cannot be digested or absorbed by humans. They support the growth of bifidobacteria, one of the beneficial bacterial strains (Williams et al., 1994). Due to this effect, some doctors recommend that patients taking bifidobacteria also supplement with fructo-oligosac-charides. Several trials have used 8 g/d. However, a review of the research has suggested that 4 g/d appears to be enough to significantly increase the amount of bifidobacteria in the gut (Gibson, 1998). Acidophilus and bifidobacteria also have the ability to manufacture B-vitamins, including niacin, folic acid, biotin, and vitamin B_6 (Bengmark, 1996).

Other health benefits associated with consumption of probiotic microorganisms in fermented foods are stabilization of Crohn's disease, stimulation of intestinal peristalsis, decreased fecal enzyme activity, activity against *Helicobacter pylori*, inhibition of some cancers, reduction of serum cholesterol, and reduction in hyper-tension (Hartley et al., 2001). Because of their health benefits, probiotics have therefore been postulated to have either a preventive or a palliative role in several health conditions, including diarrhea, vaginitis, yeast infection, canker sores, Crohn's disease (*Saccharomyces boulardii*), eczema, food allergies, HIV/AIDS (*Saccharo-myces boulardii*), immune function, infections, ulcerative colitis, and chronic can-didiasis (Hartley et al., 2001).

8.10.6 USE OF ADDITIVES

Additives are organic or inorganic compounds added to fruits and vegetables either whole or minimally processed to maintain the wholesomeness and palatability of the produce (Salunkhe, 1974; Salunkhe et al., 1991). Additives are classified into approximately seven groups. Theses include (1) colors; (2) preservatives; (3) acids, antioxidants, and mineral salts; (4) vegetable gums, emulsifiers, and stabilizers; (5) mineral salts and anti-caking agents; (6) flavor enhancers; and (7) disinfectants. Table 8.5 summarizes the common additives used in fruits and vegetables. Many people feel that additives are sometimes used when there is no real need for them, but most additives have a useful role. In fruits and vegetables, for example, preser-vatives help to prevent spoilage and prolong the shelf life of the produce. Antioxi-dants inhibit oxidative degradation of essential nutrients such as vitamin C, while antimicrobials (antibiotics, fungicides, sulfite and sulfur dioxide, chlorine, chlorine dioxide, sodium thiosulfate, trisodium phosphate, surfactants, humectants, and lactic and acetic acids) inhibit the growth of bacteria and fungi. Proteolytic enzymes help to inhibit enzymatic browning.

Apart from protecting the produce from physical deterioration, some additives have shown beneficial effects in retaining some nutrients. The blanching of cabbage in hot water containing sulfite results in a product with a higher vitamin C content than when it is blanched in plain water (Salunkhe, 1974). Sulfites also "set" the green color in green vegetables and protect them from becoming discolored when blanched with hot water (Salunkhe, 1974). Despite their benefits in preserving foods, additives have been associated with some negative effects, such as allergic reactions in some people to some of the additives, particularly sulfites. Some nutrients (e.g., thiamin) might also be lost in the presence of some additives. Therefore, the Food and Drug Administration of the United States has banned the use of sulfur-based additives in fruits and vegetables.

TABLE 8.5
Various Additives Used in Fruit and Vegetable Processing

Additive/Category	Use
Colorants	
Amaranth	Used in fruits and fruit flavored fillings; banned in some countries, including U.S.
Erythrosine	Red color used in cherries and canned fruits; banned in some countries, including the U.S.
Allura red AC	Used in fruit-flavored fillings; banned in some countries
Green S	Used in canned peas; banned in some countries, including U.S.
Plain caramel	Used in fruits and pickles
Carotenoids	Used in fruits and vegetables
Preservatives	
Sorbic acid, calcium, potassium, sodium sorbate	Used in varieties of fruits and vegetables; no adverse effect known
Sulfur dioxide	Used in raw fruits and vegetables; banned in the U.S.
Sodium sulfite, metabisulfite, hydrogen sulfite, potassium hydrogen sulfite	Used in fresh oranges and juices as a disinfectant; banned in the U.S.
Biphenyl (diphenyl)	Used in citrus fruits
Orthophenyl phenol, sodium orthophenyl phenol	Used in fruits and vegetables, especially pears, peaches, plums, prunes, carrots, sweet potatoes, citrus fruits, pineapples, tomatoes, peppers, cherries, and nectarines
Thiabendazole	Used in citrus fruits, apples, pears, potatoes, bananas, and mushrooms
Acetic acid, sodium, calcium acetate, sodium diacetate	Used in pickles and chutneys
Lactic acid	Used in fermented fruits and soft drinks
Fumaric acid	Used as antioxidant in fruits and vegetables, soft drinks

In summary, additive use should be determined on the basis of necessity and value for product enhancement. It is imperative to note that the use of additives might not always be beneficial.

8.10.7 RADURIZATION

Radurization is a processing treatment of fruits and vegetables that involves exposure of the produce to low levels of ionizing radiation ranging from 0.75 to 2.5 kGy to delay the onset of spoilage by reducing the population of microorganisms (IFT, 1983; Pauli and Tarantino, 1995). Radiation sources approved for use in foods are gamma rays (produced by the radioisotopes cobalt-60 or cesium-137), x-rays (with a maximum energy of 5 million electron volts, MeV), and electrons (with a maximum energy of 10 MeV). The low dose (< 1 kGy) of radiation energy applied to fruits and vegetables is adequate to eliminate most microorganisms, except for some viruses (Crawford and Ruff, 1996; IFT, 1983). Medium doses (1 to 10 kGy) of radiation can inhibit sprouting in potatoes and other foods and slows the ripening and spoilage process of fruits. High doses of radiation cause side effects in the produce such as an objectionable flavor, texture, and color. For this reason, a low-dose radiation process has been accepted as a useful technique to extend shelf life of fresh foods, particularly when used in conjunction with refrigeration and other auxiliary methods of preservation.

The effect of radurization on produce quality has been widely researched. In some fruits and vegetables, radurization resulted in a decline of flavor during storage. However, in others no effect on flavor was observed (Thayer, 1990; Salunkhe et al., 1991). For example, irradiation of potatoes, onions, carrots, and many other fruits and vegetables produces no detrimental effects on flavor (Salunkhe et al., 1991). Radurization cannot inactivate enzyme systems in plants. Consequently, irradiated fresh fruits and vegetables are prone to enzymic deterioration unless they are blanched prior to radurization. Also, during redurization, complex components of food such as fats, proteins, and carbohydrates (starch, sugar, inulin, cellulose, and pectin) are degraded to their simpler components. This degradation of some complex components has been attributed to a change in texture, especially the softening of produce. Radurization also destroys the green color in treated vegetables. Increasing the gamma radiation dose over 0.5 kGy results in a linear decrease in the chlorophyll content of green beans and broccoli. Irradiation has also been shown to cause 5 to 95%, 3 to 20%, and 0 to 5% destruction of carotenoids in broccoli, sweet potatoes, and carrots, respectively (Thayer, 1990; Salunkhe et al., 1991). Radurization also can lead to more than an 80% loss of vitamin C in green vegetables such as asparagus, broccoli, green beans, and spinach. The loss of some vitamin C was dependent on the irradiation dose and the variety/species of the produce. Vitamins A, C, and E are sensitive to gamma radiation, with higher doses causing more loss. Losses of vitamin C in oranges, tangerines, tomatoes, and papayas varied from 0 to 28% with an increase in radiation doses from 40 to 400 kGy (Thayer, 1990; Salunkhe et al., 1991). A review of the research related to the safety and nutritional adequacy of irradiated foods conducted by the World Health Organization (WHO) concluded that radurization is a good process for reducing or eliminating pathogens and spoilage microorganisms in foods (WHO, 1994). However, certain nutrients might be more sensitive to loss, such as vitamins A, C, and E. Radurization has potential applications in extending the storage life of fresh fruits and vegetables.

8.11 SUMMARY

Nutrient loss in fresh produce might occur at a number of different levels in the path from "farm to table." Therefore, both pre- and postharvest stages of produce have to be taken into consideration. It is also important to note that in addition to treatment of produce, or lack of it, environmental conditions, type of produce, and type of nutrient are important factors to take into consideration. Produce handling and close monitoring at all times is important to ensure that the closely interwoven physical and nutritional quality are maintained for the maximum benefit of all the key produce handlers: the consumer, producer, processor, and retailer.

REFERENCES

Aked, J. (2002), Maintaining the post-harvest quality of fruits and vegetables, in *Fruit and Vegetable Processing: Improving Quality*, Jongen, W, Ed., CRC Press, Washington, DC, pp. 119–146.

Anderson, J.W., Allgood, L.D., Lawrence, A., Altringer, L.A., Jerdack, G.R., Hengehold, D.A., and Morel, J.G. (2000), Cholesterol-lowering effects of psyllium intake adjunctive to diet therapy in men and women with hypercholesterolemia: Meta-analysis of 8 controlled trials, *Am. J. Clin. Nutr.,* 71:472479.

Animal and Plant Health Inspection Service (1985), Section III.9 and Section VI-T106. *Plant Protection and Quarantine Manual*, U.S. Dept. of Agriculture, Washington, DC.

Arunachalam, K., Gills, H.S., and Chandra, R.K. (2000), Enhancement of natural immune function by dietary consumption of *Bifidobacterium lactis* (HN019), *Eur. J. Clin. Nutr.,* 54:263–267.

Attaway, J.A (1994), Citrus juice flavonoids with anticarcinogenic and anticumor properties, in *Food Phytochemicals for Cancer Prevention in Fruits and Vegetables*, Huang, M.J., Osawa, T., Ho, C.T., and Rosen, R.T., Eds., Wiley, New York, pp. 240–248.

Augustin, J., Johnson, S.R., Teitzeol, C., Toma, R.B., Shaw, R.L., True, R.H., Hogan, J.M., and Deustch, R.M. (1978), Vitamin composition of freshly harvested and stored potatoes, *J. Food Sci.,* 43:1566–1574.

Bal, J.S., Sigh, M.P., Minhas, P.P.S., Bindra, A.S., Effect of ethephon on ripening and quality of papaya, *Acta Hortic.,* 119–122, 1992.

Bano, Z. and Rajarathnam, S. (1988), *Pleurotus* mushrooms. part II. chemical composition, nutritional value, postharvest physiology preservation and role as human food, *CRC Crit. Rev. Food Sci. Nutr.,* 27:87.

Barefoot, S.F. and Klaenhammer, T.R. (1983), Detection and activity of lactacin B, a bacteriocin produced by *Lactobacillus acidophilus, Appl. Environ. Microbiol.,* 45:1808–1815.

Battcock, M. and Azam-Ali (1998), *Fermented Fruits and Vegetables: A Global Perspective*, Food and Agriculture Organization, Rome.

Beattle, B., Wade, N. (1996), Storage ripening and handling of fruits, in *Fruit Processing* Arthey, D., and Ashurst, P.R. (Eds.), Blackie Academic and Professional, London, 130–156.

Bengmark, S. (1996), Econutrition and health maintenance: a new concept to prevent inflammation, ulceration and sepsis, *Clin. Nutr.,* 15:1–10.

Bengmark, S. (2000), Colonic foods: Pre- and probiotics, *Am. J. Gastroenterol.,* 95:S5–S7.

Biggs, M.S., Woodson, W.R., and Handa, A.K. (1988), Biochemical basis of high temperature inhibition of ethylene biosynthesis in ripening tomato fruit, *Physiol. Plant.* 72:572–578.

Bleichner, G., Blehaut, H., Mentec, H., and Moyse, D. (1997), *Saccharamyces boulardii* prevents diarrhea in critically ill tube-fed patients: A multi-center, randomized, double-blind placebo controlled trial, *Intensive Care Med.,* 23:517–527.

Bliss, D.Z., Jung, H.J., Savik, K., Lowry, A., LeMoine, M., Jensen, L., Werner, C., and Schaffer, K. (2001), Supplementation with dietary fiber improves fecal incontinence, *Nurs. Res.,* 50:203–213.

Bolin, H.R. and Huxsoll, C.C. (1989), Storage stability of minimally processed fruit, *J. Food Proc. Preserv.,* 13:218–292.

Bourne, M.C. (1987), Effect of blanch temperature on kinetics of thermal softening of carrots and green beans, *J. Food Sci.,* 52:667–690.

Bowen, P., Chen, L., Stacewics-Sapuntzakis, M., Duncan, C., Sharifi, R., Ghosh, L., Kim, H.S., Christov-Tzelkov, K., Van Breemen, R., Tomato sauce supplementation and prostate cancer: Lycopene accumulation and modulation of biomarkers of carcinogenesis, *Exp. Biol. Med,* 227:886–893, 2002.

Broekmans, W.M, Klopping-Ketelaars, I.A., Schuurman, C.R., Verhagen, H., Van Den Berg, H., Kok, F.J., and Van Poppel, G. (2000), Fruits and vegetables increase plasma carotenoids and vitamins and decrease homcysteine in humans, *J. Nutr.* 130:1578–1583.

Burke, V., Hodgson, J.M., Beilin, L.J., Giangiulioi, N., Rogers, P., and Puddey, I.B. (2001), Dietary protein and soluble fiber reduce ambulatory blood pressure in treated hypertensives, *Hypertension,* 38:821–826.

Bycroft, B., Corrigan, V., and Boulton, G. (1997), Sweetening squash with heat treatments, in *PH-1996 International Postharvest Scientific Conference*, Taupo, New Zealand, p. 148.

Chavan, J.K. and Kadam, S.S. (1989), Nutritional improvement of cereals by sprouting, *CRC Crit. Rev. Food Sci. Nutr.,* 28:40.

Clark, C.A., Weatherspoon, L., McBurney, M.I., Henry, D., and Hord, N.G. (2002), Implications of breakfast on midday meal metabolic responses in individuals with type 2 diabetes mellitus (DM), American Diabetes Association Annual Meeting, San Francisco, CA.

Colditz, G.A., Branch, L.G., Lipnick, R.J., Willett, W.C., Rosner, B., Posner, B.M. and Hennekens, C.H. (1995), Increased green and yellow vegetable intake and lowered cancer deaths in an elderly population, *Am. J. Clin. Nutr.,* 41:32–36.

Crawford, L.M. and Ruff, H.E. (1996), A review of the safety of cold pasteurization through irradiation, *Food Control,* 7:87–97.

De Simone, C., Vesely, R., and Bianchi, S.B. (1993), The role of probiotics in modulation of the immune system in man and in animals, *Int. J. Immunother.,* 9:44–48.

Dreher, M.L. (1987), *Handbook of Dietary Fiber*, Marcel Dekker, New York.

Dickerson, J.W.T. (1993), Cancer, in *Encyclopedia of Food Science, Food Technology and Nutrition*, Academic Press, London, pp 607–608.

Elkin, E.R. (1979), Nutrient content of raw and canned green beans, peaches and sweet potatoes, *Food Technol.,* 33:66–70.

Ekmer, G.W., Surawicz, C.M., and McFarland, L.V. (1996), Biotherapeutic agents, *JAMA,* 275:870–876.

Evelyn, A.M., Ueda, T.Y., Imahori, Y., Ayaki, M., L-ascorbic acid metabolism in spinach (*Spinacia oleracea* L.) during postharvest storage in light and dark, *Postharvest Biol. Technol.,* 28:47–57, 2003.

Fallik, E., Aharoni, Y., Yekutieli, O., Wiselbum, A., Regev, R., Beres, H., and Bar Lev, E. (1996), A method for simultaneously cleaning and disinfecting agricultural produce, Israel patent application 116965.

Fleischauer, A.T. and Arab, L. (2001), Garlic and cancer: A critical review of the epidemiologic literature, *J. Nutr.,* 131:525–533.

Florholmen, J., Arvidsson-Lenner, R., Jorde, R., and Burhol, P.G. (1982), The effect of Metamucil on postprandial blood glucose and plasma gastric inhibitory peptide in insulin-dependent diabetics, *Acta Med. Scand.,* 212:237–239.

Fragakis, A.S. (2003), *The Health Professional's Guide to Popular Dietary Supplements*, The American Dietetic Association, Chicago.

Gaffney, J.J. and Armstrong, J.W. (1990), High temperature, forced-air research facility for heating fruits for insect quarantine treatments, *J. Econ. Entomol.,* 83:1959–1964.

Gaffney, J.J., Hallman, G.J., and Sharp, J.L. (1990), Vapor heat research unit for insect quarantine treatments, *J. Econ. Entomol.,* 83:1965–1971.

Gaziano, J.M., Manson, J.E., Branch, L.G., Colditz, G.A., Willett. W.C., and Buring, J.E. (1995), A prospective study of consumption of carotenoids in fruits and vegetables and decreased cardiovascular mortality in the elderly, *Ann. Epidemiol.*, 5:255–265.

Gibson, G.R. (1998), Dietary modulation of the human gut microflora using probiotics, *Br. J. Nutr.*, 80(Suppl. 2):S209–S212.

Gillman. M.W., Cupples, L.A., Dagnon, D., Posner, B.M., Ellison, R.C., Castelli, W.P., and Wolf, P.A. (1995), Protective effect of fruits and vegetables on development of stroke in men, *JAMA,* 273:1113–1118.

Giovannucci, E. (1999), Tomatoes, tomato-based products, lycopene, and cancer: Review of the epidemiologic literature, *J. Natl. Cancer Inst.*, 91:317–331.

Giovannucci, E., Ascherio, A., Rimm, E.B., Stampfer, M.J., Colditz, G.A., and Willett, W.C. (1995), Intake of carotenoids and retinol in relation to risk of prostate cancer, *J. Natl. Cancer Inst.*, 87:1767–1776.

Goddard, M.S. and Matthews, R.H. (1979), Contribution of fruits and vegetables to human nutrition, *HortScience,* 14:245–247.

Golledge, C.L. and Riley, T.V. (1996), "Natural" therapy for infectious diseases, *Med. J. Austr.,* 164:94–95.

Groff, J.L., and Gropper, S.S. (2000), *Advanced Nutrition and Human Metabolism*, Wadsworth/Thomson Learning, Belmont, CA.

Hadley, C.W., Miller, E.C., Schwantz, S.J., Clinton, S.K., Tomatoes, lycopene, and prostate cancer: Progress and promise, *Exp. Biol. Med.,* 227: 869–880, 2002.

Hartley, D., Worth, F., Marcos, A., Rosado, J., Rubelgio, E., and Tannock, G. (2001), *Fermented Foods and Healthy Digestive Functions*, John Libbery Eurotext, Rome.

Hertog, M.G., Sweetnam, P.M., Fehily, A.M., Elwood, P.C., and Kromhout, D. (1997), Antioxidant flavonols and ischemic heart disease in a Welsh population of men: The Caerphilly Study, *Am. J. Clin. Nutr.*, 65:1489–1499.

Hertog, M.G.L., Ferkens, E.J.M., Kromhout, D., Hertog, M.G.L., Hollman, C.H., and Katan, M.B. (1993), Dietary anti-oxidant flavonoids and risk of coronary heart disease: The Zutphen elderly study, *Lancet,* 432:1007–1011.

Hilton, E., Isenberg, H.D., and Alperstein, P. (1992), Ingestion of yogurt containing *Lactobacillus acidophilus* as prophylactic for candidal vaginitis, *Ann. Intern. Med.*, 116: 353–357.

Holland, B., Welch, A.A., Unwin, I.D., Buss, D.H., Paul, A.A., and Southgate, D.A.T. (1992), The composition of foods, in *McCance and Widdowson's The Composition of Foods,* 5th ed. The Royal Society of Chemistry, Cambridge.

Hsing, A.W., Chokkalingam, A.P., Gao, Y., Madigan, M.P., Deng, J., Gridley, G., and Fraumeni, J.F. (2002), Allium vegetables and risk of prostate cancer: A population-based study, *J. Natl. Cancer Inst.*, 94:1648–1651.

Hugo, J.F. and Du, T. (1969), Review of literature on the health value of fruits and fruit juices, *Decid. Fruit Grower,* 19:62–75.

IFT – Institute of Food Technologists (1983), Radiation preservation of foods: a scientific status summary by the Institute of Food Technologists' Expert panel on Food Safety and Nutrition, *Food Technol.*, 37:55–61.

IFT – Institute of Food technologists (1986), Effects of food processing on nutritive values, *Food Technol.*, 12:109–116.

Jenkins, D.J., Kendall, C.W., Vuksan, V., Vidgen, E., Parker, T., Faulkner, D., Mehling, C.C., Garsetti, M., Testolin, G., Cunnane, S.C., Ryan, M.A., and Corey, P.N. (2002), Soluble fiber intake at a dose approved by the US Food and Drug Administration for a claim of health benefits: Serum lipid risk factors for cardiovascular disease assessed in a randomized controlled crossover trial, *Am. J. Clin. Nutr.,* 75:834–839.

Jongen, W.M.F., *Fruits and Vegetable Processing: Improving Quality,* CRC Press, Boca Raton, 2002, 10–118.

Kadam, A.A. and Salunkhe, D.K. (1998), Vegetables in human nutrition, in *Handbook of Vegetable Science and Technology: Production, Composition, Storage and Processing,* Salunkhe, D.K. and Kadam, A.A., Eds, Marcel Dekker, New York.

Kawase, K. (1982), Effect of nutrients on the intestinal microflora of infants, *Jpn. J. Dairy Food Sci.,* 31:A241–243.

Ketsa, S. and Pangkool, S. (1994), The effect of humidity on ripening of durians, *Postharv. Biol. Technol.,* 4:159–165.

Klein, J.D. (1989), Ethylene biosynthesis in heat-treated apples, in *Biochemical and Physiological Aspects of Ethylene Production in Lower and Higher Plants,* Clijsters, H., de Proft, M., Marcelle, R., and van Pouche, M., Eds, Kluwer Academic Press, Dordrecht.

Klein, J.D. and Lurie, S. (1992), Pre-storage heating of apple fruit for enhances postharvest quality: Interaction of time and temperature, *HortScience,* 27:326–328.

Knekt, P., Jarvinen, R., Reunanen, A., and Maatela, J. (1996), Flavonoid intake and coronary mortality in Finland: A cohort study, *Br. Med. J.,* 312:478–498.

Kucuk, O., Sarkar, F.H., Djuric, Z., Sakr, W., Pollak, M.N., Khachick, F., Banerjee, M., Bertram, J.S., Wood, D.P., Effect of lycopene supplementation in patients with localized prostate cancer, *Exp. Biol. Med.,* 227:881–885, 2002.

Lazan, H., Ali, Z.M., Mohd, A., and Nahar, F. (1987), Water stress and quality decline during storage of tropical leafy vegetables, *J. Food Sci.,* 52:1286–1292.

Louzeau, E. (1993), Can antibiotic-associated diarrhea be prevented?, *Ann. Gastroenterol. Hepatol.,* 29:15–18.

Lurie, S. (1998), Postharvest heat treatments, *Postharv. Biol. Technol.,* 14:257–269.

Lurie, S. (1999), Postharvest heat treatments, *Postharvest Biol. Technol.,* 14:257–269.

Lurie, S., Handros, A., Falllik, E., and Shapira, R. (1996), Reversible inhibition of tomato fruit gene expression at high temperature, *Plant Physiol.,* 110:1207–1214.

Lurie, S. and Klein, J.D. (1992), Ripening characteristics of tomatoes stored at 12°C and 2°C following a pre-storage heat treatment, *Sci. Hortic.,* 51:55–64.

Mahan, L.K. and Escott-Stump, S. (2000), *Krause's Food Nutrition and Diet Therapy,* 10th ed., W.B. Saunders, Philadelphia, PA.

McCarthy, M.A. and Matthews, R.H. (1994), Nutritional quality of fruits and vegetables subjected to minimal processing, in *Minimally Processed Refrigerated Fruits and Vegetables,* Wiley, R.C., Ed., Chapman and Hall, New York.

McDonough, E.E., Hitchins, A.D., and Woong, N.P. (1987), Modification of sweet acidophilus milk to improve utilization by lactose-intolerant persons, *Am. J. Clin. Nutr.,* 45:570–574.

Meydani, M., Fielding, R.A., and Fotouhi, N. (1997), Vitamin E, in *Sports Nutrition, Vitamins and Trace Elements,* Wolinsky, I. and Driskell, J.A., Eds., CRC Press, Boca Raton, FL, pp.119–135.

Middleton, E., Jr. and Kandaswami, C. (1994), Potential health promoting properties of citrus flavonoids, *Food Tech.,* 48:115–120.

Mosha, T.C. and H.E. Gaga (1999), Nutritive value and effect of blanching on the trypsin and chymotrypsin inhibitor activities of selected Tanzanian leafy vegetables, *Plant Foods Human Nutr.,* 54:271–283.

Mosha, T.C., Gaga, H.E., Pace, R.D., Laswai, H.S., and Mtebe, K. (1995), Effect of blanching on the content of anti-nutritional factors in selected vegetables, *Plant Foods Human Nutr.,* 47:361–367.

Mosha, T.C., Pace, R.D., Adeyeye, S.H., Laswai, S., and Mtebe, K. (1997), Effect of traditional processing practices on the content of total carotenoid, β-carotene, α-carotene and vitamin A activity of selected Tanzanian vegetables, *Plant Foods Human Nutr.,* 50: 189–201.

Mosha, T.C., Pace, R.D., Adeyeye, S., Mtebe, K., and Laswai, H.S. (1995), Proximate composition and mineral content of selected Tanzanian vegetables and the effect of the traditional processing on the retention of ascorbic acid, riboflavin and thiamin, *Plant Foods Human Nutr.,* 48:235–245.

National Academy of Sciences (1999), *Recommended Dietary Allowances*, 14th ed., Washington, DC.

Nguyen-The, C. and Carlin, F. (1994), The microbiology of minimally processed fresh fruits and vegetables, *Crit. Rev. Food Sci. Nutr.,* 34:371–401.

Nunes, M.C.N., Brecht, J.K., Morais, A.M.M.B., and Sargent, S.A. (1995), Physical and chemical quality characteristics of strawberries after storage are reduced by short delay to cooling, *Postharv. Bio. Technol.,* 6:17–28.

Pauli, G.H. and Tarantino, L.M. (1995), FDA regulatory aspects of food irradiation, *J. Food Protect.,* 58:209–212.

Paull, R.E. (1999), Effect of temperature and relative humidity on fresh commodity quality, *Postharv. Biol. Technol.,* 15:263–277.

Paull, R.E. (1994), Response of tropical horticultural commodities to insect disinfection treatments, *HortScience,* 29:988–996.

Paull, R.E., Nishijima, W., Reyes, M., and Cavaletto, C. (1997), Postharvest handling and losses during marketing of papaya, *Postharv. Biol. Technol.,* 11:165–179.

Picton, S. and Grierson, D. (1988), Inhibition of expression of tomato ripening genes at high temperature, *Plant Cell Environ.,* 11:265–272.

Pittia, P., Nicoli, M.C., Comi, G., and Massini, R. (1999), Shelf-life extension of fresh-like ready-to-use pear cubes, *J. Sci. Food Agric.,* 79:955–960.

Potter, N.N. (1986), *Food Science*, 4th ed., Van Nostrand Reinhold, New York.

Prakash, J. (1995), Fruits: The lesser-known health benefits, in *Proceedings of a National Seminar on Postharvest Technology of Fruits*, Bangalore, India.

Rasic, J.L. (1983), The role of dairy foods containing bifido and acidophilus bacteria in nutrition and health, *N. Eur. Dairy J.,* 4:80–88.

Reid, G., Millsap, K., Bruce, A.W. (1994), Implantation of *Lactobacillus casei* var rhamnosus into vagina, *Lancet,* 344:1229.

Riboli, E. and Norat, T. (2003), Epidemiologic evidence of the protective effect of fruit and vegetables on cancer risk, *Am. J. Clin. Nutr.,* 78:559S–569S.

Rimm, E.B., Katan, M.B., Ascherio, A., Stampfer, M.J., and Willet, W.C. (1996), Relation between intake of flavonoids and risk for coronary heart disease in male health professionals, *Ann. Int. Med.,* 125:384–389.

Ross, J.K., English, C., and Perlmutter, C.A. (1985), Dietary fiber of selected fruits and vegetables, *JAMA*, 85:1111–1116.

Roy, S.K. and Chakrabarti, A.K. (1993), Vegetables of tropical climate: Commercial and dietary importance, in *Encyclopedia of Food Science, Food Technology and Nutrition*, Academic Press, London, pp. 4715–4743.

Ryall, A.L. and Pentzer, W.T. (1974), *Handling, Transportation, and Storage of Fruits and Vegetables.* Vol. 2, The AVI Publishing Company, Westport, CT, p. 17.

Saltveit, M.E. (1999), Effect of ethylene on quality of fresh fruits and vegetables, *Postharv. Biol. Technol.,* 15:279–292.

Salunkhe, D.K. (1974), Storage, *Processing and Nutritional Quality of Fruits and Vegetables*, CRC Press, Boca Raton, FL.

Salunkhe, D.K., Bolin, H.R., and Reddy, N.R. (1991), *Storage, Processing and Nutritional Quality of Fruits and Vegetables*, 2nd ed., Vol. 1, CRC Press, Boca Raton, FL.

Salunkhe, D.K. and Kadam, S.S. (1998), *Handbook of Vegetables Science and Technology: Production, Composition, Storage and Processing*, Marcel Dekker, New York.

Salunkhe, D.K. and Kadam, S.S (1991), Introduction, in *Potato: Production, Processing and Products*, CRC Press, Boca Raton, FL, p. 1.

Sanchez, D. (1990), Keep it under an edible coat, *Agric. Res.*, 38:4–5.

Scarpignato, C. and Rampal, P. (1995), Prevention and treatment of traveler's diarrhea: A clinical pharmacological approach, *Chemotherapy*, 41:48–81.

Seddon, J.M., Ajani, U.A., Sperduto, R.D., Hiller, R., Blair, N., Burton, T.C., Farber, M.D., Gragoudas, E.S., Hiller, J., and and Miller, D.T. (1994), Dietary carotenoids, vitamins A, C, and E and advanced age-related macular degeneration: Eye disease case-control study group, *JAMA*, 272:1413–1419.

Southon, S. and Faulks, R. (2002), Health benefits of increased fruit and vegetable consumption, in *Fruits and Vegetable Processing: Improving Quality*, Jongen, W., Ed., CRC Press, New York, pp. 119–146.

Steinmetz, K.A., Kushi, L.H., Bostick, R.M., Folsom, A.R., and Potter, J.D. (1994), Vegetables, fruit, and colon cancer in the Iowa Women's Health Study, *Am. J. Epidemiol.*, 139:1–15.

Teitelbaum, J.E. and Walker, W.A. (2002), Nutritional impact of pre- and pro-biotics as protective gastrointestinal organisms, *Annu. Rev. Nutr.*, 22:107–138.

Terry, P., Giovannucci, E., Michels, K.B., Bergkvist, L., Hansen, H., Holmberg, L., and Wolk, A. (2001), Fruit, vegetables, dietary fiber, and risk of colorectal cancer, *J. Natl. Cancer Inst.*, 93:525–533.

Thayer, D.W. (1990), Food irradiation: Benefits and concerns, *J. Food Qual.*, 13:147–169.

Tudela, J.A., Espín, J.C., and Gil, M.I. (2002), Vitamin C retention in fresh-cut potatoes, *Postharv. Biol. Technol.*, 26:75–84.

USDA (1989), New edible coatings may protect fresh food, USDA news release, July 17.

Van-Duyn, M.A.S. and Pivonka, E. (2000), Overview of the health benefits of fruits and vegetable consumption for the dietetics professional: Selected literature, *J. Am. Diet. Assoc.*, 100:1511–1521.

Varoquaux, P., Mazollier, J., and Albagnac, G. (1996), The influence of raw material characteristics on the storage life of fresh-cut butterhead lettuce, *Postharv. Biol. Technol.*, 9:127–139.

Volpe, S. (2000), Vitamins and minerals for active people, in *Sports Nutrition: A Guide for the Professional Working with Active People*, Rosenbloom, C.A., Ed., The American Dietetic Association, Chicago.

Voornips, L.E., Goldbohn, R.A., Van Poppel, G., Sturnmans, F., Hermus, R.J.J., and Van den Brandt, P.A., Vegetable and fruit consumption and risks of colon and rectal cancer in a prospective cohort study: The Netherlands cohort study on diet and cancer, *Am. J. Epidemiol.*, 152(11): 1081–1092, 2000.

Wang, H., Nair, M.G., Strasburg, G.M., Chang, Y., Booren, A.M., Gray, J.I., and De Witt, D.L. (1999), Antioxidant and anti-inflammatory activities of anthocyanins and their aglycon, cyaniding from tart cherries, *J. Nat. Prod.*, 62:294–296.

Wardlaw, G.M. and Kessel, M. (2002), *Perspectives in Nutrition*, 5th ed., McGraw-Hill, Boston, p. 167.

Watada, A.E. (1987), Vitamins, in *Postharvest Physiology of Vegetables*, Weichmann, J., Ed., Marcel Dekker, New York, pp. 455–468.

Watada, A.E., Aulenbach, B.B., and Worthington, J.T. (1976), Vitamins A and C in ripe tomatoes as affected by stage of ripeness at harvest and by supplementary ethylene, *J. Food Sci.*, 41: 856–858.

Watada A.E., Ko, N.P., and Minott, D.A. (1996), Factors affecting quality of fresh-cut horticultural products, *Postharv. Biol. Technol.*, 9:115–125.

Watada, A.E. and Qi, L. (1999), Quality of fresh-cut produce, *Postharv. Biol. Technol.,* 15:201–205.

WHO (1994), *Safety and Nutritional Adequacy of Irradiated Food,* World Health Organization, Geneva.

Williams, C.H., Witherly, S.A., and Buddington, R.K. (1994), Influence of dietary neosugar on selected bacterial groups of the human fecal microbiota, *Microb. Ecol. Health Dis.,* 7:91–97.

Wills, R.B.H., Scriven, F.M., and Greenfield, H. (1983), Nutrient composition of stone fruit (*Prunus* spp) cultivar: Apricots, cherry, nectarine, peach and plum, *J. Sci. Food Agric.,* 34:1383–1389.

Winsor, G.W. and Adams, P. (1975), Changes in composition and quality of tomato fruit throughout the season, *Rep. Glasshouse Crops Res. Inst.,* 134:130–139.

Wrolstad, R.E. and Shallenberger, R.S. (1981), Free sugars and sorbitol in fruits: A compilation from literature, *J. Am. Assoc. Off. Anal. Chem.,* 64:91–103.

Yang, R.F., Cheng, T.S., and Shewfelt, R.L. (1990), The effect of high temperature and ethylene treatment on the ripening of tomatoes, *J. Plant Physiol.,* 136:368–372.

Zagory D. (1999), Effects of post-processing handling and packaging on microbial populations, *Postharv. Biol. Technol.,* 15:313–321.

Zhang, S. and Farber, J.M. (1996), The effects of various disinfectants against *Listeria monocytogenes* on fresh-cut vegetables, *Food Microb.,* 13:311–321.

9 Water and Its Relation to Fresh Produce

Mawele Shamaila
Nestlé R&D Center, Inc., Solon, OH

CONTENTS

0-8493-1902-1/05/$0.00+$1.50
© 2005 by CRC Press

TABLE 9.1
**Approximate Water Content of Some
Selected Fruits and Vegetables**

Fruit	Water (%)	Vegetable	Water (%)
Apple	85	Broccoli	91
Banana	75	Cabbage	92
Blueberry	85	Carrot	88
Cherry	86	Cauliflower	95
Grape	81	Cucumber	96
Peach	89	Lettuce	96
Pear	84	Pea	79
Raspberry	86	Potato	83
Strawberry	91	Spinach	91
Watermelon	92	Tomato	94

Source: USDA Nutrient Data Laboratory
(www.nal.usda.gov/fnic/cgi-bin/nut_search.pl).

9.1 INTRODUCTION

Water is the most important component of plant tissues, and it plays a significant role
in the quality of harvested fresh produce. It accounts for as much as 60 to 95% of the
weight of fresh produce, and carrots, lettuce, apples, and peaches contain as much as
84 to 96% of water (Table 9.1). Important plant functions, including exchange of
resources, cell expansion, physical and chemical integrity of cell walls, and cell constituents depend on water and its unique properties (Nilsen and Orcutt, 1996). Water
creates a medium in which most of the biochemical reactions necessary for plant
survival occur and participates in many important biochemical reactions. Water contributes to the quality of fresh produce and has great impact on shelf life, textural
properties, and processing potential. Fresh produce is maintained fresh due to its high
water content; however, this high water content limits the stability of its shelf life.

   The transportation and redistribution of essential life-supporting chemical components of plants is dependent on the movement of water. During plant growth, the
elaborate xylem and phloem tissue system facilitates the movement of water and
many chemical components throughout the plant. The continuous redistribution of
water is essential in the replenishment of water lost through transpiration and evaporation to the environment. This process continues in harvested fruits and vegetables
but occurs over a short distance and is much slower compared to that in living plants.
It is the loss of water that affects shelf life, quality, and marketability of fresh fruits
and vegetables. Hruschka (1977) reported weight losses of 4 to 41% in harvested
snap beans, cabbage, sweet peppers, squash, and tomatoes, resulting in shriveled
products. Maguire et al. (2001) in their review reported that excessive weight loss
from fruits can lead to a shriveled appearance; as a result of decreased turgidity this
can cause undesirable textural and flavor changes and thus make the produce unsalable.

Postharvest uptake of water may occur during washing and hydrocooling and under conditions of high humidity, but this has minimal effect on maintaining the quality of the harvested fresh produce during subsequent storage.

9.2 CHEMICAL COMPOSITION, STRUCTURE, AND PROPERTIES OF WATER

Water posses unique physical and chemical properties that enable it to be easily transportable throughout the plant's system as well as to act as a solvent. Water aids in transport of solutes across cells, cell expansion and cell turgor, temperature regulations due to the high heat of vaporization, high thermal capacity, and high conductivity (Nobel, 1999). The properties of water are mainly attributed to the polar structure of the water molecule. Thermal properties of water account for temperature control of plants by ensuring that temperatures are neither too high nor too low for the survival of the plant. Similarly, water maintains a liquid state over the wide range of temperatures at which the majority of biological reactions occur (Steinbeck, 1995).

9.2.1 POLAR NATURE OF WATER

Water is a very simple molecule composed of one oxygen atom covalently bound to two hydrogen atoms. The water molecule is V-shape with an angle of 105° between the two O-H bonds (Figure 9.1). Ruan and Chen (1998) attributed the unusual properties of water to its angled shape and the intermolecular bonds that it can form. Two orbitals of the oxygen atom participate in covalent bonds with two hydrogen atoms; the remaining two orbitals contain unpaired electrons. Due to the repulsion between the unpaired electrons and those in the covalent bonds, the orbitals arrange themselves in a tetrahedron format around the central oxygen atom.

The oxygen atom is more electronegative than hydrogen and thus it tends to attract the electrons of the covalent bond. The attraction results in a partial negative charge at the oxygen end of the molecule and a partial positive charge at each hydrogen (Figure 9.1). Although the partial charges are equal, resulting in a water molecule carrying no net charge, the combination of the separation of the partial charges and the shape of the water molecule makes the water molecule positive. This electrostatic attraction between water molecules is called the *hydrogen bond*. The energy of the hydrogen bond is about 20 kJ/mol. Although weaker than covalent or ionic bonds but stronger than the short-range, transient attractions known as Van der Waals forces (about 4 kJ/mol), it is responsible for many unique properties of water (Steinbeck, 1995). Hydrogen bonding can occur within water molecules and between neighboring water molecules, and each water molecule can form up to four hydrogen bonds.

9.2.2 PROPERTIES OF WATER

The strong hydrogen bonds within and between water molecules contribute to the unique thermal properties of water, namely, high specific heat, thermal conductivity, and high latent heat of vaporization (Cybulska and Doe, 2002). The specific heat of a substance relates to the amount of heat required to raise the temperature of 1 gram

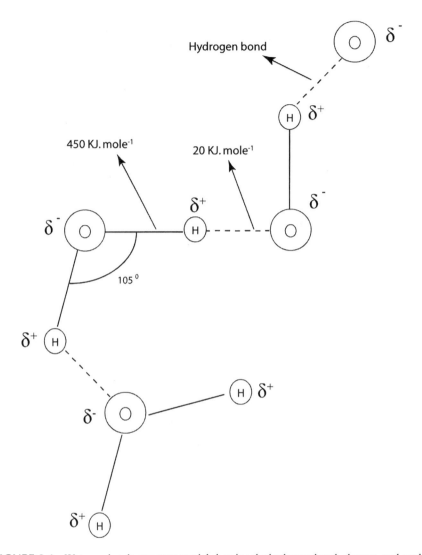

FIGURE 9.1 Water molecule structure model showing the hydrogen bond, charges, and angles.

of the substance by 1°C. The large amount of heat energy required to raise the temperature of water is due to the multiple hydrogen bonds that occur between several water molecules. As the temperature of water is raised, the water molecules vibrate, and therefore a great deal of energy must be put into the system to break the hydrogen bonds between water molecules. Taiz and Zeiger (1998) concluded that this large energy input requirement is important for plants because it helps slow potentially harmful temperature fluctuations. Steinbeck (1995) described thermal conductivity as another important property of water and attributed this to the highly ordered structure of water. Due to this high thermal conductivity, water rapidly conducts heat from the point of application and, in conjunction with specific heat, facilitates the absorption and redistribution of large amounts of heat without much temperature fluctuation. Latent heat of vaporization refers to the energy needed to separate molecules from the liquid phase into the gas phase at constant temperature, and it is important because it helps plants to cool themselves by evaporating water from leaf surfaces and other plant parts during transpiration. Specific heat of vaporization is much higher in water, and a large amount of heat energy is required to evaporate water because hydrogen bonds must be broken to permit water molecules to dissociate from one another and enter the gaseous phase.

At any airwater interface, water molecules are more strongly bound to neighboring water molecules than to the gaseous phase on the other side of the surface. On plant surfaces, a condition known as surface tension is created, and this surface tension at the evaporative surfaces of the leaves generates the physical forces that pull a stream of water through the plant's vascular system. Cohesion, which is the mutual attraction between water molecules, is another property that arises from the extensive hydrogen bonding of water molecules, whereas adhesion is the attraction of water to the solid phase. Surface tension, cohesion, and adhesion are attributed to the capillary movement of water and may be responsible for water movement in the xylem and for keeping the leaf mesophyll moistened (Steinbeck, 1995; Taiz and Zeiger, 1998).

9.3 FACTORS RESPONSIBLE FOR WATER MOVEMENT IN PLANT SYSTEMS

Energy is required to move water from one place to another, and the difference in potential energy between two locations creates an energy gradient, resulting in the movement of water from a high- to a low-energy location. Gradients due to free energy/chemical potential and concentration are important in mobility of water and solutes, with mobility occurring toward locations low in energy (Burton, 1982; Devlin and Witham, 1983; Kays, 1997). In freshly harvested produce, as water evaporates from the surface a gradient is created, and thus there is a tendency for more water to migrate toward the surface to establish equilibrium. In addition, biochemical components that tightly bind or have high affinity for water create a matrix effect that affects water mobility. Hydrophilic compounds such as proteins, cellulose, and starch tightly bind water and limit its movement. Increased temperature around the fresh produce results in increased mobility of water due to increased free energy.

9.3.1 DIFFUSION, BULK/MASS FLOW, AND OSMOTIC MOVEMENT OF WATER

Water molecules in solution are in a dynamic state and while in motion collide with one another, exchanging kinetic energy. Taiz and Zeiger (1998) define diffusion as the movement of molecules by random thermal agitation, resulting in the random but progressive movement of substances from regions of high energy to regions of low free energy. This diffusion is considered rapid over short distances but extremely slow over long distances, and it plays a significant role during loss of water vapor from leaves because the diffusion coefficient in air is much greater than that in aqueous solutions. Bulk or mass flow is the general movement of groups of water molecules in response to a pressure gradient, and this is the main mechanism responsible for long-distance movement of water in the plant through the xylem.

The movement of liquids across plant tissues is limited by the plasma membrane, which acts as a barrier. The plasma membrane is selectively permeable and allows the movement of water and other small uncharged particles across the membrane rather than large charged solutes. In osmosis, both the movements of substances down a concentration gradient and down a pressure gradient have an influence on the mobility of water as a result of increased solute dissolution. The mechanism of osmosis involves diffusion of single water molecules across the membrane bilayer and bulk flow through tiny water-filled pores of molecular dimensions; the driving force is the water/chemical potential gradient (Devlin and Witham, 1983).

9.4   WATER MIGRATION IN PLANT TISSUES

Water in liquid phase travels across membranes, through the cell's hollow channels and cell walls, and escapes as water vapor into the air spaces inside the leaf and diffuses to the environment. The gradient in water vapor concentration plays a significant role in the transpirational water loss from the leaf surfaces. Long-distance water transportation in the xylem is dependent on a pressure gradient, while that through cell layers responds to water potential gradients.

Water transportation from the roots to the leaves occurs through the apoplast and symplast routes. The symplast is an extensive interconnection of cytoplasmic cells that form a continuous system and hold as much as 70% of the water in plant tissues. Water in the symplast includes water within the plasma membrane and in the cytoplasm and the central vacuoles; 80 to 90% of the symplast water is contained in the vacuole. Water in the symplast travels from one cell to the next through the plasmodesmata as a result of the energy gradient. External to the symplast is the apoplast, which is a continous system of cell walls and intercellular air spaces in plant tissues through which water and solutes can also traverse. In the apoplast, water mainly travels through the cell wall without crossing any membranes. The plasma membrane of the cell may restrict movement of water and solutes to the apoplast. Once water is in the apoplast, mobility is toward areas of lower chemical potential. Both the symplast and the apoplast provide routes for movement of water and solutes. Water is mainly transported through the apoplast; solutes not able to diffuse across membranes are transported through the symplast. The transmembrane pathway is another route that carries water through the plant system and involves movement of water from one cell to another; water crosses at least two membranes for each cell in its path (Devlin and Witham, 1983; Steinbeck, 1995; Taiz and Zeiger, 1998).

The great majority of water transportation from the roots to the leaves occurs through the xylem, a simple pathway offering low resistance to water movement. The negative hydrostatic pressure created by leaf transpiration pulls water up the water column in the xylem (Kramer, 1995). Water movement in the xylem occurs through two types of specialized tracheary elements, the tracheids and the vessel elements. The tracheids communicate with adjacent tracheids through numerous microscopic pits, which are regions where the secondary wall is absent and the primary wall is thin and porous. The pits offer low resistance for water movement between the tracheids. Vessel elements have perforated end walls that form a perforation plate at each end of the cell and also contain pits in their walls. Vessels

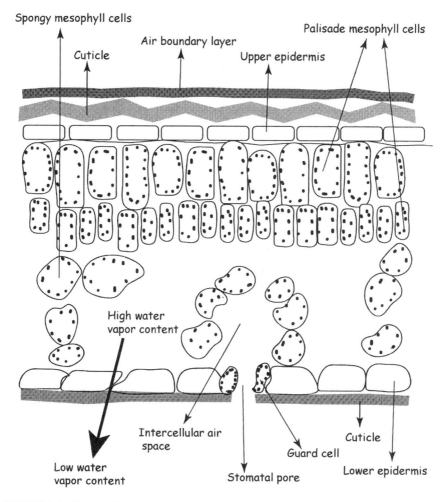

FIGURE 9.2 Sketch of a section through a leaf showing the various types of cells.

provide a very efficient, low-resistance pathway for water movement because of
their open crosswalls.

The absence of membranes in the tracheary elements and the perforations in the
walls of the vessel permit water to move freely through the water-filled capillaries
in response to a pressure gradient. Taiz and Zeiger (1998) described the movement
of water to the leaves through the xylem of the vascular bundle that terminates in a
very fine and intricate network of veins throughout the leaves. Water from the xylem
is drawn into the cells of the leaf and along the cell walls. The negative pressure
that causes water to move up through the xylem occurs at the surface of the cell
walls in the leaf, and the cell walls act like very fine capillary wicks soaked with
water. The moist mesophyll cells within the leaf are in direct contact with the
atmosphere through the intercellular air spaces (Figure 9.2). As water evaporates
from the leaf surface, more water is withdrawn from air spaces in the cells, creating
a negative pressure that drives the movement of water through the xylem system.

Nearly 90% of the water lost as water vapor to the environment occurs through the leaves from the cell walls lining the intercellular spaces adjacent to stomata. In some produce in which there are few remaining stomata, lenticels replace the function of stomata; in apples, lenticels account for up to 21% of the transpiration (Maguire et al., 2001). The outer surfaces of most common leaves are covered with cuticle, a multilayered, waxy deposit mainly composed of cutin (Figure 9.2). The waxes are hydrophobic and thus offer high resistance to diffusion by water and water vapor from underlying cells. The cuticle thus serves to restrict evaporation of water directly from the outer surfaces of leaf epidermal cells and protects both the epidermal and underlying mesophyll cells from potentially lethal desiccation. The epidermis of the leaf contains small pores called stomata, each surrounded by a pair of specialized cells called guard cells, which are responsible for the opening and closing of the stomata. The photosynthetic mesophyll cells are loosely arranged and create an interconnected system of intercellular air spaces that allows escape of gases and water vapor through the stomata.

9.5 EXTERNAL FACTORS AFFECTING WATER MIGRATION IN FRESH PRODUCE

In most harvested fresh produce, the key is to minimize the loss of water from the produce to the environment. Water loss from the produce to the environment can be replaced to a limited extent using techniques such as misting or submerging the produce in water, as is the case in the ornamental plant industry. The rate of water loss from the harvested produce is generally affected by temperature, humidity, pressure, and air movement around the product. All of these factors influence the rate of water vapor diffusion between the substomatal air chamber and the ambient atmosphere.

9.5.1 TEMPERATURE

Water loss from fresh produce is greatly increased with an increase in the temperature of the produce and the surrounding environment. The increase in temperature results in increased free energy of water molecules, which in turn increases their movement and potential for exchange. After harvest and in storage, fresh produce continues to respire and give off heat, which leads to a slight increase in the temperature of the produce and results in water loss.

Temperature has a profound effect on the postharvest quality of harvested fresh produce. Perkins-Veazie and Collins (1999) found that transferring blackberry fruit to 20°C for 2 days after low-temperature storage intervals was detrimental to fruit quality, resulting in increased weight loss, leakage, decay, and softening in 'Navaho' and 'Shawnee' blueberry cultivars. The loss in water was dependent on the fruit cultivar: 'Shawnee' fruit lost 25% of marketable fruit, while 'Navaho' had a 10% loss. In general, freshly harvested produce is held under reduced temperature to minimize respiration and weight loss. The reduction in temperature should just be above the produce's freezing-point temperature (Table 9.2) or just above its chilling threshold temperature in the case of chilling-sensitive produce. Lowering the temperature of fresh produce lowers its rate of deterioration and thus maintains quality and extends shelf life. Fresh produce subjected to 0°C or below undergoes freezing injury,

TABLE 9.2
Recommended Storage Conditions for Temperature and Relative Humidity, Approximate Shelf Life Under Optimum Conditions, and Highest Freezing Points for Some Selected Fruits and Vegetables

Product	Temperature (°C)	Relative Humidity (%)	Approximate Storage Life	Freezing Point (°C)
Fruit				
Apple	−14	90–95	1–12 months	−1.5
Apricot	0–0.6	90–95	1–3 weeks	−1.1
Blackberry	0	90–95	2–3 days	−0.8
Raspberry	0	90–95	2–3 days	−1
Strawberry	0	90–95	3–7 days	−0.8
Cherry	0	90–95	2–7 weeks	−1.8 to −1.7
Nectarine	0	90–95	2–4 weeks	−0.9
Peach	0	90–95	2–4 weeks	−0.9
Pear	−1.7 to −0.6	90–95	2–7 months	−1.6
Watermelon	10–15.6	90	2–3 weeks	−0.4
Vegetable				
Broccoli	0	95–100	10–14 days	−0.6
Cabbage	0	98–100	5–6 months	−0.9
Cantaloupe	0–2.2	95	5–14 days	−1.2
Carrot	0	98–100	7–9 months	−1.4
Sweet corn	0	95–98	5–8 days	−0.6
Cucumber	10–12.8	95	10–14 days	−0.5
Garlic	0	65–70	6–7 months	−0.8
Honeydew	7.2	90–95	3 weeks	−0.9
Lettuce	0	98–100	2–3 weeks	−0.2
Mushroom	0	95	3–4 days	−0.2
Onion	0	65–70	6–8 months	−0.9
Pea	0	95–98	1–2 weeks	−0.8
Potato	3.3–4.4	90–95	5–10 months	−0.6
Spinach	0	95–100	10–14 days	−0.3
Tomato	12.8–21.1	90–95	1–3 weeks	−0.6

Source: From Gast, K.L, *Bulletin 4135*, 1991. The University of Maine Cooperative Extension. With permission.

which involves intercellular and intracellular ice formation. Wills et al. (1998) described the physiological effects of freezing on fresh produce and how it initiates desiccation and osmotic stress of cellular structures. Expansion of water during thawing results in physical damage to the cell structure, and the affected cells rarely resume normal metabolism or regain normal texture, appearing flaccid and water-soaked. Fresh produce may also suffer chilling injury, which occurs at temperatures above the freezing point of the produce. This injury is attributed to the imbalance of metabolism and loss of cellular compartmentalization at suboptimal temperatures. Chilling injury is a function of time and temperature. Short exposure to temperatures that result in chill injury may have minimal effect, but extended periods result in irreversible damage.

9.5.2 RELATIVE HUMIDITY

Humidity is measured using dry and wet bulb thermometers and established from a psychrometer and a psychrometric chart (Gaffney, 1978). Grierson and Wardowski (1975) and Kays (1997) discussed percentage relative humidity, absolute humidity, vapor pressure, and dew point as the different ways of expressing water vapor in the environment. *Relative humidity* refers to the ratio of the quantity of water vapor present and the maximum amount possible at that temperature and pressure. Because relative humidity is affected by temperature and pressure, absolute humidity is a more precise measure of humidity. It refers to the weight of water in a given weight of dry air. Water vapor pressure difference between locations is a useful criterion for measuring potential of fresh produce to lose water to the environment. The movement of water within the produce and between the produce and its surroundings occurs in response to a gradient. Gaffney (1978) reported that moisture loss from fruits and vegetables is directly proportional to the difference between the water vapor pressure at the surface of the produce and that of the air surrounding the produce. Therefore, the water vapor deficit or the difference in water vapor pressure between fresh produce and the immediate surroundings gives an indication of the potential for the loss or gain of water by the produce (Burton, 1982; Kays, 1997; Wills et al., 1998; Thompson, 2002).

Dew point is another key parameter related to the postharvest life of fresh vegetables. It refers to the temperature at which the air is saturated with water vapor (i.e., 100% relative humidity). Water condensation occurs when the temperature is lowered below the dew point since the air can no longer hold as much water. By using a psychrometric chart, the dew point can be determined from the air temperature (dry bulb) and the relative humidity.

9.5.3 PRESSURE, AIR MOVEMENT, AND LIGHT

Although elevated pressure increases the free energy of water molecules, resulting in increased water molecule movement, this increased free energy has minimal impact on fresh produce. However, reducing pressure within the surrounding environment has an impact on water exchange in fresh produce. Reduced pressure decreases the free energy of the water molecules, resulting in an increased concentration gradient between liquid-phase molecules of water within the tissue and gaseous water molecules in the surrounding environment. The net effect of the concentration gradient is the movement of water out of the tissues to the surrounding atmosphere exterior to the product. Products stored under partial vacuum also lose water to the exterior until an equilibrium is established (Thompson, 2002).

The presence of an air boundary layer on the surface of produce reduces water loss to the surrounding environment (Ben-Yehoshua, 1987). Air movement over the produce's surface decreases the thickness of the air boundary layer. This decrease of the air boundary decreases the boundary's resistance to the exchange of water molecules between the produce and its immediate surroundings. In succulent fresh produce such as lettuce, this results in increased water loss. Air movement in closed storage systems, as is the case in refrigerated storage, can influence vapor pressure differences between the produce and its environment. Kays (1997) described the advantages that can be gained by air movement under closed refrigerated systems. The higher the volume of air circulated per unit of time over the cooling coils, the lower the differential in temperature between the return air and the delivery air. With a small temperature differential, a decreased amount of water is removed by the refrigeration coils, resulting in a decreased gradient in vapor pressure between the produce and air. Therefore, controlled, increased air movement can help maintain high humidity within the enclosed refrigeration container. Light also has an effect on the opening and closing of stomatal apertures, and loss of water in harvested products occurs through this avenue. The temperature of fresh produce can also be affected by light: water loss increases at higher light intensity.

9.6 EFFECT OF INJURY ON WATER LOSS

Any form of surface mechanical injury such as bruising or abrasion accelerates water loss from harvested fresh produce. The injury damages the surface cell tissues and thus allows migration of water through the damaged area. Greater damage occurs with cutting the produce because of the complete break of the protective surface layer and direct exposure of underlying cell tissues to the atmosphere (Bolin et al., 1991; Wills et al., 1998). In addition, any injury causes physical damage, physiological stress, and increased transpiration and enhances microbial growth (McDonald et al., 1990; Barry-Ryan and O'Beirne, 1998; Maguire et al., 2001). Hodges and Forney (2000) reported that the degree of damage during harvesting, processing, and packaging had an effect on increased electrolyte leakage in fresh-cut spinach leaves. Barry-Ryan et al. (2000) found minimal processing steps such as peeling and shredding of carrots caused physical and physiological damage and stress and enhanced microbial growth, leading to reduced shelf compared to that of whole carrots. However, these effects were reduced by storage in microporous film with CO_2 permeability of 29×10^3 mL·d^{-1}·atm^{-1}.

Fresh-cut is a large section of fresh produce where the products are intentionally sliced or cut into different sizes and shapes. The cuts can be peeling, slicing, shredding, trimming, coring, or the removal of the protective epidermis cells. Watada et al. (1996) reported that fresh-cut products (also known as lightly or minimally processed) are highly perishable because a large proportion of their surface area is without the epidermis, the outer protective layer of tissue, and they recommended that the products be held at 0 to 5°C. Barry-Ryan and O'Beirne (1998) observed that injury such as that arising from slicing of fresh produce cuts through cells, leaving large areas of internal tissues exposed, and leads to leakage of cell contents during subsequent storage. The total impact of physical damage is increased surface dehydration and total moisture loss. Osornio and Chaves (1998) reported significant exudate loss of 60% in grated beets but noted that this loss could be reduced by proper selection of packaging material. The loss in exudate was responsible for the dryness on the surface of the grated beets. Agar et al. (1999) found that peeling and slicing of kiwifruit resulted in increased mass loss compared to intact whole fruit after 3 days of storage at 2°C. Fresh-cut kiwifruits were found to have more water loss than intact fruits; this was attributed to the removal of the protective epidermal cells and the resulting increase in the surface area/mass rate. They concluded that storage temperature, degree of tissue damage, and microatmospheric gas composition were important factors for quality retention of fresh-cut kiwifruit slices in relation to water loss.

When subjected to temperatures between freezing and 12.5°C, many fruits and vegetables undergo physiological injury, called *chilling injury* since it does not involve freezing (Rolle and Chism, 1987). The injury is attributed to several factors, including impairment of membrane function, changes in cellular membrane, and effects on lipid components. Physiological damage caused by chill injury contributes to water loss and deterioration in quality of many fruits and vegetables. Purvis (1984) implicated moisture loss in grapefruit as one of the factors causing chill injury and noted that the diffusive resistance was lower on the surfaces that showed symptoms of chill injury. However, hydrocooling was found to reduce moisture loss and chilling injury due to a reduced vapor pressure gradient. Cohen et al. (1994) found a significant effect of storage temperature and chill on moisture loss of grapefruit and lemons. Although there was higher moisture and weight loss of fruit stored at 13°C compared to that held at 2°C, transfer of fruit to 20°C showed fruit previously stored at 2°C to have higher water loss. They attributed this high water loss to chill injury that caused microscopic cracks in the peel.

9.7   METHODS OF PREVENTING WATER LOSS FROM FRESH PRODUCE

Most fruits and vegetables contain the highest amount of moisture content at harvest and gain little or no moisture during storage. Any loss in water after harvest affects the quality and economic value of the produce. Table 9.3 shows the effect of water loss on the shelf life of selected fruits and vegetables and the maximum possible loss in water before the product becomes unsalable (Robinson et al., 1975). The amount of water loss that affects salability varies depending on the produce; succulent produce is greatly affected with minimal loss of water. It is therefore important to limit any further loss of water from the produce. The most ideal techniques to reduce loss of water from harvested products are to minimize the water vapor pressure deficit and the resistance to water exchange between the produce and its immediate surroundings. This can be achieved by lowering the temperature and/or increasing the relative humidity of the surrounding environment to specific levels, resulting in a reduced vapor pressure difference between the produce and its environment (Wills et al., 1998).

Both humidity and temperature are important in minimizing the difference in water vapor pressure between the harvested fresh produce and its environment. The humidity of the surrounding environment and that of the internal atmosphere of the produce should be as close to equilibrium as possible. Relative humidity of 95 to 99% is recommended for high-moisture fresh products (Table 9.2), while low humidity can be applied to products that have high resistance to water exchange (Robinson et al., 1975).

TABLE 9.3
Daily Water Loss Rate, Days of Shelf Life, and Maximum Water Loss in Some Selected Fruits and Vegetables

Produce	Water Loss of Original Fresh Weight (% loss d^{-1} mbar wvpd^{-1})	Days of Shelf Life at 20°C	Maximum Water Loss Before Becoming Unsalable (%)
Bean	1.8–3.6	4–6	6
Brussels sprout	2.8	3–5	8
Cabbage		4–11	7–10
Carrot		3–7	4–8
Cauliflower	1.9	3–4	7
Cucumber	0.4	4–7	5
Lettuce	7.5	2–3	3
Onion		7–10	10
Potato	0.05–0.5	10–15	7
Pea	1.3	3–5	5
Green Pepper		3–5	7
Spinach	11.0		
Sweet corn	1.4	2–4	7
Tomato		2–5	7
Blackberry		2–3	6
Raspberry		1–4	6
Strawberry	0.7	2–3	6

Source: Robinson, J.E. et al., *Ann. Appl. Biol.*, 81, 399, 1975. With permission.

9.7.1 TEMPERATURE

Although the relationship between humidity and water exchange is easily explained, that between temperature and water is more complex. Kays (1997) explains three thermal parameters that have a significant effect on moisture exchange in storage: (1) the actual temperature, (2) the differential in temperature between product and environment, and (3) the fluctuations in storage temperature. If the weight of water vapor in air is held constant, lowering the temperature decreases the maximum amount of water the air will hold and the relative humidity increases. Similarly, the water vapor deficit between the produce and its immediate surroundings will decrease at a given humidity with decreasing temperature. It is therefore important to lower produce temperature to decrease its rate of water loss. A produce temperature that is higher than the temperature of its environment, especially in succulent produce, results in an increased vapor pressure difference between the gas atmosphere within the produce and the surrounding cooler air (Burton, 1982). Consequently, the gradient created results in increased water loss from the produce to the environment. It is therefore strongly recommended that harvested produce be cooled as quickly as possible to minimize water loss prior to refrigeration. Table 9.4 shows

TABLE 9.4
Mature Green Bell Pepper Weight Loss and Quality in Relation to Delays to Cool at 25 or 37°C

Delay to Cool (h)	Weight Loss after Delay (%)	Weight Loss after Storage (%)	Final Weight Loss (%)	Visual Quality Score[a]	Dehydration Score[b]	Firmness (N)
0	0	2.06	0.97	8.5	1.2	19.7
At 25°C						
3	0.15	1.91	1.39	8.1	1.2	21.0
6	0.36	1.87	1.48	8.1	1.5	22.5
9	0.52	2.35	1.78	7.5	2.4	21.5
12	0.69	2.93	1.86	7.2	3.0	19.3
18	1.04	1.61	2.08	7.3	3.0	17.4
At 37°C						
3	0.25	1.62	1.19	8.6	1.2	20.9
6	0.60	1.66	1.50	8.3	1.5	19.9
9	1.33	1.67	2.30	7.5	2.3	16.6
12	1.80	2.10	2.52	6.0	2.8	14.2
18	3.25	1.64	3.07	3.6	3.4	10.7

[a] Visual score based on scale of 1 to 9; 1 = unsalable and 9 = excellent.
[b] Dehydration score based on scale of 1 to 5; 1 = none and 5 = severe dehydration.
[c] N = Newtons.

Source: Cantwell, M. et al., *Perishables Handling Q.*, 107, 17, 2001. With permission.

the effect of cooling delays on weight loss and quality of mature green bell peppers (Cantwell et al., 2001). The longer the cooling delays and the higher the produce's holding temperature prior to cooling, the greater the loss in weight, visual quality, and texture of green bell peppers. Robinson et al. (1975) reported that temperature was the single most important factor in successful storage of fresh produce, but that minimizing evaporative losses was critical. They also noted that the rate of evaporation for fresh produce was directly proportional to the water vapor pressure deficit (wvpd) of its environment, and that the amount of evaporation that can occur was limited by the amount of water the air can hold. Therefore, a reduction in water loss can be achieved by limiting the rate of evaporation and by reducing the wvpd to a minimum, or by limiting the amount of evaporation and by reducing the volume of the air in the immediate environment.

9.7.1.1 Cooling of Harvested Fresh Produce

Cooling of freshly harvested produce is one of the first steps taken to extend shelf life and quality. At harvest, produce temperature is close to ambient temperature but may rapidly change depending on the holding conditions soon after harvest. The higher the holding temperature soon after harvest, the higher the respiration rate, leading to a rapid loss in quality. Thompson et al. (2001) reported that cooling delays cause reduced product quality for three reasons: (1) they allow respiration and associated normal metabolism to continue at high rates, consuming sugars, vitamins, and other constituents; (2) they foster water loss; and (3) they increase the development of decay. Methods used to cool fresh produce include cold air circulation (room cooling, forced-air or pressure cooling), cold water treatment (hydrocooling), icing, and evaporation of water (evaporative cooling, vacuum cooling) from the produce.

Room cooling is the most common and simplest procedure used to initially cool fresh produce packaged in boxes or bulk containers. The produce is exposed to cold air in a normal cold storage environment. Thompson et al. (2002) stated that slow cooling and possibly excessive loss of water are the main disadvantages. Pressure (air-cooling) cooling, performed by forcing air through packages, is more efficient than room cooling because it cools the produce rapidly. Some of the common methods of air-cooling include tunnel-type, forced-air cooling, cold wall, serpentine cooling, forced-evaporative cooling, and container venting, as well as cooling during transportation. In general, the forced-air cooling method is much faster because it causes cold air to move through containers (Perez et al., 1998).

Water has a higher heat capacity than air, and it is used to cool fresh produce in a system referred to as *hydrocooling*. Water maintained at 0°C rapidly cools the produce as it passes under a shower or is soaked in a cold bath. Hydrocooling reduces the likelihood of water loss and may even add water to slightly wilted products, especially leafy vegetables (Wills et al., 1998; Cantwell et al., 2001; Thompson et al., 2002). Issues that arise due to hydrocooling include microbial recontamination, inability of the produce to withstand cold wetting, and chemicals such as chlorine used in water treatment.

Ice (crushed or as flakes) placed in contact with the produce is often used as a cooling agent, especially during transportation. The ice in direct contact with the product rapidly cools the product, and the melting ice contributes to high humidity, resulting in reduced water loss. Perrin and Gaye (1986) concluded that the quality of broccoli in terms of retarding yellowing, retention of chlorophyll, and retention of nutritional value was greatly retained when a bed of ice was placed beneath the broccoli heads in combination with mechanical refrigeration. Some leafy green vegetables that have a high surface area-to-volume ratio can be cooled by evaporation of water through vacuum cooling under low atmospheric pressure. Fresh produce in sealed chambers is subjected to reduced pressure, and as the water boils, the latent heat of vaporization cools the produce. Loss of water during vacuum cooling can be minimized by spraying the produce with water prior to sealing the chamber (Thompson et al., 2002). Dry air cooled by blowing across a wet surface can also be used to cool products.

9.7.2 ELEVATED RELATIVE HUMIDITY

Increasing the relative humidity of the environment in which the fresh produce is stored is one of the major steps in preventing water loss and maintaining quality of because humid environments reduce water transpiration. Watada et al. (1996) noted that due to the large surface area, especially in fresh-cut products, there was a great potential to lose a large amount of weight but that the weight loss was minimal in a controlled atmosphere containing humidified gases. The increased relative humidity of the air reduces the vapor pressure deficit between the fresh produce and the immediate air, resulting in an air-saturated environment. Gaffney (1978) reported that total moisture loss from fresh produce can be reduced by decreasing the time of exposure to a vapor pressure difference by cooling the produce surfaces close to the surrounding air temperature as quickly as possible and also maintaining the surrounding air as close to saturation as possible. van den Berg and Lentz (1978) found substantially less quality and water loss at higher humidity in cabbage, cauliflower, and potatoes, while at relative humidity of less than 95%, softening and shriveling of commodities occurred.

Relative humidity can be increased by spraying water as a mist, introduction of steam, or increasing temperature of refrigeration coils. Baldwin (1994) concluded that fruits and vegetables are often stored in a high-humidity environment (90 to 98% relative humidity) to minimize water loss and subsequent weight loss and shriveling. Automatic misting designed to spray the produce with a fine mist of water at timed intervals is commonly used to prevent dehydration, extend shelf life, and improve the appearance of fresh produce in retail display cases. Barth et al. (1990) found that misting promoted a high retention of ascorbic acid and moisture content over a 72-h storage period in broccoli. Proper control of increased relative humidity is important since very high levels of moisture in the environment promote growth of microbials.

9.7.3 Packaging of Fresh Produce

Packaging is one of the most effective methods for reducing water loss from fresh produce since it acts as physical barrier and also reduces air movement across the produce's surface. James et al. (1999) found that Muscadine grapes wrapped with polyethylene bags lost a minimal weight of 0.5 to 1.0% compared to unwrapped grapes, which lost 15% weight, and they were able to correlate the loss in weight to the decrease in fruit firmness. Collins and Perkins-Veazie (1993) packaged 'Cardinal' strawberries in plastic boxes with dome lids or polyethylene (PE) wrap and, after warming the fruit to 25°C, stored it at 1 and 5°C to simulate retail storage conditions. They found that strawberries stored in boxes covered with PE accumulated CO_2, lost less weight, and had better color retention than fruit in boxes with plastic dome lids at both 1 and 5°C. Similarly, the quality of green beans, bell peppers, and spinach packaged in polyethylene bags and stored at two different temperatures was studied by Watada et al. (1987). They found that packaging in PE reduced weight loss of green beans and spinach kept stored at 20°C, and reduced chlorophyll loss of green beans at 10°C and of spinach at 20°C. More ascorbic acid was also retained in the vegetables. Ben-Yehoshua (1985) reviewed individual seal-packaging of fruits and vegetables and noted that this technique was mainly used to extend shelf life and reduce shrinkage, weight loss, the occurrence of various blemishes, and refrigeration costs. It was further concluded that individual seal-packaging helped in securing the beneficial aspects of a water-saturated atmosphere.

Fresh produce can be packaged into bags, boxes, cartons, or mesh bags and covered with tarpaulins. Any form of close packing that restricts the passage of air around individual products reduces water loss. Perforated polymeric packaging film has long been used successfully to reduce moisture loss from produce during storage, shipment, and display by reducing the magnitude of the moisture vapor deficit between the produce and immediate in-package environment (Schlimme and Rooney, 1994). The permeability of the packaging material to water vapor migration determines water loss from fresh produce, and materials such as polyethylene films are excellent vapor barriers. Selection of packaging material and understanding the storage environment for fresh produce are important in reducing water loss. Thin plastic films are ideal in storage environments that are vapor-saturated without the associated problems of condensation. Miller et al. (1983) found that strawberries in rigid plastic baskets with solid plastic covers lost significantly less weight than those stored in mesh plastic baskets with and without covers. Presently, perforated and unperforated packaging materials are used to minimize moisture loss, reduce respiration rate, maintain product quality, and extend shelf life. Edmond et al. (1995) also found that perforated films tested on 11 horticultural products significantly reduced water loss and formation of condensation and extended the marketability of fresh fruits and vegetables in displays.

9.7.3.1 Modified Atmosphere Packaging (MAP) in Reducing Water Loss

Modified atmosphere packaging is the packaging of produce in polymeric films to establish a modified atmosphere around the produce. The produce is sealed in a polymeric bag with selected gas permeability characteristics and stored at low temperature. The relative humidity created inside the package is significant in attaining optimum shelf life and storage quality of fresh produce (Shamaila et al., 1992; Gonzalez-Aguilar et al., 2000; Bai ct al., 2001). Howard et al. (1999) found that simply placing broccoli, carrots, and green beans in Ziploc® bags at a refrigeration temperature of 4°C resulted in moisture retention. Nunes et al. (1998) used PVC film to retard water loss in strawberries and found that water loss had a great effect on the retention of ascorbic acid. Although few differences were found in wrapped and unwrapped strawberries stored at 1°C and 10°C, a much greater (fivefold) loss occurred at 20°C storage. Most packages used for MAP have low permeability to water vapor and maintain a relative humidity of greater than 85%. Relative humidity below 85% results in excessive loss of moisture, leading to wilting and shriveling (Schlimme and Rooney, 1994). However, higher relative humidity nearing 100% in the bag may lead to condensation of water on the inside of the bag and loss in produce quality in the form discoloration and decreased nutritional quality.

9.7.4 WAXING OF FRESH PRODUCE

Water loss from fresh produce occurs through any openings such as lenticels and stomata on the plant surface and can also result from damage at harvest. Therefore, any method such as waxing that seals these openings will reduce the amount of water lost from the product.

Waxing of fresh produce is carried out to restore the natural wax lost in processing, but it also aids in slowing down water loss that causes dehydration. Waxes have been found to provide a partial barrier to water vapor and gas exchange, thus delaying shrinkage of cut produce and creating a modified atmosphere around the commodity (Ben-Yehoshua et al., 1985; Ben-Yehoshua, 1987; Baldwin, 1994; Frank and McLaughlin, 1997; Amarante and Banks, 2001). The edible coatings/waxes are nontoxic and are used to enrobe some fresh produce to aid in maintaining high relative humidity and freshness and prolong shelf life (Watada et al., 1996). An edible coating is considered to be a thin layer of edible material formed naturally on the food and applied in liquid form on foods by dipping, spraying, or panning. The waxes act as sealants, cover the cuticle, and block pores on the surface, slowing down ripening and, most important, reducing water loss. Banks et al. (1993) concluded that surface coatings were important for resistance to water vapor diffusion, internal atmosphere modification, respiration, and transpiration. Baldwin et al. (1996) found that the addition of soy protein or peptides to coating material such as CMC was effective in significantly reducing water loss and gas permeability. Avena-Bustillos et al. (1997) found that edible caseinate-acetylated monoglyceride films reduced moisture loss in celery by 75% and produced a higher water vapor resistance in apples. McHugh and Senesi (2000) reported increased water barrier properties with increased lipid concentration in coatings used in fresh-cut apples; water vapor permeability varied between 69 and 325 g mm/kPa$\cdot$d$^{-1}\cdot$m^{-2}. The effectiveness of waxes in reducing water loss has been reported in pears (Meheriuk and Lau, 1988), tomatoes (Ghaouth et al., 1992; Park et al., 1994), citrus fruit (Hagenmaier and Baker, 1993), green peppers (Lerdthanangkul and Krochta, 1996), and cut apples and potatoes (Baldwin et al., 1996), and summarized by Guilbert et al. (1996). Carrillo-Lopez et al. (2000) found that the treatment of mangoes with "Semperfresh" edible coating at concentrations of 8, 16, and 24 g/L resulted in firmer fruit and less weight loss compared to untreated fruit. Similarly, Hoa et al. (2002) found that coating mangoes with formulations containing carnuba wax, shellac, zein, and cellulose derivatives retarded the loss of firmness. Waxes are commonly used on products such as apples, cucumbers, eggplants, peppers, and citrus fruits.

9.7.5 PROCESSING TREATMENTS

Fruits and vegetables that have large surface area:weight ratios are susceptible to water loss and shriveling. Although the problem can be alleviated by rapid cooling soon after harvest and storing at high relative humidity, produce that is chill-sensitive deteriorates rapidly. Gonzalez-Aguilar et al. (1999) found the combination of hot water treatment of bell peppers at 45C or 53°C and storage in low-density polyethylene bags resulted in tremendous quality improvement, including firmness, reduction in water loss, retardation of color change, and alleviation of chilling injury. Garcia et al. (1996) reported that strawberries dipped in 1 to 4% calcium chloride solutions at 25 and 45°C, followed by 3 days of storage at 18°C, retained quality characteristics. Berries treated with 2 and 4% calcium chloride solutions at 25 and 45°C lost less weight than controls. They attributed the reduced weight loss to reduced water permeability of the fruit. Bangerth et al. (1972) found that calcium chloride treatment of apples resulted in firm apples that retained more ascorbic acid and had reduced internal breakdown.

TABLE 9.5
Quality Attributes of Mature Green Bell Peppers
in Relation to Weight Loss at 20°C

Actual Weight Loss (%)	Visual Quality Score[a]	Dehydration Score[b]	Firmness (N)[c]	Gloss Meter Value[d]
0.3	8.4	1.1	23.3	6.7
2.0	7.2	2.3	16.2	5.8
2.8	5.9	2.8	12.5	4.7
3.9	4.8	3.1	7.1	3.4

[a] Visual score based on scale of 1 to 9; 1 = unsalable and 9 = excellent.
[b] Dehydration score based scale of 1 to 5; 1 = none and 5 = severe dehydration.
[c] Newtons.
[d] Gloss value: the higher the value, the glossier the produce.

Source: Cantwell, M. et al., *Perishables Handling Q.*, 107, 17, 2001. With permission.

9.7.6 AIR MOVEMENT AND AIR PRESSURE

Air movement is important in freshly harvested produce because it removes heat from the produce. The vapor pressure of the thin air layer just above the produce is generally in equilibrium with that of the produce. Air movement around the produce reduces the thickness of the boundary layer and thus increases the vapor pressure difference, resulting in increased water loss. Reduced air pressure increases the rate of water loss, but this can be minimized by misting with water or providing moisture barrier packaging (Ben-Yehoshua, 1987; Kays, 1997).

9.8 WATER LOSS AND ITS IMPACT ON PRODUCE QUALITY

Freshly harvested produce is sold based on weight; therefore, any loss in weight has great impact not only on the quality but also the economic value of the produce and its profitability. It is therefore important to take steps that minimize water loss soon after harvest and during storage. Water loss of fresh produce affects its appeal to consumers because transpiration from fresh produce induces wilting, shrinkage, loss of firmness and crispness, and succulence, all of which are components of freshness (Ben-Yehoshua, 1987). Table 9.5 shows the effect of water loss on quality in mature green bell peppers (Cantwell et al., 2001). A weight loss of 2 to 4% reduced bell peppers' firmness, gloss, and visual quality, and increased their dehydration score. In some fresh produce, water is added back by misting, water baths, and increasing humidity. However, the presence of excessive moisture may result in growth of microorganisms and decay, physical damage such as splits in fruits, undesirable growth and sprouting, and color changes. Quality of fresh produce includes a combination of visual appearance (freshness, color, absence of defects, and decay), texture (crispness, turgidity, firmness, toughness, and tissue integrity), flavor (taste and smell), nutritive value (vitamins A and C, minerals, and dietary fiber), and safety (absence of chemical residues and microbial contamination) (Thompson, 1996; Piagentini et al., 2002).

9.8.1 APPEARANCE AND COLOR

The initial attraction leading to the purchase of any fresh produce by consumers is based on its appearance. The consumer's perception of visual quality is influenced by size, shape, color, uniformity, surface texture, consistency, and visible defects and blemishes of the produce. Many physical defects of fresh produce result from water loss after harvest and during storage; some of these defects include shriveling, wilting, wrinkling, and internal drying (Hruschka, 1977; Grierson and Wardowski, 1978; Kader, 2002). Considerable loss of water from fresh produce leads to a shriveled appearance; the skin's elasticity can no longer accommodate the reduction in volume associated with weight loss. Thompson et al. (2001) reported that the two most noticeable effects of delays in cooling of produce are shriveling and the loss of fresh, glossy appearance of commodities that lose water and quickly show visible symptoms at low levels of water loss, such as leafy vegetables. They recommended a loss of no more than 1% to prevent consumers from seeing the effect of shrivelling due to water loss. Izumi et al. (1996) noted that zucchini slices stored under 0.25% O_2 lost less weight and had less browning/decay and greater shear force than those stored in air. Gonzalez-Aguilar et al. (2000) found that a combination of antibrowning agents and modified atmosphere packaging reduced browning and deterioration of fresh-cut mangoes stored at 10°C. They concluded that the high humidity created in the in-package atmosphere alleviated tissue dryness and was an important factor in the ability of the antibrowning solutions to prevent browning and decay. Baldwin et al. (1996) reported reduced weight loss and browning of cut apples treated with Nature Seal™ or ascorbic acid added to edible coating in vacuum-packed trays compared to overwrapped tray storage. Barth et al. (1992) found misting of broccoli during cabinet display resulted in significant retention of chlorophyll and green color as measured by hue angle.

9.8.2 TEXTURE

Consumers' expectation of the texture of produce is key in determining the quality of the produce being purchased. Consumers' perception of texture quality varies depending on the produce; some produce is expected to be high quality if it is firm, while others are judged to be of optimum quality if it is soft. The texture quality of fresh produce is judged by feel and sound, both of which reflect crispness and crunchiness. Plant cell constituents that are important in texture are the parenchymal cells, which have the ability to absorb water and generate hydrostatic pressure called *turgor pressure*. It is this turgor pressure that gives fresh produce its characteristic crispness and firm texture. Loss of water from cells of fresh produce results in loss of crispness and texture, which is accompanied by breakdown of cell wall components. Harker et al. (1997) in their review reported that turgor has a major function in determining tissue strength and that it is related to water loss and has a profound influence on produce's texture. Texture evaluation is also used to indicate freshness of a product, since fresh produce tends to lose its crispness during storage. James et al. (1999) found that there was minimal weight loss of 0.5 to 1.0% when Muscadine grapes were wrapped in polyethylene compared to unwrapped grapes, which lost 15% of their weight, and noted that the loss of mass was proportional to a decrease in their firmness. Arpaia et al. (1986) reported increased softening of kiwifruit kept in air or controlled atmosphere, and increased temperatures resulted in white core inclusions.

9.8.3 NUTRITIVE VALUE

Fruits and vegetables are well known for their health benefits and recently have been linked to possibly reducing the risk of chronic diseases such as cancer and other age-related diseases (Buescher et al., 1999; Kushad et al., 2003). Vitamins such as C, E, and β-carotene are important components of fruits and vegetables known to enhance the health of human beings in their role as antioxidants (Prior and Cao, 2000). The type of antioxidant varies depending on the produce, but the most common vitamin that acts as an antioxidant is ascorbic acid. The presence of any nutritive component in fresh produce is affected by processing, and water loss results in product deterioration and nutritive loss. Nunes et al. (1998) found that strawberries wrapped in PVC and stored at 1 or 10°C had less water loss and a fivefold ascorbic acid retention compared to those stored at 20°C. They concluded that water loss had a greater effect on ascorbic acid loss than temperature. Barth et al. (1990, 1992) compared the retention of ascorbic acid in misted and unmisted broccoli stored for 72 h in display cabinets and found that misting promoted the retention of ascorbic acid and significant retention of moisture content.

9.9 CONCLUSION

Water plays a significant role in the quality of fresh fruits and vegetables, and the loss of water from produce affects shelf life, quality (appearance, texture, nutritive loss), and economic value. The loss of water from fresh produce is greatly affected by temperature, relative humidity, pressure, air movement around the product, and injury to any surface or internal tissues. It is therefore important to minimize loss of water in fresh produce by lowering its temperature soon after harvest and modifying the storage environment to increase humidity with the intention of reducing the water vapor pressure deficit between the fresh produce and surrounding air. In addition, packaging fresh produce in specific films and waxing with edible coatings help reduce water loss and maintain quality and prolong shelf life.

REFERENCES

Agar, I.T., Massantini, R., Hess-Pierce, B., and Kader, A.A., Postharvest CO_2 and ethylene production and quality maintenance of fresh-cut kiwifruit slices, *J. Food Sci.*, 64, 432–440, 1999.

Amarante, C. and Banks, N.H., Postharvest physiology and quality of coated fruits and vegetables, *Hortic. Rev.* 26, 161–234, 2001.

Arpaia, M.L., Mitchell, F.G., Kader, A.A., and Mayer, G., Ethylene and temperature effects on softening and white core inclusions of kiwifruit stored in air or controlled atmospheres, *J. Am. Soc. Hortic. Sci.,* 111, 149–153, 1986.

Avena-Bustillos, R.J., Krochta, J.M., and Saltveit, M.E., Water vapor resistance of red delicious apples and celery sticks coated with edible caseinate-acetylated monoglyceride films, *J. Food Sci.,* 62, 351–354, 1997.

Bai, J.H., Saftner, R.A., Watada, A.E., and Lee, Y.S., Modified atmosphere maintains quality of fresh-cut cantaloupe (*Cucumis melo* L.), *J. Food Sci.,* 66, 1207–1211, 2001.

Baldwin, E.A., Edible coatings for fresh fruits and vegetables: Past, present, and future, in *Edible Coatings and Films to Improve Food Quality,* Krochta, J.M., Baldwin, E.A., and Nisperos-Carriedo, M., Eds., Technomic, Lancaster, 1994, Chap. 2.

Baldwin, E.A., Nisperos, M.O., Chen, X., and Hagenmaier, R.D., Improving storage life of cut apple and potato with edible coating, *Postharv. Biol. Technol.,* 9, 151–163, 1996.

Bangerth, F., Dilley, D.R., and Dewey, D.H., Effect of postharvest calcium treatment on internal breakdown and respiration of apple fruit, *J. Am. Hortic. Sci.,* 97, 679–682, 1972.

Banks, N.H., Dadzie, B.K., and Cleland, D.J., Reducing gas exchange of fruits with surface coatings, *Postharv. Biol. Technol.,* 3, 269–284, 1993.

Barry-Ryan, C. and O'Beirne, D., Quality and shelf-life of fresh cut carrot slices as affected by slicing method, *J. Food Sci.,* 63, 851–856, 1998.

Barry-Ryan, C., Pacussi, J.M., and O'Beirne, D., Quality of shredded carrot as affected by packaging film and storage temperature, *J. Food Sci.,* 65, 726–730, 2000.

Barth, M.M., Perry, A.K., Schmidt, S.J., and Klein, B.P., Misting effects on ascorbic acid retention in broccoli during cabinet display, *J. Food Sci.,* 55, 1187–1191, 1990.

Barth, M.M., Perry, A.K., Schmidt, S.J. and Klein, B.P., Misting effects on market quality and enzyme activity of broccoli during retail storage, *J. Food Sci., 57,* 954–957, 1992.

Ben-Yehoshua, S., Individual seal-packaging of fruit and vegetables in plastic film: A new postharvest technique, *HortScience,* 20, 32–37, 1985.

Ben-Yehoshua, S., Sensory quality, in *Postharvest Physiology of Vegetables,* Weichmann, J. Ed., Marcel Dekker, New York, 1987, Chap. 21.

Ben-Yehoshua, S., Transpiration, water stress and, gas exchange, in *Postharvest Physiology of Vegetables,* Weichmann, J., Ed., Marcel Dekker, New York, 1987, Chap. 6.

Ben-Yehoshua, S., Burg, S.P., and Young, R., Resistance of citrus fruit to mass transport of water vapor and other gases, *Plant Physiol.,* 79, 1048–1053, 1985.

Bolin, H.R., Stafford, A.E., King, A.D., Jr., and Huxsoll, C.C., Factors affecting the storage stability of shredded lettuce, *J. Food Sci.,* 42, 1319–1321, 1977.

Buescher, R., Howard, L., and Dexter, P., Postharvest enhancement of fruits and vegetables for improved human health*, HortScience,* 34, 1167–1170, 1999.

Burton, W.G., The physiological implications of structure: Water movement, loss and uptake, in *Postharvest Physiology of Food Crops,* Burton, W.G., Ed., Longman, New York, 1982, Chap. 3.

Cantwell, M., Thangaiah, A., and Aguiar, J., Impact of delays to cool on postharvest quality of bell peppers and eggplants, *Perishables Handling Q.,* 107, 17–20, 2001.

Carrillo-Lopez, A., Ramirez-Bustamante, F., Valdez-Toress, J.B., Rojas-Villegas, R., and Yahia, E.M., Ripening and quality changes in mango fruit as affected by coating with edible film, *J. Food Qual.,* 23, 479–486, 2000.

Cohen, E., Shapiro, B., Shalom, Y., and Klen, J.D., Water loss: A nondestructive indicator of enhance cell membrane permeability of chilling injured citrus fruit, *J. Am. Soc. Hortic. Sci.,* 119, 983–986, 1994.

Collins, J.K. and Perkins-Veazie, P., Postharvest changes in strawberry fruit stored under simulated retail display conditions, *J. Food Qual.,* 16, 133–43, 1993.

Cybulska, B. and Doe, P.E., Water and food quality, in *Chemical and Functional Properties of Food Components,* Siskorski, Z.E., Ed., CRC Press, New York, 2002, Chap. 3.

Devlin, R.M. and Witham, F.H., *Plant Physiology,* Willard Grant Press, Boston, 1983, Chaps. 2 and 3.

Emond, J., Boily, S., and Mercier, F., Reduction of water loss and consideration using perforated film packages for fresh fruits and vegetables, in *Harvest and Postharvest Technologies for Fresh Fruits and Vegetables,* Kushwaha, L., Serwatoski, R., and Brook, R., Eds., ASAE, Michigan, Proceedings of the International Conference, Guanajuato, Mexico, 1995, pp. 339–346.

Frank, T.H. and McLaughlin, E.W., *Produce: Management and Operations,* Cornell University Distance Education Program, Ithaca, NY, 1997, pp. 25, 57.

Garcia, J.M., Herrera, S., and Morilla, A., Effects of postharvest dips in calcium chloride on strawberry, *J. Agric. Food Chem.,* 44, 30–33, 1996.

Gast, K.L., *Postharvest Management of Commercial Horticultural Crops: Storage Conditions – Fruits and Vegetables,* Bulletin 4135, 1991.

Ghaouth, A.E., Ponnampalam, R., Castaigne, F., and Arul, J., Chitosan coating to extend the storage life of tomatoes, *HortScience,* 27, 1016–1018, 1992.

Gonzalez-Aguilar, G.A., Cruz, R., Baez, R., and Wang, C.Y., Storage quality of bell peppers pretreated with hot water and polyethylene packaging, *J. Food Qual.,* 22:287–299, 1999.

Gonzalez-Aguilar, G.A., Wang, C.Y., and Buta, J.G., Maintaining quality of fresh-cut mangoes using antibrowning agents and modified atmosphere packaging, *J. Agric. Food Chem.,* 48, 4204–4208, 2000.

Grierson, W. and Wardowski, W.F., Humidity in horticulture, *HortScience,* 10, 356–360, 1975.

Grierson, W. and Wardowski, W.F., Relative humidity effects on the postharvest life of fruits and vegetables, *HortScience,* 13, 570–573,1978.

Guilbert, S., Gontard, N., and Gorris, L.G.M., Prolongation of the shelf-life of perishable food products using biodegradable films and coatings, *Lebensm.-Wiss. Technol.,* 29, 10–17, 1996.

Hagenmaier, R.D. and Baker, R.A., Reduction in gas exchange of citrus fruit by wax coatings, *J. Agric. Food Chem.,* 41, 283–287, 1993.

Harker, F.R., Redgwell, R.J., Hallet, I.C., and Murray, S.H., Texture of fresh fruit, *Hortic. Rev.,* 20, 121–205, 1997.

Hoa, T.T., Ducamp, M., Lebrun, M., and Baldwin, E.A., Effect of different coating treatments on the quality of mango fruit, *J. Food Qual.,* 25, 471–486, 2002.

Hodges, D.M. and Forney, C.F., Processing line affects on storage attributes of fresh-cut spinach leaves, *HortScience,* 35, 1308–1311, 2000.

Howard, L.A., Wong, A.D., Perry, A.K., and Klein, B.P., β-Carotene and ascorbic acid retention in fresh and processed vegetables, *J. Food Sci.,* 65, 929–936, 1999.

Hruschka, H.W., Postharvest weight loss and shrivel in five fruits and vegetables, *USDA-ARS Mark. Res. Rep.,* 1059, 3–23, 1977.

Izumi, H., Watada, A.E., and Douglas, W., Low O_2 atmospheres affect storage quality of zucchini squash slices treated with calcium, *J. Food Sci.,* 61, 317–321, 1996.

James, J., Lamikanra, O., Morris, J.R., Main, G., Walker, T., and Silva, J., Interstate shipment and storage of fresh muscadine grapes, *J. Food Qual.,* 22, 605–617, 1999.

Kader, A.A., Quality and safety factors: Definition and evaluation for fresh horticultural crops, in *Postharvest Technology of Horticultural Crops,* Kader, A.A., Ed., University of California, Agriculture and Natural Resources, Publication 3311, 2002, Chap. 22.

Kays, S.J., *Postharvest Physiology of Perishable Plant Products*, Exon Press, Athens, GA, 1997, Chap. 7.

Kramer, P.J., Plants cells and water, in *Introduction to Plant Physiology*, Hopkins, W., Ed., John Wiley & Sons, New York, 1995, Chap. 3.

Kushad, M.M., Masiunas, J., Kalt, W., Eastman, K., and Smith, M.A.L., Health promoting phytochemicals in vegetables, *Hortic. Rev., 28*, 125–170, 2003.

Lerdthanangkul, S. and Krochta, J.M., Edible coating effects on postharvest quality of green bell peppers, *J. Food Sci., 61*, 176–179, 1996.

Maguire, K.M., Banks, N.H., and Opara, L.U., Factors affecting weight loss of apples, *Hortic. Rev. 25*, 197–233, 2001.

McDoanald, R.E., Risse, L.A., and Barmore, C.R., Bagging chopped lettuce in selected permeability films, *HortScience, 25*, 671–673, 1990.

McHugh, T.H. and Senesi, E., Apple wraps: A novel method to improve the quality and extend the shelf life of fresh-cut apples, *J. Food Sci., 65*, 480–485, 2000.

Meheriuk, M. and Lau, O.L., Effect of two polymeric coatings on fruit quality of 'Bartlett' and 'd'Anjou' pears, *J. Am. Soc. Hortic. Sci., 113*, 222–226, 1988.

Miller, W.R., Davis, P.L., Dow, A., and Bongers, A.J., Quality of strawberry packed in different consumer units and stored under simulated air-freight shipping conditions, *HortScience, 18*, 310–312, 1983.

Nilsen, E.T. and Orcutt, D.M., Water dynamics in plants, in *Physiology of Plants Under Stress: Biotic Factors,* John Wiley & Sons, New York, 1996, Chap. 7.

Nobel, P.S., Water, in *Physicochemical and Environmental Plant Physiology*, Academic Press, New York, 1999, Chap. 2.

Nunes, M.C.N, Brecht, J.K., Morais, A.M.M.B., and Sargent, S.A., Controlling temperature and water loss to maintain ascorbic acid levels in strawberries during postharvest handling, *J. Food Sci., 63*, 1033–1036, 1998.

Osornio, M.M.L. and Chaves, A.R., Quality changes in stored raw grated beetroots as affected by temperature and packaging film, *J. Food Sci., 63*, 327–330, 1998.

Park, H.J., Chinnan, M.S., and Shewfelt, R.L., Edible coating effects on storage life and quality of tomatoes, *J. Food Sci., 59*, 568–570, 1994.

Perez, A.G., Olias, R., Olias, J.M., and Sanz, C., Strawberry quality as a function of the high-pressure fast-cooling design, *Food Chem., 62*, 161–168, 1998.

Perkins-Veazie, P. and Collins, J.K., Shelf-life and quality of 'Navaho' and 'Shawnee' blackberry fruit stored under retail storage conditions, *J. Food Qual., 22*, 535–544, 1999.

Perrin, P.W. and Gaye, M.M., Effects of simulated retail display and overnight storage treatments on quality maintenance in fresh broccoli, *J. Food Sci., 51*, 146–149, 1986.

Piagentini, A.M., Guemes, D.R., and Pirovani, M.E., Sensory characteristics of fresh-cut spinach preserved by combined factors methodology, *J. Food Sci., 67*, 1544–1549, 2002.

Prior, L.R. and Cao, G., Antioxidant phytochemicals in fruits and vegetables: Diet and health implications, *HortScience, 35*, 588–592, 2000.

Purvis, A.C., Importance of water loss in the chilling injury of grapefruit stored at low temperature, *Sci. Hortic., 23*, 261–267, 1984.

Robinson, J.E., Browne, K.M., and Burton, W.G., Storage characteristics of some vegetables and soft fruits, *Ann. Appl. Biol., 81*, 399–408, 1975.

Rolle, R.S. and Chism, G.W., Physiological consequences of minimally processed fruits and vegetables, *J. Food Qual., 10*, 157–177, 1987.

Ruan, R.R. and Chen, P.L., Aspects of water in food and biological systems, in *Water in Foods and Biological Materials: A Nuclear Magnetic Resonance Approach*, Technomic Publishing, Lancaster, PA, 1998, Chap. 2.

Schlimme, D.V. and Rooney, M.L., Packaging of minimally processed fruits and vegetables, in *Minimally Processed Refrigerated Fruits and Vegetables*, Wiley, R.C., Ed., Chapman and Hall, London, 1994, Chap. 4.

Shamaila, M., Powrie, W.D., and Skura, B.J., Sensory evaluation of strawberry fruit stored under modified atmosphere packaging (MAP) by quantitative descriptive analysis, *J. Food Sci.,* 57, 1168–1172, 1184, 1992.

Steinbeck, J., Plants cells and water, in *Introduction to Plant Physiology*, Hopkins, W., Ed., John Wiley & Sons, New York, 1995, Chap. 3.

Taiz L. and Zeiger, E., *Plant Physiology*, 2nd ed., Sinauer Associates, Sunderland, MA, 1998, Chaps. 3 and 4.

Thompson, A.K., Quality, losses and production, in *Postharvest Technology of Fruit and Vegetables*, Blackwell Science, Cambridge, MA, 1996, Chap. 1.

Thompson, J.F., Psychrometrics and perishable commodities, in *Postharvest Technology of Horticultural Crops*, Kader, A.A., Ed., University of California, Agriculture and Natural Resources, Publication 3311, 2002, Chap. 13.

Thompson, J.F., Cantwell, M., Arpaia, M.L., Kader, A., Crisosto, C., and Smilanick, J., Effect of cooling delays on fruit and vegetable quality, *Perishables Handling Q.,* 105, 1–4, 2001.

Thompson, J.F., Mitchell, F.G., and Kasmire, R.F., Cooling horticultural commodities, in *Postharvest Technology of Horticultural Crops*, Kader, A.A., Ed., University of California, Agriculture and Natural Resources, Publication 3311, 2002, Chap. 11.

van den Berg, L. and Lentz, C.P., High humidity storage of vegetables and fruits, *HortScience,* 13, 565–569, 1978.

Watada, A.E., Kim, S.D., Kim, K.S., and Harris, T.C., Quality of green beans, bell peppers and spinach stored in polyethylene bags, *J. Food Sci.,* 52, 1637–1641, 1987.

Watada, A.E., Ko, N.P., and Minot, D.A., Factors affecting quality of fresh-cut horticultural products, *Postharv. Biol. Technol,.* 9, 115–125, 1996.

Wills, R., McGlasson, B., Graham, D., and Joyce, D., *Postharvest: An Introduction to the Physiology and Handling of Fruits, Vegetables and Ornamentals*, 4th ed., UNSW Press, Hyde Park Press, Adelaine, South Australia, 1998, Chaps. 4, 5, and 8.

10 Mechanisms of Food Additives, Treatments, and Preservation Technology

Michal Voldrich
Department of Food Preservation and Meat Technology,
FPBT, ICT, Prague, Czech Republic

CONTENTS

10.1 MECHANISMS OF SPOILAGE OF FRESH AND PROCESSED PRODUCE

Fresh fruits and vegetables and other products of plant origin are available fresh only during a brief harvest period. It is necessary to keep supplies of harvested food edible over relatively long periods of time. Empirical preservation methods have been used since ancient times; more recently, increasingly sophisticated preservation methods have been developed (Lück and Jager, 1997).This chapter is not an exhaustive review of changes that occur in produce and preservative methods to minimize them, but a brief summary that serves as an introduction to the methods of food preservation. Generally, produce degradation begins after harvest and continues during processing and storage until the product is consumed or becomes inedible or unsafe for consumption. These changes could be generally classified to four groups: physiological, enzymatic, chemical, and microbial changes.

10.1.1 PHYSIOLOGICAL CHANGES

Physiological changes are more complex and proceed in the organelles, cells, or tissues. An increase in the rate of these changes is caused by conditions that accelerate the rate of natural deterioration, such as high temperature, low atmospheric humidity, and physical injury. Abnormal physiological deterioration occurs when fresh produce is subjected to extremes of temperature, atmospheric modification, or contamination. This may cause unpalatable flavors, failure to ripen, or other changes in the living processes of the produce, making it unfit for use (Barbosa-Cánovas et al., 2003).

10.1.1.1 Mechanical Damage

Careless handling of fresh produce causes internal bruising, which results in physiological damage or splitting and skin breaks. This increases the rate of water loss and the rate of normal physiological breakdown. Skin breaks also provide sites for infection by disease organisms causing decay. The injuries can result in tissue browning, higher respiration and ethylene production rates, and some undesirable compositional changes, such as loss of ascorbic acid content and development of off-flavors. Mechanical damage can also activate the protection mechanisms of the tissues. An increase in polyphenoloxidase (PPO) activity, possibly from production of new PPO from the latent form of the enzyme, and production of quinones in the first stage of enzymatic browning reactions are examples of plant injury-induced defense responses. Similarly the biosynthesis of the steroidal glycoalcaloids saponine and chaconin occur during the storage of peeled, cut potato tubers (Glynn and Dixon, 1996; Dale et al., 1998). The highest levels of furanocoumarins were found in mechanically damaged and/or rotten impaired samples of parsnip and celeriac (Schulzova et al., 2002).

10.1.1.2 Temperature Injuries (Chilling, Freezing, and High-Temperature Injury)

Certain injuries can occur during storage under suboptimal conditions. Chilling injury usually occurs to commodities of tropical origin at temperatures below 10 to

TABLE 10.1
Susceptibility of Fruits and Vegetables to Chilling Injury at Low but Nonfreezing Temperatures

Commodity	Approximate Lowest Safe Temperature (°C)	Chilling Injury Symptoms
Aubergines	7	Surface scald, *Alternaria* rot
Avocados	5–13	Grey discoloration of flesh
Bananas (green/ripe)	12–14	Dull, gray-brown skin color
Beans (green)	7	Pitting, russeting
Cucumbers	7	Pitting, water-soaked spots, decay
Grapefruit	10	Brown scald, piking, watery breakdown
Lemons	13–15	Pitting, membrane stain, red blotch
Limes	7–10	Pitting
Mangoes	10–13	Grey skin scald, uneven ripening
Honeydew melon	7–10	Pitting, failure to ripen, decay
Watermelon	5	Pitting, bitter flavor
Okra	7	Discoloration, water-soaked areas, piking
Oranges	7	Pitting, brown stain, watery breakdown
Papaya	7	Pitting, failure to ripen, off-flavor, decay
Pineapples	7–10	Dull green color, poor flavor
Potatoes	4	Internal discoloration, sweetening
Pumpkins	10	Decay
Sweet peppers	7	Pitting, *Alternaria* rot
Sweet potatoes	13	Internal discoloration, pinking, decay
Tomatoes: Mature green	13	Water-soaked softening, decay
Tomatoes: Ripe	7–10	Poor color, abnormal ripening, *Alternaria* rot

Source: Lutz, J.M. and Hardenburg, R.E., *The Commercial Storage of Fruits, Vegetables and Florist and Nursery Stocks*, U.S. Government Printing Office, Washington, DC, 1968, p. 1.

13°C, although certain temperate commodities are also susceptible to chilling injury. The symptoms often cannot be detected at low temperatures and become visible several days after removing to warmer temperatures. The extent of damage depends on the length of exposure and temperature. The approximate lowest safe storage temperatures for selected fruits and vegetables are given in Table 10.1.

When produce is subjected to freezing temperatures between 0 and –2°C the water inside the cells starts to freeze. The ice may cause damage in tissues. Frozen produce has a water-soaked or glassy appearance. Other changes that usually occur may not be visible while the product is frozen, but during thawing the mechanical destruction of cellular membranes allows contact between enzymes and substrates. This results in enzymatic spoilage such as browning reactions and lipoxygenase oxidation. The damage is often associated with losses of juice and microorganisms present on the surface of fruits and vegetables easily contaminate exudates. Although a few commodities are tolerant to mild freezing, generally it is advisable to avoid such low temperatures because subsequent storage life is shortened. Produce that has recovered from freezing is highly susceptible to decay. When fresh produce is exposed to high temperatures the deterioration rate is increased. Exposure to solar radiation is an example; produce left in the sun after harvest may reach temperatures as high as 50°C. It will achieve a high rate of respiration and, if packed and transported without cooling or adequate ventilation, it will become unusable. Long exposure to tropical sun will cause severe water loss from thin-skinned root crops such as carrots and turnips and from leafy vegetables (Dauthy, 1995).

10.1.1.3 Loss of Water

Fresh produce continues to lose water after harvest, but unlike the growing plant it can no longer replace lost water from the soil and so must use up its water content remaining at harvest. This loss of water from fresh produce after harvest is a serious problem, causing shrinkage and loss of weight. When the harvested produce loses 5 or 10% of its fresh weight, it begins to wilt and soon becomes unusable. To extend the usable life of produce, its rate of water loss must be as low as possible.

10.1.1.4 Undesirable Effects on Respiration

Fresh produce belongs to the "respiring" group of food. Respiration, the basic reaction of all plant material, also continues after harvest until consumption, processing, or decay. The extent of gas exchange depends on various factors, such as temperature, O_2 and CO_2 concentrations, presence of ethylene, and degree of mechanical damage of plants. Conditions that accelerate respiration are usually undesirable because they usually accelerate senescence and decay of the produce. Similarly excessive reduction of the respiration rate can be also undesirable. Such an effect is caused, for example, by a reduction of the oxygen concentration in the surrounding atmosphere below the critical limits, or by an increase in the carbon dioxide concentration above tolerable levels. When the air supply is restricted and the amount of available oxygen in the environment falls to about 2% or less, there is a risk that anaerobic respiration (anoxia) and fermentation instead of respiration will occur. Fermentation breaks down sugars to ethanol and carbon dioxide, and the alcohol produced unpleasant flavors in produce and promotes premature aging (Rudell et al., 2002). Similar situations may occur when ventilation is inadequate. A restricted air supply leads to the accumulation of carbon dioxide around the produce. When the concentration of this gas rises to between 1 and 5% in the atmosphere, it will quickly ruin produce by causing bad flavors, internal breakdown, failure of fruit to ripen, and other abnormal physiological conditions (Kader et al., 1989). Maximum levels of carbon dioxide and minimum levels for oxygen for storage of selected fruits and vegetables are given in Table 10.2.

10.1.2 Enzymatic Changes

Physiological and enzymatic changes caused by endogenous enzymes are difficult to adequately classify. In this approach the physiological changes are more complex than the individual reactions previously described. Individual enzymatic reactions that are involved in postharvest senescence and spoilage belong to physiological changes with many parallel processes. Examples of undesirable enzymatic changes

TABLE 10.2
Maximum Tolerable Levels of Carbon
Dioxide and Minimum Levels of Oxygen for
Storage of Selected Fruits and Vegetables

Produce	CO_2 (%)	O_2 (%)
Apple (Golden Delicious)	2	2
Asparagus (5°C)	10	10
Avocado	5	3
Banana	5	—
Broccoli	15	1
Cabbage	5	2
Carrot	4	3
Cauliflower	5	2
Citrus fruits	—	5
Cucumber	10	3
Lettuce	1	2
Onion	10	1
Pea	7	5
Pear (Barlett)	5	2
Potato	10	10
Strawberry	20	2
Tomato	2	3

Source: Fellows, P., *Food Processing Technology, Principle and Practice*, CRC Press, Boca Raton, FL, 2000c, Part 4.

are (1) enzymatic browning reaction during peeling, cutting, mechanical damage, etc.; (2) lipoxygenase oxidation during the frozen storage of fruit and vegetables; and (3) degradation of pectin in squeezed citrus juices.

The most important endogenous enzymes and their effects on produce are summarized in Table 10.3.

In intact plant tissues enzymes are localized in the cytoplasm, while their substrates are in vacuoles (e.g., phenols), or in membranes (e.g., fatty acids). Enzymatic reactions proceed within the metabolic pathways of anabolic or catabolic processes ongoing in fruit or vegetable cells or tissues. The undesirable changes start when enzymes come in contact with substrates. The substrates are fruit or vegetable components or, in the case of oxidoreductases, oxygen. Thus, one possible way to prevent undesirable enzymatic changes is to reduce the contact of enzymes with substrates. Examples of how this is achieved involve limiting oxygen and/or minimizing the concentration of substrates using preventative processes such as using sharp knives or machine blades that make clean clean cuts without damaging plant tissue. Other general prevention methods include acidification (or changing the pH out of the enzyme's optimum range) and use of various inhibitors, especially thermal inactivation of enzymes (coagulation of protein parts of the enzyme).

TABLE 10.3
Important Endogenous Enzymes and the Effects
of Enzymatic Changes During the Processing
of Fruits and Vegetables

Endogenous Enzyme	Effect on Food
Lipoxygenase Lipase Protease	Off-flavors, off-odors
Pectolytic enzymes Cellulolytic enzymes	Texture changes (softening, sediments in citrus juices, etc.)
Polyphenoloxidase Chlorofylase Peroxidase (to some extent)	Color changes (browning, chlorophyll degradation)
Askorbase Thiaminase Polyfenoloxidase	Lowering of nutritive value (vitamin degradation, lowering of digestibility of protein, etc.)

10.1.2.1 Important Enzymes

A lipoxygenase (linoleic acid oxygen oxidoreductase, EC 1.13.11.12) enzyme occurs in many plants and also in erythrocytes and leucocytes. It catalyzes the oxidation of unsaturated fatty acids to their corresponding hydroperoxides. These hydroperoxides have the same structure as those obtained by autooxidation. In comparison with lipid autooxidation the reaction is characterized by all features of enzyme catalysis: substrate specificity, peroxidation selectivity, occurrence of a pH optimum, susceptibility to heat treatment, and a high reaction rate in the range of 0 to 20°C. Lipoxygenase oxidizes only the fatty acids with a 1-*cis*,4-*cis* pentadiene system; therefore, linoleic and linolenic acids are the preferred substrates in fruit and vegetable tissues. Hydroperoxides formed enzymatically are precursors of odorants, green-grassy or cucumberlike smelling aldehydes hexanal, 3-*cis*-hexenal (leafy aldehyde), 3-*cis*,6-*cis*-nonadienal, and others formed by subsequent enzymatic degradation catalyzed by lyases. These products are typical in fresh produce such as peeled cucumber and peeled and cut apples, but they are also undesirable in the case of frozen vegetables or fruits after some length of storage (Grosch, 1982; Whitaker, 1991).

Polyphenoloxidase (PPO) catalyzes two reactions: hydroxylation of monophenol to *o*-diphenol (E.C. 1.14.18.1, monophenol monooxygenase), which is followed by an oxidation to *o*-quinone (E.C. 1.10.3.1, *o*-diphenol: oxygenoxidoreductase). Both activities are also known as cresolase and catecholase activity. PPO is widely distributed in the plant and animal kingdom and is primarily responsible for enzymatic browning. It affects numerous plant organs that are rich in oxidizable phenols (Zawistovski et al., 1991).

Pectinolytic enzymes pectins in plant foods are attacked by a series of enzymes. Pectin esterases, which occur widely in plants and microorganisms, demethylate pectin to pectic acid. The other group of pectolytic enzymes are hydrolases and lyases, which attack the glycosidic bond in polygalacturonides. Polygalacturonases occur in plants and microorganisms; pectin and pectate lyases are only produced by microorganisms. During processing pectinolytic as well as other enzymes exude from bruised cells and may diffuse into inner tissues and contribute to softening and other modifications of texture. Pectin esterase is responsible for cloud flocculation in citrus juices. Pectinolytic enzymes and other enzymes (amylolytic and celullolytic) are produced by microorganisms (molds) during the microbial deterioration of fruit or vegetables. These enzymes are also used for pulp-enzyming before pressing of fruit and vegetable juices or to clarify the squeezed juices (Belitz and Grosch, 1999).

Peroxidase is widely present in plants and catalyzes the peroxidatic reaction:

$$ROOH + AH_2 \leftrightarrow H_2O + ROH + A.$$

Peroxidase is very important in the ripening process, although its role is not fully understood. It is believed to contribute off-flavors and off-odors in some food products (Robinson, 1991). Peroxidase is still understood as a potential source of deterioration of plant products. Peroxidase activity may result in oxidative actions that involve hydrogen donor components. For some fruits (e.g., cantaloupe melon) peroxidase activity could be related to tissue response to increased oxidative stress in cut fruit (Lamikarna and Watson, 2000). Peroxidase mainly contributes to changes in fresh produce. Residual peroxidase is often used as a measure of effectiveness of processing methods in a number of processed fruits and vegetables. In some cases, such as in frozen fruits and vegetables, residual peroxidase activity may be less important for quality assessment than residual lipoxygenases. In such cases, blanching optimization may be linked to residual lipoxygenases (Sheu and Chen, 1999).

10.1.3 CHEMICAL CHANGES

Chemical changes in fruits and vegetables occur mainly during processing and storage. At the beginning stages, after harvest, these changes are not so important compared with enzymatic or physiological deterioration. Two principal chemical changes occur during the processing and storage of fruit and vegetables that lead to a deterioration in sensory quality: lipid oxidation and nonenzymatic browning (within those also hydrolytic reactions). Chemical reactions are also responsible for changes in the color and flavor of foods during processing and storage. Oxidation of polyunsaturated lipids is responsible for production of off-flavors and off-odors. The course of the reaction is influenced by light, oxygen concentration, high temperature, the presence of catalysts (generally, transition metals such as iron and copper), and water activity. Fruit and vegetable products containing seeds or processed products in which the oil is in contact with oxygen, such as dried, unpeeled fruits and dried strawberries containing seeds, are quite susceptible to lipid oxidation during storage. Nonenzymatic browning is also a major cause of deterioration during storage of dried and concentrated foods. Nonenzymatic browning includes the Maillard reaction (reaction of reducing sugars with α-aminocompound, e.g., aminoacid), which can be divided into three stages: (1) early Maillard reactions, which are chemically well-defined steps without browning; (2) advanced Maillard reactions, which lead to the formation of volatile or soluble substances; and (3) final Maillard reactions, which lead to insoluble brown polymers. The Maillard reaction is usually a limited part of a series of very complex chemical reactions of the compounds present in the food products. Other nonenzymatic browning reactions include acid degradation of sugars (inversion of sucrose, formation of 5-hydroxymethyl-2-furancarbaldehyde (HMF) and 2-furancarbaldehyde (F) from hexoses and pentoses). The nonenzymatic browning reaction is important during processes such as frying, drying, concentration, and long-term storage of preserves at ambient or elevated temperatures, and these reactions contribute to the overall deterioration of the products (color changes, off-flavors, off-odors, usually as a contribution to damaging the fresh appearance and flavor of the processed fruit or vegetables).

Color changes, pigment modifications in foods are mainly physiological. These include the development of color in fruits such as tomatoes and bananas during ripening after harvest, or physiological degradation of chlorophyll and other pigments at the end of senescence and in early phases of decay. Chlorophylls are green plant products containing two types of chlorophyll pigment, a and b, in the ratio of about 3 to 1. Almost any type of food processing or storage causes some deterioration of the chlorophyll pigments.

Chlorophyll decomposition is a two-step process. In the first step, magnesium is removed from the molecule, producing dull, olive-brown pheophytin (phenophytinization); in the second step, pheophytins are converted to pheophorbids (Schwarts et al., 1981). Pheophytination is the major color change during processing and storage of green plant materials. Heat accelerates the rate of the change and it is acid-catalyzed. López-Ayerra et al. (1998) observed the degradation of chlorophylls in raw, frozen, and canned spinach and found that about 15.9% of chlorophylls a and b were lost during the freezing process and 99.9% after canning as a consequence of the heating used in industrial processing. Magnesium in the chlorophyll molecule can be displaced by other metal ions (e.g., copper or zinc); the product is more stable and has a different hue. The reaction with contaminating copper ions during the processing is a potential cause of undesirable color changes. Different stability of chlorophylls a and b is used for the evaluation of age of the product (frozen storage, etc.).

Anthocyanins are a group of more than 150 reddish, water-soluble pigments that are very widespread in the plant kingdom. The rate of anthocyanin destruction is pH-dependent and is greater at higher pH values. For example, according to Wesche-Ebeling et al. (1996) preserved plums after 90 days of storage at room temperature contained 77% of their original anthocyanin at pH 2.95 but only 29% at pH 3.45 and 8% at pH 3.95 as a result of anthocyanin degradation. The reaction proceeds in two steps. The glycosides anthocyanins are hydrolyzed to anthocyanidins and sugar; production of aglycons can increase the intensity of color, but aglycons are less stable and undergo other degradation reactions. Some anthocyanins are able to form complexes with metals such as Al, Fe, Cu, and Sn and are important for packaging.

These complexes usually differ in color from the original anthocyanins, and they can indicate contamination of products with metal ions (e.g., from the packaging and/or processing equipment). Products of nonenzymatic browning reactions accelerate destruction of anthocynanins (Clydesdale, 1997; Belitz and Grosch, 1999). Betalaines are nitrogenous red or yellow glycosidic, water-soluble pigments present in vegetables (red beets and other beets). Unlike the anthocyanins, the hue of the betalaines is not dependent on pH, but they are more sensitive to heat and oxygen (Clydesdale, 1997).

The carotenoids are a group of mainly lipid-soluble compounds responsible for many of the yellow and red colors of plant and animal products. Carotenoids are more stable than other pigments. During heat treatment they undergo *cis/trans* isomeration. *Cis* isomers are less stable to oxidation. Oxidation is usually a complex reaction with mechanisms similar to autoxidation of polyunsaturated lipids. Isomerization and oxidation of carotenoid pigments are accelerated by the products of other degradation processes (e.g., nonenzymatic browning reactions). The degradation of carotenoids contributes to off-flavor formation and to undesirable color changes (e.g., during the production and storage of tomato paste or tomato ketchup) (Clydesdale, 1997). Undesirable production of new pigments may occur, particularly when fruits and vegetables are exposed to light. The formation of brown or black pigments by the reaction of quinones produced by the enzymatic browning reaction or the formation of dark melanoidines at the final stages of nonenzymatic browning were mentioned above. Other examples of undesirable color changes include the formation of pink pigment in pickled, cut white cabbage, cauliflower in acid brine, onion purées, postcooking blackening of potatoes, and formation of green-blue pigments in garlic paste and other products. The blackening that may occur after cooking potatoes is the formation of a complex between chlorogenic acid present in the peridermal layers of tubers and ferric ions during cooking. These complexes are oxidized, and based on the pH of the medium they form green, blue-grey, or black products (Hugghes and Swain, 1962; Friedman, 1997). Mechanisms of color changes during the processing of garlic are not sufficiently clear. The tendency to turn green is higher when the cloves are processed in their physiological active stages (i.e., after harvest or at the end of the dormancy). The rate of greening depends on the activity of allianase and content of propenylcysteine sulfoxide (Velíšek, 1999; Ahmed and Shivhare, 2001). The other mentioned changes have not been explained yet but also correspond with physiological activities at the time of processing and are dependent on the concentration of potential precursors, which are formed in processed tissues by physiological or enzymatic reactions. The aim is to optimize processing in accordance with the physiological state of the produce and minimize undesirable enzymatic changes in processed tissues.

Softening and texture changes, cutting, peeling, and, especially, slicing result in dramatic losses in firmness of fruit tissues. The tendency to soften is dependent on the physiological state. Softening itself is a complex process that includes physiological, enzymatic, chemical, and physical changes in the tissue. Pectolytic and proteolytic enzymes exude from the bruised cells and diffuse into inner tissues (Varoquaux et al., 1990). Protopectin is transformed to water-soluble pectin, cellulose crystallinity decreases, and cell walls become thinner. Sugars and other components diffuse to the intercellular spaces. This causes loss of turgor and softening of the tissues (Poovaiah, 1986; Bolin and Huxsoll, 1989; King and Bolin, 1989). In processed fruit and vegetables softening is also caused by hydrolytic changes of macromolecules under acidic conditions, especially during heating.

10.1.4 MICROBIOLOGICAL CHANGES

Fresh fruit and vegetables contain microorganisms coming from soil, water, air, and other environmental sources. These may also include some pathogens. Vegetables are fairly rich in carbohydrates (about 5% or more), low in proteins (about 1 to 2%), and, with the exception of tomatoes, have high pH. Fruits are richer in carbohydrates (generally 10% and more) and very low in proteins (usually less than 1%); the majority of fruits have a pH value lower than 4.0. Any damage reduces the natural protective mechanisms of plant tissues and allows subsequent microbial spoilage. Different types of molds start the spoilage. High water activity as well as yeasts and bacteria also favor spoilage. Spoilage of fruits and vegetables is usually caused by various types of molds from the genera *Penicillium, Phytophtora, Alternaria, Botrytis, Rhyzopus, Aspergillus,* and others. Yeasts from the genera *Saccharomyces, Candida, Torulopsis,* and *Hansenula* have been associated with fermentation of some fruits such as apples, strawberries, citrus fruits, and dates. Bacterial spoilage associated with the souring of berries and figs has been attributed to the growth of lactic acid bacteria. Low-acid vegetables and fruits (melon, watermelon, papaya, and avocado) are often spoiled by the bacterial genera: *Pseudomonas, Erwinia, Bacillus,* and C*lostridium* and others, including other pathogens. The ability of some pathogenic bacteria such as *Salmonella* and *Shigella* to grow on sliced apples, papaya, watermelon, cantaloupe, and honeydew, especially at ambient temperature, was proven by several authors (Golden et al., 1993; Leverentz et al., 2001). Microbial spoilage of fresh fruits and vegetables is generally described by the common term *rot*, which is accompanied by changes in appearance, such as with black rot, grey rot, pink rot, soft rot, and stem-end rot. In addition to changes of color, microbial rot causes loss of texture and formation of off-flavors. In the case of toxinogenic molds and bacteria (some species of the genera *Bacillus* and *Clostridium*), mycotoxins or bacterial toxins could be produced in the product. In fruits with acidic pH, sporulating bacteria cannot germinate. If such products are to be sterilized, heat treatment below 100°C is sufficient, because there is no need to inactivate the thermoresistant bacterial spores. In 1984 a new type of spoilage bacterium, *Alicyclobacillus acidoterrestris,* in aseptically packaged apple juice was reported (Cerny et al., 1984). Due to its heat resistance (some of its published characteristics are: D-value at 90°C is about 23 min in apple juice of pH 3.5 and soluble solids 11.4°Brix, z = 7.7°C) (Splittstoesser et al., 1994) this bacterium should be considered when products that could be contaminated with this microorganism are to be pasteurized (Silva and Gibbs, 2001).

Microbial changes of processed fruit and vegetables (pastes, jams, marmalades, juices, etc.) depend on the properties of the product (i.e., on the conditions of growth of spoilage microorganisms present). The important factors affecting the safety and shelf life of such products are:

TABLE 10.4
The Most Important Hurdles

Hurdle	Requirement
Initial microbe count	As low as possible
Storage temperature	As low as possible
pH	Some bacteria are inhibited by a low pH alone; sporulating bacteria (including toxinogenic) cannot grow in acid food with pH below 4.0[a] (4.5)[b]. The lower the pH value the more efficient heat treatment, more efficient preservatives, etc.
Water activity	As low as possible
Oxygen ingress	As low as possible (vacuum or inert gas packaged low-acid vegetable products are good substrate for strictly anaerobic *Clostridium*)
Degree of heating	As high as possible (the process of reduction of microorganisms occurs according to first-order kinetics; therefore heat treatment should be at appropriate combination of temperature and duration of heating for an efficient inactivation effect)
Preservatives	Present in adequate concentration; should be added at lag phase of microbial growth

[a] From Kyzlink, V., Principles of Food Preservation, Elsevier, Amsterdam, 1990, Part 4.
[b] From Bibek, R., Fundamental Food Microbiology, CRC Press, Boca Raton, FL, 1996, Section 6.

1. The initial count and the composition of contaminating microflora,
2. The conditions of processing with regards to the possibilities of cross-contamination and contamination from the environment, workers, premises, vessels, tools, etc.,
3. The conditions during manipulation, processing, and storage with regards to the growth of the microorganisms present in the product, and
4. The count of surviving microorganisms after the antimicrobial treatment (efficiency of the antimicrobial treatment).

Storability and safety of fresh-cut or minimally processed produce are based on a combination of several preservation steps that individually may not be sufficient but in combination make the product stable for its declared shelf life. It is the principle of "hurdle technology" (Leistner, 1995). The most important hurdles are summarized in Table 10.4.

The safety of fresh and fresh-cut fruit and vegetables has received considerable attention lately. While the consumption of raw fruits and vegetables often causes outbreaks of human diseases, the number of confirmed cases of illness associated with raw plant products is lower than that associated with food of animal origin but is important, especially in developing countries. Raw fruits and vegetables have been known as vehicles of human disease since the end of the 19th century, when it became understood that microorganisms are a cause of disease. In 1903 an outbreak of typhoid fever was attributed to eating watercress grown in soil fertilized with sewage; in 1913 other cases of typhoid resulted from uncooked rhubarb grown in soil fertilized with typhoid excreta; and in 1912 it was demonstrated that lettuce and radishes grown in soil containing *Bacillus typhosa* (now *Salmonella typhi*) contained the microorganism on their surface for 31 days (Beuchat, 1998). Fruits and vegetables can be contaminated with pathogenic microflora during cultivation, harvesting, transport, distribution, marketing, and manipulation at home. Bacteria such as *Clostridium botulinum, Bacillus cereus,* and *Listeria monocytogenes* are frequent in soil. *Salmonella, Escherichia coli,* and *Campylobacter* reside in the intestinal tracts of animals, including humans, and can contaminate raw fruits and vegetables that come in contact with feces, sewage, untreated irrigation water, or surface water (Bibek, 1996; Beuchat, 1998). To minimize safety concerns, the principles of good hygiene practice and the preventive HACCP concept should be applied to avoid or minimize the contamination of products and to improve effectiveness of preservative methods.

10.2 METHODS FOR PRESERVING FRESH AND PROCESSED PRODUCE

After harvest, deterioration of fruits and vegetables commences, mainly by physiological and enzymatic changes in the tissues. The preservation methods used to extend the shelf life of fresh produce involve optimizing or slowing down of ripening and senescence.

10.2.1 Prevention of Physiological Changes

Undesirable physiological changes of fresh produce can be prevented by optimization of environmental conditions. Manipulation of environmental conditions is usually performed to lower respiration of products and to reduce the growth of decay organisms without inducing physiological injury. Temperature and relative humidity are the principle factors. The optimal conditions for manipulation and storage of fresh produce have been summarized in several sources (Flores Gutiérrez, 2000).

10.2.1.1 Temperature

Temperature is a major factor in the control of decay organisms, respiration, and transpiration. Quick cooling after harvest is critical, especially in the case of rapidly respiring tissues such as spinach or broccoli (Shewfelt, 1986). Rapid precooling of produce can be achieved by room cooling, contact or package icing, hydrocooling, forced-air cooling, hydraircooling (fine-mist spray combined with forced-air cooling), vacuum cooling, and cryogenic cooling. Recent developments in precooling techniques and their comparison have been reviewed by Brosnan and Sun (2000).

During handling and storage, chilling or freezing injury should be avoided. If the shelf life requirements are too short compared to the time needed to develop low-temperature disorders, such as overnight holding prior to processing, chilling injury may not be a major concern. The most common prevention of chilling injury is to store chilling-susceptible produce at or above the temperature at which no injury occurs or to use various postharvest techniques to alleviate chilling injury. These methods include, for example, preconditioning, such as holding produce at moderate temperatures for a period prior to low-temperature storage, which is effective for some fruits such as grapefruit (Hatton and Cubbebedge, 1983). The other methods are intermittent warming, heat treatment, controlled atmosphere storage, treatment with calcium or other chemicals, waxing, film packaging, and application of ethylene, abscisic acid, methyl jasmonate, polyamines, or other natural compounds. Low-temperature conditioning and intermittent warming maintains high levels of phospholipids, increases the degree of unsaturation of fatty acids and the amount of spermidine and spermine, and stimulates the activities of free radical scavenging enzymes. Heat treatment induces heat-shock proteins, suppresses oxidative activity, and maintains membrane stability. Methyl jasmonate can activate lipoxygenase gene expression and induce synthesis of abscisic acid and polyamines. Polyamines may act as free radical scavengers and membrane stabilizers. All of these processes can enhance chilling tolerance of tissues and alleviate chilling injury of fruits and vegetables (Wang, 2000).

In addition to chilling stress, some commodities are susceptible to heat stress. High-temperature injury has been more of a concern during the growth and development of plants than in postharvest handling, because it should not occur during the normal handling of fruits and vegetables (Shewfelt, 1986).

10.2.1.2 Atmosphere

The effect of storage or packaging atmosphere composition is described above and also in other chapters of this book. Prevention of undesirable changes necessitates storing produce within its "safe" range below or above the critical limits for carbon dioxide and oxygen, respectively. Reducing the concentration of oxygen and/or increasing the concentration of carbon dioxide in the storage atmosphere surrounding fresh produce reduces the rate of respiration of fresh fruits and vegetables and also inhibits microbial and insect growth. It is usually combined with chilling. The composition of the surrounding atmosphere can be controlled either during the storage of the produce or within its packaging. These methods are used for storage and packaging of the both respiring and nonrespiring produce. Modified atmosphere storage (MAS) and modified atmosphere packaging (MAP) use gases to replace the air around produce (or also involve sealing packages of fresh produce without gas modification) without further control of the produce. When respiring produce is stored or packaged under these conditions, gas equilibrium is established due to the properties of the packaging material. Changes in gas composition during storage depend on the respiration rate of fresh produce and hence on the temperature of storage, and, in the case of the MAP, also on the permeability of the packaging material to gases and water vapor, the internal and external humidity, which affect the permeability of some films, and on the surface area of the packaging in relation to the amount of food it contains. For MAP of fresh produce the terms *equilibrium-modified atmosphere* (EMA) or *passive atmosphere modification* (PAM) are used. In controlled atmosphere storage (CAS) or controlled atmosphere packaging (CAP) the composition of gas around respiring produce is monitored and constantly controlled. Recent advances in both systems now make the distinction between MAP and CAP less clear (Fellows, 2000c; Gorris et al., 1994). The packages for CAS or

TABLE 10.5
Examples of Refrigerated (3.5°C) Storage
Atmospheres Composition of CAS

Product	O_2 (%)	CO_2 (%)	N_2 (%)	Shelf life (months)
Apples, Bramley's seedling	13	8	79	5
Apples, Cox's Orange Pippin	3	5	92	5
Apples, Cox's Orange Pippin	1	1	98	8
Winter white cabbage	3	5	92	10

Source: Adapted from Fellows, P., Food Processing Technology: Principle and Practice, 2nd ed., CRC Press, Boca Raton, FL, 2000c, Part 4.

MAS are made airtight and the concentration of oxygen, carbon dioxide, and ethylene is monitored and regulated. The optimal compositions of gases are close to those described for MAP below and differ according to the type of fruits or vegetables, and also according to cultivar. Examples of atmospheres for CAS are given in Table 10.5.

Another way to reduce the oxygen content in the surrounding atmosphere is through pressure reduction. Hypobaric storage, patented by Burg in the late 1960s, is still used with various modifications. One of the systems is called *moderate vacuum packaging* (MVP). In this system, fresh produce is packed in a rigid, airtight container under 40 kPa of atmospheric pressure and stored at refrigerated temperature (4 to 7°C). The effect of hypobaric storage on the physiological changes of fresh produce is similar to the effect of MAP. Experiments have been conducted that confirm beneficial effects against spoilage and pathogenic microflora (Day, 1990). When gas flushing is used with MAP, the composition of the gas mixture is usually 3 to 10% CO_2, 2 to 10% O_2, and 80 to 95% N_2. The aim is always to create optimal gas equilibrium to reduce respiration to a minimal level, but the concentrations of carbon dioxide and oxygen must not be detrimental to the produce. The equilibrium concentrations are 2 to 5% CO_2 and 2 to 5% O_2 and the rest is nitrogen (Kader et al., 1989; Powrie and Skura, 1991). It is usually difficult to achieve these conditions in ready-to-eat or ready-to-use fresh fruit or vegetable products with extended shelf life. The main problem is the lack of packaging materials that are permeable enough to match the respiration of fresh produce (Day, 1994). The most permeable packaging films, such as polyethylene and oriented polypropylene are not permeable enough to allow longer shelf life of packed produce, especially when the produce has high respiration. Respiration can be reduced storing produce at a low temperature. One of the solutions used is to make microholes of defined sizes and defined quantity in the packaging material to avoid anaerobic respiration (Keteleer and Tobback, 1994). The search for plastic films of higher permeability continues. New materials being proposed include laminated films with ethylene vinyl acetate, oriented polypropylene and low-density polyethylene, combinations of ceramic materials with polyethylene, or the incorporation of finely dispersed minerals in the packaging film. These new films posses higher permeability than polyethylene or oriented polypropylene. The shelf life of shredded cabbage and grated carrots packaged in these materials is 2 to 3 days longer (7 to 8 d at 5°C) than in those packaged in oriented polypropylene, which is generally used for this purpose (Ahvenainen, 2000). The application of these "new" types of packaging films (microperforated or perforated films) can be unsuitable for produce that is sensitive to CO_2. The rate of gas movement through a perforated film is a sum of gas diffusion through the perforation and gas permeation through the polymeric film. The flow rate of CO_2 and O_2 through perforations is similar and the concentration of CO_2 inside the package can reach the critical limit (Mir and Beaudry, 1999). New developments also include films that change permeability to moisture and gases under specified temperatures designed to match the respiration rate of fresh produce. An example is the recent development of adjustable "temperature switch point" films that contain long-chain fatty alcohol-based polymeric chains. At a predetermined temperature-switch point, these chains are in a crystalline state, providing a gas barrier. When this point is reached, the side chains melt to a gas–permeable, amorphous state. Another system consists of film made from two different thicknesses of the same material, both layers containing minute cuts. When temperature rises or falls, the layers expand at different rates. As the temperature rises, the film at the cut edge retracts and curls upward, the holes enlarge, and permeability significantly increases (Ahvenainen, 2000). In addition to these "smart" packaging films other active packaging systems can be used for MAP of fresh produce (Vermeiren et al., 1999). Oxygen scavengers that are important in packaging of nonrespiring foods are not effective because of the usual lack of oxygen inside the package, but ethylene scavengers can be used. For MAP purposes, an absorbing system similar to that used for controlled storage can be employed, especially C_2H_4 scavengers based on potassium permanganate ($KMnO_4$), which oxidizes ethylene to acetate and ethanol (Zagory, 1995). Another scavenging system is based on the adsorption and subsequent breakdown of ethylene on activated carbon, charcoal containing PdCl as a metal catalyst. This is effective at 20°C for some types of fresh produce (kiwifruits, bananas, and spinach) (Abe and Watada, 1991). The system based on the incorporation of finely dispersed minerals such as zeolites, clays, or Japanese oya into the packaging films incorporates minerals that can absorb ethylene and affect the permeability of the films (Suslow, 1997). Ethylene scavengers are not yet very successful in MAP, probably because of insufficient adsorbing capacity. They are more frequently used in controlled atmosphere storage and in the case of MAP will become more important after the principal problems with internal atmosphere composition have been solved. Scavenging systems for CO_2 could be useful for reduction of internal CO_2 in the package. Various systems based on the absorption of CO_2 by the reaction with $Ca(OH)_2$, such as those used for packaging roasted coffee, have not yet been applied to fresh produce (Vermeiren et al., 1999).

10.2.1.3 Humidity

Relative humidity is another very important environmental factor in the shelf life of fresh produce. A low-humidity environment reduces growth of decay microorganisms, but it can increase moisture loss. Moisture loss is associated with structural damage in fruit and vegetable tissues. Moisture losses are undesirable in various products such as salads and packed fresh-cut produce but they can be considered as positive in some end-product applications such as drying and canning.

Relative humidity is also an important factor in the case of MAP of respiring products. The majority of plastic films are impermeable to water vapor. Due to respiration, the internal humidity inside the package increases and vapor condenses on the film and also on the surface of the packaged produce. The condensed water increases the risk of microbial decay, worsens the appearance of packaged products, and also affects the barrier properties of the plastic films. The conditions of the MAF when a passive MAF is used are also affected. The increase of humidity inside the package and subsequent condensation of water vapor decreases the permeability of permanent gases, but the more polar plastic films can soften in contact with water and increase the permeability for gases. The humidity inside the package can be controlled by the use of water absorbers and the condensation of water vapor (but not the increase of humidity inside the package) can be avoided by the use of hydrophobic plastic films (Vermeiren et al., 1999).

10.2.1.4 Light

Light is one of the factors that could affect the shelf life of fresh fruits and vegetables. It is required for anthocyanin synthesis in apples. Access to light enhances the color development of tomatoes, when chlorophyll degradation is accelerated by the red component of the spectrum and the blue component increases the biosynthesis of carotenoids (Jen, 1974). Light (including artificial lighting) can initiate chlorophyll formation in potato tubers, which is accompanied by the production of toxic steroidal glycoalakaloids (chaconine and solanine) (Salunkhe et al., 1974). Light should also be taken in consideration in the case of storage of packaged vegetables. Light may affect the gas composition in the packaging by inducing photosynthesis in green vegetables (Gorris, 1996). Access to light, especially UV light, is important for processed fruits and vegetables. It can trigger the degradation of sensitive compounds such as vitamins and pigments and initiate lipid oxidation. It is recommended that fruits and vegetables be stored in darkness (Ahvenainen, 2000).

10.2.1.5 Mechanical Wounding

Mechanical damage of plant tissues considerably affects physiological and enzymatic changes. Wounding stress results in metabolic activation, which is apparent with increased respiration rate and, in some cases, ethylene production (Varoquaux and Wiley, 1997). The wound stress sensitivity of fruits is also highly influenced by their state of maturity. Ripeness usually increases susceptibility of fruit to wounding during handling and processing (Gorny et al., 2000). The optimal stage of processing to minimize cutting damage also varies greatly depending on the species, cultivar, cultivation, and harvest and postharvest conditions (Solomos, 1997). Damaged parts must be removed before subsequent processing. However, a "mechanical damage" is a usual processing step in the production of minimally processed fruits and vegetables, which need peeling, cutting, or shredding. Peeling can done on an industrial scale in several ways (Fellows, 2000b):

- *Flash steam peeling.* The surface of the plant material is heated by high-pressure steam (1.5 MPa) in a rotating pressure vessel for 15 to 30 sec. Then the pressure is released, which causes steam to form under the skin and the surface of the food "flashes off."
- *Knife peeling.* The skin is removed by pressing the surface of fruits or vegetables against blades. Various systems use rotating fruits and vegetables and stationary blades or rotating blades and stationary plant materials.
- *Abrasion peeling.* Fruits or vegetables are fed onto carborundum rollers or into rotating bowls lined with carborundum. The abrasive surface removes the skin, which is washed away by water.
- *Caustic peeling.* Fruits or vegetables are treated with a heated solution of sodium hydroxide (1 to 2% at 100 to 120°C), then the softened skin is removed by high-pressure water sprays. When dray-caustic peeling is used, the food is dipped in 10% sodium hydroxide solution and the skin is removed with rubber discs or rollers.
- *Flame peeling.* A conveyor belt carries and rotates onions through a furnace heated to 1,000°C; the outer layers and root hairs are burned off and charred skin is removed by high-pressure water sprays.

The ideal peeling method would be gentle hand-peeling with a sharp knife. Industrial-scale knife peeling is also gentler than other methods, and losses during high-pressure peeling can also be relatively low when the regimen is sufficiently optimized. Product losses during the flash peeling of vegetables are about 8 to 18%, compared to about 25% in the case of abrasion peeling (Laurila and Ahvenainen, 2002). The choice of the peeling method also has other consequences. Abrasion-peeled potatoes should usually be treated with browning inhibitors, whereas water washing is enough for knife-peeled potatoes. Abrasion, steam peeling, and caustic peeling disturb the cell walls of vegetables, thus enhancing the possibility of microbial growth and enzymatic changes. Peeling often affects the appearance and the shelf life of the produce and is not suitable for minimally processed products.

Minimizing of mechanical damage is also important for cutting and shredding. These must be performed with knives or blades as sharp as possible and made from stainless steel to avoid dissolving metals in processed material that can affect enzymatic and chemical reactions, and subsequently the product quality. Similarly, quality retention depends on the degree of cell breakage and release of tissue fluid. Slicing machines must be installed solidly to avoid vibrations. All parts of the machine in contact with plant material should be regularly cleaned and disinfected (Fellows, 2000b).

The effect of wounding may be minimized by careful selection of appropriately mature fruit. Recently, the beneficial effect of 1-methylcyclopronene (1-MCP) treatment on climacteric fruit was described (Jiang and Joyce, 2002). 1-MCP applied on whole fruit or after cutting can block ethylene binding because of their similar chemical structure. When applied before coring and slicing of apples, ethylene production was reduced during 10 days of storage; firmness was retained and superficial scald reduced.

Wounding stress during peeling, cutting, or shredding is also affected by the distribution of different types of cells in fruit or other parts of the processed plant (root, leaf, etc.). In fruits such as apples or pears the core and adjacent tissues have higher susceptibility to the enzymatic browning reaction due to the higher concentration of substrates and enzymes; therefore, these tissues must be completely removed during peeling. The mass losses after peeling or other types of technological processing also depend on the type of plant tissue. Agar et al. (1999) observed higher mass losses in peeled and sliced kiwi fruits than in unpeeled, sliced fruits, and these changes did not correspond with the effect of wounding on ethylene and CO_2 production rates in tissues. The wounding stress expressed as the rate of ethylene and CO_2 was much higher in the unpeeled, sliced kiwi fruits.

10.2.3 PREVENTION OF ENZYMATIC REACTIONS

10.2.3.1 Application of Additives

Browning reactions occur principally as a result of cell damage during peeling, cutting, slicing, and other treatments that produce new surfaces from one or more layers of injured cells. The prevention of undesirable enzymatic and chemical changes consists usually of surface treatments. This may involve rinsing out enzymes and substrates released from the damaged cells, as well as reducing access to exogenous substrates such as oxygen. Various chemical additives that are capable of diffusion through the damaged cells may also be used. Surface treatments are the most commonly used during processing of fresh produce. These are applied by dipping, spraying, or rinsing with browning inhibitors, firming agents, growing inhibitors or stimulators, disinfectants, or antimicrobial agents. The groups of anti-browning additives are summarized in Table 10.6 and a comparison of the inhibitory effects of selected additives is shown in Figure 10.1. Dipping times normally range from 1 to 5 min (Soliva and Martin-Belloso, 2003). Increased temperature of a dipping solution can accelerate the diffusion process. Lowering the pH value of the dipping solution is recommended to minimize growth of microorganisms, but in some cases the pH value may need to be modified to a value that is close to neutral pH. For example, antibrowning treatments that involve cystein require relatively higher pH values when used to minimize the risk of formation of pinkish-red compounds in foods (Sapers and Miller, 1998). Drying of wet surfaces to avoid microbial decay should follow a dipping treatment. Some of the postdip drying methods involve draining, gentle spinning, or drying with cheesecloth (Bett et al., 2001; Gorny et al., 1999).

Intensity of browning is influenced by the activity of enzymes and by the amount of phenolic substrate. In climacteric fruits, partial ripeness corresponds to a lower tendency for browning. Sapers and Miller (1998) found minimal color changes in slightly underripe Anjou and Barlett fresh-cut pears preserved with a combination antibrowning treatment and MAP. The enzymatic browning caused by polyphenoloxidase depends also on the access to oxygen and can be reduced by MAP, vacuum packaging, and/or waxing. In some fruits and vegetables, enzymatic color changes are also caused by peroxidase enzymes (POD). Peroxidase reduces hydrogen peroxides in fruit or vegetable tissues and oxidizes any hydrogen donor. The reaction does not necessarily require oxygen (Robinson, 1991).

POD color-induced changes are probably marginal relative to PPO browning reactions. The POD browning reactions usually proceed within the complex of

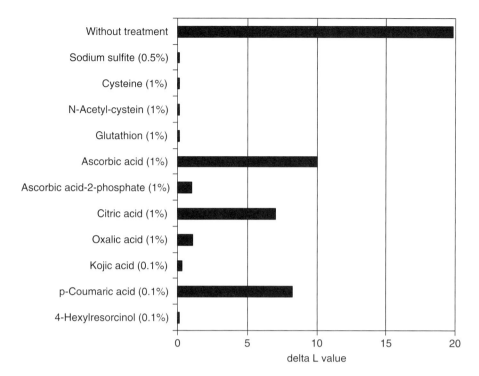

FIGURE 10.1 Browning of apple slices 3 h after treatment by dipping in solution of inhibitor for 3 min. The degree of browning is expressed as a L value (initial L value – L value after 3 h). (Adapted from Son, S.M. et al., *Food Chem.*, 73, 23, 2001.)

enzymatic changes that occur in fruits and vegetables and are inhibited by general antibrowning treatments. PPO browning is usually more severe and most treatments to prevent browning involve inactivating PPO or reducing the enzymatic activity.

In theory, PPO-catalyzed browning of vegetables and fruits can be prevented by:

- Selection of raw material to minimize enzymatic activity based on factors such as maturity level and environmetal factors
- Genetic modification
  - blocking the gene expression of PPO in plant tissues by antisense RNA (Martinez and Whitaker, 1995)
- Exclusion or removal of one or both substrates
  - removing polyphenol by sorption on insoluble poly(vinyl polypyrroli-done), poly(ethylene glycol), or other sorbents (Osuga et al., 1994)
  - exclusion of oxygen by vacuum or MAF packaging, coating (edible) and wraps, processing in an inert gas atmosphere, treatment in water or brine, efficient deaeration, etc.
- Inactivation or inhibition of enzymes
  - heat inactivation of enzymes (blanching)
  - treatment with browning inhibitors

TABLE 10.6
Antibrowning Additives

Antibrowning Additive	Mechanisms of Action (Proposed)	Limitations of Use
SO$_2$, sulfites, bisulfites	Direct inactivating effect on PPO (binding on protein, limitation of substrate access to the active sites, etc.); reduction of o-quinones; formation of colorless conjugated compounds with o-quinones and blocking of their subsequent reactions	Health risks, dangerous especially for asthmatics. Regulation of residual concentration in food products
Sulfhydryl-containing amino acids and peptides: Cystein, N-acetylcysteine, Glutathion	Similar to sulphites: Direct inactivation of PPO (probably formation of stable complexes with copper); reduction of o-quinones to o-phenols; formation of colorless conjugated compounds with o-quinones and blocking of their subsequent reactions	Cystein and glutathione cause the unpleasant odor in fruit and vegetables. High price of treatment compared with sulphites (N-acetylcysteine is most expensive); cysteine when used in lower pH values can cause the formation of colored products (pinkish-red compounds in cauliflower or cabbage)
Ascorbic acid and derivatives: Ascorbic acid and salts (Ca, Fe, Mg,Na), Ascorbic acid-2-phosphate, Ascorbic acid-2-triphosphate, Erythrobic acid and salts (Na)	Reduction of o-quinones to o-phenols; very gentle acidification	Antibrowning effect depends on the concentration of ascorbic acid in the reaction; when all ascorbic acid is oxidized the browning reaction continues
Carboxylic acids: Citric, Gluconic, Lactic, Malic, Malonic, Pyruvic, Oxalic, Oxalacetic, Tartaric	Chelating effect binding the copper (hydroxy acid only); acidification	Limited efficiency when acids without chelating effect are used
EDTA	Chelating effect binding the copper; weak acidification	

TABLE 10.6
Antibrowning Additives (continued)

Antibrowning Additive	Mechanisms of Action (Proposed)	Limitations of Use
Aromatic and phenolic acids: Benzoic Kojic Coumaric Ferulic	Competitive inhibition of PPO (binding to active sites)	With exception of kojic acid, very weak effect
4-Hexylresorcinol and other substituted 4-resorcinols	Competitive inhibition of PPO	Effective inhibitor of black-spot formation in shrimps, effective for mushrooms PPO, very low or no effect to apple and grape PPO
NaCl, NaF, CaCl$_2$	Chelating effect	
Phosphoric acids, polyphosphates	Chelating effect; acidification	
Sodium or calcium hypochlorite	Inhibition of enzyme (partial denaturation of protein)	
Proteases Isolated natural proteases or juices of natural sources: Ficin or fig extracts Papain or papaya juice Aktinidin or kiwi juice Bromelain or pineapple juice, etc.	Inhibition of enzyme	

Sources: Soliva, R.C. and Martin-Belloso, O., *Trends Food Sci. Technol.*, 14, 341, 2003; Martinez, M. and Whitaker, J.R., *Trends Food Sci. Technol.*, 6, 195, 1995; Osuga, D. et al., in *Protein Structure-Function Relationships in Foods*, Blackie, 1994, p. 62; and Son, S.M. et al., *Food Chem.*, 73, 23, 2001.

Removal of polyphenols could be used during the processing of clear juices, which could be filtered through adsorbents or adsorbents could be added to the juice and removed by filtration. Polyphenols form complexes with β-cyclodextrins. Cyclodextrines when added to juice form complexes with polyphenols that cannot be oxidized by PPO (Osuga et al., 1994). The production-scale application of such treatment is, however, limited because of price and efficiency. Exclusion or removal of oxygen is widely used during the processing of fresh produce to prevent browning and other undesirable reactions described in other parts of this review. Heat inactivation of enzymes, including the PPO (blanching), is very often used as the first step in the technological processing of fruits and vegetables. This important treatment is discussed below.

Antibrowning inhibitors are usually applied to cut fruits, leaves, roots, or other pieces of fresh produce by dipping or spraying. Sulfur dioxide can also be applied in gaseous form, usually produced by burning sulfur, but its application can be limited by the dosage of the dioxide in treated products. Individual inhibitors differ in their ability to permeate into tissues. Sulfur dioxide and disulfites or sulfites are most efficient and widely used. The inhibitory effect of sulfites is enhanced by their ability to rapidly difuse into plant tissues. Health concerns have limited the use of sulfites. It is permitted in most countries, with limitations. The potential induction of allergic and pseudoallergic reactions in humans (Simon and Stevenson, 1998) caused the Food and Drug Administration (FDA) to ban the use of sulfites in salad bars in 1995 (Martinez and Whitaker, 1995). The presence of sulfite contents above 10 µg/kg is usually required to be declared and the additive can only be used for selected foods. The potential health risk of sulfur dioxide and sulfites initiated the development of alternative browning inhibitors, but while many other PPO inhibitors are known, only a few of them have been considered as potential alternatives to sulfites. Most antibrowning agents should be used in combination with other methods, particularly with handling and packaging methods that limit exposure to oxygen. Most of the commercially available sulfite-free antibrowning agents are combinations of various compounds and ascorbic acid (McEvily et al., 1992). Such combinations usually include a chemical reductant (e.g., ascorbic acid), acidulant (e.g., citric acid), and chelating agent (e.g., EDTA or phosphoric acid or pyrophosphate). Commercial mixtures often contain calcium as a firming agent. According to Son et al. (1975), the use of oxalic acid as a browning inhibitor seems to be promising, especially due to the synergistic effect when it is used in combination with ascorbic, erythrobic, or citric acid. These commercial mixtures are applied by direct addition to homogenated produce, dipping, or spraying, but recently some authors have proposed the use of edible coating and wraps containing antibrowning additives (Baldwin et al., 1995; McHugh and Senesi, 2000). An example of an edible film is a combination that includes 61% apple puree, 23% beeswax, 7% pectin, 7% glycerol, 1% ascorbic acid, and 1% citric acid. Such a coating prevented the color changes of apple pieces stored for 12 days (McHugh and Senesi, 2000).

10.2.2.2 Blanching

Blanching is a thermal process to inactivate enzymes and remove intercellular gases. The process usually utilizes temperatures in the range of 75 to 95°C for times of about 1 to 10 min, depending on product requirements. Blanching is a necessary pretreatment for many processed fruits and vegetables to achieve satisfactory quality in dehydrated, canned, and frozen produce. It is also needed as a food processing pretreatment process where there is a risk of a delay in enzyme inactivation, such as in canning and freeze-drying. Elimination of intercellular gases is important for the reduction of oxidation and for achieving suitable vacuum conditions within a jar or can. Blanching can also contribute to microbial count reduction. Plant matter tends to shrink because of loss of turgor, but texture can be improved. Undesirable losses of nutrients may be caused during water blanching due to leaching (Selman, 1986).

10.2.2.2.1   Methods of Blanching

The most frequent commercial methods of blanching are steam and water blanching. Water blanching is probably the most widely used method. The blanchers consist of a simple water bath or various continual machines such as tubular blanchers, rotary screw blanchers, rotary drum blanchers, and thermoscrew blanchers. The loss of water-soluble nutrients is one of the limitations of water blanching. This may be reduced by maximizing the produce-to-water ratio, by recycling the blanching water, or by the use of a salt or sugar solution to reach an isotonic state (Selman et al., 1983). Recycling of blanching water represents a risk of contamination and growth of thermophilic microflora. The simplest steam blancher consists of a mesh conveyor belt that carries food through a steam atmosphere in a tunnel. The advantage of steam blanching is lower leaching losses relative to water blanching. Steam blanching is therefore used for blanching foods with large area of cut surface. Conventional steam blanching does not allow the uniform heating of multiple layers of food. In order to reach the required inactivation effect there is a risk of overheating the edges and producing changes that affect flavor and texture. This problem was overcome by the development of steam blanchers in the 1970s that utilize the Individual Quick Blanch (IQB) system (Lazar et al., 1971). IQB consists of two stages of heating. In the first stage the food is heated in a single layer to a sufficiently high temperature to inactivate enzymes. In the second stage (adiabatic holding) a deep bed of food is held for a sufficient time to allow the temperature at the center of each piece to increase to reach a combination of temperature and time needed for enzyme inactivation. The heating time is reduced (e.g., for diced carrots it is 25 sec of exposure to steam, then 50 sec in a deep bed to allow the equilibration of temperature throughout the carrot pieces. This reduction results in an improvement in the efficiency of energy consumption of 86 to 91% compared to conventional water blanching (Cummings et al., 1984).

Other methods of heating that have been tested for blanching include microwave blanching. This seems to be the most promising (Dietrich et al., 1970; Devece et al., 1999; Ramesh et al., 2002) because of the low cost, but it is not currently used on a commercial scale. Thermoinactivation of enzymes can also be achieved by other methods of heating. Hot-gas heaters have also been tested. Losses of water-soluble nutrients were comparable to those obtained by water or steam blanching, but surface drying and oxidation processes when hot air was used were disadvantages (Fellows, 2000b).

10.2.2.2.2   Effects of Blanching

Enzyme activity increases with temperature until the denaturation temperature is reached, then dramatically falls zero. The blanching process should be sufficiently fast (high temperature, short time (HTST)) to protect the labile nutrients and prevent acceleration of enzymatic activity that takes place at temperatures before denaturation is reached. It is not important for polyphenoloxidases when oxygen is limited during the process, but it could be important for processes such as the pasteurization of stewed kernel fruits, products that when heated by a slow process contain more hydrogencyanide (Voldřich and Kyzlink, 1992). Low-temperature blanching (LTB) can be also used to improve the texture of produce. Moisture heating during the

TABLE 10.7
Temperature Stability of Some Enzymes in Plant Materials

Enzyme	Plant	Z Value (°C)	F Value (min at 82°C)
Ascorbate oxidase	Peach	59	2
	Vegetables		
Catalase	Vegetables	28	6
Chlorophylase	Spinach	22	2
Lipoxygenase	Peas	16	<0.1
Pectin esterase	Citrus juice	1.4	43
Peroxidase	Peas	48	60
Phosphatase	Orange juice	9	—
Polygalactorunase	Citrus juice	16	12
	Papaya	11	23
Polyphenol oxidase	Fruits	12	1.1

Source: Adapted from Williams, D.C. et al., Food Technol., 40, 1986, 130, 1986.

blanching process leads to a swelling of the cell walls and individual cells begin to separate. LTB could, however, improve the firmness of the produce. The effects of pectin methyl esterase, which can be activated at low temperatures and inactivated at higher temperatures, cause the increase in firmness. Low-temperature heating activates demethylation of pectin, which is then cross-linked by calcium either naturally present in the tissues or added (Quintero-Ramos et al., 2002). When the blanching process is a pretreatment followed by other enzyme deactivating processes such as pasteurization, total inactivation of enzymes is not required during blanching. The enzymes present in fresh produce have different resistance levels to the heating (Table 10.7). Peroxidaxe enzymes are often the most thermoresistant enzymes, hence their use as a blanching efficiency indicator. Recent work has shown that a significant but not well-defined proportion of active peroxidase can be left in some vegetables without negatively affecting their shelf life (Selman, 2000). For various plant products it has been also concluded that the inactivation of lipoxygenase is most important for the prevention of flavor changes during the subsequent storage of the products (Williams et al., 1986).

The course of heat inactivation of enzymes depends on a number of conditions. Osmotic pressure and the presence of substances reacting with protein, such as sulfites and *o*-quinones, are among the factors to be considered. pH value is also very important. Enzymes are most thermoresistant near their isoelectric point. For example, PPO enzymes are generally most stable at about pH 6.0, and their thermoresistance decreases with changes of pH in both directions (Kyzlink, 1990). The blanching process may cause losses of nutrients. These are mainly lost in exudates leaking from the treated products. Heating also accelerates degradation changes of labile compounds such as ascorbic acid. The total losses of nutrients depend on the conditions of blanching. Losses of nutrients can be reduced, for example, by the use of microwave blanching (Ramesh et al., 2002). Exudation is undesirable in the case of nutrients but it can be beneficial for reduction of toxic constituents; for example, water blanching can reduce nitrate content from 339 mg/kg in raw carrots to 165 mg/kg after the washing and blanching process (95°C) (Selman, 2000). Blanching treatment also reduces microbial contamination. The microorganisms on the surface of produce are rinsed out or inactivated by heat. The reduction rate depends on the conditions of the process and the properties of the microorganisms.

10.2.2.2.3 Prevention of Texture Softening by Dipping

Slicing, cutting, or shredding usually results in losses of firmness of fruit or vegetable tissues due to enzymatic pectinolysis and other changes in damaged cells and adjacent layers. Softening is mainly caused by changes in pectins. Calcium salts are used to decrease softening in a number of minimally processed fruits and vegetables. The ions infiltrate the tissue and bind to the cell wall and middle lamellae (Glenn and Poovaiah, 1990). Calcium ions interact with pectin polymers to form a cross-linked polymer network that increases the mechanical strength of the tissues. A dipping solution of calcium salts is relatively common in the processing of pome fruits. Dipping in a solution of 0.1 up to 1% $CaCl_2$ is used (Soliva and Martin-Belloso, 2003). Calcium salts can also be added to the wash water or to brine. This is frequently used for vegetables such as pickled peppers and cucumbers. The direct addition of $CaCl_2$ can affect taste because of the bitter taste of calcium chloride. For this reason, the use of nonchloride salts of calcium, such as phosphates and lactates, is common (Kyzlink, 1990). When dipping is used as an antisoftening treatment, it is usually combined with other treatments: dipping solutions usually have added antibrowning agents, preservatives, or disinfectants.

10.2.3 Protection Against Microbial Spoilage

To explain the mechanisms of various preservation treatments of fruits and vegetables the following general rule can be used: R = microbe count and virulence/resistance of the environment. The intensity of food spoilage (R) is proportional to the count of microorganisms present on or in the food products and to the virulence of the microflora (ability to survive and grow) and it is inversely proportional to the resistance of the environment. The preservation methods should reduce or eliminate the microbial contamination (disinfection, cleaning, removing the microorganisms from the food products; killing the microorganisms or reducing their virulence to render them harmless) or increase the resistance of the environment (hurdle principle) (Kyzlink, 1990; Leistner, 1995; Bibek, 1996).

10.2.3.1 Removal of Microorganisms from Food Products

10.2.3.1.1 Washing and Cleaning

Raw fruits and vegetables covered with soil, mud, or sand should be carefully cleaned before processing. A second wash is usually done after peeling or cutting (Laurila and Ahvenainen, 2002). Any washing removes microorganisms and also tissue fluids containing sugars, organic acids, and other nutrients liberated from the damaged cells that can facilitate microbial growth. To reduce microbial contamination, wash water must not be contaminated and should be able to rinse out the superficial impurities from the products. Washing in flowing water or air-bubbling water is preferable to dipping into still water (Ohta and Sugarawa, 1987). The temperature of wash water should be as low as possible, preferably below 5°C. The required amount of wash water depends on the type of produce and on the extent of pollution and varies between 5 and 10 L/kg of produce. In the case of washing after peeling it is about 3 L/kg (Laurila and Ahvenainen, 2002). Wash water should be removed from the produce. A centrifuge is usually used, but the conditions of centrifugation (time and rate) should be optimized according to the produce to remove free water without damaging cells. Recently, gentle drying with cheescloth has been proposed (Zomorodi, 1990; Bolin and Huxsoll, 1991; Gorny et al., 2002).

10.2.3.1.2 *Disinfection and Decontamination*

The effectiveness of washing can be increased by addition of disinfectants or preservatives to the wash water. The use of sanitizers at concentrations higher than usual residual levels of chlorine or other disinfectants permitted for drinking water treatment is regulated. A list of the U.S. regulations dealing with disinfection of raw fruits and vegetables is given by Beuchat (2000). Generally, it is possible to use any sanitizers that are used for disinfection of food products. The use of some of these is, however, limited or regulated in some countries for fruit and vegetable handling and processing.

Chlorine is widely used for the disinfection of drinking water. It is also used for the disinfection of whole and cut fruits and vegetables as well as the surfaces that are in contact with produce during handling. Chlorine is commonly used at 200 mg/kg (available chlorine) and at pH below 8 (usually at 6.0 to 7.5 for effective disinfection without damaging equipment surfaces), with contact times ranging from 1 to 2 min. It is applied in the form of liquid chlorine (Cl_2), sodium hypochlorite, or calcium hypochlorite. When these substances are added to water, they undergo the following reaction:

$$Cl_2 + H_2O \rightarrow HOCl + H^+ + Cl^-$$

$$NaOCl + H_2O \rightarrow NaOH + HOCl$$

$$Ca(OCl_2) + 2\ H_2O \rightarrow Ca(OH)_2 + 2\ HOCl$$

$$HOCl \rightarrow H^+ + OCl^-$$

Microbicidal activity is controlled by the amount of free available chlorine (as hypochlorous acid, HOCl) in the water that comes in contact with cells (Beuchat, 1998). The dissociation of HOCl depends on pH (Table 10.8).

The antimicrobial action of chlorine is based on its powerful oxidizing effect and rapid linkage of proteins, including the cell membrane proteins. N-chloro compounds are formed and interfere with glucose oxidation or oxidation of sulfhydryl groups. Chlorine impairs membrane permeability and transport of extracellular nutrients (Camper and McFecters, 1979; Beuchat, 2000). Chlorinated water is widely used for the disinfection of whole fruits and vegetables as well as freshly peeled,

TABLE 10.8
Concentration of Undissociated HOCl (%)

pH	0°C	20°C
4	100	100
6	98.2	96.8
7	83.2	75.2
8	32.2	23.2
9	4.5	2.9

Source: Adapted from Beuchat, L.R., in *Minimally Processed Fruits and Vegetables*, Aspen Publishing, Gaithersburg, MD, 2000, p. 63.

cut, or shredded produce. Lowering pH, increasing temperature, or adding surfactants can increase the efficiency of treatment. Toxic chlorine gas is liberated from solutions of hypochlorite at pH below 4.0. The use of warm wash water is not usually applicable for raw fruits and vegetables. This accelerates microbial growth and increases the risk of infiltration of microorganisms from the water due a negative temperature differential between the water and the fruits or vegetables. The use of surfactants can adversely affect the sensorial properties of the products (Adams et al., 1989; Beuchat, 2000). The disinfection effect of chlorine also depends on the produce and on the contaminating microflora. The efficiency of chlorine and other disinfectants is limited by the neutralization effect of fruit and vegetable tissue components on the surface and also by inaccessibility of disinfectant to the microbial cells in creases, crevices, pockets, and natural openings in the skin (Nguen-The and Carlin, 1994). The application is therefore more efficient for aerobic microflora in lettuce than in root vegetables and cabbages (Garg et al., 1990; Ahvenainen and Hurme, 1994). Chlorine has very broad antimicrobial spectrum. It acts against molds, yeasts, bacteria, algae, protozoa, and many viruses, but there are differences in sensitivity of individual groups of microorganisms and of the strains. Several authors have referred to the limited effectiveness of chlorine in suppressing growth of *Listeria monocytogenes* in lettuce and cabbage (Skytta et al., 1996; Francis and O'Beirne, 1997). In addition to the health concerns related to the use of chlorine previously mentioned, limitations of its use in fruits and vegetables are related to its potential reactions with some organic compounds that have been observed during chlorination of drinking water. The amounts used during treatment of fruits or vegetables are similar or higher than those in drinking water, but the concentration of various organic compounds in tissue juice is incomparably higher. Trichlormethan and chloroacetic acid as well as various disinfection byproducts (DBPs) produced by chlorination of amino acids such as iso-butyraldehyde from valine, isovaleraldehyde from leucine, 2-methylbutyraldehyde from isoleucine, and phenylacetaldehyde from phenylalanine were identified in chlorinated drinking water. Some of these, especially the aldehydes, often cause unpleasant odors (Froese et al., 1999).

Chlorine dioxide, ClO_2, represents a valuable alternative to chlorine. It is also often used as a disinfectant for fruits and vegetables. The advantages compared to chlorine include lower reactivity with organic matter and insensitivity of the disinfection effect to pH. Chlorine dioxide is unstable and must be prepared on-site. It can be explosive at higher concentrations. Its oxidizing power is about 2.5 higher than that of chlorine. The mechanism of antimicrobial activity is similar to that of chlorine in terms of disruption of cell protein synthesis and membrane permeability control (Beuchat, 1998). Its use is also regulated. The FDA permits the use of ClO_2 for sanitizing equipment at a maximum of 200 mg/L. The maximum permitted concentrations for the treatment of fruits and vegetables with intact cuticles and for peeled products (e.g., potatoes) are 5 mg/L and 1 mg/L, respectively (Beuchat, 2000).

Peroxyacetic acid is a powerful oxidizing agent that decomposes on contact with organic matter to form acetic acid. Peroxyacetic acid is the component of sanitation preparations used for disinfection of surfaces in food production. Its disinfection effect is comparable to that of chlorine (Winniczuk and Parish, 1997). The use of peroxyacetic acid is also regulated, but it has been tested for sanitation of fresh-cut fruit and vegetable surfaces by itself (Masson, 1990) and in combination with octanoic acid (Hilgren and Salvedra, 2000).

Hydrogen peroxide is also one of the promising alternatives to chlorine. According to Sapers and Simmons (1998), treatment with H_2O_2 was effective in reducing the microbial contamination of whole cantaloupes, table grapes, prunes, raisins, walnuts, and pistachios. Exposure to H_2O_2 vapors caused bleaching of anthocyanins in strawberries and raspberries, and dipping freshly cut green bell peppers, cucumbers, zucchinis, cantaloupes, and honeydew melons in a H_2O_2 solution had no effect on sensory quality but induced severe browning in shredded lettuce (Beuchat, 2000). Residual H_2O_2 might be eliminated by endogenous catalase, but in tissues with low activity of catalase the residual peroxide can cause undesirable oxidation changes.

Application of organic acids as food additives was described as a method to inhibit PPO activity. When organic acids are used for washing of fruits and vegetables they should not be followed by rinsing with drinking water that will increase the surface pH. The mode of action of organic acid is attributed to direct pH reduction in microbial cells by dissociation of the undissociated acid molecule. Organic acids can also disrupt the substrate transport by alteration of cell membrane permeability and can inhibit NADH oxidation (Beuchat, 2000). Washing or dipping in solutions of lemon juice, citric acid, acetic acid, or lactic acid, also in combination with other disinfectants (chlorine), is used in early stages of fruit and vegetable processing.

Bromine has been used as a sanitizer in drinking water treatment, usually in combination with chlorine. The combined use of chlorine and bromine results in a synergistic effect (Beuchat, 1998).

Iodine compounds are often used as sanitizers for food processing equipment and surfaces. Free elemental iodine and hypoiodous acid have microbicidal effects. Iodine sanitation preparations are ethanol iodine solution, aqueous iodine solutions, or iodophors (combination of elemental iodine and nonionic surfactants with a polymeric carrier).

Trisodium phosphate (alkali) disinfectant, such as is used to reduce *Salmonella* in poultry and red meat, has been tested for reduction of *Salmonella* on the surface of whole and cut tomatoes. Significant reductions for whole and cut fruit were obtained by dipping for 15 sec in a 1% solution and 4 to 15% solution, respectively (Zhuang and Beuchat, 1996). In treatment of *L. monocytogenes* on shredded lettuce, however, a 2% solution has almost no effect, and a more efficient treatment with a 10% solution affected the sensory quality of the lettuce (Zhang and Farber, 1996).

Cationic surfactants (quaternary ammonium compounds) are largely used for sanitation of floors, walls, drains, equipment, and other food contact surfaces. The addition of disinfectants to wash water can increase the safety and the shelf life of produce. Washing with drinking water should not follow disinfection. The choice of disinfectant depends on the legal requirements and also on the produce. The disinfection treatment leads to the killing of a part of the surface microflora, but the mode of action is not fully understood. A comparison of the efficiency of some disinfectants against selected microflora of acidic foods is given in Table 10.9.

10.2.3.2   Direct Inactivation of Microorganisms: Abiosis

The microorganisms present in many food products are inactivated (killed) by preservation treatment to the extent that prevents spoilage during the expected shelf life (this is the concept of commercial or practical sterility).

10.2.3.2.1   Thermoinactivation
Thermal methods are extensively used for food preservation, including inactivation of enzymes as previously discussed. The principle of the preservation effect is the thermal denaturation of microbial proteins, which destroys the enzymes and enzyme-controlled metabolism in microorganisms' cells. Thermal treatment also leads to other desirable changes such as protein coagulation, starch swelling, textural softening, and formation of aroma components. However, undesirable changes such as

---

TABLE 10.9
Comparison of Efficiency of Various Disinfectants and Preservatives Against Selected Contamination Microflora of Acidic Food

Chemical Agent	Saccharomyces	Lactobacillus	Leuconostoc	Gluconobacter
Chlorine dioxide	0.0014	0.0031	0.0048	0.0031
Citric acid	>10	>10	>10	6.3
Hypochlorite (Na)	0.027	0.03	0.034	0.031
Iodophor	0.0012	0.0039	0.0098	0.0039
Lactic acid	>10	5	5.3	1.6
Peracetic acid	0.026	0.011	0.011	0.011
Quaternary ammonium compound	0.0012	<0.0001	<0.0001	0.0004
Dimethyl dicarbonate (DMDC)*	0.052	>0.1	>0.1	>0.1

* The application of DMDC is discussed in the "Abiosis" chapter (Chemosterilation) below

Note: Values are chemical concentrations (percentage weight per volume, aqueous) necessary to produce an inactivation rate of 6 log cfu mL^1 min^1 for the indicated microorganism.

Source: Adapted from Winniczuk, P.P. and Parish, M.E., *Food Microbiol.*, 14, 373, 1997.

loss of vitamins and minerals, formation of thermal reaction components, and, especially, the loss of fresh appearance, flavor, and texture could occur (Kyzlink, 1990; Ohlsson, 2002). The rate of destruction is a first-order reaction, dependent on the log of surviving microorganisms (or spores) and the time of heating at constant temperature (*death rate curve*). The time needed to kill 90% of surviving microorganism (1 log reduction) is decimal reduction time D. Destruction of microorganisms is temperature-dependent. The empirical dependence of D on temperature is the *thermal death time curve* (TDT). The slope z of the TDT curve is defined as the number of degrees Celsius required to bring about a 10-fold change in decimal reduction time. D and z values of selected microorganisms with similar data for the thermodestruction of selected food components are given in Table 10.10. There are two important consequences from the above principles (Kyzlink, 1990): (1) the higher the number of microorganisms in raw material, the longer the duration of heating necessary to reduce the contamination to a desired level, and (2) because of the logarithm of concentration in the kinetic equation, it is theoretically possible to destroy all present cells in infinite time.

The former consequence should be especially considered during processing. Handling processes that improve conditions for the growth of microorganisms must be eliminated. Efficient washing, disinfection, and other pretreatments leading to the reduction of microbial counts can allow the use of less intensive heat treatment with a lower negative effect on the processed matter.

The efficiency of heat treatment depends on various factors (Fellows, 2000b):

1. *Contaminating microflora*. Different species and strains differ in their heat resistance. Molds, yeasts, and vegetative cells are relatively sensitive. They are more resistant, but bacterial spores are more heat resistant than the cells.
2. *Incubation conditions*. An increase of heat resistance can be developed by selection and growing of resistant individuals surviving inside the equipment due to insufficient sanitation treatment (e.g., resistant *Lactobacillus* strains surviving in ketchup-producing equipment). Generally, cells or spores produced or held at sublethal temperatures are more heat-resistant. The heat resistance is also affected by the age of cultures (Bibek, 1996).
3. *Conditions during heat treatment*. Lowering the pH value increases the efficiency of heat treatment and also the sensitivity of bacterial spores. The effect of organic acids, including chemical preservatives such as sorbic and benzoic acids, on the heat resistance of ascospores of *Neosartoria fisheri* has been studied by Rajashekhara et al. (2000). Palop et al. (1999) described the effect of pH on heat resistance of *Bacillus coagulans* spores in various food items. Lowering of water activity a_w by addition of osmoactive compounds or fat generally increases the heat resistance of microrganisms. Food preservatives can increase the sensitivity of microorganisms to heat treatment.

TABLE 10.10
Thermodestruction/Thermoinactivation Parameters of Selected Microorganisms and Food Components

Food Component or Microorganism	Source	pH	z (°C)	D_t min (t ref. °C)	Temperature Range (°C)
Bacillus stearothermophilus spores	Various	>4.5	7–12	4.0–5.0 (121)	110 and more
Clostridium thermosaccharolyticum spores			8.9–12.3	3.0–4.0 (121)	
Cl. botulinum type A and B spores			7.8–10.0	0.1–0.21 (121)	
Salmonella (except for Seftenberg)			4.4–5.5	0.02–0.25 (65.5)	
Staphylococcus aureus			4.4–6.7	0.20–2.0 (65.5)	
Alicyclobacillus acidoterrestris spores	Orange juice	3.5 (11.7 °Brix)	7.8	11.9 (91)	
Spoilage microflora bacteria, yeasts, molds			4.4–6.7	0.5–3.0	
Peroxidase	Pea purée	Natural	37.2	3.0 (121)	110–138
Thiamin	Carrot purée	5.9	25	158 (121)	109–149
Chlorophyll a	Spinach	6.5	51	13 (121)	127–149

Sources: Fellows, P., *Food Processing Technology: Principle and Practice,* CRC Press, Boca Raton, FL, 2000b, Part 3; Kyzlink, V., *Principles of Food Preservation,* Elsevier, Amsterdam, 1990, Section 6; Bibek. R., *Fundamental Food Microbiology,* CRC Press, Boca Raton, FL, 1996, Section 6; and Silva, F.M. et al., *Int. J. Food Microbiol.,* 51, 95, 1999.

Efforts to optimize thermoinactivation efficiency have continued for many years. New methods of heating have been developed and used. Ohlsson (2002) reviewed the recent developments in thermal processes in minimal treatment of fruit and vegetables and food. One of the principles is the application of HTST (high temperature, short time). Inactivation of microorganisms is more dependent on temperature. With increasing temperature the rate of killing increases faster than the rate of thermodestruction of food components (see z values in Table 10.10). When produce is processed in the package, to achieve the HSTS condition the package size and the degree of convection inside the package could be optimized. More recent developments in the heat treatment of packaged food are aseptic or semi-aseptic filling. Heating equipment used includes direct steam injection and steam infusion or indirect systems that involve the use of plate, tubular, or scraped surface heat exchangers. The trend in thermoinactivation technologies will probably continue in the area of sophisticated systems such as Twintherm, in which particulate food is heated by direct steam injection in a pressurized, horizontal, cylindrical vessel that rotates slowly. Once the particles have achieved the required inactivation effect (F value), they are cooled under lower pressure (Ohlsson, 1992). Similar systems for pasteurization of vegetables with vacuum steam heating followed by vacuum cooling or a vacuum/steam/vacuum surface intervention process to reduce bacteria on the surface of fruits and vegetables (Kozempel et al., 2002) have been developed.

Fruits and, especially, vegetables are also processed using the sous-vide processing process. Raw material is vacuum-packaged in multilayer plastic film and cooked at temperatures below 100°C, followed by cooling to 3°C. As is in the case of sous-vide processing of meat, vegetables maintain nutrients compared to traditional cooking and the sous-vide produce is juicier (Schellekens, 1996).

Thermoinactivation also involves direct methods of heating, in which heat is generated inside the products such as in ohmic heating and dielectric heating using radiofrequency electric field and microwaves. In ohmic heating processes foods are made a part of an electric circuit by flowing alternating current and the resulting heat generated due to the electrical resistance of the foods through them. Ohmic heating allows a fast increase of temperature (1°C/s) with an absence of temperature gradients. The technology is used for processing liquid foods or liquid foods with small particles (Ruan et al., 2001).

When radiofrequency electric field or microwaves are used, dipolar molecules in water and other ionic components attempt to orient themselves to the field. The rapid oscillation creates frictional heat. Brody (1992) reviewed the microwave pasteurization process. Usually, microwaves can be used in blanching, thawing, and tempering of frozen products and in dehydration processes (Fellows, 2000c). Indirect heating methods have no effect on fruit and vegetable components (other than those caused by heating). Compared to traditional heating processes, this could be shorter and therefore more gentle with reduced loss of nutrients.

10.2.3.2.2 *Other Physical Methods of Killing Microorganisms*

Nonthermal preservation methods of foods include traditional technologies such as irradiation, UV radiation, ultrasound treatment, and also more modern emerging technologies such as high-pressure treatment and pulsed electric field treatment.

TABLE 10.11
Application of Irradiation in Processing of Produce

Application	Dose range (kGy)	Product
Sterilization	7–10	Herbs, spices
Control of molds	2–5	Fresh fruits
Extension of chill life (up to 1 month)	2–5	Fresh soft fruits (strawberries, etc.)
Disinfestations	0.1–2	Fruits, grains, cocoa beans
Inhibition of sprouting	0.1–0.2	Potatoes, garlic, onions

Source: Fellows, P., *Food Processing Technology: Principle and Practice,* CRC Press, Boca Raton, FL, 2000c, Part 4.

TABLE 10.12
Application of High Pressure in Processing of Produce

Products	Process Conditions
Jams, fruit dressings, fruit sauce topping, fruit jelly	400 MPa, 10–30 min, 20°C
Mandarin juice	120–400 MPa, 2–20 min, 20°C (and additional heat treatment)
Avocado	700 MPa, 600–800 L/h
Orange juice	500 MPa, 5- or 10-min cycles including 1 min hold

Source: Ohlsson, T., in *Minimal Processing Technologies in the Food Industry,* Woodhead Publishing Limited and CRC Press, Boca Raton, FL, 2002, p. 4.

Several authors have reviewed the nonthermal processes and their applications in food preservation (Kader, 1986; Barbosa-Cánovas et al., 1998; Gunes et al., 2001).

Gamma rays and, to a lesser extent, x-rays and electron emission are used for destruction of microorganisms or inhibition of physiological changes in food. Some examples of commercial application of irradiation in produce processing are given in Table 10.11. Promising results were found in delaying ripening of whole fruits by irradiation (Kader, 1986), but irradiation of cut fruit could induce undesirable changes in texture (Gunes et al., 2001).

Food products, usually in flexible packaging, can be subjected to high hydrostatic pressure in the range 300 to 1,000 MPa for several minutes. The method permits inactivation of microorganisms and enzymes without the degradation of flavor (Cheftel, 1995). Recent applications in fruit and vegetable processing summarized by Ohlsson (2002) are shown in Table 10.12. The application of high pressure has several limitations. High pressure could disrupt the structure of whole or cut fruit or vegetables. Some enzymes such as polyphenoloxidase, peroxidase, and lipoxygenase may be resistant to high pressure, and in some cases enzymatic activity could increase due to the changes in protein conformation and bacterial spores that are resistant to the treatment. Nevertheless, the technology is promising and especially suitable for fruits and vegetables because of its ability to preserve the flavor of fresh produce.

When a high-intensity pulsed electric field (HIPEF) is applied to a food in short pulses (1 to 100 μs), an external electric field induces an electric potential over the cell membranes. When this potential greatly exceeds a critical value (for vegetative cells, 15 kV/cm) irreversible pores are formed, membranes are destroyed, and cells die (Fellows, 2000c). The degree of inactivation depends on the intensity of the electric field and the number and shape of pulses and also on conditions in treated food such as temperature, pH, ionic strength, and electrical conductivity (Vega-Mercado et al., 1999). According to studies with inoculated food, HIPEF processing reduces the number of target microorganisms of up to 6 log cycles. Treatment has no effect on bacterial spores and only limited effect on enzymes (Barbosa-Cánovas et al., 1998). The technology is not commercially used for food preservation, but it is used for recovery of edible oils and fats (Fellows, 2000c) and seems to be promising pretreatment for improving drying of fruit or vegetables (Ade-Omowaye et al., 2001).

Sound waves (18 to 500 MHz) of high intensity (10 to 1,000 W/cm^2) produce in food cycles of compression and expansion; this phenomenon is known as *cavitation*. The implosion of bubbles generates spots with very high pressures and temperatures that disrupt cells (Kader, 1986). The lethal effect is relatively low and is affected by properties of treated product. Clear liquids of low viscosity are better suited than mashed vegetables with a high content of insoluble solids. Ultrasound is therefore used mainly for disinfection of product surfaces to increase the effect of disinfectants (Mason et al., 1996). Recently ultrasound treatment was observed to increase lethality of heat treatment, probably due to enhancement of the heat transfer. D values decrease 10- or 20-fold when heating is combined with ultrasound compared with heating alone (Leadly and Williams, 2001).

Ultraviolet light and high-intensity pulsed white light (25% UV, 45% visible light, 30% infrared) are used for treating produce and equipment surfaces. Produce such as whole fruits and vegetables without flat surfaces are more difficult to treat with ultraviolet radiation and the preservation effect is lower (2 to 3 log cycles reduction).

10.2.3.2.3 Chemosterilation

Chemosterilation refers to the addition of chemicals that kill microorganisms. They usually leave residues in food. Chemosterilation can be the application of peroxyacetic acid, hydrogen peroxide, or other disinfectants during the dipping or washing treatment of fresh-cut produce. Chemosterilation is applied in drinking water treatment (chlorination, ozonization). Dimethyl dicarbonate (DMDC) is commonly used for the chemical sterilization of soft drinks and wine (Ough, 1993; Dunn, 1997). DMDC is usually added to the drinks before filling. It hydrolyzes on contact with water. In pH about 2.8 and temperatures in the range of 10 to 30°C it is broken down to the detection limit within 65 to 260 min to produce methanol and carbon dioxide. It kills the microorganisms faster than conventional preservatives; after the addition of DMDC the number of microorganisms is reduced. DMDC acts primarily against yeasts; its effect against bacteria, and especially against molds, is weaker or nonexistent (Genth, 1979). The effect of DMDC on the selected microflora of acidic fruit juices is given in Table 10.9.

10.2.3.3 Indirect Preservation Methods: Anabiosis

With indirect methods of preservation the microorganisms present in food products are not inactivated, but the conditions in food products are changed or preservatives are added to inhibit their growth and development. Anabiotic treatment prolongs the lag phase of microbial growth.

10.2.3.3.1 Osmoanabiosis

Preserving food by osmoanabiosis involves the reduction of its free water content. Free water is the water available for spoilage microflora. It is related to water activity a_w and can be removed from fruits or vegetables by dehydration using drying or concentration methods or reduced by addition of solutes such as sugar, salt, and honey. The general effect of water activity on the stability of various food products is summarized in Table 10.13. Fellows (2000a,b) and Vega-Mercado et al. (2001) reviewed the development of drying equipment.

Vega-Mercado et al. (2001) divided the development of drying equipment into four generations:

1. Dryers for the processing of solid materials such as grains, sliced fruits, and vegetables or chunked products. These are cabinet and bed-type dryers such as kilns, trays, truck trays, rotary flow conveyors, and tunnels that involve hot air flowing over the produce to remove water from the surface.

2. Dryers intended for dehydration of slurries and purees. These are spray and drum dryers. During spray drying the fluid state of the food is transformed to droplets and into dried particles by spraying a hot drying medium. In drum dryers the surface of a slowly rotating, internally heated steel drum is covered with a uniform layer of food by dipping, spraying, or spreading. Before completing one revolution (20 sec to 30 min) the dried food is removed by blades in contact with the drum surface.

3. Freeze dehydration and osmotic dehydration that minimize losses of flavor and aroma compounds. The freeze drying process consists of two steps. The products are initially frozen, and this is followed by removal of water by sublimation under reduced pressure. Osmotic dehydration is based on water removal by means of osmotic pressure. Fruits or vegetables are dipped in hypertonic solutions such as sugar, salt, sorbitol, or glycerol. The efficiency of the process can be increased by the use of vacuum.

4. The fourth generation of drying equipment involves the latest developments in this technology, such as high vacuum, fluidization, heating by microwaves, radio frequency (RD), and refractance window. Microwave or RD allows heat to be brought directly to excess water in partially dried products, hence reducing the potential of overheating relative to the more traditional techniques. Refractance windows heating is a new technology that is used to transmit heat to the products. It is applied to the surface of a plastic conveyor belt floating on hot water.

TABLE 10.13
Importance of Water Activity in Food

a_w	Phenomenon	Degree of Protection Required
1.00		
0.95	Pseudomonads, bacillus, *Clostridium perfringens,* and some yeasts inhibited	
0.90	Lower limit for bacterial growth (general), salmonella, *Vibrio parahemolyticus, Clostridium botulinum,* lactobacillus, and some yeasts and fungi inhibited	Intermediate-moisture foods ($a_w = 0.90$–0.55)
0.85	Many yeasts inhibited	Other preservation treatments/other hurdles are necessary
0.80	Lower limit for enzyme activity and growth of most fungi; *Staphlococcus aureus* inhibited	
0.75	Lower limit for halophilic bacteria	
0.70	Lower limit for growth of most xerophilic fungi	
0.65	Maximum velocity of Maillard reactions	
0.60	Lower limit for growth of osmophilic or xerophilic yeasts and fungi	Minimum protection or no packaging required; depending the storage conditions, if the relative humidity is higher than the corresponding water activity of the product the prevention of water uptake is necessary
0.55	Lower limit for life	
0.50		
0.40	Minimum oxidation velocity	
0.30		
0.25	Maximum resistance of bacterial spores	
0.20		

Sources: Adapted from Bibek, R., *Fundamental Food Microbiology,* CRC Press, Boca Raton, FL, 1996, Section 6; Lund, B.M., in *The Microbiological Safety and Quality of Food,* Vol. 1, Aspen Publishing, Gaithersburg, MD, 2000, p. 122; Archer, D.L., *Int. J. Food Microbiol.,* 90, 127, 2004; and Li Bing, J. *Food Eng.,* 54, 175, 2002.

Water activity in fruit or vegetable juices or purees is also reduced by concentration, which is also used to preconcentrate foods prior drying or other processing to reduce the volume. Liquid food can be concentrated by evaporation, reverse osmosis, or freeze concentration. In spite of the development of evaporators and optimization of the process, evaporation causes changes in nutritional and sensory properties of concentrated food because of the severe heat treatment. Reverse osmosis and freeze concentration, however, are less severe and allow achievement of a higher degree of concentration.

The addition of solutes such as sugar or salt is the principal technological operation in the production of fruit jams, spreads, toppings, and syrups. After concentration, these products are classified as intermediate moisture food and their stability must be achieved in combination with additional treatments (Kyzlink, 1990).

10.2.3.3.2 Chilling and Freezing

Metabolic and enzymatic activities, as well as the rate of microbial growth, generally decrease when the temperature is lowered. Normally, the generation time for microorganisms is doubled for every 10°C reduction of temperature. The sensitivity of various microorganisms to temperature is different. The majority of spoilage and pathogenic microflora do not grow below 5°C. However, some psychrotrophic pathogens such as *Listeria monocytogenes, Yersinia enterocolitica, Clostridum botulinum* types A and E, *Aeromonas hydrophyla,* (enterotoxigenic), and *E. coli* are able to multiply slowly in refrigerated foods. Refrigeration thus appears to be the only hurdle in the preservation of high-moisture or intermediate-moisture food products, unlike fruit and vegetable products that require more than one treatment (Bibek, 1996). When the temperature of food is reduced below –2°C, free water in fruit or vegetables starts freezing and forming ice crystals. The preservation effect of freezing also results in subsequent damage of the microbial cells by extra- and intracellular ice formation and concentration of extra- and intracellular solutes (Lund, 2000). Frozen fruit and vegetable products are more stable than refrigerated products for long periods of time if the texture is unaffected by the very low temperature. Frozen foods are relatively safe. The few outbreaks of food-borne illness associated with frozen foods indicate that commercial freezing processes kill some, but not all, human pathogens. Archer (2004) evaluated freezing preservation methods. Freezing is more likely to damage the texture of the produce than is refrigeration. Compared to refrigeration, the freezing process damages the structure and processes in the plant tissues. Upon thawing of frozen food items, microbiological, enzymatic, and chemical changes are accelerated. Blanching prior to freezing will reduce the extent of enzymatic changes that occur in thawed frozen foods. Formation of ice crystals and subsequent damage of cells also decreases the microbial stability of the thawed products. The microorganisms present on the surface of frozen fruit or vegetables grow in the released juice without any barrier of the former intact fruits. The extent of undesirable changes depends on the number and size of ice crystals. The ice formation should be as fast as possible to minimize the food's structural damage. Undesirable changes of fruits and vegetables can be prevented by high-pressure freezing, the use of cryoprotectant and cryostabilizers, dehydrofreezing, and applications of antifreeze protein and ice nucleation protein.

The thawing process can be optimized by high-pressure thawing, microwave thawing, ohmic thawing, and acoustic thawing to shorten thawing time, reduce drip loss, and improve product quality (Li Bing, 2002). These methods reduce the time required for frozen materials to thaw and the risk of overheating. High-pressure freezing is a process that involves subjecting the material to be frozen under pressure and cooled below the freezing point. When the product is sufficiently cool, the pressure is released and the product immediately freezes, allowing formation of a high number of very small ice crystals with minimal damage to texture. During the rapid phase change the bubbles of air are trapped in tissues. Pressure-frozen fruits have softer textures and could be eaten without thawing (LeBail et al., 2002). During storage the structure of frozen products may change due to recrystallization, especially when the storage temperature is not constant. High-pressure thawing of fruit can reduce the juice release, and when it is followed by immersion in sugar solution the treatment can also increase the sugar uptake of fruit tissues (Eshtiaghi and Knorr, 1996). Vacuum infusion with cryprotectants (sugar or fruit juice concentrates) and cryostabilizers (e.g., pectin) prior freezing can reduce ice crystal damage in frozen fruits by notable reduction of freezable water (Torreggiani and Bertolo, 2001). The effect is also similar in the dehydrofreezing process. In this process, fruits or vegetables are dehydrated to the desired moisture level and then frozen (Spiazzi et al., 1998). Applications of antifreeze protein can lower the freezing temperature and retard recrystallization during frozen storage, while ice-nucleation proteins raise the temperatures of ice nucleation and reduce the degree of supercooling (Li and Lee, 1998). Bacteria-induced ice nucleation has been recognized as a major contributing factor to frost injury in plant products. The effect of antifreeze protein in fish is also known, although practical applications are seldom reported. A potential application of antifreeze protein could be inhibition of recrystallization of ice in ice creams, but the potential exists for the concept to be developed for a number of agricultural products.

10.2.3.3.3 Chemoanabiosis

The mechanism of chemoanabiosis is the application of a chemical agent with antimicrobial effects. Chemosterilation agents, relative to disinfection agents, do not cause rapid inactivation. They, however, act more through their microbistatic properties. When they are applied the lag phase of microbial growth is extended. Chemoanabiosis process may consist of a series of steps (Kyzlink, 1990), including addition of preservatives, addition of antibiotics (or preparations containing antibiotics), addition of natural preservatives (phytoncides), and traditional processing of food that includes a chemoanabiotic effect (smoking, curing, brining, marinating, etc.).

Preservatives that are used may include various fungicides that are applied after harvest. Fruits and vegetables to be stored may be treated with one or more fungicides. There are about 20 types of fungicides approved for use on fresh produce (Eckert and Ogawa, 1990). The list varies from one country to another. Benomyl, thiabendazol, thiophthanate methyl, and imazalil are more commonly used. In some countries these products are permitted for use as preservatives by legislation (Velíšek, 1999). The fungicides are applied by dipping or spraying, through incorporation into waxes for surface application on fruits, or in fungicide-impregnated paper. A number of antimicrobial agents also act as disinfectants, microbistatic agents, acidulants, and antioxidants (Bibek, 1996). Benzoic acid, sorbic acid, parabens, and sulfur dioxide are used most frequently for the preservation of fruit and vegetable products and are permitted in the majority countries.

Benzoic acid is used as an acid or sodium salt at a concentration of 500 to 2000 mg/kg for preservation of acidic products such as beverages, pickles, salad dressings, and mustard. The undissociated acid is more efficient, but its effectiveness decreases with increasing pH. Benzoic acid is more effective against yeasts and molds than against bacteria (Bibek, 1996). The solubility of benzoic acid is relatively low (1.8 g/L at 4°C; 2.7 g/L at 18°C; 22 g/L at 100°C). The solubility becomes more critical when a concentrated solution of sodium benzoate is added to acidic products or during heat treatment to concentrate and reduce product volume (Kyzlink, 1990).

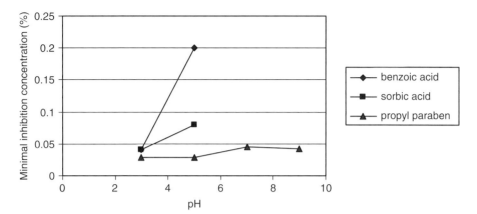

FIGURE 10.2 Effect of pH on minimal inhibition concentration of selected preservatives. (Adapted from Belitz, H.D. and Grosch, W., *Food Chemistry,* Springer, Berlin, 1999, Chap. 2.)

Sorbic acid has properties similar to those of benzoic acid. It is used as acid or salts of sodium, potassium, or calcium at concentrations that range from 500 to 2,000 mg/kg. It is more effective against yeasts and molds than against bacteria. Its inhibitory effect decreases with dissociation, but its pK value is higher than that of benzoic acid. Sorbic acid can be used in neutral pH solutions. The solubility of sorbic acid in water is also low (1.6 g/L at 20°C, 39 g/L at 100°C) and sorbic acid has technological consequences similar to those of benzoic acid (Bibek, 1996).

Parabens are alkyl esters of 4-hydroxybenzoic acid. They are used as methyl, ethyl, propyl, or butyl paraben. Parabens are broad-spectrum preservatives effective against bacteria, yeasts, and molds. They are used at concentrations from 100 to 1000 mg/kg for various acidic and low-acid food products. The antimicrobial effect of individual parabens increases with the length of alkyl. The efficiency is almost unaffected by pH (Figure 10.2). The solubilities of methyl paraben, ethyl paraben, and propyl paraben in water at 25°C are 2.5, 1.7, and 0.5 g/L, respectively (Davidson, 1993).

Sulfur dioxide and sulfites have been used for many centuries as fumigants, especially as wine preservatives. The antioxidative effect of SO_2 and sulfites was previously discussed. SO_2 and its various sulfites dissolve in water, and at low pH levels yield sulfurous acid, bisulfite, and sulfite ions. At pH values less than 4.0 the antimicrobial activity reaches its maximum. The antimicrobial action of SO_2 against yeasts, molds, and bacteria is selective, and some species are more resistant than others. SO_2 and sulfites are used in the preservation of a variety of food products. Because of their flavor and negative health effects they are used for preservation as an intermediate process in which subsequent processing such as boiling or acidification removes the additive.

Organic acids are naturally present in plant material or formed during fermentation processes. Some of these, such as citric, malic, tartaric, acetic, lactic, and propionic acid, are very often used as chemical preservatives in the treatment and processing of fresh fruits and vegetables. Usually, solutions containing combinations of organic acids or an acid and its salt are applied to reduce the undesirable flavor that could result from using too much of a single organic acid. Citric acid is often used in dipping solutions for its antioxidative and pH-lowering effect in the treatment of sliced or peeled fruits and vegetables (Soliva and Martin-Belloso, 2003). Inhibition of microorganisms by organic acids depends upon several factors, including reduction in pH, the ratio of undissociated and dissociated species of the acid, chain length, cell physiology, and metabolism. Two or more of these elements may interact to create either a microstatic or a microbicidal effect. Microorganisms differ in their sensitivity to various acids. For example, lactic acid bacteria are not only tolerant to weak lipophilic acids, but also produce them as metabolites. Acetic acid also has a microbicidal effect against lactobacilli but only a bacteristatic effect against bacilli (Doores, 1993). Acidification also increases the sensitivity of microorganism to heat treatment and to other preservation treatments. Various synergistic effects are known to occur among organic acids. Examples are the effects of mixtures of ascorbic and sorbic acids, and with lactic acid (Doores, 1993; Bibek, 1996).

Some plants naturally contain compounds with antimicrobial properties. Essential oils of spices (vanillin, cinnamol, eugenol, carvacol, thymol, and essential oils of oregano, thyme, and rosemary); humulone and lupulone in hops; allicin and alliin in garlic, onion, and horseradish; allyl isothiocyanate in mustard and other plants; and tomatidine in unripe tomatoes and other substances are known to inhibit the growth of microorganisms. The use of natural preservatives is usually limited by the flavor of the compounds or extracts and also by their potential toxicity (Bibek, 1996; Cherry, 1999; Velíšek, 1999). An exception can be the application of hexanal, which is naturally produced by enzymatic oxidation of fatty acids after peeling or cutting in damaged cells. When used in the package atmosphere of apple slices stored at 4°C, the treatment caused strong inhibition of mesophilic bacteria, molds, and yeasts and prolongation of the lag phase of psychrotrophic bacteria (Lanciotti et al., 1999).

The addition of antibiotics (natamycin, tylosin, and bacteriocins — antibiotics of Gram-positive bacteria, which include nisin, pediocin, sakacin, and subtalin) is not very common in preservation of fruit products. They may be used as one of the additives for the preservation of canned low-acid vegetables to control anaerobic spore-forming microorganisms (tylosin). Bacteriocins act against Gram-positive bacteria (streptococci, bacilli, clostridia, and other anaerobic spore-forming bacteria) (Hurst and Hoover, 1993).

The new trend in application of preservatives is antimicrobial packaging, which can take several forms, including:

- Addition of sachets or pads containing volatile antimicrobial agents into packages.
- Incorporation of volatile and nonvolatile antimicrobial agents directly into polymers.
- Coating or adsorbing antimicrobials onto polymer surfaces.
- Immobilization of antimicrobials to polymers by ion or covalent linkages.
- Use of polymers that are inherently antimicrobial.

Active packaging with various types of plastic films is used for modified atmosphere packaging of fruits and vegetables, as previously discussed, but antimicrobial packaging is not only used for respiring products, but also for processed fruit or vegetable products (Appendini and Hotchkiss, 2002).

10.2.3.3.4   Cenoanabiosis

Cenoanabiosis is essentially a traditional preservation method. In fermented products the original principle was to achieve food products of characteristic flavor rather than to improve products' shelf life. Lactic acid fermentation using homofermentative lactic acid bacteria is a typical way of processing food products such as cabbage (sauerkraut) and cucumbers (fermented cucumbers). In Balkan countries lactic acid fermentation is also used for preservation of fruits. The fermentation process proceeds either spontaneously or under controlled conditions. The fermented products have limited shelf life and another preservation treatment is commonly applied. For example, sauerkraut is usually pasteurized after finishing the fermentation process (Kyzlink, 1990). Recently the use of microbial cultures has been proposed for the preservation of unfermented foods, including produce (Rodgers, 2001). These utilize protective cultures that are food-grade bacteria (usually lactic acid bacteria) that may or may not be strains naturally present in the food type. They are selected for their ability to grow in a product and inhibit poisoning or spoilage microorganisms rather than to produce a desired texture or flavor. Under normal storage conditions protective cultures should not affect sensorial qualities of produce (Leroi et al., 1996). The principal advantage of preservation using protective cultures is the temperature-responsive action. Applications of protective cultures include improving the safety of minimally processed fruits and vegetables (Breidt and Fleming, 1997), salads, and dressings (Hutton et al., 1991; Nederland, 1998).

REFERENCES

Abe, K. and Watada, A.E., Ethylene absorbent to maintain quality of lightly processed fruits and vegetables, *J. Food Sci.,* 56, 1589, 1991.

Adams, M.R., Hartley, A.D. and Cox, L.J., Factors affecting the efficiency of washing procedures used in the production of prepared salads, *Food Microbiol.,* 6, 69, 1989.

Ade-Omowaye, B.I.O. et al., Use of pulsed electric field pretreatment to improve dehydration characteristics of plant based foods, *Trends Food Sci. Technol.,* 12, 285, 2001.

Agar, I.T. et al., Postharvest CO_2 and ethylene production and quality maintenance of fresh-cut kiwifruit slices, *J. Food Sci.,* 64, 433, 1999.

Ahmed, J. and Shivhare, U.S., Thermal kinetics of color change, rheology, and storage characteristics of garlic puree/paste, *J. Food Sci.,* 66, 754, 2001.

Ahvenainen, R., Minimal processing of fresh produce, in *Minimally Processed Fruits and Vegetables,* Alzamiora, S.M., Tapia, M.S., and López-Malo, A., Eds., Aspen Publishing, Gaithersburg, MD, 2000, p. 277.

Ahvenainen, R. and Hurme, E., Minimal processing of vegetables, in *Minimal Processing of Foods,* Ahvenainen, R., Mattila-Sandholm, T., and Ohlsson, T., Eds., VTT Espoo, Finland, 1994, p. 17.

Appendini, P. and Hotchkiss, J. H., Review of antimicrobial food packaging, *Innovat. Food Sci. Emerg. Technol.,* 3, 2002, 113.

Archer, D.L., Review article: freezing: an underutilized food safety technology?, *Int. J. Food Microbiol.* 90, 127, 2004.

Baldwin, E.A., Nisperos-Carriedo, M.O., and Baker, R.A., Use of edible coatings to preserve quality of lighty (and slighty) processed products, *Crit. Rev. Food Sci. Nutr.*, 35, 509, 1995.

Barbosa-Cánovas, G.V. et al., *Handling and Preservation of Fruits and Vegetables by Combined Methods for Rural Are*as, Technical Manual, FAO Agricultural Services Bulletin 149, Rome, 2003.

Barbosa-Cánovas, G.V. et al., *Nonthermal Preservation of Foods*, Marcel Dekker, New York, 1998.

Belitz, H.D. and Grosch, W., *Food Chemistry*, 2nd ed., Springer, Berlin, 1999, Chap. 2.

Belitz, H.D. and Grosch, W., *Food Chemistry*, 2nd ed., Springer, Berlin, 1999, Chaps. 17, 18.

Bett, K.L. et al., Flavor of fresh-cut gala apples in barrier film packaging as affected by storage time, *J. Food Qual.*, 24, 141, 2001.

Beuchat, L.R., Surface decontamination of fruits and vegetables eaten raw: a review. *Food Safety Issues, Food Safety Unit, World Health Organization,* WHO/FSF/FOS/98.2, 1998, p. 1.

Beuchat, L.R., Use of sanitizers in raw fruit and vegetable processing, in *Minimally Processed Fruits and Vegetables*, Alzamiora, S.M., Tapia, M.S., and López-Malo, A., Eds., Aspen Publishing, Gaithersburg, MD, 2000, p. 63.

Bibek, R., *Fundamental Food Microbiology*, CRC Press, Boca Raton, FL, 1996, Section 6.

Bolin, H.R. and Huxsoll, C.C., Effect of preparation procedures and storage parameters on quality retention of salad-cut lettuce, *J. Food Sci.*, 56, 60, 1991.

Bolin, H.R. and Huxsoll, C.C., Storage stability of minimally processed fruit, *J. Food Process. Preserv.*, 13, 281, 1989.

Breidt, F. and Fleming, P., Using lactic acid bacteria to improve the safety of minimally processed fruits and vegetables, *Food Technol.*, 51, 44, 1997.

Brody, A.L., Microvawe food pasteurization, sterilization and packaging, in *Food Technology International Europe*, Turner, A., Ed., Sterling Publications International, London, 1992, p. 67.

Brosnan, T. and Da-Wen Sun, Precooling techniques for horticultural products: a review, *Int. J. Refrig.*, 24, 154, 2000.

Camper, A.K. and McFecters, G.A., Chlorine injury and enumeration of waterborne coliform bacteria, *Appl. Environ. Microbiol.*, 37, 633, 1979.

Cerny, G., Hennlich, W., and Poralla, K., Fruchtsaftverderb durch bacillen: isolierung und characterisierung des verderbserregers, *Z. Lebensmitt. Unters. Forch.*, 179, 224, 1984.

Cheftel, J.C., Review: high pressure, microbial inactivation and food preservation, *Food Sci. Technol. Int.*, 1, 75, 1995.

Cherry, J.P., Improving the safety of fresh produce with antimicrobials, *Food Technol.*, 53, 54, 1999.

Clysdesdale, F.M., Color: origin, stability, measurement, and quality, in *Food Storage Stability*, Taub, I.A. and Singh, R.P., Eds., CRC Press, Boca Raton, FL, 1997, p. 175.

Cummings, D.B. et al., A new blanching system for the food industry, II, Commercial design and testing, *J. Food of Process. Preserv.*, 8, 137, 1984.

Dale, M. et al., Effect of bruising on the total glycoalkaloid and chlorogenic acid content of potato (*Solanum tuberosum*) tubers of five cultivars, *J. Sci. Food Agric.*, 77, 499, 1998.

Dauthy, M.E., Fruit and Vegetable Processing, *FAO Agricultural Services Bulletin* No. 119, Rome, 1995.

Davidson, P.M., Parabens and penolic compounds, in *Antimicrobials in Foods*, Davidson, P.M. and Branen, A.L., Eds., Marcel Dekker, New York, 1993, p. 263.

Day, B.P.F., Chilled food packaging, in *Chilled Foods: A Comprehensive Guide*, Dennis, C. and Stringer M., Eds., Ellis Horwood, London, 1990, p. 147.

Day, B.P.F., Modified atmosphere packaging and active packaging of fruit and vegetables, in *Minimal Processing of Foods*, Ahvenainen, R., Mattila-Sandholm, T., and Olsson, T., Eds., VTT Symposium Espoo, Finland, 1994, p. 173.

Devece, C. et al., Enzyme inactivation analysis for industrial blanching applications: comparison of microwave, conventional, and combination heat treatments on mushroom polyphenoloxidase activity, *J. Agric. Food Chem.*, 47, 4506, 1999.

Dietrich, W.C., Huxsoll, C.C., and Guadagni, D.C., Comparison of microwave, conventional and combination blanching of brussels sprouts for frozen storage, *Food Technol.*, 24, 613, 1970.

Doores, S., Organic acids, in *Antimicrobials in Foods*, Davidson, P.M. and Branen, A.L., Eds., Marcel Dekker, New York, 1993, p. 95.

Dunn, A.J., Developments in European Union legislation on food preservatives, *Food Chem.*, 60, 187, 1997.

Eckert, J.W., and Ogawa, J.M., Recent developments in the chemical control of post-harvest diseases, *Acta Hortic.*, 269, 477, 1990.

Eshtiaghi, M., and Knorr, D., High hydrostatic pressure thawing for the processing of fruit preparations from frozen strawberries, *Food Technol.*, 10, 143, 1996.

Fellows, P., Dehydration, in *Encyclopedia of Food Science and Technology*, 2nd ed., Francis, F.J., Ed., John Wiley & Sons, New York, 2000a, p. 480.

Fellows P., *Food Processing Technology: Principle and Practice*, 2nd ed., CRC Press, Boca Raton, FL, 2000b, Part 3.

Fellows P., *Food Processing Technology: Principle and Practice*, 2nd ed., CRC Press, Boca Raton, FL, 2000c, Part 4.

Flores Gutiérrez., A.A., *Manejo Postcosecha de Frutas y Hortalizas en Venezuela: Experiencias y Recomendaciones*, 2nd ed., UNELLEZ, San Carlos, Cojedes, Venezuela, 2000, p. 86.

Francis, G.A. and O'Beirne, D., Effects of gas atmosphere, antimicrobial dip and temperature on the fate of *Listeria inocua* and *Listeria monocytogenes* on minimally processed lettuce, *Int. J. Food Sci. Technol.*, 32, 141, 1997.

Friedman, M., Chemistry, biochemistry, and dietary role of potato polyphenols: a review, *J. Agric. Food Chem.*, 45, 1523, 1997.

Froese, K.L., Wolanski, A., and Hrudey, S.E., Factors governing odorous aldehyde formation as disinfection byproducts in drinking water, *Wat. Res.*, 33, 1355, 1999.

Garg, N., Churey, J.J., and Splittstoesser, D.F., Effect of processing conditions on the microflora of fresh-cut vegetables, *J. Food Protect.*, 53, 701, 1990.

Genth, H., Dimethyldicarbonat: ein neuer Verschwindestoff für alkoholfreie, fruchtsafthaltige Erfrischungsgetränke, *Erfrischungsgetränk* 13, 262, 1979.

Glenn, G.M. and Poovaiah, B.W., Calcium-mediated post-harvest changes in texture and cell wall structure and composition in "Golden Delicious" apple, *J. Am. Soc. Hortic. Sci.*, 115, 962, 1990.

Glynn, P. and Dixon, G.R., Glycoalkaloid concentrations in aerial tubers of potato (*Solanum tuberosum* L.), *J. Sci. Food Agric.*, 70, 439, 1996.

Golden, D.A., Rhodehamel, E.J., and Kautter, D.A., Growth of *Salmonella spp.* in cantaloupe, watermelon, and honeydew melons, *J. Food Protect.*, 56, 194, 1993.

Gorny, J.R. et al., Quality changes in fresh-cut pear slices as affected by cultivar, ripeness stage, fruit size and storage regime, *J. Food Sci.*, 65, 541, 2000.

Gorny, J.R. et al., Quality changes in fresh-cut pear slices as affected by controlled atmosphere and chemical preservatives, *Postharv. Biol. Technol.*, 24, 271, 2002.

Gorny, J.R., Hess-Pierce, B., and Kader, A.A., Quality changes in fresh-cut peach and nectarine slices as affected by cultivar storage atmosphere and chemical treatments, *J. Food Sci.*, 64, 429, 1999.

Gorris, L.G.M., de Witte, Y., and Bennik, M.H.J., Refrigerated storage under moderate vacuum, *ZFL Focus Int.*, 45, 63, 1994.

Gorris, L.G.M., Safety and quality of ready-to-use fruit and vegetables (AIR I-CT92-0125), in *EU Research,* Research Results Ready for Application (RETUER) 21 May 1996, Dublin, Ireland, 1996.

Grosch, W., Lipid degradation products and flavours, in *Food Flavours*, Morton, A. and MacLeod, J., Eds., Elsevier Scientific, Amsterdam, 1982, p. 325.

Gunes, G., Hotchkiss, J.H., and Watkins, C.B., Effect of gamma irradiation on the texture of minimally processed apple slices, *J. Food Sci.*, 66, 63, 2001.

Hatton, T.T. and Cubbebedge, R.H., Preferred temperature for prestorage conditioning of Marsh grapefruit to prevent chilling injury at low temperatures, *HortScience*, 18, 721, 1983.

Hilgren, J.D. and Salvedra, J.A., Antimicrobial efficacy of a peroxyacetic/octanoic acid mixture in fresh cut vegetables process waters*, J. Food Sci.,* 65, 1376, 2000.

Hugghes, J.C. and Swain, T., After-cooking blackening in potatoes. III Examination of the interaction of factors by in vitro experiments, *J. Sci. Food. Agric.*, 13, 358, 1962.

Hurst, A. and Hoover, D., Nisin, in *Antimicrobials in Foods*, Davidson, P.M. and Branen, A.L., Eds., Marcel Dekker, New York, 1993, p. 369.

Hutton, M.T., Chehak, P.A., and Hanlin, J.H., Inhibition of botulism toxin production by *Pediococcus acidilactici* in temperature abused refrigerated foods*, J. Food Safety*, 11, 255, 1991.

Jen, J.J., Influence of spectral quality of light on pigment systems of ripening tomatoes, *J. Food Sci.,* 39, 907, 1974.

Jiang, Y. and Joyce, D.C., 1-Methylcyclopropene treatment effects on intact and fresh-cut apple, *J. Hortic. Sci. Biotechnol.*, 77, 19, 2002.

Kader, A.A., Potential application of ionizing radiation in postharvest handling of fresh fruits and vegetables, *Food Technol.*, 40, 117, 1986.

Kader, A.A., Zagory, D., and Kerbel, E.L., Modified atmosphere packaging of fruits and vegetables, *Crit. Rev. Food Sci. Nutr.*, 28, 1, 1989.

Keteleer, A. and Tobback, P.P., Modified atmosphere packaging of respirating produce, in *Food Preservation by Combined Processes*, Leistner, L. and Gorris, L.G.M., Eds., FLAIR Concerted Action No. 7, EUR 15776 En, 1994, p. 59.

King, A.D. and Bolin, H.R., Physiological and microbiological storage stability of minimally processed fruits and vegetables, *Food Technol.*, 43, 132, 1989.

Kozempel, M. et al., Application of the vacuum steam vacuum surface intervention process to reduce bacteria on the surface of fruits and vegetables, *Innovat. Food Sci. Emerg. Technol.,* 3, 63, 2002.

Kyzlink, V., *Principles of Food Preservation*, Elsevier, Amsterdam 1990, Part IV.

Lamikarna, O. and Watson, M.A., Cantaloupe melon peroxidase: characterization and effects of food additives on activity, *Nahrung*, 44, 168, 2000.

Lanciotti, R., Corbo, M.R., Gardini, F., Sinigaglia, M., and Guerzoni, M.E., Effect of hexanal on the shelf life of fresh apple slices, *J. Agric. Food Chem.*, 47, 4769, 1999.

Laurila, E. and Ahvenainen, R., Minimal processing of fresh fruits and vegetables, in *Fruit and Vegetables Processing: Improving Duality*, Jongen W., Ed., Woodhead Publishing, Cambridge, UK, 2002, p. 294.

Lazar, M.E., Lund, D.B., and Dietrich, W.C., IQB: a new concept in blanching, *Food Technol.*, 25, 684, 1971.

Leadly, C. and Williams, A., Current and potential applications for power ultrasound in the food industry, *New Food*, 3, 23, 2001.

LeBail, A. et al., High pressure freezing and thawing of foods: a review, *Int. J. Refrig.*, 25, 504, 2002.

Leistner, L., Principles and applications of hurdle technology, in *New Methods of Food Preservation,* Gould, G.W., Ed., Blackie, London, 1995, p. 1.

Leroi, F. et al., Effect of incubation with lactic acid bacteria on extending the shelf-life of vacuum-packed cold smoked salmon, *Int. J. Food Sci. Technol.*, 31, 497, 1996.

Leverentz, B. et al., Examination of bacteriophage as biocontrol method for *Salmonella* on fresh-cut fruit: a model study, *J. Food Protect.*, 64, 116, 2001.

Li Bing, Sun Da-Wen, Novel methods for rapid freezing and thawing of foods: a review, *J. Food Eng.*, 54, 175, 2002.

Li, J. and Lee, T. C., Bacterial extracellular ice nucleator effects on freezing of foods, *J. Food Sci.*, 63, 375, 1998.

López-Ayerra, B.M., Murcia, A., and Garcia-Carmona, F., Lipid peroxidation and chlorophyll levels in spinach during refrigerated storage and after industrial processing, *Food Chem.*, 61, 113, 1998.

Lück, E. and Jager, M., *Antimicrobial Food Additives*, 2nd ed., Springer Verlag, Berlin, 1997, Chap. 12.

Lund, B.M., Freezing, in *The Microbiological Safety and Quality of Food,* Vol. I, Lund, B.M., Baird Parker, T.C., and Gould, G.W., Eds., Aspen Publishing, Gaithersburg, MD, 2000, p. 122.

Lutz, J.M. and Hardenburg, R.E., *The Commercial Storage of Fruits, Vegetables and Florist and Nursery Stocks*, U.S. Government Printing Office, Washington, DC, 1968, p. 1.

Martinez, M. and Whitaker, J.R., The biochemistry and control of enzymatic browning, *Trends Food Sci. Technol.*, 6, 195, 1995.

Mason, T.J., Paniwnyk, L., and Lorimer, J.P., The uses of ultrasound in food technology, *Ultrason. Sonochem.*, 3, 253, 1996.

Masson, R.B., Recherche de nouveaux disinfectants pour les produits de 4eme gamme, in *Proc. Congress Produits de 4eme gamme et de 5eme gamme,* C.E.R.I.A.,Brussels, 1990, p. 101.

McEvily, A.J., Iyengar, R., and Otwell, W.S., Inhibition of enzymatic browning in foods and beverages, *Crit. Rev. Food Sci. Nutr.*, 32, 253, 1992.

McHugh, T.H. and Senesi, E., Apple wraps: a novel method to improve the quality and extend the shelf life of fresh-cut apples, *J. Food Sci.*, 65, 480, 2000.

Mir, N. and Beaudry R.M., Modified atmosphere packaging, available at http://www.ba.ars. usda.gov/hb66/014modified.pdf.

Nederland, J., Preserving food, especially salad or dressing, by inoculation with lactic acid bacteria, Patent PII: SO956-7135 (98) 00143-1.

Nguen-The, C. and Carlin, F., The microbiology of minimally processed fresh fruits and vegetables, *Crit. Rev. Food Sci. Nutr.*, 34, 371, 1994.

Ohlsson, T., Minimal processing of foods with thermal methods, in *Minimal Processing Technologies in the Food Industry,* Ohlsson, T. and Bengtsson, N., Eds., Woodhead Publishing Limited and CRC Press, Boca Raton, FL, 2002, p. 4.

Ohlsson, T., R and D in aseptic particulate processing technology, in *Food Technology International Europe,* Turner, A., Ed., Sterling, London, 1992, p. 49.

Ohta, H. and Sugarawa, W., Influence of processing and storage conditions on quality stability of shredded lettuce, *Nippon Shokuhin Kogyo Gakkaishi*, 34, 432, 1987.

Osuga, D., van der Schaaf, A., and Whitaker, J.R., Polyphenol oxidases, in *Protein Structure-Function Relationships in Foods*, Yada, R.Y., Jackman, R.L., and Smith, J.L., Eds., Blackie, London, 1994, p. 62.

Ough, C.S., Dimethyl dicarbonate and diethyl dicarbonate, in *Antimicrobials in Foods*, Davidson, P.M. and Branen, A.L., Eds., Marcel Dekker, New York, 1993, p. 343.

Palop, A. et al., Influence of pH on heat resistance of spores of *Bacillus coagulans* in buffer and homogenized foods, *Int. J. Food Microbiol.*, 46, 243, 1999.

Poovaiah, B.W., Role of calcium in prolonging storage life of fruits and vegetables, *Food Technol.*, 40, 86, 1986.

Powrie, W.D. and Skura, B.J., Modified atmosphere packaging of fruit and vegetables, in *Modified Atmosphere Packaging of Food*, Ooraikul, B. and Stiles, M.E., Eds., Ellis Horwood, Chichester, 1991, p. 169.

Quintero-Ramos, A. et al., Low temperature blanching of frozen carrots with calcium chloride solutions at different holding times on texture of frozen carrots, *J. Food Process. Preserv.*, 26, 361, 2002.

Rajashekhara, E., Suresh, E. R., and Ethiraj, S., Modulation of thermal resistence of ascospores of *Neosartorya fischeri* by acidulants and preservatives in mango and grape juice, *Food Microbiol.*, 17, 269, 2000.

Ramesh, M. N. et al., Microwave blanching of vegetables, *J. Food Sci.*, 67, 390, 2002.

Robinson, D.S., Peroxidases and catalases in foods, in *Oxidative Enzymes in Food*, Robinson, D.S. and Eskin, N.A.M, Eds., Elsevier Applied Science, London, 1991, p. 1.

Rodgers, S., Preserving non-fermented refrigerated foods with microbial cultures: a review, *Trends Food Sci. Technol.*, 12, 276, 2001.

Ruan, R., Ye., X. et al., Ohmic heating, in *Thermal Technologies in Food Processing*, Richardson, P., Ed., Woodhead Publishing, Cambridge, UK, 2001, p. 241.

Rudell, D.R. et al., Investigations of aroma volatile biosynthesis under anoxic conditions and in different tissues of 'Redchief Delicious' apple fruit (*Malus domestica* Borkh.), *J. Agric. Food Chem.*, 50, 2627, 2002.

Salunkhe, D.K., Wu, M.T., and Jadhav, S.J., Effect of light and temperature on the formation of solanine in potato slices, *J. Food Sci.*, 37, 969, 1974.

Sapers, G.M. and Miller, R.L., Browning inhibition in fresh-cut pears, *J. Food Sci.*, 63, 342, 1998.

Sapers, G.M. and Simmons, G. F., Hydrogen peroxide disinfection of minimally proceed fruits and vegetables, *Food Technol.*, 52, 48, 1998.

Schellekens, M., Sous vide cooking: state of the art, in *Proceedings of Second European Symposium on Sous Vide*, Leuven, Belgium, 1996, Chap. 2.

Schulzova, V., Peroutka, R., and Hajslova, J., Levels of furanocoumarins in vegetables from organic and conventional farming, *P. J. Food Nutr. Sci.*, 11, 25, 2002.

Schwarts, S.J., Woo, S.L., and Elbe, J.H., High-performance liquid chromatography of chlorophylls and their derivatives in fresh and processed spinach, *J. Agric. Food. Chem.*, 29, 533, 1981.

Selman, J.D., The blanching process, in *Developments in Food Preservation*, Thorne, S., Ed., Vol. 4, Elsevier Applied Science, London, 1986, p. 205.

Selman, J.D., Blanching, in *Encyclopedia of Food Science and Technology*, 2nd ed., Francis, F.J., Ed., John Wiley & Sons, New York, 2000, p. 190.

Selman, J.D., Rice, P., and Abdul-Rezzak, R.K., A study of apparent diffusion coefficient for solute losses from carrot tissue during blanching in water, *J. Food Technol.*, 18, 427, 1983.

Sheu, S.C. and Chen, O., Lipoxygenase as blanching index for frozen vegetable soybeans, *J. Food Sci.*, 56, 448, 1999.

Shewfelt, R.L., Postharvest treatment for extending the shelf life of fruits and vegetables, *Food Technol.*, 40, 70, 1986.

Silva, F. M. et al., Thermal inactivation of *Alicyclobacillus acidoterrestris* spores under different temperature, soluble solids and pH conditions for the design of fruit processes, *Int. J. Food Microbiol.*, 51, 95, 1999.

Silva, F.V.M. and Gibbs, P., *Alicyclobacillus acidoterrestris* spores in fruit and design of pasteurization processes, *Trends Food Sci. Technol.*, 12, 68, 2001.

Simon, R. and Stevenson, D., Adverse reaction to sulfites, in *Allergy: Principles and Practice*, Middleton A. et al., Eds., 3rd ed., Vol. II, Mosby, St. Louis, 1988, Chap. 67.

Skytta, E. et al., Growth risk of *Listearia Listeria monocytiogenbes* in minimally processed vegetables, in *Proceedings of Food 2000 Conference on Integrating Processing, Packaging and Consumer Research,* Natick, M.A., Ed., Science and Technology Corporation, Vol. II, 1996, p. 785.

Soliva, R.C. and Martin-Belloso, O., New advances in extending the shelf-life of fresh-cut fruits: a review, *Trends Food Sci. Technol.*, 14, 341, 2003.

Solomos, T., Principles underlying modified atmosphere packaging, in *Minimally Processed Refrigerated Fruits and Vegetables,* Wiley, R.C., Eds., Chapman and Hall, New York, 1997, p. 183.

Son, S.M., Moon, K.D., and Lee, C.Y., Inhibitory effects of various antibrowning agents on apple slices, *Food Chem.*, 73, 23, 2001.

Spiazzi, E. A., Raggio, Z. I., Bignone, K. A., and Mascheroni, R. H., Experiments on dehydrofreezing of fruits and vegetables: mass transfer and quality factors, in *Advances in the Refrigeration Systems, Food Technologies and Cold Chain*, IIF/IIR, 6, 1998, p. 401.

Splittstoesser, D.F., Churey, J.J., and Lee, C.Y., Growth characteristics of aciduric sporeforming bacilli isolated from fruit juices, *J. Food Protect.*, 57, 1080, 1994.

Suslow, T., Performance of zeolite based products in ethylene removal, *Perishables Handling Q.,* 92, 32, 1997.

Torreggiani, D. and Bertolo, G., Osmotic pre-treatments in fruit processing: chemical, physical and structural effects. *J. Food Eng.* 49, 247, 2001.

Varoquaux, P. and Wiley, R.C., Biological and biochemical changes in minimally processed refrigerated fruits and vegetables, in *Minimally Processed Refrigerated Fruits and Vegetables,* Wiley, R.C., Eds., Chapman and Hall, New York, 1997, p. 226.

Varoquaux, P., Lecendre, I., and Varoquaux, F., Changes in firmness of kiwifruit after slicing, *Sci. Aliments*, 10, 127, 1990.

Vega-Mercado, H. et al., Non-thermal preservation of liquid foods using pulsed electric fields, in *Handbook of Food Preservation*, Rahman, M.S., Ed., Marcel Dekker, New York, 1999, p. 487.

Vega-Mercado, H., Góngora-Nieto, M.M., and Barbosa-Cánovas, G.V., Advances in dehydration of foods, *J. Food Eng.*, 49, 271, 2001.

Velíšek, J., *Chemie potravin*, 1st ed., Ossis, Tábor, 1999, Vol. 2, Chap. 8.

Vermeiren, L. et al., Developments in the active packaging of foods, *Trends Food Sci. Technol.*, 10, 77, 1999.

Voldřich, M. and Kyzlink, V., Cyanogenesis in canned stone fruits, *J. Food Sci.*, 57, 161, 1992.

Wang, C.Y., Postharvest techniques for reducing low temperature injury in chilling-sensitive commodities, in *Improving Postharvest Technologies of Fruits, Vegetables and Ornamentals*, Artes, F., Gil, M.I., and Conesa, M.A., Eds., International Institute of Refrigeration, Paris, 2000, p. 467.

Wesche-Ebeling, P.A. et al., Preservation factors and processing effects on anthocyanin pigments in plums, *Food Chem.,* 57, 399, 1996.

Whitaker, J.R., Lipoxygenases, in *Oxidative Enzymes in Foods*, Robinson, D.S. and Eskin, N.A.M., Eds., Elsevier Aplied Science, London, 1991, p. 175.

Williams, D.C. et al., Blanching of vegetables for freezing: which indicator enzyme to choose, *Food Technol.*, 1986, 130, 1986.

Winniczuk, P. P. and Parish M. E., Minimum inhibitory concentrations of antimicrobials against micro-organisms related to citrus juice, *Food Microbiol.*, 14, 373, 1997.

Zagory, D., Ethylene removing packaging, in *Active Food Packaging*, Rooney, M.L., Ed., Blackie Academic and Professional, London, 1995, p. 38.

Zawistovski, J., Biliarderis, C.G., and Eskin, N.A.M., Polyphenol oxidases, in *Oxidative Enzymes in Food*, Robinson, D.S. and Eskin, N.A.M., Eds., Elsevier Applied Science, London, 1991, p. 217.

Zhang, S. and Farber, J.M., The effect of various disinfectants against *Listeria monocytogenes* on fresh-cut vegetables, *Food Microbiol.*, 13, 311, 1996.

Zhuang, R.Y. and Beuchat, L.R., Effectiveness of trisodium phosphate for killing *Salmonella montevideo* on tomatoes, *Lett. Appl. Microbiol.,* 22, 97, 1996.

Zomorodi, B., The technology of processed/prepacked product produce preparing the product for modified atmosphere packaging (MAP), in *Proceedings of the 5th International Conference on Controlled/Modified Atmosphere/Vacuum Packaging*, San Jose, CA, Scotland Business Research, Princeton, CA, 1990, p. 301.

11 Role of Pesticides in Produce Production, Preservation, Quality, and Safety

Kateřina Maštovská

U.S. Department of Agriculture, Agricultural Research Service, Eastern Regional Research Center, Wyndmoor, PA

CONTENTS

0-8493-1902-1/05/$0.00+$1.50
© 2005 by CRC Press

11.1 INTRODUCTION

Produce, and crop plants in general, can be attacked by numerous pests, mainly insects, nematodes, and fungi. Moreover, various weeds may compete with the crop plants for moisture, nutrients, and light. Ever since the dawn of agriculture, human-kind has constantly struggled with pests and weeds to increase crop yields and provide an adequate supply of food. The idea of combating them using chemicals can be traced back to ancient Greece and Rome, where burning sulfur helped to avert diseases and insects. Other examples can be found throughout history [1,2]. However, it was not until the 20th century when the era of mass-produced and applied synthetic organic pesticides began and chemical pest control became an integral part of modern agriculture that we were able to sustain a rapidly growing world population with land capacity to spare.

Nevertheless, every coin has two sides, and the use of pesticides is also associated with certain risks. Real and perceived concerns about the impact of pesticides on human health and the environment have led to strict regulation of their application and pesticide residue levels in food and water supplies. These concerns are also reflected in the development of new pesticides and more effective ways of applying them with the aim of lowering their risks and tipping the balance in favor of their benefits. At the same time, scientists seek alternative approaches to chemical pest control, but their potential risks also need to be properly evaluated.

The main part of this chapter discusses the positive role of different groups of pesticides in produce production, preservation, quality, and safety. To make the picture complete, potential risks associated with pesticide use and strategies to reduce them are also briefly discussed at the end of the chapter.

11.2 POSITIVE IMPACTS OF PESTICIDE USE IN PRODUCE PRODUCTION

The scope of the term *pesticide* is very well expressed in this internationally adopted definition by the Food and Agriculture Organization (FAO) of the United Nations:

> Pesticide means any substance or mixture of substances intended for preventing, destroying, attracting, repelling or controlling any pest including unwanted species of plants or animals during the production, storage, transport, distribution and processing of food, agricultural commodities, or animal feeds or which may be administered to animals for the control of ectoparasites. The term includes substances intended for use as a plant growth regulator, defoliant, desiccant, fruit thinning agent, or sprouting inhibitor and substances applied to crops either before or after transport. The term normally excludes fertilizers, plant and animals nutrients, food additives and animal drugs.

Table 11.1 lists significant pesticide groups together with the pests they combat and/or problems they address in produce production. There are numerous sources of information on the properties, biological effects, modes of action, applications, toxicology, regulation, and monitoring of pesticides in the scientific literature, and recently the Internet has become a valuable and easily accessible tool. For instance, the Web site of the U.S. Department of Agriculture (USDA) National Agricultural

TABLE 11.1
Significant Pesticide Groups Together with Target Pests and/or Addressed Problems

Pesticide Group	Pest/Problem
Insecticides	Insects
Acaricides	Acari (mites and ticks)
Nematicides	Nematodes (eelworms)
Molluscicides	Mollusks (land slugs and snails)
Fungicides	Fungi
Bactericides	Bacteria
Herbicides	Weeds
Rodenticides	Vertebrate pests, mainly rodents (e.g., rats, mice, rabbits)
Avicides	Birds
Fumigants	Various pests in soil and enclosed spaces (e.g., greenhouses, food stores, warehouses)
Plant growth regulators	Plant growth control (e.g., growth stimulation, control of flowering, setting of fruit, leaves falling)
Fruit thinning agents	Excessive fruits
Sprouting inhibitors	Postharvest sprouting of root, tuber, and bulb crops
Defoliants	Unwanted leaves
Desiccants	Unwanted moisture (e.g., preharvest drying of potato vines)

Statistic Service (NASS) provides reports concerning agricultural chemical usage on fruits, vegetables, and other crops [3], and the National Scientific Foundation (NSF) Center for Integrated Pest Management (IPM) developed (in cooperation with the USDA) crop profiles, also accessible through the Internet, that include information on cultural practices, crop pests and diseases, and their chemical control [4]. In print, *The Pesticide Manual* [5] and the *Crop Protection Handbook* (formerly the *Farm Chemicals Handbook*) [6] serve as essential reference books on pesticide chemistry. These and other sources of information were used to compile the following sections, which discuss in detail positive effects of the application of the pesticide groups listed in Table 11.1 on produce yields, preservation, quality, and safety. Furthermore, these sections also describe basic traits of biological effects and modes of action of the discussed pesticides and provide important examples of pesticide/pest/produce combinations.

However, before going into details, let us focus on the common benefits resulting from pesticide use. Basically, the major benefits include increased produce yields and reduced spoilage during the storage, transport, distribution, and processing of produce. Increased yields and decreased postharvest spoilage translate not only into economic but also into health and environmental benefits [7]. The use of pesticides results in year-round availability and general affordability of fruits and vegetables, which are excellent sources of important vitamins, minerals, and fiber. Their increased consumption improves consumers' health, including a reduced risk of cancer [8,9]. Of course, the question remains whether pesticides potentially increase health risks more than does a lower consumption of fruits and vegetables, but let us save this discussion for later. In terms of environmental benefits, pesticides allow food to be produced on less land, and thus more land can be devoted to wildlife habitat, environmentally fragile lands can idle, and less water is needed for irrigation.

Increased domestic production results not only in lower produce prices, and consequently higher consumption, but also in other economic benefits, such as increased export and the ability to spend more money on items other than food. Thus, without practical, effective, and economically viable alternatives, the elimination of pesticide use would probably result in serious problems. For instance, an economic study from 1995 [10] estimates that complete elimination of pesticides used in the U.S. on fruits and vegetables would increase wholesale prices by 45% (retail prices by 27%) based on a 75% increase in unit production costs. The U.S. production would be reduced by 16%, exports would decrease by 27%, and imports would increase by 7%. Consequently, produce consumption would decrease by 11% and acreage required for production of fruits and vegetables in the U.S. would increase by about 44%.

Another important aspect associated with pesticide use is improved produce quality, yielding larger products with better nutritious quality (e.g., due to a lower competition from weeds), and also fruits and vegetables free from insects, surface blemishes, scars, punctures, or other damage caused by pests. Apart from the less-attractive appearance of damaged products, surface (cuticle) damages open the door to further degradative changes, moisture loss, and also bacterial and fungal colonization, resulting in microbial spoilage of produce and a potential health risk to consumers due to the possible presence of pathogenic microorganisms and toxins they may produce. Moreover, many pests, such as insects, nematodes, and mollusks, serve as vectors for plant and food-borne diseases. Thus, pesticides play an important role in produce preservation and safety by preventing or reducing primary and secondary microbial contamination. Other chapters in this book describe further degradative changes associated with mechanical injuries of produce, mechanisms of microbial spoilage, and attachment of human pathogens to fruits and vegetables. The main focus of this chapter relates to the role of pesticides in preventing these adverse consequences of pest attacks and in addressing other problems involved with produce production.

With respect to the positive effect of pesticides on food safety, it has also been suggested that pesticides may reduce the levels of naturally occurring fungal and plant toxins that may actually pose greater cancer risk than synthetic pesticides [11,12]. In the case of mycotoxins, the link between pesticide use and a decrease in fungal toxin production is relatively clear. Fungicides can control toxin-producing fungi, such as *Aspergillus flavus* or *Aspergillus parasiticus* in the case of aflatoxins, thus reducing mycotoxin levels [13,14]. Also, the application of insecticides or nematicides may indirectly contribute to decreased mycotoxin production by reducing crop damage that predisposes plant tissues to fungal invasion and/or through control of insects that spread the fungal infestations [15,16].

Natural plant toxins, also called phytoalexins (*phyton* = plant, *alexin* = defend) are considered a means of natural defense against pests. Notable phytoalexins include the glycoalkaloids produced by potatoes (solanine and chaconine) and tomatoes (tomatine) that are inhibitors of acetylcholinesterase [17], thus presumably providing insect resistance but also posing risk to humans. Furanocoumarines (e.g., psoralen, bergapten, and xanthotoxin) represent another important group of natural toxins and probable human carcinogens produced mainly by umbelliferous plants, such as celery or parsley. These phytoalexins are often present in plant tissues at relatively high levels (10 to 1,000 mg/kg) compared to synthetic pesticide residues [18]. Various stress factors (stimuli) have been shown to induce phytoalexin synthesis, such as mechanical injury, amount of light, attack by pests, or pathogen infection [19,20]. Thus, by reducing plant stress from insect and other pest attack, weed competition, and/or plant pathogen invasion, pesticides may decrease phytoalexin production. More research studies should be done in this area to better evaluate this premise and also to eliminate the possibility that pesticide application is one of the potential stress factors [21].

11.2.1 INSECTICIDES

Of more than 800,000 of the existing insect species, about 10,000 species attack crops and, when uncontrolled, can cause extensive damage and economic losses [2]. As the word implies, insecticides are designed to kill insect pests. Other chemicals used in insect control include ovicides (e.g., petroleum oil), which are intended for killing eggs of insects and mites; insect growth regulators (for disrupting molting and other life processes such as chitin synthesis); insecticide synergists (such as piperonyl butoxide for blocking insects' natural detoxification system); pheromones (for disrupting insect mating behavior); and attractants (for luring insects to a trap) and repellents.

11.2.1.1 Major Classes of Insecticides and Their Modes of Action

Table 11.2 presents major chemical classes of insecticides along with their modes of action and representative examples. Organochlorine and pyrethroid insecticides are nonsystemic (contact pesticides), which means that they do not appreciably penetrate plant tissues and therefore are not translocated within the plant. Neonicotinoids are systemic pesticides, and some of the organophosphates and carbamates possess systemic and some nonsystemic properties [5].

The mode of action of most insecticides involves the stimulation or inhibition of the transmission of nervous impulses [1,5]. The organophosphorus and carbamate insecticides inhibit the enzyme acetylcholinesterase by its phosphorylation or carbamoylation, respectively. Acetylcholinesterase catalyzes the hydrolysis of the major excitatory neurotransmitter acetylcholine to its inactive form choline, which ceases the stimulation of the receptors. Thus, inactivation of this enzyme leads to accumulation of acetylcholine at the nerve synapse and continuous transmission of nerve impulses. Nicotine and neonicotinoids can mimic acetylcholine and act as its antagonists at the postsynaptic nicotinic acetylcholine receptors, causing overstimulation, which leads to the insect's twitching, convulsing, and, finally, dying. Pyrethroids interfere with the operation of sodium channels that are involved in the propagation of action potentials along nerves [22]. Most of the organochlorine insecticides act

TABLE 11.2
Major Chemical Classes of Insecticides

Chemical Class	Mode of Action	Representatives
Carbamate	Cholinesterase inhibitors	Aldicarb, carbaryl, carbofuran, fenoxycarb, formetanate, methiocarb, methomyl, oxamyl, pirimicarb, thiodicarb
Organochlorine	Antagonists of the GABA receptor-chloride channel complex	Chlordane, dieldrin, endosulfan, endrin, heptachlor, lindane, methoxychlor
Organophosphorus	Cholinesterase inhibitors	Acephate, azinphos-methyl, chlorpyrifos, diazinon, dimethoate, disulfoton, malathion, methidathion, naled, phorate, phosmet
Neonicotinoid	Antagonists of the postsynaptic acetylcholine receptors	Acetamipirid, imidacloprid, thiamethoxam, clothianidin, dinotefuran, nitenpyram
Pyrethroid	Sodium channel modulators	Bifenthrin, cyfluthrin, cyhalothrin-lambda, cypermethrin, deltamethrin, esfenvalerate, fenpropanthrin, permethrin

Source: Tomlin, C.D.S., Ed., *The Pesticide Manual*, 12th ed., British Crop Protection Council, Surrey, UK, 2000.

as antagonists of the inhibitory neurotransmitter gamma-aminobutyric acid (GABA), thus causing hyperexcitation of the nervous system. GABA binds to the chloride channel receptor, increasing the chloride ion permeability of neutrons, which has a dampening effect on nerve impulse firing.

The environmental hazard associated with the use of certain persistent, bioaccumulative organochlorine pesticides, such as DDT (dichlorodiphenyltrichloroethane), aldrin, or dieldrin, led to the cancellation of these pesticides for crop application (DDT is still used as a mosquito vector control for the control of malaria in some tropical countries). Some more biodegradable organochlorines (endosulfan and methoxychlor) are still employed; however, there is a growing trend toward low-persistent pesticides that are either naturally occurring compounds with insecticial properties, such as abamectin or spinosad, or synthetic products derived from naturally occurring insecticides (pyrethroids and neonicotinoids) [23,24].

Abamectin and spinosad are macrocyclic lactones isolated from soil microorganisms *Streptomycetes avermitilis* and *Saccharopolyspora spinosa,* respectively. Abamectin (avermectin) stimulates the release of GABA [25], thus causing paralysis, whereas spinosad activates the nicotinic acetylcholine receptor, but at a different site than neonicotinoids [5]. Subspecies of *Bacillus thuringiensis* represent another group of relatively popular biopesticides (bacterial insecticides) [26]. At sporulation, this Gram-positive bacterium forms crystals of protein delta-endotoxins, which cause damage to the epithelial cells of the insect gut; insects stop feeding and eventually starve to death [27].

11.2.1.2 Problems Caused by Insects in Produce Production and Their Control by Insecticides

The types of potential insect pests and the degree of damage depend on many factors, such as the climate, agricultural region, crop species and their pest resistance, and natural predators [2]. Table 11.3 and Table 11.4 give examples of common insect pests on fruits and vegetables in the U.S. [4]. These tables also present some of the major problems associated with insect infestation in produce production and provide lists of insecticides that are typically used for their control.

Insects, mainly their larvae, may feed on various plant parts, such as leaves, tender twigs, stalks, buds, flowers, fruits, or roots, which often leads to severe consequences if not controlled. The loss of leaves or their damage generally results in reduced photosynthetic capacity of the plant. Thus, for most produce types, foliage loss reduces product (fruit, root, bulb, tuber, etc.) size and quality by retarding sugar development and, ultimately, leads to the plant's death if the damage is extensive. In the case of leafy and many of the cruciferous (cole) vegetables, it certainly means crop losses, significantly reduced quality, and also increased susceptibility to microbial spoilage [28,29]. Leaf-eating and damaging insects, such as leafrollers on apple trees [30] or leafhoppers on grapevines [31], may also cause lower fruit set in the next season.

Fruit trees infested with insects feeding on tender twigs, shoots, or branches, such as with the oriental fruit moth or San Jose scale [32], generally show a decrease in vigor, growth, and productivity. Similarly, most vegetables may suffer from an attack of sap-sucking insects (mainly aphids) that also lowers plant productivity and retards growth, leading to stunted plants with deformed stalks and curled leaves [28].

Insect-feeding on buds, flowers, and developing fruits translates directly into decreased fruit yields or their dramatically reduced quality. For instance, the feeding injury caused by lygus bugs (including the tarnished plant bug) on developing strawberries results in unmarketable, small, and deformed fruits that have either a characteristic "cat-faced" appearance (due to the cessation of development in the area surrounding the feeding site) or a woody texture that fails to mature ("button berry") [33].

Feeding on fruits leads to surface damage or more serious deep tunneling, but in any event, an economic loss results because the affected fruit is usually discarded during postharvest sorting. Moreover, the insect damage often provides an entry point for microbial spoilage. Similarly, larval feeding on roots renders root vegetables (e.g., carrots, parsley, or celery in the case of carrot weevil larvae infestation [34]) unacceptable for both fresh and processing markets and generally makes the roots more susceptible to diseases caused by soil-borne pathogenic microorganisms.

Fruit surface damages may appear as stings or scars, such as those caused by the codling moth or leafrollers on apples, respectively [30]. On citrus fruits, citrus thrips puncture epidermal cells, leaving scabby, grayish, or silvery scars on the rind. Second instar larvae are the most damaging stage of the citrus thrips because they feed mainly under the sepals of young fruit and are larger than first instars (larvae between hatching and the first molt). As the affected fruit grows, damaged rind tissue moves outward from beneath the sepals and appears as a conspicuous ring of scarred tissue [35].

TABLE 11.3
Examples of Fruit Insect Pests in the U.S., Major Problems They Cause, and Insecticides Used for Their Control

Commodity	Insects	Problems	Insecticides
Apples	Codling moth *Cydia pomonella*	Surface damage ("stings") of fruits and deep inner tunneling of larvae ("wormy apple") resulting in internal breakdown and possible fruit abortion	Azinphos-methyl, chlorpyrifos, esfenvalerate, permethrin, phosmet
	Leafrollers *Pandemis pyrusana, Choristoneura rosaceana*	Larvae feeding on leaves, buds, and mainly fruits, resulting in irregular, shallow scars on fruit surface	Chlorpyrifos, carbaryl, spinosad, *Bacillus thuringiensis*
Bananas	Banana aphid *Pentalonia nigronevosa*	A vector of bunchy top virus causing no fruit production and stunted leaves	Diazinon, malathion
Citrus fruits[a]	Citrus thrips *Scirtothrips citri*	Feeding on fruits leaving scabby, grayish or silvery scars or conspicuous rings of scarred tissue on the rind	Abamectin, cyfluthrin, dimethoate, fenpropanthrin, formetanate
Cherries	Cherry fruit flies *Rhagoletis cingulata Rhagoletis fausta*	Larvae feeding on fruits resulting in maggot-infested, often shrunken and deformed fruits that ripen earlier than surrounding fruits	Azinphos-methyl, carbaryl, permethrin, phosmet
Grapes	Leafhoppers *Erythroneura elegantula, Erythroneura variabilis*	Feeding on leaves and spotting of fruits with honeydew, which serves as a growth medium for fungi	Imidacloprid
	Grape berry moth *Endopiza viteana*	Larvae feeding on berries resulting in their spoilage; damage of leaves before pupating	Azinphos-methyl, carbaryl, chlorpyrifos, diazinon, endosulfan, malathion, methoxychlor, phosmet
Peaches, nectarines	Oriental fruit moth *Grapholita molesta*	Larvae feeding on tender twigs and fruits resulting in wormy fruit and rapid breakdown	Azinphos-methyl, diazinon, phosmet, esfenvalerate, permethrin, methomyl
Pears	Pear psylla *Cacopsylla pyricola*	Honeydew killing the leaves and serving as a growth medium for fungi; a vector of mycoplasma causing pear decline	Abamectin, amitraz, esfenvalerate, permethrin
Plums, prunes	San Jose scale *Quadraspidiotus perniciosus*	Feeding on branches, shoots, leaves, and fruits, potentially causing permanent injury and death to mature trees	Carbaryl, chlorpyrifos, diazinon, methidathion, phosmet

TABLE 11.3
**Examples of Fruit Insect Pests in the U.S., Major Problems They Cause,
and Insecticides Used for Their Control (continued)**

Commodity	Insects	Problems	Insecticides
Strawberries	Lygus/tarnished plant bug *Lygus lineolaris* *Lygus hesperus*	Feeding on flowers and developing fruits resulting in fruit deformations and reduced size	Azinphos-methyl, bifenthrin, endosulfan, fenpropathrin, malathion, methomyl, naled

[a] Grapefruits, lemons, limes, oranges, and tangerines

Source: National Scientific Foundation Center for Integrated Pest Management and USDA Crop Profiles Web site, available at http://pestdata.ncsu.edu/cropprofiles/.

Deep inner tunneling from larvae in the fruit is responsible for the proverbial "wormy fruit," contamination with larval feces, rapid internal degradation, and possible abortion of the fruit. Infestation of apples, pears, peaches, or nectarines by larvae of the codling or oriental fruit moth [30,36] serves as a typical example. In cherries, larvae of cherry fruit flies may also separate the pit from the pulp, resulting in shrunken and deformed fruits that ripen earlier than the uninfested cherries [37]. Moreover, brown rot (*Monilinia* spp.) usually starts in wormy fruits and then spreads to other cherries. Similarly, wounds caused by the grape berry moth on damaged grape berries provide entry ports for numerous fungal pathogens, such as powdery mildew, downy mildew, or grapevine sour rot, which accelerate grape spoilage [38].

Another source of fungal contamination is the honeydew (excess carbohydrates) excreted by certain insects, which serves as an excellent growth medium for black sooty mold fungus. Honeydew drips on fruits and foliage, making spots and causing their discoloration; this is a problem associated mainly with aphid infestation on both fruits and vegetables [39]. Other important insect pests producing honeydew include leafhoppers (problematic, for example, on grapes) or pear psylla. Under bright sunlight and dry conditions, the honeydew produced by pear psylla can kill the leaf tissue and produce a symptom called "psylla scorch" [40]. A more serious problem related to pear psylla is its ability to vector a mycoplasma that causes the disease "pear decline," which reduces tree vigor and causes poor fruit set and small fruit size and may lead to the death of the tree (a similar effect called "psylla shock" may also induce excessive contamination of the tree tissue by pear psylla toxin) [40].

Transmission of plant diseases represents an additional problem associated with insects and certain other pests that actually may be even worse than the insect-feeding itself. Among these, aphids are the most important group of viral vectors [41]. For instance, the banana aphid transmits banana bunchy top virus, resulting in no fruit production and stunted leaves ("bunchy top") [42], and numerous other aphids, such as green peach, pea, or bean aphids, are vectors for other viruses (lettuce mosaic virus, pea streak virus, etc.) that seriously affect many plants [43]. Important

TABLE 11.4
Examples of Vegetable Insect Pests in the U.S., Major Problems They Cause, and Insecticides Used for Their Control

Commodity	Insects	Problems	Insecticides
Beans, peas	Bean and pea aphids *Aphis fabae* *Acyrthosiphon pisum*	Sucking the plant sap leading to curled leaves and stunted plants; excreting honeydew (a fungi growth medium); vectors of plant virus diseases	Acephate, disulfoton, dimethoate, endosulfan, esfenvalerate, malathion, methomyl
Carrots, celery, parsley	Carrot weevil *Listronotus oregonensis*	Larvae feeding on roots and stalks causing damage	Cyfluthrin, esfenvalerate, imidacloprid, oxamyl, parathion-methyl, permethrin
Cole vegetables[a], spinach	Cabbage looper *Trichoplusia ni*; diamondback moth *Plutella xylostella*	Feeding on leaves and depositing fecal matter; damaging and contaminating broccoli and cauliflower heads	Cyhalothrin-lambda, cypermethrin, endosulfan, esfenvalerate, permethrin, spinosad, thiodicarb
Cucurbits[b]	Cucumber beetles *Acalymma vittata* *Diabrotica undecimpunctata howardi*	Feeding on young seedlings, stems, foliage and soft fruits, larvae feeding on roots; vectors of bacterial wilt disease and mosaic virus	Carbaryl, carbofuran, endosulfan, esfenvalerate, malathion, permethrin
lettuce	Aphids *Myzus persicae*, *Nasonovia ribis-nigri*, *Macrosiphum euphorbiae*	Sucking the plant sap leading to stunted plants; excreting honeydew; vectors of plant virus diseases (lettuce mosaic virus)	Acephate, diazinon, dimethoate, endosulfan, imidacloprid, oxydemeton-methyl
Onions, garlic	Onion and western flower thrips *Thrips tabaci* *Frankliniella occidentalis*	Sucking the plant sap causing stunted and deformed plants, leading to decreased yields and quality	Cyhalothrin-lambda, cypermethrin, diazinon, malathion, methomyl, permethrin,
Peppers, sweet corn	European corn borer *Ostriania nubilalis*	Feeding on leaves; larvae tunneling into pepper fruit or corn stalks and ear shanks and feeding on kernels in the ear	Bifenthrin, carbaryl, cyhalothrin-lambda, cyfluthrin, esfenvalerate, methomyl, permethrin
Potatoes, tomatoes, eggplants	Colorado potato beetle *Leptinotarsa decemlineata*	Feeding on leaves leading to decreased yields and quality and ultimately to the plant's death	Aldicarb, carbaryl, endosulfan, imidacloprid, permethrin, phorate, *Bacillus thuringiensis*

TABLE 11.4
Examples of Vegetable Insect Pests in the U.S., Major Problems They Cause, and Insecticides Used for Their Control (continued)

Commodity	Insects	Problems	Insecticides
	Green peach and potato aphids *Myzus persicae, Macrosiphum euphoribae*	Sucking the plant sap causing the leaves to curl; vectors of virus plant diseases (mosaic virus)	Cyfluthrin, cyhalothrin-lambda, diazinon, dimethoate, esfenvalerate, imidacloprid, phorate

[a] *Brassicaceae* including, e.g., broccoli, brussels sprouts, cabbage, Chinese cabbage, cauliflower, kale, kohlrabi.

[b] Cucumbers, zucchini, squash, gourds, pumpkins, watermelons, etc.

Source: National Scientific Foundation Center for Integrated Pest Management and USDA Crop Profiles Web site, available at http://pestdata.ncsu.edu/cropprofiles/.

examples of bacterial vectors include striped and spotted cucumber beetles that transmit *Erwinia tracheiphila* to cucumbers and other cucurbits causing bacterial wilt disease leading to plant death [44]. Basically, it is very difficult and mostly impossible to cure plant diseases and save the affected plants. Thus, insecticides play an important role in plant disease management by killing the disease vectors [41]. Moreover, the use of insecticides is still the most effective tool in combating insects transmitting grave diseases to humans, such as mosquitoes *Anopheles* spp., which serve as the vectors of malaria.

11.2.2 ACARICIDES

Acaricides (also called miticides) are designed to kill acari, which include ticks, and, more important for our discussion, mites. The differences in the biochemistry of acari and insects enable application of compounds highly toxic to mites and relatively innocuous to insects, although some insecticides (e.g., most organophosphates) also possess acaricidal properties [1]. In addition to interference with nervous system functions, the modes of action of acaricides involve, for instance, inhibition of oxidative phosphorylation (disruption of adenosine triphosphate, ATP, formation) or electron transport in mitochondria. Some acaricides are actually ovicides or mite growth regulators and, in several cases, the mode of action is not known [1,5]. Important acaricides include clofentezine, fenbutatin oxide, hexythiazox, propargite, and the bridged diphenyl derivates (closely related to DDT) dicofol and bromopropylate [4].

Mites are very unpleasant pests on many crops and usually become a serious problem if their natural predators are destroyed from insecticide applications, allowing mite populations to flare. Mites can pierce individual leaf cells with their mouthparts and ingest the cell contents, including the chlorophyll [45]. This results in mottled, off-color foliage, which may later appear gray or bronzed and often drop prematurely. The loss of chlorophyll means reduced photosynthetic capacity of the plant, which may lead to the same adverse consequences as infestation by leaf-eating or damaging insects.

In the U.S., the important examples of mite pests on fruit trees and other crops include the two-spotted spider mite (*Tetranychus urticae*), the European red mite (*Panonychus ulmi*), the Pacific spider mite (*T. pacificus*), and the McDaniel spider mite (*T. mcdanieli*) [4]. In the case of citrus fruits, the citrus rust mite *(Phyllocoptruta oleivora)* and some other mites may feed not only on foliage but also on stems and fruits, causing epidermal cells to die [46]. When rust mite injury occurs prior to fruit maturity, the destruction of epidermal cells will generally result in smaller fruit. Further growth leads to breaking up of the dead epidermis and a wound periderm forms over the newly formed epidermis ("russeting"). Rust mite injury to mature fruit causes epidermal cells to die and become a brownish-black color with no periderm formation. This condition is known as "bronzing." Primary effects of fruit damage caused by rust and some other epidermis-destroying mites are a reduction in citrus fruit grade and size, increased water loss, fruit drop, and reduced juice quality.

11.2.3 NEMATICIDES

Nematicides are pesticides intended for killing nematodes, which are microscopic, unsegmented eelworms, most of which live in the soil and feed on plant roots. From about 3000 known nematodes, almost one-third commonly attack crop plants [2]. Nematodes may feed on plant tissues from the inside (endoparasitic nematodes) or from the outside (ectoparasitic nematodes), and their feeding activity may interfere with water and nutrient uptake or stimulate physiological changes that disrupt healthy plant functions, leading to reduced plant vigor (decrease in growth and productivity) or even death if the infestation is enormous. Nematode-feeding on roots, tubers, and bulbs intended for consumption (such is the case with carrots, potatoes, or onions) translates directly to their lower quality and susceptibility to microbial spoilage. Furthermore, nematodes may serve as vectors of serious plant diseases.

Nematodes can overwinter in a dormant state in the soil. They are not able to travel through the soil to a large extent but are rapidly spread by running water and contaminated equipment, seeds, transplants, sets, and bulbs. Thus, the practices for effective nematode prevention mainly include the use of nematode-free seeds, uncontaminated water for irrigation, and nematicides prior to planting to eliminate their spread by farm equipment. Preplanting treatment mainly involves soil fumigation, which is usually done to combat a combination of problems associated with nematodes, weed seeds, and soil-dwelling insects and fungi.

Fumigants are vapor-active chemicals, and thus the fumigation process basically requires an enclosed space; otherwise, the active ingredient will be lost to the atmosphere. For soil treatment, the surface of the soil must be covered with a plastic tarp [1]. The most widely applied soil fumigants effective as nematicides include metam-sodium, methyl bromide, chloropicrin, and 1,3-dichloropropene [4]. Metam-sodium is a very broad-spectrum, water-dispersible, crystalline compound that is generally applied through irrigation systems. In soil, it slowly decomposes to the highly toxic fumigating agent methyl isothiocyanate [5], which reacts with nucleophilic centers, such as thiol groups, in vital nematode or other pest enzymes [1].

Methyl bromide, chloropicrin, and 1,3-dichloropropene are volatile alkyl halides that also attack nucleophilic sites ($-OH$, $-SH$, or $-NH_2$ groups) in important enzymes. Due to the environmental concerns associated with the application of methyl bromide (depletion of ozone), this fumigant is currently scheduled to be banned in the U.S. by the year 2005, and its present use is under restriction [47].

Essentially, insects and nematodes have similar nervous systems, and thus many insecticides, such as organophosphates and carbamates, can also act as nematicides. Organophosphorus insecticides, though, tend to rapidly degrade in soil, and therefore only the systemic ones (present in the plant roots), such as dimethoate or phorate, are of some practical use [1]. An example of a specially developed and widely applied organophoshorus nematicide is fenamiphos, which is a systemic compound absorbed by roots and is sufficiently persistent in soil. From the carbamate group, aldicarb, oxamyl, and carbofuran are the three most frequently used systemic car-bamates for nematode control in the U.S. [4]. A recently introduced toxin produced by fermentation of the fungus *Myrothecium verrucaria* has been also used on several fruits and vegetables (e.g., grapes) to effectively control both juvenile and adult nematodes and also egg/cyst hatching [4,5].

In the U.S., the most widespread and economically important nematode species attacking all sorts of fruit and vegetable plants include the root-knot *(Meloidogyne spp.)*, root lesion (*Pratylenchus* spp.), dagger (*Xiphinema americanum*), stubby root (*Paratrucgidirys* spp. and *Trichodorus* spp.), ring (*Criconemella xenoplax),* and sting (*Belonolaimus longicaudatus*) nematodes [4].

Presumably the most damaging nematode infestation is that of potatoes [48]. Root-knot nematodes are the major nematode pests on potatoes; they are found in abundance, especially in sandy soils. Females feeding in the tubers and the development of live young cause enlargement or bumps in the outer layers of the tubers, rendering them useless for either fresh or processing markets. Specific symptoms caused by root-knot nematodes include swellings on the roots called "galls," which may contain one to several adult root-knot females [49]. These nematodes cause field damage that is localized, usually in patches of various sizes, or that may be spread throughout an entire field, and plants become chlorotic and stunted. Damaged roots are not able to obtain soil nutrients; above ground, symptoms appear as nitrogen or micronutrient deficiencies. Plants may wilt easily, especially in warm weather, due to root damage, even though soil moisture may be adequate.

Other nematode pests on potatoes are the ectoparasitic root lesion and stubby root nematodes [49]. The root lesion nematode reduces yield indirectly by weakening the plants and increasing their stress, which causes the potato plants to be more susceptible to fungal and bacterial diseases. Stubby root nematodes also do not cause much direct damage, but they transmit the tobacco rattle virus. This virus causes a disease of potato tubers called corky ringspot, resulting in rusty brown, irregularly shaped lesions that have a corky texture.

11.2.4 MOLLUSCICIDES

Molluscicides control snails and slugs, both members of the mollusk phylum and similar in structure and biology, except that snails have external spiral shells. The most common snail and slug pests include the brown garden snail *(Helix aspersa)*, the gray garden slug *(Agriolimax reticulatus)*, the banded slug *(Limax marginatusi)*, the tawny slug *(Limax flavus)*, and the greenhouse slug *(Milax gagates)* [50].

Snails and slugs feed on a variety of living plants and also on decaying plant material [50,51]. On plants, they can chew irregular holes in leaves and flowers and also feed on fruits and young plant bark. Slugs and snails prefer succulent foliage or flowers, and thus they are primarily pests of seedlings and leafy and cruciferous vegetables. However, they can also feed on ripening fruits in close proximity to the ground, such as strawberries, artichokes, or tomatoes, and also attack foliage and fruit of some trees (citrus fruits are especially susceptible to this kind of damage). Furthermore, slugs and snails may serve as vectors of plant and animal diseases.

The control of snails and slugs mainly involves prevention (mostly removal of decaying plant matter and their diurnal hiding places), hand-picking, trapping (under boards, inverted melon rinds, or in baits with beer or other fermented food), installing barriers (e.g., copper foil), and using baits with molluscicides designed to attract and kill snails and slugs in the target area [50,51].

Metaldehyde is the most common active ingredient in the baits containing a molluscicide [50,52]. Metaldehyde is toxic to slugs and snails both by ingestion and by absorption by the mollusk's foot. It does not kill snails and slugs directly unless they eat a substantial amount, but it stimulates the secretion of large quantities of slime, causing immobilization and eventual death by loss of water (dessication) [1,5]. A recently registered mollusk bait containing ferric phosphate is much safer to mammals and other domestic and wildlife animals than is metaldehyde. Ingestion of ferric phosphate bait causes snails and slugs to cease feeding and eventually starve to death [50]. Several carbamate insecticides, such as methiocarb, thiocarb, or trimethacarb are also very effective molluscicides when formulated as baits [1].

11.2.5 FUNGICIDES

Fungicides prevent or destroy the growth of fungi, thus protecting growing plants against fungal diseases and produce against pre- and postharvest spoilage caused by fungal pathogens. Chapter 16 describes in detail mechanisms of microbial spoilage of fruits and vegetables, including the most important produce categories and spoilage fungi combinations. Thus, this discussion is restricted to the use and mode of action of fungicides serving as preventive, curative, or eradicative means in the combat with fungi and provides a general description of problems caused by fungal pests in produce production.

11.2.5.1 Classification of Fungicides, Their Application, and Mode of Action

Fungicides are mainly applied as sprays or dusts, and their postharvest application also includes dipping in fungicide baths, impregnation of fruit wrappers, or incorporation into waxes for surface application on fruits [53]. Various seed dressing techniques can be used for seed treatment to control seed- and soil-borne diseases, and fumigants may also play an important role in preplanting control of soil fungi as well as postharvest treatment of certain produce types (e.g., fumigation of potatoes with 2-aminobutane or grapes with sulfur dioxide).

Based on their ability to penetrate the plant's cuticle and translocate within the plant, fungicides can be divided into two main groups: nonsystemic (surface) and systemic. Systemic fungicides are absorbed by the plant via roots, leaves, or seeds and are redistributed within the plant, whereas nonsystemic ones stay on the surface and thus can be removed by rainfall or evaporation and do not protect newly developed or unsprayed plant parts. Moreover, most pathogenic fungi penetrate the plant's cuticle and ramify through the plant's tissues; thus, nonsystemic fungicides are applied mainly as a preventive measure (a protectant) before the fungal spores reach the plant [1]. Protectants act against spore germination to early infection (penetration of host tissues), whereas curative fungicides combat fungi at the postinfection, presymptomatic stage. Eradicants are capable of stopping fungal colonization after the symptoms develop. Most systemic fungicides combine protective, curative, and eradicative properties.

Table 11.5 presents the classification of nonsystemic and systemic fungicides based on their modes of action and chemical structures, including the most important fungicide classes along with representative examples. Generally, nonsystemic fungicides prevent or inhibit spore germination and penetration of host tissue and usually have a multisite mode of action, which makes their use less susceptible to the buildup of resistant fungi (due to mutations) compared to systemic fungicides, which often have a specific single-site action [54].

An important group of nonsystemic fungicides, including alkylenebis(dithiocarbamates), phthalimides, or sulfamides, can react with various vital cellular thiols (e.g., enzymes with a thiol group). Chlorothalonil conjugates with thiols (particularly glutathione) from germinating fungal cells and causes their depletion, leading to disruption of glycolysis and energy production [5]. Dimethyldithiocarbamates owe their fungitoxicity mainly to their ability to chelate with copper ions, thus inhibiting enzymes containing copper [1]. Chlorophenyl fungicides may interfere with miscellaneous biosynthesis pathways including protein or phospholipid biosynthesis, leading to inhibition of germination and growth of fungal mycelium. Cationic surfactants, such as dodine, can alter the permeability of fungus cell walls, thus causing the loss of vital cellular components (e.g., amino acids) [1].

Many newer fungicides have systemic properties; however, most of them are only locally systemic, which means that, after the absorption by leaves or shoots, they are moved only short distances within the transpiration stream (generally toward the leaf margin) or between plant cells [55]. Demethylation inhibitors (DMIs) may serve as a typical example. The newer strobilurin fungicides, such as azoxystrobin or kresoxim-methyl, have a slightly different distribution, called translaminar (also meso- or quasi-systemic) [56]. These compounds move into and through the leaf but do not move in the transpiration stream. Some systemic fungicides, such as metalaxyl, can be absorbed by plant roots and translocated throughout the plant, although the translocation occurs only acropetally (upwardly), with the transpiration stream (transport in xylem driven by the evaporation of water from the leaf surface) [57]. Currently, fosetyl-aluminum is the only truly systemic fungicide because it is

TABLE 11.5
Classification of Fungicides Based on Their Mode of Action and Chemical Structure

Mode of Action	Chemical Class	Representatives
Nonsystemic Fungicides (mostly protectants)		
Nonspecific thiol reactants, inhibiting respiration	Alkylenebis (dithiocarbamate)	Mancozeb, maneb, metiram, zineb
	Phthalimide	Captan, folpet
	Sulphamide	Dichlofluanid, tolylfluanid
	Inorganic	Copper hydroxide, sulfur
Inhibitors of enzymes containing copper ions or sulfhydryl groups	Dimethyldithio-carbamate	Ferbam, thiram, ziram
Inhibitors of spore germination and growth of fungal mycelium	Dicarboximide	Iprodione, procymidone, vinclozolin
	Biphenyl	o-phenylphenol, biphenyl
Inhibitors of protein or phospholipid synthesis	Chlorophenyl	Dicloran, quintozene, tecnazene, tolclofos-methyl
Disruptor of glycolysis and energy production	Chloronitrile	Chlorothalonil
Cationic surfactants altering the permeability of fungus cell wall	Guanidine	Dodine, guazatine
Systemic Fungicides (protective, curative and eradication functions)		
Demethylation inhibitors (DMIs), i.e., ergosterol biosynthesis inhibitors	Imidazole	Imazalil, prochloraz, triflumizole
	Pyrimidine	Fenarimol, nuarimol
	Triazole	Fenbuconazole, flusilazole, myclobutanil, propiconazole, tebuconazole, triadimefon
Steroid reduction inhibitor, i.e., ergosterol biosynthesis inhibitors	Morpholine	Fenpropimorph, tridemorph
Inhibitors of mitosis	Benzimidazole	Benomyl, carbendazim, thiabendazole, thiophanate-methyl
Inhibitors of methionine biosynthesis and fungal hydrolytic enzyme secretion	Anilinopyrimidine	Cyprodinil, mepanipyrim, pyrimethalin
Inhibitors of protein synthesis	Phenylamide	Benalaxyl, metalaxyl, mefenoxam
Inhibitors of mitochondrial respiration by blocking electron transfer	Strobilurin	Azoxystrobin, kresoxim-methyl, trifloxystrobin

Source: Tomlin, C.D.S., Ed., *The Pesticide Manual*, 12th ed., British Crop Protection Council, Surrey, UK, 2000.

translocated both basipetally (downward) and acropetally whether applied to roots or leaves (transport in phloem governed by osmotic pressure) [5,55].

A large group of systemic fungicides (sterol inhibitors) shares a common biochemical target: the biosynthesis of ergosterol. Ergosterol is a major sterol in fungi and it plays a vital role in membrane structure and function. The biosynthesis of ergosterol involves many steps. DMI-type fungicides (imidazoles, pyrimidines, and triazoles) block a steroid demethylation step by binding to the active site of the catalyzing enzyme (a cytochrome P450), whereas morpholine fungicides act at later stages of ergosterol biosynthesis by inhibition of isomerase and reductase enzymes [58].

Benzimidazoles inhibit nuclear division, particularly mitosis, by binding to beta-tubulin, which interferes with the microtubule assembly [55]. Cyprodinil and other anilinopyrimidines act as inhibitors of methionine biosynthesis, causing inhibition of the secretion of fungal hydrolytic enzymes that are necessary for infection [5]. Phenylamides, such as metalaxyl and mefenoxam (metalaxyl-M), interfere with ribosomal RNA synthesis, thus inhibiting protein synthesis, which would ultimately kill the fungus [55]. Strobilurin fungicides inhibit mitochondrial respiration by blocking electron transfer between cytochrome b and c_1 (at the ubiquinol oxidizing site) [5].

The growth of pathogenic fungi can also be controlled using their fungal competitors or hyperparasites (biological fungicides from the group of biopesticides). *Ampelomyces quisqualis* represents an example of a naturally occurring hyperparasite of powdery mildews (an extremely broad group of plant pathogens). It infects and forms pycnidia (fruiting bodies) within powdery mildew hyphae, conidiophores (specialized spore-producing hyphae), and cleistothecia (the closed fruiting bodies of powdery mildews). This parasitism reduces growth and may eventually kill the mildew colony [59].

Certain microbial metabolic products, antibiotics, can also be effective fungicides, but their cost is usually rather prohibitive. Moreover, their application may result in the transfer of potential antibiotic resistance to human pathogens and thus reduced efficacy of medical treatments for humans. As an example of a wider use of antibiotic fungicides, blasticidin-S or kasugamycin (protein synthesis inhibitors produced by fermentation of actinomycetes *Streptomyces griseochromogenes* or *S. kasugaensis*, respectively) is used to control the economically important disease rice blast (*Pyricularia oryzae*) in Japan [60].

11.2.5.2 Problems Caused by Fungi in Produce Production and Their Control by Fungicides

Fungi can attack different parts of the plant and cause various fungal diseases that lead to economic losses (lower yields or even plant death), decreased produce quality, and pre- and postharvest produce spoilage. Moreover, certain fungal metabolites may pose a risk to human health and negatively affect produce safety and quality in general (see Chapter 18). There are numerous factors that either favor or hamper fungal infection. These factors include mainly climatic or storage conditions (moisture usually facilitates fungal growth, whereas low temperatures suppress it), plant or produce resistance, and damage by other pests. Only a small number of fungal pathogens are capable of direct penetration of undamaged skin of most produce. Wounds caused by insects, nematodes, birds, rodents, slugs, or snails create excellent entry points for fungal and other forms of infection; thus, combat of fungal diseases and prevention of produce spoilage caused by fungi should also involve control of these pests. Table 11.6 and Table 11.7 give examples of economically important

TABLE 11.6
Examples of Fungal Diseases on Fruits and Fungicides Used for Their Control in the U.S.

Commodity	Disease/Fungus	Infected Plant Parts	Fungicides
Apples	Powdery mildew *Podosphaera leucotricha*	Buds, shoots, blossoms, developing fruits	Benomyl, cyprodinil, fenarimol, kresoxim-methyl, mancozeb, metiram, myclobutanil, sulfur, thiophanate-methyl, triadimefon, trifloxystrobin, triflumizole, ziram
	Apple scab *Venturia inaequalis*	Leaves, fruits	Benomyl, captan, cyprodinil, dodine, fenarimol, ferbam, kresoxim-methyl, mancozeb, maneb, metiram, myclobutanil, thiophanate-methyl, thiram, trifloxystrobin, triflumizole, ziram
Bananas	Black leaf streak *Mycosphaerella fijiensis*	Leaves	Azoxystrobin, benomyl, fenbuconazole, flusilazole, mancozeb, maneb, propiconazole, tebuconazole, tridemorph
Citrus fruits[a]	Brown rot *Phytophthora* spp.	Fruits	Benomyl, coppers, fosetyl-Al, mefenoxam; postharvest: benomyl, biphenyl, *o*-phenylphenol, thiabendazole
Grapes	Powdery mildew *Uncinula necator*	Leaves, stems, and berries	Azoxystrobin, captan, fenarimol, kresoxim-methyl, myclobutanil, triadimefon, trifloxystrobin, triflumizole, *Ampelomyces quisqualis*
	Downy mildew *Plasmopara viticola*	Leaves, stems, blossoms, and berries	Azoxystrobin, captan, ferbam, kresoxim-methyl, mancozeb, mefenoxam, triflumizole, ziram
	Black rot *Guignardia bidwelli*	Leaves, stems, and berries	Azoxystrobin, captan, fenarimol, ferbam, kresoxim-methyl, mancozeb, myclobutanil, tebuconazole, triadimefon, trifloxystrobin, triflumizole, ziram
Pears	Pear scab *Venturia pirina*	Leaves, fruits, twigs	Benomyl, fenarimol, ferbam, mancozeb, sulfur, triflumizole, ziram
	Gray mold *Botrytis cinerea*	Fruits (postharvest)	post-harvest: *o*-phenylphenol, sodium hypochlorite, thiabendazole
Stone fruits[b]	Brown rot *Monilinia fructicola*	Blossoms, fruits, twigs	Benomyl, captan, chlorothalonil, fenbuconazole, ferbam, iprodione, myclobutanil, propiconazole, sulfur, tebuconazole, thiophanate-methyl
Strawberries	Gray mold (botrytis fruit rot) *Botrytis cinerea*	Petals, flower stalks, fruit caps, and fruits	Benomyl, captan, cyprodinil, fenhexamid, fludioxonil, iprodione, thiophanate-methyl, thiram, vinclozolin

[a] Grapefruits, lemons, limes, oranges, and tangerines.
[b] Cherries, peaches, nectarines, plums, and prunes.

Source: National Scientific Foundation Center for Integrated Pest Management and USDA Crop Profiles Web site, available at http://pestdata.ncsu.edu/cropprofiles/.

TABLE 11.7
Examples of Fungal Diseases on Vegetables and Fungicides Used for Their Control in the U.S.

Commodity	Disease/Fungus	Infected Plant Parts	Fungicides
Beans, peas	White mold *Sclerotinia sclerotiorum*	Blossoms, stems, leaves, pods	Benomyl, chlorothalonil, dicloran, iprodione, maneb, thiophanate-methyl
Carrots, parsley	Alternaria leaf blight *Alternaria dauci*	Seedlings, leaves	Azoxystrobin, benomyl, chlorothalonil, iprodione
Celery	Late blight *Septoria apiicola* Early blight *Cercospora apii*	Leaves	Benomyl, chlorothalonil, maneb, propiconazole
Cole vegetables[a]	Alternaria leaf spot *Alternaria brassicae*	Seedlings, leaves	Chlorothalonil, maneb
	Downy mildew *Peronospora parasitica*	Seedlings, leaves	Chlorothalonil, copper hydroxide, fosetyl-Al, maneb, mefenoxam
Cucurbits[b]	Gummy stem blight/black rot *Didymella bryoniae* Phoma cucurbitacearum	Seedlings, leaves, stems, fruits	Azoxystrobin, benomyl, chlorothalonil, copper hydroxide, mancozeb, thiophanate-methyl
Lettuce	Bottom rot *Rhizoctonia solani*	Leaves	Iprodione, vinclozolin
Onions	Downy mildew *Peronospora destructor*	Leaves	Azoxystrobin, chlorothalonil, copper sulfate, fosetyl-Al, mancozeb, maneb, mefenoxam
Peppers	Phytophthora blight *Phytophthora capsici*	Seedlings, stems, leaves, fruits	Copper hydroxide, maneb, mefenoxam metalaxyl
Potatoes, tomatoes, eggplants	Late blight *Phytophthora infestans* Early blight *Alternaria solani*	Leaves, stems, tubers, fruits	Azoxystrobin, chlorothalonil, cymoxanil, dimethomorph, mancozeb, maneb, metiram, propamocarb HCl
Spinach	Downy mildew (blue mold) *Peronospora effusa* *P. farinosa, P. spinaciae*	Seedlings, leaves	Copper hydroxide, fosetyl-Al, maneb, mefenoxam
Sweet corn	Common rust *Puccinia sorghi*	Leaves, ears	Chlorothalonil, mancozeb, maneb, propiconazole

[a] *Brassicaceae* including, e.g., broccoli, brussels sprouts, cabbage, Chinese cabbage, cauliflower, kale, kohlrabi.

[b] Cucumbers, zucchini, squash, gourds, pumpkins, watermelons, etc.

Source: National Scientific Foundation Center for Integrated Pest Management and USDA Crop Profiles Web site, available at http://pestdata.ncsu.edu/cropprofiles/.

fungal diseases for fruits and vegetables in the U.S. along with fungicides used for their control [4]. These tables also indicate which plant parts or stages are usually afflicted.

Soil-borne fungi, such as *Pythium, Phytophthora,* or *Rhizoctonia*, and also some seed-borne fungi (mainly *Alternaria*), may attack germinating seeds or young seedlings and cause "damping-off" disease, which is a general term for a sudden plant death in the seedling stage caused by fungi [61]. To minimize potential losses, the seeds can be treated with protectant fungicides, mainly with thiram or captan. Also, the soil can be sterilized prior to planting using fumigants, such as methyl bromide, metam-sodium, or chloropicrin. Infection of older plants does not usually lead directly to their death, but they often develop stem and root rots, which result in decreased plant vigor and productivity and may lead to further spread of the disease to leaves, blossoms, and fruits [43]. Important examples include the devastating carrot and parsley disease alternaria leaf blight (*Alternaria dauci*) [62], alternaria leaf spot (*Alternaria brassicae*) on cole vegetables [63], and phytophthora blight (*Phytophthora capsici*), which afflicts peppers and many other vegetables [64].

Soil-borne fungi can also directly attack plant parts in close proximity to the soil, which is the case with most vegetables and some fruits (strawberries, grapes, some citrus fruits) that have leaves and fruits close to the ground or creep on it, such as cucurbits. In the case of fruit trees, their fruits and leaves are mainly affected by air-borne fungi, such as *Venturia*, which causes pear or apple scab [65], or various fungi responsible for powdery or downy mildew.

Fungal damage on leaves basically leads to the same consequences as insect or mite feeding: reduced produce quality and size (and also reduced plant vigor, ultimately leading to plant death) due to lowered photosynthetic capacity and produce loss in the case of leafy or cole (cruciferous) vegetables. Infection of produce products, such as fruits, roots, tubers, or bulbs, translates directly into yield loss, which may occur both pre- and postharvest.

Often, presymptomatic stages of fungal infections are very difficult to detect and the affected products may not be discarded during postharvest sorting, and thus the disease may spread during transit and storage. In fruits, fungi can infect unripe fruit and remain quiescent until conditions become favorable for growth. As the fruit ripens, quiescence is broken and the fungus colonizes fruit tissues [66]. Fungal pathogens on tropical fruits, such as *Colletotrichum gloeosporioides* on mango and papaya or *Colletotrichum musae* on bananas, may serve as typical examples of this behavior [67].

Postharvest application of fungicides can prevent the potential disaster caused by disease spreading from infected to sound products and provide protection from further fungal attack from another source. The most widely applied postharvest fungicides for the control of various fungi in many produce types are thiabendazole and *o*-phenylphenol [4,5]. Other examples include biphenyl (a vapor-phase fungicide impregnated in paper wraps), used for postharvest protection of citrus fruits, and diphenylamine (applied as a postharvest dip or in wax formulations), which serves as a fungicide protectant and scald inhibitor (antiscalding agent) for pome fruits [5,68]. Superficial scald is a skin disorder of certain apple cultivars (mainly Granny Smith, Rome Beauty, Delicious, Winesap, and Yellow Newtown) that may develop during storage due to the oxidation of alpha-farnesene [69]. Alpha-farnesene is a naturally occurring volatile terpene in the apple fruit that can be oxidized to a variety of products (conjugated trienes). These oxidation products cause injury to the cell membranes, eventually leading to cell death in the outermost cell layers of the fruit.

11.2.6 BACTERICIDES

Numerous bacterial pathogens can cause serious plant diseases; however, their chemical control by bactericides is rather limited and the main effort concentrates on their prevention. This consists, among other methods, of killing pests that serve as their vectors or facilitate bacterial colonization by providing entry points. Thus, many other pesticide groups play an important, indirect role in the control of bacterial infections. Moreover, some fungicides (e.g., copper compounds or dodine) also possess bactericidal properties.

One of the basic roles of bactericides in produce production lies in soil sterilization as a control of soil-borne diseases using compounds such as 8-hydroxyquinoline sulfate and the pre- or postharvest fumigants (mainly in greenhouses) methyl bromide and formaldehyde [5]. A rather unique bactericide used for soil treatment is nitrapyrin which controls *Nitrosomonas* spp., the bacteria oxidizing ammonium ions in the soil. Thus, nitrapyrin acts as a nitrification inhibitor, preventing loss of nitrogen in soil [70].

Isolated instances of the use of bactericides in produce production comprise cases in which bacterial diseases would cause serious economic losses, such as death of fruit trees or extensive damage in orchards (that is, if a chemical control is available). Important examples include the treatment of fire blight (caused by *Erwinia amylovora*) on pear and apple trees by copper compounds and the antibiotics streptomycin and oxytetracycline, or the application of oxytetracycline (in addition to the copper compounds or dodine) as a control of bacterial leaf spot on peaches and nectarines caused by *Xanthomonas campestris* pv. *pruni* [4].

The postharvest protection of produce against bacterial spoilage usually involves means other than treatment with antibiotics (see Chapter 11) because their use for this purpose is not accepted in many countries [68]. Generally, fungi constitute more serious spoilage microorganisms than bacteria. This is particularly true for fruits, in which the relatively acidic conditions tend to suppress bacterial growth. On the other hand, vegetables with a higher pH (e.g., cole vegetables) may suffer significant losses from bacterial infection. Mechanisms of bacterial infiltration and spoilage of fruits and vegetables (including examples of key spoilage bacteria) and related topics are discussed elsewhere (Chapters 13 through 18).

11.2.7 HERBICIDES

Herbicides are designed to kill weeds and other plants that grow where they are not wanted. From approximately 250,000 species of plants worldwide, about 8000 species are considered weeds, and 200 to 250 of them pose a serious concern in agriculture [71]. In addition to chemical treatment with herbicides, basic methods of weed control include mechanical operations (burial, cultivation, or mowing) and crop rotation.

Weeds do not attack fruit and vegetable plants directly; however, their presence may also have detrimental impacts on produce yields, quality, and safety. They compete with the crop plants for water, nutrients, carbon dioxide, light, space, and pollinating bees, thus potentially reducing plant productivity and product size and quality and also decreasing plant vigor, which may ultimately lead to death. Weeds may also serve as hosts for various pests (mainly insects, mites, and fungi) and plant diseases and provide a habitat for rodents. In orchards, weeds increase humidity, creating an ideal environment for the development of fungal diseases. Furthermore, dense weed stands lower orchard temperatures, increasing the risk of frost damage in the spring. Produce contamination with weed parts at harvest may be an additional problem, such as the contamination of peas with berries of eastern black nightshade (*Solanum ptycanthum*) or with flower buds of Canada thistle (*Cirsium arvense*) [72].

11.2.7.1   Classification of Herbicides, Their Application, and Mode of Action

Herbicides can be classified into groups based on several factors: (1) main point of entry (roots or leaves), and thus their principal place of application (soil-applied vs. foliage-applied); (2) translocation abilities (systemic vs. contact); (3) application time (preemergence vs. postemergence); (4) toxic or application selectivity (selective vs. nonselective); and (5) chemical structure.

Table 11.8 and Table 11.9 present major chemical classes of soil- and foliage-applied herbicides along with their modes of action and representative examples. Soil-applied herbicides are absorbed mainly by roots or hypocotyls and coleoptiles of germinating seeds and, to be effective, they must be moved into soil (usually by rain or irrigation) and be relatively persistent in it. Soil-applied herbicides are systemic by their nature, whereas foliage-applied herbicides can be both systemic and contact. Contact herbicides are absorbed by leaves (and also stems or shoots), but their translocation is very limited. Foliage-applied systemic herbicides are absorbed mainly by leaves and to a lesser extent by roots. An example of a herbicide that is strictly foliage-applied is the very popular herbicide glyphosate (and its salts, such as sulfosate), which becomes inactivated on contact with soil due to strong ion exchange interactions with soil colloids. Regardless of the entry point, most systemic herbicides are translocated in xylem (acropetally) and some also in phloem (both acropetally and basipetally).

Preemergence and postemergence application may refer to both crops and weeds [73]. Table 11.10 gives examples of pre- and postemergence herbicides used in produce production in the U.S. [3,4]. In this respect, preemergence herbicides are applied on or in soil before weed germination and prevent weed seedlings from becoming established. A large group of preemergence, soil-applied herbicides acts as germination inhibitors, affecting either cell division (microtubule assembly) or cell growth and development [1,5]. In the latter group, thiocarbamates and bensulide interfere with lipid biosynthesis, which leads to the growth inhibition in the meristematic region and reduced production of cuticular waxes. Benzonitriles (dichlobenil and its precursor chlorthiamid) and benzamide isoxaben inhibit biosynthesis of cellulose and consequently of cell walls, preventing the incorporation of glucose

TABLE 11.8
Classification of Soil-Applied Herbicides

Mode of Action = Inhibition of:	Chemical Class	Representatives
Cell division:	Alkanamide	Diphenamid, napropamide
(microtubule assembly)	Benzamide	Propyzamide (pronamide), tebutam (butam)
	Benzenedicarboxylic acid	DCPA (chlorthal-dimethyl)
	Carbamate	Chlorpropham, propham
	Chloroacetamide	Acetochlor, alachlor, dimethenamid, metazachlor, S-metolachlor, propachlor
	Dinitroaniline	Benefin, ethalfluralin, oryzalin, pendimethalin, trifluralin
	Pyridine	Dithiopyr, thiazopyr
Cell growth and development:		
Lipid synthesis	Thiocarbamate	Butylate, cycloate, EPTC, molinate,
(not acetyl CoA carboxylase)		pebulate, thiobencarb, tri-allate, vernolate
	Phosphorodithioate	Bensulide
Cellulose biosynthesis	Benzamide	Isoxaben
	Benzonitrile	Chlorthiamid, dichlobenil
Indolylacetic acid (auxin)	Phthalamate	Naptalam
transport	Semi-carbazone	Diflufenzopyr
Photosynthesis:		
Photosynthetic electron transport	Triazine and	Atrazine, cyanazine, metribuzin, simazine
(Hill reaction)	triazinone	
	Uracil	Bromacil, lenacil, terbacil
	Urea	Chlorotoluron, diuron, isoproturon, linuron
Protoporphyrinogen oxidase	Diphenyl ether	Acifluorfen, bifenox, fomesafen,
(blocking chlorophyll synthesis		oxyfluorfen
and leading to membrane lipid	Oxadiazole	Oxadiargyl, oxadiazon
destruction)	Triazolinone	Amicarbazone, sulfentrazone
Carotenoid biosynthesis	Isoxazolidinone	Clomazone
(leading to destruction of	Pyridazinone	Norflurazon
chlorophyll and membrane lipids)		

Sources: Cremlyn, R.J., *Agrochemicals: Preparation and Mode of Action,* John Wiley & Sons LTD., Chichester, 1991 and Tomlin, C.D.S., Ed. *The Pesticide Manual*, 12th ed., British Crop Protection Council, Surrey, UK, 2000.

into the cell wall glucans. Naptalam and diflufenzopyr interfere with cell growth and development by affecting indolyl-3-acetic acid (IAA or auxin) transport. Auxin is a natural plant growth regulator that promotes cell elongation in plants.

The mode of action of many other soil-applied, preemergence herbicides consists of direct or indirect inhibition of photosynthesis of emerging weed seedlings [1,5].

TABLE 11.9
Classification of Foliage-Applied Herbicides

Mode of Action	Chemical Class	Representatives
Foliage-applied, systemic herbicides:		
Inhibitors of branched-chain amino acid synthesis	Imidazolinone	Imazamox, imazapyr, imazethapyr
	Sulfonylurea	Chlorsulfuron, rimsulfuron, sulfosulfuron
Inhibitors of aromatic amino acid synthesis	Glycine derivative	Glyphosate and its salts (sulfosate)
Inhibitors of fatty acid synthesis (acetyl CoA carboxylase inhibitors)	Aryloxyphenoxypropionate	Cyhalofop-butyl, fluazifop-P-butyl, haloxyfop, quizalop-P-ethyl
	Cyclohexanedione oxime	Clethodim, cycloxydim, sethoxydim
Synthetic auxins causing abnormal growth	Aryloxyalkanoic acid	2,4-D, 2,4-DB, dichlorprop, MCPA, MCPB, mecoprop
	Benzoic acid	Chloramben, dicamba, 2,3,6-TBA
	Pyridinecarboxylic acid	Clopyralid, fluroxypyr, picloram
Foliage-applied, contact herbicides:		
Cell disrupters	Bipyridylium	Diquat, paraquat
Inhibitors of glutamine synthetase (photosynthesis inhibitors)	Phosphinic acid	Bilanofos, glufosinate-ammonium
Inhibitors of photosynthetic electron transport (Hill reaction)	Benzothiadiazinone	Bentazon
	Hydroxybenzonitrile	Bromoxynil, ioxynil
	Phenylpyridazine	Pyridate

Sources: Cremlyn, R.J., *Agrochemicals: Preparation and Mode of Action,* John Wiley & Sons Ltd., Chichester, 1991 and Tomlin, C.D.S., Ed. *The Pesticide Manual,* 12th ed., British Crop Protection Council, Surrey, UK, 2000.

TABLE 11.10
Herbicides Used in Weed Control in Produce Production in the U.S.

Commodity	Weed Preemergence Herbicides	Weed Postemergence Herbicides
Fruits:		
Apples	Diuron, norflurazon, oryzalin, oxyfluorfen, pendimethalin, simazine, terbacil	2,4-D, glufosinate-ammonium, **glyphosate**, paraquat, sethoxydim, sulfosate
Cherries	Napropamide, norflurazon, oryzalin, oxyfluorfen, pendimethalin, simazine	2,4-D, **glyphosate**, paraquat, sulfosate
Grapes	Diuron, napropamide, norflurazon, oryzalin, oxyfluorfen, pendimethalin, simazine, trifluralin	2,4-D, glufosinate-ammonium, **glyphosate**, paraquat, sethoxydim, sulfosate
Oranges	Bromacil, diuron, norflurazon, oxyfluorfen, simazine, thiazopyr	2,4-D, **glyphosate**, paraquat, sethoxydim, sulfosate

TABLE 11.10
Herbicides Used in Weed Control in Produce Production in the U.S. (continued)

Commodity	Weed Preemergence Herbicides	Weed Postemergence Herbicides
Peaches	Diuron, napropamide, norflurazon, oryzalin, oxyfluorfen, pendimethalin, simazine, terbacil	2,4-D, **glyphosate**, paraquat
Pears	Diuron, norflurazon, oxyfluorfen, simazine	2,4-D, **glyphosate**, paraquat, sulfosate
Strawberries	**Napropamide**, simazine, sulfentrazone, terbacil	Clethodim, glyphosate, paraquat, sethoxydim
Vegetables:		
Broccoli	Bensulide, **DCPA**, napropamide, oxyfluorfen, trifluralin	Sethoxydim
Carrots	**Linuron**, trifluralin	Fluazifop-P-butyl
Cucumbers	Bensulide, clomazone, ethalfluralin, S-metolachlor, naptalam, pendimethalin	Glyphosate, naptalam, **paraquat**, sethoxydim
Lettuce	Benefin, **bensulide**, pronamide	Glyphosate, paraquat, sethoxydim
Onions	Bensulide, **DCPA**, dimethenamid, S-metolachlor, oxyfluorfen, pendimethalin, trifluralin	Bromoxynil, clethodim, fluazifop-P-butyl, glyphosate, paraquat, sethoxydim
Peas	Clomazone, imazethapyr, S-metolachlor, pendimethalin, tri-allate, trifluralin	Bentazon, MPCA, MPCB, paraquat, quizalofop-P-ethyl, sethoxydim
Peppers	Clomazone, S-metolachlor, napropamide, oxyfluorfen, **trifluralin**	Glyphosate, paraquat
Potatoes	EPTC, linuron, S-metolachlor, metribuzin, pendimethalin (+ diquat, endothal, paraquat or sulfuric acid as desiccants)	Clethodim, EPTC, glyphosate, **metribuzin**, rimsulfuron, sethoxydim
Sweet corn	Acetochlor, alachlor, **atrazine**, butylate, cyanazine, dimethenamid, dimethenamid-P, S-metolachlor, pendimethalin, simazine	Bentazon, 2,4-D, glyphosate
Tomatoes	S-metolachlor, **metribuzin**, napropamide, oxyfluorfen, pendimethalin, trifluralin	Glyphosate, **metribuzin**, paraquat, sethoxydim

Note: The most widely applied herbicides in 2001 on fruits and in 2002 on vegetables are highlighted in bold.

Sources: U.S. Department of Agriculture National Agricultural Statistic Service, Agricultural Chemical Usage Reports: 2001 Fruit Summary (August 2002) and 2002 Vegetables Summary (July 2003), available at http://www.usda.gov/nass/pubs/pubs.htm and National Scientific Foundation Center for Integrated Pest Management and USDA Crop Profiles Web site, available at http://pestdata.ncsu.edu/cropprofiles/.

This group of herbicides with predominant uptake by roots can also be effective in postemergence weed control. The primary action of triazines, triazinones, uracils, and ureas is inhibition of the Hill reaction (the light reaction II consisting of water photolysis) of photosynthetic electron transport, thus depriving the weed of its energy supply. Diphenyl ethers, oxadiazoles, and triazolines inhibit the mitochondrial membrane-bound enzyme protoporphyrinogen oxidase, thus interfering with chlorophyll synthesis and causing accumulation of porphyrins in susceptible plants [74]. The photosensitizing action of porphyrins causes membrane lipid peroxidation, which results in irreversible damage to the membrane's function and structure. Other cell membrane disrupters are clomazone and norflurazon. They block carotenoid biosynthesis, leading to peroxidation and destruction of chlorophyll (chlorosis) and membrane lipids (cell leakage) by singlet oxygen, because the function of carotenoids consists of dissipation (quenching) of oxidative energy of singlet oxygen produced during photosynthesis [5,75].

Postemergence herbicides are applied mainly on the weed foliage after seeds have germinated, thus controlling already emerged and growing plants. Foliage-applied, systemic herbicides may inhibit protein or lipid synthesis or be synthetic auxins [1,5]. Various carboxylic acidic compounds, such as (2,4-dichlorophenoxy)acetic acid (2,4-D), may act like auxin, causing abnormal growth, which results in the plant's death. Sulfonylureas and imidazolinones inhibit synthesis of the branched-chain amino acids valine, leucine, and isoleucine, and glyphosate and its salts prevent synthesis of aromatic amino acids, leading to disruption of protein and DNA synthesis. Aryloxyphenoxypropionates act as acetyl CoA inhibitors, thus blocking the biosynthesis of fatty acids.

Foliage-applied, contact herbicides generally interfere with photosynthesis [1,5]. Bentazone, pyridate, and the hydroxybenzonitriles bromoxynil and ionoxynil are inhibitors of the Hill reaction, whereas glufosinate-ammonium and bilanofos inhibit glutamine synthetase, leading to accumulation of ammonia and consequently inhibition of photophosphorylation in photosynthesis. Popular postemergence herbicides and preharvest desiccants, the quartery ammonium (bipyridylium) compounds diquat and paraquat, need light and oxygen for their action. During photosynthesis, they can be reduced (by electrons from the photosystem I) to relatively stable radical cations. In the presence of oxygen, these free radicals are reorganized to the original ions and superoxide anions are formed, leading to disruption of cell membranes and cytoplasm (the ultimate toxicant responsible for peroxidation of membrane lipids is probably hydrogen peroxide originating from the superoxide anion) [76]. Similarly to glyphosate, the cationic nature of diquat and paraquat causes their inactivation on contact with soil due to strong interactions with soil colloids.

Herbicides can be applied pre- and postplanting; postplanting treatments may be termed post- or preemergence depending on whether the crop plant has already emerged [73]. In all crop planting stages, pre- or postemergence herbicides can be used depending on the stage of the weed's growth. For instance, if the crop has emerged but no weeds are present, the herbicide application is postemergence to the crop but preemergence to the weeds and herbicides would be applied to the soil surface. If both the crop and the weeds have emerged, the application is postemergence to both weeds and crops and would usually be directed to the weed foliage or a selective herbicide would be used to avoid crop losses. With regards to preplanting treatments, soil fumigation (e.g., with metam-sodium) is a very effective preemergence weed control that prevents weed seed germination and also kills other soil-dwelling pests.

11.2.7.2   Herbicide Selectivity

Herbicide selectivity is the susceptibility or tolerance of different plants to herbicide application. Nonselective herbicides kill all plants, whereas selective herbicides kill only certain plants (preferably only the weeds and not the crops). Selectivity may be based on various factors including mainly biochemical and morphological differences between the weed and crop plants and also physicochemical properties of the herbicide [1].

Selectivity based on biochemical differences relies mainly on different enzymatic systems (metabolic insensitivity) or detoxification rates of herbicides (metabolic ability) of crops and weeds [71]. For instance, the synthetic auxin 4-(2,4-dichlorophenoxy)butyric acid (2,4-DB) needs the enzyme beta-oxidase for its conversion to the active herbicide 2,4-D [5]. Several species of leguminous plants are resistant to 2,4-DB because they largely lack beta-oxidases, thus, this herbicide can be safely used for selective postemergence weed control of these crops. Another example is the tolerance of corn plants to triazines, chloroacetamides, and thiocarbamates, which is attributed to their rapid detoxification by corn glutathione S-transferase (i.e., their fast conversion to innocuous metabolites by conjugation with glutathione). This detoxification can be accelerated using so-called herbicide safeners (also considered as pesticides), such as benoxacor or flurazole, which enhance metabolism of certain herbicides mainly by inducing the detoxification enzyme production. The possibilities of biochemical selectivity may be significantly extended using modern biotechnology methods that can alter the herbicide's action site. For example, "Roundup Ready" soybeans produce an excess of 5-enolpyruvylshikimate-3-phosphate synthase (EPSPS), the enzyme that is inhibited by glyphosate (Roundup), and therefore an appropriate glyphosate application does not affect these genetically modified soybeans [71].

Morphological differences between crops and weeds mainly include the size and orientation of the leaves, character of the leaf surface, and rooting depth. Dicotyledonous plants (e.g., broad-leaved weeds) have a larger surface area and the meristematic tissue is exposed to the herbicidal spray, whereas the leaves of monocotyledonous plants (e.g., grasses) are narrow and in upright positions, which reduces the application area and access to the meristematic regions. This difference accounts, for example, for the selective toxicity of synthetic auxins (2,4-D and others) toward broad-leaved weeds [1]. The character of the leaf surface determines the herbicide's penetration ability. Generally, the more waxy and hairy the surface, the more difficult the absorption of a foliage-applied herbicide. Similarly, the more deeply rooted the plant is, the more difficult the uptake of a soil-applied herbicide.

Physicochemical properties of herbicides may also play an important role, particularly water solubility and mobility in soil, in the case of uptake by the roots. For instance, the selectivity of triazines is a result of their low water solubility combined with a fairly high degree of adsorption onto soil colloids. Thus, they penetrate only a few centimeters downward in the soil, so deep-rooted plants are not affected, whereas shallow-rooted or germinating weeds are killed [1]. On the other hand, the foliage-applied bipyridylium herbicides (diquat and paraquat) and glyphosate strongly bind to soil colloids, and they are useful in directed applications because only those plants hit by the spray are affected and there is no uptake from the soil.

11.2.8 RODENTICIDES AND AVICIDES

Rodenticides are intended to kill vertebrate pests, mainly rodents, which may cause significant damage in produce production and storage and also pose a great risk to human and animal health. The application of rodenticides must be carefully monitored because they are also highly toxic to humans and domestic and other nontarget animals. In special situations, fumigants, such as hydrogen cyanide, phosphine, sulfur dioxide, or methyl bromide are used to control rodents (and many other pests) in enclosed spaces (warehouses, etc.) or their burrows [1]. However, rodenticides are more generally applied as baits; the most popular active ingredients include zinc phosphide, strychnine, sodium fluoroacetate, fluoroacetamide, chloralose, and various coumarin or indandione anticoagulants [1,5].

Zinc phosphide reacts with stomach acids to liberate toxic phosphine, which enters the bloodstream and affects liver, kidney, and heart functions. Strychnine is one of the oldest rodenticides that is still being used. This naturally occurring alkaloid (extracted from seeds of *Strychnos* spp., Loganiaceae family) acts as an antagonist of the neurotransmitter glycine. The toxicity of the sodium salt or amide of fluoroacetic acid (sodium fluoroacetate or fluoroacetamide) arises from its role in the biosynthesis of fluorocitrate, which inhibits the enzyme aconitase in the tricarboxylic acid (Krebs) cycle. Anticoagulants, such as brodifacoum, bromadiolone, chlorophacinone, difenacoum, or warfarin, interfere with the action of vitamin K, thus blocking the formation of the blood clotting factor prothrombin, so that a minor injury can cause a fatal hemorrhage (the indandione anticoagulant chlorophacinone also uncouples oxidative phosphorylation). Chloralose is primarily a sleep-inducing narcotic drug that acts by retarding metabolism and lowering body temperature to a fatal level (causing hypothermia). Chloralose can also be used as a relatively low-hazard avicide, enabling pest birds to be caught and killed by other means.

Pest birds represent another problem in produce production because they may lower yields and quality and cause surface damages. Moreover, their droppings can pose a health risk to consumers. Because of environmental and wildlife concerns, bird control using avicides and bird repellants should be avoided and other means, such as frightening and restriction devices, preferably should be used [77]. Thus, avicides are more often applied around public buildings in cities than in crop fields. In addition to chloralose, 4-aminopyridine (Avitrol) is another compound used in bird control; in essence it serves as a bird repellant because it may affect only a few members of a flock, causing them to become hyperactive, thus frightening away other birds [78].

11.2.9 PLANT GROWTH REGULATORS

Plant growth regulators (PGRs) are substances that can control (accelerate or retard) plant growth; initiate flowering; induce fruit setting; control fruit maturation; stimulate root development; cause blossoms, fruits, and leaves to fall; or delay fruit drop. Thus, PGRs may significantly improve produce yields and quality (size, shape, color, etc.). They can also be used as herbicides, sprouting inhibitors, defoliants, and fruit thinning agents.

TABLE 11.11
Examples of Compounds Used As Plant Growth Regulators (PGRs)

PGR Group	Representatives
Auxins	Indolyl-3-acetic acid (IAA), indolyl-3-butyric acid (IBA), 1-naphthylacetic acid (NAA), 2-(1-naphthyl)acetamide (NAD or NAAm), aryloxyalkanoic acids (2,4-D, dichlorprop, etc.)
Gibberellins	Gibberellic acid, gibberellins
Cytokinin	6-Benzylaminopurine, kinetin
Ethylene releasers	Ethephon
Defoliants	Dimethipin, endothal, thidiazuron, tribufos
Growth inhibitors	Maleic hydrazide, chlorpropham, propham, tecnazene
Growth retardants	Chlormequat, trinexapac-ethyl

Table 11.11 gives examples of compounds used as PGRs. PGRs may be naturally occurring plant hormones (phytohormones) or synthetic organic compounds often derived from them [79]. There are five major types of phytohormones: auxins, gibberellins, cytokinins, ethylene, and abscisic acid [80]. The synthetic PGRs may mimic natural phytohormones, may interfere with them, or may affect plant growth in some other way.

The term *auxin* is derived from the Greek word *auxein*, which means to grow. Compounds are generally considered auxins if they are able to induce cell elongation in stems and otherwise resemble physiological activity of the main naturally occurring auxin indolyl-3-acetic acid (IAA) biosynthesized from the amino acid tryptophan [81]. In addition to elongation of stem cells (internode elongation), auxins usually affect other processes, such as differentiation of phloem and xylem; root, leaf, and fruit growth; suppression of lateral buds below the apical bud (apical dominance); bending in response to gravity or light (gravitropism and phototropism); fruit setting and ripening; or fruit and leaf abscission.

IAA and its more stable analog indolyl-3-butyric acid (IBA) are used to induce rooting of cuttings. Examples of synthetic auxins include aryloxyalkanoic acids (e.g., 2,4-D), which are applied mainly as herbicides; however, their sublethal doses may prevent fruit drop or promote fruit set. Other synthetic auxins, 1-naphthylacetic acid (NAA) and 2-(1-naphthyl)acetamide (NAD or NAAm), are used to improve fruit setting and prevent premature flower and fruit drop in apples and many other fruits, to induce flowering in pineapples, or as fruit thinning agents [5].

Fruit thinning means the removal of a portion of the crop before it matures. It is usually necessary for apples, pears, peaches, nectarines, and plums. The benefits of fruit thinning include improvements in size and quality (color, flavor, and sugar content) of the remaining fruit, promotion of return blooms, and reduction of limb breakage from excessive crop loads. Hand thinning is accurate and selective but also laborious and very difficult on large trees. Chemical thinning represents a viable alternative, although it lacks the ability to selectively position fruit and may result in removal of too many or too few fruit. The most frequently applied fruit thinning agents include the mentioned synthetic auxins NAA and NAD, which induce formation of an abscission zone in peduncle [5], and carbamate insecticide carbaryl, which also possesses growth regulation properties [82], but its use may create potential mite problems by killing natural mite predators.

Gibberellins stimulate cell division and elongation. Gibberellic acid (GA) was first isolated in 1926 from the fungus *Gibberella fujikuroi*, and since then more than 70 gibberellins have been found in a wide variety of plant species [5,80]. GA is produced in the seed embryo, where it initiates growth, possibly by stimulating RNA and protein synthesis and increasing the transport of carbohydrates from the endosperm to the embryo [1]. The most significant effect of gibberellin treatment consists of stem growth stimulation. GA has a wide variety of applications, such as to improve fruit setting in pears, blueberries, and some other fruits; to control fruit maturity by delaying the yellow color in lemons; to loosen and elongate clusters and increase berry size in grapes; to break dormancy and stimulate sprouting in seed potatoes; to produce brighter-colored, firmer, and larger cherries; and to advance flowering and increase the yield of strawberries, among other purposes [1,4,5].

Cytokinins are derivatives of the purine adenine, and they are essential for the cell division process [80,83]. The first identified cytokinin was zeatin, isolated from maize grain in 1964. Examples of synthetic cytokinins include kinetin and 6-benzylaminopurine (6-benzaldenine). Their application range is also relatively wide. For instance, cytokinins delay senescence and thus can be used to prolong shelf life of fresh vegetables; they also stimulate fruit set, flower bud formation, and regular bearing in fruit trees or promote seed germination [1,5].

Ethylene is the simplest plant hormone and affects various plant growth processes [84]. It has been demonstrated that IAA promotes ethylene production in plants, which may account for some of the effects caused by auxins [1]. Ethylene releasers, such as ethephon, can be sprayed onto plants; they penetrate into the plant tissues and decompose to ethylene. Depending on the time of application, ethylene (from ethephon) may control different phases of plant development. Ethylene is used to promote preharvest ripening in apples, tomatoes, citrus fruits, etc.; to accelerate postharvest ripening in bananas, mango, and citrus fruits; to induce flowering and regulate ripening in pineapples; to increase fruit setting and yield in cucumbers; or to act as a fruit thinning agent [1,5].

Abscisic acid (ABA) is the best-known natural plant growth inhibitor that induces bud dormancy and probably also leaf fall in deciduous trees [1,85]. Examples of important synthetic growth inhibitors or retardants include compounds that act as sprouting inhibitors, such as maleic hydrazide, defoliants (dimethipin, endothal, tribufos, or thidiazuron), or agents that (among other functions) prevent premature fruit drop, such as chlormequat or trinexapac-ethyl. Chlormequat and trinexapac-ethyl inhibit gibberellin biosynthesis, which results in inhibition of cell elongation, mainly shortens and strengthens the stem and produces a sturdier plant [86]. Thus, these PGRs are also used to prevent lodging in cereals (wheat, rye, oats, and triticale) [5].

Sprouting inhibitors control postharvest sprouting in root (carrots, beets), tuber (potatoes), and bulb (onions) crops [68]. The most important sprout suppressants include preharvest-applied maleic hydrazide or postharvest-applied chlorpropham, propham, or tecnazene. Maleic hydrazide inhibits cell division in meristematic regions but does not affect cell elongation. It is an isomer of the pyrimidine base uracil, and thus it may interfere with DNA or RNA synthesis [1]. Chlorpropham and propham are carbamate herbicides that also inhibit cell division (microtubule organization). Tecnazene is a fungicide used for control of *Fusarium* rot of potatoes, so the suppression of sprouting is an additional benefit of this pesticide [5,68].

11.2.10 DESICCANTS AND DEFOLIANTS

Desiccants are compounds that accelerate drying of plant tissues. The most widely used desiccants are the contact herbicides diquat and paraquat, and also sulfuric acid or sodium chlorate. Preharvest destruction of potato haulms (vines) is probably the most important example of the application of desiccants in produce production. The main reason for vine desiccation is to promote tuber maturity, which, apart from other things, involves skin thickening and hardening, called skin set [87]. A good skin set reduces bruising of tubers during harvest and handling, water loss during storage, and susceptibility to infection by spoilage microorganisms.

Defoliants are substances intended for causing the leaves (foliage) of a plant to abscise, or fall off. In agriculture, defoliants are mainly applied to cotton and hops to facilitate mechanical harvest. Sodium chlorate was among the first chemicals used as a defoliant (and also desiccant), and this nonselective herbicide is still extensively utilized for this purpose. Modern defoliants from the group of PGRs, such as tribufos or thidiazuron, are absorbed by leaves and stimulate formation of an abscission layer between the plant stem and the leaf petioles, causing the dropping of entire green leaves [5].

A defoliant known as "Agent Orange," consisting of herbicides 2,4-D and 2,4,5-T (2,4,5-trichlorophenoxyacetic acid), was infamously used in the Vietnam War to remove the plant cover and was later shown to cause postexposure carcinogenic and teratogenic effects in humans due to contamination by the dioxin TCDD (2,3,7,8-tetrachlorodibenzo-*p*-dioxin) [88].

11.3 RISKS ASSOCIATED WITH PESTICIDES AND STRATEGIES TO REDUCE THEM

Potential risks associated with pesticide use relate mainly to their impact on human health and the environment. The concerns about pesticides may be legitimate, based on the results of scientific studies [89], or perceived. Unfortunately, a thorough scientific evaluation of the risks is practically impossible due to the high number of variables involved [90,91]. Toxicological studies can determine acute toxicity and other effects induced in tested animals, but chronic toxicity from pesticide exposure at actual concentrations found in food is unknown. Analytical chemists provide valuable data about pesticide levels in the human diet and about their fate in the environment. Epidemiological and environmental studies try to correlate pesticide use with the observed incidence of diseases or other adverse effects in humans and wildlife. Those are the main sources of information currently available for assessments that may indicate potential risks, but they hardly provide all the answers, especially in terms of the long-term effects involving cancer and other chronic diseases; impacts on reproduction, immune, and nervous systems; and ecological balance.

The perceived concerns may originate from unfortunate experiences with some of the early synthetic organic pesticides [92], lack of education or relevant information, general misconceptions (not only among the lay public [93]) or be caused by unbalanced, exaggerated, or even unsupported news releases. Generally, the public becomes particularly sensitive if the concerns relate to infants and small children (the "Alar affair" in 1989 may serve as an example [94]). Recent studies show that infants and children may be more sensitive than adults to pesticides in their diet [89]. In many countries, these findings (supported by sociological factors) led to stricter regulations of pesticide residues in baby foods and other products, impositions of more stringent requirements for newly registered pesticides, and reevaluations of already registered pesticides. For instance, the European Commission Directive 1999/50/EC requires baby foods to contain less than 0.01 mg/kg of any pesticide residue [95]. In the U.S., the Food Quality Protection Act (FQPA) of 1996 established a new safety standard for pesticide residues in food with special consideration given to assessing potential risks to infants and children by including an additional 10-fold safety factor [96]. Based on this requirement, the U.S. Environmental Protection Agency (EPA) is currently reassessing pesticide tolerance levels and reevaluating pesticide registrations. As a consequence, the use of a number of pesticides has been curtailed or banned, and others are scheduled for complete or partial (commodity-dependent) withdrawal.

In general, regulation of pesticide levels in food and strict enforcement play an important role in reducing consumers' exposure as a means of compliance with requirements for pesticide applications (generally with good agricultural practice, GAP). The implementation of principles of integrated pest management (IPM) has also been proved to reduce potential risks associated with pesticide use [97]. The FAO defines IPM as "a pest management system that, in the context of the associated environment and the population dynamics of the pest species, utilizes all suitable techniques and methods in as compatible a manner as possible and maintains the pest population at levels below those causing economically unacceptable damage or loss" [98]. The USDA uses the definition "IPM is a management approach that encourages natural control of pest populations by anticipating pest problems and preventing pests from reaching economically damaging levels. All appropriate techniques are used such as enhancing natural enemies, planting pest-resistant crops, adapting cultural management, and using pesticides judiciously" [99]. Thus, contrary to organic ("pesticide-free") production, IPM does not exclude pesticide use, but approaches pesticides more like "prescription drugs" rather than a "first aid medicine."

In addition to chemical control using pesticides, the list of tools available for IPM includes: (1) cultural control (cultivation, sanitation, crop rotation, destruction of pests' habitat, etc.); (2) biological control (biopesticides and natural enemies, such as parasites, predators, or pathogens); (3) genetic control (e.g., pest-resistant crops or application of sterile male technique in insect control); or (4) legal measures (quarantines, inspections) [100]. Despite the availability of alternative methods, the use of pesticides is and certainly will remain an important part of IPM systems, because many of these approaches are still too expensive or not sufficiently effective

[97]. However, the pesticides should be applied at the right moment using modern application methods, such as remote sensing, precision application, or controlled-release formulations, and the compounds used should be efficient at low rates, possess selective toxicity to pests and low toxicity to beneficial organisms and humans, be highly biodegradable, and have a low mobility in water.

11.4 SUMMARY

Pesticides have many important functions in modern produce production, increasing yields, preventing spoilage, and generally improving produce quality and safety. However, they should be used judiciously; the adage "moderation in all things" also holds true in this case.

REFERENCES

1. Cremlyn, R.J., *Agrochemicals: Preparation and Mode of Action*, John Wiley & Sons Ltd., Chichester, 1991.
2. Hajšlová, J., Pesticides, in *Environmental Contaminants in Food*, Moffat, C.F. and Whittle K.J., Eds., Sheffield Academic Press, 1999, Chap. 7.
3. U.S. Department of Agriculture (USDA) National Agricultural Statistic Service (NASS), Agricultural Chemical Usage Reports: 2001 Fruit Summary (August 2002) and 2002 Vegetables Summary (July 2003), available at http://www.usda.gov/nass/pubs/pubs.htm.
4. National Scientific Foundation (NSF) Center for Integrated Pest Management (IPM) and USDA Crop Profiles, available at http://pestdata.ncsu.edu/cropprofiles/.
5. Tomlin, C.D.S., Ed. *The Pesticide Manual,* 12th ed., British Crop Protection Council, Surrey, UK, 2000.
6. *Crop Protection Handbook 2003*, Meister Publishing, Willoughby, 2003.
7. Rawlins, S., The economic, health, and environmental benefits of pesticide use, in *Pesticides: Managing Risks and Optimizing Benefits*, Ragsdale, N.N. and Seiber, J.N., Eds., American Chemical Society, Washington, DC, 1999, Chap. 14.
8. National Research Council, *Diet and Health: Implications for Reducing Chronic Disease Risk*, National Academy Press, Washington, DC, 1989.
9. Block, G., Patterson, B., and Subar, A.F., Fruit, vegetables, and cancer prevention: a review of the epidemiological evidence, *Nutr. Cancer*, 18, 1, 1992.
10. Taylor, C.R., Economic impacts and environmental and food safety tradeoffs of pesticide use reduction on fruit and vegetable, Study ES95-1, Auburn University, June 1995.
11. National Research Council, *Carcinogens and Anticarcinogens in the Human Diet*, National Academy Press, Washington, DC, 1996.
12. Gold, L.S., Slone, T.H., Stern, B.R., Manley, N.B., and Ames, B.N., Rodent carcinogens: setting priorities, *Science*, 258, 261, 1992.
13. Arino, A.A. and Bullerman, L.B., Growth and aflatoxin production by *Aspergillus parasiticus* NRRL 2999 as affected by the fungicide iprodione. *J. Food Prot.*, 56, 718, 1993.
14. Calori-Domingues, M.A. and Fonseca, H., Laboratory evaluation of chemical control of aflatoxin production in unshelled peanuts (Arachis hypogaea L.). *Food Addit. Contam.*, 12, 347, 1995.

15. Windstrom, N.W., The role of insects and other plant pests in aflatoxin contamination of corn, cotton and peanuts: a review, *J. Environ. Qual.*, 8, 5, 1979.
16. El-Morshedy, M.M.F. and Aziz, N.H., Effects of fenamiphos, carbofuran and aldicarb on zearalenone production by toxigenic Fusarium spp. contaminating roots and fruits of tomato, *Bull. Environ. Contam. Toxicol.,* 54, 514, 1995.
17. Wierenga, J.M. and Hollingworth, R.M., Inhibition of insect acetylcholinesterase by the potato glycoalkaloid alpha-chaconine. *Nat. Toxins,* 1, 96, 1992.
18. Ministry of Agriculture, Food and Fisheries (MAFF), *Annual Report of the Steering Group on Chemical Aspects of Food Surveillance*, Food Surveillance Paper 1, MAFF Publications, London, 1996.
19. Hammond-Kosack, K. and Jones, J.D.G., Responses to plant pathogens, in *Biochemistry & Molecular Biology of Plants*, Buchanan, B.B., Gruissem, W., and Russell, L.J., Eds., American Society of Plant Physiologists, Rockville, MD, 2000, Chap. 21.
20. Bray, E.A., Bailey-Serres, J., and Weretilnyk, E., Responses to abiotic stresses, in *Biochemistry & Molecular Biology of Plants*, Buchanan, B.B., Gruissem, W, and Russell, L.J., Eds., American Society of Plant Physiologists, Rockville, MD, 2000, Chap. 22.
21. Winter, C.K., Pesticides and human health: the influence of pesticides on levels of naturally-occurring plant and fungal toxins, in *Pesticides: Managing Risks and Optimizing Benefits*, Ragsdale, N.N. and Seiber, J.N., Eds., American Chemical Society, Washington, DC, 1999, Chap. 12.
22. Bloomquist, J.R., Neuroreceptor mechanisms in pyrethroid mode of action and resistance, *Rev. Pestic. Tox.*, 2, 185, 1993.
23. Addor, R.W., Insecticides, in *Agrochemicals from Natural Products*, Godfrey, C.R.A., Ed., Marcel Dekker, New York, 1995, Chap. 1.
24. Henrick, C.A., Pyrethroids, in *Agrochemicals from Natural Products*, Godfrey, C.R.A., Ed., Marcel Dekker, New York, 1995, Chap. 2.
25. Kornis, G.I., Avermectins and milbemycins, in *Agrochemicals from Natural Products*, Godfrey, C.R.A., Ed., Marcel Dekker, New York, 1995, Chap. 4.
26. Faull, J.L. and Powell K.A., Biological control agents, in *Agrochemicals from Natural Products*, Godfrey, C.R.A., Ed., Marcel Dekker, New York 1995, Chap. 9.
27. Gill, S.S., Cowles, E.A., and Pietrantonio, P.V., The mode of action of *Bacillus thuringiensis* endotoxins, *Annu. Rev. Ent.*, 37, 615, 1992.
28. Capinera, J.L., *Handbook of Vegetable Pests*, Academic Press, San Diego, CA, 2001.
29. Boucher, T.J. and Adams, R.G., Integrated pest management for Connecticut cole, Cooperative Extension Publication 93-19, Integrated Pest Management, University of Connecticut, Storrs, CT, 1993.
30. Cranshaw, W.S. and Zimmerman, R.J., Apple and pear insects, Fact Sheet No. 5.519, The Colorado State University Cooperative Extension, Fort Collins, CO, 2002.
31. Flaherty, D.L., Christensen, L.P., Lanini, W.T., Marois, J.J., Phillips, P.A., and Wilson, L.T., Eds., *Grape Pest Management*, 2nd ed., University of California Division of Agriculture and Natural Resources Publication 3343, 1992.
32. Welty, C., San Jose scale on fruit trees, Ohio State University Extension Fact Sheet HYG-2039-92, Department of Horticulture and Crop Science, Ohio State University, Columbus, OH 1992.
33. Cermak, P. and Walker, G.M., Tarnished plant bug: a major pest of strawberry, Fact Sheet 92-108, Ministry of Agriculture and Food, Ontario, Canada, 1992.
34. Torres, A.N., Hoy, C.W., and Welty, C., An integrated pest management program for carrot weevil in parsley, Extension Fact Sheet CV-1001-02, The Ohio State University, Columbus, OH, 2002.

35. Kerns, D.L. and Tellez, T., Susceptibility of lemons to citrus thrips scarring based on fruit size, in *1998 Citrus and Deciduous Fruit and Nut Research Report*, Publication AZ1051, College of Agriculture, The University of Arizona, Tucson, AZ 1998.

36. Howitt, A., Oriental fruit moth, in *Common Fruit Tree Pests*, Fruit IPM Fact Sheet NCR 63, Michigan State University Extension, Paw Paw, MI, 1993.

37. Riedl, H. and Kuhn, E., Cherry fruit fly and black cherry fruit fly, Tree Fruit Fact Sheet 102GFSTF-I 15, New York State Integrated Pest Management Program, New York State College of Agriculture and Life Sciences, Cornell University, Ithaca, NY, 1988.

38. Oliva, S., Navarro, S., Navarro, G., Camara, M.A., and Barba, A., Integrated control of grape berry moth, powdery mildew (*Uncinula necator*), downy mildew (*Plasmopara viticola*) and grapevine sour rot (*Acetobacter* spp.), *Crop Prot.*, 18, 581, 1999.

39. Shetlar D.J., Aphids on trees and shrubs, Ohio State University Extension Fact Sheet HYG-2031-90, Department of Horticulture and Crop Science, Ohio State University, Columbus, OH, 1990.

40. Leeper, J. and Tette, J., Pear psylla, Tree Fruit Fact Sheet 102GFSTF-I1, New York State Integrated Pest Management Program, New York State College of Agriculture and Life Sciences, Cornell University, Ithaca, NY, 1978.

41. Fry, W.E., *Principles of Plant Disease Management*, Academic Press, London, 1982.

42. Ferreira, S.A., Trujillo, E.E., and Ogata, D.Y., Banana bunchy top virus, Plant Disease Fact Sheet PD-12, Cooperative Extension Service, University of Hawaii, Manoa, HI, 1997.

43. Sherf, A.F. and MacNab, A.A., *Vegetable Diseases and Their Control*, 2nd ed., John Wiley & Sons, New York, 1986.

44. Venette, J.R., Smith, R.C., Lamey, H.A., and McBride, D.K., Bacterial wilt of cucurbits, Fact Sheet PP-747, North Dakota State University of Agriculture and Applied Science, Fargo, ND, 1996.

45. Shetlar D.J., Spider mites and their control, Ohio State University Extension Fact Sheet HYG-2012-92, Department of Horticulture and Crop Science, Ohio State University, Columbus, OH, 1992.

46. Childers, C.C., Hall, D.G., Knapp, J.L., McCoy, C.W., Rogers, J.S., and Stansly, P.A., Citrus rust mites, in *2000 Florida Citrus Pest Management Guide*, Institute for Food and Agricultural Sciences and Florida Cooperative Extension Service Document ENY-603, University of Florida, Gainesville, FL, 2000.

47. U.S. Environmental Protection Agency (EPA), U.S. government nominates critical use exemption for methyl bromide, Environmental News, available at www.epa.gov/newsroom, 2003.

48. Nickle, W.R., Nematode parasites of vegetable crops, in *Plant and Insect Nematodes*, Marcel Dekker, New York, 1984, Chap. 9.

49. Noling, J.W., Nematode management in potatoes (Irish or white), Institute for Food and Agricultural Sciences and Florida Cooperative Extension Service Document ENY-29, University of Florida, Gainesville, FL, 2003.

50. Flint, M.L., Slugs and snails, UC Pest Management Guidelines, UC ANR Publication 7427, University of California, Davis, CA, 2003.

51. Shetlar D.J., Slugs and their management, Ohio State University Extension Fact Sheet HYG-2010-95, Department of Horticulture and Crop Science, Ohio State University, Columbus, OH, 1995.

52. Price, J.F., Nagle, C., McCord, E., and Webb, S.E., Insecticides, miticides, and molluscicides for management of insect, mite, snail and slug pests of Florida strawberry, Institute for Food and Agricultural Sciences and Florida Cooperative Extension Service Document ENY-657, University of Florida, Gainesville, FL, 2001.

53. Eckert, J.W. and Ogawa J.M., The chemical control of post-harvest diseases: deciduous fruits, berries, vegetables and root/tuber crops. *Annu. Rev. Phytopathol.*, 26, 433, 1988.

54. Geoghiou, G.H. and Saito, T., Eds., *Pest Resistance to Pesticides*, Plenum Press, London, 1983.

55. Lyr, H., Ed., *Modern Selective Fungicides: Properties, Applications, Mechanisms of Action*, 2nd ed., Gustav Fischer Verlag, Villengang, Germany, 1995.

56. Bartlett, D.W., Clough, J.M., Godwin, J.R., Hall, A.A., Hamer, M., and Parr-Dobrzanski, B., The strobilurin fungicides (a review), *Pest Manag. Sci.*, 58, 649, 2002.

57. Corbett, J.R., Wright, K., and Baillie, A.C., *The Biochemical Mode of Action of Pesticides,* Academic Press, London, 1984.

58. Berg, D., Biochemical mode of action of fungicides, in *Human Welfare and the Environment*, Vol. 1, Pergamon Press, Oxford, UK, 1983, p. 55.

59. Falk, S.P., Gadoury, D.M., Pearson, R.C., and Seem, R.C., Partial control of grape powdery mildew by the mycoparasite *Ampelomyces quisqualis, Plant Dis.*, 79, 483, 1995.

60. Godfrey, C.R.A, Fungicides and bactericides, in *Agrochemicals from Natural Products*, Godfrey, C.R.A., Ed., Marcel Dekker, New York, 1995, Chap. 7.

61. Pfleger, F.L. and Gould, S.L., Damping-off of seedlings, Extension Service Fact Sheet FS-01167, Department of Plant Pathology, University of Minnesota, Minneapolis, MN, 1994.

62. Dillard, H.R., Carrot leaf blight, Vegetable Crops Fact Sheet 739, Department of Plant Pathology, Cooperative Extension, Cornell University, Ithaca, NY, 1988.

63. Pfleger, F.L. and Gould, S.L., Diseases of cole crops, Extension Service Fact Sheet FS-01169, Department of Plant Pathology, University of Minnesota, Minneapolis, MN, 1999.

64. Roberts, P.D., McGovern, R.J., Kucharek, T.A., and Mitchell, D.J., Vegetable diseases caused by *Phytophthora capsici* in Florida, Plant Pathology Fact Sheet SP-159, Plant Pathology Department, University of Florida, Gainesville, FL, 2001.

65. Biggs, A.R, Apple scab, in *Compendium of Apple and Pear Diseases*, Jones, A.L. and Aldwinckle, H.S., Eds., APS Press, St. Paul, MN, 1990, p. 6.

66. Swinburne, T.R., Quiescent infections in post-harvest diseases, in *Post-harvest Pathology of Fruits and Vegetables*, Dennis, C., Ed., Academic Press, New York, 1983, Chap. 1.

67. Jeffries, P., Dodd, J.C., Jeger, M.J., and Plumbley, R.A. The biology and control of *Colletotrichum* species on tropical fruit, *Plant Pathol.*, 39, 353, 1990.

68. Aked, J., Fruits and vegetables, in *The Stability and Shelf-Life of Food*, Kilcast, D. and Subramaniam, P., Eds., CRC Press, Boca Raton, FL, 2000, Chap. 11.

69. Ingle, M. and D'Souza, M.C., Physiology and control of superficial scald of apples: a review, *Hortic. Sci.*, 24, 28, 1989.

70. Johnson, J.W., Nitrification inhibitors potential use in Ohio, Agronomy Fact Sheet AGF-201-95, Department of Horticulture and Crop Science, Ohio State University Extension, Columbus, OH, 1995.

71. Lingenfelter, D.D., Introduction to weeds and herbicides, Publication No. CAT UC175, Agricultural Research and Cooperative Extension, Pennsylvania State University, University Park, PA, 2002.

72. Crotser, M.P. and Masiunas, J.B, Eastern black nightshade (*Solanum ptycanthum* Dun.) competition with processing pea (*Pisum sativum* L.), *Hortic. Sci.*, 31, 88, 1997.

73. Tredaway Ducar, J. and MacDonald, G.E., Principles of weed management, Agronomy Department Document SS-AGR-100, Florida Cooperative Extension Service, Institute of Food and Agricultural Sciences, University of Florida, Gainesville, FL, 2003.

74. Matringe, M., Camadro, J.M., Labbe, P., and Scalla, R., Protoporphyrinogen oxidase as a molecular target for diphenyl ether herbicides, *Biochem. J.*, 260, 231, 1989.

75. Cantrell, A., McGarvey, D.J., Truscott, T.G., Rancan, F., and Bohm, F., Singlet oxygen quenching by dietary carotenoids in a model membrane environment, *Arch. Biochem. Biophys.*, 412, 47, 2003.

76. Summers, L.A., *The Bipyridinium Herbicides*, Academic Press, London, 1980.

77. Arnold, K.A., Environmental control of birds, in *CRC Handbook of Pest Management in Agriculture*, Pimentel, D., Ed., Vol. 1, CRC Press, Boca Raton, FL, 1981, p. 499.

78. Extension Toxicology Network EXTOXNET, Pesticide information profile: 4-aminopyridine, available at http://pmep.cce.cornell.edu/profiles/extoxnet/, Cornell University, Ithaca, NY, 1993.

79. Stonard, R.J. and Miller-Wideman, M.A., Herbicides and plant growth regulators, in *Agrochemicals from Natural Products*, Godfrey, C.R.A., Ed., Marcel Dekker, New York, 1995, Chap. 6.

80. Crozier, A., Kamiya, Y., Bishop, G., and Yokota, T., Biosynthesis of hormones and elicitor molecules, in *Biochemistry & Molecular Biology of Plants*, Buchanan, B.B., Gruissem, W., and Russell, L.J., Eds., American Society of Plant Physiologists, Rockville, MD, 2000, Chap. 17.

81. Grauslund, J., Chemical thinning of the apple cultivar 'Summer red' with NAA and carbaryl, *Acta Hortic.*, 120, 77, 1981.

82. Hedden, P. and Kamiya, Y., Gibberellin biosynthesis: enzymes, genes and their regulation, *Annu. Rev. Plant Physiol. Plant Mol. Biol.*, 48, 431, 1997.

83. Moc, D.W.S. and Mok, M.C., Eds., *Cytokinins: Chemistry, Action and Function*, CRC Press, Boca Raton, FL, 1994.

84. Kannelis, A.K., Chang, C., Kende, H., and Grierson, D., Eds., *Biology and Biotechnology of the Plant Hormone Ethylene*, NATO ASI Series, Kluwer Academic Publishers, Dordrecht, The Netherlands, 1997.

85. Walton, D.C. and Li, Y., Abscisic acid biosynthesis and metabolism, in *Plant Hormones: Physiology, Biochemistry, and Molecular Biology*, Davies, P.J., Ed., Kluwer Academic Publishers, Dordrecht, The Netherlands, 1995, p. 140.

86. Rademacher, W., Growth retardants: effect on gibberellin biosynthesis and other metabolic pathways, *Annu. Rev. Plant Physiol. Plant Mol. Biol.*, 51, 501, 2000.

87. Pavlista, A.D. and Ojala, J.C, Potatoes: chip and french fry processing, in *Processing Vegetables: Science and Technology*, Smith, D.S., Cash, J.N., Nip, W.-K., and Hui, Y.H., Eds., Technomic Publishing Company, Lancaster, PA, 1997, Chap. 10.

88. Schecter, A., Quynh, H.T., Pavuk, M., Päpke, O., Malisch, R., and Constable J.D., Food as a source of dioxin exposure in the residents of Bien Hoa City, Vietnam, *J. Occup. Environ. Med.*, 45, 781, 2003.

89. National Research Council (NRC), *Pesticides in the Diets of Infants and Children*, National Academy Press, Washington, DC, 1993.

90. Seiber, J.N. and Ragsdale, N.N., Examining risks and benefits associated with pesticide use: an overview, in *Pesticides: Managing Risks and Optimizing Benefits*, Ragsdale, N.N. and Seiber, J.N., Eds., American Chemical Society, Washington, DC, 1999, Chap. 1.

91. Winter, C.K., Dietary pesticide risk assessment, *Rev. Environ. Contam. Toxicol.*, 127, 23, 1992.
92. Carson, R., *Silent Spring*, Houghton Mifflin, Boston, MA, 1962.
93. Ames, N.A., Science and the environment: facts vs. phantoms. *Priorities for Health*, 5, American Council on Science and Health, 1993.
94. Sewell, B.H., Whyatt, R., Hathaway, J., and Mott, L., *Intolerable Risk: Pesticides in Our Children's Food,* Natural Resources Defense Council, New York, 1989.
95. Commission Directive 1999/50/EC amending Directive 91/321/EEC on infant formulae and follow-on formulae, *Off. J. European Com.*, L 139, 29, 1999.
96. *Food Quality Protection Act of 1996*, U.S. Public Law 104-170, 104th Congress, Washington, DC, 1996.
97. Forney, D.R., Importance of pesticides in integrated management, in *Pesticides: Managing Risks and Optimizing Benefits*, Ragsdale, N.N. and Seiber, J.N., Eds., American Chemical Society, Washington, DC, 1999, Chap. 13.
98. Food and Agricultural Organization (FAO) of the United Nations (UN), *International Code of Conduct on the Distribution and Use of Pesticides*, Rome, Italy, 1990, Article 2.
99. U.S. Department of Agriculture (USDA), Agricultural Research Service (ARS), USDA programs related to integrated pest management, USDA Program Aid 1506, Beltsville, MD, 1993.
100. U.S. Department of Agriculture (USDA), Adoption of integrated pest management in U.S. agriculture, *Agriculture Information Bulletin* 707, 1994.

12 Microbial Ecology of Spoilage

M. L. Bari, Y. Sabina, S. Kawamoto and K. Isshiki
Food Hygiene Research Team, National Food Research Institute, 2-1-12, Kannondai, Tsukuba, 305-8642, Japan

CONTENTS

12.1 INTRODUCTION

As soon as vegetables and fruits are harvested physiological changes occur, and some of these lead to a loss in quality. Respiratory activity involving the breakdown of carbohydrates by the plant enzymes continues and the changes induced, whether advantageous or deleterious, are markedly influenced by the maturity of the plant when harvested; thus, plants can usually be stored for lengthy periods with little change in quality if they are harvested at the right time. Many fleshy fruits such as bananas are harvested before maturation and ripening continues thereafter, but citrus fruits only ripen satisfactorily on the tree [1]. However, although spoilage can be induced by autolytic enzymes, it is caused more usually by the activities of microorganisms.

The number of microorganisms on vegetables received from the field is highly variable. All green plants posses microflora, which normally subsist on the surface of vegetables. Soil, water, air, insects, and animals all contribute to the microflora of vegetables. The relative importance of these sources differs with the structural entity of the plant (e.g., leaves have greater exposure to air, whereas root crops have greater exposure to soil). High aerobic plate counts occur in vegetables that are in contact with soil, such as garlic, and on vegetables grown above the ground, such as spinach. The activities of humans have important effects. For example, the use of pesticides to control insects often limits the spread of microorganisms. Similarly, cultivation, either by hand or mechanically, introduces or distributes microorganisms into ecological niches from which previously they were absent. Finally, the introduction of human or other animal waste material into the water or soil has an obvious impact on the flora of vegetables [2].

The microbial contamination of raw vegetables usually occurs on exposed surfaces, while the internal tissues remain essentially free of microorganisms. The means by which microorganisms penetrate the tissues has not been established clearly, but their presence usually is not deleterious to the growing plant. An equilibrium of coexistence exists, although it can be broken, and spoilage can develop under certain circumstances. However, harvesting often injures produce, and plant tissues can rupture. As a result, nutrients that enhance microbial growth are released and create an entrance to internal tissues. Consequently, the harvested vegetable is more prone to support microbial growth than is the growing plant. [3]

Most of the microorganisms present on fresh vegetables are saprophytes such as coryneforms, lactic acid bacteria, spore-formers, coliforms, micrococci, and pseudomonads derived from soil, water, and air. Fungi including *Aureobasidium*, *Fusarium*, and *Alternaria* often are present but in relatively lower numbers than bacteria. To date, the strictly anaerobic organisms that also may be present have not been well characterized except for certain heat-resistant spore-formers important in spoilage of canned vegetables.

The predominant microorganisms on healthy, raw vegetables are usually bacteria, although significant numbers of molds and yeasts may be present. Spoilage organisms may be on the plant in the field or may be introduced during harvesting and transport. Most spoilage is caused by fungi, chiefly members of the genera *Penicillium*, *Sclerotinia*, *Botrytis*, and *Rhizopus*. It has also been reported that *Sclerotinia sclerotium*, the cause of pink rot of celery, produces photoxins, which cause a blistering cutaneous reaction in field workers who handle the produce [4].

Bacteria are responsible for approximately one-third of the total microbial spoilage loss of vegetables. Such spoilage may be due to bacteria that cause soft rots and other rots, spots, blights, and wilts. Soft rots, occurring during transport and storage, are usually caused by coliforms, *Erwinia carotovora*, and certain pseudomonads, such as *Pseudomonas fluorescens (marginalis)*. Organisms causing rots other than soft rot include corynebacteria, xanthomonads, and pseudomonads. Often infection can occur in the field, thereby permitting invasion of the plant tissue by soft-rot organisms as a result of trauma induced during subsequent transport and storage [5]. Soft rot of potatoes by *Clostridium* has also been reported.

12.2 NATURE OF MICROORGANISMS ON FRUITS AND VEGETABLES

12.2.1 SAPROPHYTIC MICROORGANISMS

The microbial species that prevail on fruits and vegetables are commonly found on plants in the field or after harvest and probably originated from the epiphytic microflora of the raw materials. *P. fluorescens*, *E. herbicola*, and *E. agglomerans* are major components of the epiphytic microflora of many vegetables [6–10]. An average *Leuconostoc* spp. population of 2.5×10^4 CFU/g was found on plants by Mundt et al. [11], although other studies indicate lower frequencies [12] (1 to 33% in various crops). Yeast species have been identified from raw fruits and vegetables [6, 13–15]. Similarly, pectinolytic *P. fluorescens*, pectinolytic *Xanthomonas* spp., *Cytophaga* spp., and *Flavobacterium* spp. have been isolated from various unprocessed vegetables sampled in retail outlets [16–18]. More generally, pectinolytic fluorescent pseudomonds are well-known agents of soft-rot diseases of leafy vegetables [19–25] and can account for an important fraction of epiphytic microflora: 10^4 pectinolytic *Pseudomonas* cells/cm^{-2} have been counted on white cabbage leaves [6]. Counts as high as 10^9 CFU/g^1 have been noted in fresh soil [26], although lower values have also been reported [27]. It is significant that the coliform population on vegetables was reduced by disinfection of irrigation water, whereas the number of pectinolytic bacteria was unaffected [28]. The presence of pseudomonads on vegetables presumably does not depend on external contamination, because they are likely endemic.

In contrast, *Erwinia* spp., a major cause of soft-rot diseases of vegetables [29], have only been sporadically isolated from minimally processed fresh vegetables (MPF) (*E. carotovora* was found in only few samples by Brocklehurst et al. [30]). Similarly, *P. cichorii*, an important disease agent of lettuce and chicory [31], has not been reported in MPF vegetables. Processing, sorting, and trimming of raw material to remove all decayed or diseased parts could explain the low frequency of such plant pathogens in processed products. In addition, soft-rot *Erwinia* are rarely present in high numbers in the environment [29], are not isolated from soil samples by direct plating on selective media [26], and are usually recovered only after enrichment [32].

12.2.2 PECTINOLYTIC MICROORGANISMS

By isolating representative samples of colonies from count plates and testing the isolates for pectinolysis, 10 to 20% of isolates among the mesophilic bacteria on shredded lettuce were found to be pectinolytic [14]. A high proportion of *Pseudomonas* (20 to 60%) was found to be pectinolytic in many samples of shredded carrots and shredded chicory salads [33,34]. In contrast, Brocklehurst et al. [30] counted only 10^6 to 10^7 pectinolytic bacteria/g[1] by direct plating on Hankin media, for a total mesophilic flora count of 10^8 to 10^9 CFU/g. Pectinolytic isolates were usually identified as *Pseudomonas fluorescens* of different biovers, *P. paucimobilis*, *P. viridiflava*, *P. luteola*, *Xanthomonas maltophila*, *Flavobacterium* spp., *Cytophaga* spp., or *Vibrio fluvialis* [33–36]. Some pectinolytic fungi (*Mucor* spp. and *Sclerotinia sclerotiorum*) and yeasts (*Trichosporon* spp.) have also been isolated from shredded carrots [33,37].

12.2.3 FOODBORNE PATHOGENS

Raw fruits and vegetables may harbor many potential foodborne pathogens [38–41]. *Listeria monocytogenes* has been isolated in lettuce heads [42] in England (9% of samples contaminated [43]) and Spain (7.8% of sample contaminated [44]) and in potatoes and radishes in the U.S. (25.8 and 30.3% of samples contaminated, respectively [45]). Vegetables were found to carry 10^2 to 10^4 CFU/g of cytotoxic and hemolytic *Aeromonas* spp. in the U.S. [46]. Fresh fruits and vegetables can also play a role in the transmission of *Salmonella* spp. [47]. An investigation in the U.S. after an outbreak of *Salmonella poona* found that 1% of melon rinds from fruits imported from Mexico were contaminated with the bacteria [48]. Similarly, in an outbreak of *Salmonella* caused by the consumption of bean sprouts in England, bacteria were isolated on the bean seeds used by the producer [49]. Vegetables sampled in the field or in retail outlets were contaminated with *Salmonella* spp. at frequencies of 7.5% in Spain [50] and 8 to 63% in the Netherlands [51]. In Egypt, vegetables and salads were found to be contaminated with *Shigella* spp., *Salmonella* spp., and *S. aureus* (10^3 CFU/g, [52]). *C. botulinum* spores were isolated from 13.6% of cabbage samples [53] and from onion skin [54], garlic [55], and many other vegetables, as reviewed by Notermans [56]. *B. cereus* was isolated from vegetable seeds and from seed sprouts [57].

12.2.4 YEAST AND MOLD

The low pH (< 4.5) of most fruits means that spoilage is caused mainly by fungi. On the other hand, the pH range of most vegetables varies between 5.0 and 7.0 and thus spoilage may be caused by either fungi or bacteria, although the former are the most important group [1]. In terms of their spoilage characteristics fungi are often somewhat arbitrarily divided into two groups: the plant pathogens, which infect plants before harvesting, and the saprophytic fungi, which attack commodities after harvesting. An important property of most spoilage organisms, both fungal and bacterial, is their ability to secrete pectolytic enzymes, which soften and disintegrate plant tissues. Thus, the growth of fungi on fruits and vegetables usually results in

TABLE 12.1
Microbial Spoilage of Different Vegetables

Spoilage Type	Microorganisms Involved	Vegetables Affected
Alternaria leaf spot	*Alternaria brassicae* and *Alternaria oleracea*	All cruciferous leafy vegetables
Bacterial soft rot	*Erwinia carotovora* and others	Celery, all cruciferous leafy vegetables, beets and chards, lettuce, spinach, etc.
Bacterial leaf spot	*Pseudomonas maculicola*	Broccoli and cauliflower
Black rot	*Xanthomonas campestris*	All cruciferous leafy vegetables
Bacterial zonate spot	*Pseudomonas cichorii*	Cabbage primarily
Brown spot	*Cephalosporium apii*	Celery
Brown rot	*Alternaria brassicae*	Cauliflower (curd)
Big vein	*Olipidium brassicae*	Lettuce
Cercospora leaf spot	*Cercospora beticola*	Beets and chards
Downy mildew	*Peronospora parasitica*	Cabbage and cauliflower
Downy mildew	*Bremia lactucae* and *Peronospora effusa*	Lettuce/spinach
Gray mold rot	*Botrytis cinerea*	Celery, broccoli, brussels sprouts, cabbage and cauliflower, artichokes, and lettuce, etc.
Late blight	*Septoria apiicola*	Celery
Rhizoctonia head rot	*Rhizoctonia solani*	Cabbage mainly
Rhizopus soft rot	*Rhizopus stolonifer*	Brussels sprouts, cabbage, and cauliflower
Ring spot	*Mycosphaerella brassicicola*	Mainly cauliflower; occasionally brussels sprouts, cabbage, and kale
Watery soft rot	Various *Sclerotinia* spp.	Celery and lettuce

Source: From Rubatzky, V.E. and Yamaguchi, M., *World Vegetables: Principles, Production and Nutritive Values*, 2nd ed., Aspen Publishing, Gaithersburg, MD, 1999.

severe tissue breakdown, causing mushy areas; this spoilage is termed "rot." The names given to the different rots indicate the appearance of the food when it is spoiled [1]. The most common forms of bacterial and fungal spoilage are listed in Table 12.1 and Table 12.2.

An important cause of spoilage is *Penicillium*, many species of which are able to attack fruits; perhaps as much as 30% of all fruit decay can be attributed to this genus. Many fruits and vegetables, such as tomatoes, cucumbers, potatoes, and beets, are susceptible. *Penicillium* has a velvety colony that has blue-green centers with pale to bright yellow or yellow exudate. It has a fruity odor and has been isolated from decaying cabbage and barley plants, soils, stored seeds of cereals, grapes, nuts, dried fruits, and fruit juices. Another important disease is *Rhizopus* soft rot, which affects a wide range of fruits and vegetables, particularly during transit under poor refrigeration. Harvested strawberries and potatoes are often attacked and spoiled as indicated by soft, mushy areas with grayish mycelium evident in the affected areas [58].

Fusarium is a common soil fungus that sporulates in warm, wet weather and is widely found on grasses and other plants. It is found regularly on banana roots and other fruits and vegetables such as tomatoes and watermelons.

TABLE 12.2
Microbial Spoilage of Different Unripe and Ripe Fruits

Spoilage Type	Microorganisms Involved	Fruits Affected
Anthracnose	*Colletotrichum lindemuthianum*	Watermelon, tomato, cucumber, beans
Alternaria rot	*Alternaria tenuis*	Pumpkin, winter squashes, persians, cantaloupe, honeydew, crenshaw, casabas, bell pepper, eggplant, etc.
Bacterial soft rot	*Erwinia carotovora* and others	Tomato, cantaloupe, honeydew, persians, crenshaw, asparagus, casabas, bell pepper, eggplant, etc.
Bacterial spot	*Pseudomonas* spp. and *Xanthomonas* spp.	Tomato, cucumbers, bell pepper
Black rot	*Mycosphaerella citrullina*	Watermelon, cucumber
Blue mold rot	*Penicillium* spp.	Cantaloupes and Honeydews mainly
Buckeye rot	Various *Phytophthora* spp.	Tomato
Cottony leak	*Pythium butleri*	Cucumbers, beans
Cladosporium rot	*Cladosporuim* spp.	Tomato, cucumber, cantaloupe, honeydew, bell pepper, etc.
Fusarium rot	Various *Fisarium* spp.	Cantaloupe, honeydews, persians, crenshaw, casabas, asparagus, etc.
Gray mold rot	*Botrytis cinerea*	Tomato, bell pepper, peas
Late blight rot	*Phytophthora infestans*	Watermelon, tomato
Phompsis rot	*Phompsis vexans*	Eggplant
Phytophthora rot	Various *Phytophthora* spp.	Asparagus
Pleospora rot	*Pleospora lycopersici*	Tomato
Rhizopus rot	*Rhizopus stolonifer*	Cantaloupe, honeydews, persians, crenshaw, tomato, bell pepper, cucumber, casabas, etc.
Soil rot	*Rhizoctonia solani*	Tomato, beans
Stem-end rot	*Diplodia natalensis*	Watermelon
Watery soft Rot	Various *Sclerotinia* spp.	Beans, peas

Source: From Rubatzky, V.E. and Yamaguchi, M., *World Vegetables: Principles, Production and Nutritive Values*, 2nd ed., Aspen Publishing, Gaithersburg, MD, 1999.

Downy mildew disease of plants, especially in cool, humid regions, is caused by several fungi, including species of *Basidiophora, Bremia, Peronospora, Phytophthora, Plasmopara, Pseudoperonospora,* and *Sclerospora.* White, gray, bluish, or violet downy patches of mildew form mostly on the undersides of leaves in damp weather. Pale-green to yellow or brown areas usually develop on the upper leaf surface opposite the downy growth. Affected leaves often wilt, wither, and die early. Stems, flowers, and fruits are sometimes infected. Garden plants, bush fruits, vegetables, and certain trees, shrubs, field crops, and weeds are susceptible. The black spots seen on tomatoes are usually caused by *Alternaria*, which appears when the weather is warm and is often found on window frames where condensation has occurred [58].

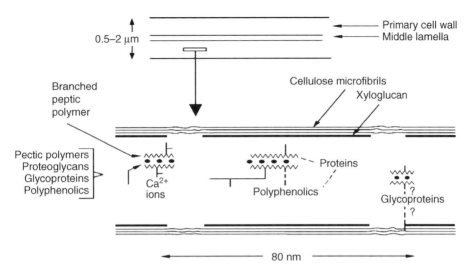

FIGURE 12.1 Pectic polymers in the primary wall of dicotyledonous and monocotyledonous plants. (From Selvendren, R.R., *J. Cell Sci.*, 2 (Suppl.), 51, 1985. With permission.)

12.3 GROWTH OF BACTERIA ON VEGETABLES

12.3.1 DECAY OF VEGETABLES BY BACTERIA

Bacteria are a major cause of disease and decay of vegetables in the field as well as after harvest. The ability of bacteria to degrade plant tissue is attributed to their ability to degrade pectic substances, which are important components of the middle lamella and the primary cell walls of the parenchymatous tissue of dicotyledonous and monocotyledonous plants. (Figure 12.1, Selvendren [59]).

The major constituents of pectic substances are rhamnogalacturonans. Those in which a proportion of the galacturonic acid residues are present as methyl esters are designated pectinic acids or pectins; those without methyl ester groups are called pectic acids.

According to Selvendren [59] there is still debate about the localization of pectins within the walls of parenchymatous tissues. The middle lamella of parenchymatous tissue is thought to consist principally of the calcium salts of pectins. The properties of pectin substances in plant cell walls have been reviewed by Selvendren [59].

The ability of bacteria to cause rotting of plant tissue results mainly from their production of an array of enzymes that attack the rhamnogalacturonan polymer by hydrolysis (hydrolases) or by elimination (lyases) [60,61]. The bacteria that are mainly responsible for rotting of vegetables after harvest are *E. carotovora* and certain fluorescent pseudomonads (*Pseudomonas marginalis*) [3,17,18]. They also contribute to the spoilage of prepared, ready-to-use, fresh vegetables [34]. Pectic enzymes are formed by many plant pathogenic bacteria, including species of *Erwinia, Pseudomonas,* and *Xanthomonas.* Other bacteria that form pectic enzymes and cause softening of plant tissue include *A. liquefaciens*, associated with the softening of ripe olives [62,63], *Cytophaga* spp. [16], *Bacillus* spp., *Clostridium* spp.

[3], *Leuc. mesenteroides* [64], and *Agrobacterium tumefaciens* biover 3 [65]. Bacteria reported to form pectic enzymes but not shown to cause marked softening of plant tissue include *Flavobacterium* spp. [66], *Klebsiella pneumoniae* and *Y. enterocolitica* [67–69], *Kl. oxytoca*, *Rhizobium* spp., *Arthrobacter* spp., and *Bacteriocides* spp.

The type and amount of pectic enzymes formed by soft-rot bacteria enable them to degrade the tissue of vegetables. To what extent the formation of more limited types and quantities of pectic enzymes by other bacteria facilitates their growth in association with plant tissue is not known.

12.3.2 Factors That Influence Microbial Growth

12.3.2.1 Introduction

Certain conditions favor the development of specific types of microflora and inhibit the development of others. These can be divided into intrinsic and extrinsic factors. Intrinsic factors are those that are characteristic of the produce itself; extrinsic factors are those that refer to the environment surrounding the produce.

12.3.2.2 Intrinsic Factors

12.3.2.2.1 Moisture Content

Microorganisms need water in an available form to grow in produce. The control of the moisture content in foods is one of the oldest exploited preservation strategies. Food microbiologists generally describe the water requirements of microorganisms in terms of the water activity (a_w) of the food or environment. Water activity is defined as the ratio of water vapor pressure of the food substrate to the vapor pressure of pure water at the same temperature [2]: $a_w = p/p_o$, where p = vapor pressure of the solution and p_o = vapor pressure of the solvent (usually water). The a_w of pure water is 1.00 and the a_w of a completely dehydrated food is 0.00. The a_w of a food on this scale from 0.00 to 1.00 is related to the equilibrium relative humidity above the food on a scale of 0 to 100%. Thus, % Equilibrium Relative Humidity (ERH) = $a_w \times 100$. The a_w of a food describes the degree to which water is "bound" in the food, its availability to participate in chemical biochemical reactions, and its availability to facilitate growth of microorganisms.

Most fresh fruits and vegetables have a_w values that are close to the optimum growth level of most microorganisms (0.97 to 0.99). Microorganisms respond differently to a_w depending on a number of factors. Microbial growth, and, in some cases, the production of microbial metabolites, may be particularly sensitive to alterations in a_w. Microorganisms generally have optimum and minimum levels of a_w for growth depending on other growth factors in their environments. One indicator of microbial response is their taxonomic classification. For example, Gram-negative bacteria are generally more sensitive to low a_w than are Gram-positive bacteria. Table 12.3 lists the approximate minimum a_w values for the growth of selected microorganisms relevant to food. It should be noted that many bacterial pathogens are controlled at water activities well above 0.86 and only *S. aureus* can grow and produce toxin below a_w 0.90. It must be emphasized that these are approximate values because solutes can vary in their ability to inhibit microorganisms at the same a_w value.

TABLE 12.3
Approximate Minimum a_w Values for Growth of Selected Microorganism at Optimal Temperature

Microorganisms	a_w
Bacteria	
Bacillus cereus	0.95
B. subtilis	0.90
Campylobacter spp.	0.98
Clostridium botulinum type E	0.97
Clostridium botulinum types A and B	0.94
Clostridium perfringens	0.94
Enterohemorrhagic *Escherichia coli*	0.95
Enterobacter aerogens	0.94
Listeria monocytogenes	0.92
Microbacterium spp.	0.94
Salmonella spp.	0.94
Shigella spp.	0.97
Staphylococcus aureus growth	0.83
Staphylococcus aureus toxin	0.88
Vibrio vulnificus	0.96
Vibrio parahaemolyticus	0.94
Yersinia enterocolitica	0.97
Mold	
Alternaria citri	0.84
Aspergillus candidus	0.75
A. niger	0.77
Botrytis cinerea	0.93
Chrysosporium fastidium	0.69
Pencillium brevicompactum	0.81
Rhizopus nigricans	0.93
Yeast	
Debaryomyces hansenii	0.83
Saccharomyces bailii	0.80
S. cerevisiae	0.90
S. rouxii	0.62

Sources: Troller, J.A. and Christian, J.H.B., *Water Activity and Food*, Academic Press, New York, 1978 and ICMSF, *Microbial Ecology of Foods*, Vol. 1, *Factors Affecting Life and Death of Microorganisms*, Academic Press, Orlando, FL, 1980.

12.3.2.2.2   pH and Acidity

Increasing the acidity of foods, either through fermentation or the addition of weak acids, has been used as a preservation method since ancient times. In their natural state, vegetables are slightly acidic, while most fruits are moderately acidic. Table 12.4 lists the pH ranges of some common fruits and vegetables.

TABLE 12.4
pH Ranges of Some Common Fruits and Vegetables

Food	pH	Ref.
Fruits		
Apples	2.9–3.3	1
Apple cider	3.6–3.8	7
Bananas	4.5–4.7	1, 3
Figs	4.6	1, 7
Grapefruit (juice)	3.0	1
Grapes	3.4–4.5	1, 7
Limes	1.8–2.0	1
Honeydew melons	6.3–6.7	7
Oranges (juice)	3.6–4.3	1, 7
Plums	2.8–4.6	1, 3
Watermelons	5.2–5.6	1
Vegetables		
Asparagus (buds and stalks)	5.7–6.1	1, 3
Beans (string and lima)	4.6–6.5	1, 3, 6
Beets (sugar)	4.2–4.4	3, 7
Broccoli	6.5	1,7
Brussels sprouts	6.3–6.6	1, 3
Cabbage (green)	5.4–6.0	1, 3, 6
Carrots	4.9–5.2; 6.0	1, 3
Cauliflower	5.6	3
Celery	5.7–6.0	1, 3, 6
Corn (sweet)	7.3	1, 3, 6, 7
Cucumbers	3.8	7
Eggplant	4.5	1, 7
Lettuce	6.0–6.4	1, 3
Olives (green)	3.6–3.8	1, 7
Onions (red)	5.3–5.8	1, 3, 6
Parsley	5.7–6.0	1, 6, 7
Potatoes (tubers and sweet)	5.3–5.6	2, 6, 7
Pumpkin	4.8–5.2	1, 7
Rhubarb	3.1–3.4	1, 7
Spinach	5.5–6.0	1, 3
Squash	5.0–5.4	1, 7
Tomatoes (whole)	4.2–4.3	1, 3, 7
Turnips	5.2–5.5	1, 3

Sources: Jay, J.M., *Modern Food Microbiology*, 6th ed., Aspen Publishing, Gaithersburg, MD, 2000; Burton, W.G., *The Potato*, 2nd ed., Veenman, Wageningen, Holland, 1966; Banwart, G.J., *Basic Food Microbiology*, 2nd ed., Van Nostrand Reinhold, New York, 1989; Daeschel, M.A. et al., *J. Food Sci.*, 55, 186, 1990; Sapers, G.M. et al., *J. Food Sci.*, 49, 233, 1984; Wolf, I.D. et al., *J. Food Sci.*, 44, 1008, 1979; and ICMSF, *Microbial Ecology of Foods*, Vol. 1, *Factors Affecting Life and Death of Microorganisms*, Academic Press, Orlando, FL, 1980.

It is well known that groups of microorganisms have optimum, minimum, and maximum pH for growth in produce. As with other factors, pH usually interacts with other parameters in the produce to inhibit growth. The pH can interact with factors such as a_w, salt, temperature, redox potential, and preservatives to inhibit growth of pathogens and other organisms. The pH of the produce also significantly affects the lethality of heat treatment of the plant food. Less heat is needed to inactivate microbes as the pH is reduced [72]. Some produce, such as fully ripe tomatoes, is in a pH range (3.9 to 4.5) that prevents or retards growth of enteric pathogens such as *Shigella* and *E. coli* O157:H7. The pH of many vegetables, melons, and soft fruits is 4.6 or higher, which is suitable for the growth of pathogenic bacteria. The growth and survival of human pathogens could be affected by the presence of plant pathogens such as *Botrytis cinerea* or *Penicillium* spp. [73]. Growth of post-harvest fungi in subsurface tissues can alter the pH of plant tissues, allowing the growth of pathogenic bacteria. Populations of *L. monocytogenes* inoculated into decayed apple tissue increased on fruit infected by *Glomerella cingulata* but not by *Penicillium expansum* [73–78]. This difference was attributed, in part, to the increase in pH of the infected tissues from 4.7 to 7.0 as a result of infection by *G. cingulata* compared to a decrease in pH from 4.7 to 3.7 as a result of infection by *P. expansum* [73]. Similar results were obtained with *E. coli* O157:H7 when it was coinoculated with *G. cingulata* or *P. expansum* on apples [75,77].

12.3.2.2.3 Nutrient Content

Microorganisms require certain basic nutrients for growth and maintenance of metabolic functions. The amount and type of nutrients required range widely depending on the microorganism. These nutrients include water, a source of energy, nitrogen, vitamins, and minerals [84–86]. Varying amounts of these nutrients are present in foods. Plant foods have high concentrations of different types of carbohydrates and varying levels of proteins, minerals, and vitamins.

Foodborne microorganisms can derive energy from carbohydrates, alcohols, and amino acids. Most microorganisms will metabolize simple sugars such as glucose. Others can metabolize more complex carbohydrates, such as starch or cellulose found in plant foods. Amino acids serve as a source of nitrogen and energy and are utilized by most microorganisms. Some microorganisms are able to metabolize peptides and more complex proteins. Other sources of nitrogen include, for example, urea, ammonia, creatinine, and methylamines. Examples of minerals required for microbial growth include phosphorus, iron, magnesium, sulfur, manganese, calcium, and potassium. In general, small amounts of these minerals are required; thus, a wide range of foods can serve as good sources of minerals [84–86].

In general, the Gram-positive bacteria are more fastidious in their nutritional requirements and thus are not able to synthesize certain nutrients required for growth [86]. For example, the Gram-positive foodborne pathogen *S. aureus* requires amino acids, thiamine, and nicotinic acid for growth [86]. Fruits and vegetables that are deficient in B vitamins do not effectively support the growth of these microorganisms. The Gram-negative bacteria are generally able to derive their basic nutritional requirements from the existing carbohydrates, proteins, minerals, and vitamins that are found in a wide range of food [86].

The microorganisms that usually predominate in foods are those that can most easily utilize the nutrients present. Generally, the simple carbohydrates and amino acids are utilized first, followed by the more complex forms of these nutrients. The complexity of foods in general is such that several microorganisms can be growing in a food at the same time. The rate of growth is limited by the availability of essential nutrients. The abundance of nutrients in most foods is sufficient to support the growth of a wide range of foodborne pathogens. Thus, it is very difficult and impractical to predict pathogen growth or toxin production based on the nutrient composition of the food.

12.3.2.2.4 Biological Structure

Plant foods, especially in the raw state, have biological structures that may prevent the entry and growth of microorganisms. Examples of such physical barriers include testa of seeds, skin of fruits and vegetables, and shells of nuts. Plant foods may have pathogenic microorganisms attached to the surface or trapped within surface folds or crevices. Intact biological structures thus can be important in preventing entry and subsequent growth of microorganisms. Several factors may influence penetration of these barriers. The maturity of plant foods influences the effectiveness of their protective barriers. Physical damage due to handling during harvest, transport, or storage, as well as invasion of insects, can allow the penetration of microorganisms [84,86]. During the preparation of foods, processes such as slicing, chopping, grinding, and shucking will destroy the physical barriers. Thus, the interior of the food can become contaminated and growth can occur depending on the intrinsic properties of the food. For example, *Salmonella* spp. have been shown to grow on the interior of portions of cut cantaloupe, watermelon, honeydew melons [87], and tomatoes [88], given sufficient time and temperature.

Fruits are an example of the potential of pathogenic microorganisms to penetrate intact barriers. After harvest, pathogens will survive but usually not grow on the outer surface of fresh fruits and vegetables. Growth on intact surfaces is not common because foodborne pathogens do not produce the enzymes necessary to break down the protective outer barriers on most produce. This outer barrier restricts the availability of nutrients and moisture. One exception is the reported growth of *E. coli* O157:H7 on the surface of watermelon and cantaloupe rinds [89]. Survival of foodborne pathogens on produce is significantly enhanced once the protective epidermal barrier has been broken either by physical damage, such as punctures or bruising, or by degradation by plant pathogens (bacteria or fungi). These conditions can also promote the multiplication of pathogens, especially at higher temperatures. Infiltration of fruit was predicted and described by Bartz and Showalter [90] based on the general gas law, which states that any change in pressure of an ideal gas in a closed container of constant volume is directly proportional to a change in the temperature of the gas. In their work, Bartz and Showalter described a tomato; however, any fruit, such as an apple, can be considered a container that is not completely closed. As the container or fruit cools, the decrease in internal gas pressure results in a partial vacuum inside the fruit, which then results in an influx from the external environment. For example, an influx of pathogens from the fruit's surface or cooling water could occur as a result of a increase in external pressure due to immersing warm fruit in cool water. Internalization of bacteria into fruits and vegetables could also occur due to breaks in the tissues or through morphological structures in the fruit itself, such as the calyx or stem scar. Although infiltration was considered a possible scenario, there is insufficient epidemiological evidence to require refrigeration of intact fruit.

12.3.2.2.5 Redox Potential

The oxidation–reduction, or redox, potential of a substance is defined in terms of the ratio of the total oxidizing (electron accepting) power to the total reducing (electron donating) power of the substance. In effect, redox potential is a measurement of the ease by which a substance gains or loses electrons. The redox potential (Eh) is measured in terms of millivolts. A fully oxidized standard oxygen electrode will have an Eh of +810 mV at pH 7.0, 30°C (86°F), and under the same conditions, a completely reduced standard hydrogen electrode will have an Eh of –420 mV. The Eh is dependent on the pH of the substrate; normally, the Eh is taken at pH 7.0 [86, pp. 45–47].

The major groups of microorganisms based on their relationship to Eh for growth are aerobes, anaerobes, facultative aerobes, and microaerophiles. Examples of foodborne pathogens for each of these classifications include *Aeromonas hydrophila, Clostridium botulinum, Escherichia coli* O157:H7, and *Campylobacter jejuni,* respectively. Generally, the range at which different microorganisms can grow are as follows: aerobes +500 to +300 mV; facultative anaerobes +300 to –100 mV; and anaerobes +100 to less than –250 mV ([88], pp. 69–70). For example, *C. botulinum* is a strict anaerobe that requires an Eh of less than +60 mV for growth; however, slower growth can occur at higher Eh values. The measured Eh values of various foods can be highly variable depending on changes in the pH of the food, microbial growth, packaging, the partial pressure of oxygen in the storage environment, and ingredients and composition (protein, ascorbic acid, reducing sugars, oxidation level of cations, and so on). Another important factor is the poising capacity of the food, which is analogous to buffering capacity and relates to the extent to which a food resists external affected changes in Eh. The poising capacity of the food will be affected by oxidizing and reducing constituents in the food as well as by the presence of active respiratory enzyme systems. Fresh fruits and vegetables continue to respire; thus, low Eh values can result [91].

12.3.2.2.6 Naturally Occurring Antimicrobials

Some foods intrinsically contain naturally occurring antimicrobial compounds that convey some level of microbiological stability to them. There are a number of plant-based antimicrobial constituents, including many essential oils, tannins, glycosides, and resins, that can be found in certain foods. Specific examples include eugenol in cloves, allicin in garlic, cinnamic aldehyde and eugenol in cinnamon, allyl isothiocyanate in mustard, eugenol and thymol in sage, and carvacrol (isothymol) and thymol in oregano [86, pp. 266–267]. Other plant-derived antimicrobial constituents include the phytoalexins and the lectins. Lectins are proteins that can specifically bind to a variety of polysaccharides, including the glycoproteins of cell surfaces [84, pp. 175–214]. Through this binding, lectins can exert a slight antimicrobial effect. The usual concentration of these compounds in formulated foods is relatively low, so the antimicrobial effect alone is slight. However, these compounds may produce greater stability in combination with other factors in the formulation.

Some types of fermentations can result in the natural production of antimicrobial substances, including bacteriocins, antibiotics, and other related inhibitors. Bacteriocins are proteins or peptides that are produced by certain strains of bacteria that inactivate other, usually closely related, bacteria [92]. The most commonly characterized bacteriocins are those produced by the lactic acid bacteria (e.g., fermentation of Chinese cabbage for the production of kimchi [93]).

12.3.2.2.7 Competitive Microflora

The potential for microbial growth of pathogens in temperature-sensitive foods depends on the combination of the intrinsic and extrinsic factors and the processing technologies that have been applied. Within the microbial flora in a food there are many important biological attributes of individual organisms that influence the species that predominates. These include the individual growth rates of the microbial strains and the mutual interactions or influences among species in mixed populations [71, pp. 221–231].

12.3.2.2.7.1 Growth

In a food environment, an organism grows in a characteristic manner and at a characteristic rate. The length of the lag phase, generation time, and total cell yield are determined by genetic factors. Accumulation of metabolic products may limit the growth of particular species. If the limiting metabolic product can be used as a substrate by other species, these may take over (partly or wholly), creating an association or succession [71, p. 222]. Due to the complex of continuing interactions between environmental factors and microorganisms, a food at any one point in time has characteristic flora, known as its association. The microbial profile changes continuously and one association succeeds another in what is called "succession." Many examples of this phenomenon have been observed in th microbial deterioration and spoilage of foods [71, p. 226]. As long as metabolically active organisms remain, they continue to interact, so that dominance in the flora occurs as a dynamic process. Based on their growth-enhancing or -inhibiting nature, these interactions are either antagonistic or synergistic.

12.3.2.2.7.2 Competition

In food systems, antagonistic processes usually include competition for nutrients, competition for attachment/adhesion sites (space), unfavorable alterations of the environment, and a combination of these factors. Early studies demonstrated that the natural biota of frozen pot pies inhibited inoculated cells of *S. aureus*, *E. coli*, and *Salmonella typhimurium* [84, p. 52]. Even though *S. aureus* is often found in low numbers in this product, staphylococcal enterotoxin is not produced. The reason is that the *Pseudomonas-Acinetobacter-Moraxella* association that is always present in this food grows at a higher rate, outgrowing the staphylococci [71, p. 222].

Organisms of high metabolic activity may consume required nutrients, selectively reducing these substances and inhibiting the growth of other organisms. Depletion of oxygen or accumulation of carbon dioxide favors facultative obligate anaerobes that occur in vacuum-packaged fresh meats held under refrigeration [71, p. 222].

Staphylococci are particularly sensitive to nutrient depletion. Coliforms and *Pseudomonas* spp. may utilize amino acids necessary for staphylococcal growth and make them unavailable. Other genera of Micrococcaceae can utilize nutrients more rapidly than staphylococci. Streptococci inhibit staphylococci by exhausting the supply of nicotinamide or niacin and biotin [71, p. 222]. *Staphylococcus aureus* is a poor competitor in both fresh and frozen foods. At temperatures that favor staphylococcal growth, the normal food saprophytic biota offers protection against staphylococcal growth through antagonism, competition for nutrients, and modification of the environment to conditions less favorable to *S. aureus* [84, p. 455). Changes in the composition of the food, as well as changes in intrinsic or extrinsic factors, may either stimulate or decrease competitive effects.

12.3.2.2.7.3   Effects on Growth Inhibition
Changes in growth stimulation have been reported among several foodborne organisms, including yeasts, micrococci, streptococci, lactobacilli, and Enterobacteriaceae [71, p. 224]. Growth-stimulating mechanisms can have a significant influence on the buildup of typical flora. There are several of these mechanisms [71, p. 224]. Metabolic products from one organism can be absorbed and utilized by other organisms. Changes in pH may promote the growth of certain microorganisms. An example is natural fermentations, in which acid production establishes the dominance of acid-tolerant organisms such as the lactic acid bacteria. Growth of molds on high-acid foods has been found to raise the pH, thus stimulating the growth of *C. botulinum*. Changes in Eh or a_w in the food can influence symbiosis. There are some associations where maximum growth and normal metabolic activity are not developed unless both organisms are present. This information can be used to control microorganisms in foods.

12.3.2.3   Extrinsic Factors

12.3.2.3.1   Types of Packaging/Atmospheres
Many scientific studies have demonstrated the antimicrobial activity of gases at ambient and subambient pressures on microorganisms important in foods [94]. Gases inhibit microorganisms by two mechanisms. First, they can have a direct toxic effect that can inhibit growth and proliferation. Carbon dioxide (CO_2), ozone (O_3), and oxygen (O_2) are gases that are directly toxic to certain microorganisms. This inhibitory mechanism is dependent upon the chemical and physical properties of the gas and its interaction with the aqueous and lipid phases of the food. Oxidizing radicals generated by O_3 and O_2 are highly toxic to anaerobic bacteria and can have an inhibitory effect on aerobes depending on their concentration. Carbon dioxide is effective against obligate aerobes and at high levels can deter other microorganisms. A second inhibitory mechanism is achieved by modifying the gas composition, which has indirect inhibitory effects by altering the ecology of the microbial environment. When the atmosphere is altered, the competitive environment is also altered. Atmospheres that have a negative effect on the growth of one particular microorganism

may promote the growth of another. This effect may have positive or negative consequences depending upon the native pathogenic microflora and their substrate. Nitrogen replacement of oxygen is an example of this indirect antimicrobial activity [94].

Controlled atmosphere and modified atmosphere packaging (MAP) of certain foods can dramatically extend their shelf life. The use of CO_2, N_2, and ethanol are examples of MAP applications. In general, the inhibitory effects of CO_2 increase with decreasing temperature due to the increased solubility of CO_2 at lower temperatures [84, p. 286]. Carbon dioxide dissolves in the food and lowers the pH of the food. Nitrogen, being an inert gas, has no direct antimicrobial properties. It is typically used to displace oxygen in the food package either alone or in combination with CO_2, thus having an indirect inhibitory effect on aerobic microorganisms [94]. This principle of antimicrobial atmospheres has been applied to fruits and vegetables and a variety of prepared, ready-to-eat foods.

12.3.2.3.2   Effect of Time/Temperature Conditions on Microbial Growth

12.3.2.3.2.1   Impact of Time

When considering growth rates of microbial pathogens, in addition to temperature, time is a critical consideration. Food producers or manufacturers address the concept of time as it relates to microbial growth when a product's shelf life is determined. Shelf life is the time period from when the product is produced until the time it is intended to be consumed or used. Several factors are used to determine a product's shelf life, ranging from organoleptic qualities to microbiological safety. The shelf life of a perishable food product is expressed in terms of a "sell by" date [95]. The "sell by" date must incorporate the shelf life of the product plus a reasonable period for consumption that consists of at least one-third of the approximate total shelf life of the perishable food product.

Under certain circumstances, time alone at ambient temperatures can be used to control product safety. When time alone is used as a control, the duration should be equal to or less than the lag phase of the pathogen(s) of concern in the product in question. For refrigerated food products, the shelf life or use-period required for safety may vary depending on the temperature at which the product is stored. For example, Mossel and Thomas [96] report that the lag time for growth of *L. monocytogenes* at 10°C (50°F) is 1.5 days, while at 1°C (34°F) the lag time is ~3.3 days. Likewise, they reported that at 10°C (50°F) the generation time for the same organism is 5 to 8 h, while at 1°C (34°F), the generation time is between 62 and 131 h. However, according to the USDA Pathogen Micromodel Program (version 5.1) [97], at 2% NaCl concentration and a_w of 0.989, a temperature shift from 10°C (50°F) to 25°C (77°F) decreases the lag time of *L. monocytogenes* from 60 to 10 h. In a similar manner, a pH increase from 4.5 to 6.5 decreases the lag time from 60 to 5 h.

Therefore, the safety of a product during its shelf life may differ, depending upon other conditions such as temperature of storage, pH of the product, and so on. Mossel and Thomas [96], along with numerous others, illustrated that various time/temperature combinations can be used to control product safety depending on the product's intended use.

12.3.2.3.2.2   Impact of Temperature

All microorganisms have a defined temperature range in which they grow, with a minimum, maximum, and optimum. An understanding of the interplay between time, temperature, and other intrinsic and extrinsic factors is crucial to selecting the proper storage conditions for a food product. Temperature has dramatic impact on both the generation time of an organism and its lag period. Over a defined temperature range, the growth rate of an organism is classically defined as an Arrhenius relationship [84, pp. 79–80]. The log growth rate constant is found to be proportional to the reciprocal of the absolute temperature:

$$G = -\mu/2.303 \, RT$$

where G = log growth rate constant, μ= temperature characteristic (constant for a particular microbe), R = gas constant, and T = temperature (°K).

The above relationship holds over the linear portion of the Arrhenius plot. However, when temperatures approach the maximal for a specific microorganism, the growth rate declines more rapidly than when temperatures approach the minimal for that same microorganism. A relationship that more accurately predicts growth rates of microorganisms at low temperatures follows [86, p. 51]:

$$r = b(T - To),$$

where r = growth rate, b = slope of the regression line, T = temperature (°K), and To = conceptual temperature of no metabolic significance.

At low temperatures, two factors govern the point at which growth stops: (1) reaction rates for the individual enzymes in the organism become much slower and (2) low temperatures reduce the fluidity of the cytoplasmic membrane, thus interfering with transport mechanisms [84, pp. 79–80]. At high temperatures, structural cell components become denatured and inactivation of heat-sensitive enzymes occurs. While the growth rate increases with increasing temperature, the rate tends to decline rapidly thereafter, until the temperature maximum is reached.

The relationship between temperature and growth rate constant varies significantly across groups of microorganisms. Four major groups of microorganisms have been described based on their temperature ranges for growth: thermophiles, mesophiles, psychrophiles, and psychrotrophs. The optimum temperature for growth of thermophiles is between 55 and 65°C (131 to 149°F) with the maximum as high as 90°C (194°F) and a minimum of around 40°C (104°F). Mesophiles, which include virtually all human pathogens, have an optimum growth range of between 30°C (86°F) and 45°C (113°F) and a minimum growth temperature ranging from 5 to 10°C (41 to 50°F). Psychrophilic organisms have an optimum growth range of 12°C (54°F) to 15°C (59°F) with a maximum range of 15°C (59°F) to 20°C (68°F). Psychrotrophs such as *L. monocytogenes* and *C. botulinum* type E are capable of growing at low temperatures (minimum of –0.4°C [31°F] and 3.3°C [38°F], respectively, to 5°C [41°F]) but have a higher growth optimum range (37°C [99°F] and 30°C [86°F], respectively) than true psychrophiles [98–100]. Psychrotrophic organisms are much more relevant to food and include spoilage bacteria, spoilage yeast and molds, and certain foodborne pathogens.

Growth temperature is known to regulate the expression of virulence genes in certain foodborne pathogens [101]. For example, the expression of proteins governed by the *Yersinia enterocolitica* virulence plasmid is high at 37°C (99°F), low at 22°C (72°F), and not detectable at 4°C (39°F). Growth temperature also affects an organism's thermal sensitivity. It must be emphasized that the lag period and growth rate of a microorganism are influenced not only by temperature but also by other intrinsic and extrinsic factors as well. For example, the growth rate of *Clostridium perfringens* is significantly lower at pH 5.8 vs. pH 7.2 across a wide range of temperatures [71, p. 10]. Salmonellae do not grow at temperatures below 5.2°C (41°F).

The intrinsic factors of the food product, however, have been shown to affect the ability of salmonellae to grow at low temperatures. *Salmonella senftenberg, S. enteritidis*, and *S. manhattan* were not able to grow in ham salad held at 10°C (50°F) but were able to grow in chicken à la king held at 7°C (45°F) [71, p. 9]. *Staphylococcus aureus* has been shown to grow at temperatures as low as 7°C (45°F), but the lower limit for enterotoxin production has been shown to be 10°C (50°F). In general, toxin production below about 20°C (68°F) is slow. For example, in laboratory media at pH 7, the time to produce detectable levels of enterotoxin ranged from 78 to 98 h at 19°C (66°F) to 14 to 16 h at 26°C (79°F) [71, p. 10].

12.3.2.3.3 Storage/Holding Conditions

When considering growth rates of microbial pathogens, time and temperature are integral and must be considered together. As has been stated previously in this chapter, increases in storage or display temperature will decrease the shelf life of refrigerated foods since the higher the temperature the more permissive conditions are for growth. Generation times as short as 8 min have been reported in certain foods under optimal conditions [98]. Thus time/temperature management is essential for product safety.

The literature is replete with examples of outbreaks of foodborne illness that have resulted from cooling food too slowly, a practice that may permit growth of pathogenic bacteria. Of primary concern in this regard are the spore-forming pathogens that have relatively short lag times and the ability to grow rapidly and/or that may normally be present in large numbers. Organisms that possess such characteristics include *C. perfringens* and *Bacillus cereus*. As with *C. perfringens,* foodborne illness caused by *B. cereus* is typically associated with consumption of food that has supported growth of the organism to relatively high numbers. The FDA "Bad Bug Book" notes that "The presence of large numbers of *B. cereus* (greater than 10^6 organisms/g) in a food is indicative of active growth and proliferation of the organism and is consistent with a potential hazard to health" [102]. In this case, the time and temperature (cooling rate) of certain foods must be addressed to ensure rapid cooling for safety.

The effect of the relative humidity of the storage environment on the safety of foods is somewhat more nebulous. The effect may or may not alter the a_w of the food. Such changes are product-dependent. The earlier discussion on a_w and its effect on microorganisms in foods provides some background information. In addition, the possibility of surface evaporation or condensation of moisture on a surface should be considered.

Generally, foods that depend on a certain a_w for safety or shelf life considerations will need to be stored such that the environment does not markedly change this characteristic. Foods will eventually come to moisture equilibrium with their surroundings. Thus, processors and distributors need to provide for appropriate storage conditions to account for this fact.

Packaging plays a major role in the vulnerability of the food to the influence of relative humidity. But even within a sealed container, moisture migration and the phenomenon of environmental temperature fluctuation may play a role. It has been observed that certain foods with low a_w can be subject to moisture condensing on the surface due to wide environmental temperature shifts. This surface water will result in microenvironments favorable to growth of spoilage, and possibly pathogenic, microorganisms. As a general guideline, the product should be held such that environmental moisture, including that within the package, does not have an opportunity to alter the a_w of the product in an unfavorable way.

The ability of bacteria to multiply on vegetables is influenced strongly by environmental factors including the storage temperature, the presence of free water, the relative humidity, and the gaseous environment. There has been considerable interest in the possible use of modified atmospheres to extend the storage life of vegetables [103–106]. Controlled atmospheres usually contain a lower concentration of oxygen and a higher concentration of carbon dioxide than is found in air. This tends to slow the respiration of the vegetable, retard ripening, and maintain quality for a longer time than during storage in air. The minimum oxygen concentration that will avoid injury to produce is approximately 2% in most cases. Below this level anaerobic respiration may occur [106]. Controlled or modified atmospheres are combined with refrigeration for greatest effectiveness.

A study of the effect of controlled atmospheres on the growth of *L. monocytogenes* on asparagus, broccoli, and cauliflower and on the shelf life of these vegetables was conducted by Berrang et al. [107]. A high relative humidity was maintained during the experiments. The controlled atmospheres extended the storage time at 4 or 15°C for which the vegetables remained acceptable for consumption but did not affect the growth of *L. monocytogenes*. Similar results were obtained with *A. hydrophila* [108]. Controlled atmosphere has similar effects on the sensory qualities and the natural microflora of broccoli [109] and bell peppers [110].

Modified, but not controlled, atmospheres may be used for packaging vegetables but respiration of the produce can result in high concentrations of carbon dioxide and depletion of oxygen to concentrations below 1 to 2%. Spoilage of such products is associated with high counts of facultatively anaerobic bacteria and usually occurs before any significant multiplication of *Clostridium* spp. If severe depletion of oxygen occurred in packs, then it would be expected that a decrease in quality of the product would make it unsaleable before growth of *Clostridium* spp. had occurred. In some cases, however, microenvironments have become established in vegetable products other than canned foods and have resulted in growth of *C. botulinum* and outbreaks of botulism.

12.3.3 MICROENVIRONMENTS IN VEGETABLES

Some outbreaks of botulism have resulted from the use of partially processed vegetable products in which the risk had not been foreseen but in which the microenvironments allowed the growth of *C. botulinum*.

Potato salad has been implicated in three outbreaks of type A botulism in the U.S. One in Colorado in 1969 involved six people [111,112], one in New Mexico in 1978 affected 34 people [113], and one in Colorado in 1978 affected 8 people [114]. In at least two of these incidents the salad was prepared from potatoes that had been baked in aluminum foil and kept for several days at room temperature. Potatoes that were surfaced or stab-inoculated with spores of type A *C. botulinum*, wrapped in aluminum foil, baked, then left at 22°C became toxic in 6 to 7 days when they were not always overtly spoiled [115]. Boiled potatoes inoculated with spores used immediately to prepare potato salad with a pH of 5.2 and incubated anaerobically at room temperature for 9 d failed to become toxic [114]. Plain potatoes treated similarly were toxic in 24 h. Both proteolytic and nonproteolytic strains of *C. botulinum* have been reported to survive the commercial cooking process for vacuum-packed potatoes and to produce toxin during storage, sometimes before obvious spoilage [116–118].

In 1983 an outbreak of botulism occurred in Peoria, Illinois, in which 28 people were hospitalized and 1 died. Type A toxin was involved and sautéed onions were identified as the probable food vehicle [54]. It was claimed that the sautéed onions were prepared freshly each morning and that any leftovers were discarded at the end of the day. Strains of *C. botulinum* type A were isolated from patients and the skins of onions, and spores that were inoculated into onions had been sautéed in margarine and cooled, which were then incubated at 35°C to simulate the holding temperature in the restaurant [54]. As few as two spores per gram of these strains in sautéed onions resulted in growth and toxin formation within 48 h at 35°C, while the appearance of the food was normal. It was concluded that the margarine probably created an anaerobic environment that allowed growth of the organism.

An outbreak of Type A botulism occurred in Japan in 1984 and involved 36 people, of whom 11 died [119]. The causative food was vacuum-packed, deep-fried, mustard-stuffed lotus root. After the stuffing had been placed in the lotus root and the food was refrigerated. The pH and the a_w of the stuffing increased to levels that allowed growth of *C. botulinum*. Growth of this organism and toxin formation occurred, but only if the product was subjected to a heat treatment, which was part of the manufacturing process. Whether the heat treatment killed competing bacteria or altered the food so as to provide more favorable conditions for growth was not determined. The author considered that vacuum packaging did not enhance growth of *C. botulinum* in the product. It may have contributed to the outbreak in that people assumed that the vacuum-packaged food had a longer shelf life than products not so packed.

Between July and September 1985, 36 cases of type B botulism were caused by food served in a restaurant in Vancouver, British Colombia [120,121]. Commercially prepared bottled, chopped garlic in soybean oil was implicated epidemiologically. Although the product was labeled "keep refrigerated" in very small print, the jar of garlic in the restaurant had been kept at room temperature (25 to 32°C). The chopped garlic in soybean was produced by rehydration of dry, chopped garlic to which soybean oil was added. The product was filled into 8-ounce glass jars. Each jar contained 125 g of garlic and 90 mL of water-in-oil emulsion, pH 5.8 to 5.9. Inoculation studies showed that type A and proteolytic type B strains multiplied and formed toxin in the product during incubation at 35°C [55]. The product remained organoleptically acceptable even when highly toxic. Toxin was also formed when the product was incubated at room temperature. (It is of interest that a type B strain from a patient involved in the outbreak produced a relatively high level of toxin in the product.) One or two nonproteolytic type B strains grew and formed toxin in the chopped garlic in oil incubated at room temperature. Thus, both onions and garlic in an oil menstruum supported the growth of and toxin formation by *C. botulinum* types A and B from a minimal inoculum of one to five spores per gram while remaining organoleptically acceptable. In 1989 three cases of type A botulism were attributed to the consumption of a chopped garlic-in-oil product [122]. The remains of the products contained high concentrations of the organism and the toxin. After this second outbreak the U.S. Food and Drug Administration ordered companies to stop making any garlic-in-oil mixes that are protected only by refrigeration. Such products must now contain specific labels of microbial inhibitors or the acidifying agents, such as phosphoric or citric acid.

12.4 INTERACTIONS BETWEEN SPOILING BACTERIA

Microbial food spoilage is a process involving growth of microorganisms to numbers (10^7 to 10^9 cfu/g) at which the microorganisms also must be assumed to interact and influence the growth of one another [123]. The interactions between microorganisms may be classified on the basis of their effects as being detrimental or beneficial [124]. Several types of interactions have been studied in food ecosystems, including both antagonistic and coordinated behavior and interactions where growth or a particular metabolism of one organism is favored by the growth of another organism.

12.4.1 ANTAGONISM

Changes in environmental conditions (e.g., by lowering of pH) can be a powerful way for a microorganism to antagonize other bacteria and create a selective advantage. Also, competition for nutrients may select for the organisms best capable of scavenging the limiting compounds. Several microorganisms important in food spoilage have such antagonistic abilities. Thus, the lactic acid bacteria cause a lowering of pH and may produce antibacterial peptides (bacteriocins) [125]. The spoilage reactions of certain Gram-negative bacteria may produce NH_3 and trimethyl amine, which are toxic to a number of other bacteria and sometimes to the producing organism itself. *Pseudomonas* spp., in particular the fluorescent group, produce a range of antibacterial and antifungul compounds such as antibiotics and cyanide, and at the same time they compete very efficiently for iron [126]. Iron is essential for most microorganisms and is used in bacterial respiration (as electron shuttler) and in redox enzymes. Due to the high oxidative power of Fe^{+3}, iron is mostly bound in insoluble complexes in the environment and in mammals and plants. Most microorganisms have therefore developed highly specific iron-chelating systems, and often they produce siderophores, which are iron-chelators secreted by the cell. Upon binding of iron the siderophore-iron complex is taken up by the microbial cells and iron is liberated internally [127]. In particular, pseudomonads are prominent producers of siderophores with high iron binding constants. The iron-chelating ability has been of particular interest in the rhizosphere, where the use of pseudomonads as biocontrol agents against fungal diseases has been attributed in part to their competitive advantage vis-à-vis iron [126,128]. Bacteriocinogenic strains of *Pediococcus* and *Enterococcus* have recently been shown to control the growth of *L.monocytogenes* on mung bean sprouts [129].

Several antagonistic microorganisms have been discovered to reduce postharvest fungal decay of apples and other pome fruits. Strains of *Pseudomonas syringae* Van Hall are effective in controlling blue mold and gray mold, caused by *Penicillium* spp. and *Botrytis* spp., respectively, of citrus and pome fruits [130,131]. Other bacteria such as *Pseudomonas cepacia* Burkholder [132], *Pseudomonas gladioli* Severini [133], *Bacillus pumilus*, and *Bacillus amyloliquefaciens* [134] also have been reported to reduce blue mold and/or gray mold on apples and pears. The yeast *Candida oleophila* (Aspire™) is effective for the control of blue mold and gray mold on citrus and pome fruits [135,136]. Species of *Cryptococcus*, including *C. laurentii*, *C. albidus*, and *C. flavus*, reduced gray mold in pears [137]. *Cryptococcus laurentii* and *Rhodotorula glutinis* were effective in controlling blue mold on apples and pears [75]. The yeasts *Picia anomala* and *Candida sake* have controlled both blue mold and gray mold on apples and pears [73,138]. *P. syringae* [139] are also effective in controlling blue mold and gray mold on apples. Besides the control of blue mold and gray mold, these *P. syringae* isolates have been effective for control of peach brown rot (*M. fructicola*) and rhizopus rot (*Rhizopus* spp.) [139] and apple scab (*V. inaequalis*) [140]. However, these *P. syringae* isolates performed poorly when tested against *M. fructicola* on sweet cherries.

12.4.2 METABIOSIS

Innumerable interdependencies exists between different organisms. The term *metabiosis* describes the reliance by an organism on another to produce a favorable environment. This can be the removal of oxygen by Gram-negative microflora, allowing anaerobic organisms such as *Clostridium botulinum* to grow [141], or it can be situations where one organism provides nutrients that enhance the growth of another. Several studies have shown that despite the inhibitory activity of pseudomonads describe above, their presence may also enhance growth of some microorganisms. Preinoculation with different Gram-negative psychrotrophic bacteria subsequently yielded higher growth and more acid from lactic acid bacteria [142], and growth of *Staphylococcus aureus* may also be stimulated by *Pseudomonas* spp. [143]. Such nutrient interdependency may also play a role in food spoilage [144,145].

12.4.3 ACYLATED HOMOSERINE LACTONE (AHL)-BASED COMMUNICATION

In recent years, it has been recognized that bacteria can behave not only as individual cells but, under appropriate conditions and when their numbers reach a critical level, they can modify their behavior to act as a multicellular group. This is achieved by cell-to-cell communication. Bacteria secrete chemicals into the surrounding environment, and the concentration of the chemicals that accumulates is dependent on population density. By detecting and reacting to these chemicals, individual cells can sense how many cells surround them, and whether there are enough bacteria (a "quorum") to initiate the change toward acting in a multicellular fashion. This is known as *quorum sensing* [146].

Changes in behavior result from switching on specific genes in response to the signal. The purpose of the change in behavior is that a population of bacteria can cooperate to exploit their environment in ways that individual cells cannot. For example, a single pathogenic bacterium attempting to invade its host has little chance of overcoming the plant's defense systems. Such a pathogen benefits from delaying expression of virulence factors until there are sufficient numbers of bacteria present to ensure success. Similarly, beneficial bacteria responsible for nitrogen fixation may use quorum sensing to optimize nodule formation on plant roots. Many bacteria produce substances that kill, or inhibit the proliferation of other, disease-causing microorganisms, and can therefore be useful as "biocontrol" agents. However, these substances may only be produced when the bacteria reach a critical population density.

One set of signals that these diverse types of bacteria use in quorum sensing is a family of structurally related chemicals. These chemicals are based on a modified amino acid (homoserine lactone) carrying a variable acyl chain substituent and are called acyl homoserine lactones (AHLs). The diversity in the acyl chain (chain length, degree of oxidation and saturation) can confer some specificity on the communications system. Nevertheless, it is likely that there is some cross-talk between bacterial genera: plant-growth-promoting bacteria can influence the quorum sensing systems of plant pathogens. Some of this cross-talk may represent a way for bacteria to acquire information on the total bacterial population. This could permit a response to competitors, or potential associates, or a method of direct competition if a particular AHL has a detrimental effect on other species. Optimizing the beneficial traits of plant-growth-promoting bacteria, therefore, requires an understanding of the cell-to-cell communication that occurs between members of one species, as well as cross-talk with other bacterial types [141,147].

The production of AHLs and regulation of different phenotypic traits have been reported in many Gram-negative bacteria [146]. Classical examples include regulation of symbiotic behavior (e.g., bioluminescence in *V. fischeri*) or virulence factors (e.g., elastase in *Pseudomonas aeruginosa* [148], antibiotic production in *Erwinia carotovora* [149], and Ti-plasmid transfer in *Agrobacterium tumefaciens* [150], but also more complex behavior such as surface motility and colonization of *S. liquefaciens* [151,152] and biofilm formation of *P. aeruginosa* [147] and *Burkholderia cepacia* [153] have been noted. Cell density-dependent regulations of gene expression probably reflect the need for the invading pathogen to reach a critical population density sufficient to overwhelm host defenses and thus establish infection. For example, transgenic plants producing *N*-oxoacyl-honoserine lactone (OHL) showed increased resistance to infecting *E. carotovora*. The plant-originating signal molecules force the *E. carotovora* to switch on production of virulence factors at low bacterial population density. Production of virulence factors elicits a plant defense that, due to the low and insufficient bacterial density, pushes the balance of the plant–pathogen interaction in the direction of infection abortion [154].

12.5 ECOLOGICAL FACTORS INFLUENCING SURVIVAL AND GROWTH OF PATHOGENS ON FRUITS AND VEGETABLES

Outbreaks of human infections associated with consumption of raw fruits and vegetables have occurred with increased frequency during the past decade. Factors contributing to this increase may include changes in agronomic and processing practices, an increase in per capita consumption of raw or minimally processed fruits and vegetables, increased international trade and distribution, and an increase in the number of immunocompromised consumers. A general lack of efficacy of sanitizers in removing or killing pathogens on raw fruits and vegetables has been attributed, in part, to their inaccessibility to locations within structures and tissues that may harbor pathogens. Understanding the ecology of pathogens and naturally occurring microorganisms is essential before interventions for elimination or control of their growth can be devised [155]. However, changes in dietary habits, methods of fruit and vegetable production and processing, sources of produce, and the emergence of pathogens previously not recognized for their association with raw produce have enhanced the potential for outbreaks [156,157].

While much is known about the ecology of microbial pathogens in foods of animal origin, the behavior of pathogens in association with naturally occurring microflora on fruits and vegetables is ill-defined. Tremendous differences in surface morphology, internal tissue composition, and metabolic activities of leaves, stems, florets, fruits, roots, and tubers provide a wide range of diverse ecological niches selective for specific species or groups of microorganisms.

Bruised and cut surface tissues exude fluids containing nutrients and numerous phytoalexins and other antimicrobials that may enhance or retard the growth of naturally occurring microflora and pathogens [158]. The presence of soil or fecal material on the surface of produce that may permeate cut tissues may alter the ecological environment and, perhaps, also the behavior of pathogens and other microflora. The growth of molds in these environments may result in increased pH, thus enhancing the probability of growth of pathogenic bacteria. Colonization and biofilm development may ensue, resulting in conditions that would protect against death of pathogens or promote growth of spoilage or pathogenic microorganisms. The viability of parasites as affected by extrinsic and intrinsic factors unique to fruits and vegetables is unknown.

12.5.1 BEHAVIOR OF FOODBORNE PATHOGENS ON FRUITS AND VEGETABLES

Although spoilage bacteria, yeasts, and molds dominate the microflora on raw fruits and vegetables, the occasional presence of pathogenic bacteria, parasites, and viruses capable of causing human infections has also been documented [103,159–163]. All types of produce have potential to harbor pathogens [164], but *Shigella* spp., *Salmonella*, enterotoxigenic and enteroheamorrhagic *Escherichia coli*, *Campylobacter* spp., *Listeria monocytogenes*, *Yersinia enterocolitica*, *Bacillus cereus*, *Clostridium botulinum*, viruses, and parasites such as *Giardia lamblia*, *Cyclospora cayetanensis*, and *Cryptosporidium parvum* are of greatest public health concern [158,159,165,167]. Fruits and vegetables can become contaminated with pathogenic organisms while growing in fields, orchards, vineyards, or greenhouses, or during harvesting, postharvest handling, processing, distribution, and preparation in food service or home settings. Each vegetable possesses a unique set of intrinsic factors that can influence the survival and growth of human pathogenic microorganisms.

A wide range of fresh fruits and vegetables, as well as unpasteurized apple juices, has been implicated in outbreaks of infections. Examples are listed in Table 12.5. This is not a comprehensive list but does illustrate the diversity of types of produce potentially capable of serving as vehicles for human infection. The survival and growth of a pathogen on or in raw produce or unpasteurized produce products are dictated by its metabolic capabilities. However, manifestation of these capabilities can be greatly influenced by intrinsic and extrinsic ecological factors naturally present in produce or imposed at one or more points during the entire system of production, processing, distribution, and preparation at the site of consumption.

12.5.2 UNDERSTANDING THE ECOSYSTEM OF PATHOGENS IN FRESH PRODUCE

A better understanding of microbial ecosystems on the surface of raw fruits and vegetables would be extremely useful when developing interventions to minimize contamination, prevent the growth of pathogens, and kill or remove pathogens at various stages of production, processing, marketing, and preparation for consumption. These ecosystems are extremely diverse and complex. The presence and number of bacteria, yeasts, molds, parasites, and viruses differ, depending on the type of produce, agronomic practices, geographical area of production, and weather conditions prior to harvest [41,155,164,205]. Microbial ecosystems unique to various type of produce after harvesting can be greatly influenced by handling and storage conditions as well as conditions of processing, packaging, distribution, and marketing.

Pathogens, along with spoilage microorganisms, may contaminate fruits and vegetables via several different routes and at several points throughout the pre- and postharvest system. Potential preharvest sources of microorganisms include soil, feces, irrigation water, water used to apply fungicide and insecticides, dust, insects, inadequately composted manure, wild and domestic animals, and human handling. Postharvest sources includes feces, human handling, harvesting equipment, transport, containers, wild and domestic animals, agronomic practices, and geographical area of production [155]. Figure 12.2 illustrates the cycle of infection and contamination of fresh produce.

12.5.3 INTERACTIONS WITH EPIPHYTIC MICROORGANISMS

Interactions between foodborne pathogens and background microflora have been studied extensively in meat and dairy products. Experiments conducted with monoxenic beef minces showed that *L. monocytogenes* did not grow on sterile meat, grew when coinoculated with *P. fluroscens*, and decreased with *Lactobacillus plantarum*. More

TABLE 12.5
Example of Outbreaks of Infections Epidemiologically Associated with Raw Fruits and Vegetables

Microorganism	Year	Country	Produce	Ref.
Bacteria				
Bacillus cereus	1973	U.S.	Seed sprouts	167
Clostridium botulinum	1987	U.S.	Cabbage	53
E. coli O157:H7	1995	U.S.	Lettuce	168, 169
	1997	Japan	Radish sprouts	170
	1997	U.S.	Alfalfa sprouts	171
	2001	Japan	Fermented cabbage	172
	2002	Japan	Cucumber	173
E.coli (enterotoxigenic)	1993	U.S.	Carrots	174
Listeria monocytogenes	1979	U.S.	Celery, lettuce, tomato	175
	1981	Canada	Cabbage	176
Salmonella				
Minami	1954	U.S.	Watermelon	177
Typhimurium	1974	U.S.	Apple cider	178
Saint-Paul	1988	UK	Mungbean sprouts	49
Poona	1991	U.S./Canada	Cantaloupes	48
Montevideo	1993	U.S.	Tomatoes	157
Bovismorbificans	1994	Sweden/Finland	Alfalfa sprouts	179
Stanley	1995	U.S.	Alfalfa sprouts	180
Typhi	1998–99	U.S.	Mamey	181
Mbandaka	1999	U.S.	Alfalfa sprouts	182
Shigella flexneri	1998	UK	Fruit salad	183
S. Sonnei	1986	U.S.	Lettuce	184
	1994	Norway	Lettuce	185
	1995	U.S.	Scallions	186
	1998	U.S.	Parsley	187
Vibrio cholerae	1970	Israel	Vegetables	188
Viruses				
Calicivirus	1998	Finland	Raspberries (frozen)	189
Hepatitis A	1983	UK	Raspberries (frozen)	190
	1988	U.S.	Lettuce	191
	1994	U.S.	Tomato	192
	1997	U.S.	Strawberries	193
Norwalk and Norwalk-like	1987	UK	Melon	194
	1991	U.S.	Celery	195
Parasites				
Cyclospora cayetanensis	1996–97	U.S.-Canada	Raspberries	196–200
	1997	U.S.	Lettuce	201
	1997	U.S.	Basil	202
	1997	Peru	Raw vegetables	165
Cryptosporidium parvum	1995	U.S.	Mixed salad	203
	1997	Peru	Raw vegetables	165
Giardia lamblia	1992	U.S.	Raw vegetables	204

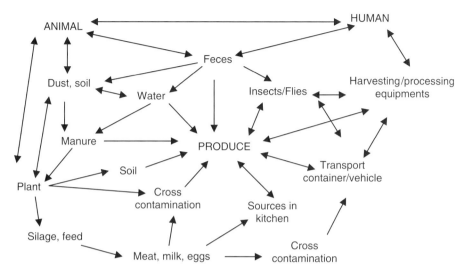

FIGURE 12.2 Cycle of infection and contamination of fresh produce. (Modified from Beuchat, L.R., *J. Food Prot.*, 59, 204, 1996.)

generally, lactic acid starter cultures have a protective effect against foodborne pathogens in milk and in meat products [206,207] and their antagonistic properties have been studied extensively and reviewed recently [208,209]. Lactic acid bacteria, among which *Leuconostoc* spp. predominate, can reach high counts on minimally processed vegetables after a few days of storage, and more particularly in products that contain high carbon dioxide concentrations [33,210,211]. It is worth noting that a strain of *Leuconostoc* spp. produces bacteriocins active against *L. monocytogenes* [212].

It has been observed that fluorescent pseudomonads can activate growth of *L. monocytogenes* in pasteurized milk as well as on meat, although other workers found that pseudomonads slightly inactivated *L. monocytogenes* in pasteurized milk after the exponential growth phase [213]. Presumably this occurred because pseudomonads hydrolyzed milk proteins, whereas *L. monocytogenes* alone did not [214,215]. There are no similar results for vegetable products, but fluorescent pseudomonads are the main component of their microflora and can actively participate in the breakdown of plant tissues. Their role in releasing potential nutrients for pathogenic microorganisms is worth investigating further.

12.5.4 COHABITATION WITH OTHER MICROORGANISMS MAY AFFECT SURVIVAL AND GROWTH OF PATHOGENS

While the pH of many vegetables is in a range suitable for growth of pathogenic bacteria, some, for example, fully ripe tomatoes, are in a pH range (3.9 to 4.4) that prevents or retards growth. Yeasts and molds, on the other hand, have a competitive advantage over bacteria that may access bruised tissues of acidic vegetables and many fruits, because they are able to grow at a lower pH range (2.2 to 5.0) characteristic of much of this produce. Spoilage of fruits is often caused by specific molds or groups of molds and yeasts [216]. When surface tissues of fruits are punctured or broken by insects or mechanical abuse, yeasts and molds naturally present on the skin surface can rapidly grow in the abundance of nutrients available in the released cell fluids.

Spoilage by yeasts usually results from fermentative activity. Molds, many of which can utilize ethanol and simple sugars as sources of energy, then grow and eventually degrade structural polysaccharides. Many molds produce ammonia and other alkaline by-products during the course of metabolizing substrate nutrients. Some molds and yeasts utilize organic acids, leading to reduced acidity and increased pH. At least two reports show that growth of molds on the surface of tomato juice (pH 4.2) increases the pH to between 6.3 to 7.6, allowing *C. botulinum* to grow and produce toxin [217,218]. Of the 58 species representing 21 genera of mold examined by Mundt [219], all except 2 raised the pH of tomato juice (pH 4.1) to a range of 4.9 to 9.0.

Insects such as the lesser mealworm and house fly have been shown to carry *E. coli* O157:H7 [220,221]. Fruits flies have been shown to transmit *E. coli* O157:H7 to apples [222]. Recognizing the outbreaks of *E. coli* O157:H7 infections associated with apple cider may have been due, in part, to amplification of the pathogen within bruised apple tissue. Dingman [223] investigated survival and growth of *E. coli* O157:H7 in tissue of five apple cultivars. While growth occurred in bruised tissue of all five cultivars, initiation of growth varied from 2 to 6 d after inoculation and was influenced by the time elapsed between picking and inoculating apples. Bacteria other than *E. coli* O157:H7 were not isolated from inoculated bruised tissue. Yeast and mold populations were not determined; however, the pH of bruised apples was significantly higher than the pH of undamaged apples, suggesting that mold growth may have occurred. In any case, the increased pH would favor survival and growth of *E. coli* O157:H7 and other pathogens.

In a survey of 401 samples of raw fruits and vegetables collected in retail markets, 66% affected by bacterial soft rot were positive for presumptive colonies of *Salmonella* [224]. Thirty percent of 166 representative isolates from 20 different commodities, including cantaloupe and tomato, were confirmed to be *Salmonella*. Coinoculation of potatoes, carrots, and peppers with a soft-rot bacterium and with *Salmonella typhimurium*, followed by incubation for 24 h, resulted in 10-fold higher counts of the pathogen compared to those in vegetables inoculated with *Salmonella* alone. Vegetables coinoculated with *Pseudomonas viridiflava* and *S. typhimurium* contained *Salmonella* populations approximately three times higher than vegetables inoculated with *Salmonella* alone. Jainisiewicz et al. [225], on the other hand, reported that inoculation of *Pseudomonas syringae* into wounds in apples prevented *E. coli* O157:H7 from growing. Populations of yeasts and molds to test produce were not reported in these studies [224,225].

With the exception of some types of melons, such as cantaloupe (pH 6.2 to 6.9) and watermelon (pH 5.2 to 5.7), which are recognized as good substrates for growth of *Salmonella* [87] and *E. coli* O157:H7 [89], fruits and fruit juices with pH less than 4.0 are generally not considered as substrates to support the growth of pathogenic bacteria. However, the development of a pH gradient surrounding mycelial growth in bruised tissues or as a mat on the surface of juice could provide conditions for growth of incident cells of pathogenic bacteria.

Most of the natural microflora on the surface of fresh produce do not exert a deleterious effect on sensory qualities. However, when spoilage does occur, *Pseudomonas*, *Xanthomonas*, *Erwinia*, *Bacillus*, *Clostridium*, and several genera of yeasts and molds are commonly involved [89]. Some naturally occurring microorganisms may have a lethal or antagonistic effect on bacteria capable of causing human diseases. Bacteriocinogenic strains of *Pediococcus* and *Enterococcus*, for example, have recently been shown to control the growth of *L. monocytogenes* on mung bean sprouts [129].

12.5.5 BEHAVIOR OF PATHOGENS IN BIOFILMS

Pathogens contaminate fresh produce by several means, including environmental sources in the field or orchard, or contact with harvesting equipment and containers used to transport produce from the field to the marketplace, and perhaps in food service and home settings. Exopolysaccharides secreted by bacteria can form a bound capsule layer when associated with the cell wall or released by the cell to create a matrix structure [226]. Microbial aggregates that have been shown to harbor bacteria, yeasts, and molds within this matrix have been observed on plant surfaces, and these structures are referred to as *biofilms* [227].

Colonization of spoilage and nonspoilage microorganisms of fruits, vegetables, and postharvest contact surfaces can provide a protective environment for pathogens, reducing the effectiveness of sanitizers and other inhibitory agents [228]. *L. monocytogenes*, in a multispecies biofilm containing *Pseudomonas fragi* and *Staphylococcus xylosus*, has been reported to be essentially unaffected by treatment with 500 ppm free chlorine [229]. Fett [230] examined the cotyledons, hypocotyls, and roots of alfalfa, broccoli, cloves, and sunflower sprouts. Biofilms were observed on plant parts. He concluded that naturally occurring biofilms on sprouts might afford protected colonization sites for human pathogens such as *Salmonella* and *E. coli* O157:H7, making their elimination with antimicrobial compounds difficult. The formation of biofilms on leaf surfaces of spinach, lettuce, Chinese cabbage, celery, leeks, basil, parsley, and endive has been demonstrated [227]. Estimates of biofilm abundance in phyllosphere communities show that bacteria in biofilms constitute 10 to 40% of the bacterial population on broad-leaf endive and parsley [231].

Containers used to harvest, transport, and display raw fruits and vegetables are often not effectively cleaned and sanitized, which often leads to the development of biofilms [232–234]. Even single-use containers may hold produce for a sufficient time to allow the formation of biofilms. Contamination of fresh produce with pathogens may result from contact with surfaces harboring these biofilms. If pathogens attach to biofilms during transport or processing, their survival and growth may be enhanced [235–237]. Growth of pathogens incorporated into biofilms would increase the probability of cross-contamination of produce. Jeong and Frank [238,239] determined that *L. monocytogenes* grows in multispecies biofilms containing microflora from meat and dairy plants. No information is available on behavior of *L. monocytogenes* or other pathogenic bacteria in biofilms formed by microflora associated with raw fruits and vegetables. Predominant microorganisms in biofilms on surfaces of containers and equipment used in the fresh fruit and vegetable industry would likely differ greatly from those on containers and equipment used in meat and dairy industries. Even within the produce industry, microflora in biofilms on various container and equipment surfaces would be predicted to differ greatly, depending on the type of produce being harvested and processed. Survival and growth characteristics of pathogens would also likely be influenced by these differences.

During growth and maturation of fruits and vegetables as well as during harvesting, transport, processing, and storage after processing, opportunities arise for the development of biofilms. These biofilms may provide protection against sanitizers. Growth of *L. monocytogenes* in a multispecies biofilm, with concurrent development of resistance to sodium hypochlorite, has been demonstrated [237]. A model system needs to be developed to stimulate produce biofilms for the purpose of determining the behavior of pathogens incorporated into them. The ability of pathogens to survive in biofilms subjected to dehydration and treatment with sanitizers needs to be determined.

REFERENCES

1. Hayes, P.R., *Food Microbiology and Hygiene*, Elsevier Applied Science, London, 1985, p. 113.
2. Jay, J.M., *Modern Food Mocrobiology,* 6th ed., Aspen Publishing, Gaithersburg, MD, 2000, p. 679.
3. Lund, B.M., Bacterial spoilage, in *Post Harvest Pathology of Fruits and Vegetables*, Dennis, C., Ed., Academic Press, London, 1983, p. 219.
4. Garrett, C.M.E., Bacterial diseases of food plants: an overview, in *Bacteria and plants*, Rhodes-Roberts, M.E and Skinner, F.A., Eds., Academic Press, London, 1982.
5. Lund, B.M., Bacterial spoilage of vegetables and certain fruits, *J. Appl. Bacteriol.*, 34, 9, 1971.
6. Geeson, J.D., The fungul and bacterial flora of stored white cabbage, *J. Appl. Bacteriol.*, 46, 189, 1979.
7. Khan, M.R., Saha, M.L., and Kibria, H.M.G., A bacteriological profile of salad vegetables in Bangladesh with special reference to coliforms, *Lett. Appl. Microbiol.*, 14, 88, 1992.
8. Senter, S.D., Bailey, J.S., and Cox, N.A., Aerobic microflora of commercially harvested, transported and cryogenically processed collards (*Brassica oleracea*), *J. Food Sci.*, 52, 1021, 1987.
9. Senter, S.D. et al., Microbiological changes in fresh market tomatoes during packing operations, *J. Food Sci.*, 50, 254, 1985.
10. Senter, S.D., Cox, N.A., Bailey, J. S., and Meredith, F.I., Effect of harvesting, transportation and cryogenic processing on the microflora of southern peas, *J. Food Sci.*, 49, 1410, 1984.
11. Mundt, J.O., Wanda, F.G., and McCarty, I.E., Spherical lactic acid-producing bacteria of southern grown raw and processed vegetables, *Appl. Microbiol.*, 15, 1303, 1967.
12. Lin, C., Bolsen, K.K., and Fung, D.Y.C., Epiphytic lactic acid bacteria succession during the preensiling periods of alfalfa and maize, *J. Appl. Bacteriol.*, 73, 375, 1992.
13. Deak, T., Foodborne yeasts*, Adv. Appl. Microbiol.* 36,179,1991.
14. Magnusson, J.A., King, A.D., Jr., and Torok, T., Microflora of partially processed lettuce, *Appl. Environ. Microbiol.*, 56, 3851, 1990.

15. Torok, T., and King, A.D., Jr., Comparative study on the identification of foodborne yeasts, *Appl. Environ. Microbiol.*, 57, 1207, 1991.
16. Liao, C.H. and Wells, J.M., Properties of *Cytophaga johnsonae* strains causing spoilage of fresh produce at food markets, *Appl. Environ. Microbiol.*, 52, 1261, 1986.
17. Liao, C.H. and Wells, J.M., Association of pectolytic strains of *Xanthomonas campestris* with soft rots of fruits and vegetables at retail markets, *Phytopathology*, 77, 418, 1987.
18. Liao, C.H. and Wells, J.M., Diversity of pectolytic, fluorescent pseudomonads causing soft rots of fresh vegetables at produce markets, *Phytopathology*, 77, 673, 1987.
19. Beraha, L., and Kwolek, W.K., Prevalence and extent of eight market disorders of western-grown head lettuce during 1973 and 1974 in the greater Chicago, Illinois area, *Plant Dis Rep.*, 59, 1001, 1975.
20. Brocklehurst, T.F. and Lund, B.M., Properties of pseudomands causing spoilage of vegetables stored at low temperature, *J. Appl. Bacteriol.*, 50, 259, 1981.
21. Ohata, K.T., Tsuchiya, Y., and Shirata, A., Difference in kinds of pathogenic bacteria causing head rot of lettuce of different cropping types, *Ann. Phytopathogen. Soc. Japan*, 45, 333, 1979.
22. Pieczarka, K.J. and Lorbeer, J.W., Microorganisms associated with bottom rot of lettuce grown on organic soil in New York State, *Phytopathology*, 65, 16, 1975.
23. Sellwood, J.E., Ewart, J.M., and Brucker, E., Vascular blackening of chicory caused by pectolytic isolate of *Pseudomonas fluorescens. Plant Pathol.*, 30, 179, 1981.
24. Tsuchiya, Y. et al., Identification of causal bacteria of head lettuce, *Bull. Nat. Inst. Agric. Sci.*, C, 33, 77, 1979.
25. Vantomme, R. et al., Bacterial soft rot of witloof chicory caused by strains of *Erwinia* and *Pseudomonas*: symptoms, isolation and characterization, *J. Phytopathol.*, 124, 337, 1989.
26. Cuppels, D. and Kelman, A., Evaluation of selective media for isolation of soft-rot bacteria from soil and plant tissue, *Phytopathology*, 64, 468, 1974.
27. Sands, D.C. and Hankin, L., Ecology and physiology of fluorescent pectolytic pseudomonads, *Phytopathology*, 65, 921, 1975.
28. Robinson, I. And Adams, R.P., Ultraviolet treatment of contaminated irrigation water and its effect on the bacteriological quality of celery at harvest, *J. Appl. Bacteriol.*, 45, 83, 1978.
29. Peombelon, M.C.M. and Kelman, A., Ecology of the soft rot Erwinias, *Annu. Rev. Phytopathol.*, 18, 361, 1980.
30. Brocklehurst, T.F., Zaman-Wong, C.M., and Lund, B.M., A note on the microbiology of retail packs of salad vegetables, *J. Appl. Bacteriol.*, 63, 409, 1987.
31. Grogan, R.G. et al., Varnish spot, destructive disease of lettuce in California caused by *Psudomonas cichorii*, *Phytopathology*, 67, 957, 1977.
32. McCarter-Zorner, N.J. et al., Soft rot *Erwinia* bacteria in the rhizosphere of weeds and crops plants in Colorado, United States and Scotland, *J. Appl. Bacteriol.*, 59, 357, 1985.
33. Carlin, F. et al., Microbiological spoilage of fresh ready-to-use grated carrots, *Sci. Aliments*, 9, 371, 1989.
34. Nguyen-The, C. and Prunier, J.P., Involvement of pseudomonads in the deterioration of "ready-to-use" salads, *Int. J. Food Sci. Technol.*, 24, 47, 1989.
35. Barriga, M.I. et al., Microbial changes in shredded iceberg lettuce stored under controlled atmospheres, *J. Food Sci.*, 56, 1586, 1991.
36. Denis, C. and Picoche, B., Microbiologie des legumes frais predecoupes, *Ind. Agr. Alim.*, 103, 547, 1986.

37. Babic, I. et al., The yeast flora of stored ready to use carrots and their role in spoilage, *Int. J. Food Sci. Technol.*, 27, 473, 1992.
38. Doyle, M.P., Fruits and vegetable safety-microbiological considerations, *Hortscience*, 25, 1478, 1990.
39. Lund, B.M., Anaerobes in relation to foods of plant origin, in *Anaerobic Bacteria in Habitats Other than Man*, Barnes, E.M. and Mead, G.C., Eds., Blackwell Scientific, Oxford, 1986, p. 351.
40. Lund, B.M., Bacterial contamination of food crops, *Asp. Appl. Biol.*, 17, 71, 1988.
41. Lund, B.M., Ecosystems in vegetable foods, *J. Appl. Bacteriol.*, 73 (Suppl.), 115S, 1992.
42. Steinbruegge, E.G., Maxcy, R.B., and Liewen, M.B., Fate of *Listeria monocytogenes* on ready to serve lettuce, *J. Food Prot.*, 51, 596, 1988.
43. McLauchlin, J. and Gillbert, R.J., Listeria in food: report from the PHLS Committee on *Listeria* and listeriosis, *PHLS Microbiol. Digest*, 7, 54, 1990.
44. De Simon, M., Tarrago, C., and Ferrer, M.D., Incidence of *Listeria monocytogenes* in fresh foods in Barcelona (Spain), *Int. J. Food Microbiol.*, 16, 153, 1992.
45. Heisick, J.E. et al., *Listeria* spp. found on fresh market produce, *Appl. Environ. Microbiol.*, 55, 1925, 1989.
46. Callister, S.M. and Agger, W.A., Enumeration and characterization of *Aeromonas hydrophila* and *Aeromonas caviae* isolated from grocery store produce, *Appl. Environ. Microbiol.*, 53, 249, 1987.
47. Oosterom, J., Epidemiological studies and proposed preventive measures in the fight against human salmonellosis, *Int. J. Food Microbiol.*, 12, 41, 1991.
48. CDC, Multistate outbreak of *Salmonella Poona* infections: United states and Canada, 1991, *NMWR*, 40, 549, 1991.
49. O'Mahony, M. et al., An outbreak of *Salmonella saint-paul* infection associated with beansprouts, *Epidemiol. Infect.*, 104, 229, 1990.
50. Garcia-Villanova Ruiz, B., Galvez Vargas, R., and Garcia-Villaniva, R., Contamination on fresh vegetables during cultivation and marketing, *Int. J. Food Microbiol.*, 4, 285, 1987.
51. Tamminga, S.K., Beumer, R.R., and Kampelmacher, E.H., The hygienic quality of vegetables grown in or imported into the Netherland: a tentative survey, *J. Hyg. Camb.*, 80, 143, 1978.
52. Saddik, M.F., El-Sherbeeny, M.R., and Bryan, F.L., Microbiological profiles of Egyptian raw vegetables and salads, *J. Food Prot.*, 48, 883, 1985.
53. Solomon, H.M., Kautter, D.A., Lilly, T., and Rhodehamel, E.J., Outgrowth of *Clostridium botulinum* in shredded cabbage at room temperature under modified atmosphere, *J. Food Prot.*, 53, 831, 1990.
54. Solomon, H.M. and Kautter, D.A., Growth and toxin production by *Clostridium botulinum* in sauted onions, *J. Food Prot.*, 49, 618, 1986.
55. Solomon, H.M. and Kautter, D.A., Outgrowth and toxin production by *Clostridium botulinum* in bottled chopped garlic, *J. Food Prot.*, 51, 862, 1988.
56. Notermans, S.H.W., Control in fruits and vegetables, in *Clostridium botulinum: Ecology and Control in Foods*, Hauschild, A.H.W. and Dodds, K.L., Eds., Marcel Dekker, New York, 1993, p. 233.
57. Harmon, S.M., Kautter, D.A., and Solomon, H.M., *Bacillus cereus* contamination of seeds and vegetable sprouts grown in a home sprouting kit, *J. Food Prot.*, 50, 62, 1987.
58. Rubatzky, V.E. and Yamaguchi M., *World Vegetables: Principals, Production and Nutritive Values*, 2nd ed., Aspen Publishers, Gaithersburg, MD, 1999, p. 105.

59. Selvendren, R.R., Developments in chemistry and biochemistry of pectic and hemi-cellulosic polymers, *J. Cell Sci.*, 2 (Suppl.), 51, 1985.

60. Collmer, A. and Keen, N.T., The role of pectic enzymes in plant pathogenicity, *Annu. Rev. Phytopathol.*, 24, 383, 1986.

61. Kotoujansky, A., Molecular genetics of pathogenesis by soft rot *Erwinias*, *Annu. Rev. Phytopathol.*, 25, 405, 1987.

62. Vaughn, R.H. et al., Gram-negative bacteria associated with sloughing, a softening of California ripe olives, *J. Food Sci.*, 34, 224, 1969.

63. Hsu, E.J. and Vaughn, R.H., Production and catabolite repression of the constitutive polygalacturonic acid transeliminase of *Aeromonas liquefaciens*, *J. Bacteriol.*, 98, 172, 1969.

64. Juven, B.J., Lindner, P., and Weisslowicz, H., Pectin degradation in plant material by *Leuconostoc mesenteroides*, *J. Appl. Bacteriol.*, 58, 533, 1985.

65. McGuire, R.G. et al., Polygalacturonase production by *Agrobacterium tumefaciens* Biovar 3, *Appl. Environ. Microbiol.*, 57, 660, 1991.

66. Lund, B.M., Properties of some pectolytic, yellow pigmented, gram-negative bacteria isolated from fresh cauliflowers, *J. Appl. Bacteriol.*, 32, 60, 1969.

67. Starr, M.P. et al., Enzymatic degradation of polygalacturonic acid by *Yersinia* and *Klebsiella* in relation to clinical laboratory procedures, *J. Clin. Microbiol.*, 6, 379, 1977.

68. Chatterjee, A.K. et al., Synthesis and excretion of polygalacturonic trans-eliminase in *Erwinia*, *Yersinia* and *Klebsiella* species, *Can. J. Microbiol.*, 25, 94, 1979.

69. Bagley, S.T. and Starr, M.P., Characterization of intracellular polygalacturonic acid trans-eliminase from *Klebsiella oxytoca, Yersinia enterocolitica* and *Erwinia chrysanthemi*, *Curr. Microbiol.*, 2, 381, 1979.

70. Troller, J.A. and Christian, J.H.B., *Water Activity and Food,* Academic Press, New York, 1978.

71. [ICMSF] International Commission on Microbiological Specification for Foods, *Microbial Ecology of Foods.* Vol. 1, Factors Affecting Life and Death of Microorganisms, Academic Press, Orlando, FL, 1980, p. 311.

72. Mossel, D.A.A. et al., *Essentials of the Microbiology of Foods*: *A Textbook for Advanced Studie*s, John Wiley & Sons, Chichester, 1995, p. 699.

73. Jijakli, M.H. and Lepoivre, P., Biological control of postharvest *Botrytis cinerea* and *Penicillium expansum* on apples, in IOBC/WPRS bulletin: *Biological Control of Foliar and Post-Harvest Diseases*, 16th ed., 1993, pp. 106–110.

74. Ippolito, A. et al., Control of postharvest decay of apple fruit by *Aureobasidium pullulans* and induction of defense responses, *Postharv. Biol. Technol.*, 19, 265–272, 2000.

75. Chand-Goyal, T. and Spotts, R.A., Biological control of postharvest diseases of apple and pear under semi-commercial and commercial conditions using three saprophytic yeast, *Biol. Control*, 10, 199–206, 1997.

76. Filonow, A.B., et al., Biological control of *Botrytis cinerea* in apple by yeast from various habitats and their putative mechanisms of antagonism, *Biol. Control*, 7, 212–220, 1996.

77. Beuchat, L.R., Surface disinfection of raw produce, *Dairy Food Environ. Sanit.*,12, 6–9, 1992.

78. Janisiewicz, W.J. et al., Postharvest biological control of blue mold on apples, *Phytopathology*, 77, 481–485, 1987.

79. Burton, W.G., *The Potato*, 2nd ed., Veenman, Wageningen, Holland, 1966, p. 162.

80. Banwart, G.J., *Basic Food Microbiology*, 2nd ed., Van Nostrand Reinhold, New York, 1989, p. 117.

81. Daeschel, M.A., Fleming, H.P., and Pharr, D.M., Acidification of brined cherry peppers, *J. Food Sci.*, 55, 186, 1990.

82. Sapers, G.M., Phillips, J.G., and Divito, A.M., Correlation between pH and consumption of foods comprising mixture of tomatoes and low-acid ingredients, *J. Food Sci.*, 49, 233, 1984.

83. Wolf, I.D. et al., The pH of 107 varieties of Minnesota-grown tomatoes, *J. Food Sci.*, 44, 1008, 1979.

84. Mossel, D.A.A. et al., *Essentials of the Microbiology of Foods: A Textbook for Advanced Studies,* John Wiley & Sons, Chichester, 1995, p. 47.

85. Ray, B., *Fundamental Food Microbiology*, CRC Press, Boca Raton, FL, 1996, p. 62.

86. Jay, J.M., *Modern Food Microbiology*, 6th ed., Aspen Publishing, Gaithersburg, MD, 2000, p. 47.

87. Golden, D.A., Rhodehamel, E.J., and Kautter, D.A., Growth of *Salmonella* spp. in cantaloupe, watermelon, and honeydew melons, *J. Food Prot.* 56, 194, 1993.

88. Lin, C.M. and Wei, C.I., Transfer of *Salmonella montevideo* onto the interior surfaces of tomatoes by cutting, *J. Food Prot.*, 60, 858, 1997.

89. Del Rosario, B.A. and Beuchat, L.R., Survival and growth of enterohemorrhagic *Escherichia coli* 0157:H7 in cantaloupe and watermelon, *J. Food Prot.* 58, 105, 1995.

90. Bartz, J.A. and Showalter, R.K., Infiltration of tomatoes by aqueous bacterial suspensions, *Phytopathology*, 71, 515, 1981.

91. Morris, J.G., The effect of redox potential, in: Lund B.L., Baird-Parker, T.C., and Gould, G.W., Eds., *The Microbiological Safety and Quality of Food*, Vol. 1, Aspen Publishing, Gaithersburg, MD, 2000, p. 235.

92. Luck, E. and Jager, M., *Antimicrobial Food Additives: Characteristics, Uses, Effects*, Springer, Berlin, 1997, p. 251.

93. Cheigh, H.-S. and Park, K.-Y., Biochemical, microbiological, and nutritional aspects of kimchi (Korean fermented vegetable products), *Crit. Rev. Food Sci. Nutr.*, 34, 175, 1994.

94. Loss, C.R. and Hotchkiss, J.H., Inhibition of microbial growth by low-pressure and ambient pressure gasses, in *Control of Foodborne Microorganisms*, Juneja, V.K. and Sofos, J.N., Eds., Marcel Dekker, New York, 2002, p. 245.

95. [NIST] National Institute of Standards and Technology, Uniform laws and regulations in the areas of legal metrology and engine fuel quality [as adopted by the 84th National Conference on Weights and Measures 1999]. (2000 ed.) U.S. Dept. of Commerce, Technology Administration, National Institute of Standards and Technology, Gaithersburg, MD, Uniform open dating regulation, pp. 117–122. (NIST Handbook 130), 2000.

96. Mossel, D.A.A. and Thomas, G., Securite microbioligique des plats prepares refrigeres: recommendations en matiere d'analyse des risques, conception et surveillance du processus de fabrication, *Microbiol. Aliements Nutr.*, 6, 289, 1988.

97. [USDA] U.S. Dept. of Agriculture, Agricultural Research Service, Eastern Regional Laboratory. USDA Pathogen Modeling Program Version 5.1.

98. Roberts, T.A., Baird-Parker, A.C., and Tompkin, R.B., Eds., *Microorganisms in Foods*, Vol. 5, *Characteristics of Microbial Pathogens*, Blackie Academic & Professional, London, 1996, p. 513.

99. Doyle, M.P., Beuchat, L.R., and Montville, T.J., *Food Microbiology: Fundamentals and Frontiers*, 2nd ed., Doyle, M.P., Beuchat, L.R., and Montville, T.J., Eds., American Society for Microbiology, Washington, DC, 2001.

100. Lund, B.M. and Snowdon, A.L., Fresh and processed fruits, in *The Microbiological Safety and Quality of Food*, Vol. 1, Lund, B.M., Baird-Parker, T.C., and Gould, G.W., Eds., Aspen Publishing, Gaithersburg, MD, 2000, p. 738.

101. Montville, T.J. and Matthews, K.R., Principles which influence microbial growth, survival, and death in foods, in *Food Microbiology: Fundamentals and Frontiers*, 2nd ed., Doyle, M.P., Beuchat, L.R., and Montville, T.J., Eds., American Society for Microbiology, Washington, DC, 2001, Chaps. 2, 13.

102. [FDA] Food and Drug Administration, Center for Food Safety and Applied Nutrition. The "Bad Bug Book" *Foodborne Pathogenic Microorganisms and Natural Toxins Handbook*], available at http://www.cfsan.fda.gov/~mow/intro.html, accessed Dec. 10, 2001.

103. Lioutas, T.S., Challenges of controlled and modified atmosphere packaging: a food company's perspective, *Food Technol.*, 42, 78, 1988.

104. Zagory, D. and Kader, A.A., Modified atmosphere packaging of fresh produce, *Food Technol.*, 42, 70, 1988.

105. Huxsoll, C.C. and Bolin, H.R., Processing and distribution alternatives for minimally processed fruits and vegetables. *Food Technol.*, 43, 124, 1989.

106. Mayers, R.A., Packaging considerations for minimally processed fruits and vegetables, *Food Technol.*, 43, 129, 1989.

107. Berrang, M.E., Brackett, R.E., and Beuchat, L.R., Growth of *Listeria monocytogenes* on fresh vegetables stored under controlled atmosphere, *J. Food Prot.*, 52, 702, 1989.

108. Berrang, M.E., Brackett, R.E., and Beuchat, L.R., Growth of *Aeromonas hydrophilia* on fresh vegetables stored under controlled atmosphere, *Appl. Environ. Microbiol.*, 55, 2167, 1989.

109. Brackett, J., Changes in the microflora of packaged fresh broccoli, *J. Food Quality*, 12, 169, 1989.

110. Brackett, J., Influence of modified atmosphere packaging on the microflora and quality of fresh bell peppers, *J. Food Prot.*, 51, 829, 1988.

111. Ryan, D.W. and Cherrington, M., Human type A botulism, *J. Am. Med. Assoc.*, 216, 513, 1971.

112. Cherrington, M., Botulism: ten years experience, *Arch. Neuro. Chicago*, 30, 432, 1974.

113. CDC, Botulism: New Mexico, *MMWR*, 27, 138, 1978.

114. Seals, J.E. et al., Restaurant-associated type A botulism: transmission by potato salad, *Am. J. Epidemiol.*, 113, 436, 1981.

115. Sugiyama, H. et al., Production of botulinum toxin in inoculated pack studies of foil wrapped baked potatoes, *J. Food Prot.*, 44, 896, 1981.

116. Tamminga, S.K. et al., Microbial spoilage and development of food poisoning bacteria in peeled, completely or partly cooked, vacuum-packed potatoes, *Archiv. Lebensmittel hygiene*, 29, 215, 1978.

117. Notermans, S., Dufrenne, J., and Keijbets, M.H.J., Vacuum-packed, cooked potatoes: toxin production by *Clostridium botulinum* and shelf-life, *J. Food Prot.*, 44, 572, 1981.

118. Lund, B.M., Graham, A.F., and George, S.M., Growth and formation of toxin by *Clostridium botulinum* in peeled, inoculated, vacuum-packed potatoes after a double pasteurization and storage at 25°C. *J. Appl. Bacteriol.*, 64, 241, 1988.

119. Hayashi, K., Sakaguchi, S., and Sakaguchi, G., Primary multiplication of *Clostridium botulinum* type A in mustard-miso stuffing of 'karashi-renkon'(deep fried mustard-stuffed lotus roots), *Int. J. Food Microbiol.*, 3, 311, 1986.

120. CDC, International outbreak of restaurant-associated botulism: Vancouver, British Colombia, Canada, *MMWR*, 34, 643, 1985.

121. St Louis, M.E. et al., Botulism from chopped garlic: delayed recognition of major outbreak, *Ann. Intern. Med.*, 108, 363, 1988.

122. Morse, D.L. et al., Garlic-in-oil associated botulism: episode leads to product modification, *Am. J. Pub. Health*, 80, 1372, 1990.

123. Boddy, L. and Wimpenny, J.W.T., Ecological concepts in food microbiology, *J. Appl. Bacteriol. Symp. Suppl.* 73, 23S, 1992.

124. Fredrickson, A.G., Behavior of mixed cultures of microorganisms, *Annu. Rev. Microbiol.*, 31, 63, 1977.

125. Adams, M.R. and Nicolaides, L., Review of the sensitivity of different foodborne pathogens to fermentation, *Food Control*, 8, 227, 1997.

126. Ellis, J., Timms-Wilson, T.M., and Bailey, M.J., Identification of conserved traits in fluorescent pseudomonads with antifungul activity. *Environ. Microbiol.*, 2, 274, 2000.

127. Corsa, J.H., Signal transduction and transcriptional and post-translational control of iron regulated genes in bacteria, *Microbiol. Mol. Rev.*, 61, 319, 1997.

128. O'Sullivan, D.J. and O'Gara, F., Traits of fluorescent pseudomonas spp. involved in suppression of plant root pathogens, *Microbiol. Rev.*, 56, 662, 1992.

129. Bennik, M.H.J., van Overbeek, W., and Gorris, L.G.M., Biopreservation in modified atmosphere stored mung bean sprouts: the use of vegetable associated bacteriocinogenic lactic acid bacteriato control the growth of *Listeria monocytogenes*, *Lett. Appl. Microbiol.*, 28, 226, 1999.

130. Janisiewicz, W.J. and Marchi, A., Control of storage rots on various pear cultivars with a saprophytic strain of *Pseudomonas syringae*, *Plant Dis.* 76, 555, 1992.

131. Janisiewicz, W.J. and S.N. Jeffers, Efficacy of commercial formulation of two biofungicides for control of blue mold and gray mold of apples in cold storage, *Crop Prot.*, 16, 629, 1997.

132. Janisiewicz, W.J. and Roitman, J., Biological control of blue mold and gray mold on apple and pear with *Pseudomonas cepacia*, *Phytopathology*, 78, 1697, 1988.

133. Mao, G.H. and Cappellina, R.A., Postharvest biocontrol of gray mold of pear by *Pseudomonas gladioli*, *Plant Pathol.*, 79, 1153, 1989.

134. Mari, M. et al., Bioassays of glucoinolate-derived isothiocyanates against postharvest pear pathogens, *Plant Pathol.* 45, 753, 1996.

135. Wilson, C.L. et al., Potential on induced resistance to control postharvest diseases of fruits and vegetables, *Plant Dis.*, 78, 837, 1994.

136. Lurie, S., Droby, S., Chalupowicz, L., and Chalutz, E., Efficacy of *Candida oleophila* strain 182 in preventing *Penicillium expansum* infection of nectarine fruits, *Phytoparasitica*, 23, 231, 1995.

137. Roberts, R.G., Biological control of gray mold of apple by *Cryptoccus laurentii*, *Phytopathology*, 80, 526, 1990.

138. Teixido, N., Usall, J., and Vinas, I., Efficacy of preharvest and postharvest *Candida sake* biocontrol treatments to prevent blue mould on apples during cold storage, *Int. J. Food Microbiol.*, 50, 203, 1999.

139. Zhou, T., Northover, J., and Schneider, K.E., Biological control of postharvest diseases of peach with phyllosphere isolates of *Pseudomonas syringae*, *Can. J. Plant Pathol.*, 21, 375, 1999.

140. Zhou, T. and DeYoung, R., Control of apple scab with applications of phyllosphere microorganisms, in *Advances in Biocontrol of Plant Diseases*, Tang, W., Cook, R.J., and Rovira, A., Eds., Beijing China Agricultural University Press, Beijing, 1996, p. 369.

141. Gram, L. et al., Food spoilage: interactions between food spoilage bacteria, *Int. J. Food Microbiol.*, 78, 79, 2002.

142. Cousin, M.A. and Marth, E.H., Lactic acid production by *Streptococcus thermophilus* and *Lactobacillus bulgaricus* in milk precultured with psychrotrophic bacteria, *J. Food Prot.*, 40, 475, 1977.

143. Seminiano, E.N. and Frazier, W.C., Effect of pseudomonads and Achromobacteraceae on growth of *Staphylococcus aureus*, *J. Milk Food Technol.*, 29, 161, 1966.

144. Dainty, R.H. et al., Bacterial sources of putrescine and cadaverine in chill stored vacuum- packaged beef, *J. Appl. Bacteriol.*, 61, 117, 1986.

145. Jorgensen, L.V., Huss, H.H., and Dalgaard, P., The effect of biogenic amine production by single bacterial cultures and Metabiosis on cold-smoked salmon, *J. Appl. Microbiol.*, 89, 920, 2000.

146. Whitehead, N.A. et al., Quorum-sensing in Gram-negative bacteria, *FEMS Microbiol. Rev.*, 25, 365, 2001.

147. Davies, D.G. et al., The involvement of cell-to-cell signals in the development of a bacterial biofilms, *Science*, 280, 295, 1998.

148. Passador, L. et al., Expression of *Pseudomonas aeruginosa* virulence gene requires cell-to-cell communication, *Science*, 260, 1127, 2001.

149. Bainton, N.J. et al., N-(-3-oxohexanoyl)-L-homoserine lactone regulates carbapenem antibiotic production in *Erwinia carotovora*, *Biochem. J.*, 288, 997, 1992.

150. Zhang, L. et al., Agrobacterium conjugation and gene regulation by *N*-acyl-homoserine lactones, *Nature*, 362, 446, 1993.

151. Eberl, L. et al., Differentiation of *Serratia liquefaciens* into swarm cells is controlled by the expression of the *flhD* master operon, *J. Bacteriol.*, 178, 554, 1996.

152. Eberl, L., Molin, S., and Givskov, M., Surface motility in *Serratia liquefaciens*, *J. Bacteriol.*, 181, 1703, 1999.

153. Huber, B. et al., The cep quorum-sensing system of *Burkholderia cepacia* H111 controls biofilms formation and swarming motility, *Microbiology*, 147, 2517, 2001.

154. Mae, A. et al., Transgenic plants producing the bacterial pheromone *N*-acyl-homoserine lactone exhibit enhanced resistance to the bacterial phytopathogen *Erwinia carotovora*, *Mol. Plant-Microb. Interact.*, 14, 1035, 2001.

155. Beuchat, L.R., Ecological factors influencing survival and growth of pathogens on fruits and vegetables, *Microbes Infect.*, 4, 413, 2002.

156. Beuchat, L.R. and Ryu, J.H., Produce handling and processing practices, *Emerg. Infect. Dis.*, 3, 459, 1997.

157. Hedberg, C.W., MacDonald, K.L., and Osterholm, M.T., Changing epidemiology of foodborne disease: a Minnesota perspective, *Clin. Infect. Dis.*, 18, 671, 1994.

158. Sofos, J.N. et al., Naturally occurring antimicrobial agents in food, Task Force Rep. 132, *Council for Agricultural Science and Technology,* Ames, IA, 132, 103, 1998.

159. Beuchat, L.R., Pathogenic microorganisms associated with fresh produce, *J. Food Prot.*, 59, 204, 1996.

160. Beuchat, L.R., *Listeria monocytogenes*: incidence on vegetables, *Food Control*, 7, 223, 1996.

161. Beuchat, L.R., Surface decontamination of fruits and vegetables eaten raw: a review, Food Safety Unit, *World Health Organization,* WHO/FSF/98.2, 42, 1998.

162. Francis, G.A., Thomas, C., and O'Beirne, D., The microbiological safety of minimally processed vegetables, *Int. J. Food Sci. Technol.*, 34, 1, 1999.

163. Nguyen-The, C. and Carlin, F., Fresh and processed vegetables, in *The Microbiological Safety and Quality of Food*, Vol. 1, Lund, B.M., Baird-Parker, T.C., and Gould, G.W., Eds., Aspen Publishing, Gaithersburg, MD, 2000, p. 620.

164. Brackett, R.E., Incidence, contributing factors, and control of bacterial pathogens on produce, *Postharv. Biol. Technol.*, 15, 305, 1999.

165. Ortega, Y.R. et al., Isolation of *Cryptosporidium parvum* and *Cyclospora cayetanensis* from vegetables collected in markets of an endemic region in Peru, *Am. J. Trop. Med. Hyg.*, 57, 683, 1997.

166. Sterling, C.R. and Ortega, Y.R., *Cyclospora*: an enigma worth unraveling, *Emerg. Infect. Dis.*, 5, 48, 1999.

167. Portney, B.L., Goepfert, J.M., and Harmon, S.M., An outbreak of *Bacillus cereus* food poisoning resulting from contaminated sprouts, *Am. J. Epidemiol.*, 103, 589, 1976.

168. CDC, Outbreak of *E.coli* O157:H7, Northwestern Montana, *EPI-AID*, 1995, p. 95.

169. CDC, Outbreak of *Escherichia coli* O157:H7 infections among boy scouts, Maine, September 1995, *EPI-AID*, 1995, p. 93.

170. Ministry of health and Welfare of Japan, Vero-toxin-producing *Escherichia coli* (enteroheamorrhagic *E. coli*) infection, Japan, 1996-June, 1997, *Infect. Agents Surveillance Rep.*, 18, 153, 1997.

171. CDC, Outbreaks of *Escherichia coli* O157:H7 associated with eating alfalfa sprouts-Michigan and Virginia, June-July, 1997, *MMWR*, 46, 741, 1997.

172. Tanaka, H. et al, Epidemiological and bacteriological investigation of enteroheamorrhagic *Escherichia coli* infection in the Chugaku-Shikoku Area, *J. Jap. Assoc. Infect. Dis.*, 76, 449, 2002.

173. IASR, An outbreak of *Escherichia coli* O157:H7 caused by lightly fermented cucumber in a kindergarten in Japan, *Infect. Agent Surveillance Rep.*, 24, 132, 2002.

174. CDC, Foodborne outbreaks of enterotoxigenic *Escherichia coli*: Rhode Island and New Hampshire, 1993, *MMWR*, 43, 81, 1994.

175. Ho, J.L. et al., An outbreak of type 4b *Listeria monocytogenes* infection involving patients from eight Boston hospitals, *Arch. Intern. Med.* 146, 520, 1986.

176. Schlech, W.F. et al., Epidemic listeriosis: evidence for transmission in food, *N. Engl. J. Med.*, 308, 203, 1983.

177. Gaylor, G.E., MacCready, J.P., Reardon, J.P., and McKernan, B.F., An outbreak of salmonellosis traced to watermelon, *Publ. Health Rep.*, 70, 311, 1995.

178. Parish, M.E., Public health of nonpasteurized fruit juices, *Crit. Rev. Microbiol.*, 23, 109, 1997.

179. Ponka, A. et al., *Salmonella* in alfalfa sprouts, *Lancet*, 345, 462, 1995.

180. Mahon, B.E. et al., An international outbreak of *Salmonella* infection caused by alfalfa sprouts grown from contaminated seeds, *J. Infect. Dis.*, 175, 876, 1997.

181. FDA, FDA warns consumers about frozen mamey, U.S. FDA talk paper T99-11, U.S. *Food and Drug Administration,* Rockville, MD, February 20, 1999.

182. National Advisory Committee on Microbiological Criteria for foods, Microbiological safety evaluations and recommendations on sprouted seeds, *Int. J. Food Microbiol.*, 52, 123, 1999.

183. O'Brien, S., *Shigella flexneri* outbreak in southwest England, *Euro Surveillance Weekly*, 34, 2, 1998.

184. Davis, H. et al, A shigellosis outbreak traced to commercially shredded lettuce, *Am. J. Epidemiol.*, 128, 1312, 1988.

185. Kapperund, G. et al., Outbreak of *Shigella sonnei* infection traced to imported iceberg lettuce, *J. Clin. Microbiol.*, 33, 609, 1995.

186. CDC, Outbreaks of *Shigella sonnei* infection associated with eating fresh parsley–Minnesota, Massachusetts, California, Florida, and Canada, July–August, 1998, *MMWR*, 48, 285, 1999.

187. Cook, K.A. et al., Scallions and shigellosis: a multistate outbreak traced to green onions, Prog. Abstr. 44th *Annu. Epidemic Intelligence Service Conf.*, Atlanta, GA, 1995, p. 35.

188. Shuval, H.I., Yekutiel, P., and Fattal, B., Epidemiological evidence for helminth and cholera transmission by vegetables irrigated with wastewater: Jerusalem — a case study, *Water Sci. Technol.*, 17, 433, 1984.

189. Ponka, A. et al., Outbreak of calicivirus gastroenteritis associated with eating frozen raspberries, *Euro Surveillance Weekly*, 4, 66, 1999.

190. Reid, T.M.S. and Robinson, H.G., Frozen raspberries and hepatitis A, *Epidemiol. Infect.* 98, 109, 1987.

191. Rosenblum, L.S. et al., A multifocal outbreak of hepatitis A traced to commercially distributed lettuce, *Am. J. Publ. Health*, 80, 1075, 1990.

192. Williams, I.T. et al., Foodborne outbreak of hepatitis A, Arkansas, Prog. Abst. 40th *Annu. Epidemic Intelligence Service Conf.*, Atlanta, GA, 1995, p. 19.

193. CDC, Hepatitis A associated with consumption with frozen strawberries–Michigan, March, 1997, *MMWR*, 46, 295, 1997.

194. Iverson, A.M., Gill, M., and Bartlett, C.L.R., Two outbreaks of foodborne gastroenteritis caused by a small round structure virus: evidence of prolonged infectivity in a food handler, *Lancet*, II, 556, 1987.

195. Warner, R.D. et al., A Large nontypical outbreak of Norwalk like virus gastroenteritis associated with exposing celery to nonporable water and with *Citrobacter freundii*, *Arch. Intern. Med.*, 151, 2419, 1991.

196. CDC, Outbreak of *Cyclospora cayetanensis* infection — United States, 1996, *MMWR*, 45, 549, 1996.

197. CDC, Update: outbreak of *Cyclospora cayetanensis* infection — United States and Canada, 1996, *MMWR*, 45, 611, 1996.

198. CDC, Outbreak of cyclosporiasis — United States, 1997, *MMWR*, 46, 451, 1997.

199. CDC, Update: outbreak of cyclosporiasis — United States, 1997, *MMWR*, 46, 461, 1997.

200. Herwaldt, B.L. and Ackers, M.L., An outbreak in 1996 of cyclosporiasis associated with imported raspberries, *N. Engl. J. Med.*, 336, 1548, 1997.

201. CDC, Update: outbreak of cyclosporiasis — United States and Canada, 1997, *MMWR*, 46, 521, 1997.

202. CDC, Update: outbreak of cyclosporiasis — Northern Virginia, Washington, DC, Baltimore, Maryland, Metropolitan Area, 1997, *MMWR*, 46, 689, 1997.

203. Besser-Wiek, J.W., Forfang, J., and Hedberg, C.W., Foodborne outbreak of diarrheal illness associated with *Cryptosporidium parvum* — Minnesota, 1995, *MMWR*, 45, 783, 1996.

204. Mintz, E.D. et al., Foodborne giardiasis in a corporated office setting, *J. Infect. Dis.*, 167, 250, 1993.

205. Nguyen-The, C and Carlin, F., The microbiology of minimally processed fresh fruits and vegetables, *Crit. Rev. Food Sci. Nutr.*, 34, 371, 1994.

206. El-Gazzar, F.E., Bohner, H.F., and Marth, E.H., Antagonism between *Listeria monocytogenes* and *Lactococci* during fermentation of products from ultrafiltered skim milk, *J. Dairy Sci.*, 75, 43, 1991.

207. Schillinger, U. and Lucke, F.-K., Lactic acid bacteria as protective cultures in meat products, *Fleischwirtschaft*, 70, 1296, 1990.

208. Piard, J.C. and Desmazeaud, M., Inhibiting factors produced by lactic acid bacteria. I. Oxygen metabolites and catabolism end-products, *Lait*, 71, 525, 1991.

209. Piard, J.C. and Desmazeaud, M., Inhibiting factors produced by lactic acid bacteria. II. Bacteriocins and other antibacterial substances, *Lait*, 72, 113, 1992.

210. Carlin, F. et al., Effects of controlled atmospheres on microbial spoilage, electrolyte leakage and suger content on fresh, "ready-to-use", grated carrots, *Int. J. Food Sci. Technol.*, 25, 110, 1990.

211. Carlin, F. et al., Modified atmosphere packaging of fresh "ready-to-use" grated carrots in polymeric films, *J. Food Sci.*, 55, 1033, 1990.

212. Harris, J.L. et al., Antimicrobial activity of lactic acid bacteria against *Listeria mono-cytogenes*, *J. Food Prot.*, 52, 3784, 1989.

213. Farrag, S.A. and Marth, E.H., Variation in initial populations of *Pseudomonas fluorescens* affects behavior of *Listeria monocytogenes* in skim milk at 7 or 13°C, *Milchqissenschaft*, 46, 718, 1991.

214. Marshall, D.L. and Schmidt, R.H., Growth of *Listeria monocytogenes* at 10°C in milk preincubated with selected pseudomonads, *J. Food Prot.*, 51, 277, 1988.

215. Marshall, D.L. and Schmidt, R.H., Physiological evaluation of stimulated growth of *Listeria monocytogenes* by *Pseudomonas* species in milk, *Can. J. Microbiol.*, 37, 594, 1991.

216. Splittstoesser, D.F., Fruits and fruits products, in *Food and Beverage Mycology*, Beuchat, L.R., Ed., Van Nostrand Reinhold, New York, 1987, p. 101.

217. Huhtanen, C.N. et al., Growth and toxin production by *Clostridium botulinum* in moldy tomato juice, *Appl. Environ. Microbiol.*, 32, 711, 1976.

218. Odlaug, T.E. and Pflug, I.J., *Clostridium botulinum* growth and toxin production in tomato juice containing *Aspergillus gracilis*, *Appl. Environ. Microbiol.*, 37, 456, 1979.

219. Mundt, J.O., Effect of mold growth on the pH of tomato juice, *J. Food Prot.*, 41, 267, 1978.

220. Iwasa, M. et al., Detection of *Escherichia coli* O157:H7 from *Muisca domestica* (Diptera: Muscidae) at a cattle farm in Japan, *J. Med. Entomol.*, 36, 108, 1999.

221. McAllister, J.C. et al., Reservoir competence of *Alphitobus diaperinus* (Coleptera: tenebrionidae) for *Escherichia coli* (Eubacteriales: Enterobacteriaceae), *J. Med. Entomol.*, 33, 983, 1996.

222. Janisiewicz, W.J. et al., Fate of *Escherichia coli* O157:H7 on fresh cut apple tissue and its potential for transmission by fruit flies, *Appl. Environ. Microbiol.*, 65, 1, 1999.

223. Dingman, D.W., Growth of *Escherichia coli* O157:H7 in bruised apple (*Malus domestica*) tissue as influenced by cultivar, date of harvest, and source, *Appl. Environ. Microbiol.*, 66, 1077, 2000.

224. Wells, J.M. and Butterfield, J.E., *Salmonella* contamination associated with bacterial soft rot of fresh fruits and vegetables in the market place, *Plant Dis.*, 81, 867, 1997.

225. Janisiewicz, W.J., Conway, W.S., and Leverentz, B., Biological control of postharvest decays of apple can prevent growth of *Escherichia coli* O157:H7 in apple wounds, *J. Food Prot.*, 62, 1372, 1999.

226. Leigh, J.A. and Coplin, D.L., Exopolysaccarides in plant-bacterial interaction, *Annu. Rev. Microbiol.*, 46, 307, 1992.

227. Morris, C.E., Monier, J.E., and Jacques, M.A., Methods for observing microbial biofilms directly on leaf surfaces and recovering them for isolation of culturable microorganisms, *Appl. Environ. Microbiol.*, 63, 1570, 1997.

228. Carmichael, I. et al., Bacterial colonization and biofilms development on minimally processed vegetables, *J. Appl. Microbiol.*, 85, 45S, 1999.

229. Norwood, D.E. and Gilmour, A., The growth and resistance to sodium hypochlorite of *Listeria monocytogenes* in a steady-state multispecies biofilms, *J. Appl. Microbiol.*, 88, 512, 2000.

230. Fett, W.F., Naturally occurring biofilms on alfalfa and other types of sprouts, *J. Food Prot.*, 63, 625, 2000.

231. Morris, C.E., Monier, J.E., and Jacques, M.A., A technique to quantify the population size and composition of the biofilms component in communities of bacteria in the phyllosphere, *Appl. Environ. Microbiol.*, 64, 4789, 1998.

232. Blackman, I.C. and Frank, J.F., Growth of *Listeria monocytogenes* as a biofilms on various food processing surfaces, *J. Food Prot.*, 59, 827, 1996.

233. Costerton, J.W. et al., Bacterial biofilms in nature and disease, *Annu. Rev. Microbiol.*, 41, 435, 1987.

234. Gabis, D. and Faust, R.E., Controlling microbial growth in food processing environments, *Food Technol.*, 42, 81, 1988.

235. Dewanti, R. and Wong, A.C.L., Influence of culture conditions on biofilms formation by *Escherichia coli* O157:H7, *Int. J. Food Microbiol.*, 26, 147, 1995.

236. Farrell, B.L., Ronner, A.B., and Wong, A.C.L., Attachment of *Escherichia coli* O157:H7 in ground beef to meat grinders and survival after sanitation with chlorine and peroxyacetic acid, *J. Food Prot.*, 61, 817, 1998.

237. Helke, D.M. and Wong, A.C.L., Survival and growth characteristics of *Listeria monocytogenes* and *Salmonella* Typhimurium on stainless steel and bunan rubber, *J. Food Prot.*, 57, 963, 1994.

238. Jeong, D.K. and Frank, J.F., Growth of *Listeria monocytogenes* at 10°C in biofilms with microorganisms isolated from meat and dairy processing environments, *J. Food Prot.*, 57, 576, 1994.

239. Jeong, D.K. and Frank, J.F., Growth of *Listeria monocytogenes* at 21°C in biofilms with microorganisms isolated from meat and dairy processing environments, *Lebensm. Wiss. Technol.*, 27, 415, 1994.

13 Attachment of Bacterial Human Pathogens on Fruit and Vegetable Surfaces

Dike O. Ukuku, Ching-Hsing Liao, and Shirley Gembeh
U.S. Department of Agriculture[1], Agricultural Research Service, Eastern Regional Research Center, Wyndmoor, PA

CONTENTS

13.1 INTRODUCTION

The microbiological safety of fresh and minimally processed fruits and vegetables has been questioned as a result of recent outbreaks of foodborne illness associated with unpasteurized juices, sprouts, melons, lettuce, berries, and other commodities.

---

[1] Mention of a brand or firm name does not constitute an endorsement by the U.S. Department of Agriculture over others of similar nature not mentioned.

The presence and survival of human pathogens in these commodities have been demonstrated. Fruits and vegetables are frequently in contact with soil, insects, animals, or humans during growing or harvesting [1] and in the processing plant. Thus, their surfaces are exposed to natural contaminants, and by the time they reach the packing house, most fresh produce retains populations of 10^4 to 10^6 microorganisms/g [1,2]. Many vegetables, including bean sprouts, cabbage, cucumber, potatoes, and radishes, have been found to be contaminated with *Listeria monocytogenes* [1,3–6]. This microorganism has been isolated from soil, sewage sludge, vegetation, and water [3,4] and therefore has the potential to contaminate produce surfaces. Despite several guides to the produce and fresh-cut industry on how to reduce microbial food safety hazards for fresh-cut fruits and vegetables [7,8], the incidence of salmonellosis is frequently reported. *Salmonella* is among the most frequently reported cause of foodborne outbreaks of gastroenteritis in the U.S. [9,10].

The ability of pathogenic and spoilage-causing bacteria to adhere to surfaces of fruits and vegetables continues to be a potential food safety problem of great concern to the produce industry. Surface structure and biochemical characteristics of bacteria and of a substratum, in this case fruits and vegetables, play a major role in how and where bacteria may attach [11].

13.2 OUTBREAKS OF FOODBORNE ILLNESS ASSOCIATED WITH PRODUCE

The number of documented outbreaks of human infections associated with the consumption of raw fruits and vegetables has increased in recent years. In the U.S. the number of reported produce-related outbreaks per year doubled between the period from 1973 to 1987 and 1988 to 1992 [10,12]. Five (1990,1991, 2000, 2001, 2002) multistate outbreaks of salmonellosis have been associated epidemiologically with cantaloupes. The first involved *Salmonella* Chester, which affected 245 individuals (two deaths) in 30 states. The second involved more than 400 laboratory-confirmed *Salmonella* Poona infections and occurred in 23 states and Canada [10]. The most recent (April/May 2002) outbreak was due to *Salmonella* Poona associated with 43 illnesses [13]. Other human pathogens including *E. coli* O157:H7 and *Shigella* are capable of growth on melon flesh [14,15]. The recent FDA survey of imported fresh produce reported an incidence of 5.3% positives for *Salmonella* and 2% for *Shigella* for 151 samples of cantaloupe. All contaminated melons originated in Mexico, Costa Rica, and Guatemala [16]. In a survey of domestic fresh produce [17], of 115 samples of cantaloupes, 2.6% were positive for *Salmonella* and 0.9% were positive for *Shigella*.

Among the greatest concerns with human pathogens on fresh fruits and vegetables are enteric pathogens (e.g., *E. coli* O157:H7, *Listeria monocytogenes*, and *Salmonella*) that have the potential for growth prior to consumption or have a low infectious dose. More recently, outbreaks of salmonellosis have been linked to tomatoes, seed sprouts, cantaloupe, mamey, apple juice, and orange juice [18]. *Escherichia coli* O157:H7 infection has been associated with lettuce, sprouts, and apple juice, and enterotoxigenic *E. coli* has been linked to carrots [19,20]. Documented associations of shigellosis with lettuce, scallions, and parsley; cholera with strawberries; parasitic diseases with raspberries, basil, and apple cider; hepatitis A virus with lettuce, raspberries, and frozen strawberries; and Norwalk/Norwalk-like virus with melon, salad, and celery have been made [21,22]. A better understanding of bacterial adhesion to fruits and vegetables is needed for the development of more effective washing treatments to control microorganisms on fresh-cut produce.

13.3 SURFACES OF FRUITS AND VEGETABLES

13.3.1 Surface Characteristics

The surfaces of fruits and vegetables show a large diversity in structure and composition; the epidermis is covered by an epicuticular wax on aerial organs (leaves, stem, flowers, and fruits) or periderm on roots and tubers. Stomata, lenticels, broken trichomes, and scars from detached organs represent natural ways of entry for microorganisms. Cracks in the surface of vegetables and fruits may occur in certain growing conditions [23,24], and postharvest handling may cause injuries and bruising.

Unlike fruits with smooth surfaces, the outer surface (rind) of a cantaloupe presents a variety of surfaces to which a bacterium may bind. The epidermal cell surface is ruptured with a meshwork of raised tissue (the net). This net consists of lenticels and phellum (cork) cells. These cells have hydrophobic suberized walls to reduce water loss and protect against pathogen ingress. Also imparting a hydrophobic nature to the outer surface of cantaloupe is the cuticle composed of waxes and cutin that covers the epidermal cells [25]. Hydrophilic components of plant cell walls and middle lamella may also be exposed to bacterial invasion due to cuticular cracks and injuries to the epidermal surface.

13.3.2 Native Microflora of Fruits and Vegetables

There is great variation in the number and type of native microflora among fruits and vegetables due in part to their surface structures, chemical composition, type of fruits or vegetables, growth and harvesting condition, including processing treatments before, during, or after storage. It has been reported that 40 to 70% of the total flora of peas, snap beans, and corn consist of leuconostocs, streptococci, and corynebacteria, a Gram-positive rod [26]. Others are groups of bacteria that cause soft rot. Fruits contain less water and more carbohydrates and therefore would support growth of bacteria, yeast, and mold. However, the pH of fruits is below the level that generally favors bacterial growth [27].

13.3.3 Types of Spoilage Microflora

Microorganisms responsible for postharvest diseases are not necessarily dominant on the surface of sound vegetables. On cabbage leaves *Botrytis cinerea* represented less than 0.1% of the total microflora [28], and on witloof chicory only 3% of the epiphytic isolates caused spoilage upon inoculation [29]. Some spoilage microorganisms may be specific to a few vegetable species, whereas others, such as *Botrytis*

TABLE 13.1
Resuscitation of Acid-Injured *Erwinia carotovora* subsp. *carotovora* (Ecc) Cells on Cut Surfaces of Cucumber and Apple Fruits

Incubation Time (h)	Plate Count (log cfu/membrane) on Cucumber			Plate Count (log cfu/membrane) on Apple		
	BHIA	VRBA	% Injury	BHIA	VRBA	% Injury
0	5.3[a] ± 0.2[b]	3.9 ± 0.1[b]	96[b]	4.9 ± 0.3[b]	3.8 ± 0.4[b]	93[b]
4	5.3 ± 0.1[b]	4.2 ± 0.2[c]	91[c]	4.7 ± 0.1[b]	3.7± 0.2[b]	94[b]
8	5.3 ± 0.3[b]	4.9 ± 0.3[d]	59[d]	4.3 ± 0.2[c]	3.0± 0.3[c]	95[b]
16	7.0 ± 0.4[c]	6.5 ± 0.2[e]	20[e]	4.5 ± 0.1[c]	2.9 ± 0.4[c]	97[b]

Note: Filter membranes containing acetic acid-treated Ecc (strain SR319 cells (0.3%, 6 min) were placed on cut surfaces of cucumber or apple fruits. The changes in the number of injured Ecc cells in the population were monitored after incubating the membranes containing acid–treated cells on either fruit for 4, 8, and 16 h. BHIA = brain heart infusion agar; VRBA = violet red bile agar. % Injury = (plate count on BHIA – plate count on VRBA)/plate count on BHIA] × 100%.

[a] The value represents the average of three experiments, two duplicates in each experiment ± standard deviation.

[b,c,d,e] Within a column, the numbers not followed by the same letter are significantly different ($P < 0.05$) by the Bonferroni LSD separation technique.

cinerea, Sclerotinia species, *Sclerotium rolfsii, Rhizopus stolonifer*, and soft-rot bacteria, may attack a wide range of products. Many strains of *Alternaria alternata* and *Fusarium* spp. produce mycotoxins in decaying tomatoes [30] or potatoes [31,32]. However, a recent report concluded that vegetable products do not represent a significant hazard with regard to mycotoxins [33,34].

The relative importance of spoilage organisms for a given vegetable may differ in different countries and climates. *Mycocentrospora acerina* and *Phythophthora megasperma* were the main cause of spoilage of stored carrots in Normandy, France [35], whereas in England only *B. cinerea* and *Rhizoctonia carotae* were significant agents of spoilage [36]. In Denmark *R. carotae* was the major cause of postharvest decay [37]. On freshly harvested cabbage, *Alternaria tenuis* was the main cause of spoilage, *B. cinerea* prevailed on cabbage stored under air, and *Fusarium roseum* was an important cause of spoilage for cabbage stored under modified atmosphere [38]. At times injury may occur to bacterial cells on fruit and vegetable surfaces, probably due to adverse environmental conditions at the farm or during poststorage. Acid-injured *Erwinia carotovora* subsp. *carotovora* cells survived up to 16 h on cut surfaces of cucumber and apple fruits [39,40] (Table 13.1), and injured *Salmonella* Mbandaka or *Salmonella* Typhimurium also survived up to 16 h on fresh-cut apple disks (Table 13.2). Some microbial species do not cause spoilage of raw vegetables but may have an important impact on the quality of processed vegetables. Many postharvest disease agents colonize the mother plant, produce inoculum, and contaminate the vegetables before harvest [41]. This is the case with *Botrytis* spp., *Fusarium* spp., and *S. rolfsii* on various vegetable crops, *Phytophthora* spp. on

TABLE 13.2
Differential Responses of *Salmonella* Mbandaka and *S. Typhimurium* on Fresh-Cut Surfaces of Apple Disks to Water and Acetic Acid Treatment (2.4%, 5 min)

Washed with:	Log Cells Killed (CFU/disk)[a] on the Disks Inoculated with:		Log (%) Cells Injured on the Disks Inoculated with:	
	S. Mbandaka	*S. Typhimurium*	*S. Mbandaka*	*S. Typhimurium*
None	0.00 [f]	0.00[f]	0.02 (4)[c f]	0.07 (14)[e]
Water	0.60[b e]	0.63[e]	0.03 (7)[ef]	0.08 (17)[e]
2.4% acetic acid	1.13[d]	1.41[d]	0.26 (45)[de]	0.29 (45)[d]
Water, then 2.4% acetic acid	1.17[d]	1.41[d]	0.32 (51)[d]	0.26 (45)[d]

[a] Initial cell numbers of *S. mbandaka* or *S. typhimurium* on the disk were in the range of 7.30 to 7.34 log cfu/disk.

[b] Values represent the average of three experiments, two duplication for each experiment.

[c] Numbers in parentheses represent the rate of injury in percentage; injury was determined based on the counts on XLT4.

[d,e,f] Within a column, the means followed by the same letter are not significantly different ($P < 0.05$) by the Bonferroni LSD separation technique.

solanaceous vegetables, *Alternaria* spp. on carrots, and soft-rot erwinias on potatoes [42]. The plant or its immediate environment may also provide conditions for survival of microorganisms not commonly found in the field. For example, the soft-rot bacterium *E. carotovora* survived for a longer period in the rhizosphere of crops than in the soil; pectolytic fluorescent pseudomonads were not found in soil [43] but were abundant in a field where carrots had suffered extensive soft rot [44]. *Listeria monocytogenes* is also suspected to survive better in the rhizosphere of plants than in nonrhizosphere soil [45]. Pathogenic bacteria [46–50] viruses [21,22,48,51] and helminths [53] declined on cultivated plants, but in some studies pathogens survived until harvest.

13.3.4 SOURCES OF BACTERIAL PATHOGEN CONTAMINATION

It is very difficult to determine the primary source of contamination that leads to an outbreak, especially for produce. Determining the primary source of the pathogen will help in devising strategies and interventions to minimize risks of future outbreaks. For example, only 2 of 27 outbreaks during a particular time period of investigations of fresh produce were clearly identified to a point of contamination [54]. Bacterial pathogens may contaminate fruits and vegetables at any point throughout the production system. Potential preharvest sources of contamination include soil, feces, irrigation water, water used to apply fungicides and insecticides, dust, insects, inadequately composted manure, wild and domestic animals, and human handling [1]. *Salmonella*, *Escherichia coli* O157:H7, and *Listeria monocytogenes* can be found in animal feces. Transmission of *E. coli* O157:H7 from manure-contaminated soil and irrigation water to lettuce plants and its migration throughout the plant were recently reported [55,56]. Evidence of an association of salmonellae with stems and leaves of tomato plants grown hydroponically in inoculated solution has been presented [57]. Postharvest sources of contamination include feces, human handling, harvesting equipment, transport containers, wild and domestic animals, insects, dust, rinse water, ice, transport vehicles, and processing equipment [58].

13.4 ATTACHMENT OF HUMAN BACTERIAL PATHOGENS ON FRUITS AND VEGETABLES

The mechanism of attachment of bacterial cells to plant surfaces has been studied most extensively for plant pathogens and symbionts [59,60]. According to Fletcher [61], bacterial adhesion occurs in three steps: reversible adsorption, primary adhesion, and colonization. During the reversible adsorption phase, the bacterium is at a distance of greater than 50 nm and is affected by van der Waal interactions with the substratum. This means that the bacteria can be easily washed off at this stage. At the primary adhesion stage, the distance between the bacteria and the substratum ranges from 10 to 20 nm and the type of force affecting adhesion is electrostatic unless the opposing surface has a net surface charge, then attractive forces will come into play. The colonization step is the final phase and biofilms may be formed. According to Buscher et al. [62], once bacteria overcome the water barrier and a separation distance of about less than 1.0 nm, additional adhesion interactions such as hydrogen bonding, cation bridging, and receptor-ligand interactions between bacteria and plant surfaces will occur. At this stage the bacteria are very difficult to remove. Flagella, fimbriae, outer membrane proteins, and extracellular polysaccharides have all been implicated in attachment. Cellulose production and the presence of curli may allow for strong attachment of *Salmonella* to produce surfaces. Under natural conditions when contamination occurs in the field the production of cellulose and curli by *Salmonella* may allow for the bacterium to strongly bind to the plant surface and be highly resistant to removal by rain or by washing steps during processing. Attachment of bacterial human pathogens to fruits and vegetables has not been fully investigated.

Bacterial attachment to fruits with primarily smooth surfaces (apples, tomatoes, pears, and honeydew melons) involves less surface area than bacterial attachment to fruits with greater surface roughness. Bacterial attachment on the surface of apple fruits was greater in the vicinity of the calyx and stem, as compared to the remaining skin surface, with more than 94% recovery from the stem and calyx areas [63]. When a green pepper disk was inoculated, greater than 84% of the attached bacteria were on the injured surface, 15% on the inner skin, and less than 1% on the outer skin. Bacterial attachment was enhanced when the surfaces of apples or bell peppers were punctured or intentionally cut [14,63]. The adhesion of bacteria on the surface of cucumbers in wash water was less extensive at lower temperatures and shorter exposure times [64]. In this study the authors reported that various species of bacteria were adsorbed to cucumber surfaces in the following order: *Salmonella* Typhimurium > *Staphylococcus aureus* >*Lactobacillus plantarum* >*Listeria monocytogenes*. Cells were adsorbed at all temperatures tested (5, 15, 25, and 35°C) at levels that depended on incubation time, but the numbers of cells adsorbed were larger at higher incubation temperatures. Levels of adhesion of bacteria to dewaxed cucumber were higher for *L. monocytogenes* and lower for *Salmonella* Typhimurium, *L. plantarum,* and *S. aureus* than were levels of adhesion to waxed cucumbers.

Some strains of *E. coli* O157:H7 are also known to produce fimbriae and bacterial exopolysaccharides [65,66], and these may function as plant surface adhesins. Studies using confocal scanning laser microscopy indicated that *Salmonella, E. coli* O157:H7 and *L. monocytogenes* can attach to intact plant surfaces, trichomes, and be present in substomatal chambers, but are found most often at wounded surfaces and within cuticular cracks [66–71]. Ukuku and Fett [11] reported higher initial adhesion of *E. coli* O157:H7 on melon surfaces compared to *Salmonella* and *L. monocytogenes*; however, *E. coli* O157:H7 were more easily removed than the other two pathogens by washing treatments. Barak et al. [13] reported that *Salmonella* binds more strongly on alfalfa sprouts than *E. coli* O157:H7.

13.4.1 Factors That Influence Bacterial Attachment

Most bacteria are readily suspended in aqueous medium because of the polar, hydrophilic moieties that abound on bacteria cell surfaces [72]. Hydrophilic sites consist of charged moities such as carboxyl, phosphate, amino, and guanidyl groups as well as neutral hydroxyl groups. Hydrophobic sites consist of lipids and lipopolysaccharides [73]. Bacterial surfaces are heterogeneous, with physicochemical properties determined primarily by teichoic acid (Gram-positive strains) or other polysaccharides (Gram-negative strains) along with proteinaceous appendages (fimbriae) [59,74,75]. The chemistry of teichoic acid or polysaccharides confers regions of hydrophobic or hydrophilic properties on bacterial surfaces, which aids in their attachment to surfaces. Bacterial attachment to surfaces is influenced not only by cell surface charge [76,77] and hydrophobicity [11,78–80] but also by the presence of particular surface appendages such as flagella and fimbriae as well as extracellular polysaccharides [81,82]. Flagella, fimbriae (pili), outer membrane proteins, and extracellular polysaccharide may influence bacterial attachment to plant surfaces [59]. Plant surfaces and microbes both have negative surface potential, which results in electrostatic repulsion between the two surfaces. Surface appendages such as pili already present on microbes prior to or induced by the presence of a plant surface or other favorable conditions are used to bridge the gap exerted by electrostatic repulsion [53]. It is difficult to predict the surface properties of bacterial human pathogens when the pathogens are first exposed to a plant surface because environmental conditions can significantly affect bacterial surface properties, including charge and hydrophobicity [66,78,83]. Specific interactions between complementary moieties such as bacterial carbohydrate polymers with plant lectins or fimbriae with plant carbohydrate-containing moieties may also play a role [84,85], especially in attachment to exposed plant cell wall materials and damaged tissues.

Recently, *Salmonella* was demonstrated to produce the extracellular carbohydrate polymer cellulose; this, along with curli (aggregative fimbriae), the two principle components of the extracellular matrix, is thought to be responsible for biofilm formation [86,87]. Interestingly, for the plant pathogen *Agrobacterium* and the plant

TABLE 13.3
Bacterial Hydrophobicity and Strength
of Attachment to Cantaloupe Rind
24 Hours Postinoculation

Bacteria[a]	HIC[b]	S_R-value[c]
Salmonella spp.	0.484 ± 0.118	0.934 + 0.010
Escherichia coli O157:H7	0.220 ± 0.018	0.751 + 0.051
Listeria monocytogenes	0.281 ± 0.063	0.816 + 0.036

[a] Inoculum for *Salmonella* spp. contained a cocktail of the following bacteria: *Salmonella* Stanley (3.43×10^8 CFU/mL), *Salmonella* Poona (2.65×10^8 CFU/mL), and *Salmonella* Saphra (2.46×10^8 CFU/mL).

Inoculum for *E. coli* cocktail consisted of ATCC 25922 (2.70×10^8 CFU/mL); O157:H7-Odwala outbreak strain (2.38×10^8 CFU/mL); O157:H7-Oklahoma outbreak strain (2.30×10^8 CFU/mL). *Listeria monocytogenes* inoculum consisted of 2.32×10^8 CFU/mL Scott A; 2.02×10^8 CFU/mL ATCC 15313; 2.30×10^8 CFU/mL CCR1-L-G; 2.32×10^8 CFU/mL H7778.

[b] HIC = hydrophobic interaction chromatography.
[c] S_R-value = strength of attachment.

symbiont *Rhizobium*, cellulose production plays an important role in the firm attachment of the bacteria to plants and in the formation of bacterial aggregates at the plant surface [53]. After initial attachment, *Agrobacterium* synthesizes cellulose fibrils that bind the bacteria very tightly to the host cell surface and to each other, and the bacterial cells can only be removed by digesting the bacterial or host cell wall [84]. Fibrils of unknown nature have been observed for *Pseudomonas putida* and *P. tolaasi* binding to the surface of *Agaricus bisporus* mycelium [89], *P. syringae* pv. *syringae* binding to apple tissues [90], and *Azospirillum brasilense* binding to tomato, cotton, and pepper roots [91]. Ukuku and Fett [11] reported that *Salmonella* had the highest S_R-value, followed by *L. monocytogenes* and then *E. coli* (Table 13.3). Higher S_R-values indicate stronger bacterial attachment to surface of melons, as indicated by the relative inability of washing treatments to detach the pathogen from the melons' surface using water. Also, surface hydrophobicity of *Salmonella* was higher than that of *E. coli* and *L. monocytogenes* (Table 13.3).

The overall cell surface charge for *Salmonella* varied among serovars; two isolates originating from cantaloupe (serovars Poona and Saphra) had higher cell surface charges than the serovar (Stanley) from alfalfa sprouts (Table 13.4). Strength of attachment on cantaloupe surfaces increased slightly for *E. coli* over 7 d of storage but decreased for *L. monocytogenes* [11]. There was no difference between cell surface charge for *E. coli* ATCC 25922 and *E. coli* O157:H7 strains (Table 13.5), and the cell surface charge for *L. monocytogenes* strains was very similar (Table 13.6). Among all the bacterial human pathogens tested, *Salmonella* had higher

TABLE 13.4
Relative Surface Charge for *Salmonella* spp. Determined by Electrostatic Interaction Chromatography

Serovar/strain	(–)	(+)	Source
Stanley HO558	21.48	4.10	Alfalfa sprout
Poona RM2350	33.71	1.82	Cantaloupe
Saphra 97A3312	50.00	6.08	Cantaloupe

Note: For experimental details see Ukuku, D.O. and Fett, W.F., *J. Food Prot.*, 65, 1093, 2002.

TABLE 13.5
Relative Surface Charge for *Escherichia coli* spp. Determined by Electrostatic Interaction Chromatography

Strain	(–)	(+)	Source
ATCC 25922	1.62	0.12	Type strain
O157:H7 (Odwala)	1.48	0.18	Apple juice
O157:H7 (Oklahoma)	1.50	0.16	Apple juice

Note: For experimental details see Ukuku, D.O. and Fett, W.F., *J. Food Prot.*, 65, 1093, 2002.

TABLE 13.6
Relative Surface Charge for *Listeria monocytogenes* Determined by Electrostatic Interaction Chromatography

Strain	(–)	(+)	Source
Scott A	38.06	0.40	Clinical isolate
CCR1-L-G	37.68	0.20	Food isolate
ATCC 151313	38.11	0.32	Type strain
H7778	37.47	0.50	Food isolate

Note: For experimental details see Ukuku, D.O. and Fett, W.F., *J. Food Prot.*, 65, 1093, 2002.

surface hydrophobicity and relative negative charges than *E. coli* and *Listeria mono-cytogenes*. However, *Listeria monocytogenes* had a greater significant relative cell surface charge than did *E. coli*. This heterogeneity may help explain the differences observed in bacterial hydrophobicity or cell surface charge in relation to their attachment to vegetable and fruit surfaces, especially the cantaloupe surface.

13.4.2 Factors Limiting Detachment of Microorganisms

Irregularities such as roughness, crevices, and pits have been shown to increase bacterial adherence by increasing cell attachment and reducing the ability to remove cells [92,93]. However, preventive mechanisms should be geared towards physical or chemical treatments to prevent bacterial transfer from the surfaces of the produce to the interior flesh. The effectiveness of chlorination of wash water in reducing the population of bacteria on produce is dependent on the interval between contamination and application of the washing treatment [70,94–96]. If bacterial attachment occurs more than 24 h prior to washing, detachment or inactivation using chlorine or hydrogen peroxide treatments was shown to be less effective, and the difference between the two treatments diminished. It is likely that the limited ability of washing to remove established bacterial populations from the surface of fresh produce is due in part to biofilm formation, microbial infiltration, and internalization. At the retail level or at food establishments, produce is usually washed only using potable water, and the fresh-cut pieces may not always be prepared using clean and sanitized utensils. Thus, fresh-cut fruits and vegetables may not be adequately sanitized and protected from cross-contamination. However, because the time of contamination is not generally known and may precede washing by many days, more effective means of decontaminating produce are needed.

13.4.3 Biofilm Formation on Produce Surfaces

The ability of bacteria to form biofilms on food contact surfaces, which increases their resistance to cleaning and to antimicrobial agents, is well known [15]. However, relatively little is known about biofilm formation on fruit and vegetable surfaces. Babic et al. [97] described biofilm-like structures associated with bacteria within spinach leaf tissue. Carmichael et al. [98] observed native bacterial biofilms on the surface of lettuce. Large differences in surface morphology and metabolic functions of different plant organs (e.g., fruits, flowers, leaves, and roots) provide a wide range of diverse ecological niches that could be selective for specific species or communities of microorganisms. Microbial growth on raw fruits and vegetables can result in the formation of biofilms by spoilage and nonspoilage microorganisms. These biofilms can provide a protective environment for pathogens and reduce the effectiveness of sanitizers and other inhibitory agents. A number of *Pseudomonas* species associated with plants can produce exopolysaccharides characteristic of biofilms [74,99]. Human pathogens including *Campylobacter jejuni*, *E. coli* O157:H7, *L. monocytogenes* and *Salmonella* Typhimurium also are able to form biofilms on inert surfaces [15,100].

13.5 BACTERIAL DEGRADATION OF FRUIT AND VEGETABLE SURFACES

So far there is no reported evidence that suggests direct degradation of produce by human bacterial pathogens; rather, they may coexist with the spoilage organisms associated with the produce. Antagonistic relationships between pseudomonads and *Listeria monocytogenes* on potato slices have been reported [101]. *Salmonella* attached in greater numbers and survived washing with sanitizers to a greater extent on cut surfaces of pepper disks, compared to natural external or internal surfaces. Similarly, *Salmonella* attachment and survival during washing were greater in the calyx and stem areas and on cut surfaces of apples, compared to the unbroken skin surface [63].

Also, respiration, transpiration, and enzymatic activity of living tissue after harvest can cause fruit and vegetable deterioration. Consumer evaluation of processed vegetables and fruits is based on the absence of discoloration, resulting from enzymatic browning of cut surfaces, and the yellowing of green vegetables [101–104]. The surfaces of fruits and vegetables are covered with epicular wax and cuticle that function as a hydrophobic barrier to water and gas exchange, and the amount of this cuticle and epicular wax varies from species to species; therefore, the postharvest enzymatic activity varies. The enzymes of particular interest are the pectolytic enzymes, which gradually cause the ripening of the fruits, making them softer, and chlorophyllase, which catalyzes the cleavage of phytol from chlorophyll to form chlorophyllides, which leads to the formation of pheophorbides. The browning or the darkening of fruits and vegetables is a significant problem, and several research efforts are currently underway to prevent these actions [102,105]. The surface matrix of fruits and vegetables is made up of cellulose, polyuronic acids, proteins, and phenolic compounds and waxes [105]. Ethylene promotes many changes in fruits and vegetables. The decrease of ethylene leads to the loss of green color in green peppers.

13.5.1 PREVENTION OF MICROBIAL CONTAMINATION AND DEGRADATION OF PRODUCE

There are no clear-cut answers as to how bacterial attachment to produce surfaces can be avoided or eliminated. Fruits and vegetables are frequently in contact with soil, insects, animals, or humans during growing or harvesting [106]. A major factor limiting the efficacy of conventional sanitizing treatments for apples and other commodities is the inaccessibility of attached bacteria. Several studies have demonstrated the presence of enteric bacteria, including Enterobacteriaceae and Pseudomonadaceae, within fresh tomatoes and cucumbers [43,107]. An evaluation of mature apples for the presence of *Erwinia amylovora* (cause of fire blight) could not detect this organism but did reveal the presence of internalized Enterobacteriaceae in about 5% of the samples examined [108]. Internalization of *E. coli* O157:H7 has been reported in lettuce [109,110] and radish sprouts [19,111].

The interventions aimed at reducing microbial spoilage on produce surfaces should be applied at preharvest, during harvesting, and postharvest. At preharvest, intervention applications should consider cultural practices such as crop rotation, pruning of produce and destruction of crop debris. During harvesting, emphasis should be on hygiene (i.e., hygienic condition of the environment, the harvesters, and containers used for harvesting and transporting the produce to the storage facility). At postharvest, intervention technologies may include physical or chemical treatments geared toward reducing surface contaminants. Fumigation of fruits and vegetables during storage will reduce the potential for decay [112–114]. The efficacy of sanitizing treatments in decontaminating fruits and vegetables has been investigated [11,70,71,115–124], and population reductions of 2.6 to 3.8 $\log_{10}$ CFU/g have been reported for *Salmonella* and *E. coli* 0157:H7 [70,71,122–124]. However, some studies have shown that substantially smaller reductions are obtained when the targeted bacteria have been on the fruit surfaces for more than a few days [11,70,71]. Limited bactericidal action (1 to 3 log reductions) of chlorine on produce, including whole melons, has been reported [2,11,20,96,125–128]. Depending upon the fruit or vegetable and whether it is whole or cut, up to 200 to 300 ppm chlorine is usually recommended as a sanitizer in wash water [20,129]. The lack of an effective antimicrobial treatment at any step from planting to consumption means that pathogens introduced at any point may be present on the final food product. Washing and rinsing some types of fruits and vegetables prolong shelf life by reducing the number of microorganisms on their surfaces. However, only a portion of pathogenic microorganisms may be removed with this simple treatment. Use of a disinfectant can enhance efficiency of removal up to 100-fold, but chemical treatments administered to whole and cut produce typically will not reduce populations of pathogens by more than 2 to 3 $\log_{10}$ CFU/g [129].

Pathogens also vary in their sensitivity to sanitizers. For example, *L. monocytogenes* is generally more resistant to chlorine than are *Salmonella* and *E. coli* 0157:H7 [111,129,130]. The general lack of efficacy of sanitizers on raw fruits and vegetables can be attributed, in part, to their inaccessibility to locations within structures and tissues that harbor pathogens. Pathogenic bacteria are able to infiltrate cracks, crevices, and intercellular spaces of seeds and produce. Infiltration is dependent on temperature, time, and pressure and occurs when the water pressure on the produce surface overcomes internal gas pressure and the hydrophobic nature of the surface of the produce [18,62]. Infiltration may also be enhanced by the presence of surfactants and when the temperature of the fruit or vegetable is higher than the temperature of contaminated wash water. The protective mechanism of these sites is not well understood but the concept that hydrophobicity of microbial cells aids in their protection by inhibiting penetration of the disinfectants has been proposed. Recently, Ukuku et al. [131] reported reduced microbial populations of fresh-cut melon prepared from whole melon treated with hot water. In this study they concluded that decontamination of whole cantaloupes designated for fresh-cut processing with hot water could have major advantages over the use of sanitizers, including a significant reduction or elimination of vegetative cells of pathogenic bacteria on melon surfaces, thus reducing the probability of potential transfer of pathogenic bacteria from the rind to the interior tissue during cutting. Although application of sanitizer and heat treatment has been used to kill *Salmonella* Stanley inoculated on alfalfa seeds [132] and vegetables [133], this type treatment was able to reduce the population of native microflora and inoculated *Salmonella* on whole melon surfaces without compromising the quality of the fresh-cut pieces [132,133]. Alternatively, lactic acid bacteria can be used to improve the microbial safety of minimally processed fruits and vegetables [42,134–138]. As with infiltration, preharvest internalization of human pathogens within fruits and vegetables would greatly limit the efficacy of washing as a means of decontamination.

REFERENCES

1. Beuchat, L., Pathogenic microorganisms associated with fresh produce, *J. Food Prot.,* 59, 204, 1996.
2. Brackett R.E., Shelf stability and safety of fresh produce as influenced by sanitation and disinfection, *J. Food Prot.,* 55, 808, 1992.
3. Farber, J.M., and Peterkin, P.I., *Listeria monocytogenes*, a food-borne pathogen, *Microbiol. Rev.,* 55, 476, 1991.
4. Faber, J.M., Sanders, G.W., and Johnston, M.A., A survey of various foods for the presence of *Listeria* species, *J. Food Prot.,* 52, 456, 1989.
5. Heisick, J.E., Wagner, D.E., Nierman, M.L., and Peeler, J.T., *Listeria* spp. found on fresh market produce, *Appl. Environ. Microbiol.,* 55, 1925, 1989.
6. Heisick, J.E., Rosas-Marty, L., and Tatini, S.R., Enumeration of viable *Listeria* species and *Listeria monocytogenes* in foods, *J. Food Prot.,* 58, 733, 1995.
7. FDA, Guidance for industry: guide to minimize microbial food safety hazards for fresh fruits and vegetables, Food and Drug Administration, U.S. Department of Agriculture, Centers for Disease Control and Prevention, Oct. 26, 1998, available at http://www.foodsafety.gov.
8. FDA, Guidance for industry: reducing microbial food safety hazards for sprouted seeds and guidance for industry: sampling and microbial testing of spent irrigation water during sprout production, *Fed. Reg.,* 64, 57893, 1999.
9. Centers for Disease Control, Multistate outbreak of *Salmonella poona* infections — United States and Canada, *Morbid. Mortal. Weekly Rep.,* 40,, 1991, pp. 549–552.
10. Mead, P.S., Slutsker, L., Dietz, V., McGaig, L.F., Bresee, J.S., Shapiro, C., Griffin, P.M., and Tauxe, R.V., Food-related illness and death in the United States, *Emerg. Infect. Dis.,* 5, 607, 1999.
11. Ukuku, D.O. and Fett, W.F., Relationship of cell surface charge and hydrophobicity to strength of attachment of bacteria to cantaloupe rind, *J. Food Prot.,* 65, 1093, 2002.
12. Bean, N.H., Goulding, J.S., Daniels, M.T., and Angulo, F.J., Surveillance for food-borne disease outbreaks: United States, 1988–1992, *J. Food Prot.,* 60, 1265, 1997.
13. Barak, J.D., Whitehand, L.C., and Charkowski, A.O., Differences in attachment of *Salmonella* enterica Serovars and *Escherichia coli* O157:H7 to alfalfa sprouts, *Appl. Environ. Microbiol.,* 68, 4758, 2002.
14. Liao, C.-H. and Cooke, P.H., Response to trisodium phosphate treatment of *Salmonella* Chester to fresh-cut green pepper slices, *Can. J. Microbiol.,* 47, 25, 2001.
15. Zottola, E.A., Microbial attachment and biofilm formation: a new problem for the food industry?, *Food Technol.,* 48, 107, 1994.
16. FDA, FDA survey of imported fresh produce, U.S. Food and Drug Administration, Center for Food Safety and Applied Nutrition, 2001, available at http://vm.cfsan.fda.gov/~dms/.

17. FDA, FDA survey of domestic fresh produce, U.S. Food and Drug Administration, Center for Food Safety and Applied Nutrition, 2001, available at http://www.cfsan.fda.gov/~dms/.

18. Beuchat, L.R., Ecological factors influencing survival and growth of human pathogens on raw fruits and vegetables, *Microbes Infect.,* 4, 413, 2002.

19. Itoh, Y., Sugita-Konishi, Y., Kasuga, F., Iwaki, M., Hara-Kudo, Y., Saito, N., Noguchi, Y., Konuma, H., and Kumagai, S., Enterohemorrhagic *Escherichia coli* O157:H7 present in radish sprouts, *Appl. Environ. Microbiol.,* 64, 1532, 1998.

20. Beuchat, L.R., Pathogenic microorganisms associated with fresh produce, *J. Food Prot.,* 59, 204, 1995.

21. Bagdasaryan, G.A., Survival of viruses of the enterovirus group (poliomyelitis, echo, coxsackie) in soil and on vegetables, *J. Hyg. Epidemiol. Microbiol. Immunol.,* 8, 497, 1964.

22. Buck, J.W., Walcott, R.R. and Beauchat, L.R., Recent trends in microbiological safety of fruits and vegetables, *Plant Health Prog.,* Jan./Feb. 2003.

23. Ehret, D.L., Helmer, T., and Hall, J.W., Cuticle cracking in tomato fruits, *J. Hortic. Sci.,* 68, 195, 1993.

24. Peet, M.M., Fruit cracking in tomato, *Hortic. Technol.,* 2, 216, 1992.

25. Webster, B.D. and Craig, M.E., Net morphogenesis and characteristics of the surface of muskmelon fruits, *J. Am. Soc. Hortic. Sci.,* 101, 412, 1976.

26. Splittstoesser, D.F. and Gadjo, I., The groups of micro-organism composing the total count population in frozen vegetables, *J. Food Prot.,* 31, 234, 1966.

27. James, J., Incidence and types of microorganisms in food, in *Modern Food Microbiology,* Van Nostrand Reinhold, New York, 1992, p. 63.

28. Geeson, J.D., The fungal and bacterial flora of stored white cabbage, *J. Appl. Bacteriol.,* 46, 189, 1979.

29. Varaquaux, P., Unit operations of fresh-cut produce, in *Proceedings of the Second International Conference on Fresh Cut Produce,* Campden and Chorleywood Food Research Association Group, Chipping Campden, UK, 2001.

30. Stinson, E.E., Osman, S.F., and Heisler, E.G., Mycotoxins production in whole tomatoes, apples, oranges and lemons, *J. Agric. Food Chem.,* 29, 790, 1981.

31. Kim, J.-C, and Lee, Y.-W., Sambutoxin, a new mycotoxin produced by toxic *Fusarium* isolates from rotted potato tubers, *Appl. Environ. Microbiol.,* 60, 4380, 1994.

32. Kim, J.-C., Lee, Y.-W., and Yu, S.-H., Sambutoxin-producing isolates of Fusarium species and occurance of sambutoxin in rotten potato tubers, *Appl. Environ. Microbiol.,* 61, 3750, 1995.

33. Smith, J.E., Lewis, C.W., Anderson, J.G., and Solomons, G.L., *Mycotoxins in Human Nutrition and Health,* European Commission, Directorate-General XII, Brussells, 1994, p. 300.

34. Snowdon, A.L., *A Colour Atlas of Post-harvest Disease and Disorders of Fruits and Vegetables.* Vol. 2, *Vegetables,* Wolf Scientific, London, 1991, p. 416.

35. Le Cam, B., Rouxel, F., and Villeneuve, F., Analyse de la flore fongique de la carotte conservée au froid: prépondéreance de mycocentrospora acerina (Hartig) Deighton, *Agronomy* 13, 125, 1993.

36. Geeson, J.D., Browne, K.M., and Everson, H.P., Storage diseases of carrots in East-Anglia 1978–82, and the effects of some pre- and post-harvest factors, *Ann. Appl. Biol.,* 112, 503, 1988.

37. Jensen, A., Storage diseases of carrots, especially *Rhizoctonia* crater rot, *Acta Hortic.* 20, 125, 1971.

38. Adair, C.N., Influence of controlled-atmosphere storage condition on cabbage post-harvest decay fungi, *Plant Dis. Rep.*, 55, 864, 1971.
39. Liao, C.-H. and Shollenberger, L.M., Lethal and sublethal action of acetic acid on *Salmonella in vitro* and on cut surfaces of apple slices, *J. Food Sci.*, 68, 2793, 2003.
40. Liao, C.-H. and Shollenberger, L.M., Enumeration, resuscitation, and infectivity of the sublethally injured *Erwinia* cells induced by mild acid treatment, *Phytopathology*, 94, 76, 2004.
41. Lund, B.M., Bacterial spoilage, in *Post-Harvest Pathology of Fruits and Vegetables*, Denis, C., Ed., Academic Press, London, 1983, p. 219.
42. Torriani, S., Orsi, C., and Vescovo, M., Potential of *Lactobacillus casei*, culture permeate, and lactic acid to control microorganisms in ready-to-use vegetables, *J. Food Prot.*, 60,1564, 1997.
43. Sands, D.C., Hankin, L., and Zucker, M., Ecology and physiology of fluorescent pectolytic pseudomonads, *Phytopathology*, 65, 921, 1975.
44. Cuppels, D. and Kelman, A., Evaluation of selective media for isolation of soft-rot bacteria from soil and plant tissue, *Phytopathology*, 64, 468, 1974.
45. Van Renterghem, B., Huysman, F., Rygole, R., and Verstarte, W., Detection and prevalence of *Listeria monocytogenes* in the agricultural ecosystem, *J. Appl. Bacteriol.*, 71, 211, 1991.
46. Al-ghazali, M.R. and Al-azawi, S.K., *Listeria monocytogenes* contamination of crops grown on soil treated with sewage sludge cake, *J. Appl. Bacteriol.*, 69, 642, 1990.
47. Dunlop, S.G. and Wang, W.-L.L., Studies on the use of sewage effluent for irrigation of truck crops, *J. Milk Food Technol.*, 24, 44, 1961.
48. Grigoreva, L.V., Garodetsky, A.S., Omel'Yanets, T.G., and Bogdanenko, L.A., Survival of bacteria and viruses on vegetables crops irrigated with infected water, *Hyg. Sanit.*, 30, 357, 1965.
49. Nichols, A.A., Davis, P.A., and King, K.P., Contamination of lettuce irrigated with sewage effluent, *J. Hortic. Sci.*, 46, 425, 1971.
50. Rudolfs, W., Falk, L.L., and Ragotskie, R.A., Contamination of vegetables grown in polluted soil, *Sew. Ind. Wastes*, 23, 253, 1951.
51. Kott, H. and Fishelson, L., Survival of enteroviruses on vegetables irrigated with chlorinated oxidation pond effluents, *Isr. J. Technol.*, 12, 290, 1974.
52. Samish, Z., Etinger-Tulczynsky, R., and Bick, M., The microflora within the tissue of fruits and vegetables, *J. Food Sci.*, 28, 259, 1963.
53. Romantschuk, M., Roine, E., Bjorklof, K., Ojanen, T., Nurmiaho-Lassila, E.-L., and Haahtela, K., Microbial attachment to plant aerial surfaces, in *Aerial Plant Surface Microbiology*, Morris, C.E., Nicot, P.C., and Nguyen-The, C., Eds., Plenum Press, New York, 1996, p. 43.
54. National Advisory Committee on Microbiological Criteria for Foods (NACMCF), Microbiological safety evaluations and recommendations on fresh produce, *Food Control*, 10, 117, 1999.
55. Solomon, E.B., Yaron, S., and Matthews, K.R., Transmission of *Escherichia coli* O157:H7 from contaminated manure and irrigation water to lettuce plant tissue and its subsequent internalization, *Appl. Environ. Microbiol.*, 68, 397, 2002.
56. Wachtel, M.R., Whitehand, L.C., and Mandrell, R.E., Association of *Escherichia coli* O157:H7 with preharvest leaf lettuce upon exposure to contaminated irrigation water, *J. Food Prot.*, 65, 8, 2002.
57. Guo, X., van Iersel, M.W., Chen, J., Brackett, R.E., and Beuchat, L.R., Evidence of association of salmonellae with tomato plants grown hydroponically in inoculated nutrient solution, *Appl. Environ. Microbiol.*, 68, 3639, 2002.

58. Burnett, S.L. and Beuchat, L.R., Human pathogens associated with raw produce and unpasteurized juices, and difficulties in contamination, *J. Indust. Microbiol. Biotechnol.*, 27, 104, 2001.

59. Romantschuk, M., Attachment of plant pathogenic bacteria to plant surfaces, *Annu. Rev. Phytopathol.*, 30, 225, 1992.

60. Sadorski, A.Y., Fattal, B., and Goldberg, D., High levels of microbial contamination of vegetables irrigated with wastewater by the drip method, *Appl. Environ. Microbiol.*, 36, 824, 1978.

61. Fletcher, M., Bacterial attachment in aquatic environments: a diversity of surfaces and adhesion strategies, in *Bacterial Adhesion: Molecular and Ecological Diversity*, Fletcher, M., Ed., John Wiley & Sons, New York, 1996, p. 1.

62. Busscher, H.J., Sjollema, J., and Van der Mei, H.C., Relative importance of surface free energy as a measure of hydrophobicity in bacterial adhesion to solid surfaces, in *Microbial Cell Surface Hydrophobicity*, Doyle, R.J. and Rosenberg, M., Eds., 1990, p. 335.

63. Liao, C.-H. and Sapers, G.M., Attachment and growth of *Salmonella* Chester on apple fruits and *in vivo* response of attached bacteria to sanitizer treatments, *J. Food Prot.*, 63, 876, 2000.

64. Reina, L.D. Fleming, H.P., and Breidt, F., Jr., Bacterial contamination of cucumber fruit through adhesion, *J. Food Prot.*, 65, 1881, 2002.

65. Fratamico, P.M., Bhaduri, S., and Buchanan, R.L., Studies on *Escherichia coli* serotype O157:H7 strains containing a 60-Mda plasmid and on 60-Mda plasmid-cured derivatives, *J. Med. Microbiol.*, 39, 371, 1993.

66. Sherman, P., Soni, R., Petric, M., and Karmali, M., Surface properties of Vero cytoxin-producing *Escherichia coli* O157:H7, *Infect. Immun.*, 55, 1824, 1987.

67. Burnett, S.L., Chen, J., and Beuchat, L.R., Attachment of *Escherichia coli* O157:H7 to surfaces and internal structures of apples as detected by confocal scanning laser microscopy, *Appl. Environ. Microbiol.*, 66, 4679, 2000.

68. Kenney, S.J., Burnett, S.L., and Beuchat, L.R., Location of *Escherichia coli* O157:H7 on and in apples as affected by brushing, washing, and rubbing, *J. Food Prot.*, 64, 1328, 2001.

69. Takeuchi, K., Matute, C.M., Hassan, N.A., and Frank, J.F., Comparison of attachment of *Escherichia coli* O157:H7, *Listeria monocytogenes*, *Salmonella typhimurium*, and *Pseudomonas fluorescens* to lettuce leaves, *J. Food Prot.*, 63, 1433, 2000.

70. Ukuku, D.O., Pilizota, A.V., and Sapers, G.M., Bioluminescence AT assay for estimating total plate counts of surface microflora of whole cantaloupe and determining efficacy of washing treatments, *J. Food Prot.*, 64, 813, 2001.

71. Ukuku, D.O., Pilizota, A.V. and Sapers, G.M., Influence of washing treatment on native microflora and *Escherichia coli* population of inoculated cantaloupes, *J. Food Safety*, 21, 31, 2001.

72. Mafu, A.A., Roy, D., Savoie, L., and Goulet, J., Bioluminescence assay for estimating the hydrophobic properties of bacteria as revealed by hydrophobic interaction chromatography, *Appl. Environ. Microbiol.*, 57, 1640, 1991.

73. Noda, Y. and Kanemasa, Y., Determination of hydrophobicity on bacterial surfaces by nonionic surfactants, *J. Bacteriol.*, 167, 1016, 1986.

74. Fett, W.F., Cescutti, P., and Wijey, C., Exopolysaccharides of the plant pathogens *Pseudomonas corrugata* and *Ps. flavescens* and the saprophyte *Ps. chloroaphis*, *J. Appl. Bacteriol.*, 81, 181, 1996.

75. Pringle, J.H. and Fletcher, M., The influence of substratum hydration and adsorbed macromolecules on bacterial attachment to surfaces, *Appl. Environ. Microbiol.*, 51, 1321, 1986.

76. Fletcher, M., and Floodgate, G.D., An electron-microscopic demonstration of an acidic polysaccharide involved in the adhesion of a marine bacterium to solid surfaces, *J. Gen. Microbiol.*, 74, 325, 1973.

77. Gilbert, P.D., Evans, D.J., Duguid, I.G., and Brown, M.R.W., Surface characteristics and adhesion of *Escherichia coli* and *Staphylococcus epidermidis*, *J. Appl. Bacteriol.*, 71, 72, 1991.

78. Dewanti, R. and Wong, A.C.L., Influence of culture conditions on biofilm formation by *Escherichia coli* O157:H7, *Int. J. Food. Microbiol.*, 26, 147, 1995.

79. Van loosdrecht, M.C.M., Lyklema, J., Norde, W., Scharaa, G., and Zehnder, A.J.B., The role of bacterial cell wall hydrophobicity in adhesion, *Appl. Environ. Microbiol.*, 53, 1893, 1987.

80. Van Der Mei, H.C., Rosenberg, M., and Busscher, H.J., Assessment of microbial cell surface hydrophobicity, in *Microbial Cell Surface Analysis*, Mozes, N., Handley, P.S., Busscher, H.J., and Rouxhet, P.G., Eds., VCH, New York, 1991, p. 263.

81. Fletcher, M. and Loeb, G.I., Influence of substratum characteristics on the attachment of marine pseudomonad to solid surfaces, *Appl. Environ. Microbiol.*, 37:67, 1979.

82. Frank, J.F., Microbial attachment to food and food contact surfaces, *Adv. Food Nutr. Res.*, 43, 320, 2000.

83. Briandet, R., Meylheuc, T., Maher, C., and Bellon-Fontaine, M.N., *Listeria monocytogenes* Scott A: cell surface charge, hydrophobicity, and electron donor and acceptor characteristics under different environmental growth conditions, *Appl. Enivron. Microbiol.*, 65, 5328, 1999.

84. Matthysse, A.G., Adhesion in the *Rhizophere*, in *Bacterial Adhesion: Molecular and Ecological Diversity*, Fletcher, M., Ed., John Wiley and Sons, New York, 1996, p. 129.

85. Matthysse, A.G., Initial interactions of *Agrobacterium tumefaciens* with plant host cells, *CRC Crit. Rev. Microbiol.*, 13, 281, 1986.

86. Brown, P.K., Dozois, C.M., Nickerson, C.A., Zuppardo, A., Terlonge, J., and Curtiss, R., III, MirA, a novel regulator of curli (AgF) and extracellular matrix synthesis by *Escherichia coli* and *Salmonella enterica* serovar Typhimurium, *Mol. Microbiol.*, 41, 349, 2001.

87. Zogaj, X., Nimtz, M., Rohde, M., Bokranz, W., and Romling, U., The multicellular morphotypes of *Salmonella typhimurium* and *Escherichia coli* produce cellulose as the second component of the extracellular matrix, *Mol. Microbiol.*, 39, 1452, 2001.

88. Magnusson, K.E., Hydrophobic interaction: a mechanism of bacterial adhesion, *Scand. J. Infect. Dis. Suppl.*, 33, 32, 1982.

89. Rainey, P.B., Phenotypic variation of *Pseudomonas putida* and *P. tolaasi* affects attachment to *Agaricus bisporus* mycelium, *J. Gen. Microbiol.*, 137, 2768, 1991.

90. Mansvelt, E.L. and Hatting, M.J., Scanning electron microscopy of invasion of apple leaves and blossoms by *Pseudomonas syringae* pv. *syringae*, *Appl. Environ. Microbiol.*, 55, 535, 1989.

91. Bashan, Y., Levanony, H., and Whitmoyer, R.E., Root surface colonization of non cereal crop plants by pleomorphic *Azospirillum brasilense* Cd, *J. Gen. Microbiol.*, 137, 187, 1991.

92. Frank, J.F. and Koffi, R.A., Surface adherent growth of *Listeria monocytogenes* is associated with increased resistance to surfactant sanitizer and heat, *J. Food Prot.*, 53, 550, 1990.

93. International Commission on Microbiological Specifications for Foods (ICMS), Factors affecting life and death of microorganisms, in *Microbial Ecology of Foods*, Vol. 1, Academic Press, New York, 1980.

94. Sapers, G.M., Miller, R.L., and Mattrazzo, A.M., Effectiveness of sanitizing agents in inactivating *Escherichia coli* in Golden Delicious apples, *J. Food Sci.*, 64, 734, 1999.

95. Sapers, G.M., Miller, R.L., Pilizota, V., and Mattrazzo, A.M., Antimicrobial treatments for minimally processed cantaloupe melon, *J. Food Sci.*, 66, 345, 2001.

96. Ukuku, D.O., and Sapers, G.M., Effect of sanitizer treatment on *Salmonella* Stanley attached to the surface of cantaloupe and cell transfer to fresh-cut tissues during cutting practices, *J. Food Prot.*, 64, 1286–1291, 2001.

97. Babic, I., Roy, S., Watada, A.E., and Wergin, W.P., Changes in microbial populations on fresh cut spinach, *Int. J. Food Microbiol.*, 31, 107, 1996.

98. Carmichael, I., Harper, I.S., Coventry, M.J., Taylor, P.W.J., Wan, J., and Hickey, M.W., Bacterial colonization and biofilm development on minimally processed vegetables, *J. Appl. Microbiol. Symp. Suppl.*, 85, 45S, 1999.

99. Bennett, R.A. and Billing, E., Origin of the polysaccharide ooze from plants infected with *Erwinia amylovora*, *J. Gen. Microbiol.*, 116, 341, 1980.

100. Strom, M.S. and Lory, S., Structure-function and biogenesis of the type IV pili, *Annu. Rev. Microbiol.*, 47, 565, 1993.

101. Liao, C.-H., and Sapers, G.M., Influence of soft rot bacteria on growth of *Listeria monocytogenes* on potato tuber slices, *J. Food Prot.*, 62, 343, 1999.

102. Bolin, H. and Huxsoll, C., Effect of preparation and storage parameters on quality retention of salad-lettuce, *J. Food Sci.*, 56, 60, 1991.

103. Watada, A. and Qi, L., Quality of fresh-cut produce, *Postharv. Biol. Technol.*, 15, 201–205, 1999.

104. Wei, C.I., Huang, T.S., Kim, J.M., Lin, W.F., Tamplin, M.L., and Bartz, J.A., Growth and survival of *Salmonella montevideo* on tomatoes and disinfection with chlorinated water, *J. Food Prot.*, 58, 829, 1995.

105. Lauchli, A., Apoplasmic transport in tissues, in Transport in plants 11. Part B. Tissues and organs, *Encyclopedia of Plant Physiology*, Luttge, U. and Pittman, M.G., Eds., Springer-Verlag, Berlin, 1976, p. 3.

106. Shewfelt, R.L., Quality of minimally processed fruits and vegetables, *J. Food Qual.*, 10, 143, 1987.

107. Meneley, J.C. and Stanghellini, M.E., Detection of enteric bacteria within locular tissue of healthy cucumbers, *J. Food Sci.*, 39, 1267, 1974.

108. Roberts, R.G., Reymond, S.T., and Mclaughlin, R.J., Evaluation of mature apple fruit from Washington State for the presence of *Erwinia amylovora*, *Plant Dis.*, 73, 917, 1989.

109. Seo, K.H. and Frank, J.F., Attachment of *Escherichia coli* O157:H7 to lettuce leaf surface and bacterial viability in response to chlorine treatment as demonstrated by using confocal scanning laser microscopy, *J. Food Prot.*, 62, 3, 1999.

110. Somers, E.B., Schoeni, J.L., and Wong, A.C.L., Effect of trisodium phosphate on biofilm and planktonic cells of *Campylobacter jejuni*, *Escherichia coli* O157:H7, *Listeria monocytogenes* and *Salmonella typhimurium*, *Int. J. Food Microbiol.*, 22, 269, 1994.

111. Brackett, R.E., Incidence, contributing factors, and control of bacterial pathogens in produce, *Postharv. Biol. Technol.*, 15, 305, 1999.

112. Sholberg, P.L., Delaquis, P., and Moyls, A.L., Use of acetic acid fumigation to reduce the potential for decay in harvested crops, in *Recent Research Developments in Plant Pathology*, *Research Signpost* 2, 31, 1998.

113. Sholberg, P.L. and Gaunce, A.P., Fumigation of fruit with acetic acid to prevent postharvest decay, *Hortic. Sci.*, 30, 1271, 1995.
114. Sholberg, P.L. and Gaunce, A.P., Fumigation of stonefruit with acetic acid to control postharvest decay, *Crop. Prot.*,15, 681, 1996.
115. Costilow, R., Uebersax, M.A., and Ward, P.J., Use of chlorine dioxide for controlling microorganisms during the handling and storage of fresh cucumbers, *J. Food Sci.*, 49, 396, 1984.
116. Delaquis, P., Graham, H.S., and Hocking, R., Shelf-life of coleslaw made from shredded cabbage fumigated with gaseous acetic acid, *J. Food Proc. Preserv.*, 21, 129, 1997.
117. Delaquis, P., Sholberg, P.L., and Stanich, K., Disinfection of mung bean seed with vaporized acetic acid, *J. Food Prot.*, 62, 953, 1999.
118. Roberts, R.G. and Reymond, S.T., Chlorine dioxide for reduction of postharvest pathogen inoculum during handling of tree fruits, *Appl. Environ. Microbiol.*, 60, 2864, 1994.
119. Wei, C.-I., Cook, D.I. and Kirk, R.J., Jr., Use of chlorine compounds in the food industry, *Food Technol.*, 1, 107, 1985.
120. Wickramanayake, G.B., Disinfection and sterilization by ozone, in *Disinfection, Sterilization, and Preservation*, 4th ed., Block, S.S, Ed., Lea & Febiger, Philadelphia, PA, 1991, p. 182.
121. Wright, J.R., Sumner, S.S., Hackney, C.R., Pierson, M.D., and Zoecklein, B.W., Reduction of *Escherichia coli* O157:H7 on apples using wash and chemical sanitizer treatments, *Dairy, Food Environ. Sanit.*, 20, 120, 2000.
122. Zhang, S. and Farber, J.M., The effects of various disinfectants against *Listeria monocytogenes* on fresh-cut vegetables, *Food Microbiol.*, 13, 311, 1996.
123. Zhuang, R.-Y. and Beuchat, L.R., Effectiveness of trisodium phosphate for killing *Salmonella montevideo* on tomatoes, *Lett. Appl. Microbiol.*, 232, 97, 1996.
124. Zhuang, R.-Y., Beuchat, L.R., and Angulo, F.J., Fate of *Salmonella montevideo* on and in raw tomatoes as affected by temperature and treatment with chlorine, *Appl. Environ. Microbiol.*, 61, 2127, 1995.
125. Adams, M.R., Hartley, A.D. and Cox, L.J., Factors affecting the efficacy of washing procedures used in the production of prepared salads, *Food Microbiol.*, 6, 69, 1989.
126. Ayhan, Z., Chism, G.W. and Richter, E.R., The shelf-life of minimally processed fresh cut melons, *J. Food Quality*, 21, 29, 1998.
127. Green, D.E. and Stumpe, P.K., The mode of action of chlorine, *J. Am. Water Works Assoc.*, 38, 1301–1305, 1946.
128. Nguyen–The, C. and Carlin, F., Fresh and processed vegetables, in *The Microbiological Safety and Quality of Food*, Vol. 1, Lund, B.M., Baird-Parker, T.C., and Gould, G.W., Eds., Aspen Publishing, Gaithersburg, MD, 2000, p. 620.
129. Beuchat, L.R., Surface decontamination of fruits and vegetables eaten raw: a review, Food Safety Unit, World Health Organization. WHO/FSF/FOS/98.2, 1998.
130. Brackett, R.E., Antimicrobial effect of chlorine on *Listeria monocytogenes*, *J. Food Prot.*, 50, 999, 1987.
131. Ukuku, D.O., Pilizota, V., and Sapers, G.M., Effect of hot water treatments on survival of *Salmonella* and microbial quality of whole and fresh-cut cantaloupe, *J. Food Prot.*, 67, 432, 2004.
132. Jaquette, C.B., Beuchat, L.R., and Mahon, B.E., Efficacy of chlorine and heat treatment in killing *Salmonella* Stanley inoculated onto alfalfa seeds and growth and survival of the pathogen during sprouting and storage, *Appl. Environ. Microbiol.*, 62, 2212, 1996.

133. Spotts, R.A. and Cervantes, L.A., Effect of ozonated water on postharvest pathogens of pear in laboratory and packinghouse tests, *Plant Dis.,* 76, 256, 1992.
134. Breidt, F. and Fleming, H.P., Using lactic acid bacteria to improve the safety of minimally processed fruits and vegetables, *Food Technol.,* 51, 44, 1997.
135. Janisiewicz, W.J., Conway, W.S., and Leverentz, B., Biological control of postharvest decays of apples can prevent growth of *Escherichia coli* O157:H7 in apple wounds, *J. Food Prot.,* 62, 1372, 1999.
136. Janisiewicz, W.J. and Jeffers, S.N., Efficacy of commercial formulation of two bio-fungicides for control of blue mold and gray mold of apples in storage, *Crop Prot.,* 16, 629, 1997.
137. Leibinger, W., Breuker, B., Hahn, M., and Mendgen, K., Control of postharvest pathogens and colonization of the apple surface by antagonistic microorganisms in the field, *Phytopathology,* 87, 1103, 1997.
138. Vescovo, M., Torriani, S., Orsi, C., Macchiarolo, F., and Scolari, G., Application of antimicrobial-producing lactic acid bacteria to control pathogens in ready-to-use vegetables, *J. Appl. Bacteriol.,* 81, 113, 1996.

14 Bacterial Infiltration and Internalization in Fruits and Vegetables

Aubrey Mendonca
Department of Food Science and Human Nutrition, Iowa State University, Ames, IA

CONTENTS

14.1 INTRODUCTION

The surfaces of fruits and vegetables growing in orchards and fields inevitably become contaminated with microorganisms from a variety of sources in the natural environment. In addition to epiphytic bacteria that multiply on leaves and fruits, human enteric pathogens can be transiently present on the surface of fresh produce. Bacteria can become internalized in fruits and vegetables by penetrating deeper tissues through damaged sites on the surface of these products. Internalization may occur via infiltration with contaminated water in the fields or during postharvest processing. Internalized bacteria can increase postharvest losses or compromise the microbial safety of fruits and vegetables.

In response to public health concerns with microbial safety of minimally processed fresh fruits and vegetables, various methods for destroying microbial contaminants on the surface of these popular food products are actively being studied. Unfortunately, surface decontamination methods are ineffective in eliminating pathogenic bacteria that infiltrate fresh produce and become internalized in deeper subsurface areas. Internalized bacteria are inaccessible to chemical sanitizers due to physical protection from subsurface tissue in which they are embedded. Therefore, prevention of infiltration is crucial for controlling internalization of bacteria and improving the microbial shelf life and safety of fruits and vegetables.

This chapter provides an overview of the occurrence of bacteria and bacterial adhesion on fresh produce followed by a review of current published research in the following areas: (1) bacterial infiltration and internalization in fruits and vegetables; (2) factors affecting infiltration and internalization; (3) control of bacterial infiltration; and (4) survival and growth of internalized bacteria. Additionally, recommendations for future research will be provided.

14.2 OCCURRENCE OF BACTERIA ON FRUITS AND VEGETABLES

The surfaces of fruits and vegetables growing in orchards and fields provide a habitat for a variety of microorganisms including bacteria, yeast, and molds. Among these microbial groups bacteria constitute a major part of the microflora on leaf surfaces and the surface of fruits. In fact, epiphytic bacteria are the first to populate newly formed plant tissues, including buds. Bacterial colonization of newly formed buds occurs from small populations of resident epiphytic bacteria within the bud tissue or from a variety of environmental sources or vectors (Andrews and Hirano, 1991; Beattie and Lindlow, 1994; Beattie, 2002). Sources of bacterial contamination in fresh produce include soil, water, windblown dust, insects, animals (wild or domesticated), animal manure used as fertilizer, humans, harvesting equipment, transport vehicles, and processing equipment (Centers for Disease Control and Prevention, 1997; Beuchat, 1998). Of these sources inadequately composted animal manure is a major contributor to contamination of fresh produce with human pathogens (U.S. FDA, CFSAN, 1998; De Roever, 1999; Himathongkham et al., 1999). Apart from epiphytic bacteria that multiply on undamaged fruit and vegetable surfaces, other bacteria including human enteric pathogens can be transiently present on plant surfaces (Leben, 1965). Contamination of tree fruit with human enteric pathogens such as *Salmonella* and enterohemorrhagic *Escherichia coli* may come from the feces of birds; Wallace et al. (1997) isolated vero cytotoxin-producing *Escherichia coli* O157:H7 from wild birds.

Compared to epiphytic bacteria, certain human enteric bacteria can survive and grow in a more limited set of conditions created by bruises and wounds on fruits and vegetables. Several reports have highlighted the ability of *Salmonella* and *E. coli* O157:H7 to survive on wounded or seemingly undamaged fresh produce (Beuchat, 1996; Kakiomenou et al., 1998; Porto and Eiora, 2001). A variety of vegetables and relatively low-acid fruits can support rapid proliferation of *Salmonella* and *E. coli* O157:H7 within the temperature range of 15 to 25°C that is usually encountered during handling of fresh produce. Postharvest proliferation of these pathogens on fresh produce, particularly on various types of lettuce, can occur under temperature and relative humidity conditions that permit bacterial growth, thus increasing the risk of foodborne disease (Abdul Raouf et al., 1993; Diaz and Hotchkiss, 1996). The potential for contamination of fresh produce with pathogenic bacteria from the environment and the limited efficacy of mere washing to remove pathogens pose a food safety risk if leafy vegetables are eaten without cooking. This risk can be further increased when postharvest processing conditions permit infiltration of pathogenic bacteria in fresh produce and render them inaccessible to the killing effects of chemical sanitizers applied to the surface of fruits and vegetables.

14.3 BACTERIAL ADHESION TO FRUITS AND VEGETABLES

14.3.1 BACTERIAL ADHESION

Bacterial adhesion to or contact with damaged or intact plant surfaces precedes entry of these organisms into fresh produce and is facilitated by several factors, including cell surface charge (Fletcher and Loeb, 1979; Ukuku and Fett, 2002), hydrophobicity (Van Loosdrecht et al., 1987; Van der Mei et al., 1991; Ukuku and Fett, 2002), and extracellular polysaccharides (Fletcher and Floodgate, 1973; Frank, 2000). As found in some fungi, the process of bacterial adhesion to plant surfaces may occur in two stages: a rapid stage that occurs soon after bacteria contact the plant surface and a slower stage that increases during colonization of the surface. The concept of a rapid stage of bacterial adhesion is supported by Hass and Rotem (1976). Those researchers demonstrated that washing of cucumber leaves immediately after inoculating them with *Pseudomonas lachrymans* only removed about 17% of this organism. This percentage of *P. lachrymans* cells removed by washing remained relatively constant irrespective of whether a low (10^4 cells/mL) or high (10^{11} cells/mL) level of inoculum was used.

Few studies have addressed the effects of differences in types of foodborne bacteria on the extent of bacterial adhesion to the surface of fresh and minimally processed vegetables. Some foodborne bacteria such as *Pseudomonas fluorescens* have been shown to preferentially adhere to intact leaf surfaces compared to cut surfaces (Seo and Frank, 1999; Takeuchi et al., 2000). Seo and Frank (1999) reported that viable *E. coli* O157:H7 cells did not adhere to intact surfaces of lettuce leaves but were found attached to the interior surfaces of stomata following chlorine treatment of the lettuce. Takeuchi et al. (2000) studied the extent of attachment of *E. coli* O157:H7, *Listeria monocytogenes*, *Salmonella* Typhimurium, and *Pseudomonas fluorescens* on iceberg lettuce using confocal scanning laser microscopy (CSLM). These researchers observed that a significantly larger amount *E. coli* O157:H7 and *L. monocytogenes* cells adhered to cut edges, whereas *Pseudomonas fluorescens* adhered preferentially to intact surfaces. *Salmonella* Typhimurium adhered equally to both intact and cut surfaces. On the intact leaf surface the extent of adhesion of each organism in descending order was *Pseudomonas fluorescens* followed by *E. coli* O157:H7, *L. monocytogenes,* and *S.* Typhimurium. Some cells of each organism adhered to and subsequently penetrated subsurface tissues through the cut edges of the lettuce leaves.

Because leaf surfaces are generally hydrophobic due to the presence of a waxy cuticle (Romberger et al., 1993), it can be assumed that the hydrophobicity of bacterial cells helps them to adhere to leaves or other hydrophobic surfaces. *Pseudomonas fluorescens* readily adhered to the intact surface of lettuce leaves compared to the edges of cut surfaces (Seo and Frank, 1999). Spores of *Pasteuria* species have been reported to adhere rapidly to the waxy cuticle of plants and nematodes (Mendoza de Gives et al., 1999). Both hydrophobic and electrostatic interactions seemed to be involved in the mechanism of rapid attachment of *Pasteuria penetrans* spores (Afolabi et al., 1995; Davies et al., 1996). The hydrophobicity of plant surfaces including the surfaces of leaves can be evaluated by measuring the "contact angle" of water droplets on the leaf surface. The contact angle is the angle formed between the advancing edge of a water droplet and a solid surface (Rentschler, 1971). There is an inverse relationship between the contact angle and the extent of attachment between a liquid and a solid surface; therefore, the lower the extent of attachment between liquid droplet and the surface (larger contact angle) the greater the hydrophobicity. Generally, the leaves of many plants are hydrophobic and have contact angles larger than 80 or 100° (Neinhuis and Barthlott, 1997). In this regard, Beattie (2002) suggested that even the most hydrophobic of bacterial cell surfaces are unlikely to be sufficiently hydrophobic to promote rapid adhesion to leaf surfaces.

14.3.2 INVOLVEMENT OF BACTERIAL APPENDAGES

Although bacterial cell surface charge and hydrophobicity may contribute to bacterial adhesion to plant surfaces, bacterial appendages such as pili or other protein structures help to facilitate this process. Generally, the surfaces of both plants and bacterial cells are negatively charged. Therefore, electrostatic repulsion will occur once these surfaces come in close proximity to each other. The rapid attachment of bacterial cells on leaves may be mediated by pili or via carbohydrates or outer membrane proteins on the bacterial cell surface. These cell surface adhesions bridge the gap created by electrostatic repulsion (Romantschuk et al., 1996). The involvement of pili in relatively rapid attachment of *Pseudomonas syringae* and *Pseudomonas phaseolicola* to leaf surfaces has been demonstrated (Romantschuk et al., 1993; Suoniemi et al., 1995).

14.3.3 EXTRACELLULAR POLYSACCHARIDES

The second stage of bacterial adhesion to plant surfaces seems to involve the production of extracellular polysaccharides. This process occurs within hours or days after contact between bacteria and plant surfaces. Increases in bacterial populations that become firmly bound to leaves over time have been reported (Beattie and Lindlow, 1994; Wilson et al., 1999). Such increases in the extent of bacterial adhesion may be attributed to colonization of internal leaf sites, especially by plant pathogens (Wilson et al., 1999), and production of extracellular polysaccharides (Takahashi and Doke, 1984). Published reports describing amorphous material around bacteria on leaf surfaces (Beattie and Lindlow, 1999) and around clumps of bacterial cells detached from leaves (Morris et al., 1997) provide ample evidence in support of bacterial production of extracellular polysaccharides on leaves. Davies et al. (1993) reported the induction of some bacterial genes linked to the production of extracellular polysaccharides via bacterial contact with a solid surface. The presence of extracellular polysaccharides on surfaces usually results in the formation of biofilms, which enhance bacterial colonization and survival on leaves (Morris et al., 1997; Beattie and Lindlow, 1999). Injury of the leaf cuticle and release of juices from damaged tissue could result in multiplication of bacteria in biofilms and subsequent spread of bacterial cells to internal leaf tissue. A comprehensive account of bacterial attachment and biofilms on fruits and vegetables is provided in Chapter 13 of this book.

14.4 INTERNALIZATION OF BACTERIA IN FRUITS AND VEGETABLES

14.4.1 FACTORS THAT FACILITATE INTERNALIZATION

14.4.1.1 Natural Openings

Most microorganisms on the surface of intact fresh produce are prevented from entering subsurface tissues by the cuticular layer that covers the epidermis of leaves, stems, and fruits (Nguyen-The and Carlin, 2000). However, natural openings on the surface of fruits and vegetables can provide channels through which bacteria can enter these food products. For example, Leben (1972) found bacterial cells inside more than 60% of the buds of field-grown crops, including certain legumes (soybean and red clover), turnips, grapes, and cucumbers. Itoh et al. (1998) reported that *E. coli* O157:H7 were found in the stomata of radish sprouts grown in contaminated water. Bartz and Showalter (1981) demonstrated that *Serratia marcescens* became internalized in tomatoes that were immersed in cell suspensions of the organism. Populations of this organism were more concentrated in tissues just below the stem scar area of tomatoes. Similar studies involving immersion of apples (Buchanan et al., 1999), oranges, and grapefruit (Merker et al., 1999) in a dye solution indicated the potential for pathogenic bacteria to infiltrate natural openings and become internalized in the core areas of these fruits. Major routes of bacterial entry were the blossom end and stem scar area of apples and citrus fruits, respectively. Bacteria have been shown to enter leaves of plants via the stomata and into certain fruits via the calyx or stem scar (Samish and Etinger-Tulczynska, 1963; Samish et al., 1963; Zhuang et al., 1995). Seo and Frank (1999) demonstrated that *E. coli* O157:H7 entered lettuce leaves through the stomata during dipping the lettuce in a cell suspension of this pathogen. Burnett et al. (2000) reported that *E. coli* O157:H7 were sporadically present in lenticels of intact mature apples immersed in suspensions of this organism. The development of lenticels on mature apples follows formation of cutin within damaged trichomes or stomata, whose guard cells are permanently opened as a result of growth-induced stretching. Stomata facilitate gas exchange in young fruits; however, as lenticels form and loosely packed wax platelets develop in mature fruits, the lenticels usually close and become impervious to liquids and gases (Clements, 1935). Interestingly, about 5% of the lenticels of Red Delicious apples remain open (Clements, 1935).

14.4.1.2 Microbial Damage

Microbial damage to the leaf cuticle will facilitate the entry of bacteria into subsurface areas of the leaf tissue. The mechanisms of bacterial entry may involve physical force or breakdown of the cuticle. These actions are believed to be associated with fungal penetration of leaves (Kolattukudy et al., 1995). Many filamentous fungi such as *Aspergillus*, *Fusarium*, *Penicillium*, and *Trichoderma* can produce a broad spectrum of enzymes including amylases, cellulases, and pectinases. Also, some yeasts can produce pectinases (Blanco et al., 1999). Pectinases are well recognized for their involvement in softening of fruits and vegetables and the properties of these enzymes have been comprehensively reviewed (Bateman and Miller, 1996). The involvement of pectinases in plant–fungi interactions was proposed by De Bary in 1886 (Lang and Dornenburg, 2000). Breakdown of plant tissue by fungal pectinases and cellulases increases the potential for bacterial penetration into deeper tissues of fruits and vegetables.

Apart from certain fungal enzymes that degrade fresh produce, some bacterial enzymes also initiate damage. Bacterial enzymes (cutinases) that degrade the leaf cuticle have been found in *Pseudomonas aeruginosa*, *Pseudomonas putida*, and *Pseudomonas syringae* (Sebastian et al., 1987; Fett et al., 1992; Liu and Kolattukudy, 1998). Degradation of seemingly intact fruits and vegetables by certain bacteria such as *Erwinia carotovora* causes soft rot. Soft rot lesions in fruits and vegetables can facilitate entry of other bacteria, including human enteric pathogens such as *Salmonella*, which may be unable to degrade plant tissue. Vegetables such as tomatoes, peppers, celery, and potatoes are very susceptible to bacterial soft rot (Lund and Kelmann, 1977; Bartz and Kelman, 1986; Robbs et al., 1996). Interestingly, *Salmonella* spp. were isolated from 18 to 20% of vegetables including lettuce, tomatoes, carrots, sprouts, beans, cantaloupe, and broccoli that had signs of soft rot. This occurrence of *Salmonella* spp. was approximately twice that (9 to 10%) which was found on intact, unblemished samples of those same vegetables (Wells and Butterfield, 1997). The incidence of bacterial soft rot increased following certain packinghouse procedures including fluming and use of dump tanks (Lund and Kelmann, 1977; Segall et al., 1977).

14.4.1.3 Damage by Insects and Birds

Insects and birds are a source of microorganisms (including pathogenic bacteria) in fresh fruits and vegetables. Microorganisms transferred to fresh produce during contact with insects or birds can enter areas of preexisting damage (Beuchat, 1996). In addition, these pests may facilitate entry of microorganisms into fresh produce via damage caused when they feed on leaves or fruits. Insects and birds have been reported to be carriers of *E. coli* O157:H7 (Wallace et al., 1997; Shere et al., 1998; Iwasa et al., 1999; Janisiewicz et al., 1999; Rahn et al., 1997). Vinegar flies and nitidulid beetles contaminated 75 to 100% of damaged peaches and nectarines with plant pathogens (Michailides and Spotts, 1999). Houseflies can carry about 100 different pathogenic microorganisms and transmit about 65% of these pathogens (Kettle, 1982). Human enteric pathogens found in houseflies include *Shigella* spp., *Salmonella* Typhimurium, *Campylobacter jejuni*, *Entamoeba histolytica*, *Vibrio cholerae* O139, and pathogenic *Escherichia coli* (Olsen, 1998). Iwasa et al. (1999) reported that wild house flies on a cattle farm were persistently contaminated with *E. coli* O157:H7. The presence of this pathogen on these flies was most likely linked to the flies' frequent contact with cow manure. A high incidence of *E. coli* O157:H7 in cuts and bruises in whole apples was linked to microbial contamination from fruit flies (Janisiewicz et al., 1999).

14.4.1.4 Physical Damage

14.4.1.4.1 Environmental Factors

Environmental factors such as strong winds, hail, and frost can cause physical damage to fresh produce in the form of bruises, cuts, punctures, and splits. The action of strong winds and hail can result in damage to the protective cuticle of leaves (Van Gardingen et al., 1991; Rogge et al., 1993), including rupture or removal of cuticular waxes by bending, rubbing, particle impaction, or flexing action. Generally, the cuticular layer of plant leaves offers protection from invasion by bacteria and other microorganisms by helping the leaves to repel water. Water repellency is a common property of leaves and is linked to an intact layer of epicuticular waxes (Neinhuis and Barthlott, 1997). Bermandinger-Stabentheiner (1994) suggested that the action of strong winds may even melt waxes via frictional heat. The removal of cuticular waxes certainly decreases the water repellency of leaves and increases their wettability, predisposing them to invasion by microorganisms. Damage to the cuticular layer can permit microbial proliferation in cellular fluids and moisture released from the damaged sites. Sugars in released juices from damaged tissue attract insects, which can further injure fresh produce and facilitate entry of microorganisms (Heard, 2002).

14.4.1.4.2 Maturation of Fruits

Maturation of certain fruits increases their susceptibility to mechanical damage. During maturation, some fruits such as citrus endure folding or creasing of the rind and degradation in the rind at the stem end (Ryall and Pentzer, 1982). Such types of age-induced changes in the rind of citrus fruits can cause damage, including splitting during harvesting, washing, or packaging, and permit the entry of bacteria into these fruits. Almed et al. (1973) reported that damage to citrus fruits during harvesting allowed the entry of spoilage microorganisms.

14.4.1.4.3 Harvesting and Postharvest Operations

Mechanically harvested fruits are relatively more susceptible to splitting, puncturing, and bruising compared to fruits harvested manually. For example, the mechanical harvesting of citrus produces a higher incidence of fruits with attached stems (more than 1 cm long) that can inflict puncture wounds on other fruits during handling. Manual harvesting of citrus fruits results in more fruits that endure "plugging" or removal of part of the peel at the stem (Almed et al., 1973). Carballo et al. (1994) noted that field-packed bell peppers sustained fewer bruises than peppers packaged in a packing house. Forceful contact of fruits with hard surfaces during harvesting, high-pressure spray washing, processing, packaging, and storage may cause bacteria to become embedded in sites just beyond the surface of the fruits. Petracek et al. (1998) suggested that high-pressure spray washing of citrus does not generally result in visible damage to sound fruit; however, this process will rupture fruit in areas that were previously injured.

When fruits endure pressure on their surfaces via rubbing or other types of cleaning actions such as brushing, bacterial cells may be forced into the layer of natural wax platelets and in cracks on the skin surface. Kenney et al. (2001) used CSLM to observe the location of *E. coli* O157:H7 cells on the surface of artificially inoculated, bruised Red Delicious apples. The inoculated apples were washed and/or rubbed with a polyester cloth then checked to ascertain whether the treatments introduced cells into areas beyond the surface of the skin. The researchers found that there was no significant difference in the location of *E. coli* O157:H7 cells on or in undamaged and bruised apples that were not rubbed or washed. Cells adhered preferentially to the edges of cuticular wax platelets and in gaps between the platelets. However, cells on the surface of rubbed apples were sealed in the naturally occurring cracks on the wax platelets. These cells most likely became embedded in the wax platelets when pressure was applied to the surface of apples during rubbing.

14.5 INFILTRATION OF BACTERIA IN FRUITS AND VEGETABLES

Bacteria cells suspended in water may infiltrate the internal tissues of fruits and vegetables during contact with the surfaces of these products. Research conducted as early as the 1960s demonstrated the internalization of bacteria in otherwise healthy tissues of vegetables (Samish, et al., 1961, 1963). In the natural environment bacterial infiltration may occur when dew, rainwater, or irrigation water accumulates on the surface of fruits and vegetables. Also, contact between fallen fruit and ground water could facilitate bacterial infiltration. Several studies have attempted to determine whether bacteria infiltrated fruits and vegetables under natural environmental conditions. Solomon et al. (2002) demonstrated that when lettuce was grown in soil with artificially contaminated manure or irrigation water, *E. coli* O157:H7 can infiltrate the root system and migrate throughout the edible portion of the plant. Seeman et al. (2003) investigated the internalization of *E. coli* ATCC 25922 in whole apples in a controlled outdoor setting. Apples of three varieties (Red Delicious,

Golden Delicious, and Rome Beauty) were placed on soil that was artificially inoculated with *E. coli* to simulate contamination of drop or windfall apples. *E. coli* was detected in samples of the inner core and flesh of all apple varieties. The presence of *Salmonella* in the stem scar area of whole tomatoes and in aseptically cut internal tissues from fruits that were in contact with moist, artificially inoculated soils indicated that bacteria can infiltrate tomatoes through the stem scar and into the pulp (Guo et al., 2002). The results of these studies underscore the potential for bacterial infiltration into fresh fruit outside of laboratory conditions.

Based on the results of dye uptake studies conducted using citrus fruits (Merker et al., 1999) and apples (Buchanan et al., 1999), it is conceivable that bacteria can become internalized in deeper tissues of fruits via infiltration with water. The uptake of dye occurs mainly through natural structures such as the stem scar and blossom end; however, Merker et al. (1999) reported that lesions such as older puncture wounds that seemed to be "healed" provide other routes for ingress of dye into citrus fruits. Buchanan et al. (1999) observed that apples were infiltrated through the blossom end and those apples that took up a substantial amount of dye had clearly recognizable open channels into the core region. Dye uptake was also observed through the skin, especially at sites where the apples sustained bruises or punctures.

In studies involving immersion of apples in bacterial cell suspensions, *E. coli* O157:H7 cells have been shown to infiltrate the inner core of sound Red Delicious apples. The organism spread and attached to the cartilaginous pericarp of the ventral cavity and seed locules of the fruit. Infiltration started at the blossom end of the calyx and moved up the floral tube into the core area of the apple (Burnett et al., 2000). Buchanan et al. (1999) recovered *E. coli* O157:H7 in greater numbers from the outer core region (compared to the skin) of intact apples following immersion in a suspension of the pathogen. These studies support observations that the blossom end of apples allows infiltration of bacteria into these fruits. Bacterial infiltration leads to internalization of bacteria in deeper tissues, which can protect the organisms from contact with sanitizers applied to the surface of the fruit.

14.5.1 FACTORS AFFECTING BACTERIAL INFILTRATION

14.5.1.1 Temperature

During postharvest processing of fruits and vegetables the wash and flume water used for cleaning these products may permit infiltration of bacteria. This problem is further exacerbated when there is a negative temperature differential (warm product, cold wash water). Previous research involving tomatoes (Bartz and Showalter, 1981; Bartz, 1982, 1999), apples (Buchanan et al., 1999; Burnett et al., 2000), and citrus fruit (Merker et al., 1999) have demonstrated that immersion of warm fruit in cold water produces a transient pressure difference that can lead to infiltration of bacterial cells. Washing of warm, fresh produce with cold water has been shown to allow uptake of water (Bartz and Showalter, 1981; Bartz, 1982; Zhuang et al., 1995; Buchanan et al., 1999; Merker et al., 1999). Burnett et al. (2000) reported that infiltration of *E. coli* O157:H7 through the floral tube of apples occurred regardless of the temperature differential during inoculation of the apples. The researchers suggested that infiltration by this pathogen may be partly linked to capillary action, which pulled the inoculum into the apple core over an extended period of time (18 h). Therefore, if the wash water for fresh produce is contaminated with pathogenic bacteria, the internal tissues of these products will most likely become contaminated following ingress of water.

The infiltration of bacteria with water taken up by warm fresh produce immersed in cold wash water can be explained by the general gas law. As the temperature of fruits and vegetables is lowered the gases trapped in their tissues contract and create a reduced internal pressure. This reduction in pressure forces the combined atmospheric and hydrostatic pressure on the produce to equilibrate with the internal pressure and thus permits infiltration of water (Bartz, 1982). Zhuang et al. (1995) demonstrated that significantly higher populations of *Salmonella* Montevideo infiltrated the core tissue of tomatoes (25°C) immersed in a cell suspension of the organism at 10°C compared with tomatoes immersed at 25 or 37°C. Relatively warm (26 to 40°C) tomatoes submerged for 10 min or longer in cool (20 to 22°C) suspensions of *Serratia marcescens*, *Erwinia carotovora*, *Pseudomonas aeruginosa*, or *Pseudomonas marginalis* were infiltrated by these bacteria (Bartz and Showalter, 1981). Infiltration of bacteria was attributed to a negative temperature differential between the water and the tomatoes. The creation of a positive temperature differential (water temperature higher than that of the immersed product) can decrease the extent of bacterial infiltration in fruits (Bartz, 1981, 1991; Zhang et al., 1995; Burnett et al., 2000).

14.5.1.2 Hydrostatic Pressure

Fruits and vegetables deposited in dump tanks can endure various levels of hydrostatic pressure depending how deeply they are submerged in the wash water. Bartz (1982) demonstrated that certain levels of hydrostatic pressure increased infiltration of *E. carotovora* in tomatoes. Tomatoes were immersed in water or in *E. carotovora* cell suspensions in a 19-L pressure cooker at different pressure levels of air entering the sealed vessel to simulate various depths in water. Tomatoes immersed in the aqueous cell suspension with no hydrostatic pressure did not gain weight or develop decay (soft rot). Tomatoes subjected to a hydrostatic pressure of 24 inches for 2 min exhibited no detectable increase in weight but had a 20% incidence in decay, indicating infiltration of *E. carotovora*. Based on this result even a slight infiltration of water into fruits could lead to bacterial internalization. A hydrostatic pressure of 24 inches for 10 min resulted in an average weight gain of 0.1 g/fruit and a 40% incidence of decay. In contrast, an increase in hydrostatic pressure to 48 inches for only 1 sec resulted in a weight gain of less than 0.1 g/fruit but there was a 70% incidence of decay. It is likely that internalization of bacteria is initiated almost instantaneously when hydrostatic pressure is applied to the surfaces of fruit (Bartz, 1999). The impact of hydrostatic pressure on bacterial infiltration in fruits is relevant to situations in which packing houses may overfill dump tanks to maintain efficient operation of the packing line. In overfilling dump tanks some of the fruits will be submerged under many layers of fruit and remain in the water for extended periods of time. Bartz (1999) suggested that such situations are likely to lead to infiltration of wash water.

14.5.1.3 Surfactants

The addition of surfactants to enhance the cleaning efficacy of wash water can affect the extent of bacterial infiltration into fruits and vegetables. The influence of added surfactants in wash water on the infiltration of tomatoes with *E. carotovora* was studied (Bartz, 1982). Surfactants such as Triton X-100 and Tergitol-NPX were added to water or to cell suspensions of *E. carotovora* to give concentrations ranging from 0.001 to 1.0% weight/volume. Percentage weight increases in tomatoes were 0.23, 0.35, 0.38, 1.02, and 0.68 for surfactant concentrations of 0.0 (control), 0.001, 0.01, 0.1, and 1.0%, respectively. Weight increases in tomatoes were highly correlated to surfactant concentration. Correlations were 0.96 and 0.99, respectively, for two separate tests. The incidence of decay in tomatoes increased with an increase in water uptake. Surprisingly, less infiltration resulted from using 1.0% surfactant compared to a concentration of 0.1%. Tomatoes treated with water that contained 1.0% surfactant appeared to exude liquid from the stem scars following removal of the fruits from the wash water (Bartz, 1982). This loss of liquid might have contributed to the relatively lower percentage weight increase in tomatoes treated with the highest concentration of surfactant (1%) used in the study. Apart from infiltration and subsequent decay of tomatoes by *E. carotovora*, tomatoes submerged in tap water with added surfactant developed decay from naturally occurring spoilage organisms on the fruit surface (Bartz, 1982).

The extent of bacterial infiltration into tomatoes is enhanced in the presence of surfactant and hydrostatic pressure. Tomatoes immersed at a depth of 12 inches in cell suspensions of *E. carotovora* without added surfactant were infiltrated by the organism if exposure time was 10 min or more. Weight gain in the fruits submerged for 0, 2, or 5 min was not significantly different. However, when tomatoes were immersed at a depth of 27 inches in cell suspensions with added surfactant, infiltration of water and *E. carotovora* occurred within 2 min and increased dramatically by 5 min (Bartz, 1982).

14.5.1.4 Type of Cooling System

The type of system used to cool fruits and vegetables may affect the extent of bacterial infiltration into these products. Merker et al. (1999) suggested that the potential for infiltration of bacteria (from contaminated water) into intact citrus fruits may be increased when the fruits are placed in a dump tank or hydrocooler. Vigneault et al. (2000) demonstrated that a laboratory-scale shower hydrocooler reduced the temperature of tomatoes from 35 to 15°C within 13.3 min, whereas a flume cooler gave the same temperature reduction in 10.5 min. In both cooling systems the tomatoes gained weight due to infiltration of water. However, tomatoes cooled in the shower hydrocooler exhibited larger increases in weight than those cooled in the flume. Interestingly, significantly larger weight increases occurred in tomatoes with an upward orientation of the stem scars under the shower. The authors suggested that this increase in weight was due water continuously flooding the pores of the stem scar. Increased uptake of water by tomatoes in a shower cooler could lead to increased infiltration of bacteria from contaminated cooling water into the fruit.

14.5.2 CONTROL OF BACTERIAL INFILTRATION

Control of bacterial infiltration in fresh produce is important for ensuring microbial safety and reducing postharvest losses. From a food safety perspective, surface sanitizing treatments aimed at enhancing the microbial safety of fruits and vegetables can be ineffective when human enteric pathogens such as *E. coli* O157:H7 and *Salmonella* infiltrate these products. In fact, bacteria that become internalized in subsurface areas of fresh produce are more difficult to destroy than those located on the surface. This is most likely due to the inaccessibility of the internalized organisms to sanitizing treatments applied to the surface of fresh produce (Adams et al., 1989; Parish, 1997; Beuchat, 1998; Senkel et al., 1999). Based on the findings of previous studies (Bartz et al., 1975; Bartz, 1981; Bartz and Showalter, 1981), inadvertent microbial inoculation of fruit via bacterial infiltration can cause more extensive postharvest losses than mere contamination of the fruit surface. During infiltration bacteria may become internalized over a relatively wide area within certain fruits such as tomatoes, whereas in surface contamination bacteria initiate proliferation in a limited area (Bartz and Showalter, 1981).

14.5.2.1 Barriers to Effective Control of Infiltration

Certain factors related to the natural environment, irrigation of fields, inherent properties of fresh produce, or postharvest processing of these products, make it challenging to control bacterial infiltration. As previously mentioned, environmental factors such as strong winds, hail, and frost can inflict physical damage (e.g., cuts, bruises, splits, and punctures) to fresh produce. Damaged fruits with surface lesions are more prone to bacterial infiltration than are sound fruits (Buchanan et al., 1999; Merker et al., 1999). Also, preharvest contact of fruits and vegetables with contaminated irrigation water or soil makes postharvest control of pathogens more challenging due to the increased populations of microorganisms on fresh produce entering packing houses (Beuchat, 1998). Samish et al. (1963) observed that healthy tomatoes from farms that used overhead irrigation carried more bacteria than tomatoes from farms that employed furrow irrigation. It is likely that overhead irrigation systems create a high-humidity microclimate that favors attachment and growth of certain bacteria on tomatoes.

Another important factor that contributes to bacterial infiltration and further exacerbates this problem is the inherent variation in porosity in fresh produce. Porosity and size of natural openings vary widely among types of fruits and vegetables or within one cultivar of fruit or vegetable. Variations in porosity have been observed among stem scars of tomatoes within a given lot (Bartz, 1982) and also among different cultivars (Bartz, personal communication). Buchanan et al. (1999) observed that among the Golden Delicious apples immersed in a dye solution, those apples that were infiltrated to a greater extent by the dye had relatively larger open channels into the core region.

During postharvest processing of fresh fruits and vegetables, particularly during washing, certain factors have been shown to contribute to bacterial infiltration. Two important factors are negative temperature differential (warm product, cold water) and level of hydrostatic pressure. When submerged in water fresh produce may endure bacterial infiltration mainly because of the interaction of these two factors (Bartz, 1999). The combined effect of temperature differential and hydrostatic pressure on fresh produce cannot be evaluated by simple calculations. This statement is supported by research data indicating that higher temperatures in immersion depth studies resulted in increased infiltration (Bartz, 1982). More importantly, previous research has shown that infiltration of water can occur almost instantaneously under certain conditions of temperature and hydrostatic pressure (Bartz and Showalter, 1981). However, when the temperature of wash water and hydrostatic pressure are controlled, produce have to be immersed in water long enough during cooling to facilitate infiltration (Bartz and Showalter, 1981; Bartz, 1999).

14.5.2.2 Prevention of Infiltration

Based on current knowledge of bacterial infiltration in fruits and vegetables, interventions to prevent this problem are needed from cultivation to processing. Preventive measures should be applied at production steps wherever factors that contribute to bacterial infiltration are identified. Certainly, it might be particularly difficult to control this problem in fields and orchards. However, interventions applied at key points in the production of fresh produce can provide an extra margin of safety against bacterial infiltration. Also, the use of a sanitizer combined with conditions to prevent bacterial infiltration during washing of fresh produce is likely to prove more effective than either intervention used separately.

A multifaceted approach to preventing bacterial infiltration in fresh fruits and vegetables should include good agricultural practices, thorough screening of produce for visible damage, and a combination of bactericidal and infiltration control interventions during washing. Previous studies have demonstrated the capability of human enteric pathogens such as *Salmonella* and *E. coli* O157:H7 to infiltrate fruits and vegetables in contact with contaminated moist soil or manure (Guo et al., 2002; Solomon et al., 2002; Seeman et al., 2003). Therefore, the use of good manufacturing practices, including avoiding the use of animal manure or application of thoroughly composted manure, and irrigation of crops with potable water, will minimize contamination of fruits and vegetables with pathogenic bacteria.

It is well recognized that food crops may sustain physical damage in the natural environment. For example, dropped fruit may develop cuts or bruises upon impact with the ground. Such damage will predispose fruits to bacterial infiltration. Also, the impact with which fruits hit the ground may force bacteria into protected sites on the fruit surface. This problem is further exacerbated if the soil is contaminated with animal feces that carry human pathogens. Dropped apples used to make apple cider were linked to at least two outbreaks of *E. coli* O157:H7 infections (Besser et al., 1993; Health Canada, 1999). Dingman (1999) reported that 8 of 11 orchard owners in Connecticut admitted to using dropped apples to produce cider that was incriminated in another outbreak. It is therefore important to avoid using dropped fruit in the production of juices and apple cider that may not be subjected to bactericidal interventions to eliminate human pathogens. Also, stringent culling of fruits and vegetables with visible damage is important to reduce the incidence of bacterial infiltration during washing. Fruits with cuts, bruises, or punctures are relatively more susceptible to bacterial infiltration (Buchanan et al., 1999; Merker et al., 1999). Flume and wash waters have been reported to be possible contributors to bacterial contamination in fresh produce (Bartz and Showalter, 1981; Bartz, 1999; Buchanan et al., 1999; Merker et al., 1999; Walderhaug et al., 1999; Burnett et al., 2000).

Control of wash water temperature and the depth at which fresh produce is immersed is an additional intervention to prevent bacterial infiltration. Bartz (1982) recommended that the water used in handling or washing fresh produce should be warmer than the incoming fruits and vegetables except for the following conditions: (1) the fruit or vegetable is not susceptible to infiltration (2) the fruit or vegetable and the wash water are not contaminated with undesirable microorganisms or substances; and/or (3) the shelf-life and quality if the fruit or vegetable would not be substantially affected by infiltration. As fruits and vegetables are immersed to greater depths in water the increase in hydrostatic pressure on their surfaces can facilitate bacterial infiltration (Bartz, 1982, 1999). When tomatoes were subjected to pressure to simulate an immersion depth of 24 inches in water, the incidence of microbial spoilage increased from 20 to 40% following exposure times of 2 and 10 min, respectively. When the pressure on tomatoes was increased to simulate a depth of 48 inches, exposure for only 1 sec caused a 70% incidence of spoilage (Bartz, 1982). Based on these findings, overloading of dump tanks and prolonged immersion of produce in water should be avoided to prevent bacterial infiltration.

The fact that infiltration can occur rapidly under relatively high hydrostatic pressure justifies the need for adequate concentrations of sanitizer to inactivate bacteria immediately when produce is immersed in the dump tank. Also, adequate amounts of sanitizer are necessary to prevent cross-contamination. Research findings indicate that if chlorination of water used for washing apples is inadequate to prevent cross-contamination, then uncontaminated apples are likely to be infiltrated by *E. coli* O157:H7 (Buchanan et al., 1999). Dipping surface-inoculated tomatoes in water with a free chlorine concentration of 60 to 110 ppm significantly ($P < 0.05$) reduced populations of *Salmonella* Montevideo; however, increased concentrations of up to 320 ppm did not produce a significant reduction compared to dipping in water with 110 ppm free chlorine (Zhuang et al., 1995) The incidence of postharvest disease linked to infiltration of tomatoes with contaminated water was decreased but not eliminated by 50 to 100 ppm free chlorine in wash water (Bartz, 1988). These results indicate that treatment of fresh produce with chlorine may not be fully effective in destroying various types of microorganisms that could be present on tomatoes in the packing house (Senter et al., 1985). Zhuang et al. (1995) recommended that tomato packing houses maintain their dip tanks at a higher temperature than that of the tomatoes and at a free chlorine concentration of 200 ppm. As with any other sanitizer, careful monitoring of the concentration of chlorine, in the form of hypochlorous acid, is necessary to maintain adequate levels of the sanitizer for destroying microorganisms in produce wash water. Maintaining adequate levels of hypochlorous acid to effect microbial destruction in wash water could be difficult since chlorine readily reacts with organic matter. In this regard, debris of harvested produce and exudate from damaged tissue of fresh produce may neutralize some of the chlorine. However, the use of chlorination in combination with conditions to prevent infiltration into fresh produce can offer a margin of safety against bacterial infiltration that either intervention is unable to provide separately.

14.6 SURVIVAL AND GROWTH OF INTERNALIZED BACTERIA

Results of previous research indicated the potential for survival and growth of bacteria internalized in fresh produce. Hill and Faville (1951) inoculated oranges, still attached to the tree, with *Aerobacter*, *Achromobacter*, and *Xanthomonas*. The researchers reported a 3-log increase in bacterial numbers during 5 weeks following inoculation. In a market survey of fresh produce bacteria were frequently isolated from inside whole tomatoes, cucumbers, beans, and peas and less frequently in bananas and melons. Bacteria were rarely isolated from inside grapes, olives, peaches, and citrus fruit (Samish et al., 1963). From a food safety perspective, human pathogenic bacteria, internalized within fruits and vegetables, have to survive in these products to pose a health hazard to consumers. For certain pathogens even low numbers of internalized cells surviving in fresh produce might be important. For example, under natural conditions, *E. coli* O157:H7 could pose a significant health risk because the infective dose is less than 1,000 cells (Ackers et al., 1998). *E. coli* O157:H7 can grow within damaged or decayed apples (Riordan et al., 2000) and in areas of the fruit that are not accessible to washing (Sapers et al., 2000; Annous et al., 2001).

The survival and growth of bacteria internalized in fruits and vegetables may be influenced by several factors, including type of bacteria, physical and chemical characteristics of the fruit or vegetable, and postharvest processing and storage conditions. Walderhaug et al. (1999) studied the potential for survival and growth of *Salmonella* Hartford and *E. coli* O157:H7 in California Valencia oranges artificially inoculated with these pathogens. The organisms were introduced in whole oranges via injection into the core region, the albedo (white part of the orange) and the orange section portion. The oranges were also inoculated through simulated wounds produced by the tips of 1-mL plastic pipettes. The pathogens were deposited in the oranges at two depths: shallow (4 to 5 mm) into the albedo and deep (10 to 11 mm) into the orange section portion. The oranges were held at 4°C or at 21°C for 5 d.

Results of this study indicated some differences in survival of internalized pathogens based on type of bacteria tested and in storage temperature for the oranges. Survival of *E. coli* O157:H7 decreased after 5 d at 4°C irrespective of the internal location of the organism. At 21°C an increase (2-log) in growth of *E. coli* O157:H7 was observed in the section portions that were inoculated via simulated wounds. No growth of this pathogen occurred when it was injected directly into the alberdo; however, almost a 1-log increase in numbers occurred when the organism was injected into the core or introduced to the alberdo through a simulated wound.

Generally, *Salmonella* survived better than *E. coli* O157:H7 in oranges at 4°C; populations of the pathogen remained relatively constant in all injected oranges. At 21°C *Salmonella* increased in all experimental treatment samples; increases ranged from 0.2 to 3.9 $\log_{10}$ CFU/mL. Other researchers have demonstrated that populations of *Salmonella* Montevideo in the core tissue of tomatoes remained constant during storage of the tomatoes at 10°C for 8 d. However, populations of the pathogen in tomatoes increased significantly during storage at 20°C (Zhuang et al., 1995). As observed with *E. coli* O157:H7, the extent of growth was greater in the section portions of the oranges. Growth of both pathogens in the section portions of oranges at 21°C may be attributed to the location of the organisms at internal sites, which are separated from the acid juice. The juice of oranges is retained within segregated vesicles. Therefore, internalized bacteria may be able to grow in areas outside of intact vesicles.

The results of these studies involving oranges underscored the importance of low temperatures for preventing growth of certain pathogens in fresh produce. Additionally, pH microenvironments that are conducive to bacterial survival and growth may exist in intact fruit. The results also confirm the findings of previous research that demonstrate the growth of human enteric pathogens in tomatoes (Asplund and Nurmi, 1991; Zhuang et al., 1995), melons (Golden et al., 1993), and apples (Janisiewicz et al., 1999).

14.7 CONCLUSION

The internalization of bacteria in fresh fruits and vegetables may compromise the shelf life and microbial safety of these products. Bacteria adhere to fresh fruit and vegetables or infiltrate them with the ingress of water through natural openings or at damaged sites on the produce surface. Bacterial infiltration is affected by the temperature of water in contact with fresh produce, hydrostatic pressure, surfactants, and type of cooling system used in postharvest processing. Infiltration of bacteria into intact fruit occurs most often when warm fruit is washed in cold water. This negative temperature differential makes the tissues of the fruit contract and drives the uptake of water by the fruit. Bacterial infiltration can occur relatively rapidly when intact fruits endure hydrostatic pressure depending on the depth of immersion in water. Intact fruits may have pH microenvironments that are conducive to survival of internalized bacteria. A multifaceted approach to prevent bacterial infiltration in fresh produce should include good agricultural practices, thorough screening of produce for visible damage, and a combination of interventions to destroy bacteria and control temperature and hydrostatic pressure during postharvest processing.

14.8 FUTURE RESEARCH

Published studies on internalization of bacteria in fruits and vegetables have provided some evidence on how bacteria infiltrate these products as well as survive and grow within them. However, gaps in current knowledge on bacterial infiltration and internalization in fresh produce warrant further research in the following areas: (1) natural occurrence of human enteric pathogens internalized in intact fruit and conditions that facilitate internalization; (2) assessment of risk of foodborne illness from fresh produce that are infiltrated by human pathogens; (3) optimum conditions of temperature and hydrostatic pressure for preventing infiltration in various types of fruits and vegetables; and (4) extent of survival of human enteric pathogens in various types of fresh produce.

REFERENCES

Abdul Raouf, U.M., Beuchat, L.R., and Ammar, M.S., Survival and growth of *Escherichia coli* O157:H7 on salad vegetables, *Appl. Environ. Microbiol.*, 59, 1999–2006, 1993.

Ackers, M.L., Mahon, B.E., Leahy, E., Goode, B., Damrow, T., Hayes, P.S., Bibb, W.F., Rice, D.H., Barrett, T.J., Hutwagner, L., Griffin, P.M., and Slutsker, L., An outbreak of *Escherichia coli* O157:H7 infections associated with leaf lettuce consumption, *J. Infect. Dis.*, 177, 1588–1593, 1998.

Adams, M.R., Hartley, A.D., and Cox, L.J., Factors affecting the efficacy of washing procedures used in the production of prepared salads, *Food Microbiol.*, 6, 69–77, 1989.

Afolabi, P., Davies, K.G., and O'Shea, P.S., The electrostatic nature of the spore of *Pasteuria penetrans*, the bacterial parasite of root-knot nematodes, *J. Appl. Bacteriol.*, 79, 244–249, 1995.

Almed, E.M., Martin, F.G., and Fluck, R.C., Damaging stresses to fresh and irradiated citrus fruit, *J. Food Sci.*, 38, 230–233, 1973.

Andrews, J.H. and Hirano, S.S., *Microbial Ecology of Leaves*, Springer-Verlag, New York, 1991.

Annous, B.A., Sapers, G.M., Mattrazzo, A.M., and Riordan, D.C.R., Efficacy of washing with a commercial flatbed brush washer, using conventional and experimental washing agents, in reducing populations of *Escherichia coli* on artificially inoculated apples, *J. Food Prot.*, 64, 159–163, 2001.

Asplund, K. and Nurmi, E., The growth of salmonellae in tomatoes, *Int. J. Food Microbiol.*, 13, 177–182, 1991.

Bartz, J.A., Potential for post harvest disease in tomato fruit infiltrated with chlorinated water, *Plant Dis.*, 72, 9–13, 1988.

Bartz, J.A., Washing fresh fruits and vegetables: lessons from treatment of tomatoes and potatoes with water, *Dairy Food Environ. Sanit.*, 19, 853–864, 1999.

Bartz, J.A., Infiltration of tomatoes immersed at different temperatures to different depths in suspensions of *Erwinia carotovora* subsp *carotovora*, *Plant Dis.*, 66, 302–305, 1982.

Bartz, J.A. and Kelman, A., Reducing the potential for bacterial soft rot in potato tubers by chemical treatments and drying, *Am. Potato J.*, 63, 481–493, 1986.

Bartz, J.A. and Showalter, R.K., Infiltration of tomatoes by aqueous bacterial suspensions, *Phytopathology*, 71, 515–518, 1981.

Bateman, D.F. and Miller, R.L., Pectic enzymes in tissue degradation, *Annu. Rev. Phytopathol.*, 49, 119–146, 1996.

Beattie, G.A., Leaf surface waxes and the process of leaf colonization by microorganisms, in *Phyllosphere Microbiology*, Lindlow, S.E., Ed., APS Press, St. Paul, MN, 2002, pp. 3–23.

Beattie, G.A. and Lindlow S.E., Bacterial colonization of leaves: a spectrum of strategies, *Phytopathology*, 89, 353–359, 1999.

Beattie, G.A. and Lindlow S.E., Comparison of the behavior of epiphytic fitness mutants of *Pseudomonas syringae* under controlled and field conditions, *Appl. Environ. Microbiol.*, 60, 3799–3808, 1994.

Bermadinger-Stabentheiner, E., Problems in interpreting effects of air pollutants on spruce epicuticular waxes, in *Air Pollutants and the Leaf Cuticle*, Percy, K.E., Cape, J.N., and Jagels, R., Eds., Springer-Verlag, Berlin, 1994.

Besser, R.E., Lett, S.M., Weber, J.T., Doyle, M.P., Barrett, T.J., Wells. J.G., and Griffith, P.M., An outbreak of diarrhea and hemolytic uremic syndrome from *Escherichia coli* O157:H7 in fresh pressed apple cider, *J. Am. Med. Assoc.*, 269, 2217–2220, 1993.

Beuchat, L.R., Pathogenic microorganisms associated with fresh produce, *J. Food Prot.*, 59, 204–216, 1996.

Beuchat, L.R., Surface decontamination of fruits and vegetables eaten raw: a review, *Food Safety Issues,* World Health Organization, 1998.

Blanco, O., Sieiro, C. and Villa, G., Production of pectic enzymes in yeasts, *FEMS Microbiol. Lett.,* 175, 1–9, 1999.

Buchanan, R.L., Edelson, S.G., Miller, R.L., and Sapers, G.M., Contamination of intact apples after immersion in an aqueous environment containing *Escherichia coli* O157:H7, *J. Food Prot.,* 62, 444–450, 1999.

Burnett, S.L., Chen, J., and Beuchat, L.R., Attachment of *Escherichia coli* O157:H7 to the surface and internal structures of apples as demonstrated by confocal scanning laser microscopy, *Appl. Environ. Microbiol.,* 66, 4679–4687, 2000.

Carballo, S.J., Blankenship, S.M., Sanders, D.C., Ritchie, D.F., and Boyette, M.D., Comparison of packing systems for injury and bacterial soft rot on bell pepper fruit, *Hortic. Tech.,* 4, 269–272, 1994.

Centers for Disease Control and Prevention, The control of non-point source microbial indicators and pathogens using best management practices: a literature review and evaluation, Centers for Disease Control and Prevention, National Center for Infectious Diseases, Division of Parasitic Disease, Atlanta, GA, 1997.

Clements, H.F., Morphology and physiology of the pome lenticels of *Pyrus malus, Bot. Gaz.,* 97, 101–117, 1935.

Davies, D.G., Afloabi, P., and O'Shea, P., Adhesion of *Pasteuria penetrans* to the cuticle of root-knot nematodes *(Meloidogyne* spp.) inhibited by fibronectin: a study of electrostatic and hydrophobic interactions, *Parasitology,* 112, 553–559, 1996.

Davies, D.G., Chakrabarty, A.M., and Geesey, G.G., Exopolysaccharide production in biofilms: substratum activation of alginate gene expression by *Pseudomonas aeruginosa, Appl. Environ. Microbiol.,* 59, 1181–1186, 1993.

Diaz, C. and Hotchkiss, J.H., Comparative growth of *Escherichia coli* O157:H7, spoilage organisms and shelf-life of shredded iceberg lettuce stored under modified atmospheres, *J. Sci. Food Agric.,* 70, 433–438, 1996.

De Roever, C., Microbiological safety evaluations and recommendations on fresh produce, *Food Control,* 10, 117–143, 1999.

Dingman, D.W., Prevalence of *Escherichia coli* O157:H7 in apple cider manufactured in Connecticut, *J. Food Prot.,* 62, 567–573, 1999.

Fett, W.F., Gerard, H.C., Moreau, R.A., Osman, S.F., and Jones, L.E., Cutinase production by *Streptomyses* spp., *Curr. Microbiol.,* 25, 165–171, 1992.

Fletcher, M. and Floodgate, G.D., An electron microscopic demonstration of an acidic polysaccharide involved in the adhesion of a marine bacterium to solid surfaces, *J. Gen. Microbiol.,* 74, 325–334, 1973.

Fletcher, M. and Loeb, G.I., Influence of substratum characteristics on attachment of marine pseudomonad to solid surfaces, *Appl. Environ. Microbiol.,* 37, 67–72, 1979.

Frank, J.F., Microbial attachment to food and food contact surfaces, *Adv. Food Nutr. Res.,* 43, 320–370, 2000.

Golden, D.A., Rhodehamel, E.J., and Kautter, D.A., Growth of *Salmonella* spp. in cantaloupe, watermelon, and honeydew melons, *J. Food Prot.,* 56, 194–196, 1993.

Guo, X., Chen, J., Brackett, R.E., and Beuchat, L.R., Survival of *Salmonella* on tomatoes stored at high relative humidity, in soil, and on tomatoes in contact with soil, *J. Food Prot.,* 65, 274–279, 2002.

Haas, J.H. and Rotem, J., *Pseudomonas lachrymans* adsorption, survival, and infectivity following precision inoculation of leaves (cucumbers), *Phytopathology,* 13, 992–997, 1976.

Health Canada, An outbreak of *Escherichia coli* O157:H7 infection associated with unpasteurized, non-commercial, custom-pressed apple cider: Ontario 1998, *Can. Communicable Dis. Rep.*, 25, 13–20, 1999.

Heard, G.M., Microbiology of fresh-cut produce, in *Fresh-Cut Fruits and Vegetables*, Lamikanra, O., Ed., CRC Press, Boca Raton, FL., 2002, pp. 187–248.

Hill, E.C. and Faville, L.W., Studies on the artificial infection of oranges with acid tolerant bacteria, *Proc. Florida State Hortic. Soc.*, 64, 174–177, 1951.

Himathongkham, S., Baharw, S., Rieman, H., and Cliver, D. Survival of *Escherichia coli* O157:H7 and *Salmonella typhimurium* in cow manure slurry, *FEMS Microbiol. Lett.*, 178, 251–257, 1999.

Itoh, Y., Sugita-Konishi, Y., Kasuga, F., Iwaki, M., Hara-Kudo, Y., Saito, N., Nogouchi, Y., Konuma, H., and Kumagai, S., Enterohemorrhagic *Escherichia coli* O157:H7 present in radish sprouts, *Appl. Environ. Microbiol.*, 64, 1532–1535, 1998.

Iwasa, M., Sou-Ichi, M., Asakura, H., Hideaski, K., and Morimoto, Y., Detection of *Escherichia coli* O157:H7 from *Musca domestica (Diptera: Muscidae)* at a cattle farm in Japan, *J. Med. Entomol.*, 36, 108–112, 1999.

Janisiewicz, W.J., Conway, W.S., Brown, N.W., Sapers, G.M., Fratamico, P., and Buchanan, R.L., Fate of *Escherichia coli* O157:H7 on fresh-cut apple tissue and its potential for transmission by fruit flies, *Appl. Environ. Microbiol.*, 65, 1–5, 1999.

Kakiomenou, K., Tassou, C., and Nychas, G.J., Survival of *Salmonella enteritidis* and *Listeria monocytogenes* on salad vegetables, *World J. Microbiol. Biotechnol.*, 3, 383–387, 1998.

Kenney, S.J., Burnett, S.L., and Beuchat, L.R., Location of *Escherichia coli* O157:H7 on and in apples as affected by bruising, washing and rubbing, *J. Food Prot.*, 64, 1328-1333, 2001.

Kettle, D.S., *Muscidae* (houseflies, stableflies), in *Medical and Veterinary Entomology*, John Wiley & Sons, New York, 1982.

Kolattukudy, P.E., Li, D., Hwang, C.S., and Flaishman, M.A., Host signals in fungal gene expression involved in penetration into the host, *Can. J. Bot.*, 73, S1160–S1168, 1995.

Lang, C. and Dornenburg, H., Perspectives in biological function and the technological application of polygalacturonases, *Appl. Microbiol. Biotechnol.*, 53, 366–375 2000.

Leben, C., Epiphytic microorganisms in relation to plant disease, *Annu. Rev. Phytopathol.*, 3, 209–230, 1965.

Leben, C., Microorganisms associated with plant buds, *J. Gen. Microbiol.*, 71, 327–331, 1972.

Liu, Z.M. and Kolattukudy, P.E., Identification of a gene product induced by hard-surface contact of *Colletotrichum gloeosporoides conidia* as a ubiquitin-conjugating enzyme by yeast complementation, *J. Bacteriol.*, 180, 3592–3597, 1998.

Lund, B.M. and Kelman, A., Determination of the potential for development of bacterial soft rot of potatoes, *Am. Potato J.*, 54, 211–255, 1977.

Mendoza de Gives, P., Davies, K.G., Morgan, M., and Behnke, J.M., Attachment tests of *Pasteuria penetrans* to the cuticle of plant and animal parasitic nematodes, free living nematodes and srf mutants of *Caenorhabditis elegans*, *J. Helminthol.*, 73, 67–71, 1999.

Merker, R., Edelson-Mammel, S.G., Davis, V., and Buchanan, R.L., Preliminary experiments on the effect of temperature differences on dye uptake by oranges and grapefruit, U.S. Food and Drug Administration, Center for Food Safety and Applied Nutrition, Washington, DC, 1999.

Michailides, T.J. and Spotts, R.A., Transmission of *Mucor pyriformis* to fruit of *Prunus persica* by *Carpophilus* spp. and *Drosophilia melanogaster*, *Plant Dis.*, 74, 287–291, 1990.

Morris, C.E., Monier, J.M., and Jacques, M.A., Methods for observing microbial biofilms directly on leaf surfaces and recovering them for isolation of culturable microorganisms, *Appl. Environ. Microbiol.*, 63, 1570–1576, 1997.

Neinhuis, C. and Barthlott, W., Characterization and distribution of water-repellent, self-cleaning plant surfaces, *Ann. Bot.*, 79, 667–677, 1997.

Nguyen-The, C. and Carlin, F., Fresh and processed vegetables, in *The Microbiological Safety and Quality of Food*, Vol. 1, Lund, B.M., Baird-Parker, T.C., and Gould, G.W., Eds., Aspen Publishing, Gaithersburg, MD, 2000, pp. 620–684.

Olsen, A.R., Regulatory action criteria for filth and other extraneous materials III. Review of flies and foodborne enteric disease, *Regulat. Toxicol. Pharmacol.*, 28, 199–211, 1998.

Parish, M.E., Public health and unpasteurized fruit juices, *Crit. Rev. Microbiol.*, 23, 109–119, 1997.

Petracek, P.D., Kelsey, D.F., and Davis, C., Response of citrus fruit to high pressure washing, *J. Am. Soc. Hortic. Sci.*, 123, 661–667, 1998.

Porto, E. and Eiora, M., Occurrence of *Listeria monocytogenes* in vegetables, *Dairy Food Environ. Sanit.*, 21, 282–286, 2001.

Rahn, K., Renwick, S.A., Johnson, R.P., Wilson, J.B., Clarke, R.C., Alves, D., McEwen, S., Lior, H., and Spika, J., Persistence of *Escherichia coli* O157:H7 in dairy cattle and the dairy farm environment, *Epidemiol. Infect.*, 119, 251–259, 1997.

Rentschler, I., The wettability of leaf surfaces and the submicroscopic structure, *Planta*, 96, 119–135, 1971.

Riordan, D.C.R., Sapers, G.M., and Annous, B.A., The survival of *Escherichia coli* O157:H7 in the presence of *Penicillium expansum* and *Glomerella cingulata* in wounds on apple surfaces, *J. Food Prot.*, 63, 1637–1642, 2000.

Robbs, P.G., Bartz, J.A., McFie, G., and Hodge, M.C., Causes of decay of fresh-cut celery, *J. Food Sci.*, 61, 444–448, 1996.

Rogge, W.F., Hildermann, L.M., Mazurek, M.A., Cass, G.R., and Simoneit, B.R.T., Sources of fine organic aerosol. 4. particulate abrasion products from leaf surfaces of urban plants, *Environ. Sci. Technol.*, 27, 2700–2711, 1993.

Romantschuk, M., Nurmiaho-Lassila, E.L., Roine, E., and Suoniem, A., Pilus-mediated adsorption of *Pseudomonas syringae* to the surface of host and non-host plant leaves, *J. Gen. Microbiol.*, 139, 2251–2260, 1993.

Romantschuk, M., Roine, E., Bjorklof, K., Ojanen, T., Nurmiaho-Lassila, E.-L. and Haahtela, K., Microbial attachment to plant aerial surfaces, in *Aerial Plant Surface Microbiology*, Morris, C.E., Nicot, P.C., and Nguyen-The, C., Eds., Plenum Press, New York, 1996.

Romberger, J.A., Hejnowicz, Z., and Hill, J.F., *Plant Structure: Function And Development,* Springer-Verlag, Berlin, 1993.

Ryall, A.L. and Pentzer, W.T., *Handling, Transportation and Storage of Fruits and Vegetables*, Vol. 2, *Fruits and Tree Nuts*, 2nd ed., AVI Publishing Company, Westport, CT, 1982.

Samish, Z. and Etinger-Tulczynska, R., Distribution of bacteria within the tissue of healthy tomatoes, *Appl. Microbiol.*, 11, 7–10, 1963.

Samish, Z., Etinger-Tulczynska, R., and M. Bick., Microflora within healthy tomatoes, *Appl. Environ. Microbiol.*, 9, 20–25, 1961.

Samish, Z., Etinger-Tulczynska, R., and Bick, M., The microflora within the tissue of fruits and vegetables, *J. Food Sci.*, 28, 259–266, 1963.

Sapers, G.M., Miller, R.L., Jantschke, M. and Mattrazzo, A.M., Factors limiting the efficacy of hydrogen peroxide washes for decomposition of apples containing *Escherichia coli. J. Food Sci.,* 65, 529–532, 2000.

Sebastian, J., Chandra, A.K., and Kolattukudy, P.E., Discovery of a cutinase-producing *Pseudomonas* sp. cohabiting with an apparently nitrogen-fixing *Corynebacterium* sp. in the phyllosphere, *J. Bacteriol.*, 169, 131–136, 1987.

Seeman, B.K., Sumner, S.S., Marini, R., and Kniel, K.E., Internalization of *Escherichia coli* in apples under natural conditions, *Dairy Food Environ. Sanit.*, 22, 667–673, 2003.

Segall, R.H., Henry, F.E., and Dow, A.T., Effect of dump tank water temperature on the incidence of bacterial soft rot of tomatoes, *Proc. Florida State Hortic. Soc.*, 90, 204–205, 1997.

Senkel, I.A., Henderson, R.A., Jolbitato, B., and Meng, J., Use of hazard analysis critical control point and alternative treatments in the production of apple cider, *J. Food Prot.*, 62, 778–785, 1999.

Senter, S.D., Cox, N.A., Bailey, J.S., and Forbes, W.R., Microbiological changes in fresh market tomatoes during packing operations, *J. Food Sci.*, 50, 254–255, 1985.

Seo, K.H. and Frank, J.F., Attachment of *Escherichia coli* O157:H7 to lettuce leaf surface and bacterial viability in response to chlorine treatment as demonstrated confocal scanning laser microscopy, *J. Food Prot.*, 62, 3–9, 1999.

Shere, J.A., Bartlett, K.J. and Kaspar, C.W. Longitudinal study of *Escherichia coli* 0157:H7 dissemination on four dairy farms in Wisconsin. *Appl. Environ. Microbiol.*, 64, 1390–1399, 1998.

Solomon, E.B., Yaron, S., and Matthews, K.R., Transmission and internalization of *Escherichia coli* O157:H7 from contaminated manure and irrigation water into lettuce plant tissue, *Appl. Environ. Microbiol.*, 68, 397–400, 2002.

Suoniemi, A., Bjorklof, K., Haahtela, K., and Romantschuk, M., Pili of *Pseudomonas syringae* pathovar *syringae* enhance initiation of bacterial epiphytic colonization of bean, *Microbiology*, 141, 497–503, 1995.

Takahashi, T. and Doke, N., A role of extracellular polysaccharides of *Xanthomonas campestris* pv. *Citri* in bacterial adhesion to citrus leaf tissues in preinfectious stage, *Ann. Phytopathol. Soc. Jap.*, 50, 565–573, 1984.

Takeuchi, K., Matute, C.M., Hasan, A.N., and Frank, J.F., Comparison of the attachment of *Escherichia coli* O157:H7, *Listeria monocytogenes, Salmonella Typhimurium,* and *Pseudomonas fluorescens* to lettuce leaves, *J. Food Prot.*, 63, 1433–1437, 2000.

Ukuku, D.O. and Fett, W.F., Relationship of cell surface charge and hydrophobicity to strength of attachment of bacteria to cantaloupe rind, *J. Food Prot.*, 65, 1093–1099, 2002.

U.S. FDA, CFSAN, *Guide to Minimize Microbial Food Safety Hazards for Fresh Fruits and Vegetables*, U.S. Food and Drug Administration, October 1998, pp. 1–40.

Van der Mei, H.C., Rosenburg, M., and Brusscher, H.J., Assessment of microbial cell surface hydrophobicity, in *Microbial Cell Surface Analysis*, Mozes, H.J., Handley, P.S., and Busscher, H.J., VCH, New York, 1991.

Van Gardingen, P.R., Grace, J., and Jeffree, C.E., Abrasive damage by wind to the needle surfaces of *Picea sitchensis* (Bong.) Carr. and *Pinus sylvestris* L., *Plant Cell Environ.*, 14, 185–193, 1991.

Van Loosdrrecht, M.C.M., Lyklema, J., Norde, W., Schara, G., and Zehnder, A.J.B., The role of bacterial cell wall hydrophobicity in adhesion, *Appl. Environ. Microbiol.*, 53, 1893–1897, 1987.

Vigneault, C., Bartz, J.A., and Sargent, S.A., Postharvest decay risk associated with hydro-cooling tomatoes, *Plant Dis.*, 84, 1314–1318, 2000.

Walderhaug, M.O., Edelson-Mammel, S.G., DeJesus, A.J., Eblen, B.S., Miller, A.J., and Buchanan, R.L., Preliminary studies on the potential for infiltration, growth and survival of *Salmonella enterica* serovar *Hartford* and *Escherichia coli* O157:H7 within oranges, U.S. Food and Drug Administration, Center for Food Safety and Applied Nutrition, Washington, DC, 1999.

Wallace, J.S., Cheasty, T., and Jones, K., Isolation of vero-cytotoxin producing *Escherichia coli* O157 from wild birds, *J. Appl. Microbiol.*, 82, 399–404, 1997.

Wells, J.M. and Butterfield, J.E., *Salmonella* contamination associated with bacterial soft rot of fresh fruits and vegetables in the marketplace., *Plant Dis.*, 81, 867-872, 1997.

Wilson, M., Hirano, S.S., and Lindlow, S.E., Location and survival of leaf associated bacteria in relation to pathogenicity and potential for growth within the leaf, *Appl. Environ. Microbiol.*, 65, 1435–1443, 1999.

Zhuang, R.Y., Beuchat, L.R., and Angulo, F.J., Fate of *Salmonella* Montevideo on and in raw tomatoes as affected by temperature and treatment with chlorine, *Appl. Environ. Microbiol.*, 61, 2127–2131, 1995.

15 Mechanisms of Microbial Spoilage of Fruits and Vegetables

Brendan A. Niemira, Christopher H. Sommers, and Dike O. Ukuku
United States Department of Agriculture,[1] Agricultural Research Service, Eastern Regional Research Center, Wyndmoor, PA

CONTENTS

---

[1] Mention of trade names or commercial products in this publication is solely for the purpose of providing specific information and does not imply recommendation or endorsement by the U.S. Department of Agriculture.

15.1   INTRODUCTION

In the various stages of shipment, transshipment, and storage that separate the producers of fruits and vegetables from wholesalers, distributors, and retailers, accumulated losses due to spoilage can, depending on the commodity in question, destroy 25 to 80% of fresh produce before it reaches the consumer (Baldwin, 2001). In recent decades, the technologies available to the fresh produce industries to reduce spoilage have become increasingly sophisticated. Improved understanding of produce storage physiology and response to modified atmospheres, microprocessor controlled atmospheric and temperature sensors and control systems, and novel approaches to controlling spoilage organisms have improved the overall efficiency of the fresh produce handling infrastructure, but new challenges are emerging. These advances are being applied to a globalized produce distribution network that is offering an increasingly diverse selection of fruits and vegetables, in addition to an expanding range of complex products such as ready-to-eat salads and mixed vegetables (Garrett, 2002). The diversity of the fresh fruits and vegetables available, coupled with the logistical complexity of a globalized network of produce growers, distributors and retailers, make the issue of spoilage a significant economic factor. This chapter will provide an overview of the key issues surrounding microbial spoilage of fresh produce, including the mechanisms by which produce may become infected, the types of microorganisms that cause spoilage of produce, and a presentation of case studies of microbial spoilage of archetypal fresh fruits and vegetables.

15.2   DISEASE VS. SPOILAGE

15.2.1   ABIOTIC SPOILAGE AND ITS RELATIONSHIP TO BIOTIC SPOILAGE

Spoilage is a general term that describes a loss of marketable quality of fresh produce (Brackett, 1997). This spoilage may be the loss of appealing qualities, such as aroma (intensity, complexity, etc.), texture (firmness, crunch, mouth feel, etc.), taste (the balance of sweetness and acidity), or appearance (color, evenness, etc.). Spoilage of this type can generally be ascribed to abiotic (or, more precisely, apathogenic) factors, such as the physiological age of the produce, the temperature at which the produce is stored, or the atmosphere mix used in storage. These circumstances alter the physiology of the produce such that marketable quality is lost through chill injury, water loss or some other mechanism. Fruits and vegetables that continue to ripen after having been picked (climacteric fruits) are especially susceptible to this kind of spoilage and have a relatively short shelf life before they become overripe and lose marketable value (e.g., banana, avocado, tomato). Control of atmosphere or use of special edible coatings may help to extend the shelf life of these types of fruits and vegetables (Baldwin, 2001).

It should be noted that fresh produce that is shipped across great distances is vulnerable to spoilage during transit, and greater distances and longer transit times become increasingly more problematic. Produce that is shipped internationally must also withstand required disinfestation procedures between ports of export and import. To accommodate these regulatory and economic factors, loss of a certain amount of marketable quality may be unavoidable. The loss of positive attributes may result in a lower price premium for the produce but does not necessarily result in removal of the produce from the marketplace.

Spoilage may also take the form of the development of undesirable qualities, including off-aromas, off-flavors, or textural and appearance changes such as sliminess, moldiness, blemishes, etc. This type of spoilage typically results from the action of microorganisms such as bacteria and fungi. While the abiotic or physiological spoilage can, in the eyes of the consumer, make "appealing" produce merely "acceptable," microbial spoilage renders produce "unacceptable," with concomitantly greater economic loss to the producer, wholesaler and retailer of fresh produce. In some cases, microbial spoilage can result in vegetable food products that are not merely unacceptable, but harmful, as in the case of mycotoxins such as aflatoxin on grains and patulin on apples.

15.2.2 PREHARVEST VS. POSTHARVEST

Phytopathogens are organisms that cause injury or disease to plants. In a discussion of spoilage, a distinction should be drawn between preharvest diseases of living plants and postharvest diseases of the plant organs that are of economic importance as fresh produce, even though, in many cases, the same phytopathogens cause both classes of disease. A classic example of this is potato late blight disease (*Phytophthora infestans*), the cause of the Irish potato famine of the 1840s. *Ph. infestans* attacks potato leaves and shoots, degrading and decaying the tissues, reducing or eliminating the photosynthetic capacity of the crop. It also causes the harvested potatoes to rot in storage and during shipment and is therefore considered to be a spoilage concern (Niemira et al., 1999). In presenting the distinction between *field* pathogens and *storage* or *spoilage* pathogens (the provinces of *plant pathology* and *food microbiology*, respectively), it should be acknowledged that this is a somewhat arbitrary separation based on the technical factors of produce production and distribution, rather than a distinction that arises from the biology or ecology of plants and their pathogens. Indeed, in many cases spoilage organisms are introduced to the produce during its time in the field but do not cause damage until the conditions of storage and shipment allow them to proliferate. In some cases, field infections with pathogens or symbionts stimulate an induced resistance response that renders the resulting produce less susceptible to spoilage (Tuzun and Kloepper, 1995; Niemira et al., 1996).

Field diseases of crop plants are frequently treated using cultural or agrochemical interventions to exclude, contain, or eradicate the phytopathogens that cause them (Agrios, 1997). Living plants interact with their environment; disease-resistant varieties are able to mount a complex range of native defenses against phytopathogens

(Niemira et al., 1999). In contrast, diseases that occur on the harvested produce, are, by definition, attacking an isolated plant organ. Fresh produce is physiologically alive, in that it is able to exchange water and gas with its environment, mount physiological defenses to attack, and undergo changes in metabolism and physiology. However, the specialized nature of the tissue in question means that the harvested produce can only draw on a more limited range of physiological options, and a more limited metabolic reserve, in responding to attack by phytopathogens (Brackett, 1997). As with field disease resistance, resistance of the produce to spoilage organisms is incorporated in the varieties during the breeding process as much as possible (Niemira et al., 1999). However, unlike field diseases, which act on plants growing outdoors, storage diseases occur under controlled conditions (e.g., storage facility, the packing house, the shipping container, and the wholesale redistribution center). In this context, although varietal- and chemical-based interventions are used, it is the manipulation and control of the environment that becomes the predominant means of preserving produce quality.

15.2.3 SPOILAGE ORGANISMS

Postharvest spoilage can be caused by a wide variety of different organisms. The proverbial "worm in the apple" is an obvious example of the kind of damage caused by invertebrates such as arthropods and nematodes. Although not the subject of the present discussion, invertebrates such as the carrot root knot nematode (*Meloidogyne hapla*), the green peach aphid (*Myzus persicae*) and the Mediterranean fruit fly (*Ceratitis capitata*) are only a few of the thousands of pests that can lead to spoilage diseases (van Emden et al., 1969; White and Elson-Harris, 1994; Agrios, 1997).

Phytopathogenic fungi and bacteria are varied, physiologically and taxonomically, and often closely related to species that are nonpathogenic, or are pathogenic to organisms other than plants. Of the microorganisms that cause plant diseases, a smaller subset are responsible for microbial spoilage of fruits and vegetables. An abbreviated fungal taxonomy (adapted from Agrios, 1997) is presented below:

1. Kingdom: Protozoa (phagotrophic psuedofungi)
   a. Phylum: Plasmodiophoromycota
      i. Class: Plasmodiophoromycetes (endoparasitic slime molds).
         Key spoilage genera: *Plasmodiophora, Spongospora*
2. Kingdom: Chromista (phototrophic psuedofungi)
   a. Phylum: Oomycota
      i. Class: Oomycetes (water molds, white rusts, downy mildews).
         Key spoilage genera: *Pythium, Phytophthora, Plasmopara, Bremia, Psuedoperonospora*
3. Kingdom: Fungi (true fungi)
   a. Phylum: Chitridiomycota
      i. Class: Chitridiomycetes. Key spoilage genus: *Synchytrium*
   b. Phylum: Zygomycota
      i. Class: Zygomycetes (bread molds). Key spoilage genera: *Rhizopus, Mucor*

c. Phylum: Ascomycota
   i. Class: Pyrenomycetes. Key spoilage genus: *Glomerella*
   ii. Class: Loculoascomycetes. Key spoilage genera: *Elsinoe, Venturia*
   iii. Class: Discomycetes. Key spoilage genera: *Monilinia, Sclerotinia, Sclerotium*
   iv. Class: Dueteromycetes: Key spoilage genera: *Penicillium, Aspergillus, Fusarium, Alternaria, Botrytis, Rhizoctonia*

Fungi have been broadly grouped into four classes, according to their phyto-pathogenicity (Prell and Day, 2000): pure saprophytes (nonpathogenic, lives on dead tissue), opportunistically phytopathogenic saprophytes (attack weakened, stressed, or senescent plants, live on freshly digested tissue), primary pathogens/necrotrophs (attack healthy plants, live on freshly digested tissue), and biotrophs (feed parasitically or symbiotically on living tissue). Of the more than 100,000 known species of fungi, roughly 10,000 species can cause disease in plants; nearly the same proportion of the approximately 1,600 species of bacteria, about 100 species, are phytopathogenic (Agrios, 1997). A relative handful of genera are responsible for the most serious bacterial plant diseases, and a subset of these are most significant from the standpoint of postharvest spoilage of produce: *Clavibacter, Erwinia, Pseudomonas, Xanthomonas,* and *Streptomyces.* Among the species within each of these genera, there is great variation of biochemistry, ecology, and pathogenic significance. For example, some of these, such as the *E. carotovora* group, are capable of producing the pectolytic enzymes that allow them to penetrate plant epidermal tissue, while others, such as the *E. amylovora* group, do not; this is the type of difference that separates primary from secondary or opportunistic pathogens.

In order to improve the clarity and utility of the nomenclature of the phytopathogenic bacteria, the taxonomy has recently been revised to regroup the pectolytic members of *Erwinia* under a new genus, *Pectobacterium*, and also to elevate certain subspecies to species status (Hauben et al., 1998; Gardan et al., 2003). Thus, *E. carotovora* is now more properly referred to as *Pe. carotovorum*; for the purposes of clarity and ease of reference to the literature, the older nomenclature will be used in the remainder of this chapter when referring to the pectolytic members of *Erwinia*.

15.3 EARLY SPOILAGE PATHOGENESIS EVENTS: OVERVIEW

15.3.1 PLANT DEFENSES

Spoilage disease occurs at the intersection of the factors that make up the disease triangle (Figure 15.1). The first element of the triangle is the host plant, or, more specifically, the plant part that is of value as a produce commodity. Plants use a variety of mechanisms to defend against pathogen attack. These complex and highly evolved defenses may be a preexisting part of the plant's normal state, such as the anatomical leaf structure that isolates the cells from the environment (Figure 15.2). The outer, waxy layers of the epidermis form the bulk of the cuticle and are known as cutin (Figure 15.3). The complex waxes of the cutin and the additional pectin, cellulose, and hemicellulose structures are resistant to physical or chemical degradation, although

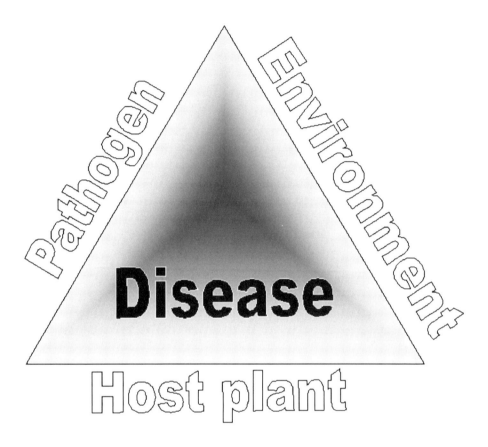

FIGURE 15.1 The plant pathology "disease triangle." Disease occurs when a compatible pathogen and host are together under the proper environmental conditions.

enzymatic degradation by capable phytopathogens is a weakness. Other forms of defense may consist of constitutively produced biochemicals, or responses that are activated upon pathogen attack, such as hypersensitivity responses and induction of phytoalexins and/or pathogenesis-related proteins (PRP). Induced resistance responses can, in some cases, persist and act as a form of "immunization" of the plant against subsequent pathogen attack in the field and/or in storage (Tuzun and Kloepper, 1995; Niemira et al., 1996). Disease-resistant varieties have been bred to possess systemic and/or inducible defenses against a particular pathogen. This varietal resistance may be of the broad/horizontal type (somewhat resistant against all races/genotypes of a particular pathogen), or it may be narrow/vertical (completely resistant to certain races/genotypes of the pathogen, completely susceptible to others). Unfortunately, the biochemical details of disease resistance are often only poorly understood, and the ability of one plant organ (e.g., leaves) to resist pathogen attack does not necessarily correlate with the disease resistance of another (e.g., tubers) (Niemira et al., 1999).

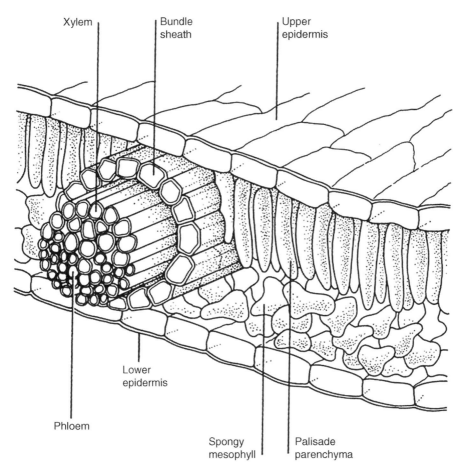

FIGURE 15.2 Cross-sectional anatomy of leaf tissue. (From Mauseth, 1988. With permission.)

15.3.2 PATHOGEN ATTACK

The second element in the progression of microbial spoilage is the introduction of a compatible pathogen to the produce. Bacterial and fungal phytopathogens can be introduced to fruits and vegetables at almost any point during the growth, maturation, harvest, storage, or shipment of the produce by exposure to contaminated water, dust, mechanical equipment, etc. (Beuchat, 1995). As previously indicated, the surfaces of fruits and vegetables have evolved to present as impervious a barrier as feasible, with multiple layers of chemical and mechanical defenses. Thus, while the first step in spoilage is getting the pathogen *onto* the produce, the second, more important step in spoilage is getting the pathogen *into* the produce. Obligate phyto-pathogens have evolved a suite of tools that allows them to penetrate the plants' defenses. These may take the form of specialized physical structures, such as fungal appressoria (Figure 15.4), in addition to specialized enzymes that can degrade the

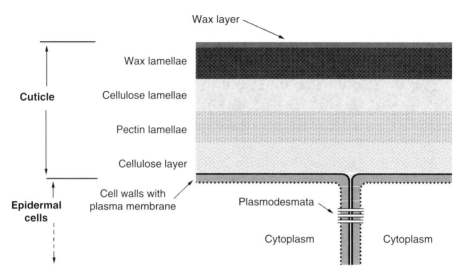

FIGURE 15.3 Diagrammatic representation of foliar epidermal cells and associated cuticle layer.

cellulose, hemicellulose, pectin, and cutin of the epidermal and endodermal tissues (Figure 15.5). While cellulases are of notable importance with regard to field diseases, they are considered to be relatively less significant than pectinases with regard to storage diseases (Brackett, 1997).

Many damaging forms of spoilage are caused by organisms that, by themselves, are unable to penetrate the epidermis of the produce. These organisms, often referred to as secondary or opportunistic pathogens, rely on a breach of the surface integrity to gain entry to the inner tissues of the leaf, fruit, tuber, etc. This opening may be a naturally occurring anatomical structure opening such as a stomate, hydathode, stem end scar, etc., or it may be a breach caused by an obligate pathogen, a pest (nematode, insect, etc.), or some other biotic agent (Figure 15.4).

Frequently, the breach is of abiotic origin, such as a puncture, fracture, abrasion, or some other wound resulting from mishandling at some stage of the production cycle. Poorly designed or maintained equipment can cause these wounds at a variety of stages during the growth, harvest, washing, packing, or shipping of the produce. Produce that has been harvested by cutting or has been prepared by cutting, sectioning, peeling, chopping, or otherwise treated in some way that breaches the epidermis is therefore especially vulnerable to spoilage. The complexity of the physiology and microbiology of processed fruit and vegetable products, including those with more than one vegetable component, warrants a fuller discussion than can be presented herein. Many of the food quality and food safety issues related to minimally processed and/or fresh-cut fruits and vegetables have been recently reviewed (Garrett, 2002; Zhuang et al., 2003).

Two of the most commonly destructive bacteria of stored produce are *E. carotovora* pv. *carotovora* and *Ps. fluorescens* (Agrios, 1997). These are responsible for the soft rot of a wide variety of fruits and vegetables in storage; these pathogens

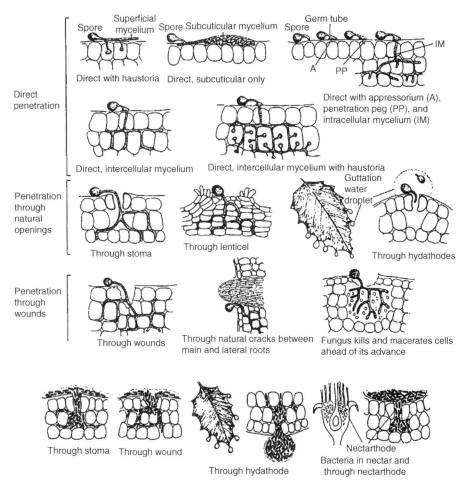

FIGURE 15.4 Various methods of penetration of plant tissue by phytopathogenic fungi (top three rows) and bacteria (bottom row). (From Agrios, 1997. With permission.)

will be cited repeatedly later in this chapter, as part of the case studies of key fruits and vegetables. Stored produce that is clean from the field may still become infected by contact with contaminated surfaces (equipment, hands, detritus, etc.), and via infiltration or absorption of contaminated wash water. Wounds provide common entry points, and the bacteria multiply in intercellular spaces. Produce that is under attack by soft-rotting bacteria can quickly degrade, such that by the time a soft-rot infection is apparent by appearance or by a characteristic smell the shelf life of the produce can be counted in days, if not hours. In the initial phase of infection by *Erwinia*, the bacteria reproduces on the surface and internally without digesting the polysaccharide matrix of the cell wall; when the population density reaches a critical level, a quorum-sensing pathway is initiated. Acyl-homoserine lactone (AHL) is a signal molecule that initiates the production of pectolytic enzymes that release oligo- and monosaccharides on which the bacteria feed, degrading the plant tissues in the

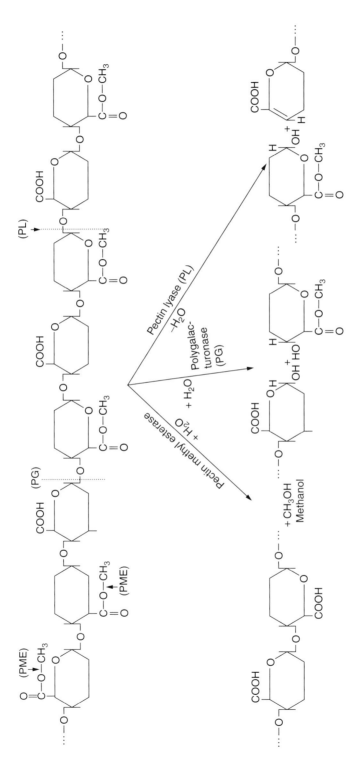

FIGURE 15.5 Pectin degradation by pectin methyl esterase, polygalacturonase, and pectin lyase. (From Agrios, 1997. With permission.)

process (Dong et al., 2001; Leadbetter, 2001). The bacteria multiply rapidly in the sugar- and nutrient-rich medium of the degraded tissues, and, because the bacteria are already at elevated population levels before the tissue degradation triggers plant defense responses, these responses can be quickly overwhelmed. Given the speed and destructive power of bacterial soft rots, it is generally accepted that the control of bacterial soft rot is best accomplished by exclusion and sanitation. The ultimate goals are to prevent the initial infection of the produce, and, through closely controlled conditions in storage, to prevent the growth of the bacteria.

15.3.3 DISEASE-CONDUCIVE CONDITIONS

The final element of the disease triangle is the environment in which the host and pathogen interact. When speaking of phytopathogenesis in general, it is difficult, if not impossible, to make sweeping statements regarding the conditions under which phytopathogens can potentially cause disease, given the diverse nature of the pathogens, their plant hosts, and environmental variations they experience. However, in a discussion concerned more narrowly with postharvest spoilage, the only relevant microclimate conditions are predetermined by (1) the produce in question and (2) market forces of time, geography, and economics. According to the (arbitrary) distinction previously described, a pathogen that is introduced to the produce is relevant to this discussion only if it begins to cause, or continues to cause, disease in the postharvest environment. Therefore, the effects of environment are more easily addressed, because they are artificially created and maintained. Modern storage and shipment/transshipment facilities maintain produce in conditions that are designed to preserve freshness and marketability. Typically, this means circulating air maintained at 5 to 7°C, with relatively high humidity (90%+). Air circulation helps to prevent standing water droplets or films on the surface of the produce and also assists in removing the heat generated by the respiring produce; this is especially important for produce that will be stored for extended periods of time.

15.4 CASE STUDIES OF KEY PATHOGENS

15.4.1 CATEGORIES OF PRODUCE

With regard to potential for spoilage, the thousands of commercially available varieties of fresh fruits and vegetables may be arbitrarily grouped in a number of ways. A strict botanical grouping would separate fruits (e.g., melon, tomato) from leaves (e.g., lettuce, cabbage) from roots (e.g., carrot, sweet potato) from tubers (e.g., potato), etc. A physiological grouping might draw a distinction between climacteric (e.g., banana, tomato) and nonclimacteric produce (e.g., strawberry, orange). A focus on the potential for field exposure to soil-borne phytopathogens might distinguish arboreal (e.g., apple, tomato) from terrestrial (e.g., squash, melon) from subterranean (e.g., carrot, potato). The potential economic impact of spoilage may prompt a grouping based on market value, separating relatively low value (e.g., apple, carrot, potato) from relatively high value (e.g., blueberry, tomato, asparagus). Given the potential for transmission of unwanted phytopathogens presented by workers and equipment, a distinction can be drawn based on hand-harvested (e.g., melon, tomato) and semimechanized (e.g., cherry, orange) and mechanized (e.g., carrot, potato). Historically, seasonal variation in storage condition influenced the spoilage of spring (e.g., peas, rhubarb) vs. summer (e.g., blueberry, cucumber) vs. fall (e.g., apple, potato) crops. A distinction could be drawn between mechanized, high-input agriculture, and low-input or organic practices. The geographical source of the produce has a significant impact on the potential for spoilage; locally grown produce is handled less and spends less time in storage than produce grown regionally, shipped from across a continent or imported from distant locales. Of the various factors that could be considered, the issue most significant in a globalized market, with produce shipped long distances, the relative amount of time spent in storage, and therefore the length of the window of opportunity for a spoilage microorganism to act, will be the initial grouping factor used herein. Relatively perishable commodities, those that spend relatively short periods of time in storage, will be grouped separately from relatively storable commodities. Within these groups, arbitrary subgroups based on gross structure will be established.

Each of the fruits and vegetables that will be considered can fall victim to a number of phytopathogens, and, conversely, a given pathogen may have a wide host range. Certain common themes will become apparent in the ways in which spoilage develops among the various commodity/pathogen combinations, yet each disease progression has unique aspects that are instructive to consider. Note that with regard to nomenclature many diseases are referred to by the genus of the responsible pathogen (e.g., "Fusarium rot" is caused by *Fusarium* spp., while in other cases the genus may be a synonym for another common name of the disease, e.g., "*Botrytis*" is equivalent to gray mold, caused by *Botrytis cinerea)*.

15.4.2 PERISHABLE PRODUCE

15.4.2.1 Leafy Produce

Botrytis cinerea is a fungus that attacks a wide range of fruits and vegetables and is the cause of gray mold of leafy vegetables such as lettuce, endive, and cabbage (Anon., 2000). *Botrytis* establishes itself in the inner leaves of the head, and the mycelium can spread to encompass the entire leaf. Under conducive conditions, the inner core of the head may be completely engulfed in mycelium with little external sign of disease. Initial infection typically occurs when the ambient temperature in the leaf canopy is less than 25°C; thereafter, fungal growth can continue at temperatures ranging from 0 to 35°C. The fungus can grow under refrigeration temperatures (0 to 10°C), and therefore presents a particular problem in storage for produce that was infected relatively close to harvest. When infecting leaves, *Botrytis cinerea* can express cutinases, pectinases, and cellulases to digest the leaf tissue to component oligo- and monosaccharides (Carlile et al., 2001). The degraded leaf tissue resulting from the growth of *Botrytis* can lead to secondary infection by other fungi or by bacterial pathogens, such as *E. carotovora*, which accelerates the extent of spoilage.

Downy mildew of lettuce, caused by the oomycte *Bremia lactucae*, is similar to *Botrytis* in that, following inoculation in the field, subsequent spoilage during storage or shipment can approach 100% (Raid and Datnoff, 1992). This pathogen is typically most active under conditions of high relative humidity, and when a thin film of water is persistent on the leaf surface. Sporangia are deposited on the leaf surface, and a germ tube grows across the leaf, penetrating stomata. In susceptible varieties, initial infections are visible as small spots, typically on the lower leaf surface. While these lesions can expand into the inner leaves, the outer leaves are most susceptible to expanding chlorosis, browning and necrotic streaking, with the necrotic tissue serving as a gateway for other pathogens such as *Botrytis* or bacterial pathogens.

15.4.2.2 Cruciferous Produce

The fungal pathogen *Alternaria brassicae* attacks cabbage, broccoli, and cauliflower. Infection of mature plants typically occurs in the field, where conidia are dispersed by wind and water, although postharvest infection can also be a problem. High levels of humidity or water films on produce are required for infection (Humperson-Jones and Phelps, 1989). Older, senescing tissue is typically the part of mature plants most susceptible to *A. brassicae*. After germination, fungal hyphae can penetrate through stomata or wounds and pervade the vascular system, causing streaking, spotting and wilting throughout the plant. Early in the season, this can result in loss of the entire plant as the stem rots; in terms of postharvest spoilage, the damaged outer leaves of cabbage are no longer marketable and must be removed, while the florets of broccoli and cauliflower turn brown, reducing market value. While *A. brassicae* infections are readily managed by fungicide applications and sanitation controls, this pathogen is ubiquitous and a perennial problem for growers and shippers. The diseased tissue can be a serious problem, in that it can provide a breeding ground for other, more aggressive and less readily controlled fungal or bacterial pathogens.

15.4.2.3 Fibrous Produce

Asparagus spears are harvested by cutting in the field; additional trimming is performed in packing sheds and storage facilities. Trimmed spear ends and other plant detritus can accumulate on the packing line, providing a breeding ground for bacteria; contaminated knives, wash water and cooling water can be vectors of soft-rot bacteria such as *E. carotovora*. Asparagus is therefore at a greater risk from cross-contamination than is produce that is harvested whole and intact. Asparagus is also a relatively high-value commodity, which increases the economic impact of superficial or cosmetic loss of quality. Soft-rotting bacteria such as *E. carotovora* and *Pseudomonas* spp. can cause darkened, slimy, sunken spots to appear at the cut base end or at the tips by enzymatic degradation of the tissues. Modified atmosphere packaging that uses an elevated CO_2 level can suppress bacterial soft rots in refrigerated storage (Anderson and Tong, 1993).

Spoilage of celery is frequently caused by grey mold (*Botrytis*), watery soft rot (*Sclerotinia* spp.) and/or bacterial soft rot (*Erwinia* and *Pseudomonas*) (Gross et al., 2002). *Sc. sclerotiorum* and *Sc. minor* can survive in field soils in infected plant detritus for multiple seasons, leading to soil-, water- and wind-borne contamination of stems (Laemmlen, 2001). *Sclerotinia* establishes itself on senescent tissue and spreads mycelium into succulent tissues, using a complex suite of enzymes, including several different polygalacturonases, to break down pectins as it grows (Martel et al., 1996). As the celery stalk tissue is consumed, cottony mycelium may appear on the surface, and ascospore containing sclerotia form. Initial infection is favored by relatively warm conditions (~25°C), but once underway, growth of the fungus can continue at refrigeration temperatures (Agrios, 1997). As with many spoilage conditions previously discussed, the digested and partially digested plant tissue can encourage the growth of other phytopathogenic fungi or bacteria.

15.4.2.4   Soft-Skinned Produce

The growth of *Botrytis* on grape clusters leads to gray mold, also know as bunch rot. Infection frequently occurs in the field, leading to the establishment of a fungal culture within the bunch, in the relatively high-humidity center. When the grape bunches are harvested and stored, the fungal mycelium penetrates the mature grapes and spreads throughout the bunch, leading to a cottony mass that pervades all or part of the bunch (Cappellini et al., 1986). Following the enzymatic degradation of the bunched grapes, secondary fungal or bacterial pathogens may also grow on the stressed and damaged fruit, leading to further discoloration and loss of turgor. The high humidity within a closed plastic bag provides a good environment for Botrytis to grow rapidly; current practice is to use open or perforated bags for table grapes.

As the grape skins lose integrity, water loss may cause them to shrivel. For this reason, *Botrytis* may, in some cases, be intentionally allowed to persist in the field to infect grapes that are intended for processing into some sweet wines. The water loss resulting from the *Botrytis* infection (in this context sometimes referred to as "the noble rot") concentrates the sugars and acids of the grape, resulting in a complex and valuable wine. However, the infection can grow out of control under the wrong environmental conditions, and *Botrytis*-infested detritus can be a source of inoculum for subsequent or adjacent crops (Bachman, 2001). This risk, and the increased problems associated with spoilage of the *Botrytis*-infested bunches before processing, or of cross-contamination of other grapes, particularly table grapes, via contaminated workers and equipment make this a somewhat inconsistently effecacious management strategy.

The tomato is susceptible to a wide variety of spoilage pathogens, including (but not limited to) fungal rots caused by *Alternaria alternata, Colletotrichum phomoides, Fusarium* spp., *Phytophthora* spp., and *Sclerotinia* spp., as well as bacterial soft rots caused by *Erwinia, Pseudomonas* and *Xanthomonas campestris* (Gross et al., 2002). Infection of the tomato fruit with spoilage organisms can occur in the field but also results from improper handling during harvest and shipment. Vine-ripened tomatoes are easily damaged and must be handled with care to avoid inoculation with phytopathogens. Although the flavor and desirability (and therefore the market value) of vine-ripened tomatoes is typically higher than that of tomatoes that are picked unripe, the latter are firmer and more resistant to bruising and are better able to withstand handling and can therefore be shipped and packaged more easily. However, the relative durability of the unripe fruit means that it is more likely to receive rougher treatment, which can crack, puncture, crush, or otherwise damage the skin.

Accumulated detritus from damaged fruit on packing and washing equipment, shipping containers, and line workers can introduce spoilage organisms to the produce. Tomato anthracnose, also called ripe rot, is caused by *Colletotrichum phomoides*. The growth of this fungus is favored under relatively warm conditions with high humidity, such as those found late in the season. It can lead to severe postharvest storage losses in most tomato varieties, but particularly in canning varieties (Agrios, 1997). As the mycelium penetrates the tomato tissue, small, sunken, water-soaked spots appear on the surface. These blemishes enlarge to cover an appreciable part of the tomato surface, merge, and ultimately lead to a watery softening and collapse of the fruit as the mycelium grows throughout the tomato fruit. Secondary infection by bacterial pathogens is common once the blemishes begin to merge, rapidly accelerating the spoilage of the produce. Control of this disease centers around exclusion and sanitation.

15.4.2.5 Firm-Skinned Produce

Citrus fruits such as oranges, lemons, and limes are susceptible to blue molds and green molds caused by *Penicillium* spp. Fractured, punctured, or otherwise wounded fruit is most vulnerable to infection, but otherwise healthy fruit may become infected via direct hyphal growth if left in contact with diseased fruit (Agrios, 1997). *Penicillium* rots on citrus fruits first appear as small, water-soaked discolorations. The fungus produces pectinases and cellulases to digest and invade the plant tissue (Ribon et al., 2002). A visible, cottony white mold spreads across the surface of the fruit, and sporulation takes place at the center of the infection, giving the characteristic blue or green color.

Injury, such as that resulting from mishandling, can lead to a variety of infections of firm-skinned melons such as cantaloupe, watermelon, and honeydew. Even in the absence of any injury, however, healthy melons are susceptible to fungal rots caused by *Fusarium* spp. and *Phytophthora* spp. and associated bacterial infections by *Erwinia* and *Pseudomonas* (Roberts et al., 2001). Ripe watermelons can become infected with *Phytophthora capsici* from excessively moist soils in the field or by contact with contaminated equipment or personnel. The fungus can survive in the field in the absence of a host for several seasons. Infectious soils can be carried into previously clean fields on agricultural equipment. *Ph. capsici* fruit rot begins as a small brownish discoloration, circular or oval in shape. These lesions expand rapidly, leading to a darkened, water-soaked appearance of the fruit. As the mycelium invades and digests the rind and interior of the melon, the center of the lesion can become visibly moldy as the fungus sporulates. Once the rot is established, the entire fruit may be consumed within a matter of a few days by the *Phytophthora* alone or in conjunction with other fungal or bacterial pathogens.

15.4.2.6 Berries

Strawberries are picked nearly fully ripe and are easily damaged. The warm, moist conditions of the harvest period, coupled with the growth habit of the strawberry (low to the ground), make soil and water contamination key risk factors. The most common spoilage pathogens on strawberries are grey mold (*Botrytis*) and Rhizopus rot (*Rhizopus stolonifer*) (Gross et al., 2002). The growth and development of these molds on strawberries are typically exacerbated by the easily damaged nature of the product. Strawberries are packed in small containers to avoid crushing and bruising to the greatest extent possible, and although these containers are typically perforated or vented, even slight damage to the berry surface can lead to excessive moisture and/or free fluid accumulation in the container. Once established at warmer temperatures (~25°C), *Botrytis* can continue to grow at refrigeration temperatures, penetrating the fruit and digesting the sugar-rich tissues (Sommer et al., 1973). The close contact of one fruit with another makes possible direct hyphal growth, ultimately leading to a cottony mycelial mass that encompasses the available fruit. As with the development of this pathogen on other commodities, the partially digested fruit tissue supports the growth of secondary spoilage pathogens.

Raspberries are one of the most fragile and easily damaged fruits on the market. They are handled and packaged much the way strawberries are (although with an even greater degree of care) and are similarly susceptible to *Botrytis* and *Rhizopus stolonifer* (Gross et al., 2002). As with *Botrytis*, *Rhizopus* is frequently inoculated onto the fruit while in the field and can lead to many of the same symptoms: extensive hyphal growth and penetration of the fruit, enzymatic digestion of the tissues, envelopment of the available fruit in a cottony mycelial mass, and opportunistic infection by other fungi and bacteria. However, while *Botrytis* can continue to grow (albeit slowly) at temperatures as low as 0°C, *Rhizopus* ceases development below 5°C. The primary means of control of this fungus is rapid chilling of the fruit to < 5°C and consistent temperature control throughout the handling chain.

15.4.3 STORABLE PRODUCE AND KEY SPOILAGE FUNGI AND BACTERIA

15.4.3.1 Roots and Tubers

The major storage pathogens of carrots are, by now, familiar: *Botrytis, Erwinia, Pseudomonas,* and *Sc. sclerotiorum* (Gross et al., 2002). Unlike the commodities considered to this point, carrots are a subterranean crop and are therefore presented with a combination of factors that serve to increase the risks of inoculation and subsequent spoilage. Crops that suffer from a high degree of field disease are unlikely to be harvested or brought into storage. Therefore, from the standpoint of postharvest spoilage, the greatest spoilage risk factor is the commingling of slightly diseased produce with healthy produce in the storage facility. Carrots and other subterranean crops are more exposed to soil-borne pathogens than are surface-growing or aerial crops. Harvesting of subterranean crops necessarily requires digging, pulling, lifting, or otherwise extracting the produce from the soil, with all of the potential for equipment-related injury. In preparation for market, carrots are topped, that is, the greens are removed; in general, procedures that involve cutting the produce provide an additional avenue for spoilage. Also, the need to wash soil from the carrot requires a more physically aggressive series of handling steps than is required for other commodities. Thus, while the potential for cross contamination via wash water is a common theme for many fruits and vegetables, the intensive scrub process that carrots undergo can lead to greater damage to the surface, providing wounds as entry points for the spoilage organisms. Careful handling during harvest, and exclusion of the pathogens from the growing and packing environment, are the only effective means of control for these pathogens.

Major storage diseases of potatoes include Fusarium rot (*Fusarium* spp.), pink rot (*P. erythroseptica*) and, most notably, late blight (*P. infestans*). *Phytophthora* spp. are field pathogens of most major crop plants. *P. infestans* is the most famous and widely studied oomycete in the history of plant pathology (Kamoun, 2003). This aggressive fungus, responsible for the Irish potato famine of the 1840s, has reemerged in recent years to become the single most devastating pathogen of potatoes in the field and in storage, causing losses as high as US$5 billion annually (Kamoun, 2003). Spores are typically soil- or water-borne but may also be spread via air or by contact with diseased detritus in the field, on equipment, or in storage. Upon germination, fungal hyphae invade and digest tuber tissue, entering via wounds or stomata on the tuber surface. A significant alternate route of infection is via the stolon by which the tuber is connected to the parent plant in the field. Lesions appear as dark brown or purplish blotches, which soon turn sunken and water-soaked (Niemira et al., 1999). The fungus pervades the outer vascular ring of the tuber first, and the entire surface of the tuber may be mottled and diseased before the fungus extends into the interior. Infection with *P. infestans* is typically soon accompanied by growth of secondary soft-rotting pathogens; the disease complex of fungus and bacteria quickly destroys the original diseased tuber and any previously healthy tubers that it contacts, producing an unforgettably putrid characteristic odor. During the harvest process, the need to wash soil from the potato tubers can lead to wounding of the surface and cross contamination among the tubers as they are placed in storage via wash water or contaminated equipment. Intensive chemical applications are required to control the development of *P. infestans* on foliage in the field; sanitation and exclusion are the only means to control this pathogen in storage.

15.4.3.2 Bulbs

Bulb crops such as onion and garlic suffer from spoilage by blue molds (*Penicillium* spp.) and black molds (*Aspergillus niger*), and, more so in the case of onion, concomitant or subsequent growth of *Botrytis* and bacterial soft rots. These pathogens colonize and grow best under conditions of high moisture and/or free water on the bulbs; the growth of these pathogens can be effectively suppressed by reduction of the relative humidity to 60 to 70% (Gross et al., 2002). For many of the commodities considered to this point, low humidity causes an unacceptable degree of wilting, cracking, shrinkage, etc. and loss of marketable quality. However, the bulbous nature of onion and garlic allow them to tolerate these levels with little economic impact, providing a relatively straight-forward means of controlling the spoilage (i.e., harvest procedures that keep the bulbs as dry as possible, and low humidity in storage).

15.4.3.3 Soft-Skinned Produce

Apples are most commonly spoiled by *Penicillium* molds (chiefly *Penicillium expansum*), with significant spoilage also caused by gray mold (*B. cinerea*) and bitter rot

(*Colleototrichum* spp.) (Pierson et al., 1971; Sanderson and Spotts, 1995). *Colleototrichum* is an imperfect fungus that causes anthracnose diseases of numerous crops (Agrios, 1997). *Colleototrichum* typically infects apples late in the season, causing loss of fruit in the field and spoilage in storage. Fungal conidia germinate on the apple surface and the hyphae invade the fruit tissue. The circular areas of infected tissue appear sunken and brownish, and within these areas form cushion-like masses of developing fungal conidia called acervuli. The rot spreads toward the center of the fruit, making the apple bitter. The diseased spots grow until they fuse, and, unlike the watery rot that this fungus causes in tomatoes, apples become mummified and infectious with fresh conidia (Agrios, 1997). As with tomatoes, the only practical control for *Colleototrichum* is exclusion of infected fruit from the storage facility.

15.4.3.4 Firm-Skinned Produce

Despite the physically tough, thick, and waxy rind that characterizes pumpkins and other squash such as the acorn, butternut, crookneck, Hubbard, and winter squash, these cucurbits are vulnerable to a number of spoilage fungi, with the risk of damage proportional to the time in storage. Fungi such as *Aspergillus, Colleototrichum orbiculare, Rhizopus, Sclerotinia sclerotiorum,* and *Fusarium* are able to infect the squash via natural openings, such as the stem scar and blossom end, but wounds are the most significant means of entry (Zitter, 1992). Infection of the squash typically occurs in the field, prior to harvest, or during the harvest and storing process. Mechanical damage to the rind can arise from rough handling, but also from improperly stacking squash too high, resulting in bruising and cracking. Following germination of the spores of *Fusarium* on the surface or within a wound, the hyphae invade the plant tissues in a familiar sequence of events; the resulting rot tends to be drier and more corky in appearance than many other fruit rots. Brown, sunken spots on the surface of the fruit spread more slowly than the internal growth of the fungus suggests. Extensive Fusarium rot manifests as a whitish pink or yellow mold on the surface of the fruit, and colonization of the rotted fruit by secondary bacterial pathogens is likely (Agrios, 1997).

15.5 CONCLUSIONS AND SUMMARY

The spoilage organisms examined in this chapter form a complex, overlapping ecology. For each of the commodities discussed, multiple pathogens operate in concert to attack, feed, and grow, helping each other to accelerate the spoilage of the stored produce. For many pathogen complexes, a reasonably accurate, though simplistic, summary is that a fungus opens the door and *Erwinia* walks in. Exclusion of the pathogens from the field, harvest, and storage is the most straightforward means of controlling these pathogens. For most crops, this means adherence to crop rotations, fungicide application programs during the growing season, and use of resistant varieties. Specific agricultural practices can further reduce the risk of establishment and development of spoilage pathogens. Examples of these would be minimizing the use of wash water as produce is going into storage, to reduce the presence of free water and water films; proper care and handling to avoid mechanical damage to the produce; and close control of temperature and humidity to suppress pathogen development while preserving product quality. The single most important factor is sanitation of the harvest and packing equipment and environment. The removal of pathogen refugia such as plant detritus, and avoiding cross-contamination vectors such as recycling wash waster, poorly designed equipment or poorly trained personnel are the first lines of defense against spoilage microorganisms.

ACKNOWLEDGMENTS

The authors would like to thank R. Flores and Y. Karaibrahimoglu for their thoughtful reviews of this manuscript.

REFERENCES

Agrios, G.A., *Plant Pathology*, 4th ed., Academic Press, San Diego, CA, 1997.

Anderson, L. and Tong, C., Commercial postharvest handling of fresh market asparagus (*Asparagus officinalis*), University of Minnesota Extension Service Bulletin FS-06236, Minneapolis, MN, 1993.

Anon., Gray-mold rot or Botrytis blight of vegetables, Extension RPD 942, University of Illinois, Urbana-Champaign, IL, 2000.

Bachman, H., France's sweet winemakers vow to stick together, *Wine Spectator*, Oct. 15, 2001, available at http://www.winespectator.com/Wine/Archives/Show_Article/0,1275,3398,00.html.

Baldwin, E., New coating formulations for the conservation of tropical fruits, 2001, available at http://technofruits2001.cirad.fr/en/baldwin_en.htm.

Beuchat, L.R., Pathogenic microorganisms associated with fresh produce. *J. Food Prot.*, 59: 204–216, 1996.

Brackett, R.E., Fruits, vegetables and grains, in *Food Microbiology: Fundamentals and Frontiers*, Doyle, M.P., L.R. Beuchat and T.J. Montville, Eds., American Society of Microbiology, Washington, DC, 1997, pp. 117–128.

Cappellini, R.A., M.J. Ceponis, and G.W. Lightener, Disorders in table grape shipments to the New York market, 1972–1984, *Plant Dis.*, 70, 1075–1079, 1986.

Carlile, M.J., S.C. Watkinson, and G.W. Gooday, Parasites and mutualistic symbionts, *The Fungi*, Bath Press, Avon, UK, 2001, pp. 363–452.

Dong, Y.-H., L-H. Wang, J.-L. Xu, H.-B. Zhang, X.-F. Zhang, and L.-H. Zhang, Quenching quorum-sensing dependent bacterial infection by an N-acyl homoserine lactonase, *Nature*, 411, 813–817, 2001.

Gardan, L., C. Gouy, R. Christen, and R. Samson, Elevation of three subspecies of *Pectobacterium carotovorum* to species level: *Pectobacterium atrosepticum* sp. nov., *Pectobacterium betavasculorum* sp. nov. and *Pectobacterium wasabiae* sp. nov., *Int. J. Syst. Evol. Microbiol.*, 53, 381–391, 2003.

Garrett, E.H., Fresh-cut produce: tracks and trends, *Fresh-Cut Fruits and Vegetables: Science, Technology, and Market*, CRC Press, Boca Raton, FL, 2002, pp. 1–10.

Gross, K.C., C.Y. Wang, and M. Saltveit, *The Commercial Storage of Fruits, Vegetables, and Florist and Nursery Crops*, Agriculture Handbook 66, USDA Agricultural Research Service, Beltsville, MD, Nov. 8, 2002, available at http://www.ba.ars.usda.gov/hb66/index.html.

Hauben, L., Moore, E.R.B., Vauterin, L., Steenackers, M., Mergaert, J., Verdonck, L. and Swings, J., Phylogenetic position of phytopathogens within the Enterobacteriaceae, *Syst. Appl. Microbiol.*, 21, 384–397, 1998.

Humperson-Jones, F.M. and K. Phelps, Climactic factors influencing spore production in *Alternaria brassicae* and *Alternaria brassicola*, *Ann. Appl. Biol.*, 114, 449–458, 1989.

Kamoun, S., Molecular genetics of pathogenic oomycetes, *Eukaryotic Cell*, 2, 191–199, 2003.

Laemmlen, F., Sclerotinia diseases, University of California, Agricultural and Natural Resources Pub. 8042, 2001.

Leadbetter, J.R., Plant microbiology: quieting the raucous crowd, *Nature*, 411, 748–749, 2001.

Martel, M.-B., R. Letoublon, and M. Fevre, Purification of endo polygalacturonases from *Sclerotinia sclerotiorum*: multiplicity of the complex enzyme system, *Curr. Microbiol.*, 33, 243–248, 1996.

Mauseth, J.D., *Plant Anatomy*, Pearson Education, Glenview, IL, 1988.

Niemira, B.A., R. Hammerschmidt, and G.R. Safir, Post-harvest suppression of potato dry rot (*Fusarium sambucinum*) in prenuclear minitubers by arbuscular mycorrhizal fungal inoculum, *Am. Potato J.*, 73, 509–515, 1996.

Niemira, B.A., W.W. Kirk, and J.M. Stein, Screening for late blight susceptibility in potato tubers by digital analysis of cut tuber surfaces, *Plant Dis.*, 83, 469–473, 1999.

Pierson, C.F., M.J. Ceponis, and L.P. McColloch, Market diseases of apples, pears and quinces, USDA Agricultural Handbook 376, Washington, DC, 1971.

Prell, H.H. and P.R. Day, *Plant-Fungal Pathogen Interactions*, Springer, New York, 2000.

Raid, R.N. and L.E. Datnoff, Downy mildew of lettuce. Fact Sheet HS-147, Florida Cooperative Extension Service, University of Florida, Gainesville, FL, 1992.

Ribon, A.O.B., V.V. Queiroz, and E.F. de Araujo, Structural organization of polygalacturonase-encoding genes from *Penicillium griseoroseum*, *Genet. Mol. Biol.*, 25, 489–493, 2002.

Roberts, P.D., R.J. McGovern, T.A. Kucharek, and D.J. Mitchell, Vegetable diseases caused by *Phytophthora capsici* in Florida, Florida Cooperative Extension Service Document PP-176, Institute of Food and Agricultural Sciences, University of Florida, Gainsville, FL, 2001.

Sanderson, P.G. and R.A. Spotts, Postharvest decay of winter pear and apple fruit caused by species of *Penicillium*, *Phytopathology*, 85, 103–110, 1995.

Sommer, N.F., R.F. Fortlage, F.G. Mitchell, and E.C. Maxie, Reduction of postharvest losses of strawberry fruits from gray mold, *J. Am. Soc. Hortic. Sci.*, 98, 285–288, 1973.

Tuzun, S. and J. Kloepper, Practical application and implementation of induced resistance, in *Induced Resistance to Disease in Plants*, Hammerschmidt, R. and J. Kuc, Eds., Kluwer Academic Publishers, Boston, MA, 1995, pp. 152–168.

van Emden, H.F., V.F. Eastop, R.D. Hughes, and M.J. Way, The ecology of *Myzus persicae*, *Annu. Rev. Entomol.*, 14, 197–270, 1969.

White, I.M., and M.M. Elson-Harris, *Fruit Flies of Economic Significance: Their Identification and Bionomics*, CAB International, Oxon, UK, 1994.

Zhuang, H., M.M. Barth, and T.R. Hankinson, Microbial safety, quality and sensory aspects of fresh-cut fruits and vegetables, in *Microbial Safety of Minimally Processed Foods*, Novak, J.S., G.M. Sapers, and V.K. Juneja, Eds., CRC Press, Boca Raton, FL, 2003, pp. 255–278.

Zitter, T.A., Vegetable crops: fruit rots of squash and pumpkins, Cooperative Extension Fact Sheet Page 732, Cornell University, Ithaca, NY, 1992.

16 Role of Fluorescent Pseudomonads and Their Pectolytic Enzymes in Spoilage of Fresh and Fresh-Cut Produce

Ching-Hsing Liao and Dike Ukuku
Eastern Regional Research Center, U.S. Department of
Agriculture[1], Agricultural Research Service, Wyndmoor, PA

CONTENTS

[1] Mention of trade names or commercial products in this chapter is solely for the purpose of providing specific information and does not imply recommendation or endorsement by the U.S. Department of Agriculture.

16.1 INTRODUCTION

Fresh and fresh-cut produce have become the fastest growing food category in the supermarket during the last two decades. A recent USDA survey [1] showed that the demand for fresh produce increased by more than 12% in the last decade and per capita consumption jumped from 284 pounds in 1987 to 318 pounds in 1997. To meet the market expansion, new strategies are required to improve the production of these commodities on the farm and to reduce the losses caused by physical, physiological, and microbiological disorders after harvest. An estimated 10 to 30% of fresh fruits and vegetables produced in the U.S. are wasted after they are harvested [2]. A large part of these losses are due to spoilage caused by bacteria, fungi, or yeasts [3–6]. The spoilage of acidic fruits such as apples, oranges, and strawberries is usually caused by molds, yeasts, or lactic acid bacteria (LAB) [5,6]. However, the spoilage of fresh produce with neutral pH, including edible roots/tubers and salad vegetables, is often the result of pectolytic bacteria causing a form of "soft rot" [3,4]. Results from a series of USDA surveys show that "bacterial soft rot" accounts for a very large proportion of postharvest disorders in potato, tomato, lettuce, bell pepper, and cucumber shipments at wholesale produce markets in New York [7–9]. Apart from the economic impact, soft-rotted produce more often harbors human pathogens such as *Salmonella* than their healthy counterparts [10] and has become an important food safety concern [11].

Bacterial soft rot is commonly known to be caused by three groups of *Erwinia*, including *E. carotovora* subsp. *carotovora* (Ecc), *E. carotovora* subsp. *atroseptica* (Eca), and *E. chrysanthemi* (Ech). However, pectolytic bacteria in at least six other genera including *Pseudomonas, Xanthomonas, Cytophaga, Flavobacterium, Bacillus,* and *Clostridium* can be involved [4,5]. A series of studies conducted in our laboratory during the 1980s [12–14] showed that over 40% of the rotted fruits and vegetables collected at retail and wholesale markets were likely caused by non-*Erwinia* soft-rotting bacteria including *Cytophaga, Xanthomonas,* and pectolytic fluorescent (PF) pseudomonads. Bartz [15] also found that PF pseudomonads accounted for almost one-third of the soft-rotting bacteria isolated from decayed tomatoes. The assumption that PF pseudomonads consisting of *P. fluorescens* and *P. viridiflava* are more likely to be involved in spoilage of refrigerated produce than other pectolytic bacteria can be attributed to at least two reasons. First of all, these pseudomonads are nutritionally versatile and capable of growing in a simple salt solution containing only four minerals and one of several utilizable carbon sources [14]. Second, both *P. fluorescens* and *P. viridiflava* are psychotropic and capable of inducing soft rot of fresh produce that is stored at 10°C or below [14,16,17]. Other genera of soft-rot bacteria usually grow very poorly and are unable to induce tissue maceration at low temperatures. Possibly because of these mentioned capabilities, fluorescent pseudomonads, including pectolytic strains, are commonly found on the surfaces and often constitute a major component of native microflora on fresh and minimally processed produce [16–25].

The soft-rot symptoms caused by PF pseudomonads are in general similar to those caused by Ecc, Eca, and Ech. However, under the most favorable conditions PF pseudomonads are less virulent than the erwinias. It is now generally believed that the ability of soft-rot erwinias and pseudomonads to macerate plant tissue results mainly from their ability to degrade plant cell walls by producing an array of pectin-degrading enzymes [26]. The enzymatic and molecular mechanism by which pectolytic erwinias cause soft-rot disease in plants has been extensively investigated and reviewed [26–29]. However, the role of PF pseudomonads and their pectic enzymes in spoilage of fresh and fresh-cut produce has not yet been studied to the same extent as the soft-rot erwinia systems. The subjects to be discussed in this chapter include: (1) the distribution and relationship of PF pseudomonads to spoilage of fresh produce; (2) pectic and other depolymerizing enzymes produced by PF pseudomonads; (3) biochemical and molecular genetic evidence that a single alkaline pectate lyase (PL) is required for induction of soft rot, (4) interactions between fluorescent pseudomonads, native microflora, and human pathogens (*Listeria monocytogenes, Escherichia coli* O157:H7, and *Salmonella*) on fresh produce; and (5) potential postharvest treatments for inactivation of unwanted microorganisms on fresh produce.

16.2 PF PSEUDOMONADS AS A MAJOR CAUSE OF PRODUCE SPOILAGE

16.2.1 Physiological Diversity of PF Pseudomonads

PF pseudomonads represent a very heterogeneous taxonomic group mainly consisting of *P. viridiflava* [30] and five biovars of *P. fluorescens* [31]. Soft-rotting strains of *P. fluorescens*, often referred to as *P. marginalis* in the plant pathology literature [32], were the first among PF pseudomonads to be recognized as a soft-rotting pathogen of head lettuce in the field and after harvest [33,34]. With the exception of this disease-causing ability, *P. marginalis* is indistinguishable from other strains of *P. fluorescens* genetically and physiologically [31,33]. Until now, the description of *P. marginalis* continues to be used for those soft-rotting pseudomonads that are fluorescent and positive in oxidase and arginine dehydrogenase activities [32,34]. Although the type strains of *P. marginalis* previously available from culture collections were identified as *P. fluorescens* biovar II [31], soft-rotting strains belonging to the four other biovars of *P. fluorescens* have been isolated. Following the characterization of 55 strains of PF pseudomonads isolated from naturally rotted specimens, Liao and Wells [14] found that only 19 of those strains exhibited the characteristics typical of biovar II. The remaining 36 strains were identified as *P. fluorescens* biovar IV (9 strains), *P. fluorescens* biovar V (11 strains), and *P. viridiflava* (16 strains). Brocklehurst and Lund [16] reported that *P. fluorescens* strains belonging to biovars I and III were possibly involved in the spoilage of cabbage stored in the cold. Diverse biovars of *P. fluorescens* have also been shown to be associated with spoilage of endive leaves and ready-to-use salad vegetables [23,35]. These reports and others not cited suggest that PF pseudomonads are likely responsible for a substantial proportion of postharvest rot of fresh and fresh-cut produce that is stored at low temperatures.

It should be noted, however, that other species of fluorescent pseudomonads including *P. aeruginosa* [36] and *P. cichorii* [7] have also been reported to be associated with spoilage of potato tubers and leafy vegetables. To the best of our knowledge, neither *P. aeruginosa* nor *P. cichorii* has ever been shown to produce pectolytic enzymes required for induction of soft rot as discussed below. In contrast, certain *P. syringae* pathovars including pv. *lachrylmans,* although exhibiting strong pectolytic activity *in vitro,* are unable to cause soft rot on potato tuber slices or cucumber fruits [33]. *P. syringae* pathovars are very closely related to *P. viridiflava* genetically and physiologically. So far, there is no report showing the involvement of pectolytic *P. syringae* in spoilage of fresh produce. Therefore, PF pseudomonads to be discussed in this chapter are limited to the soft-rotting strains of fluorescent pseudomonads, mainly consisting of *P. fluorescens* and *P. viridiflava.*

16.2.2 OCCURRENCE OF PF PSEUDOMONADS ON FRESH PRODUCE

Unlike soft-rot erwinias, PF pseudomonads are generally considered weak and opportunistic pathogens that do not cause large-scale disease outbreaks in the field. However, these bacteria are widespread in nature and can be isolated from very diverse environments and plant sources including soil, water, root rhizospheres, and surfaces of fruit and vegetables. As common epiphytes, they can become a major component (up to 40%) of the native microflora on potato tubers [37,38], collards [39], peas [40], tomatoes [41], spinach [42], lettuce [43–45], cabbage [46], and salad vegetables [47–49]. Although the direct involvement of these pseudomonads in spoilage is difficult to demonstrate, it is generally assumed that reduction in shelf life of fresh produce as a result of spoilage may be caused by complex interactions between PF pseudomonads and nonpathogenic microflora including lactic acid bacteria (LAB), yeasts, or fungi. Because of their ubiquity and potential to induce tissue maceration at low temperatures, PF pseudomonads are expected to play a very critical role in the quality and safety of fresh produce.

16.3 BIOCHEMICAL CHARACTERIZATION OF PECTATE LYASE

16.3.1 PRODUCTION OF PECTIC ENZYMES AND OTHER DEPOLYMERASES BY PF PSEUDOMONADS

PF pseudomonads produce at least four types of pectinases: polygalacturonase (PG), pectate lyase (PL), pectin lyase (PNL), and pectin methyl esterase (PME). PG is a glycosidase that cleaves α-1,4 glycosidic bonds between uronic acid residues in polygalacturonic acid (PGA) by hydrolysis. PL is a lytic enzyme that cleaves α-1,4 glycosidic bonds between uronic acid residues in PGA or low-methylated pectin by *trans*-elimination. PNL is also a lytic enzyme that cleaves uronic acid residues in highly (> 91%) methylated pectin. PME is a saponifying enzyme that causes the hydrolysis of methyl ester groups in highly methylated pectin with the production of methanol and PGA. The method for analyzing the activity of each enzyme [50–54] has been developed mainly based on detection of unsaturated oligogalacturonate products generated by PL or PNL or detection of saturated digalacturonates or methanol products generated, respectively, by PG and PME. Based on the data obtained so far, production of PG and PME by PF pseudomonads appears to be rare and has been demonstrated only in a few strains of *P. fluorescens* so far examined [53,54]. Production of PNL was detected only in cultures of certain *P. fluorescens* strains that had been exposed to DNA-damaging agents such as mitomycin C, UV irradiation, and nalidixic acid [55,56]. As will be discussed in more detail, almost all of the PF pseudomonads so far examined produce PL in culture either constitutively or inducibly. Production of PL therefore represents a common feature among soft-rotting pseudomonads, suggesting that PL may be the primary enzyme required for induction of soft rot [57]. On the contrary, production of PME, PNL, and PG appears not to be essential for induction of soft rot but may aid the survival and growth of these pseudomonads in plants or other environments.

In addition to pectin-degrading enzymes, PF pseudomonads also produce proteases (Prt) [58] and cellulases (Cel) [59], but not lipases and amylases [60]. Production of lipases, however, has been detected in at least two *P. fluorescens* strains used for biological control [61]. Production of Prt and Cel by pseudomonads is possibly for the purpose of catabolic or nutritional functions. Both enzymes degrade polymeric substrates (protein or cellulose) readily available in host plants or environments into monomeric end products that can be utilized by bacteria as energy or carbon sources. When inoculated onto plants, purified Prt or Cel is unable to cause visible disintegration of plant tissues. It has yet to be determined, however, whether the combined action of PL and other depolymerases such as Prt, Cel, or lipase may augment the extent of spoilage or tissue maceration. Although the biological functions of all the diverse extracellular enzymes produced by PF pseudomonads are not clear, Prt produced by *P. fluorescens* has been extensively investigated. Analysis of the concentrated culture supernatant of *P. fluorescens* (CY091) by isoelectric focusing (IEF) electrophoresis and overlay enzyme-activity staining revealed the presence of at least two Prts [62]. Two biocontrol strains (B52 and A506) of *P. fluorescens* have also been shown to produce more than one Prt [61; J. Loper, personal communication]. The predominant Prt produced by *P. fluorescens* CY091, designated AprX, has been characterized and the gene operon (in 7.3-kb genomic fragment) encoding the structural enzyme protein and its secretory apparatus has been cloned and sequenced [58]. The Prt AprX has an estimated molecular mass of 50 KDa and is a zinc-metalloprotease requiring Ca^{+2} for activity. Production and secretion of AprX by strain CY091 is dependent on Ca^{+2} or Sr^{+2}, and two conserved sequence domains associated with Ca^{+2} or Sr^{+2} binding have been identified. As an extracellular alkaline enzyme, AprX exhibits 50 to 60% identity in amino acid sequence to related proteases produced by *P. aeruginosa* [63] and *E. chrysanthemi* [64].

PF pseudomonads can also produce other types of depolymerases or secondary metabolites to enhance their ecological fitness or pathogenesis requirement. Production of a peptidolipid biosurfactant, viscosin, by a pectolytic strain of *P. fluorescens* has been shown to facilitate the initiation and spread of soft rot [65]. Although the health benefit of a diet rich in fresh produce is well known, fruits and vegetables have been shown to contain a potentially hazardous compound named "rutin" [66]. Rutin is a flavonol glycoside consisting of the mutagenic aglycone quercetin and the disaccharide rutinoside. Certain PF pseudomonads are able to produce a glycosidase to degrade rutin and thereby minimize the safety hazard associated with this compound [67].

16.3.2 DESCRIPTION OF THE *ERWINIA* PECTIC ENZYME SYSTEM

Much of our knowledge about the enzymatic mechanism of soft-rot pathogenesis was derived from studies with *Erwinia*. Soft-rotting *Erwinia* are well known for their ability to produce a wide variety of pectic enzymes including PL, PNL, PME, PG, and PAE (pectin acetylesterase). Each pectinase cleaves a preferred substrate (pectin or pectate) by hydrolysis, *trans*-elimination, or saponification. Although the pathological function of each pectinase is not fully understood, PL is generally believed to be the principal enzyme involved in tissue maceration, electrolyte loss, and cell death [68]. PLs produced by *Erwinia* species are unique for their occurrence as multiple (greater than five) isozymes with isoelectric points (pIs) ranging from 4.0 to 10.0 [69]. It has been demonstrated that the alkaline PL (pI > 9.0) is more efficient than neutral or acidic PLs in inducing tissue maceration [70], and that an alkaline PL by itself is sufficient to cause tissue maceration even in the absence of live bacteria [71]. The biochemical basis for the difference in tissue-macerating ability among PL isozymes has not yet been determined. In addition, the pathological basis for producing more PLs than are required for induction of soft rot by *Erwinia* is not fully understood. Whether production of multiple PL isozymes is required for attacking different host plant species or different organs within the same species needs to be further investigated [72]. Because of their pathological and biotechnological importance, the molecular mechanisms by which soft-rot *Erwinia* mediate the synthesis and secretion of various depolymerases possibly in response to environmental changes or stresses have been extensively investigated and reviewed [27–29].

16.3.3 DESCRIPTION OF THE *PSEUDOMONAS* PECTIC ENZYME SYSTEM

Unlike the complex range of pectinases produced by soft-rot erwinias, the pectic enzyme system of PF pseudomonads is much simpler. So far, only a few strains of *P. fluorescens* and *P. viridiflava* have been shown to produce PME, PG, and PNL. However, all but one strain of PF pseudomonads examined in our laboratory produce PL [57]. In order to determine whether PF pseudomonads also produce multiple PL isozymes, the IEF gel electrophoresis and overlay enzyme-activity staining techniques [69] have been applied to analyze the IEF profile of PLs produced by 18 strains of PF pseudomonads. All 8 strains of *P. viridiflava* and all 10 strains of *P. fluorescens* investigated in our laboratory produce a single PL with an approximate pI of 9.7 to 10.0 [57]. The IEF-overlay enzyme-activity staining techniques have also been used to analyze the IEF profiles of PLs produced by other non-*Erwinia* pectolytic bacteria. Results obtained so far suggest that production of a single alkaline PL is a common feature among non-*Erwinia* pectolytic bacteria including *Cytophaga johnsonae* [57,73], *Xanthomonas campestris* [74,75], *Bacillus subtilis* [76–79], *Clostridium* spp. [80,81], and possibly *E. rubrifaciens* [82].

16.3.4 PURIFICATION, ENZYMATIC PROPERTIES, AND TISSUE-MACERATING ABILITY OF *PSEUDOMONAS* PLs

Because of the simplicity of the pectic enzyme system, PLs produced by non-*Erwinia* soft-rotting bacteria including *P. fluorescens*, *P. viridiflava*, *C. johnsonae*, and *X. campestris* can be easily purified from culture filtrates by two simple steps including ammonium sulfate precipitation and anion-exchange chromatography [57]. Analysis of PL samples by SDS-polyacrylamide gel electrophoresis showed that the enzymes had been purified to near homogeneity following the two purification steps. Molecular weights (M_r) of PLs from *P. fluorescens* (CY091) and *P. viridiflava* (SF312) were estimated to be 41 and 42 KDa, respectively, based on their electrophoretic mobility in SDS-polyacrylamide gels. Further analysis of purified PLs by IEF gel electrophoresis confirmed the alkaline nature of the enzymes (pI = 9.7 to 10). However, *Pseudomonas* PLs appear to migrate in SDS-polyacrylamide gels at a rate slightly slower than expected for the sizes of proteins predicted from the genes cloned [83–85], possibly due to the unique β-helix protein structure similar to that demonstrated for *E. chrysanthemi* PLc [86]. In addition to a slight difference in M_r and pI, the PLs from *P. fluorescens* and *P. viridiflava* can be distinguished by differences in other biochemical properties including Km, Vmax, and optimal pH and temperature for activity [87]. For both PLs, the optimal Ca^{+2} concentration for activity is 0.5 mmol/L, the optimal pH for activity is 8.5 to 9.0, and they are stable at low temperatures (25°C or below) for at least 30 d. However, at 37°C, the activity decreased 50% in 36 h. Thermostability of both enzymes at elevated temperatures (48°C or higher) increases in the presence of $CaCl_2$ or a positively charged molecule such as polylysine and decreases in the presence of a negatively charged molecule such as heparin. Both PLs exhibit differential degrees of sensitivity to group-specific inhibitors such as iodoacetic acid and diethylpyrocarbonate, indicating that sulfhydryl and imidazole groups are important for their catalytic function [87]. PLs purified from culture supernatants of *C. johnsonae* and one strain each of *P. fluorescens* and *P. viridiflava* were all able to induce soft rot on potato tuber slices to different degrees. In general, PLs from the two pseudomonads are about 10-fold more efficient in inducing tissue maceration than the PL from *C. johnsonae* [57]. When inoculated onto potato tuber slices, a minute amount (less than one unit of activity; one unit of activity being the amount of enzyme required to release 1 μmol of unsaturated uronides) of purified PL from either pseudomonad is sufficient to induce maceration of plant tissue even in the absence of live bacteria [57,87].

16.4 MOLECULAR GENETIC ANALYSIS OF PL PRODUCTION BY PF PSEUDOMONADS

16.4.1 ANALYSIS OF TRANSPOSON (TN5) MUTANTS DEFICIENT IN PL PRODUCTION AND SECRETION

Two classes of *P. viridiflava* and *P. fluorescens* mutants defective in pectolytic activity and designated as Pel⁻ and Rep⁻, respectively, have been isolated using transposon Tn5 mutagenesis [71]. The Pel⁻ mutation resulted from the insertion of Tn5 into the structural *pel* gene, whereas Rep⁻ mutation resulted from the insertion of Tn5 into one of two regulatory genes, designated *repA* (= *gacS* = *lemA*) and *repB* (= *gacA*). The Rep⁻ mutants exhibit pleotrophic phenotypic changes including the loss of the ability to synthesize PL, Prt, exopolysaccharides, and fluorescent siderophores. Since the loss of pectolytic activity in Pel⁻ and Rep⁻ mutants was always accompanied by the loss of the soft-rotting ability on bell pepper fruits [88–90], production of PL is absolutely required for soft rot development. However, as discussed above, production of additional depolymerases such as Prt is not essential for induction of soft rot. Prt⁻ mutants resulting from the transposition of Tn5 into the structural *aprX* gene retain the wild-type level of tissue-macerating ability of *P. fluorescens* [89] and *P. viridiflava* [90]. Production of Prt does not appear to play a significant role in induction of soft rot.

16.4.2 CLONING AND CHARACTERIZATION OF THE *PEL* GENES FROM NON-*ERWINIA* PECTOLYTIC BACTERIA

Current knowledge about *pel* genes was derived primarily from the studies of soft-rotting *Erwinia*. A number of reviews on this subject are available in the literature (for examples, see 26–28,72). Recently, *pel* genes have been cloned from several non-*Erwinia* phytopathogens including *P. viridiflava* [84], *P. fluorescens* [85], *P. marginalis* [91], *P. s.* pv. *lachrymans* [92], *Bacillus* spp. [93–95], *X. c.* pv. *campestris* [74], *X. c.* pv. *vesicatoria* [75], and *X. c.* pv. *malvacearum* [83]. Nucleotide sequences of *pel* genes cloned from these bacteria have been determined and found to be closely related to alkaline PLs (PLd and PLe) of *E. chrysanthemi* [83]. These *pel* genes usually encode pre-Pel proteins consisting of 377 to 380 amino acid (a.a.) residues with a signal peptide consisting of 26 to 29 a.a. at the N-terminus. Four conserved sequence domains presumably involving Ca^{+2} binding, catalytic activities, and protein-export functions were revealed. Multiple sequence alignment analysis shows that PL proteins of non-*Erwinia* phytopathogens including *Xanthomonas*, *Pseudomonas*, and *Bacillus* constitute a distinct cluster that shows 20 to 43% a.a. identity to the four established PL enzyme families of *Erwinia* [83]. Two lines of evidence further confirm the alkaline PL produced by non-*Erwinia* soft-rotters as the single factor responsible for tissue maceration. *Escherichia coli* cells carrying a *Pseudomonas pel* gene were able to induce tissue maceration inside potato tubers and under anaerobic conditions [85]. Restoration of soft-rotting ability in Pel⁻ mutants could be accomplished by transferring the functional *pel* gene into the mutants [84]. These results further confirm the earlier conclusion that non-*Erwinia* soft-rotting bacteria including PF pseudomonads produce a single PL for induction of soft rot as compared to multiple PLs produced by *Erwinia*.

16.4.3 MECHANISMS REGULATING THE PRODUCTION AND SECRETION OF PL

Although much is known about the molecular genetic mechanisms by which soft-rot *Erwinia* regulates the production of pectic enzymes [26–28], very little is presently known about the mechanism by which PF pseudomonads mediate the synthesis of PL. Pleotropic mutants of *P. fluroescens* and *P. viridiflava* displaying the simultaneous loss of pectolytic and proteolytic activities have been identified by transposon mutagenesis [88–90]. Results from Southern blot analysis using an internal fragment of Tn5 as a probe revealed that these mutants were derived from the transposition of Tn5 into one of two distinct genomic fragments. Two functional genes designated *gacS* (= *repA* or *lemA*) and *gacA* (= *repB*) in these two fragments have been identified, cloned, and confirmed by complementation studies. Following nucleotide sequence analyses, the *gacS* and *gacA* genes were predicted to encode a sensory and a regulator protein, respectively, in a two-component regulatory protein family [96–100]. The *gacS/gacA* pair thus likely acts in concert to mediate the production of PL, Prt, EPS, and siderophores [88–90], possibly in response to environmental needs or stresses. The two-component regulators GacS and GacA in a biological control strain of *P. fluorescens* have been shown to regulate the production of phospholipase C [96], lipase [61], and antibiotics [97–99] in biological control strains of *P. fluorescens*. The GacS/GacA system is also involved in the formation of disease lesions on snap beans by *Pseudomonas syringae* pv. *syringae* [100]. This system also interacts with the stationary-phase factor δˢ (encoded on *rpoS*) playing a predominant role in the regulatory cascade controlling stress responses in a biocontrol strain of *P. fluorescens* [101]. The global activator GacA of *P. aeruginosa* interacts with a quorum-sensing regulatory system (LuxR-LuxI) to control the production of the autoinducer-butyryl-homoserine lactone [102]. It has not yet been investigated whether the RpoS and autoinduction regulatory cascade as demonstrated in other strains of *P. fluorescens* or *P. aeruginosa* also operates in PF pseudomonads to control the production of tissue-macerating factor PL.

A group of *P. viridiflava* mutants failing to excrete PL and Prt across the outer membrane have been generated by transposon mutagenesis [71]. These secretion-defective mutants, designated Out⁻ mutants, resulting possibly from the insertion of

Tn*5* into a cluster of genes in the Type II secretion gene family [103,104], were unable to induce soft rot on potato tuber slices or bell pepper fruits [71]. The synthesis and secretion of PL thus represent two consecutive functions required by *P. fluorescens* and *P. viridiflava* to be efficient soft-rotting pathogens.

Production of PL in certain strains of *P. fluorescens* is induced by pectic substrates [53,105] or by plant tissue extracts [106–108]. However, in other *P. fluorescens* strains, production of PL is not affected by the type of carbon source included in the medium [53,108]. Recently, we investigated the mode of PL production in 24 strains of *P. fluorescens* and found that production of PL in 4 out of 24 strains was not induced by pectic substrates but by Ca^{+2} [109]. These four strains produce 10 times more PL in medium containing 1 m*M* $CaCl_2$ than in one containing no $CaCl_2$ supplement. Presence of $CaCl_2$ in the medium not only affects the amount but also the final destination of PL. Over 86% of total PL produced by strain CY091 in $CaCl_2$-supplemented medium was excreted into the culture fluid. By comparison, only 13% of total PL produced by this strain in $CaCl_2$-deficient medium was detected in the extracellular fraction. The effect of Ca^{+2} on PL (and also Prt) production is concentration-dependent and can be replaced by Sr^{+2}, but not by Zn^{+2}, Fe^{+2}, Mn^{+2}, Mg^{+2}, or Ba^{+2}. Because of the indispensable role of Ca^{+2} in PL production and pectic degradation, the potential of using ion-chelating agents such as EDTA for control of *Pseudomonas* rot has been investigated [109]. Treatment of potato tuber disks with 40 ppm of EDTA especially in the presence of nisin (a bacteriocin) is effective in suppressing the development of soft rot [109,110].

16.5 INTERACTIONS OF PF PSEUDOMONADS AND NATIVE MICROFLORA ON FRESH PRODUCE

The changes in microflora on fresh and fresh-cut produce as affected by processing, decontamination treatments, and storage conditions have been extensively investigated and reviewed [5,6]. The numbers and the types of microorganisms identified are variable and largely dependent on the sources of the samples analyzed. The population of mesophilic bacteria as determined on plate count agar can range from 10^3 to 10^9 colony-forming units (CFU) per gram of tissue. Very diverse groups of microflora are present on the surfaces of fresh fruits and vegetables. In addition to fluorescent pseudomonads and *Erwinia*, other genera of microflora including *Serratia*, *Klebsiella*, *Citrobacter*, *Enterobacte*, yeast, and LAB have been detected on various types of produce [17,18–20,22,40,42,43,45,47,48]. PF pseudomonads often constitute a major proportion of native flora on salad vegetables [17], shredded lettuce [45], cauliflower florets [47], endive leaves [22], spinach [42], tomatoes [41], and alfalfa seed [25], and appear to play the critical role in the development of soft rot or spoilage. As discussed above, softening and maceration of plant tissues results mainly from the action of PL or other depolymerases produced by pectolytic *Erwinia* and fluorescent pseudomonads.

The coliforms and enterobacteria present on the surfaces of fresh produce are generally considered saprophytic and nonpectolytic, although production of PL and exoPG within the cells has been detected in certain strains of *Klebsiella* and *Yersinia* [115,116]. The role of these enterobacteria in spoilage of fresh produce is unclear, and a direct correlation between the number of bacteria present and the degree of spoilage observed has not yet been consistently demonstrated. It appears that the shelf life of fresh produce is more dependent on the type of microorganisms present than on the total number of bacteria detected. Pectolytic microflora including *Erwinia* [3–6], PF pseudomonads [3–6,35], LAB [111,112], and yeasts [113,114] are more likely to cause spoilage than nonpectolytic flora such as *Klebsiella* and *Citrobacter*. Under natural conditions, spoilage of fresh and fresh-cut produce results largely from the complex interactions between pectolytic and nonpectolytic microflora present on the surfaces of produce. A nonlinear mathematical model to predict the growth of *P. marginalis* and its relationship to vegetable spoilage has been proposed [117,118]. It should be noted, however, that not all produce spoilage is microbiological in nature. Physiological spoilage of fresh produce can be caused by endogenous pectic enzyme activities or fermentative reactions inside plant tissues [24].

16.6 INTERACTIONS OF PF PSEUDOMONADS AND HUMAN PATHOGENS ON FRESH PRODUCE

The ability of pathogenic pseudomonads to infect and multiply in plants is mainly due to their ability to produce enzymes, toxins, or other virulence factors to disrupt plant cells in order to obtain nutrients for growth [119]. Recently, a number of gastrointestinal human pathogens including *E. coli* O157:H7, *Salmonella*, and *L. monocytogenes* have also been found to survive and grow on cut surfaces of fresh fruits or vegetables. It is unclear, however, as to how human pathogens acquire nutrients for growth on uninjured plants. With the exception of *Yersinia enterocolitica*, the other human pathogens including *E. coli*, *Salmonella*, and *L. monocytogenes* do not produce enzymes or toxins associated with pathogenicity on plants. Based on a recent survey [116] conducted in our laboratory, *Yersinia enterocolitica* is the only foodborne pathogen that has been shown to produce pectic enzymes, including one endolytic PL and one exolytic PG. Pectic enzymes produced by *Yersinia* and *E. coli* strains carrying the genes coding for these two enzymes are accumulated within the perplasmic space of bacterial cells and were unable to disrupt plant cells and to induce maceration of potato slices [116]. It is not known whether pectolytic flora including PF pseudomonads interfere with the colonization of plants by human pathogens. The data accumulated so far suggest that the presence of PF pseudomonads on fresh produce can exert either positive or negative effects on the survival and growth of foodborne pathogens, as discussed below.

Wells and Butterfield [10] were the first to report a higher incidence of *Salmonella* contamination on rotted produce than on apparently healthy plant counterparts (18 to 20% compared to 9 to10%, respectively). The higher incidence of *Salmonella* contamination was thought to be caused by enhanced growth of the pathogen in rotted tissue. A challenge study showed a 5- to 10-fold increase in the population of *S. typhimurium* was observed in potato disks coinoculated with *E. carotovora* or *P. viridiflava* [10]. Carlin et al. [22,23] also found that increases in the population of *L. monocytogenes* were directly correlated with the extent of spoilage of fresh endive leaves. The rotted tissue may provide the nutrients needed for the growth of human pathogens. Furthermore, the macerated tissue may also serve as a source or vehicle for dissemination of foodborne pathogens and spread of disease [120].

Rotted tissues infected with fungal pathogens are also more likely to harbor *Salmonella* than their healthy counterparts [121]. Foodborne pathogens such as *Salmonella* and *L. monocytogenes* usually do not grow, or grow very poorly, on acidic fruits (pH < 4) such as apples or oranges. The growth of postharvest rot pathogens in fruits can markedly change the pH surrounding the infected tissue. Conway et al. [122] reported that for fresh-cut apple *L. monocytogenes* grew in decayed areas infected by *Glomerella cingulata* but populations decreased in decayed areas infected by *Penicillium expansum*. The pH in tissues infected with *G. cingulata* increased from 4.7 to 7.7, whereas the pH in tissues infected with *P. expansum* decreased from 4.7 to 3.7. Similarly, Riordan et al. [123] found that the population of *E. coli* O157:H7 increased 1 to 3 logs in wounded apple tissue infected with *G. cingulata* but no change in the *E. coli* population was observed in wounded tissues infected with *P. expansum*. The pH in the former increased from 4.1 to 6.8 and the pH in the latter showed no significant increase. Increase in pH in fruit tissue infected with *G. cingulata* can therefore promote the growth of human pathogens.

Contrary to the positive effects described above, a number of studies have shown that growth of *L. monocytogenes* on potato slices [124], spinach [42], and endive [23] can be negatively affected or suppressed by the presence of fluorescent pseudomonads, possibly in part due to the production of ion-chelating siderophores [25]. As discussed above, production of acids by *P. expansum* in infected tissue also reduced the growth of *L. monocytogenes* significantly [122,123]. In fact, fluorescent pseudomonads antagonistic to foodborne pathogens can be found commonly on the surfaces of fresh produce and sprouting seeds [23,25,48]. Elimination of spoilage microorganisms such as PF pseudomonads from fresh produce may prolong the shelf life of fresh produce but at the same time may generate a less competitive environment for human pathogens to proliferate to an infectious dosage level. Although not supported by experimental data, it has been suggested [5,11] that an increase in the incidence of the association of foodborne disease outbreak with fresh produce may be in part due to the increase in postharvest treatments for elimination of indigenous microflora on fresh produce.

16.7 POSTHARVEST TREATMENTS OF FRESH PRODUCE AND THEIR EFFECTS ON PF PSEUDOMONADS

After harvest, fruits and vegetables are usually subjected to cleaning and decontamination treatments to remove soil, spoilage microorganisms, and occasional human pathogens. Due to a sharp increase in the association of fresh produce with disease outbreaks during the past two decades [11], extensive research efforts have been made to develop effective treatment methods for enhancing the microbiological safety of fresh and fresh-cut produce [125]. Primary focuses of these studies were to eliminate human pathogens presumably present sporadically at extremely low levels on the surfaces of fresh produce [125]. A number of physical, chemical, and biological intervention technologies [126], previously developed for elimination of spoilage bacteria and human pathogens on animal food products, have been modified and tested for their efficacy against harmful microorganisms on fresh and fresh-cut produce.

16.7.1 IRRADIATION

The use of ionizing irradiation (e.g., γ-rays from ^{60}Co or ^{137}Cs, accelerated electrons, or x rays) on raw fruits and vegetables has become a potential means of extending shelf life and inactivating pathogenic microorganisms on fresh produce [127]. Irradiation is measured in grays (Gy) or kilograys (kGy) to indicate a dose of irradiation energy required to kill an organism. Complex life forms with large genomic DNAs are in general more sensitive to the lethal effect of irradiation than simpler organisms with small genomes. The lethal irradiation dose for humans, insects, bacteria, and viruses has been estimated to be 0.004, 0.1, 1.5 to 4.5, and 10 to 45 kGy, respectively [128]. Spoilage of fresh produce caused by *Erwinia* or PF pseudomonads can be suppressed by irradiation at dose levels of 1 to 3 kGy without adversely affecting the sensory qualities of fruits and vegetables [127]. For fresh produce that is more sensitive to irradiation treatments, an even lower dose (0.5 kGy) can be applied, usually in combination with other treatments such as chlorination, heat, or modified atmosphere, to inactivate insect pests [129], spoilage bacteria [130], or human pathogens [131].

16.7.2 OZONE

Ozone has been approved in the U.S. as generally recognized as safe (GRAS) for treatment of bottled water but has not yet been permitted for use as a disinfectant in fresh produce processing [132]. Ozonated water at the concentration of 20 ppm is lethal to most of the bacteria pathogenic to humans or plants such as *P. aeruginosa* [133]. Exposure of *P. fluorescens* to 2.5 ppm of ozone for 40 sec reduces the population of this spoilage bacterium by 5 to 6 logs [134]. Use of ozone to disinfect fruits and vegetables including lettuce has been reported [135]. Moor et al. [136] found that ozone is in general more effective against Gram-negative than against Gram-positive bacteria and is ideal as a terminal disinfectant for food processing because of the lack of odor and residue. Application of ozone is being actively tested for its potential to improve the safety of fresh fruits and vegetables [137].

16.7.3 CHLORINE

Chlorine-based sanitizers including elemental chlorine, sodium hypochlorite (NaOCl), calcium hypochlorite (CaOCl), and chlorine dioxide are commonly used disinfectants in washing, spray, and flume waters in fresh produce processing plants. At concentrations of 50 to 200 ppm with a contact time of 1 to 2 min, chlorine is effective in removing over 99% of human pathogens and spoilage bacteria including PF pseudomonads on raw fruits and vegetables [138]. The antimicrobial activity of chlorine is pH-dependent and mainly due to the formation of hydrochlorous acid (HOCl) when dissolved in water. As the pH of the solution is reduced, the equilibrium is in favor of the formation of HOCl. Fruit and vegetable tissue components can neutralize chlorine, making it inactive against microorganisms. Therefore, the pH and active chlorine content in chlorinated water should be monitored regularly to ensure the maximal antimicrobial effect of chlorine treatment.

16.7.4 HYDROGEN PEROXIDE

H_2O_2 is classified as GRAS for use in food processing as a bleaching agent, oxidizing and reducing agent, and antimicrobial agent. Although it has not been approved for use in the fresh produce industry, the efficacy of H_2O_2 in improving the microbiological quality and extending the shelf life of minimally processed fruits and vegetables has been investigated [139]. H_2O_2 vapor treatments delayed or diminished the severity of bacterial soft rot in fresh-cut cucumber, green bell pepper, and zucchini but had no effect on the spoilage of fresh-cut broccoli, carrot, cauliflower, celery, or fresh strawberry. Similar treatments are able to delay the spoilage of mushrooms caused by *P. tolaasii* but also induce browning in the mushrooms. Dipping fresh-cut zucchini, cantaloupe, or cucumber in an H_2O_2 solution reduced the load of fluorescent pseudomonads by 90% and was similar in effectiveness to chlorine treatment [139]. The presence of H_2O_2 residues in some treated commodities and the adverse effect of treatments on produce color and flavor are the two concerns that require further investigation.

16.7.5 ORGANIC ACIDS

Organic acids including lactic acid and acetic acid (AA) have been approved for disinfection of beef, lamb, pork, and poultry carcasses. The application of organic acids to the surfaces of fresh produce for the purpose of reducing the populations of pathogenic and spoilage bacteria including PF pseudomonads has been investigated. Gastrointestinal human pathogens such as *Salmonella* and *E. coli* O157:H7 are approximately 10 to 50 times more resistant to AA treatment than plant-associated bacteria such as *Erwinia* or PF pseudomonads. The minimal concentration of AA required to kill 90% of *Salmonella*, *Erwinia,* and *P. fluorescens* within 5 min was estimated to be 2.4, 0.3, and 0.06%, respectively [Liao, unpublished]. Following exposure to AA, a very large proportion of *Erwinia* or PF pseudomonads become injured and are more susceptible to the action of other antimicrobial substances. A combination of AA and H_2O_2 was the most effective treatment against *Salmonella* and possibly spoilage bacteria among five sanitizer treatments examined.

16.7.6 MODIFIED ATMOSPHERE

The use of modified atmospheres (MA) on fresh produce packaged in polymeric film products has a significant effect on the microflora and quality of fresh produce. While small amounts of CO_2 stimulate the growth of many organisms [140], high concentrations of CO_2 (> 3%) inhibit the growth of most organisms, including PF pseudomonads [141–143]. *Pseudomonas* spp. as a group are in general more sensitive to CO_2 than native bacteria found on meat products such as *Proteus, Bacillus,* and *Micrococcus.* Thus, MA packaging provides an effective means to extend the shelf life of produce and to reduce the proliferation of spoilage pseudomonads [20]. The increase in the concentration of CO_2 inside the packaging pouches usually leads to a decrease in the *Pseudomonas* population but often leads to a drastic increase in the population of lactic acid bacteria, which is thought to play a role in spoilage of fresh produce [24].

16.8 CONCLUSION

PF pseudomonads consisting of *P. fluorescens* and *P. viridiflava* are the cause of a large proportion of postharvest rot of fresh fruits and vegetables. They are commonly found on the surfaces of fresh produce and constitute a major component of resident microflora on potato tubers and leafy vegetables. The ability of these pseudomonads to cause spoilage (often in the form of soft rot) results from their ability to produce a variety of depolymerases including pectinases (primarily PL), proteases (Prt), cellulases, and lipases. Unlike multiple PL isozymes produced by *Erwinia*, a single alkaline PL is produced by most if not all PL pseudomonads. The conclusion that a single alkaline PL is the sole or principal pectinase required for induction of soft rot is based on the results from a series of experiments including enzyme purification, isoelectric-focusing electrophoresis, transposon mutagenesis, gene cloning, and complementation studies. Two genes regulating the production and/or secretion of PL, Prt, the exopolysaccharides (alginate and levan), and fluorescent iron-chelating siderophores have been identified. These two genes, designated *gacS* and *gacA*, are members of the two-component regulatory gene family and are predicted to encode a sensory protein for receiving the external or internal signals (GacS) and an activator protein for mediating the synthesis and secretion of the aforementioned extracellular compounds (GacA). The presence of calcium is absolutely required by *P. fluorescens* and *P. viridiflava* to produce and excrete PL and Prt and is also required for catalytic activity of both enzymes. Application of ion chelators such as EDTA and organic acids such as acetic acid and citric acid thus becomes a possible approach for reducing the soft rot caused by PF pseudomonads. The potential of using irradiation, chemical sanitization, and modified atmospheres as well as biological control agents to reduce the population of spoilage and pathogenic bacteria on fresh produce needs to be further investigated.

ACKNOWLEDGMENT

We want to thank Jim Smith, Jim McEvoy, William Fett, and Vijay Juneja for their suggestions and reviews during the preparation of this chapter.

REFERENCES

1. Kaufman, P.R. et al., Understanding the dynamics of produce markets: consumption and consolidation growth, USDA, Economic Research Service, Agriculture Information Bulletin 758, available at http://www.ers.usda.gov/publications/aib758.pdf, accessed Aug. 24, 2001.

2. Harvey, J.M., Reduction of losses in fresh market fruits and vegetables, *Annu. Rev. Phytopathol.*, 16, 321, 1978.
3. Lund, B.M., The effect of bacteria on post-harvest quality of vegetables and fruits, with particular reference to spoilage, in *Bacteria and Plants*, Rhodes-Roberts, M. and Skinner, F.A., Eds., Academic Press, London, 1982.
4. Lund, B.M., Bacterial spoilage, in *Post-Harvest Pathology of Fruits and Vegetables*, Dennis, C., Ed., Academic Press, London, 1983, Chap. 9.
5. Nguyen-The, C. and Carlin F., The microbiology of processed fresh fruits and vegetables, *Crit. Rev. Food Sci. Nutr.*, 34, 37, 1994.
6. Nguyen-The, C. and Carlin F., Fresh and processed vegetables, in *The Microbiological Safety and Quality of Food*, Lund, B., Baird-Parker, T.C., and Gould, G.W., Ed., Vol. 1, Aspen Publishing, Gaithersburg, MD, 2000, Chap. 25.
7. Ceponis, M.J., Diseases of California head lettuce on the New York market during the spring and summer months, *Plant Dis.*, 54, 964, 1974.
8. Ceponis, M.J. and Butterfield, J.E., Market losses in Florida cucumbers and bell pepper in metropolitan New York, *Plant Dis.*, 58, 558, 1974.
9. Ceponis, M.J., Cappellini, R.A., and Lightner, G.W., Disorders in tomato shipments to the New York market, 1972–1984, *Plant Dis.*, 70, 261, 1986.
10. Wells, J.M. and Butterfield, J.E., *Salmonella* contamination associated with bacterial soft rot of fresh fruits and vegetables in the marketplace, *Plant Dis.*, 81, 867, 1997.
11. NACMCF [National Advisory Committee on Microbiological Criteria for Foods], Microbiological safety evaluations and recommendations on fresh produce, *Food Control*, 10, 117, 1999.
12. Liao, C.-H. and Wells, J.M., Properties of *Cytophaga johnsonae* strains causing spoilage of fresh produce at food markets, *Appl. Environ. Microbiol.*, 52, 1261, 1986.
13. Liao, C.-H. and Wells, J.M., Association of pectolytic strains of *Xanthomonas campestris* with soft rots of fruits and vegetables at retail markets, *Phytopathology*, 77, 418, 1987.
14. Liao, C.-H. and Wells, J.M., Diversity of pectolytic, fluorescent pseudomonads causing soft rot of fresh vegetables at produce markets, *Phytopathology*, 77, 673, 1987.
15. Bartz, J.A., Causes of postharvest losses in a Florida tomato shipment, *Plant Dis.*, 64, 934, 1980.
16. Brocklehurst, T.F. and Lund, B.M., Properties of pseudomonads causing spoilage of vegetables stored at low temperature, *J. Appl. Bacteriol.*, 50, 259, 1981.
17. Brocklehurst, T.F., Zaman-Wong, C.M., and Lund, B.M., A note on the microbiology of retail packs of prepared salad vegetables, *J. Appl. Bacteriol.*, 63, 409, 1987.
18. Brackett, R.E., Changes in the microflora of packaged fresh tomatoes, *J. Food Qual.*, 11, 89, 1988.
19. Brackett, R.E., Changes in the microflora of packaged fresh broccoli, *J. Food Qual.*, 12, 169, 1989.
20. Brackett, R.E., Influence of modified atmosphere packaging on the microflora and quality of fresh bell peppers, *J. Food Prot.*, 53, 255, 1990.
21. Brackett, R.E., Shelf stability and safety of fresh produce as influenced by sanitation and disinfection, *J. Food Prot.*, 55, 808, 1992.
22. Carlin, F., Nguyen-The, C., and Abreu da Silva, A., Factors affecting the growth of *Listeria monocytogenes* on minimally processed fresh endive, *J. Appl. Bacteriol.*, 78, 636, 1995.
23. Carlin, F., Nguyen-The, C., and Morris, C.E., Influence of background microflora on *Listeria monocytogenes* on minimally processed fresh broad-leaved endive (*Cinchorium endivia var. latifolia*), *J. Food Prot.*, 59, 698, 1996.

24. Carlin, F. et al. Microbiological spoilage of fresh, ready-to-use grated carrots, *Sci. Aliments*, 9, 371, 1989.
25. Liao, C.-H. and Fett, W.F., Analysis of native microflora and selection of strains antagonistic to human pathogens on fresh produce, *J. Food Prot.*, 64, 1110, 2001.
26. Collmer, A. and Keen, N.T., The role of pectic enzymes in plant pathogenesis, *Annu. Rev. Phytopathol.*, 24, 383, 1986.
27. Barras, F., Gijsegem, F., and Chatterjee, A.K., Extracellular enzymes and pathogenesis of soft-rot *Erwinia*, *Annu. Rev. Phytopathol.*, 32, 201, 1994.
28. Hugouvieux-Cotte-Pattat, N. et al., Regulation of pectinolysis in *Erwinia chrysanthemi*, *Annu. Rev. Microbiol.*, 59, 213, 1996.
29. Py, B. et al., Extracellular enzymes and their role in *Erwinia* virulence, *Methods Microbiol.*, 27, 157, 1998.
30. Billing, E., *Pseudomonas viridiflava* (Burkholder, 1930; Clara 1934), *J. Appl. Bacteriol.*, 33, 492, 1970.
31. Stanier, R.Y., Palleroni, J.J., and Doudoroff, M., The aerobic pseudomonads: a taxonomic study, *J. Gen. Microbiol.*, 43, 159, 1966.
32. Lelliott, R.A., Billing, E., and Hayward, A.C., A determinative scheme for the fluorescent plant pathogenic pseudomonads, *J. Appl. Bacteriol.*, 29, 470, 1966.
33. Sands, D.C., Schroth, M.N., and Hildebrand, D.C., Taxonomy of phytopathogenic pseudomonads, *J. Bacteriol.*, 101, 9, 1970.
34. Fahy, P.C. and Lloyd, A.B., *Pseudomonas*: the fluorescent pseudomonads, in *Plant Bacterial Diseases: A Diagnostic Guide*, Fahy, P.C. and Persley, G.J., Eds., Academic Press, Australia, 1983, Chap. 8.
35. Nguyen-The, C. and Prunier, J.P., Involvement of pseudomonads in deterioration of "ready-to-use" salads, *Int. J. Food Sci. Technol.*, 24, 47, 1989.
36. Cother, E.J., Darbyshire, B., and Brewer, J., *Pseudomonas aeruginosa*: cause of internal brown rot of onion, *Phytopathology*, 66, 828, 1976.
37. Cuppels, D. and Kelman, A., Isolation of pectolytic fluorescent pseudomonads from soil and potatoes, *Phytopathology*, 70, 1110, 1980.
38. Sampson, P.J. and Hayward, A.C., Characteristics of pectolytic bacteria associated with potato in Tasmania, *Aust. J. Biol. Sci.*, 24, 917, 1971.
39. Senter, S.D., Bailey, J.S., and Cox, N.A., Aerobic microflora of commercially harvested, transported and cryogenically processed collards (*Brassica oleracea*), *J. Food Sci.*, 52, 1020, 1987.
40. Senter, S.D. et al., Effects of harvesting, transportation, and cryogenic processing on the microflora of Southern peas, *J. Food Sci.*, 49, 1410, 1984.
41. Senter, S.D. et al., Microbiological changes in fresh market tomatoes during packing operations, *J. Food Sci.*, 50, 254, 1985.
42. Babic, I. et al., Changes in microbial populations of fresh cut spinach, *Intl. J. Food Microbiol.*, 31, 107, 1996.
43. Magnuson, J.A., King, A.D. Jr., and Török, T., Microflora of partially processed lettuce, *Appl. Environ. Microbiol.*, 56, 3851, 1990.
44. Bolin, H.R. et al., Factors affecting the storage stability of shredded lettuce, *J. Food Sci.*, 42, 1319, 1977.
45. King, A.D., Jr. et al., Microbial flora and storage quality of partially processed lettuce, *J. Food Sci.*, 56, 459, 1991.
46. Chesson, A., The fungal and bacterial flora of stored white cabbage, *J. Appl. Bacteriol.*, 46, 189, 1979.
47. Garg, N., Churey, J.J., and Splittstoesser, D.F., Effect of processing conditions on the microflora of fresh-cut vegetables, *J. Food Prot.*, 53, 701, 1990.

48. Francis, G.A. and O'Beirne, D., Effects of the indigenous microflora of minimally processed lettuce on the survival and growth of *Listeria innocua, Int. J. Food Sci. Technol.*, 33, 477, 1998.

49. Carlin, F. et al., Microbiological spoilage of fresh, ready-to-use grated carrots, *Sci. Aliments*, 9, 371, 1989.

50. Collmer, A., Ried, J.L., and Mount, M.S., Assay methods for pectic enzymes, *Methods Enzymol.*, 161, 329, 1988.

51. Oxford, A.E., Production of a soluble pectinase in a simple medium by certain plant-pathogenic bacteria belonging to the genus *Pseudomonas, Nature(London)*, 154, 271, 1944.

52. Paton, A.M., Pectin-decomposing strains of *Pseudomonas, Nature (London)*, 181, 61, 1958.

53. Nasuno, S. and Starr, M.P., Pectic enzymes of *Pseudomonas marginalis, Phytopathology*, 56, 1414, 1966.

54. Smith, W.K., A survey of the production of pectic enzymes by plant pathogenic and other bacteria, *J. Gen. Microbiol.*, 18, 33, 1958.

55. Sone, H. et al., Production and properties of pectin lyase in *Pseudomonas marginalis* induced by mitomycin C, *Agric. Biol. Chem.*, 52, 3205, 1988.

56. Schlemmer, A.F., Ware, C.F., and Keen, N.T., Purification and characterization of a pectin lyase produced by *Pseudomonas fluorescens* W51, *J. Bacteriol.*, 169, 4493, 1987.

57. Liao, C.-H., Analysis of pectate lyase produced by soft rot bacteria associated with spoilage of vegetables, *Appl. Environ. Microbiol.*, 55, 1677, 1989.

58. Liao, C.-H. and McCallus, D.E., Biochemical and genetic characterization of an extracellular protease from *Pseudomonas fluorescens* CY091, *Appl. Environ. Microbiol.*, 64, 914, 1998.

59. Gilbert, H.J. et al., Evidence for multiple carboxymethylcellulase genes in *Pseudomonas fluorescens* subsp. *cellulosa, Mol. Gen. Genet.*, 210, 551, 1987.

60. Liao, C.-H. and Wells, J.M., Properties of yellow (orange)-pigmented strains of soft rot bacteria in the genus *Xanthomonas, Pseudomonas, Cytophaga*, and *Xanthomonas*, in *Plant Pathogenic Bacteria*, Civerolo, E.L., Collmer, A., Davis, R.E. and Gillaspie, A.G., Eds., Martinus Nijhoff Publisher/Kluwer Academic Publishers Group, Boston, 1987, pp. 578-583.

61. Woods, R.G. et al., The *aprX-lipA* operon of *Pseudomonas fluorescens* B52: a molecular analysis of metalloprotease and lipase production, *Microbiology* 147, 345, 2001.

62. McCallus, D.E. and Liao, C.-H., Biochemical and genetic characterization of protease from the soft-rotting bacterium *Pseudomonas fluorescens*, in 6th Intl. Symp., Mol. Plant-Microbe Interact., Univ. Washington, Seattle, Washington, Abstr. 189, 1992.

63. Guzzo, J. et al., Cloning of the *Pseudomonas aeruginosa* alkaline protease gene and secretion of the protease into the medium by *Escherichia coli, J. Bacteriol.*, 172, 942, 1991.

64. Delepelaire, P. and Wandersman, C., 1989, Protease secretion by *Erwinia chrysanthemi, J. Biol. Chem.*, 264, 9083, 1989.

65. Laycock, M.B et al., Viscosin, a potent peptidolipid biosurfactant and phytopathogenic mediator produced by a pectolytic strain of *Pseudomonas fluorescens, J. Agric. Food Chem.*, 39, 483, 1991.

66. Brown, N.P. and Dietrich, P.S., Mutagenicity of plant flavonols in the *Salmonella*/mammalian microsome test: activation of flavonol glycosidase by mixed glycosidases from rat cecal bacteria and other sources, *Mutat. Res.*, 66, 223, 1979.

67. Hildebrand, D.C. and Caesar, A., The widespread occurrence of rutin glycosidase in fluorescent phytopathogenic pseudomonads, *Lett. Appl. Microbiol.*, 8, 117, 1989.

68. Mount, M.S., Bateman, D.F., and Basham, H.G., Induction of electrolyte loss, tissue maceration, and cellular death of potato tissue by an endopolygalacturonate trans-eliminase, *Phytopathology*, 60, 924, 1970.

69. Ried, J.L. and Collmer, A., Activity stain for rapid characterization of pectic enzymes in isoelectric focusing and sodium dodecyl sulfate-placrylamide gels, *Appl. Environ., Microbiol.*, 50, 615.

70. Payne, J.H. et al., Multiplication and virulence in plant tissue of *Escherichia coli* clones producing pectate lyase isozymes PLb and PLe at high levels and an *Erwinia chrysanthemi* mutant deficient in PLe, *Appl. Environ. Microbiol.*, 53, 2315, 1987.

71. Liao, C.-H., Hung, H.Y., and Chatterjee, A.K., An extracellular pectate lyase is the pathogenicity factor of the soft-rotting bacterium *Pseudomonas viridiflava*, *Mol. Plant-Microbe Interact.*, 1, 199, 1988.

72. Kotoujansky, A., Molecular genetics of pathogenesis by soft-rot erwinias, *Annu. Rev. Phytopathol.*, 25, 405, 1987.

73. Kurowski, W.M. and Dunleavy, J.A., Pectinase production by bacteria associated with improved preservative permeability in Sitka spruce: synthesis and secretion of polyg-alacturonate lyase by *Cytophaga johnsonii*, *J. Appl. Bacteriol.*, 41, 119, 1976.

74. Dow, J.M., et al., Molecular cloning of a polygalacturonate lyase gene from *Xanth-omonas campestris* pv. *campestris* and role of the gene product in pathogenicity, *Physiol. Mol. Plant Pathol.*, 35, 113, 1989.

75. Beaulieu, C. et al., Biochemical and genetic analysis of a pectate lyase gene from *Xanthomonas campestris* pv. *vesicatoria*, *Mol. Plant-Microbe Interact.*, 4, 446, 1991.

76. Watada, A.E. et al., Polygalacturonate lyase production by *Bacillus subtilis* and *Flavobacterium pectinovourum*, *Appl. Microbiol.*, 27, 346, 1974.

77. Chesson, A. and Codner, R.C., The maceration of vegetable tissue by a strain of *Bacillus subtilis*, *J. App. Bacteriol.*, 44, 347, 1978.

78. Karbassi, A. and Vaughn, R.H., Purification and properties of polygalacturonic acid *trans*-eliminase from *Bacillus stearothermophilus*, *Can. J. Microbiol.*, 26, 377, 1980.

79. Leary, J.V. et al., Isolation of pathogenic *Bacillus circulans* from callus cultures and healthy offshoots of date palm (*Phoenix dactylifrera* L.), *Appl. Environ. Microbiol.*, 52, 1173, 1986.

80. Lund, B.M. and Brocklehurst, T.F., Pectic enzymes of pigmented strains of *Clostrid-ium*, *J. Gen. Microbiol.*, 104, 59, 1978.

81. Macmillian, J.D. and Vaughn, R.H., Purification and properties of a polygalacturonic acid-*trans*-eliminase produced by *Clostridium multifermentans*, *Biochemistry*, 3, 564, 1964.

82. Gardner, J.B. and Kako, C.I., Polygalacturonic acid *trans*-eliminase in the osmotic shock fluid of *Erwinia rubrifaciens*: characterization of the purified enzyme and its effect on plant cells, *J. Bacteriol.*, 127, 451, 1976.

83. Liao, C.-H. et al., Cloning of a pectate lyase gene from *Xanthomonas campestris* pv. *malvacearum* and comparison of its sequence relationship with *pel* genes of soft-rot *Erwinia* and *Pseudomonas*, *Mol. Plant-Microbe Interact.*, 9, 14, 1996.

84. Liao, C.-H. et al., Cloning and characterization of a pectate lyase gene from the soft-rotting bacterium *Pseudomonas viridiflava*. *Mol. Plant-Microbe Interact.*, 5, 301, 1992.

85. Liao, C.-H., Cloning of pectate lyase gene *pel* from *Pseudomonas fluorescens* and detection of sequences homologous to *pel* in *Pseudomonas viridiflava* and *Pseudomo-nas putida*, *J. Bacteriol.*, 173, 4386, 1991.

86. Yoder, M.D., Keen, N.T., and Jurnak, F., New domain motif: the structure of pectate lyase C, a secreted plant virulence factor, *Science*, 260, 1503, 1993.

87. Liao, C.-H. et al. Biochemical characterization of pectate lyases produced by fluorescent pseudomonads associated with spoilage of fresh fruits and vegetables, *J. Appl. Microbiol.*, 83, 10, 1997.
88. Liao, C.-H., McCallus, D.E., and Fett, W.F., Molecular characterization of two gene loci required for production of the key pathogenicity factor pectate lyase in *Pseudomonas viridiflava*, *Mol. Plant-Microbe Interact.*, 7, 391, 1994.
89. Liao, C.-H. et al., Identification of gene loci controlling pectate lyase production and soft-rot pathogenicity in *Pseudomonas marginalis, Can. J. Microbiol.*, 43, 425, 1997.
90. Liao, C.-H. et al. The *repB* gene required for production of extracellular enzymes and fluorescent siderophores in *Pseudomonas viridiflava* is an analog of the *gacA* gene of *Pseudomonas syringae*, *Can. J. Microbiol.*, 42, 177, 1996.
91. Nikaidou, N., Kamio, Y., and Izaki, K., Molecular cloning and nucleotide sequence of the pectate lyase gene from *Pseudomonas marginalis* N6301, *Biosci. Biotechnol. Biochem.*, 57, 957, 1993.
92. Bauer, D.W. and Collmer, A., Molecular cloning, characterization, and mutagenesis of a *pel* gene from *Pseudomonas syringae* pv. *lachrymans* encoding a member of the *Erwinia chrysanthemi* PelADE family of pectate lyases, *Mol. Plant-Microbe Interact.* 10, 369, 1997.
93. Nagel, C.W. and Baughn, R.H., Comparison of growth and pectolytic enzyme production by *Bacillus polymyxa*, *J. Bacteriol.*, 83, 1, 1962.
94. Karbassi, A. and Vaughn, R.H., Purification and properties of polygalacturonic acid *trans*-eliminase from *Bacillus stearothermophilus*, *Can. J. Microbiol.*, 26, 377, 1980.
95. Nasser, W. et al., Pectate lyase from *Bacillus subtilis*: molecular characterization of the gene, and properties of the cloned enzyme, *FEBS Lett.*, 335, 319, 1993.
96. Sacherer, P., Défago, G., and Haas, D., Extracellular protease and phosphlipase C are controlled by the global regulatory gene *gacA* in the biocontrol strain *Pseudomonas fluorescens* CHA0, *FEMS Microbiol. Lett.*, 116, 155, 1994.
97. Laville, J. et al., Global control in *Pseudomonas fluorescens* mediating antibiotic synthesis and suppression of black root rot of tobacco, *Proc. Natl. Acad. Sci. USA*, 89, 1562, 1992.
98. Gaffney, T.D. et al., Global regulation of expression of anti-fungal factors by a *Pseudomonas fluorescens* biological control strain, *Mol. Plant-Microbe Interact.*, 7, 455, 1994.
99. Corbell, N. and Loper, J.E., A global regulator of secondary metabolite production in *Pseudomonas fluorescens* Pf-5, *J. Bacteriol.*, 177, 6230, 1995.
100. Hrabak, E.M. and Willis, D.K., The *lemA* gene required for pathogenicity of *Pseudomonas syringae* pv. syringae on bean is a member of a family of two-component regulators, *J. Bacteriol.*, 174, 3011, 1992.
101. Whistler, C.A. et al., The two-component regulators GacS and GacA influence accumulation of stationary-phase sigma factor δ^s and the stress response in *Pseudomonas fluorescens* Pf-5, *J. Bacteriol*, 180, 6635, 1998.
102. Reimmann, C. et al., The global activator GacA of *Pseudomonas aeruginosa* PAO positively controls the production of the autoinducer *N*-butyryl-homoserine lactone and the formation of the virulence factors pyocyanin, cyanide, and lipase, *Mol. Microbiol.*, 24, 309, 1997.
103. Sandkvist, M., Biology of type II secretion, *Mol. Microbiol.*, 40, 271, 2001.
104. Koster, M., Bitter, W., and Tommassen, J., Protein secretion mechanisms in Gram-negative bacteria, *Int. J. Med. Microbiol.*, 290, 325, 2000.

105. Fuchs, A., The *trans*-eliminative breakdown of Na-polygalacturonate by *Pseudomonas fluorescens*, *Antonie van Leeuwenhoek J. Microbiol. Serol.*, 31, 323, 1965.
106. Zucker, M. and Hankin, L., Regulation of pectate lyase synthesis in *Pseudomonas fluorescens* and *Erwinia carotovora*, *J. Bacteriol.*, 104, 13, 1970.
107. Zucker, M. and Hankin, L., Inducible pectate lyase synthesis and phytopathogenicity of *Pseudomonas fluorescens*, *Can. J. Microbiol.*, 17, 1313, 1971.
108. Zucker, M., Hankin, L., and Sands, D., Factors governing pectate lyase synthesis in soft rot and non-soft rot bacteria, *Physiol. Plant Pathol.*, 2, 59, 1972.
109. Liao, C.-H., McCallus, D.E., and Wells, J.M., Calcium-dependent pectate lyase production in the soft-rotting bacterium *Pseudomonas fluorescens*, *Phytopathology*, 83, 813, 1993.
110. Wells, J.M., Liao, C.-H., and Hotchkiss, A.T., *In vitro* inhibition of soft-rotting bacteria by EDTA and nisin and *in vivo* response on inoculated fresh cut carrots, *Plant Dis.*, 82, 491, 1998.
111. Juven, J., Lindner, P., and Weisslowicz, H., Pectin degradation in plant material by *Leuconostoc mesenteroides*, *J. Appl. Bacteriol.*, 58, 533, 1985.
112. Sakellaris, G., Nikolaropoulos, S., and Evangelopoulos, A.E., Purification and characterization of an extracellular polygalacturonase from *Lactobacillus plantarum* strain BA 11, *J. Appl. Bacteriol.* 67, 77, 1989.
113. Call, H.P., Harding, M., and Emeis, C.C., Screening for pectinolytic *Candida* yeasts: optimization and characterization of the enzymes, *J. Food Biochem.*, 9, 193, 1985.
114. Fellows, P.J. and Worgan, J.T., An investigation into the pectolytic activity of the yeast *Saccharomycopsis fibuliger*, *Enzyme Microbiol. Technol.*, 6, 405, 1984.
115. Starr, M.P. et al., Enzymatic degradation of polygalacturonic acid by *Yersinia* and *Klebsiella* species in relation to clinical laboratory procedures, *J. Clin. Microbiol.*, 6, 379, 1977.
116. Liao, C.-H. et al., Genetic and biochemical characterization of an exopolygalacturonase and a pectate lyase from *Yersinia enterocolitica*, *Can. J. Microbiol.*, 45, 396, 1999.
117. Membré, J.M. and Burlot, P.M., Effects of temperature, pH and NaCl on growth and pectinolytic activity of *Pseudomonas marginalis*, *Appl. Environ. Microbiol.*, 60, 2017, 1994.
118. Membré, J..M., Goubet, D., and Kubaczka, M., Influence of salad constituents on growth of *Pseudomonas marginalis*: a predictive microbiology approach, *J. Appl. Bacteriol.*, 79, 603, 1995.
119. Gross, D.C. and Cody, Y.S., Mechanisms of plant pathogenesis by *Pseudomonas* species, *Can. J. Microbiol.*, 31, 403, 1985.
120. Buck, J.W., Walcott, R.R., and Beuchat, L.R., Recent trends in microbiological safety of fruits and vegetables, *Plant Health Prog.*, available at http://www.aps-net.org/online/feature/safety/, accessed Jan. 29, 2003.
121. Wells, J.M. and Butterfield, J.E., Incidence of *Salmonella* on fresh fruits and vegetables affected by fungal rots or physical injury, *Plant Dis.*, 83, 722, 1999.
122. Conway, W.S. et al., Survival and growth of *Listeria monocytogenes* on fresh-cut apples slices and its interaction with *Glomerella cingulata* and *Penicillium expansum*, *Plant Dis.*, 84, 177, 2000.
123. Riordan, D.C.R., Sapers, G.M., and Annous, B.A., The survival of *Escherichia coli* O157:H7 in the presence of *Penicillium expansum* and *Glomerella cingulata* in wounds on apple surfaces, *J. Food Prot.*, 63, 1637, 2000.
124. Liao, C.-H. and Sapers, G.M., Influence of soft rot bacteria on growth of *Listeria monocytogenes* on potato tuber slices, *J. Food Prot.*, 62, 343, 1999.

125. FDA [U.S. Food and Drug Administration], Analysis and evaluation of preventive control measures for the control and reduction/elimination of microbial hazards on fresh and fresh-cut produce, available at http://www.cfsan.fda.gov/~comm/ift3-1.html, accessed Jan. 15, 2002.

126. Beuchat, L.R., Surface decontamination of fruits and vegetables eaten raw: a review, World Health Organization, Food Safety Unit, WHO/FSF/FOS/98.2, available at www.who.int?fsf/fos982~1.pdf, accessed Mar. 2, 2002.

127. Thayer D.W. and Rajkowski, K.T., Developments in irradiation of fresh fruits and vegetables, *Food Technol.*, 53, 62, 1999.

128. Tauxe, R.V., Food safety and irradiation: Protecting the public from foodborne infections, *Emerg. Infect. Dis.*, 7, Supplement June 2001, available at http://www.dcd.gov/ncidod/eid/vol7no3_supp/tauxe.htm, accessed 8/11/2001

129. McGuire, R.G., Response of lychee fruit to cold and gamma irradiation treatments for quarantine eradication of exotic pests, *HortScience* 32, 1255, 1997.

130. Hagenmaier, R.D. and Baker, R.A., Low-dose irradiation of cut iceberg lettuce in modified atmosphere packaging, *J. Agric. Food Chem.*, 45, 2864, 1997.

131. Prakash, A. et al., Effects of low-dose gamma irradiation and conventional treatments on shelf life and quality characteristics of diced celery, *J. Food Sci.*, 65, 1070, 2000.

132. Graham, D.M., Use of ozone for food processing, *Food Technol.* 51, 72, 1997.

133. Restaino, L. et al., Efficacy of ozonated water against various food-related microorganisms, *Appl. Environ. Microbiol.*, 61, 3471, 1995.

134. Kim, J.G. and Yousef, A.E., Inactivation kinetics of foodborne spoilage and pathogenic bacteria by ozone, *J. Food Sci.*, 65, 521, 2000.

135. Kim, J.G., Yousef, A.E., and Chism, G.W., Use of ozone to inactivate microorganisms on lettuce, *J. Food Safety*, 19, 17, 1999.

136. Moor, G., Griffith, C., and Peters, A., Bactericidal properties of ozone and its potential application as a terminal disinfectant, *J. Food Prot.*, 63, 1100, 2000.

137. Xu, L, Use of ozone to improve the safety of fresh fruits and vegetables, *Food Technol.*, 53, 58, 1999.

138. Beuchat, L.R. et al., Efficacy of spray application of chlorinated water in killing pathogenic bacteria on raw apples, tomatoes, and lettuce, *J. Food Prot.*, 61, 1305, 1998.

139. Sapers, G.M. and Simmons, G.F., Hydrogen peroxide disinfection of minimally processed fruits and vegetables, *Food Technol.*, 52, 48, 1998.

140. Gill, C.O. and Tan, K.H., Effect of carbon dioxide on growth of *Pseudomonas fluorescens*, *Appl. Environ. Microbiol.*, 38, 237, 1979.

141. Wells, J.M., Growth of *Erwinia carotovora*, *E. atroseptica*, and *Pseudomonas fluorescens* in low oxygen and high carbon dioxide atmosphere, *Phytopathology*, 64, 1012, 1974.

142. Enfors, S.-O. and Molin, S., Effect of high concentration of carbon dioxide on growth rate of *Pseudomonas fragi*, *Bacillus cereus* and *Streptococcus cremoris*, *J. Appl. Bacteriol.*, 48, 409, 1980.

143. Eyles, M.J., Moir, C.J., and Davey, J.A., The effects of modified atmospheres on the growth of psychrotrophic pseudomonads on a surface in a model system, *Int. J. Food Microbiol.*, 20, 97, 1993.

17 Microbial Metabolites in Fruits and Vegetables

Keith Warriner
Department of Food Science, University of Guelph, Guelph,
Ontario N1G 2W1, Canada

Svetlana Zivanovic
Department of Food Science and Technology,
University of Tennessee, Knoxville, TN

CONTENTS

0-8493-1902-1/05/$0.00+$1.50
© 2005 by CRC Press

17.1   INTRODUCTION

The association of microbes with plants has occurred throughout evolution. Plants have successfully harnessed the benefits derived from microbial associations through establishing symbiotic relationships with, for example, mycorrihiza. However, activity of many other microbes can result in disease (preharvest) and spoilage (postharvest). However, it must be noted that not all plant-associated microbes fall into the broad classification of beneficial or detrimental since many simply coexist.

The microbial ecology of plants is very diverse and has yet to be fully characterized. Nevertheless, a common feature is that those present on plants at harvest can significantly affect the subsequent rate and type of spoilage that occurs. Therefore, when describing the role of microbial metabolites on produce quality the interaction during plant growth needs to be considered. The complexity of microbial (bacteria, yeasts, and molds) populations present on plants depends upon several extrinsic and intrinsic factors (reviewed by Lund, 1992 and Lindow and Brandl, 2003). The surface of leaves represents a harsh environment that exposes cells to extreme variations in temperature, ultraviolet radiation, and relative humidity. The high degree of competition that exists between microbes is an additional barrier to survival along with the activity of preformed and inducible plant defenses. In this respect the enzymes, metabolic products, and other compounds produced by microbes are central to their survival and growth on vegetables and fruits. Bacteria (especially *Pseudomonas* and *Erwinia* spp.) tend to dominate the phyllosphere because they have specially adapted systems to modify the environment in a way that enhances their persistence. The production of fluorescent or pigmented compounds provides a barrier to the high levels of ultraviolet radiation encountered on the surface of leaves. The hydrophobic waxy cuticle of plants can inhibit the movement and accessibility of nutrients to bacterial cells. However, biosurfactants produced by the majority of *Pseudomonas* spp. decrease the water tension, enabling relatively free movement across the leaf surface to nutrient sources and natural openings such as stomata. *Pseudomonas* are also known to release a toxin called syringomycin that can produce holes in the plant cell membrane, allowing access to intracellular nutrients without necessarily resulting in disease symptoms. More prominent spoilage effects are observed due to the release of enzymes such as pectinases into the environment that degrade plant walls. The more subtle effects of microbial metabolites are less well understood. Many of the microbial metabolites that have a negative impact on product sensory quality are derived from secondary metabolism. In its most basic definition, secondary metabolism can cover any pathway that is not essential for cellular function or viability. Common examples of secondary metabolism include volatiles, antibiotics, toxins, and sideophores.

The microbial ecology of vegetables and fruits also plays a role in product degradation. Here the activity of one type of microbe can release compounds that are subsequently catabolized by others. As an added layer of complexity, it should also be noted that many of the metabolic products encountered in microbes can also occur in plants. This naturally makes differentiating microbial spoilage from that of autolytic action problematic. Nevertheless, this chapter aims to provide a broad overview of the impact of microbial metabolites on vegetable and fruit degradation.

This will encompass describing the significant enzymes associated with cell survival and degradation. The impact of primary and secondary metabolites on produce quality will be covered. Finally, aspects of regulation of microbial metabolite production will be discussed with specific emphasis on the role of quorum sensing.

17.2 ROLE AND FUNCTION OF MICROBIAL ENZYMES ASSOCIATED WITH FRUITS AND VEGETABLES

The enzymatic activity of microbes can have adverse effects on vegetables' or fruits' visual appearance and sensory properties. However, it must be noted that many of the spoilage enzymes in microbes are also endogenous to plants. For example, oxidative enzymes such as polyphenoloxidase and peroxidase in unprocessed vegetable and fruit products cause browning or other color changes. Taste and flavor taints are caused by lipid oxidation due to the action of the plant enzyme lipoxygenase. Hydrolytic enzymes that cause softening (i.e., pectinesterase and cellulase enzymes) and amylases that degrade starch are common in plant tissue. With this background activity it can be difficult to assess the significance of microbial enzymes in degradation reactions. However, microbial enzymes do play a key role in surmounting plant defenses and/or gaining access to nutrients.

17.2.1 SUPEROXIDE DISMUTASE, CATALASE, AND GLUTATHIONE

The early Earth was anaerobic with an atmosphere similar to that of Mars and Venus. However, through microbial evolution the first photosynthetic bacteria arose that could harbor the energy emitted by sunlight and further evolve into the complex eukaryotic cells. Critically, the side product of the photosynthetic reaction was oxygen, which accumulated in the atmosphere and enabled the evolution of aerobic life forms. The rise in oxygen concentration proved detrimental to obligate anaerobes, and these had to retreat to specialized, oxygen-free niches within the environment. The success of aerobic and facultative anaerobic microbes can be principally attributed to the energy-rich rewards of utilizing oxygen as an electron acceptor. However, early evolution had to find methods to minimize the toxic side products of oxygen, specifically hydrogen peroxide (H_2O_2) and oxygen-free radicals. Like all organisms, plants have to contend with the toxic effects oxygen radicals. Environmental stresses such as drought, salinity, UV radiation, and ozone cause production of high levels of oxygen-free radicals as a result of increased activity in the mitochondria and chloroplasts (Dat et al., 2003). However, relevant to the theme of this chapter, the plant utilizes the antimicrobial action of hydrogen peroxide as a potent defense mechanism against plant pathogen invasion. Essentially, the early interaction of pathogens with plants stimulates a plasma membrane NADPH oxidase to produce hydrogen peroxide (Breusegem et al., 2002). The main effect is to inactivate the invading pathogen in place by degrading lipid membranes, enzymes, and DNA. The plant is not immune to hydrogen peroxide and localized necrosis occurs, in addition to cross-linking of cell membranes. However, the latter effects also provide a means of localizing the sites of infection, thereby protecting the healthy plant tissue.

As a defense against the plant oxidative burst the microbial invader uses a combination of catalase and superoxide dismutase to inactivate H_2O_2 and O_2^-, respectively. Glutathione-associated enzymes and small proteins, such as thioredoxin and glutaredoxin, work in concert within the cell as antioxidants, maintaining a reducing environment. Therefore, in terms of produce degradation the role of SOD, catalase and glutathione is indirect insofar as it enables the plant pathogens to go forward and cause disease in the host plant. However, SOD, catalase, and glutathione are common features of all aerobes, so there remains uncertainty about the significance of their activity during plant interactions (Lebeda et al., 2001). This is especially true considering that many fungal pathogens produce greater concentrations of hydrogen peroxide during invasion compared to that released by the plant's hypersensitive response.

The antimicrobial properties of hydrogen peroxide could potentially be harnessed to extend the shelf life of vegetables and fruits through suppression of microbial activity. One approach is based on using lactic acid bacteria to produce hydrogen peroxide *in situ* and hence retard the growth of spoilage bacteria such as *Pseudomonas* spp. Although lactic acid bacteria are facultative anaerobes they are devoid of catalase activity, even though hydrogen peroxide is typically produced during aerobic growth. Instead, a flavoprotein, NADH peroxidase, functions to remove hydrogen peroxide, but this tends to be slower than its production and the net result is the accumulation of H_2O_2. This is one reason why the growth of lactic acid bacteria is often greater under anaerobic conditions, where hydrogen peroxide production is negligible, compared to when they are cultivated aerobically. The potential of excess hydrogen peroxide production as a preservation method has been evaluated using *Lactobacillus delbrueckii* subsp. *lactis* (Harp and Gilliland, 2003). The model systems studied were broccoli, cabbage, carrots, and lettuce inoculated with *Escherichia coli* and *Listeria monocytogenes*. The workers found that populations of *L. delbrueckii* subsp. *lactis* remained at high levels during storage but there was no noticeable antagonistic action against the pathogens. The reason for the limited efficacy of the hydrogen peroxide-producing strain was probably caused by the action of endogenous catalase activity from the vegetables.

17.2.2 LACCASESE, POLYPHENOL OXIDASES, AND LIPOXYGENASE

Laccasese catalyzes the reduction of O_2 to H_2O using a range of phenolic compounds as hydrogen donors. The enzyme shares a number of suitable substrates with monophenol monooxygenase, catachol oxidase, and polyphenol oxidase. The common substrates shared by all these enzymes can make direct identification of activity problematic. It is well established that laccasese and polyphenol oxidases are widely distributed in plants, playing a significant part in enzymic browning and flavor. The same enzymes are also widely distributed in fungi, including the human pathogen *Cryptococcus neoformans*. It is thought the oxidation of aromatic substances within the body by *C. neoformans* plays a central role in protecting the organisms against the immune system (Liu et al., 1999). How the enzymes contribute to vegetable and fruit degradation remains obscure. It is known that fungal polyphenol oxidases have an antioxidant effect in juice systems by removing residual oxygen present, thus preserving flavor and color. A more indirect effect is derived from the enzyme found in mycorrhizal fungi that is thought to contribute to nutrient recycling in the soil, thereby enhancing plant development (Burke and Cairney, 2002).

17.2.3 Pectinases

Pectic substances, commonly called pectins, are a heterogeneous group of polysaccharides composed of a partially methylated polygalacturonic backbone with possible rhamnose, arabinose, and xylose side groups. Protopectin is a water-insoluble, high-molecular-weight molecule with numerous branches. It is usually referred to as a "parent" molecule since it is the principal form that occurs in nature. Pectic substances, mainly protopectins, form middle lamella in plant tissues, which keeps cells together. Pectins also make up about one-third of the dry weight of plant cell walls, where they occur as amorphous, gel-like structures and serve as cementing material between cellulose and hemicellulose fibers. During industrial extraction of pectins, side chains of protopectin are eliminated; commercial pectins are mainly linear α-1,4-galacturonans with defined degree of esterification.

Microorganisms do not contain pectins in their cell walls regardless of the species. However, in the absence of other carbon sources, microorganisms, both saprophytes and parasites, may hydrolyze pectins and utilize them as nutrients. Furthermore, enzymatic degradation of plant cell walls is the main pathway of attack by plant pathogens.

Pectin-degrading enzymes are widespread in nature (reviewed by Gummadi and Panda, 2003). Endogenous plant pectinases have an important role in fruit growth and ripening. Depending on the reaction they catalyze, pectin-degrading enzymes are categorized as hydrolyses (e.g., polygalacturonase, polygalacturonosidase), lyases (e.g., pectin lyase, pectate lyase), and esterases (e.g., pectin methylesterase, pectin acetylesterase).

Protopectinases (PPases) are hydrolytic enzymes that solubilize protopectin. Plant endogenous protopectinases are triggered during maturation and result in softening of the fruits. Microbial protopectinases, isolated from yeast and fungal species, mainly *Aspergillus, Geotrichum,* and *Trichosporon,* exhibit two forms of action. Protopectinase A-type acts as an endoenzyme, splitting glycosidic bonds within the polygalactronic region, while B-type PPase hydrolyze side chains and cleave the link between protopectin molecule and other constituents in the wall (Sakai, 1992).

Pectinesterase (EC 3.1.1.11; pectin pectylhydrolase, pectin methylesterase; PE; PME) release methanol from methylated galacturonase units. Pectin methylesterase is a common enzyme in plants and microorganisms. However, plant PMEs remove methyl groups in blocks, leaving "pouches" of deesterified galacturonic acid residues, while fungal PMEs cause random deesterification, resulting in random distribution of unmethylated galacturonic acids along the pectin chain (Benen and Visser, 2003). This enzymatic reaction is being used in industry to improve the consistency of tomato products. After deesterification by fungal pectinesterases, carboxylic groups at galacturonic acid residues became available to bind divalent calcium ions and increase, for example, viscosity of tomato paste (Voragen and Pilnik, 1989).

Pectin depolymerizing enzymes, usually referred to as pectinases, include hydro-lases that require water as a reactant and lyases that cleave glycosidic bonds by β-elimination reactions. Both groups may act either as endo- or exoenzymes. Pectinases are usually acidophilic, with optimum pH in the range of 3.5 to 5.5, and have extensive application in the extraction and clarification of fruit juices and wine (Pretel et al., 1997). Some alkaline pectinases have been isolated from *Bacillus* and *Pseudomonas* spp. and are used in the pretreatment of waste water from vegetable processing and degumming and processing of plant fibers (Tanabe et al., 1988; Cao et al., 1992; Bruhlmann et al., 2000).

Polygalacturonase (PG) exist as endo-PG (EC 3.2.1.15; poly(1,4)-α-D-galactu-ronide glycanohydrolase) and exoPG (EC 3.2.1.67; poly(1,4)-α-D-galacturonide galacturonohydrolase). PGs cleave the α-(1,4) glycosidic bond but require demeth-ylated galacturonic acids in order to act. EndoPGs are widely produced by plants and microorganisms but not by animals (except snails). Endogenous plant endoPGs are responsible for softening during fruit maturation, while fungal growth on fruits and vegetables results in localized softening and spoilage due to rapid mercerization and liquefaction of the product. ExoPG, although it occurs in various fruits and vegetables, has only been detected in a few fungi and bacteria such as *Clostridium thermosaccharolyticum*, *Erwinia chrysanthenu*, and *Ralstonia solancearum* (Huang and Allen, 1997; Shevchik et al., 1999; van Rijssel et al., 1993). ExoPGs attack the pectin molecule from the nonreducing end, liberating one galacturonic acid molecule at a time.

Pectic lyases (*trans*-eliminases) cleave α-1,4 glycosidic bonds but do not involve water. The catalyzed reaction occurs by the mechanism of β-elimination. Pectin lyases (EC 4.2.2.10; PL) act as endoenzymes, preferentially splitting highly esterified pectin, while pectate lyases (EC 4.2.2.2; PAL) act on nonesterified or low esterified pectate and exist in both endo and exo forms. Lyases have not been found in plants but are common in microorganisms. Pectin lyases are predominantly of fungal (e.g., *Asperigillus* spp.) and pectate lyases of bacterial (e.g., *Bacillus* and *Erwinia* spp.) origin. Since both polygalacturonases and lyases result in softening of fruits and vegetables or decreased viscosity of pectin-rich products, determination of the 4,5 unsaturated end products of β-elimination indicate microbiological sources of the enzyme.

17.2.4 CELLULOSE- AND HEMICELLULOSE-DEGRADING ENZYMES

Cellulose is a major structural component in plant cell walls. It is a large, linear polysaccharide made exclusively of β-1,4 linked glucose units. Long cellulose mol-ecules, being inherently rigid and extended, arrange themselves in a parallel mode and develop extensive hydrogen bonds. This results in formation of crystalline microfibrils with about 70 cellulose chains per fibril. However, small, noncrystalline regions in the fibrils may exist due to the presence of other sugar residues entangled within cellulose chains. In plant cell walls, the cellulose fibers are embedded in an amorphous complex of branched and linear mannans, xylans, arabans, and galactans, called hemicellulose, which provide support for the structural network. This structure is particularly important since highly crystalline fibrils resist hydrolytic enzymes and solubilization in common solvents, while noncrystalline regions are easily accessible and represent the point of enzymatic and chemical attack.

A group of enzymes that degrade cellulose are referred to as cellulases and enzymes that hydrolyze hemicellulose are called hemicellulases. Cellulases include numerous endo- and exoglucanases, but none of them can individually degrade crystalline cellulose fibers. The degradation starts with the C_1-enzyme, which "opens up" the cellulose matrix, making it more amorphous and accessible to hydrolysis by the C_x-enzymes (Reese, 1956; Mansfield et al., 1999). C_1-Cellulase is thought not to be a hydrolytic enzyme but rather to be involved in the disruption of intercellular hydrogen bonds of cellulose fibers. On the other hand, C_x-cellulases are true hydrolases with affinities towards different substrates. Endo-C_x-cellulase attacks the long-chain glucans and has no activity on disaccharide cellobiose, while exo-C_x-cellulases show greater activity towards short oligomers. Thus, exo-β-1,4 glucan glucohydrolase (glucohydrolase; EC 3.2.1.74) hydrolyzes single glucose units from the nonreducing end of soluble cellulose, while exo-β-1,4 glucan cellobiohydrolase (cellobiohydrolase; EC 3.2.1.91) catalyzes the liberation of cellobiose from the nonreducing, and possibly the reducing, end of the polymer (Wood and McCrae, 1979). Finally, β-glusosidase or cellobiase (β-1,4 oligoglucan glucohydrolase; EC 3.2.1.21) produces monomeric glucose from cellobiose. Interestingly, those cellulases exhibiting an exoglucanase mode of action generally have a tunnel-shaped molecular structure, while the more randomly acting endoglucanases have a more "cleft-shaped" catalytic domain (Warren, 1996).

Cellulases, especially C_x-cellulases, occur in numerous higher plants, and their activities increase during fruit ripening. Nevertheless, the extracts of the enzymes cannot degrade cellulose fibers and therefore endogenous cellulases are not considered to be a significant factor in tissue softening and degradation, especially compared to pectinases. However, various microorganisms, aerobic and anaerobic, produce cellulolytic enzymes. Plant pathogens, for example, secrete high levels of cellulases along with pectolytic and hemicellulolytic enzymes. This combination of enzymatic activity presents an excellent tool for degradation of plant cell walls and invasion of plant tissue. Besides degrading the support and protective structure in the plant cell, these enzymes completely hydrolyze the polysaccharides into soluble sugars, providing nutrients for all microorganisms present and resulting in extensive softening and spoilage in fruits and vegetables (De Vries and Visser, 2001). On the other hand, bacteria such as *Bacillus* and *Pseudomonas* secrete mainly endoglucanases that randomly cleave cellulose. In order to be completely utilized, produced oligomers have to be further processed by the wall-bound or intracellular enzymes in the bacteria, which, consequently, make them less available to competing microorganisms (Shallom and Shoham, 2003). Although bacteria are not considered to be as efficient cellulose decomposers as fungi, there are some exceptions. *Clostridium thermocellum* possesses several very efficient endocelluloses, exocelluloses, and hemicellulases (Beguin et al., 1988; Lamed and Bayer, 1988). However, it can grow only on cellobiose and cannot utilize xylose, arabinose, mannose, and other sugars derived from hemicellulose. One possible explanation for this is that the hemicellulases degrade surrounding polysaccharides, making cellulose more accessible to cellulose-degrading enzymes (Lamed and Bayer, 1988).

Microbial cellulases are inducible enzymes, which means that these enzymes will be produced when cellulose is present in the medium (Reese et al., 1969; Kubicek et al., 1993). However, this raises the question of how the crystalline, highly insoluble cellulose can induce enzyme synthesis, which is a strictly intracellular process, without entering the microbial cell. It is highly possible that microorganisms constantly produce low levels of cellulolyitc enzymes. These enzymes produce soluble cellobiose, cellobiono-δ-1,5-lactone, and short cellulose oligomers that can enter the cell and promote cellulase gene expression (Kubicek et al., 1993). Hemicellulases are a diverse group of enzymes and include xylanases (EC 3.2.1.8), β-mannanases (EC 3.2.1.78), α-L-arabinofuranosidases (EC 3.2.1.55), and α-L-arabinanases (EC 3.2.1.99), among others. Similar to degradation of cellulose, hydrolysis of hemicellulose is carried out by microorganisms that can be found either freely in nature, mainly plant pathogens, or as a part of the digestive tract of ruminant animals. Although cellulose- and hemicellulolose-degrading enzymes are mainly produced by microorganisms that grow and cause spoilage of fruits and vegetables, their optimum pH is generally between 4.5 and 6.5. Their optimum temperature range is between 35 and 50°C, but cellulases from many sources show thermostability, which may have detrimental effects on insufficiently processed plant tissues.

In the food industry, commercial cellulases are usually used in combination with pectinases and hemicellulases. Their applications include the extraction and clarification of fruit juices, extraction of oils from seeds, isolation of proteins from soybeans and starch from corn, improving the rehydrability of dried vegetables, removal of cell walls to facilitate the release of flavors and enzymes, and the production of soluble sugars and alcohol from cellulosic wastes (Bhat and Bhat, 1997).

17.2.5 STARCH-DEGRADING ENZYMES

Starch is the main reserve polysaccharide in plants and presents the chief carbon source for animals, plants, and microorganisms. However, having thousands of glucose units in its structure, it is too large to be transported through the cell membrane and enter the cell. When there are no other, preferable, carbon sources available in the surrounding media, microorganisms begin to produce extracellular starch-degrading enzymes to provide necessary nutrients. Depending on their activity on starch molecules, these enzymes can be classified as endo- or exoenzymes. Endoenzymes, such as α-amylase or pullulanase, hydrolyze glycosidic bonds randomly within the starch molecule, while exoenzymes, such as β-amylase or glucoamylase, cleave molecule of maltose or glucose one by one from the nonreducing end of the polysaccharide (Hyun and Zeikus, 1985). Generally, endoenzymes cause rapid loss in viscosity while exoenzymes do not affect viscosity but increase sweetness.

α-Amylase (EC 3.2.1.1; α-1,4-D-glucan glucanohydrolase) is the main enzyme in carbohydrate metabolism in microorganisms, animals, and plants. It randomly hydrolyzes α-1,4 glycosidic bonds in amylase and amylopectin but has no effect on α-1,6 branching points. The resulting molecules are α-limit dextrins of various lengths, and, if the reaction is prolonged, maltose and even glucose. Due to the rapid decrease of molecular weight of starch molecules, the viscosity of their dispersions significantly decreases.

Although most, if not all, bacterial and fungal species produce α-amylase, *Bacillus cereus, B. subtilis, B. stearothermophilus, Lactobacillus cellobiosus, Streptococcus* spp., *Clostridium butiricum, Cl. thermosaccharolyticum, Aspergillus niger, A. oryzae, Fusarium oxysporum, Candida japonica,* and *Pichia polymorpha* produce the enzyme in significant amounts (Antranikian, 1992). Industrial α-amylase is usually obtained from *Bacillus* spp. (Pariza and Johnson, 2001). It should be noted that the optimum temperature for α-amylase activity is usually higher than the optimum temperature for growth of the microorganism that produces the enzymes. Thus, some bacterial α-amylases are active at temperatures as high as 90°C (Asther and Meunier, 1990; Maesmans et al., 1994). Generally, the optimum temperature for bacterial α-amylases ranges from 20 to 90°C, while α-amylases from fungal species require only 25 to 50°C for optimal activity.

Inhibitors of α-amylase have been extracted from white kidney beans (*Phaseoulus vulgaris*) and roselle (*Hibiscus sabctarifla*). Their optimum temperature was determined to be 40°C. Interestingly, the inhibitors are effective only towards animal α-amylase and not toward enzymes of plant or microbial origin (Hansawasdi et al., 2000; Chun et al., 2001).

Similarly to α-amylase, β-amylase (EC 3.2.1.2; α-1,4-glucan maltohydrolase) hydrolyzes only α-1,4 bonds, but this enzyme eliminates one maltose molecule at a time starting from the nonreducing end of amylase or amylopectin. It is commonly found in plants and some bacteria, but it is relatively rare in fungi. Good bacterial producers of β-amylase are *Bacillus* spp. and *Cl. thermosulfurogenes* (Wang and Yu, 1994; Huang and Allen, 1997; Mohan-Reddy et al., 1999); the optimum temperature for their enzymatic activity ranges from 30 to 75°C.

Another starch-degrading exoenzyme, glucoamylase (EC 3.2.1.3; α-1,4-D-glucan glucohydrolase; amyloglucosidase), liberates glucose molecules from the nonreducing ends of starch molecules. The reaction is fast on α-1,4 linkages but significantly slower on α-1,6 bonds. Glucoamylase is often detected in fungi, especially in *Aspergillus* spp., and yeast such as *Pichia, Candida,* and *Saccharomyces*. Glucoamylase is, on the other hand, rare in bacteria but has been detected in *B. stearothermophylus* and *Cl. therosaccharolyticum.* Similarly, α-glucosidase (EC 3.2.1.20; α-D-glucoside glucohydrolase; maltase) releases one glucose molecule from the nonreducing end, with a higher affinity towards oligosaccharides than towards high-molecular-weight molecules. Contrary to glucoamylase, α-glucosidase is widely distributed not only in fungi and yeasts but in bacteria as well. Both glucoamylase and α-glucosidase have optimum temperatures above 40°C, with extremes of 70 and 115°C for *Cl. thermohydrosulfuricum* and *Pyrococcus furiosus*, respectively (Costantino et al., 1990; Antranikian, 1992). However, contrary to α- and β-amylases, which exhibit their highest activity under neutral pH conditions, glucoamylase and α-glucosidase require acidic pH.

The starch disbranching enzyme pullulanase (EC 3.2.1.41; α-dextrin-6-glucono hydrolase) hydrolyzes only amylopectin's α-1,6 bonds, resulting in the accumulation of oligomers of various lengths. This enzyme has been detected in bacteria and plants (Warner and Knutsen, 1991; Stathopoulos et al., 2000; Schindler et al., 2001).

B. acidopullulyticus has a characteristically high pullulanase activity and is used for large-scale production of the enzyme (Kusano et al., 1990).

Starch-degrading enzymes of microbial origin have an important role in the production of corn syrups, bakery products, alcohol, and alcoholic beverages such as beer and whiskey. In addition to food industry applications, thermoresistant α-amylase is a common ingredient in laundry detergents.

17.2.6 ENZYME TRANSPORT IN MICROBES

Polymeric materials of plants (e.g., starch and pectin) must be degraded into transportable units in order to be assimilated by the microbial cell. To achieve this the microbes release enzymes (exoenzymes) to attack the plant cell walls, etc. It is interesting to note that more than 50% of the protein synthesized within microbial cells is exported. This underscores the significance of enzyme secretion in cells, which, if controlled, could provide a novel method of enhancing the shelf life of produce.

Within bacteria there are three secretory systems, commonly termed Types I, II, and III. The Type I system is characterized by transferring preformed enzymes (e.g., protease) in a single step from the cytoplasm of the cell.

Type II secretion systems are relevant to plant degradation; the enzymes pectinase and cellulase are transported via this system. In contrast to the Type I system, the Type II system involves a two-step mechanism. The protein within the cytoplasm is transported to the periplasm via a sec-dependent protein transporter from which it passes through a cytoplasmic membrane spanning structure to the outside of the cell (reviewed by Sandkvist, 2003).

The enzymes are synthesized as precursors with additional amino-terminal sequences, called signal sequences, that designate the protein for export. These signal sequences are eventually cleaved via a peptidase at the sec transporter. The protein is transported into the periplasmic space, where it may undergo further modifications such as subunit assembly, before it is translocated across the outer membrane. Type II transport is an energetic process involving ATP hydrolysis via a sec ATPase.

The Type III secretion system can be best described as a tubelike structure (200 nm in length) that spans both the cytoplasmic membrane of Gram-negative phytopathogens and cell walls of plants. Interestingly, a very similar system is used in animal pathogens, which suggests a common evolutionary link (reviewed by Innes, 2003). Similar to the mechanism in animal pathogens, the main role of Type III secretion systems is to interface with the host cell so that effector molecules and enzymes (e.g., proteases) can be released into the plant to alter cellular function. For example *Pseudomonas syringae* pv. *tomato* DC3000 transfers proteins into plants that inhibit the hypersensitive response that would otherwise result in an attack on the invading pathogen. Other proteins also nullify plant defense systems, enabling the invading pathogen to become established (Buttner and Bonas, 2003).

17.3 MICROBIAL EFFECTS ON PRODUCE FLAVOR

17.3.1 MICROBIAL METABOLITES THAT ENHANCE PRODUCE QUALITY

Although many microbial metabolites can be considered detrimental to the quality of fruits and vegetables, there are examples of positive effects. Methylotrophs (e.g., *Methylobacterium*) have specifically been identified as enhancing the flavor volatiles in fruit. Due to their characteristic pink colonies they are sometimes referred to as PPFMs (pink-pigmented facultative methylotrophs). A common feature of methylotrophs is their ability to grow on methanol and methylamine as well as on a variety of C_2, C_3, and C_4 compounds. Methylotrophs can readily become established on plants by utilizing the methanol that accumulates on leaf surfaces as a consequence of plant metabolism.

Although common inhabitants of plants, they were previously considered to be neutral, having no adverse or positive effects. However, it has been established that some *Methylobacterium* strains can produce plant regulators such as cytokinins, PQQ, urease, and polyhydroxybutyrate that stimulate plant development. In addition, because of their relatively good growth on leaf surfaces, their role in suppressing phytopathogens through competitive inhibition has been observed (Romanovskaya et al., 2001). Interestingly, the interaction of *Methylobacterium extorquens* with growing strawberry plants also has a positive impact on fruit flavor. The characteristic flavor of strawberries is produced in part by the concentration of 2,5-dimethyl 4-hydroxy 2H-furan 3-one (DMHF). This compound is synthesized by the plant from 1,2-propanediol. It has been demonstrated that the activity of *Methylobacterium extorquens* significantly increases the concentration of DMHF in the subsequent fruit. Although not fully investigated, it is thought that the bacterium oxidizes 1,2-propanediol to form DMHF, in addition to other precursors, that the plant's biosynthetic apparatus converts to other flavor furanines (Zabetakis, 1997; Zabetakis et al., 1999). Therefore, enhancing fruit's flavor by introducing methylotrophs to growing plants would be a simple, environmentally acceptable means of improving produce quality.

The odorous secondary metabolites produced by a diverse range of fungi are known to include esters, terpens, alkanes, and fatty acids. Most provide fruity odors that are perceived as beneficial for fruit fermentation. For example, *Saccharomyces cerevisiae* and *Kluyveromyces marxianus* both produce low levels of 2-phenylethanol and terpenes that enhance fruity odors during fermentation. Interest in deriving bioflavors and fragrances from microbial metabolism has been expressed (reviewed by Vadamme and Soetaert, 2002), but its impact on fresh produce quality remains obscure.

The products derived from vegetable and fruit fermentation is a further example of how produce quality can be enhanced (Steinkraus, 1996). Favorable fermentation of vegetables and fruits are performed by yeasts (alcoholic fermentations) and lactic acid bacteria (Table 17.1). One of the most popular fermented vegetable products is sauerkraut, produced by the fermentation of cabbage. Typically a combination of *Leuconsotoc mesenteroides, L. plantarium, L. cucumeris, and L. pentoaceticus* is used to produce desirable product flavor. The end products of sauerkraut fermentation are lactic acid along with smaller amounts of acetic and propionic acids and carbon dioxide, in addition to a mixture of aromatic esters. The acids, in combination with alcohol, form esters, which contribute to the characteristic flavor of sauerkraut.

Vegetables can also be fermented by *Bacillus* (*B. pumilus, B. licheniformis*), but alkalization, rather than acidification, occurs. The alkaline pH is produced by the deamination of amino acids by the action of the *Bacillus* protease enzyme. Due to

TABLE 17.1
Microbes Recovered in Fermented Fruits and Vegetables

Microbe	Reaction
Bacteria	
Acetobacter aceti	Oxidizes a range of organic compounds including ethanol to acetic acid
Streptococcus faecalis	Homofermentative producing predominantly lactic acid
Leuconostoc mesenteroides	Heterofermentative producing lactate, acetate, ethanol, and carbon dioxide from sugar fermentation
Pediococcus cerevisiae	Homofermentative
Lactobacillus plantarium	Homofermentative
Lactobacillus brevis	Heterofermentative
Yeast	
Saccharomyces cervisiae	Produces ethanol under anaerobic conditions

the need for a high protein content to sustain the fermentation the vegetables used in this process are typically soybeans and other legumes.

The fermentation of fruits and vegetables is considered beneficial due to the historical acceptance of such products. However, the same fermentations by wild yeast and contaminating bacteria can lead to spoilage via production of pigmentation, secondary alcohols, and other volatile products such as diacetyl.

17.3.2 MICROBIAL METABOLITES AND VEGETABLE SPOILAGE

The spoilage of fresh (unfermented) vegetables and fruits is dependent on the type of produce and the preservation method applied. As noted previously, the main action of endogenous lytic plant enzymes plays a significant role in limiting shelf life. Indeed, the autodegradation of plant tissue typically releases nutrients upon which spoilage microflora can grow.

Vegetables stored under aerobic conditions are spoiled by the action of Gram-negative obligate aerobes (e.g., *Erwinia* and *Pseudomonas* spp). Spoilage by *Pseudomonas* spp. is often caused before significant visual damage has occurred due the accumulation of volatile substances (off-odors) such as ammonia and hydrogen cyanide. These are side products resulting from the metabolism of plant-derived nutrients and both have a very low detection threshold. Through prolonged storage the action of microbially derived pectinases, along with endogenous plant enzymes, causes visible rots.

Actinomycetes has been implicated in the spoilage of both beans and mushrooms due to the formation of 2-methylisoborneol that results in a musty off-flavor. The growth of *Actinomycetes* on vegetables is an obvious source of 2-methylisoborneol but it can also occur if contaminated processing water is used.

Rotting potatoes have a very distinctive odor which is often likened to that of pig farms. The odor is attributed to the formation of skatole, indole, and *p*-cresol. The common spoilage bacteria associated with potato rot are the aerobes *Erwinia carotovora* and *E. chrysanthemi*. Under anaerobic conditions *Clostridium scatologenes* can grow and produce significant amounts skatole, indole, and *p*-cresol. The contamination of potatoes typically occurs in the field, augmented by tissue damage and plant stress.

Modified atmosphere packaging (MAP) has been extensively employed to enhance the shelf life of fresh-cut produce. By modifying the atmosphere to contain high oxygen and carbon dioxide the activity of aerobes can be retarded. However, this in turn leads to a different form of spoilage, typically involving yeasts and lactic acid bacteria. Yeast spoilage in MAP occurs through pectinase activity, gas production, off-flavor development, and visible colony formation (Flett, 1992). In vegetables with a high sugar content (e.g., root crops) lactic acid bacteria can become established and cause spoilage. A study performed by Jacxsens et al. (2003) provides a good example of the role of different microbes on the spoilage of MAP vegetables. With celeriac and grated carrots under modified atmosphere (3% O_2 and 5% CO_2) initial spoilage events are associated with degradation of plant cell walls by yeast and other pectinase-expressing microbes. The autolytic action of endogenous plant enzymes augments the further breakdown of cell walls, releasing glucose, fructose, and sucrose. The sugars support the growth of lactic acid bacteria that typically produce polysaccharides (slime production), CO_2, and lactic and acetic acid, in addition to diacetyl.

17.3.3 MICROBIAL METABOLITES AND FRUIT SPOILAGE

The most significant spoilage agent of fruit is fungal activity. Typically, under conditions that do not favor bacterial growth (e.g., high acidity) fungi can proliferate with visible deleterious effects. Among the significant fungal diseases, blue mold caused by *Penicillium expansum*, gray mold caused by *Botrytis cinerea*, and Mucor rot caused by *Mucor piriformis* are common on fruits.

Molds such as *P. italicum* and *P. digitatum* cause spoilage of fruits via the formation of pigmented rots, in addition to volatiles such as 4-vinylguaiacol. Methyl esters of acids such as butanoic acid also lead to sulfurous and rancid odors due to a combination of mold and fruit autolytic pathways. Apples are frequently spoiled by *P. expansum*, which causes discoloration and a strong aroma due to the formation of geosmin.

Aspergillus flavus, *Eurotium oxysporum*, and *Penicillium* species have all been implicated in dried fruit taints caused by formation of 2,4,6-trichloroanisole from 2,4,6-trichlorophenol. The odor of 2,4,6-trichloroanisole can be described as phenolic, iodine, or musty and can be perceived at levels on the order of 20 ppm. However, there is debate as to whether 2,4,6-trichloroanisole recovered in fruit is derived from microbial activity, since it has been found in vine fruit devoid of both microorganisms and the chlorophenol precursor. This doubt was confirmed by the fact that dried fruit sterilized using either propylene oxide or hydrogen peroxide produced the highest concentrations of 2,4,6-trichloroanisole. This further underlines the difficulty of establishing the impact of microbial metabolites on produce quality. A similar mold metabolite that has the potential to cause a musty taint in packaged food is 2,4,6-tribromoanisole, derived from the metabolic breakdown of the fungicide 2,4,6-tribromophenol.

Xerophilic fungi (e.g., *Eurotium amstelodami, E. chevaliere, E. herbariorum,* and *P. citrinum*) have been associated with the formation of ketone odors on desiccated coconut. The ketones are a result of a modified β-oxidation of fatty acids that produces 2-pentanone, 2-heptanone, 2-nonanone, 2-undecanone, 2-heptanol, and 2-nonanol. Desiccated coconut stored under high relatively humidity enables the growth of *Baciilus subtilis*, which produces secondary metabolic products such as 2,3,5,6-tetramethylpyrazine and 2,3,5-trimethylpyrazine that give the product a pungent aroma.

Acid-tolerant lactic acid bacteria can take part in fruit spoilage by the production of volatile flavor compounds in products such as citrus juice. The formation of neutral compounds (e.g., acetoin) is favored in fruit juices due to the presence of fermentable substrates and high acidity. The low pH causes a switch from acid production to that of neutral compounds, thereby preventing autoinhibition by lactate or acetate accumulation.

Taints in fruit can be caused by the formation of halophenols such as 2, 6-dichlorophenol. Halophenols are formed by the reaction of a phenolic precursor with hydrogen peroxide catalyzed by haloperoxidase. Like many enzymes, haloperoxidase is widely distributed in plants but is also present in the acidophile spore-forming bacterium *Alicyclobacillus acidoterrestris*. The key difference of the bacterial enzyme is that it does not require cofactors or halide ions for activity. There is debate on the significance of the halophenols formed by *Ali. acidoterrestris* due to the presence of the reaction pathway in fruits. Nevertheless, it has been demonstrated that fruit juice inoculated with *Ali. acidoterrestris* accumulates 2,6-dibromophenol and 2,6-dichlorophenol, confirming its potential role in spoilage (Jensen and Whitfield, 2003).

17.3.4 ANTIMICROBIAL AGENTS FROM MICROORGANISMS

The microbial metabolites that cause adverse quality changes in fruits and vegetables are of key significance. However, it must be noted that not all associated microbes are detrimental to plants. Microbial populations on plants can provide a defense against phytopathogen attack through the production of antimicrobial compounds. This has prompted the possibility of using such beneficial microbes as biocontrol agents in pre- and postharvest operations as an alternative to chemical-based treatments. In the following section the main antimicrobial agents produced by microbes during their interaction with plants will be described. Although not directly related to produce degradation, such compounds can indirectly improve quality.

17.3.4.1 Antibiotics

Antibiotics encompass a chemically heterogeneous group of organic, low-molecular-weight compounds produced by microbes. Mass-produced antibiotics have been used in horticulture for many years to control plant disease (Anjaiah et al., 1998; McManus et al., 2002). However, it should be noted that many plant-associated microbes can also produce antibiotics via secondary metabolism. For example, numerous strains of antibiotic-producing *Pseudomonas* spp. have been isolated from roots of various plants grown in soils from diverse geographical regions (Raaijmakers

TABLE 17.2
Antibiotics Produced by Plant-Associated Bacteria

Antibiotic	Species/Strain	Origins	Ref.
DAPG	*Psuedomonas* spp.		
	Q2-87	Wheat	Vincent et al., 1991
	CHAO	Tobacco	Keel et al., 1992
	F113	Sugar beet	Shanaham et al., 1992
DDR	*Pseudomonas borealis*	Wheat	Hokeberg et al., 1998
Phenazines	*Pseudomonas* spp.		
	PNA1	Chickpea	Bruhlmann et al., 2000
	30-84	Wheat	Pierson and Thomashow, 1992
Kanosamine	*Bacillus cereus*	Alfalfa	Milner et al., 1996
AFC-BC11	*Burkholderia cepacia* BC11	Cotton	Kang et al., 1998
Oomycin A	*Pseudomonas fluorescens* Hv37a	Barley	Gutterson et al., 1986

et al., 2002). Indeed, the rhizosphere of plants is typically dominated by antibiotic-producing strains that can become dominant in such a competitive environment. Pyoluteorin, pyrrolnitrin, phenazines, and 2,4-diacetylphloroglucinol are the most significant classes of antibiotics produced by rhizosphere pseudomonads (Shanaham et al., 1992). Antibiotics produced by plant-associated microbes have a wide spectrum of antagonism toward viruses, bacteria, and fungi (Table 17.2). Among the soil-borne fungi, *Gaeumannomyces graminis* var. *tritici*, *Pythium ultimum*, and *Thielaviopsis basicola* can be controlled by *Pseudomonas* strains producing 2,4-diacetylphloroglucinol. Some strains produce a single type of antibiotic whereas others such as *P. fluorescens* CHA0 produce multiple types, thereby extending the antagonistic spectrum of the bacterium.

Antibiotic production by *Pseudomonas* is complex and dependent on a wide range of factors. For example, plant stress, nutrient availability, temperature, and signal molecules released in plant root exudates all affect antibiotic production. A most striking example of the latter is antibiotic production in *Baciilus cereus*, which increases over 300% in the presence of alfalfa extracts (Milner et al., 1996). In general, the production of antibiotics is favored in low-stress, nutrient-rich environments.

The application of antibiotic-producing strains in the field is a contentious issue since it is well established that overexploitation in the medical field has led to multi-drug-resistant strains of human pathogenic bacteria. To deliberately add antibiotic-producing strains to fresh-cut produce or fruit would probably work in the short term to extend shelf life but eventually resistant strains would likely evolve. Attempts to introduce antibiotic-producing strains into the rhizosphere have to date met with limited success. Many strains show variable levels of antibiotic production in the field environment. To be successful as a biocontrol agent the introduced bacterium must become established and express its antibiotic phenotype. Typically an introduced bacterium competes relatively poorly in the presence of an established competitive microflora and cannot colonize the roots in sufficient numbers to exert its biocontrol effect.

TABLE 17.3
Selection of Bacteriocins Produced by Bacteria

Bacteriocin	Target Organisms	Ref.
Nisin A	*Listeria monocytogenes*	Cutter and Siragusa, 1998
Pisciolin 126	*Listeria monocytogenes*	Jack et al., 1996
Pedicon AcH	*Listeria monocytogenes*	Goff et al., 1996
Enterocin	*Listeria monocytogens*	Aymerich et al., 2000
Colicins	*Coliforms*	Lazdunski, 1998
Pyrocins	*Pseudomonas aeruginosa*	Michel-Briand and Baysse, 2002
Agrocin 84	*Agrobacterium*	Penyalver and Lopez, 1999

17.3.4.2 Bacteriocins

The biopreservative effect of lactic acid bacteria is well established and has been exploited in fermentations since early civilization. Of course, the main mode of enhancing food safety and stability by fermentation is through the production of organic acids that reduce the pH of the food matrix. As discussed previously, the generation of antimicrobial oxygen species can also contribute to suppressing microbial growth of spoilage microbes. However, a number of lactic acid bacteria also produce small peptides that have antibacterial properties and are termed *bacteriocins* (Table 17.3). Characteristically, bacteriocins have a narrow spectrum of activity (compared to antibiotics) centered about homologous (closely related) species. To date the majority of studies of bacteriocins have focused on antimicrobial properties against virulent pathogens such as *Listeria monocytogenes* and *Clostridium* spp. (reviewed by Cleveland et al., 2001). The role of bacteriocins in preventing spoilage of vegetables and fruits has received less attention. This is probably based on the poor antimicrobial activity of bacteriocins produced by lactic acid bacteria on key spoilage microbes such as pseudomonads. Although the bacteriocins from lactic acid bacteria have attracted the most attention, it should be noted that Gram-negative bacteria also produce antimicrobial peptides. Indeed, the very first bacteriocins isolated were colicins from *E. coli*. The mode of action of the majority of antimicrobial peptides is via disruption of cell membranes or nucleic acids.

In terms of suppressing plant pathogens attention has been given to a group of bacteriocins (pyocins) produced by *Pseudomoas* spp. The capacity to produce pyocins is widespread in *P. aeruginosa* species, but they are only produced in significant amounts when cultures are exposed to mutagenic agents, for example, UV radiation or mitomycin C (Michel-Briand and Baysse, 2002). In broad terms there are three types of pyocin produced by *P. aeruginosa*. The R2 (R-type) and F2 (F-type) pyocins are phage tails, both of which are resistant to nuclease and protease. S-type pyocins are thought to act as nucleases and kill cells by breaking down the DNA of susceptible cells. It should be noted that pyocins produced by other pseudomonads are different from those of *P. auruginosa*, although the narrow specificity is a common feature (Parret and de Mot, 2002; Parret et al., 2003).

The physiological role of pyocins is unclear but is likely to decrease potential competition in niches such as the rhizosphere. This has promoted their use as a potential biocontrol agent for plant pathogens, although studies have only met with limited success. As with many bacteriocins the narrow specificity range is a key problem along with the generation of resistant strains. The need for DNA-damaging conditions to induce pyocin production could be a further limitation. In addition, because *P. auruginosa* is an opportunistic pathogen there would be a need to insert the genes into an alternative host. Nevertheless, the pyocins may offer a novel method to suppress the activity of spoilage pseudomonads on fresh-cut produce.

17.3.4.3 Mycotoxins

Mycotoxins are a type of antibiotic in terms of antimicrobial activity, but the key difference is their toxicity to both animals and plants. Indeed, mycotoxins produced by some molds are well-established safety hazards that cause acute, chronic, mutagenic, and teratogenic toxicity (Bennett and Klich, 2003). In plants mycotoxins frequently lead to stunted plant growth and decrease in yields (Masuda et al., 2003).

In basic terms, mycotoxins are small, secondary metabolites produced by fila-mentous fungi. In addition to the toxic effects against humans and animals the mycotoxins of some molds are also important in plant pathogenesis. Despite the significance of mycotoxins, their actual physiological role is unclear. It is known that mycotoxins have a wide range of biological activities, including antibiotic (antibacterial and antifungal) properties, that may function to suppress competitive microflora.

In terms of plant degradation the key mycotoxins of concern are fumonisins and patulin. However, it should be noted that the food safety risks of mycotoxins greatly outweigh the detrimental effects on crop quality by mold growth. Fumonsins are made by the plant phytopathogen of *Fusarium* species (Hirooka et al., 1996). Several fumonisins have so far been isolated, but only FB1, FB2, and FB3 are produced in the natural environment of the plant. The mold is typically carried on seeds of corn, and some (e.g., *Fusarium moniliforme*) can become integrated into the internal (endophytic) microflora. Interestingly, the fumonisin of *F. moniliforme* inhibits the production of aflatoxin by *Aspergillus flavus* (Hirooka et al., 1996), but this obviously does not lead to a safer product.

Patulin was initially isolated from *Penicillium patulum* and received significant attention due to its potent antibiotic activity against protozoa, bacteria, and viruses. The same secondary metabolite was also isolated from other species and given the names clavacin, claviformin, expansin, mycoin C, and penicidin. However, it soon became apparent that patulin was also a potent toxin against plants and animals, and therefore it was reclassified as a mycotoxin. Patulin is commonly associated with soft fruits (apples, pears) contaminated with *Penicillium expansum*.

Conditions that stimulate mycotoxin production are typically associated with high relative humidity in combination with temperatures on the order of 20 to 30°C. In the field this resembles drought conditions wherein the stress imposed on plants compromises their defenses, thereby enabling invasion by pathogenic fungi. It fol-lows that improper storage of grain (especially if mechanically damaged) will be conducive to mycotoxin production. Methods for controlling mycotoxins are largely preventive. They include good agricultural practice and sufficient drying of crops after harvest. The construction of fungi-resistant plant breeds and introduction of biocontrol agents are future preventative strategies.

17.4 REGULATION OF GENE EXPRESSION IN MICROBES ASSOCIATED WITH PLANTS

17.4.1 QUORUM SENSING

Quorum sensing, also equivalent to autoinduction or cell-to-cell communication, allows both Gram-positive and Gram-negative bacteria to sense each other in the local environment (reviewed by Whitehead et al., 2001). In Gram-negative bacteria the quorum-sensing system was first described as the autoinducer of the *lux* regulon, responsible for bioluminescence of the marine symbiotic bacterium *Vibrio fisheri*. In this system a *lux* I gene encodes for an autoinducer synthase and *lux* R encodes a transcriptional activator protein (R-protein). The autoinducer synthase produces an autoinducer molecule that freely passes the cytoplasmic membrane and accumulates in the microenvironment surrounding the cell. The levels of autoinducer increase with cell density and the effecter molecule begins to accumulate within cells due to the decreasing concentration gradient. At a critical concentration the autoinducer molecule associates with the R-protein and the complex subsequently binds to specific DNA sequences upstream of target genes, thereby enhancing transcription. Therefore, the quorum-sensing system enables a population of bacterial cells to coordinate gene function as opposed to acting individually. This allows for more effective and efficient invasion of the plant.

The most-studied autoinducers in Gram-negative bacteria are N-acylhomoserine lactones (AHLs), which vary in length, saturation state, and C3 substitution of the N-acyl side chain. AHLs control the production of many secondary metabolites and virulence determinants in a range of Gram-negative bacteria. In most cases, quorum sensing is related to pathogenicity traits such as conjugal transfer of the Ti plasmids from *Agrobacterium tumefaciens* (see later), the production of extracellular cell-wall-degrading enzymes in *E. carotovora*, the induction of virulence factors in *P. aeruginosa*, and antibiotic production in *P. aurefaciens*. In addition, there is growing evidence that the ability to form surface-associated, structured, and cooperative consortia (referred to as biofilms) is also controlled by quorum sensing.

Quorum sensing plays a significant role in plant–bacteria interactions. Plants (e.g., pea seedlings) release AHL molecules in root exudates that can induce or suppress quorum-controlled genes in rhizobacteria (Bauer and Teplitski, 2001). The main question yet to be answered is what physiological role plant-produced AHLs have. The inhibitory AHLs would logically prevent expression of virulent genes in pathogens, thereby acting as an effective means of defense. The role of stimulatory AHLs is less clear but possibly leads to an uncoordinated attack by pathogens and hence places less stress on the plant's defenses.

In Gram-positive bacteria quorum sensing has been shown for genetic competence in *Bacillus subtilis* and *Streptococcus aureus*. Significantly, the production of antimicrobial peptides such as bacteriocins has also been demonstrated to be under the control of quorum sensing (Strume et al., 2002). The key difference in Gram-positive quorum sensing is that peptides (specifically, autoinducing peptides) rather than AHLs act as the signaling molecule. Peptides implicated in quorum sensing are small and diverse and undergo posttransitional modification to enhance stability and functionality. Interestingly, nisin can act as an autoinducing peptide, in addition to having antimicrobial effects.

The role of quorum sensing in food spoilage is only at the early stages of investigation. It is known that several pectinolytic *Erwinia* and *Pseudomonas*, which spoil ready-to-eat vegetables such as bean sprouts, produce AHLs. Storage trials with bean sprouts have demonstrated that inoculation of sprouts with AHL-producing, pectinolytic *E. carotovora* results in faster spoilage. If quorum sensing does indeed have a significant role in regulating the spoilage capacity of bacteria associated with fruits and vegetables it would provide a novel preservation strategy. An obvious method would be to introduce AHL-degrading enzymes (or bacterial cells expressing the enzyme) onto fruits or vegetables to disrupt the cell-to-cell communication. To date such an enzyme has been recovered from *Bacillus* species and *P. aureofaciens*. An alternative approach would be to add chemicals that could block or inhibit the quorum-sensing response. One promising group of quorum-sensing inhibitors (QSI) are the halogenated furanones produced by the Australian red algae, *Delisea pulchra*. These furanones interfere with the receptor proteins and release the AHL signal. In certain examples, such as the increased production of antimicrobial peptides, the quorum-sensing stimuli could be amplified or disconnected from cell density dependency. In this way lactic acid bacteria, for example, could be made to produce bacterocins constitutively. However, as refered to previously, this would not enhance produce's shelf life.

17.4.2 MICROBIAL PHYTOHORMONES

Phytohormones (auxin, cytokinin, and gibberellins) play a major role in regulating plant growth, development, and reproduction. Phytohormones also play a role in the communication between microbes and the plant that can either be beneficial (symbiotic) or detrimental. Fungi are known to produce a wide range of phytohormones that affect plant development. Indeed, many of the phytohormones were first discovered in fungi, suggesting coevolution with plants. The production of phytohormones by microbes is considered beneficial in terms of enhancing plant development and yield. However, many bacteria associated with plants produce the plant regulator indole 3-acetic acid (IAA). IAA promotes cell wall loosening and, along with auxin, can facilitate nutrient leakage from plant cells. This not only facilitates survival of the bacterium on the plant's surface in the field but can also accelerate the degradation process.

In addition to chemicals, genetic material can also be transferred to plants from bacteria. For example, the Ti plasmid found in the soil-borne bacterium *Agrobacterium tumefaciens* can be transferred to plants, resulting in tumor induction (crown-gall). Upon infection a 30,000-kb segment separates from the plasmid and integrates into the plant's chromosomes. The inserted genes encode for proteins that stimulate auxin and cytokinin synthesis. The net result is to reprogram plant development, leading to stunted growth and visible disease symptoms. The Ti plasmid has been used in genetic engineering of plants and may offer a powerful tool in developing resistant cultivars.

17.5 CONCLUSIONS

The effect of microbial metabolites on the degradation of fruits and vegetables largely remains an unexplored area. The rots and pigmentation formed by microbial activity are all visual indications of spoilage. However, the subtle formations of volatile compounds (taints) prove a challenge to identify and characterize. This is further complicated by the confusion caused by the common enzymic processes in microbes and plants. Nevertheless, as knowledge is gained on how microbial metabolites affect produce quality more opportunities will exist for applying novel preservation technologies. In this regard the disruption of quorum sensing, implicated in fruit and vegetable spoilage, could be an interesting route to follow.

REFERENCES

Anjaiah, V., Koedman, N., Nowak-Thompson, B., Loper, J.E., Hofte, M., Tambong, J.T., and Cornelis, P., Involvement of phenazines and anthranilate in the antagonism of *Pseudomonas aeruginosa* PNA1 and Tn 5 derivatives toward *Fusarium* spp. and *Pythium* spp., *Mol. Plant-Microbe Inter.*, 11, 847, 1998.

Antranikian, C., Microbial degradation of starch, in *Microbial Degradation of Natural Products*, Winkelmann, G., Ed., VCH, New York, 1992, pp. 27–54.

Asther, M. and Meunier, J.M., Increased thermal stability of *Bacillus licheniformis* α-amylase in the presence of various additives, *Enzyme Microb. Technol.*, 12, 902 1990.

Aymerich, T., Garriga, M., Ylla, J., Vallier, J., Monfort, J.M., and Hugas, M., Application of enterocins as biopreservatives against *Listeria innocua* in meat products, *J. Food Prot.*, 63, 721, 2000.

Bauer, W.D. and Teplitski, M., Can plants manipulate quorum sensing?, *Aus. J. Plant Physiol.*, 28, 913, 2001.

Beguin, P., Millet, J., Grepinet, O., Navarro, A., and Juy, M., The cell genes of *Clostridium thermocellum*, in *Biochemistry and Genetics of Cellulose Degradation*, Aubert, J.P., Beguin, P., and Millet, J., Eds., Academic Press, London, 1988, p. 267.

Benen, J.A.E. and Visser, J., Polygalacturonases, in *Handbook of Food Enzymology*, Whitaker, J.R, Voragen, A.G.J., and Wong, D.W.S., Eds., Marcel Dekker, New York, 2003, pp. 857–867.

Bennett, J.W. and Klich, M., Mycotoxins, *Clin. Microbiol. Rev.*, 16, 497, 2003.

Bhat, M. K. and Bhat, S., Cellulose degrading enzymes and their potential industrial applications, *Biotechnol. Adv.*, 15, 583, 1997.

Breusegem, S.Y., Clegg, R.M., and Loontiens, F.G., Base-sequence specificity of Hoechst 33258 and DAPI binding to five (A/T)(4) DNA sites with kinetic evidence for more than one high-affinity Hoechst 33258-AATT complex, *J. Mol. Biol.*, 315, 1049, 2002.

Bruhlmann, F., Leupin, M., Erismann, K.H., and Fiechter, A., Enzymatic de-gumming of ramie bast fibers, *J Biotechnol.*, 76, 43, 2000.

Burke, R.M. and Cairney, J.W.G., Laccases and other polyphenol oxidases in ecto- and ericoid mycorrhizal fungi, *Mycorrhiza*, 12, 105, 2002.

Buttner, D. and Bonas, U., Common infection strategies of plant and animal pathogenic bacteria, *Curr. Opin. Plant Biol.*, 6, 312, 2003.

Cao, J., Zheng, L., and Chen, S., Screening of pectinase producer from alkalophilic bacteria and study on its potential application in de-gumming of ramie, *Enzyme Microb. Technol.*, 14, 1013, 1992.

Chun S.H, Ryu I.H, Park, S.T., and Lee, K.S., Properties of α-amylase inhibitor from white kidney beans (*Phaseoulus vulgaris*), *Food Sci. Biotechnol.*, 10, 20, 2001.

Cleveland, J., Montville, T.J., Nes, I.F., and Chikindas, M.L., Bacteriocons: safe, natural antimicrobials for food preservation, *Int. J. Food Microbiol.*, 71, 1, 2001.

Costantino, H.R., Brown, S.H., and Kelly, R.M., Purification and characterization of an α-glucosidase from a hyperthermophilic archaebacterium, *Pyrococcus furiosus*, exhibiting a temperature optimum of 105 to 115°C, *J Bacteriol.*, 172, 3654, 1990.

Cutter, A.D. and Siragusa, G.R., Incorporation of nisin into a meat binding system to inhibit bacteria on beef surfaces, *Lett. Appl. Microbiol.*, 27, 19, 1998.

Dat, J.F., Pellinen, R., Beeckman, T., Van de Cotte, B., Langebartels, C., Kangasjarvi, J., Inze, D., and Van Breusegem, F., Changes in hydrogen peroxide homeostasis trigger an active cell death process in tobacco, *Plant J.*, 33, 621, 2003.

De Vries, R.P. and Vissser, J., *Aspergillus* enzymes involved in degradation of plant cell wall polysaccharides, *Microbiol. Mol. Biol. Rev.*, 65, 497, 2001.

Flett, G., Spoilage yeasts, *Crit. Rev. Biotechnol.*, 12, 1, 1992.

Goff, J.H., Bhunia, A.K., and Johnson, M.G., Complete inhibition of low levels of *Listeria monocytogenes* on refrigerated chicken meat with Pediocin AcH bound to heat killed *Pediococcus acidilactici* cells, *J. Food Prot.*, 59, 1187, 1996.

Gummadi, S.N. and Panda, T., Purification and biochemical properties of microbial pectinases: a review, *Process Biochem.*, 38, 987, 2003.

Gutterson, N.I., Layton, T.J., Ziegle, J.S., and Warren, G.J., Molecular cloning and genetic determinants for inhibition by fungal growth by fluorescent pseudomonad, *J. Bacteriol.*, 165, 696, 1986.

Hansawasdi, C., Kawabata, J., and Kasai, T., Alpha-amylase inhibitors from roselle (*Hibiscus sabdariffa Linn.*) tea, *Biosci. Biotehnol. Biochem.*, 64, 1041, 2000.

Harp, E. and Gilliland, S.E., Evaluation of a select strain of *Lactobacillus delbrueckii* subsp *lactis* as a biological control agent for pathogens on fresh-cut vegetables stored at 7C, *J. Food Prot.*, 66, 1013, 2003.

Hirooka, E.Y., Yamaguchi, M.M., Aoyama, S., Sugiura, Y., and Ueno, Y., The natural occurrence of fumonisins in Brazilian corn kernels, *Food Add. Contam.*, 13, 173, 1996.

Hokeberg, M., Wright, S.A.I., Svensson, M., Lundgren, L.N., and Gerhardson, B., Mutants of *Pseudomonas chlororaphis* defective in production of an anti-fungal metabolite express reduced biocontrol activity, *Abstr. Proc.* ICPP98, Edinburgh, Scotland, 1998.

Huang, O. and Allen, C., An exo-poly-alpha-D-galacturonosidase, PeHB, is required for wild-type virulence of *Ralstonia solanacearum*, *J. Bacteriol.*, 179, 7369, 1997.

Hyun, H.H. and Zeikus, J.G., General biochemical characterization of thermostable pullulanase and glucoamylase from *Clostridium thermohydrosulfuricum*, *Appl. Environ. Microbiol.*, 49, 1168, 1985.

Innes, R., New effects of type III effectors, *Mol. Microbiol.*, 50, 363, 2003.

Jack, R.W. et al., Characterization of the chemical and anti-microbial properties of pisicicolin 126 a bacteriocin produced by *Carnobacterium piscicola* JG126, *Appl. Environ. Microbiol.*, 62, 2897, 1996.

Jacxsens, L., Devlieghere, F., Ragaert, P., Vanneste, E., and Debevere, J., Relation between microbiological quality, metabolic production and sensory quality of equilibrium modified atmosphere packaged fresh-cut produce, *Int. J. Food Microbiol.*, 83, 263, 2003.

Jensen, N. and Whitfield, F.B., Role of *Alicyclobacillus acidoterrestris* in the development of a disinfectant taint in shelf-stable fruit juice, *Lett. Appl. Microbiol.*, 36, 9, 2003.

Kang, Y.W., Carlson, R., Tharpe, W., and Schell, M.A., Characterization of genes involved in biosynthesis of a novel antibiotic from *Burkholderia cepacia* BC11 and their role in biocontrol of *Rhizoctonia solani*, *Appl. Environ. Microbiol.*, 64, 3939, 1998.

Keel, C. et al., Suppression of root diseases by *Pseudomonas fluorescens* CHA0: importance of the secondary metabolite 2, 4-diacetylphloroglucinol, *Mol. Plant-Microbe Int.*, 5, 4, 1992.

Kubicek, C.P., Messner, R., Gruber, F., Mach, R.L., and Kubicek-Pranz, E.M., The *Trichoderma reesei* cellulase regulatory puzzle: from the interior life of a secretory fungus, *Enzyme Microb. Technol.*, 15, 90, 1993.

Kusano, S., Takahashi S.I., Fujimoto, D., and Sakano, Y., Effects of reduced malto-oligosaccharides on the thermal stability of pullulanase from *Bacillus acidopullulyticus*, *Carbohydr. Res.*, 199, 83, 1990.

Lamed, R. and Bayer, E.A., The cellulosome of *Clostridium thermocellum*, *Adv. Appl. Microbiol.*, 33, 1, 1988.

Lazdunski, C.J., Pore-forming colicins: synthesis, exracellular release, mode of action, immunity, *Biochimie*, 70, 1291, 1998.

Lebeda, A., Luhova, L., Sedlarova, M., and Jancova, D., The role of enzymes in plant-fungal pathogens interactions, *Plant Dis. Prot.*, 108, 89, 2001.

Lindow, S.E. and Brandl, M.T., Microbiology of the phyllosphere, *Appl. Environ. Microbiol.*, 69, 1875, 2003.

Liu, L., Wakamatsu, K., Ito, S., and Williamson, P.R., Catecholamine oxidative products, but not melanin, are produced by *Cryptococcus neoformans* during neuropathogenesis in mice, *Infect. Immun.*, 67, 108, 1999.

Lund, B.M., Ecosystems in vegetable foods, *J. Appl. Bacteriol.*, 73, 115S, 1992.

Maesmans, G., Hendrickx, M., De Cordt S., Van Loey, A., Noronha, J., and Tobback, P., Evaluation of process value distribution with time temperature integrators, *Food Res. Int.*, 27, 413, 1994.

Mansfield, S.D., Mooney, C., and Saddler, J.N., Substrate and enzyme characteristics that limit cellulose hydrolysis, *Biotechnol. Prog.*, 15, 804, 1999.

Masuda, D., Yamaguchi, K., Kimura, M., Yamaguchi, I., and Nishiuchi, T., Mycotoxin-induced dwarfism in Arabidopsis plant, *Plant. Cell. Physiol.*, 44, S38, 2003.

McManus, P.S., Stockwell, V.O., Sundin, G.W., and Jones, A.L., Antibiotic use in plant agriculture, *Annu. Rev. Phytopathol.*, 40, 443, 2002.

Michel-Briand, Y. and Baysse, C., The pyocins of *Pseudomonas aurginosa*, *Biochimie*, 84, 499, 2002.

Milner, J.L., Silo-Suh, L., Lee, J.C., He, H.Y., Clardy, J., and Handelsman, J., Production of kanosamine by *Bacillus cereus* UW85, *Appl. Environ. Microbiol.*, 62, 3061, 1996.

Mohan-Reddy, R., Gopal-Reddy, P., and Seenayya, G., Enhanced production of thermostabile β-amylase and pullunase in the presence of surfactant by *Clostridium thermosulfurogenes* SV2, *Process Biochem.*, 34, 87, 1999.

Pariza, M.W. and Johnson, E.A., Evaluating the safety of microbial enzyme preparations used in food processing: update for a new century, *Reg. Toxicol. Pharmacol.*, 33, 173, 2001.

Parret, A.H.A. and de Mot, R., Bacteria killing their own kind: novel bacteriocins of *Pseudomonas* and other protobacteria, *Trends Microbiol.*, 10, 107, 2002.

Parret, A.H.A., Schoofs, G., Proost, P., and de Mot, R., Plant lactin-like bacteriocin from a rhizosphere-colonizing *Pseudomonas* isolate, *J. Bacteriol.*, 185, 897, 2003.

Penyalver, R. and Lopez, M.M., Cocolonization of the rhizosphere by pathogenic *Agrobacterium* strains and non-pathogenic strains K84 and K1206 used for crown gall biocontrol, *Appl. Environ. Microbiol.*, 65, 1936, 1999.

Pierson, L.S. and Thomashow, L.S., Cloning and heterologous expression of the phenazine biosynthetic locus from *Pseudomonas aureofacines* 3084, *Mol. Plant-Microbe Int.*, 5, 330, 1992.

Pretel, M.T., Lozano, P., Riquelme, F., and Romojaro, F., Pectic enzymes in fresh fruit processing: optimization of enzymatic peeling of oranges, *Process Biochem.*, 32, 43, 1997.

Raaijmakers, J.M., Vlami, M., and de Souza, J.T., Antibiotic production by biocontrol agents, *Antonie van Leeuwenhoek*, 81, 537, 2002.

Reese, E.T., A microbiological progress report: enzymatic hydrolysis of cellulose, *Appl. Microbiol.*, 4, 39, 1956.

Reese, E.T., Lola, J.E., and Parrish, F.W., Modified substances and modified products as inducers of carbohydrases, *J. Bacteriol.*, 100, 1151, 1969.

Romanovskaya, V.A., Stolyar, S.M., Malashenko, Y.R., and Dodatko, T.N., The ways of plant colonization by *Methylobacterium* strains and properties of these bacteria, *Microbiology*, 70, 221, 2001.

Sakai, T., Degradation of pectins, in *Microbial Degradation of Natural Products*, Winkelmann, G., Ed., VCH, New York, 1992, pp. 57–82.

Sandkvist, M., Biology of type II secretion, *Mol. Microbiol.*, 40, 271, 2003.

Schindler, I., Renz, A., Schmid, F.X., and Beck, E., Activation of spinach pullulanase by reduction results in a decrease in the number of isomeric forms, *Biochim. Biophys. Acta Protein Struct. Molec. Enzymol.*, 1548, 175, 2001.

Shallom, D. and Shoham, Y., Microbial hemicellulases, *Curr. Opin. Microbiol.*, 6, 219, 2003.

Shanaham, P., O'Sullivan, D.J., Simpson, P., Glennon, J.D., and O'Gara, F., Isolation of 2, 4-diacetylphloroglucinol from a fluorescent pseudomonad and investigation of physiological parameters influencing its production, *Appl. Environ. Microbiol.*, 58, 353, 1992.

Shevchik, V.E. et al., Characterization of the exo-pectate lyase Pe1X of *Erwinia chrysanthemi*, *J. Bacteriol.*, 181, 1652, 1999.

Stathopoulos, C., Hendrixson, D.R., Thanassi, D.G., Hultgren, S.J., St. Geme, S.T., III, and Curtiss, R, III, Secretion of virulence determinants by the general secretory pathway in Gram-negative pathogens: an evolving story, *Microbes Infect.*, 2, 1061, 2000.

Steinkraus, K.H., *Handbook of Indigenous Fermented Foods*, Steinkraus, K.H.,Ed., Marcel Dekker, New York, 1996, Chap. 4.

Strume, M.H.L., Kleerebezem., M., Nakayama, J., Akkermans, A.D.L., Vaughan, E.E., and de Vos, W.M., Cell to cell communication by autoinducing peptides in Gram-positive bacteria, *Antonie van Leeuwenhoek*, 81, 233, 2002.

Tanabe, H., Kobayashi, Y., and Akamatsu, T., Pretreatment of pectic waste water with pectate lyase from an alkalophilic *Bacillus* sp., *Agric. Biol. Chem.*, 52, 18530, 1988.

Vandamme, E.J. and Soetaert, W., Bioflavours and fragrances via fermentation and biocatalysis, *J. Chem. Technol. Biotechnol.*, 77, 1323, 2002.

van Rijssel, M., Gerwig, G.J., and Hanse, T.A., Isolation and characterization of an extracellular glycosylated protein complex form *Clostridium thermosaccharolyticum* with pectin methylesterase and polygalacturonate hydrolase activity, *Appl. Environ. Microbiol.*, 59, 828, 1993.

Vincent, M.N., Harrison, L.A., Brakin, J.M., Kovacevich, P.A., Murkerji, P., Weller, D.M., and Pearson, E.A., Genetic analysis of the anti-fungal activity of soilborne *Pseudomonas aureofaciens* strain, *Appl. Environ. Microbiol.*, 57, 2928, 1991.

Voragen, A.G.J. and Pilnik, W., Pectin-degrading enzymes in fruit and vegetable processing, in *Biocatalysis in Agricultural Biotechnology*, Whitaker, J.R. and Sonnet, P.E., Eds., American Chemical Society, Washington, DC, 1989 p. 93.

Wang, M.J. and Yu, R.C., Selective isolation of β-amylase derepressed mutants of *Bacillus acidopullulyticus*, *J. Ferment. Bioeng.*, 77, 243, 1994.

Warner, D.A. and Knutson, C.A., Isolation of α-amylases and other starch degrading enzymes from endosperm of germinating maize, *Plant Sci.*, 78, 143, 1991.

Warren, R.A.J., Microbial hydrolysis of polysaccharides, *Annu. Rev. Microbiol.*, 50, 183, 1996.

Whitehead, N.A., Barnard A.M.L., Slater H., Simpson N.J.L., and Salmond G.P.C., Quorum-sensing in Gram-negative bacteria, *FEMS Microbiol. Rev.*, 25, 365, 2001.

Wood, T.M. and McCrae, S.I., Synergism between enzymes involved in the solubilization of native cellulose, *Adv. Chem. Ser.*, 181, 181, 1979.

Zabetakis, I., Enhancement of flavor biosynthesis from strawberry (*Fragraria ananassa*) callus cultures by *Methylobacterium* species, *Plant Cell Tissue Organ Cult.*, 50, 179, 1997.

Zabetakis, I., Gramshaw, J.W. and Robinson, D.S., 2, 5-dimethyl 4-hydroxy 2H-furan 3-one and its derivatives: analysis, synthesis and biosynthesis: a review, *Food Chem.*, 65, 139, 1999.

18 Microstructure of Produce Degradation

Delilah F. Wood, Syed H. Imam, Glen P. Sabellano, Paul R. Eyerly, William J. Orts, and Gregory M. Glenn
USDA, ARS, WRRC, Albany, CA

CONTENTS

18.1 INTRODUCTION

Botanical nomenclature is based on the origin of each component during development and maturation [1]. Edible plant parts encompass a wide array of differing origins and types of tissues [2] (Figure 18.1) and include both vegetative and reproductive tissues. Vegetative tissues include green leafy vegetables (lettuce, spinach), stem vegetables (celery, rhubarb), roots (carrot, radish), tubers (potato), and bulbs (onion, garlic). Reproductive tissues consist of flower buds (broccoli florets, artichokes), fruits (bananas, squash, green beans), seeds (mature beans, almonds), and grains (wheat, corn).

This chapter covers a few selected plant materials by discussing a brief survey of the literature and by showing scanning electron micrographs of specific commodities that were purchased at the local supermarket. We hope to acquaint the reader with a general idea of plant structure and surface morphology as they relate to degradation of produce.

Fresh material was prepared for microscopy as soon as it was purchased. Aged material was purchased fresh and then allowed to age at ambient conditions for a week or more prior to preparation for microscopy. Plant material was fixed in

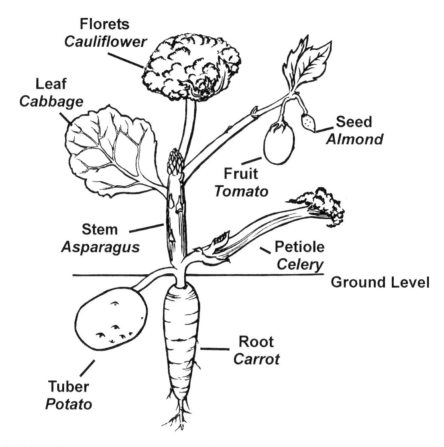

FIGURE 18.1 Schematic representation of the various parts of plants consumed as food. (Adapted from Jewell, G.G., Fruits and vegetables, in *Food Microscopy, Food Science and Technology,* Vaughn, J.G., Ed., Academic Press, New York, 1979.)

formalin-acetic acid-ethanol [3] or in 3% glutaraldehyde-2% formaldehyde and post-fixed in 1% osmium tetroxide [4] and dehydrated in a graded series of ethanol [3,4]. The samples were then either cryofractured in liquid nitrogen [5] or sliced under 100% ethanol, critical point dried in a Tousimis Autosamdri 815 (Tousimis, Rockville, MD), sputter coated with gold-palladium in a Denton Desk II sputter coating unit (Denton Vacuum, Moorestown, NJ), and observed and photographed in an Hitachi S-4700 field emission scanning electron microscope (Japan).

18.2 INHERENT PRESERVATION IN PLANT ARCHITECTURE

Intact fruits and vegetables are protected from rapid breakdown from environmental influences by their surrounding outer layers that form a skin. The outermost protective layer is a thin, continuous waxy layer, the cuticle [6], which slows dehydration and serves as an effective barrier against the entry of pathogens and insects [7] and is, therefore, the natural package giving a first layer of protection. Beneath the cuticle

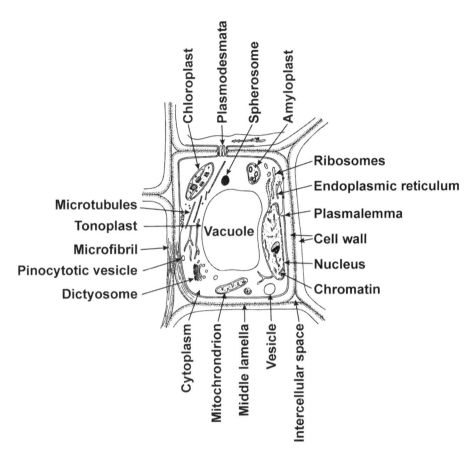

FIGURE 18.2 Schematic drawing of a plant cell. Amyloplast, starch-storing plastid; Chromatin, deoxyribonucleoprotein fibrils; Chloroplast, double-membrane-bound organelle with photosynthetic lamellae; Cell wall, rigid extracellular layer composed of microfibrils in a matrix; Cytoplasm, the fluid matrix and associated molecules between the vacuole and the cell wall; Dictyosome, membrane system of a few disc-shaped cisternae with secretory vesicles. The golgi apparatus is the cell's whole complement of dictyosomes; Endoplasmic reticulum, flattened cisternae and tubules of membrane sometimes with attached ribosomes; Intercellular space, the space formed at the junction of three or more adjacent cell walls; Mitochondrion, a cell organelle responsible for cellular respiration; Microfibril, the cellulose strands of the cell wall, having a diameter of 10 to 30 nm; Middle lamella, a more darkly staining zone between adjacent cells, rich in pectin; Microtubules, 20-nm-wide tubules that may be involved in movement in the cytoplasm; Nucleus, a double-membrane-bound compartment enclosing the chromosomes, nucleolus, and nucleoplasm. The envelope has nuclear pores; Plasmodesmata, intercellular cytoplasmic connections; Plasmalemma, cell membrane; Pinocytotic vesicle, an "infolding" of the cell membrane; Spherosome, single-membrane-bound vesicle with stainable contents; Tonoplast, vacuole membrane; Vacuole, aqueous, membrane-bound area within the cytoplasm; Vesicle, membrane-bound cytoplasmic inclusion. (Adapted from Jewell, G.G., Fruits and vegetables, in *Food Microscopy, Food Science and Technology,* Vaughn, J.G., Ed., Academic Press, New York, 1979.)

lies the epidermis consisting of cells that may possess thickened outer cell walls. The skin may also contain other additional cell layers with thickened cell walls.

Cell walls also provide inherent preservation of structure to the plant. Plant cells, unlike animal cells, which have only a membrane, are also surrounded by rigid cell walls (Figure 18.2). Cell walls are highly reliant on an adequate water potential or turgor of the cells. Cell walls are also found on other edible fresh produce (often thought of as plants), the mushrooms, which are actually fungi.

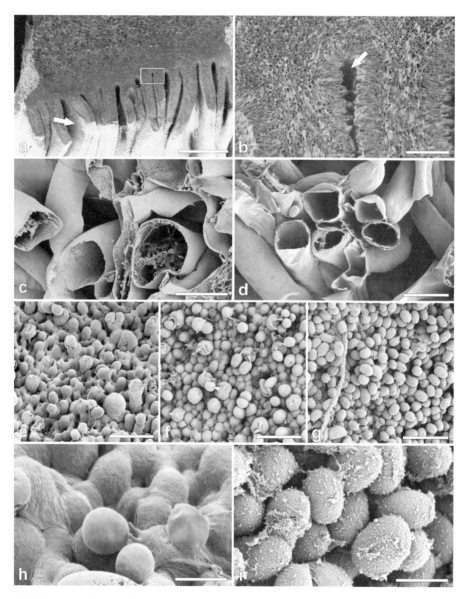

FIGURE 18.3 (See caption on opposite page)

Fungi, unlike plants, are nonphotosynthetic organisms. Edible mushrooms are essentially a mass of hyphae that has joined together to form a basidiocarp, or fruiting body. A high-quality, fresh, white button mushroom has a tightly closed cap, or pileus, the gills are covered by a membrane or velum, the stipe (stalk) is short, and the mushroom is white. Discoloration or browning is an unacceptable quality of mushrooms that might develop upon storage. Several mechanisms may be associated with the browning reaction, ranging from rough handling to the action of bacteria or mold on the mushroom tissues, causing brown spots. To inhibit browning, low-dose irradiation was shown effective as a preventive measure to browning both as an antioxidant and a microbial inhibitor. A high dose rate was found to be less effective than the low dose rate in reducing the onset of browning. The high dose rate induced changes in the pileus, including cell wall thickening, which was shown to change permeability rates in the cells [8].

Another unacceptable quality parameter has to do with normal mushroom development, or aging. The stipe grows, the velum stretches and breaks, and the pileus opens to expose and release the spores on the surfaces of the growing gill tissue. The broken velum leaves a ring of tissue on the stipe, which is then termed the annulus [9]. Part of the pileus structure is shown in Figure 18.3a and b. Spores form on the inner portions of the gills shown by the arrows (Figure 18.3a and b); the rest of the pileus and the stipe are composed of masses of hyphae. Hyphae in the fresh mushroom are tightly compacted and have smooth, plump, hydrated cell walls and evident cytoplasm (Figure 18.3c). In the aged mushroom, the cell walls are somewhat collapsed and wrinkled (Figure 18.3d), indicative of dehydration stress, and cytoplasm is no longer evident. The apparent disappearance of cytoplasm agrees with findings of Braaksma et al. [9]; growth during aging of the pileus tissue was reported to occur by vacuolar expansion. Spores in the fresh mushroom are smooth, rounded, and immature (Figure 18.3e, f, and h) and some of them are covered by a smooth membranous substance. The aged mushroom (Figure 18.3g and i) has spores that are closer to maturity, are ovoid in shape and are covered with irregular, sticky-looking deposits on their surfaces, which may be the remnants of the smooth membranous substance that has undergone digestion in the formation of the spores and dehydration.

FIGURE 18.3 (Opposite page) Scanning electron micrographs of a white button mushroom. (a) Cross section of the cap showing the cap body and the gills, on the surfaces of which spores are formed (arrow). The boxed area is shown in higher magnification in b. (b) Close view of two gill fingers; arrow indicates the gap between the gill fingers; spores form on the surfaces on either side of the gap. (c) Fresh hyphae showing the smoothness of the cell walls of the fully hydrated mushroom. (d) Aged hyphae showing collapsing and wrinkling indicative of dehydration. (e) Immature spores from a fresh sample. (f) Immature spores from a fresh sample; these look as if they are more developed than those shown in e. (g) Mature spores from an aged sample. (h) Spores from a fresh sample; it appears as if the spores are covered with a thin membranous substance. (i) Spores from an aged sample; it appears that the thin membranous substance of h has undergone digestion and dehydration and that remnants remain on the surfaces of the mature spores. Magnification bars: a, 500 μm; b, 50 μm; c, 10 μm; d–g, 20 μm; h, i, 5 μm.

Dehydration of plant cells results in a loss of cell turgor. The turgor of plant cells and cell walls is essential in providing textural "crispness" attributes to fresh vegetables such as lettuce and parsley. Figure 18.4 shows the surfaces of fresh (Figure 18.4a, c, g, and i) and aged (b, d, e, f, and h) Italian parsley leaves. The upper (adaxial) side of the leaf (Figure 18.4a) shows plump epidermal cells, indicating

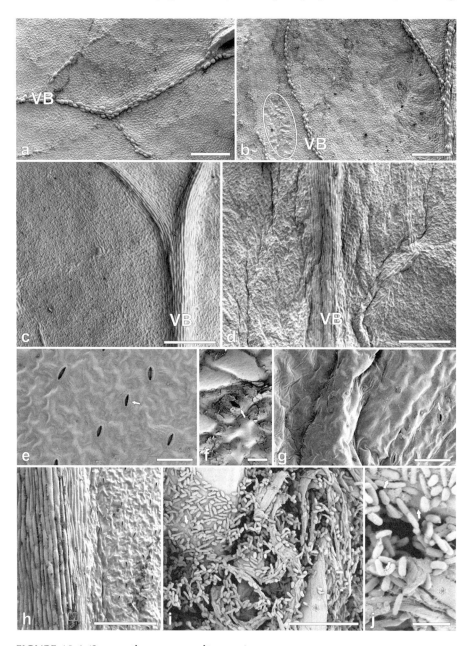

FIGURE 18.4 (See caption on opposite page)

that they are fully hydrated or have normal turgor pressure. The vascular bundles appear normal and there are no wrinkles or folds in the tissue. In contrast, the adaxial surface of the aged leaf (Figure 18.4b) has an irregular appearance, areas where epidermal cells appear to be missing, and irregular dark spots. Although the vascular bundles look normal, epidermal cells are not as plump as in the fresh sample. More pronounced changes are evident on the bottom (abaxial) leaf surface, where the fresh leaf (Figure 18.4c) shows a distinct vascular bundle, stomata and plump epidermal cells; the aged leaf (Figure 18.4d) shows a vascular bundle that blends in with the rest of the extensive folds in the epidermis. In the fresh sample, the epidermis appears flat, the stomata are nicely arranged and the epidermal cells are plump (Figure 18.4e); however, early signs of degradation are also apparent in the fresh leaf in areas where the epicuticular wax has started to peel and crack (Figure 18.4f). The aged leaf (Figure 18.4g) shows deformities of the epidermis, which has shriveled and wrinkled due to dehydration; some of the epidermal cells have collapsed and others appear to have maintained some internal turgor pressure. In the region of the vascular bundle of the aged leaf (Figure 18.4h), dark spots, which are not stomata, are evident. One of the dark spots on the outside of the vascular bundle proved to be a bacterial colony (Figure 18.4i and j). The colony was associated with a fibrous substance, probably a biofilm, which some bacteria exude in order to adhere to a substrate. Biofilms are typically composed of highly hydrated polysaccharides [10,11]. Because the polysaccharides have a high water content, they are difficult to preserve using aqueous fixation techniques typically used for scanning electron microscopy. During fixation and subsequent dehydration, portions of the polysaccharide dissolve and the remaining insoluble portions lose definition and tend to clump together, forming what looks like fibers under the microscope. The larger fibrous components are part of the parsley leaf that is disintegrating due to microbial activity (Figure 18.4i and j) that develops readily on parsley due to its short shelf life.

FIGURE 18.4 (Opposite page) Italian parsley. Scanning electron micrographs show the outer epidermal surfaces of leaves. Adaxial, upper leaf surfaces showing the vascular bundles and epidermal cells in the fresh (a) and aged (b) leaves. (a) Fresh leaf showing regular, hydrated epidermal cells. (b) Aged leaf with shrunken epidermal cells, areas where epidermal cells appear to be missing (oval) and irregular spots. Abaxial, lower, leaf surfaces show the outer epidermis of fresh (c, e, f) and aged (d, g, h, i) leaves. (c) Fresh leaf has distinct vasuclar bundles, epidermal cells and stomata. (d) Aged leaf has extensive wrinkling or folding, epidermal cells appear to have lost turgor pressure, and the vascular bundle is indistinct. (e) Fresh leaf showing open stomata (arrow) and epidermal cells that have adequate turgor. (f) Fresh leaf showing a single stomatal complex (arrow) and epidermal cells with rounded shapes indicative of adequate turgor; the "flaky" appearance is due to epicuticular wax that has started to peel and crack. (g) Aged leaf showing shrunken stomata (arrow) and folds. Arrows, stomatal complexes. (h) Aged leaf showing a distinct view of the vascular bundle and stomata; the epidermal cells have lost turgor pressure and wrinkles are apparent; darkened areas are apparent (box). (i) High magnification of the boxed area in h showing a bacterial colony associated with a thin, filmy substance, probably a biofilm (arrows) composed of polysaccharides that microorganisms exude to allow them to adhere to a substrate; fibers (*) are plant material that is decomposing. (j) Bacterial colony showing greater detail; arrows indicate possible biofilm. VB, vascular bundle. Magnification bars: a–d, 500 μm; e–g, 50 μm; h, 200 μm; i, 10 μm, j, 2 μm.

Short shelf life is common in green produce. Broccoli provides another example of short shelf life and is harvested immature when flowering heads are growing rapidly [12]. Broccoli should be consumed as the fresh product within a month if stored at 0°C or 3 d if stored under ambient conditions [13]. The floret buds (Figure 18.5a and b) tend to be more prone to decay than the rest of the stalk and quickly show dehydration and superficial mold growth (Figure 18.5a and b) upon aging. Microbial growth also accelerates undesirable changes in color due to the lowering

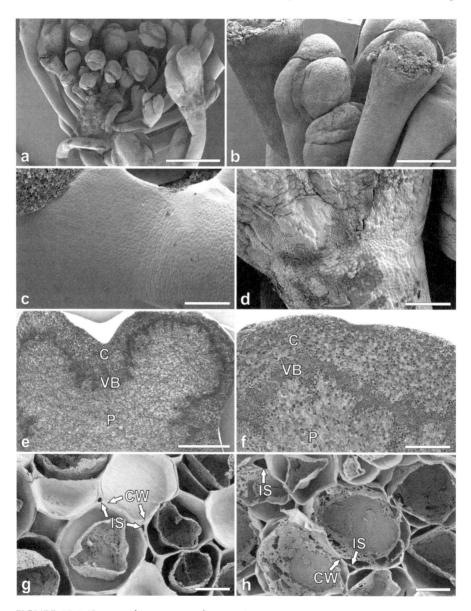

FIGURE 18.5 (See caption on opposite page)

of the pH and the conversion of chlorophyll to pheophytin [14]. The epidermis of fresh broccoli has the characteristic turgid epidermal cells and smooth appearance to the skin (Figure 18.5c). The aged epidermis shows signs of dehydration and superficial mold growth (Figure 18.5d). Stem cross sections of both fresh (Figure 18.5e) and aged (Figure 18.5f) tissue are virtually identical at low magnification. At higher magnifications, minor differences may be seen between the fresh (Figure 18.5g) and aged (Figure 18.5h) tissue. Parenchyma cell walls tend to be thicker and intercellular spaces tend to be smaller in the fresh (Figure 18.5g) than in the aged (Figure 18.5h) sample. The cells in the aged sample also seem to be less turgid and more rounded than comparable cells in the fresh sample.

Green beans, botanically fruits, are harvested and consumed in the immature, green state. Fresh green beans are bright green in color and have a characteristic "snap" when broken with the fingers. The green bean consists, in large part, of the pod, most of which is fruit coat, or pericarp. The bean pod has a number of immature seeds in the center. The bulk of the pericarp is mesocarp, which consists of thin-walled parenchyma cells (Figure 18.6a). Upon aging, mesocarp cells start to collapse in the outermost layer (Figure 18.6b). In the inner mesocarp, cell walls break and several cells or cell layers conjoin and form huge gaps (Figure 18.6b). The epidermis in the fresh bean (Figure 18.6c) is flat and covered with hairs or papillae and stomata. The unprotected green bean loses turgor or dehydrates as it ages and large folds occur (Figure 18.6d). At higher magnification, the epidermal cells in the fresh green bean (Figure 18.6e) have a linear appearance, the stomata are evident, and the fine papillae are slightly hooked on the ends. Debris can also be found on the surface of the fresh bean. Epidermal papillae are hollow, evident once they have been fractured (Figure 18.6f). The papillae of the aging bean are dehydrated and may be constricted at points. Epidermal cells have begun to pucker (Figure 18.6g) due to the effects of dehydration and fungal hyphae may be seen on the surface (Figure 18.6g) or growing into stomata (Figure 18.6h).

Cucumber is another fruit that is consumed in its immature, green state. The epidermis in the fresh fruit is covered with epicuticular wax and is completely smooth (Figure 18.7a). In the aging cucumber, the epidermis begins to have irregular "pools" of differing electron densities (Figure 18.7b). Closer inspection of the darkened "pools" reveals areas where the epicuticular wax has begun to be disrupted and the rounded epidermal cells become exposed (Figure 18.7c and d). Large areas that have

FIGURE 18.5 (Opposite page) Scanning electron micrographs of broccoli florets. (a, b) Aged broccoli flower buds showing some growth of fungi on the epidermis and dehydration. (c) Fresh floret stalk epidermis showing smooth, hydrated epidermal cells. (d) Aged floret stalk epidermis showing the effects of dehydration and fungal growth. (e) Fresh cross section of floret stalk showing the organization of the various tissues. (f) Aged cross section of floret stalk with structure similar to that of the fresh tissue shown in e (g) Fresh cells of the cortex of the floret stalk showing thickened cell walls, somewhat angular cell shapes, and small intercellular spaces. (h) Aged cells of the cortex of the floret stalk; compared to f, cells are more rounded in shape, have less apparent turgor, and have thinner cell walls and larger intercellular spaces. C, cortex; CW, cell wall; IS, intercellular space; P, pith; VB, vascular bundles. Magnification bars: a, e, 1 mm; b, 500 μm; c, d, f, 200 μm; g, h, 10 μm.

visually become soft rot have extensive growth of fungi (Figure 18.7e and f). In cross section, large gaps appear in the mesocarp adjacent to the junction of the exocarp (Figure 18.8a–d). Additionally, fungal hyphae, which have most likely degraded the parenchyma cell walls or weakened them to the extent of collapse, are

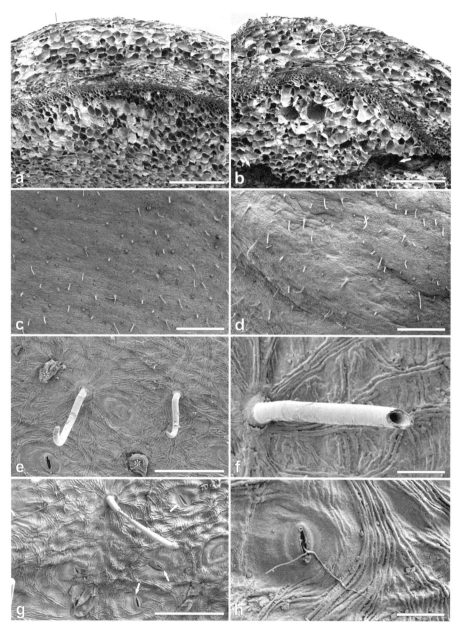

FIGURE 18.6 (See caption on opposite page)

apparent in the voided areas (Figure 18.8c and d). The exocarp, which has thick, sclerified cell walls, shows relatively little change (Figure 18.8e). The cell walls of the inner mesocarp also do not show much change in structure at this point of degradation (Figure 18.8f).

In contrast to produce with a short shelf life, the carrot, a root, has a very long shelf life unless it is grown in the presence of *Chalara elegans*, which can cause postharvest black root rot, common in the Fraser Valley of British Columbia, Canada [15]. Carrots can be harvested at any time after a carrot has formed, although older carrots tend to be "woody" and are less acceptable to consumers. Small, ready-to-eat carrots were chosen for SEM study because of their convenience to consumers. The carrots are peeled and scrubbed prior to packaging, so their surfaces differ from those that are freshly harvested. The outer surfaces of fresh or aged prepackaged carrots are similar and consist of crushed outer cortex (Figure 18.9a and b). The surfaces of carrots that have been peeled and stored often have a white, translucent appearance. The white appearance was shown to be due the shredding of superficial layers of the carrot that then dehydrate, giving a white appearance to the carrot skin [16]. Slicing of carrots causes physical damage, stress, and the increased risk of microbial growth. The risk and severity of the effects were found to be less with gentle handling and the use of extremely sharp knives for cutting [17] and were even less following vacuum packaging [18].

Both fresh (Figure 18.9c) and aged (Figure 18.9d) carrots are similar in cross section even though the aged carrot had been aged to the point of wilting. Carrots are resilient and they nearly fully recovered their original structure due to the imbibition of water during aqueous fixation for scanning electron microscopy; therefore, most of the cell walls in the aged carrot appear turgid. Some cells in the aged carrot were found to have broken cell walls and the cells had joined with nearby cells to form larger gaps in the tissue (Figure 18.9e) and some cells appeared to be collapsed (Figure 18.9f). Broken cell walls and collapsed cells were not apparent in the fresh sample (Figure 18.9g).

FIGURE 18.6 (Opposite page) Green bean. Scanning electron micrographs (SEMs) showing cross sections of pericarp of fresh (a) and aging (b) green bean. (a) Fresh green bean showing regular, thin-walled parenchyma cells of the mesocarp. (b) Aging green bean showing collapse of cells in the outer mesocarp (circle) and gaps that have formed by cell wall breakage have joined multiple cells (arrows). SEMs of the surface of the epidermis of green bean of fresh (c, e, f) and aging (d, g, h) tissue. (c) Fresh green bean showing fine hairs (papillae) and stomata and an unwrinkled surface. (d) Aging bean showing the very disrupted, wrinkled surface. (e) Fresh epidermis showing hooked papillae and stomatal complexes with linear epidermal cells and debris (*). (f) Fractured papilla of fresh green bean. (g) Aging green bean epidermis showing a dehydrated and constricted papilla, sunken stomata (arrows) and irregular, pinched epidermal cells. (h) Aging green bean epidermis showing fungal hyphae that have begun growing into a stomate. Magnification bars: a, b, 250 μm; c, d, 500 μm; e, g, 100 μm; f, g, 25 μm.

FIGURE 18.7 Cucumber. Scanning electron micrographs showing the outer surface of the epicuticular-wax-covered epidermis. (a) Fresh sample. (b) Aged sample showing "pools" of differing electron densities (i.e., dark and light areas). (c) Close view of the dark and light areas showing that the dark areas are devoid of epicuticular wax. (d) Dark area with only patchy areas that contain epicuticular wax. (e) Top of the micrograph is the edge of a soft-rot spot; fungi proliferate and the epidermis has debris and other irregularities on its surface. (f) Close view of the fungal hyphae. Magnification bars: a, e, 1 mm; b, 200 μm; c, f, 50 μm; d, 20 μm.

18.3 TEXTURAL CHARACTERISTICS

Produce texture is determined by tissue structure and physiology. Microscopy has been used to relate tissue microstructure and texture by measuring tissue failure under tension followed by observation of the areas of failure using various micros-copy methods. Texture is largely dependent on cell structure and the relationships between cells. Microstructural studies can show many of the characteristics giving rise to texture in any given produce commodity.

All fruits and vegetables are composed of cells. Cell characteristics include cell structure (composition, size, shape, type, water content); air spaces between cells; cell wall structure, composition, and thickness; and adhesion between cells. Adhesion is

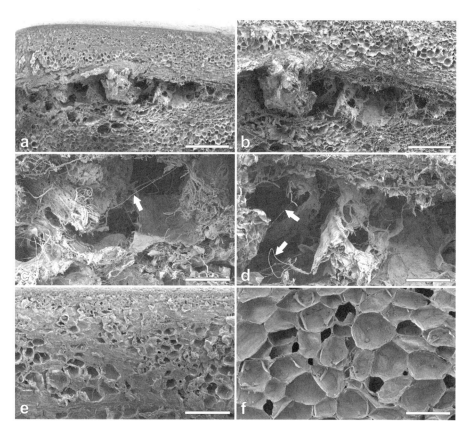

FIGURE 18.8 Cucumber. Scanning electron micrographs exhibiting cross-sectional views of the aged fruit. (a) Outermost edge shows the epidermis, exocarp, and mesocarp; the fruit has huge gaps in the mesocarp adjacent to the junction of the exocarp. (b) Close views of the degenerated area from a. (c, d) Close views of the degenerated area from a. showing the growth of fungal hyphae (arrows). (e) Exocarp showing thick-walled cells that have no change from fresh material. (f) Inner mesocarp tissue showing little or no change from fresh material. Magnification bars: a, 1 mm; b–f, 200 μm.

determined mainly by the state of the middle lamella. Fruits and vegetables are dynamic, living systems; thus, structural and textural changes occur continuously. Textural diversity at the cellular level has been studied in various tissues using tensile strength measurements and correlating the measurements with observation by microscopy. Examination of the fracture surfaces allowed the detection of differences in tissue strength and juiciness and provided insight into the cellular basis of plant texture. It also helped to identify specific cell characteristics that influence the sensory texture attributes of "hardness" and "juiciness" [19].

Soft texture is a desirable characteristic in kiwifruit. In contrast to cells of apple fruit, kiwifruit cells remain in close apposition, whereas apple cells have more intercellular spaces and have very little cell-to-cell contact. The intercellular spaces account for part of the reason the two fruits have such differing textures [20].

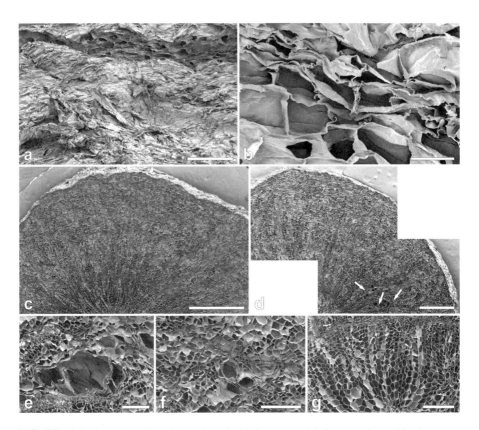

FIGURE 18.9 Prepackaged peeled and washed baby carrot. (a) Outer surface of fresh carrot. (b) Cross section of the outer layers of fresh carrot showing crushing of the cells due to initial processing steps of peeling and washing. (c) Cross section of fresh carrot. (d) Cross section of aged carrot showing cell wall breakage and cell collapse (arrows). (e) Aged carrot, close view showing the effects of cell wall breakage. (f) Aged carrot, close view showing cell wall breakage and cell collapse. (g) Fresh carrot cross section showing no cell wall breakage or collapse. Magnification bars: a, 200 μm; b, 50 μm; c, d, 1 μm; e–g, 200 μm.

Kiwifruit is harvested at the mature but unripe stage, at which time the fruit is quite firm. Softening occurs during the ripening process. Microscopy has been used to document changes occurring during ripening at the microstructural level and correlated with penetrometer tests to measure the hardness of the whole fruit. Unripe kiwifruit has angular cells and, as they ripen, the cells become more rounded and starch granules degrade.

Kiwifruit is composed of three distinct regions: outer pericarp, inner pericarp, and core. Each tissue differs in composition and rate of change in cell wall components. Cell walls thicken and undergo considerable change in all tissues of the fruit during ripening [21,22]. The thickening or swelling of cell walls in ripening kiwifruit coincides with the solubilization of pectin [23]. Pectin hydrolysis and the modification of hemicellulose were also involved in ripening and fruit softening in papaya [24]. The swollen cell walls in kiwifruit were resistant to staining with Ruthenium red, which is characteristic of pectic material, compared with those in unripe fruit.

Morphometric studies showed that the volume of intercellular air spaces increased with ripening and correlated with the disintegration of the middle lamella, suggesting that the middle lamella was replaced by air. However, intercellular spaces decreased as fruit became very ripe probably due to the loss of tissue integrity.

Changes in cellular structure are tissue-dependent. Intercellular air spaces did not increase in the locular region in the inner pericarp. Similarly, cell walls show less change in the locular region than those of other tissues, and thickening of cell walls occur minimally. In the locular wall, however, intercellular space did increase [20].

The swelling of the cell walls and weakening of the middle lamella produced a desirable soft texture in the kiwifruit [20]. The changes in kiwifruit are similar to those in other fruit; however, the changes in kiwifruit occurred prior to the climacteric (or commencement of the ethylene production), contrary to that in other climacteric fruits. Structural studies using tensile strength measurements were applied to kiwifruit at different stages of ripeness and the rupture locations were determined by scanning electron microscope observation. Cell walls ruptured in freshly harvested, unripe kiwifruit. As the tissue got softer, the stress breaks occurred around cells, indicating that the adhesive substance between cells, the middle lamella, was weaker than the cell walls [25].

The skin of kiwifruit is covered with numerous hairs or trichomes of two size classifications. The large trichomes are roughly 2.5 mm in length and the shorter ones are twisted and are about 200 μm long (Figure 18.10a–h). In addition to the elongated trichomes, platelets occur on the surface of the skin. The platelets do not really look like wax, since they are quite thick, but look more like part of the epidermis. They might, therefore, be flattened trichomes or outgrowths of epidermal cells [1]. In fresh kiwifruit, the platelets appear to be closely appressed to the surface of the fruit (Figure 18.10c and e), whereas those in the aged fruit have started peeling up from the fruit surface (Figure 18.10d and f), allowing the fungal hyphae, which are now covering the surface, to penetrate into the fruit and cause decay of the fleshy fruit.

The outer pericarp tissue of fresh kiwifruit (Figure 18.11a) contains a mixture of large and small cells. The outer cells, just beneath the skin, are radially compressed in the fresh (Figure 18.11a) as well as in the aged (Figure 18.11b) fruits. Cells further into the outer pericarp tissue of the aged fruit are compressed slightly (Figure 18.11b). Closer examination reveals that the cells in the outer region of the outer pericarp in the fresh kiwifruit are rounded just below the radially compressed cells (Figure 18.11c and e). Cells in aged tissue have begun to lose their shapes (Figure 18.11d and f) and cell walls have been damaged (Figure 18.11f).

Starch granules apparent in the fresh (Figure 18.11e and g) kiwifruit have virtually disappeared in the aged (Figure 18.11f and h) fruit, in agreement with Hallett et al. [26] in their observations of starch degradation. Cell walls in the outer pericarp in the aged fruit (Figure 18.11f) were much thicker than those in the fresh (Figure 18.11e) kiwifruit.

FIGURE 18.10 Kiwifruit. Scanning electron micrographs of the skin of kiwifruit. (a) Fresh fruit showing the long trichomes covering the skin surface. (b) Aged fruit. (c) Fresh fruit showing the short, twisted trichomes and platelets covering the skin surface; the platelets might be flat trichomes since they do not appear to be wax and appear to be outgrowths of the epidermis. (d) Aged fruit showing extensive growth of fungal hyphae and the "lifting" of the surface platelets. (e) Fresh fruit showing the long and short trichomes. (f) Aged fruit showing long and short trichomes and the extensive growth of fungi. (g) Fresh fruits showing the base of a long trichome. (h) Aged fruit showing the base of a long trichome.

18.4 ARCHITECTURAL DECAY AND MICROSTRUCTURAL CHANGES

Fruits that are usually eaten ripe are typically harvested in the unripe or "green" state to allow desirable changes (i.e., ripening) to occur during the time it takes for the product to reach the consumer. Ripening usually causes a loss of firmness in tissue resulting from the depolymerization of cell wall components and the dissolution of the middle lamella, except in regions that contain plasmodesmata [27]. Intercellular adhesion decreases with ripening and continues to decrease during aging [28]. Physiological and chemical changes occur during ripening and have been well documented [29]. Many papers on ripening have been published, and these provide insight into some of the physiological changes that continue to occur as the fruit decays.

Fruit decay can be accelerated by dehydration or water loss. Water loss (weight loss) is an important consideration in the preservation of structure and texture. Stomata and the cuticle play roles in vapor diffusion or water loss [30].

Loss of water during storage was found to be associated with an increase in the volume of intercellular space and the breakdown of the middle lamella, resulting in separation between cells that correlated well with mealy texture in apples [31]. Water loss generally occurs through cracks in the cuticle; in 'Braeburn' apples, water loss was positively correlated with microcracks in the cuticle [32]. Cracks also occurred in the cuticle of 'Golden Delicious' apples and grew larger during maturation and continued to increase during postharvest storage [33]. Thus, the presence of cracks in the cuticle may decrease firmness of fruit.

Fruit firmness is also related to cellular characteristics such as the strength of the cell walls, intracellular turgor pressure, the number and sizes of intercellular spaces, and cell-to-cell adhesion [34]. Adhesion between cells, a function of the middle lamella, plays a significant role in the texture of produce. Firm apple texture is highly dependent on an intact middle lamella with good adhesive properties and firm cell walls. Mealy texture in apple fruit is attributed to the disintegration of the middle lamella such that, when a force is applied, breaks occur between instead of through cells [35].

Differences in cell-to-cell cohesiveness was also noted in calcium-treated vs. untreated 'Golden Delicious' apples [36], as well as in 'Granny Smith' (a hard, long-term-storage cultivar) vs. 'Rubinette' (a soft-textured, mealy cultivar) apples [37], in which the calcium-treated and the hard cultivar fractured intracellularly and the untreated and the softer cultivar fractured intercellularly. Calcium has been shown to have positive effects on maintaining cell wall structure [38]. Calcium chloride treatments firm fruit tissue by reacting with pectic acid in the cell wall to form calcium pectate, which strengthens molecular bonding between constituents of the cell wall [39]. Transmission electron microscope studies of apple mesocarp tissue during storage revealed that most of the changes due to calcium treatment involved the middle lamella [40]. Calcium treatment of 'Golden Delicious' apples is common practice and is known to have positive effects during long-term storage and has been the subject of numerous investigations [41–43].

The development of senescence in 'Golden Delicious' apples was studied by microstereology and electron microscopy over a 7-month storage period. In the skin tissue, the cytoplasm decreased in volume and density, ribosomes disappeared, and the tonoplast and plasmalemma were frequently ruptured. The flesh exhibited the most changes and was already found to be aging at the time of harvest. In the tissue aged 7 months, the cytoplasm was almost nonexistent and only fragments of membranes and vesicles were apparent. The matrix of the middle lamella disappeared

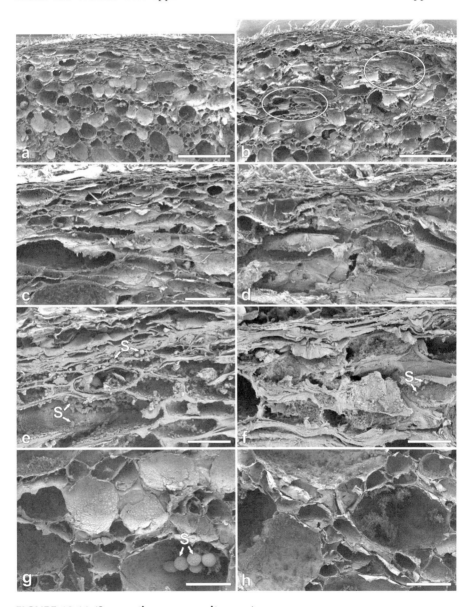

FIGURE 18.11 (See caption on opposite page)

and cellulose microfibrills, resulting from the depolymerization of the cell wall, became more apparent [28]. Thus, long-term storage can result in a number of microstructural changes in tissues.

The microstructure of pears has a tremendous influence on their storability, and different treatments to increase shelf life influence the microstructure. Calcium sprays and early harvest dates reduced disorders such as brown core, cork spot, and superficial scald on pears [44]. Irradiation and calcium treatment was found to be effective in increasing shelf life; irradiation reduced the microbial load but softened the skin and calcium reduced the amount of softening caused by irradiation treatments [45]. Neither treatment changed the microstructure of the pear skin.

De Belie et al. [46] studied the influences of cell turgor and fruit ripening on pear tissue and characteristics of the cells. Intercellular fractures occurred in firm tissues, intracellular fractures occurred in soft tissues, and both types of fractures occurred at intermediate firmnesses. Pears were incubated in solutions of differing water potential, and for firm pears a clear influence of the water potential of the incubation solution was found. There was no influence of water potential in soft pears, however. In firm pears, incubation in hypotonic solutions resulted in an increase in tensile strength and they became stress-hardened, while submersion in hypertonic solutions caused a decrease in tensile strength.

Video microscopy was employed to view the dynamics of tissue failure during tensile fracturing of tissue slices of pear [47]; this method gave the added dimension of time to texture assessment and allowed viewing of cell changes during deformation of living tissue. Time is an important concept since the sensation of texture occurs over the period of time required for mastication.

'Forcelle' is a small pear with brownish yellow skin. The fresh skin has a regular, reticulated appearance throughout plus larger, rounded cracks, or lenticels (Figure 18.12a). Extended, refrigerated storage, for a period of about a month, resulted in a soft rot at the blossom end of the fruit. Portions of the rotted skin and the tissue underneath were studied by SEM. The wax seemed to be completely destroyed and the skin had a reticulated appearance due to cracking that outlined groups of epidermal cells. The reticulations were different and much farther apart in the aged fruit (Figure 18.12b) than those on the fresh sample. A closer view revealed that the reticulations in the fresh fruit had a fibrous nature that connected the wax plates

---

FIGURE 18.11 (Opposite page) Scanning electron micrographs of cross sections of kiwifruit. (a) Fresh cross section of the outer pericarp showing a mixture of large and small cells in the flesh and the radially compressed cells in the flesh near the surface. (b) Aged cross section showing areas of further compression resulting from aging. (c) Fresh fruit periphery showing extent of compression of cells. (d) Aged fruit periphery showing more compression and cell damage. (e) Fresh fruit just beneath surface periphery showing starch granules and intact but slightly radially compressed cells. (f) Aged fruit of a similar area as in e showing more compression, damaged cells and fewer starch granules than in the fresh sample; cell wall thickening is also apparent. (g) Fresh sample; cells in the inner portion of the outer pericarp showing regular shapes and starch granules. (h) Aged sample; cells in the inner portion of the outer pericarp showing regular cell shapes and the lack of starch granules; cell walls might be slightly thicker.

(Figure 18.12c). In the aged sample, it appeared as if the wax had "bloomed" or migrated on the fruit surface since there was a roughened surface that did not appear to be due to microbial growth (Figure 18.12d). The interiors of the lenticels of the fresh pear fruit had limited fungal growth (Figure 18.12e). In the aged fruit, the fungal growth in the lenticels was excessive, the fungus had sporulated, and the underlying layers of wax and epidermal cells had been completely destroyed. Cells of the outer pericarp tissue were exposed (Figure 18.12f). The fresh sample had

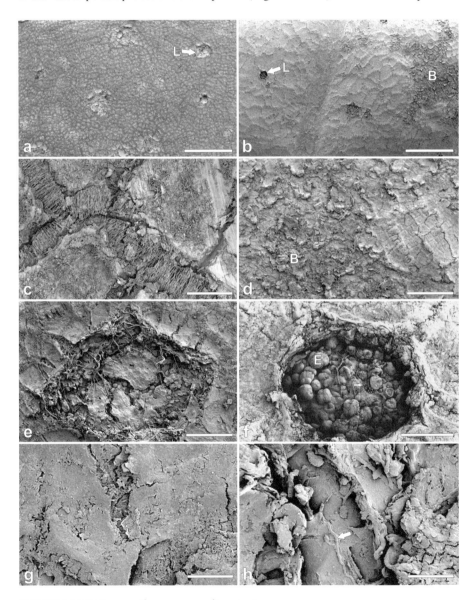

FIGURE 18.12 (See caption on opposite page)

cracks in the wax layer on the surface (Figure 18.12g); however, the cracks in the aged sample were much deeper and fungal hyphae were evident (Figure 18.12h).

A cross section of fresh pear showed angular, thin-walled parenchyma in the mesocarp and the tissue fractured through cells (Figure 18.13a). The middle lamella in the aged pear sample was completely destroyed, evidenced by the fracture plane that occurred around rather than through cells (Figure 18.13b and c). Cells had lost turgor and intercellular gaps (Figure 18.13b and c) were apparent in the aged sample. Cell walls in the inner mesocarp were angular and rigid in the fresh pear (Figure 18.13e), whereas in the aged pear the inner mesocarp cell walls were reticulated and had obviously lost turgor (Figure 18.13f). The cell walls in the aged sample may have been thicker but it was difficult to determine due to the angle of fracture, the difficulty in obtaining a clean fracture, and the depth of focus in the SEM.

18.5 CELL COLLAPSE AND DEGRADATION

Grapes, botanically berries, are consumed as fresh fruit, dried fruit, juice, and wine. There are a multitude of cultivars that make up this vast array of end products. Wine grapes are typically small and grow in tightly compacted bunches. Some cultivars are particularly susceptible to bunch rot disease, caused by *Botrytis cinerea*. The susceptibility to the disease might be related to the fact that the grape grows in tight clusters, and thus the microenvironment is ideal for the growth of the fungi [48]. Cell walls also undergo significant changes during development; for example, during ripening, changes in cell wall composition include significant modification of specific polysaccharides as well as large changes in protein composition [49,50].

Red seedless grapes, consumed as fresh berries, have a smooth, slightly dimpled skin (Figure 18.14a and b). There are no stomata on the berry surface, and therefore the cuticle plays the major role in vapor diffusion. The aged grape shows enormous changes in the skin resulting from dehydration followed by the growth of microorganisms on the surfaces (Figure 18.14c and d). Even in areas having

---

FIGURE 18.12 (Opposite page) Scanning electron micrographs of 'Forcelle' pear skin. (a) Fresh skin showing the reticulated pattern and lenticels. (b) Aged sample showing the edges of epidermal cells and a larger reticulated pattern that resulted from cracks around groups of epidermal cells. The layer of wax had been destroyed and redistributed and as a "bloom" pattern. (c) close view of the reticulated pattern in the fresh sample that appeared to be fibers holding the wax platelets together. (d) Aged sample showing part of the bloom and the tops of the epidermal cells without a wax covering. (e) Lenticel in fresh fruit; fungal hyphae are growing on the surface and the wax covering is cracked. (f) Lenticel in aged fruit showing the profuse growth of fungi, the complete destruction of the wax layer, fungal spores, and exposed epidermal cells. (g) Close view of wax layer on fresh pear. (h) Aged sample, close view of the disrupted wax in the bloom area; arrow indicates fungal hypha. B, bloom; E, epidermal cell; L, lenticel. Magnification bars: a, b, 1 mm; c, 10 μm; d, e, 200 μm; f, 100 μm; g, h, 20 μm.

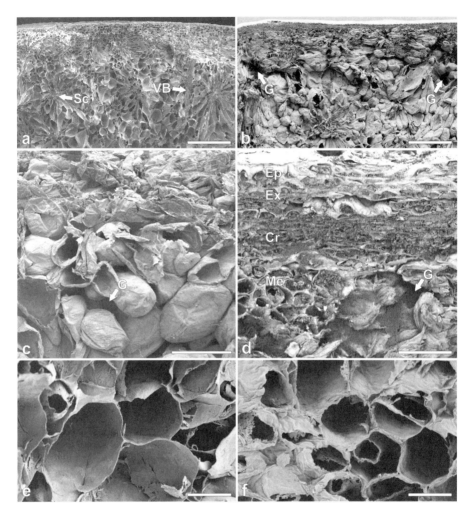

FIGURE 18.13 Scanning electron micrographs of 'Forcelle' pear cross sections. (a) Fresh surface and underlying tissues of the mesocarp showing fresh fracture plane where breaks occur through cells. (b) Aged surface and underlying tissues where the fracture occurs between cells, indicating that the middle lamella has failed. (c) Aged sample of mesocarp cells showing nonturgid cells; most breaks have occurred between cells, intercellular gaps and extensive tissue disruption toward the top of the micrograph. (d) Aged sample at the periphery showing the epidermis, exocarp, crushed peripheral mesocarp, and underlying mesocarp tissues; note the intercellular gap in the mesocarp. (e) Fresh inner mesocarp showing fully hydrated cell walls. (f) Aged inner mesocarp showing dehydrated cell walls and possible cell wall thickening. Magnification bars: a, b, 500 μm; c, 200 μm; d, 50 μm; e, f, 25 μm.

relatively minor wrinkling, more debris was found on the surfaces (Figure 18.14e), and at high magnifications the wax seems to have dissolved and redeposited in an irregular fashion (Figure 18.14f).

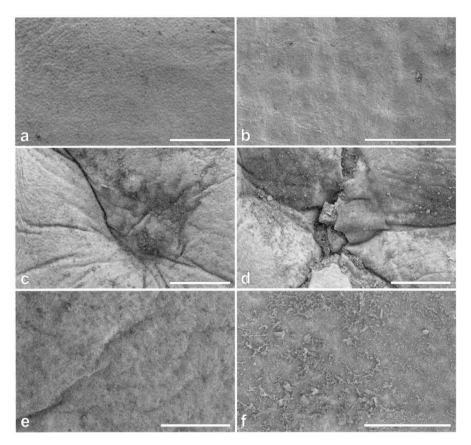

FIGURE 18.14 Scanning electron micrographs of skin of red seedless grape. (a) Fresh sample showing smooth surface. (b) Aged sample showing minor dimpling. (c) Aged sample showing major wrinkling. (d) Aged sample showing major surface wrinkling with embedded debris. (e) Aged sample; close view of a minor wrinkled area. (f) Aged sample showing disrupted cuticular wax. Magnification bars: a, c, d, 500 μm; b, f, 100 μm; e, 200 μm.

A cross section of the fresh red seedless grape reveals thin epidermal cells, small exocarp cells with slightly thickened cell walls that look as if they are filled with pigmented cytoplasm, and large, thin-walled parenchyma cells in the mesocarp (Figure 18.15a). A damaged grape (Figure 18.15b) looks as if it were pierced by a foreign object that was then removed. The object damaged the external and internal tissues and resulted in growth of callus tissue on the surface and in the interior of the tissue.

Cell walls and cells lost turgor and showed the effects of dehydration (Figure 18.15c). Cell walls in the fresh sample were angular in shape and tightly appressed (Figure 18.15d). Cell walls in the aged sample were reticulated and intercellular spaces became more apparent (Figure 18.15e). Cells in the aged sample appeared to have thickened cell walls or middle lamella, which may have been disintegrating (Figure 18.15g); this was not apparent in cells in the fresh sample (Figure 18.15f).

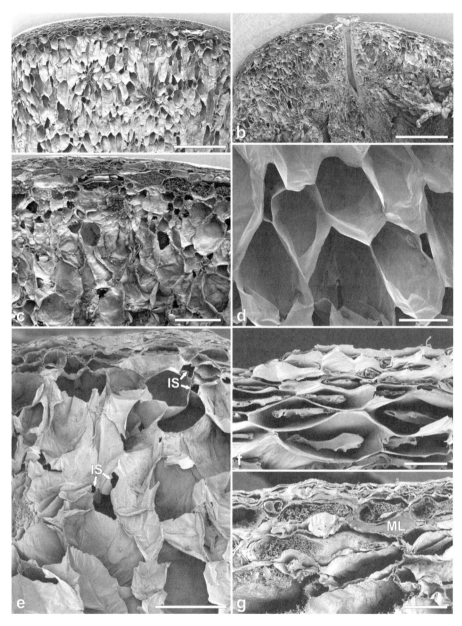

FIGURE 18.15 Scanning electron micrographs of the interior of red seedless grape. (a) Fresh external layers and mesocarp tissue. (b) Aged external layers and mesocarp showing the remnants of an object that had pierced the grape and was removed, showing callus growth on the grape surface. (c) Aged sample showing collapsed cells of the mesocarp. (d) Fresh sample showing mesocarp cell walls. (e) Aged sample showing collapse and distortion of the mesocarp cell walls. (f) Fresh section epidermis and exocarp showing thin cell walls. (g) Aged epidermis and exocarp showing thickened cell walls and some collapse of the epidermis and exocarp tissues. C, callus tissue; IS, intercellular space; ML, middle lamella. Magnification bars: a, b, 1 mm; c, 250 μm; d, 100 μm; e, 200 μm; f, g, 50 μm.

18.6 STRATEGIES FOR FRUIT PRESERVATION AND IMPROVED SHELF LIFE

Since fresh produce is shipped worldwide, the time for a product to reach the market may be considerable. Cold storage and commodity-specific packaging is a typical method of protection and preservation for many fruits and vegetables. Modified atmosphere packaging and storage under controlled atmospheres [51] are also of importance, particularly in the minimally processed food arena.

At the consumer level, water sprays to avoid dehydration combined with refrigeration is common practice in the supermarket. However, refrigeration simply slows degradative processes and is often insufficient to meet the demands of today's marketing techniques. Therefore, other postharvest treatments are needed to further increase storage life. Edible coatings, such as waxes, and those that also include antibrowning agents form a semipermeable barrier to air to control respiration and prolong shelf life [39,52]. Due to increasing concern about unacceptable residues on fruit surfaces following chemical treatments, a number of chemical-free postharvest techniques have been investigated to lengthen the storage life of fresh produce.

Postharvest heat treatment has been in commercial use for many years [53]; however, the technology is still being perfected for different commodities. Short hot water rinses were investigated as a method of reducing postharvest losses of Galia melon (cantaloupe) [54] and sweet pepper [55] fruits. Scanning electron microscopy showed that the fruit surfaces were free of debris and fungal spores and that superficial cracks in the epidermis were sealed. Observation of fresh orange bell pepper revealed superficial cracks and debris on the surface (Figure 18.16a). Aging resulted in collapse of mesocarp tissue (Figure 18.16b and c) just beneath the outer tissue layers.

Hot water immersion was effective in disinfestation and in maintaining fruit quality of Valencia oranges [56]. Light and electron microscopy showed that there was little effect on surface waxes of the orange fruit as a result of heating. Mandarin oranges were heat-treated for 3 min prior to storage at temperatures ranging from 50 to 58°C in the assessment of temperature effectiveness in storage performance. Treated fruit surfaces were compared to control fruits by SEM observation. Heat treatment of the mandarins at 50 to 54°C smoothed the waxy surface, which was granular in the control, partly removed the wax at 56°C, and completely removed the wax at 58°C [57]. Hot water brushing of organic citrus fruits reduced postharvest decay and scanning electron microscopy showed that the epicuticular waxes had been smoothed by the treatment and the cracks and stomata were sealed by the wax, minimizing the sites of pathogen entry [58].

Heat treatment of mango was ineffective for heat-sensitive cultivars, in which damage resulted from the heat treatment. Such heat-related injuries were investigated using scanning and transmission electron microscopy where cuticle and exocarp ruptured and exposed internal cells. Cell walls of the mesocarp were convoluted and thickened and starch granules still remained in the tissue, suggesting that the carbohydrate metabolic enzymes had been activated [59].

Other nontoxic shelf-life extension methods include treatment with carbon dioxide followed by refrigerated storage. Firmness was increased and decay susceptibility

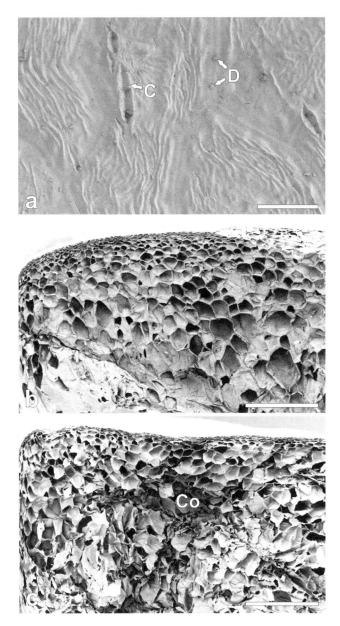

FIGURE 18.16 Scanning electron micrographs of orange bell pepper. (a) Fresh skin showing cracks in the cuticle. (b) Fresh cross section of outer layers and mesocarp. (c) Aged cross section showing outer layers and mesocarp exhibiting collapse of the outer mesocarp. C, crack; Co, collapsed mesocarp cells, D, debris. Magnification bars: a, 50 μm; b, c, 1 mm.

decreased (compared to those at harvest) in strawberries following storage at 0°C. The positive effects were enhanced by treatment with CO_2 followed by storage at 0°C. Examination of the fracture surfaces of strawberry following tensile testing indicated that the primary mode of tissue failure was cell-to-cell debonding, not cell breakage, in both air- and CO_2-treated fruit [60]. An increase in firmness over harvest is common in fruit that is stored at 0°C. The mechanism is not well understood but is theorized to be due to an increase in viscosity of pectin because no change in structure or chemistry is observed [61,62].

18.7 SPECIFIC DISORDERS AND THEIR IMPACT ON PRODUCE QUALITY

Produce shows an increase in susceptibility to postharvest diseases and infestation during prolonged storage partly due to ongoing physiological changes that enable pathogen development on or in the fruits [63]. Electron microscopy exhibited the "withered juice sac" or dry pulp disorder, where the soluble solids content is higher in the rind than in the pulp, in Ponkan mandarin orange at the cellular level. Prior to the onset of symptoms of the disorder, nuclear divisions occurred in the cells of the flavedo (the colored part of the rind). The appearance of symptoms corresponded to further changes in the flavedo cells, the enlargement of the cells and their vacuoles, and a decrease in the cytoplasm content. The tonoplast disappeared in latter stages of the disorder and only a nucleus and small amounts of cytoplasm remained. Thus, the disorder was caused by changes occurring in the rind: cell division, growth, and senescence, which caused water to be translocated to the rind due to a water potential gradient. Prestorage treatment with gibberellic acid, a plant hormone that slows senescence, delayed the onset of the disorder [64].

Water loss, determined by weight loss, was found to be a nondestructive predictor of chilling injury in grapefruit and lemon. Chilling injury appears as distinct, swollen areas with pitting and cuticular damage. Prior to the appearance of the gross symptoms of chilling injury, scanning electron microscopy revealed calcium oxalate crystals growing inside cracks that had developed around the stomata [65].

Electron microscopy and electrophoresis were useful in determining the reason for degreening inhibition of bananas. In bananas, degreening is inhibited above 24°C; ripening at higher temperatures results in retained thylakoid membranes, a delayed breakdown in chlorophyll b, and a reduced breakdown of pigment-protein complexes. Retention of thylakoid membranes is an important factor in the failure of Cavendish bananas to degreen when ripened at tropical temperatures [66]. Fruit softening in bananas is the result of the coordinated degradation of pectin, hemicellulose, and starch (Figure 18.17a and b) in banana pulp [67]. Stress-relaxation probe measurements were used to describe the changes in physical properties during softening of the banana fruit pulp.

Eggplants had a higher average gloss reading than mature green tomatoes or apples. Scanning electron microscopy showed that the epicuticular wax covering of eggplants was smoother than that on tomatoes or apples, thus, giving a more effective light scattering surface. The removal of wax from eggplants resulted in decreases in gloss. The results indicate how wax influences shininess; however, the roughness measurements of flattened peel after wax removal suggest that surface topography also influences gloss [68]. The tomato has a dimpled surface (Figure 18.18a) with

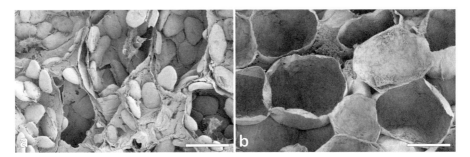

FIGURE 18.17 Scanning electron micrographs of banana. (a) Fresh banana showing starch granules. (b) Aged banana showing the same type of cells without starch granules. Magnification bars: a, b, 50 μm.

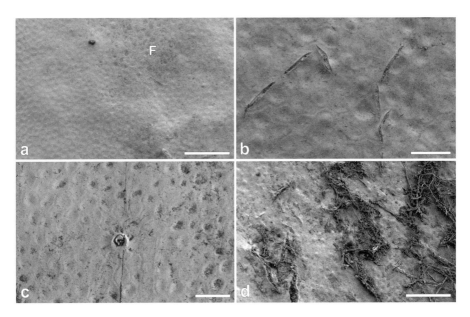

FIGURE 18.18 Scanning electron micrographs of aged tomato skin. (a) Normal dimpled surface with some bubbling and fungal growth. (b) Cracks in the cuticle and epidermis. (c) Damaged area and some cracks. (d) Fungal growth on the surface and into cracks. Magnification bars: a, d, 200 μm; b, 100 μm; c, 50 μm.

no apparent stomata. During aging, a soft rot appeared and revealed cracking (Figure 18.18b), damage (Figure 18.18c), and extensive fungal growth (Figure 18.18d).

A number of malformations may result during the growing of tomatoes from stress to the plant [69,70].The fresh tomato fruit interior has thin epidermal and exocarp layers with large parenchyma making up the mesocarp (Figure 18.19a). In the region of the soft rot, the epidermis, exocarp, and first few layers of the mesocarp are crushed (Figure 18.19b) and there is some further damage in the outer mesocarp

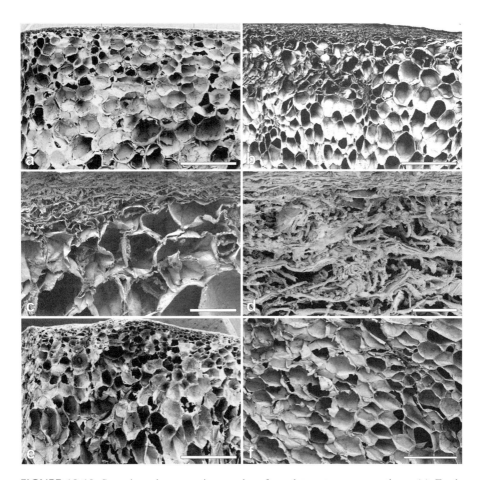

FIGURE 18.19 Scanning electron micrographs of aged tomato cross sections. (a) Fresh tomato showing thin epidermis and exocarp just beneath and large parenchyma cells of the mesocarp. (b) Aged sample in the location of soft rot showing the crushed outer layers and damage to cell walls in the mesocarp layer. (c) Aged sample, close view of part of b showing crushed outer layers and convoluted cell walls of the mesocarp. (d) Crushed outer layers. (e) Aged sample from less damaged area showing underlying tissues under one of the bubbled portions of skin with some distortion of mesocarp cell walls. (f) Aged sample showing inner mesocarp tissue; cell walls are dehydrated, notable by the collapsed cell walls. Magnification bars: a–d, 1 mm; e, 100 μm; f, 200 μm.

(Figure 18.19c). The crushed layers of the epidermis, exocarp, and mesocarp might have fungal growth but it is difficult to differentiate hyphae from the fibers of the tomato tissue. In a less damaged area of the tomato away from the region of the soft rot, the structure appears more normal with some slight deformities in the mesocarp (Figure 18.19e). The inner mesocarp from the less damaged area of the aged tomato shows minor disruption of tissue structure (Figure 18.19f).

18.8 CONCLUSION

All plant cells are surrounded by rigid cell walls. With adequate turgor and nutrients, cell walls provide support to the plant, maintain structure, and give produce commodities their specific textural characteristics. Dehydration, physical damage, and physiological change all contribute to the phenomenon of aging in plant tissues. Aging wears down the defenses of the fruit or vegetable and opens a window of opportunity for microbial or insect invasion. Specific measures can be taken to slow the aging process, such as refrigeration, commodity-specific packaging, and various postharvest treatments. None of the treatments are preventative measures; they extend shelf life, which means that we are able to ship produce commodities worldwide throughout the year.

Specific examples of produce degradation have been shown in scanning electron micrographs in this chapter. Similarities between various plants include an intolerance of dessication followed by rapid wrinkling of the outer layers and cracking of the cuticle. The mesocarp may separate from the outer layers. At advanced stages, the middle lamella degenerates and cells separate under strain. Finally, microbes start invading the unprotected tissues.

REFERENCES

1. Esau, K., *Anatomy of Seed Plants*, 2nd ed., John Wiley & Sons, New York, 1977.
2. Jewell, G.G., Fruits and vegetables, in *Food Microscopy, Food Science and Technology*, Vaughn, J.G., Ed., Academic Press, New York, 1979, p. 1.
3. Jensen, W.A., *Botanical Histochemistry*, W.H. Freeman, San Francisco, 1962.
4. Wood, D.F. and Cornish, K., Microstructure of purified rubber particles, *Int. J. Plant Sci.*, 161, 435, 2000.
5. Humphreys, W.J., Spurlock, B.O., and Johnson, J., Critical point drying of ethanol-infiltrated, cryofractured biological specimens, *Scanning Elect. Microsc.*, 1, 275, 1974.
6. Glenn, G.M. et al., Role of cuticles in produce quality and preservation, in *Produce Degradation: Pathways and Their Prevention*, Lamikanra, O. and Imam, S., Eds., CRC Press, Boca Raton, FL, 2004, Chap. 3.
7. Bukovac, M.J., Rasmussen, H.P., and Shull, V.E., The cuticle: surface structure and function, *Scanning Elect. Microsc.*, 3, 231, 1981.
8. Beaulieu, M. et al., Dose rate effect of irradiation on phenolic compounds, polyphenol oxidase, and browning of mushrooms (*Agaricus bisporus*), *J. Agric. Food Chem.*, 47, 2537, 1999.
9. Braaksma, A. et al., Morphometric analysis of ageing mushrooms (*Agaricus bisporus*) during postharvest development, *Postharvest Biol. Technol.*, 13, 71, 1998.
10. Costerton, J.W. et al., Bacterial biofilms in nature and disease, *Annu. Rev. Microbiol.*, 41, 435, 1987.
11. Costerton, J.W. et al., Microbial biofilms, *Annu. Rev. Microbiol.*, 49, 711, 1995.
12. Tian, M.S. et al., A role for ethylene in the yellowing of broccoli after harvest, *J. Am. Soc. Hortic. Sci.*, 119, 276, 1994.
13. King, G.A. and Morris, S.C., Early compositional changes during postharvest senescence of broccoli, *J. Am. Soc Hortic. Sci.*, 119, 1000, 1994.

14. Gunawan, M.I. and Barringer, S.A., Green color degradation of blanched broccoli (*Brassica oleracea*) due to acid and microbial growth, *J. Food Process. Preserv.*, 24, 253, 2000.

15. Punja, Z.K., Chittaranjan, S., and Gaye, M.M., Development of black root rot caused by *Chalara elegans* on fresh market carrots, *Can. J. Plant Pathol.*, 14, 299, 1992.

16. Tatsumi, Y., Watada, A.E., and Wergin, W.P., Scanning electron microscopy of carrot stick surface to determine cause of white translucent appearance., *J. Food Sci.*, 56, 1357, 1991.

17. Barry-Ryan, C. and O'Beirne, D., Quality and shelf-life of fresh cut carrot slices as affected by slicing method, *J. Food Sci.*, 63, 851, 1998.

18. Buick, R.K. and Damoglou, A.P., The effect of vacuum packaging on the microbial spoilage and shelf-life of 'ready-to-use' sliced carrots, *J. Sci Food. Agric.*, 38, 167, 1987.

19. Harker, F. R. et al., Texture of parenchymatous plant tissue: a comparison between tensile and other instrumental and sensory measurements of tissue strength and juiciness, *Postharv. Biol. Technol.*, 11, 63, 1997.

20. Hallett, I.C., MacRae, E.A., and Wegrzyn, T.F., Changes in kiwifruit cell wall ultra-structure and cell packing during postharvest ripening, *Int. J. Plant Sci.*, 153, 49, 1992.

21. Sutherland, P. et al., Localization of cell wall polysaccharides during kiwifruit (*Actinidia deliciosa*) ripening, *Int. J. Plant Sci.*, 160, 1099, 1999.

22. Bauchot, A.D. et al., Cell wall properties of kiwifruit affected by low temperature breakdown, *Postharv. Biol. Technol.*, 16, 245, 1999.

23. Redgwell, R.J. et al., *In vivo* and *in vitro* swelling of cell walls during fruit ripening, *Planta*, 203, 162, 1997.

24. Paull, R.E., Gross, K., and Qiu, Y., Changes in papaya cell walls during fruit ripening, *Postharv. Biol. Technol.*, 16, 79, 1999.

25. Harker, F.R. and Hallett, I.C., Physiological and mechanical properties of kiwifruit tissue associated with texture change during cool storage, *J. Am. Soc. Hortic. Sci.*, 119, 987, 1994.

26. Hallett, I.C., Wegrzyn, T.F., and MacRae, E.A., Starch degradation in kiwifruit: *in vivo* and *in vitro* ultrastructural studies, *Int. J. Plant Sci.*, 156, 471, 1995.

27. Roy, S., Watada, A.E., and Wergin, W.P., Characterization of the cell wall micro-domain surrounding plasmodesmata in apple fruit, *Plant Physiol.*, 114, 539, 1997.

28. de Barsy, T., Deltour, R., and Bronchart, R., Ultrastructrual and microstereological study of aging in Golden Delicious apples during cold storage, *Acta Hortic.*, 258, 379, 1989.

29. White, P.J., Recent advances in fruit development and ripening: an overview, *J. Exp. Bot.*, 53, 1995, 2002.

30. Possingham, J.V. et al., Cuticular transpiration and wax structure and composition of leaves and fruit of *Vitis vinifera*, *Aust. J. Biol. Sci.*, 20, 1149, 1967.

31. Tu, K. et al., Monitoring post-harvest quality of *Granny Smith* apple under simulated shelf-life conditions: destructive, non-destructive and analytical measurements, *Int. J. Food Sci. Technol.*, 31, 267, 1996.

32. Maguire, K.M. et al., Relationship between water vapour permeance of apples and micro-cracking of the cuticle, *Postharv. Biol. Technol.*, 17, 89, 1999.

33. Roy, S. et al., Changes in the ultrastructure of the epicuticular wax and postharvest calcium uptake in apples, *Hortic. Sci.*, 34, 121, 1999.

34. Harker, R.F. et al., Texture of fresh fruit, *Hortic. Rev.*, 20, 121, 1997.

35. Reeve, R.M., Relationships of histological structure to texture of fresh and process fruits and vegetables, *J. Texture Stud.*, 1, 247, 1970.
36. Glenn, G.M. and Poovaiah, B.W., Calcium-mediated postharvest changes in texture and cell wall structure and composition in Golden Delicious apples, *J. Am. Soc. Hortic. Sci.*, 115, 962, 1990.
37. Lapsley, K.G., Escher, F.E., and Hoehn, E., The cellular structure of selected apple varieties, *Food Struct.*, 11, 339, 1992.
38. Glenn, G.M., Reddy, A.S.N., and Poovaiah, B.W., Effect of calcium on cell wall structure, protein phosphorylation and protein profile in senescing apples, *Plant Cell Physiol.*, 29, 565, 1988.
39. Lee, J.Y. et al., Extending shelf-life of minimally processed apples with edible coatings and antibrowning agents, *Lebensm.-Wiss. Technol.*, 36, 323, 2003.
40. Siddiqui, S. and Bangerth, F., The effect of calcium infiltration on structural changes in cell walls of stored apples, *J. Hortic. Sci.*, 71, 703, 1996.
41. Roy, S., et al., Heat treatment affects epicuticular wax structure and postharvest calcium uptake in 'Golden Delicious' apples, *HortSci.*, 29, 1056, 1994.
42. Roy, S. et al., Surfactants affect calcium uptake from postharvest treatment of 'Golden Delicious' apples, *J. Am. Soc. Hortic. Sci.*, 121, 1179, 1996.
43. Siddiqui, S. and Bangerth, F., Studies on cell wall mediated changes during storage of calcium-infiltrated apples, *Acta Hortic.*, 326, 105, 1993.
44. Raese, J.T. and Drake. S.R., Effect of calcium sprays, time of harvest, cold storage, and ripeness on fruit quality of 'Anjou' pears, *J. Plant Nutr.*, 23, 843, 2000.
45. Kovács, E. et al., Structural and chemical changes of pear skin, *Acta Hortic.*, 368, 243, 1994.
46. De Belie, N. et al., Influence of ripening and turgor on the tensile properties of pears: A microscopic study of cellular and tissue changes, *J. Am. Soc. Hortic. Sci.*, 125, 350, 2000.
47. Hallett, I.C. and Harker, F.R., Microscopic investigations of fruit texture, *Acta Hortic.* 464, 411, 1998.
48. Comménil, P., Brunet, L., and Audran, J.-C., The development of the grape berry cuticle in relation to susceptibility to bunch rot disease, *J. Exp. Bot.*, 48, 1599, 1997.
49. Nunan, K.J. et al., Changes in cell wall composition during ripening of grape berries, *Plant Physiol.*, 118, 783, 1998.
50. Nunan, K.J. et al., Isolation and characterization of cell walls from the mesocarp of mature grape berries (*Vitus vinifera*), *Planta*, 203, 93, 1997.
51. Mattheis, J. and Fellman, J.K., Impacts of modified atmosphere packaging and controlled atmospheres on aroma, flavor, and quality of horticultural commodities, *HortTechnology*, 10, 507, 2000.
52. King, A.D. and Bolin, H.R., Physiological and microbiological storage stability of minimally processed fruits and vegetables, *Food Technol.*, 43, 132, 1989.
53. Barkai-Golan, R. and Phillips, D.J., Postharvest heat treatment of fresh fruits and vegetables for decay control, *Plant Dis.*, 75, 1085, 1991.
54. Fallik, E. et al., Reduction of postharvest losses of Galia melon by a short hot-water rinse, *Plant Pathol.*, 49, 333, 2000.
55. Fallik, E. et al., A unique rapid hot water treatment to improve storage quality of sweet pepper, *Postharv. Biol. Technol.*, 15, 25, 1999.
56. Williams, M.H. et al., Effect of postharvest heat treatments on fruit quality, surface structure, and fungal disease in Valencia oranges, *Aust. J. Exp. Agric.*, 34, 1183, 1994.
57. Schirra, M. and D'hallewin, G., Storage performance of Fortune mandarins following hot water dips, *Postharvest Biol. Technol.*, 10, 229, 1997.

58. Porat, R. et al., Reduction of postharvest decay in organic citrus fruit by a short hot water brushing treatment, *Postharv. Biol. Technol.*, 18, 151, 2000.

59. Jacobi, K.K. and Gowanlock, D., Ultrastructural studies of 'Kensington' mango (*Mangifera indica* Linn.) heat injuries, *HortScience*, 30, 102, 1995.

60. Harker, R.F. et al., Physical and mechanical changes in strawberry fruit after high carbon dioxide treatments, *Postharv. Biol. Technol.*, 19, 139, 2000.

61. Werner, R.A. and Frenkel, C., Rapid changes in the firmness of peaches an influenced by temperature, *HortScience*, 13, 470, 1978.

62. Werner, R.A., Hough, L.F., and Frenkel, C., Rehardening of peach in cold storage, *J. Am. Soc. Hortic. Sci.*, 103, 90, 1978.

63. Eckert, J.W. and Ogawa, J.M., The chemical control of postharvest diseases: deciduous fruits, berries, vegetables and root/tuber crops, *Annu. Rev. Phytopath.*, 26, 433, 1988.

64. Wang, S., Physiology and electron microscopy of the withered juice sac of Ponkan (*Citrus reticulata*) during storage, *Acta Hortic.*, 343, 45, 1993.

65. Cohen, E. et al., Water loss: a nondestructive indicator of enhanced cell membrane permeability of chilling-injured *Citrus* fruit, *J. Am. Soc. Hortic. Sci.*, 119, 983, 1994.

66. Blackbourn, H.D. et al., Inhibition of degreening in the peel of bananas ripened at tropical temperatures. III. Changes in plastid ultrastructure and chlorophyll-protein complexes accompanying ripening in bananas and plantains, *Ann. Appl. Biol.*, 117, 147, 1990.

67. Kojima, K., Sakurai, N., and Kuraishi, S., Fruit softening in banana: correlation among stress-relaxation parameters, cell wall components and starch during ripening, *Physiol. Plant.*, 90, 772, 1994.

68. Ward, G. and Nussinovitch, A., Gloss properties and surface morphology relationships of fruits, *J. Food Sci.*, 61, 973, 1996.

69. Boyle, J.S., Abnormal ripening of tomato fruit, *Plant Disease*, 78, 936, 1994.

70. Powell, R.K., Stoffella, P.J., and Powell, C.A., Internal tomato irregular ripening symptoms do not diminish upon storage, *HortScience*, 33, 157, 1998.

19 Structure and Function of Complex Carbohydrates in Produce and Their Degradation Process

Syed Imam, Justin Shey, Delilah Wood,
Greg Glenn, Bor-Sen Chiou, Maria Inglesby,
Charles Ludvik, Artur Klamczynski,
and William Orts
Bioproduct Chemistry & Engineering Research,
Western Regional Research Center, ARS-USDA, Albany, CA

CONTENTS

19.1 INTRODUCTION

Nature utilizes sophisticated and ingenious biochemical pathways in developing plants and/or fruits to synthesize simple carbohydrates such as monomeric sugars and assemble them into simple to complex polysaccharide structures to perform various physiological functions. Many of the molecules that make up the living tissue in plants are polymers, which include cellulose, complex carbohydrates of starchy foods, protein molecules, lipid, and DNA. Additionally, polysaccharides are further complexed or linked with polymers such as protein, lipids, and lignin to construct complex supramacromolecular structures and to confer tailored properties. For example, in woody plants, lignin, a polyphenolic polymer, is complexed with waxes that confer a shiny and smooth surface coating on the exterior of the plant cell walls, providing protection from infections and restricting the flow of moisture across the cell wall. In the absence of such coating, carbohydrate polymers will be destined to rapid hydrolytic degradation and the plant will quickly dehydrate. In this chapter we will only focus on carbohydrates such as starch, cellulose, hemicellulose, and pectin, which play critical roles in providing the structural integrity and important functionalities to plants, particularly in fruits and vegetables, during their growth and maturity, as well as during their ripening, storage, and other postharvest processing.

In general, polysaccharides in plants perform two main functions: they serve as a powerhouse to store energy for carrying out various metabolic functions and they provide structural integrity to plants. The simple distinction of α- vs. β-linkages between the glucose monomer units determines the function of polysaccharides in plant cells. Generally, the storage polysaccharides possess an α-linkage between the 1- and 4-positions of adjacent glucose units. In plant cells, this energy source is starch, which is present as either amylose (linear molecules of glucose units) or amylopectin (a branched glucose macromolecule). In animal cells, the energy source is a highly branched arrangement of glucose units, called glycogen. The α-linkage of storage polysaccharides imparts a helical conformation to the glucopyranosyl chain. Because the helix inhibits extensive interchain associations, it is unfit as a structural entity but excellent as an energy source because rapid degradation for energy release is not hampered by the necessity of first breaking strong intermolecular interactions.

On the other hand, structural polysaccharides provide the rigidity and elasticity needed to protect cells. Such structural polysaccharides are characterized by the β-linkages between glucopyranosyl units and, like storage polysaccharides, their macromolecular structure varies with variation in life form. Unmodified cellulose is the primary structural polysaccharide of plants. The β-linkage produces a nearly linear, extended macromolecular conformation that permits close packing of polymer chains; this close packing in turn encourages intermolecular hydrogen bonding (and crystallinity) between adjacent chains to produce the rigidity required in a structural material. The chemistry, structure, and functional properties of plant polysaccharides will be covered in greater detail in the subsequent sections. Also included is a brief discussion on enzymes, which are specific biological catalysts used in the synthesis and assembly of carbohydrate polymers that provide useful functionalities to plants.

TABLE 19.1
Possible Factors That Can Induce Changes in Postharvest Produce

Factors That Can Induce or Initiate Changes Leading to Produce Deterioration and Degradation	Associated Changes in Produce
Light (photo)	Molecular chain scission
Heat (thermal)	Depolymerization of polysaccharides
Water (hydrolysis)	Viscosity
Mechanical injury (physical)	Appearance
Chemicals (chemical)	Texture
Microbes and enzymes (bio)	Flavor
	Color

The role of enzymes in the syntheses, degradation, and modification of polysaccharides has been the subject of several excellent reviews [1–7]. In addition to enzymes, plants also synthesize and/or utilize any required cofactors and, eventually, inhibitors to terminate reactions once the polymer synthesis is complete or not needed anymore. Because nature assembles these structures, it has both the inherent ability and the capacity to degrade and depolymerize these complex structures into simple molecules or polymer building blocks (precursors) via specialized enzymes. Once the polymer degrades back to simple molecules, naturally occurring microbes and/or their enzymes can utilize them as a carbon source, recycling these materials back into the biosphere.

After the produce is harvested, it proceeds through a variety of phases that involve ripening, processing, packaging, transportation, and storage before it is finally delivered to the customers. During these phases, produce is exposed to a variety of stresses such as heat, irradiation, altered humidity conditions, mechanical injury, suffocation, infestation, etc., that can lead to postharvest deterioration of complex carbohydrates in fruits and vegetables, which alters their color, texture, flavor, and appearance (Table 19.1). This chapter focuses on the chemistry of the major structural carbohydrates generally found in produce, including their structure-function properties, interactions, and biochemical mechanisms that lead to their biodegradation and physical deterioration.

19.2 CHEMISTRY, FUNCTIONAL PROPERTIES, AND REACTION MECHANISMS OF PLANT CARBOHYDRATES

19.2.1 STARCH

Starch, a major storage product found in fruits and vegetables, is essentially a condensation polymer of glucose molecules connected by acetal linkages. Glucose sugars (Figure 19.1) in starch are the principal food molecules of the cell. All sugars contain hydroxyl groups and either an aldehyde ($_H > C = O$) or a ketone ($> C = O$) group. The hydroxyl group of one sugar can combine with the aldehyde or ketone group of a second sugar with the elimination of water to form a disaccharide (Figure 19.2).

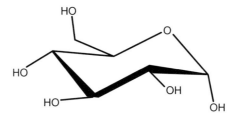

FIGURE 19.1 Chemical structure of glucose, a monosaccharide sugar.

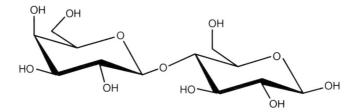

FIGURE 19.2 Chemical structure of lactose, a disaccharide sugar.

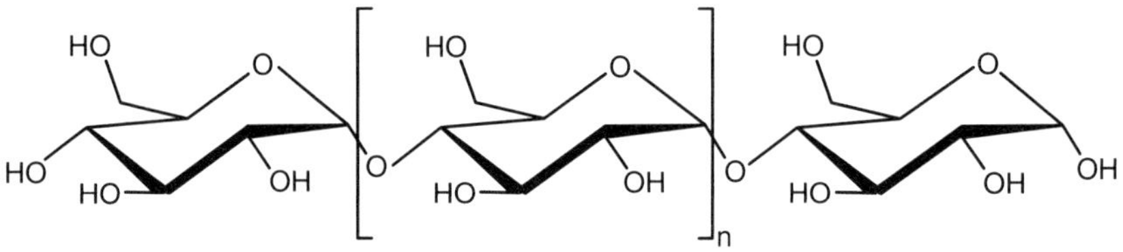

$$G—G_n—G$$

FIGURE 19.3 Linear amylose molecule.

The addition of more monosaccharides in this manner yields very large polysaccharide molecules containing thousands of monosaccharide units (residues).

Starch polymer is composed of two major components, amylose and amylopectin. The amylose is mostly composed of linear α-D-(14)-glucan (Figure 19.3), whereas amylopectin is a highly branched α-D-(14)-glucan with α-D-(16) linkages at the branch points (Figure 19.4). The linear amylose molecules comprise about 30% of common cornstarch and have molecular weights of 200,000 to 700,000, while the branched amylopectin molecules have molecular weights as high as 100 to 200 million.

Factors that influence the properties and functions of starch are briefly described here. Starch is stored in plants as granules or solid particles composed of molecules of both amylose and amylopectin. The granules vary in size from a few micrometers to more than 50 μm, depending on their botanical source. A scanning electron micrograph of starch granules recovered from corn is shown in Figure 19.5.

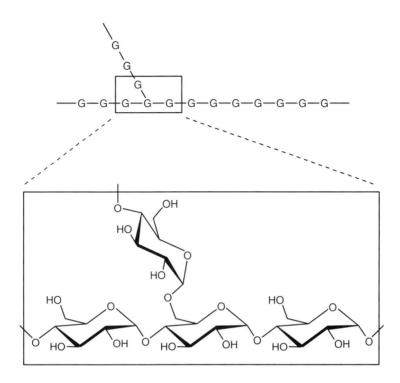

FIGURE 19.4 Branched amylopectin molecule.

Starch granules are hydrophilic, since each starch monomer unit contains three free hydroxyl groups. Consequently, starch changes moisture content as relative humidity changes. Cornstarch granules retain about 6% moisture at 0% relative humidity (RH) but contain 20% moisture when in equilibrium with about 80% RH.

Starch granules are thermally stable when heated in an open atmosphere to about 250°C. Above that temperature the starch molecules begin to decompose. Dry granules absorb moisture when immersed in water but retain their basic structure due to their crystallinity and hydrogen bonding within the granules. Native granular starch contains crystalline areas within the amylopectin (branched) component, but the linear amylose component is largely amorphous and can be mostly extracted in cold water. The granular structure is ruptured by heating in water or treating with aqueous solutions of reagents that disrupt crystalline areas and hydrogen bonding within the granules. The constituent molecules become completely soluble in water at 130 to 150°C and at lower temperatures in alkaline solutions. Starch granules that have been ruptured in aqueous media are commonly referred as gelatinized or destructurized starches. The temperature at which starch granules are completely gelatinized is known as the gelatinization temperature, which varies depending on the botanical source of the starch. Application of high pressure and shear to starch granules permits disruption of the organized structure at lower water content than is possible at atmospheric pressure.

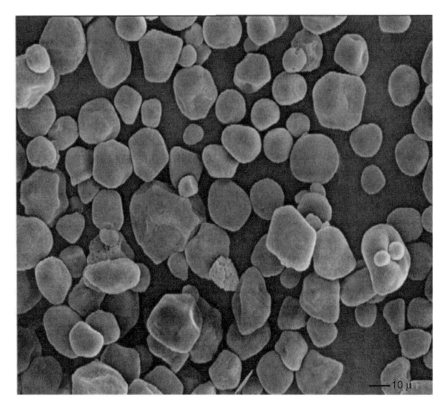

FIGURE 19.5 Scanning electron micrograph showing cornstarch granules, bar = 10 μm.

Starch solutions are unstable at low temperatures. On standing in dilute solutions, the linear amylose component crystallizes. Many branches of amylopectin may also crystallize. Rapid cooling of concentrated starch dispersions creates stiff gels, which crystallize more slowly. Amylose, and to a lesser degree the outer branches of amylopectin, can assume helical conformations that have a hydrophobic core (Figure 19.6). Each turn of the helix comprises about six monomer units. Iodine, fatty acids, lipids, alcohols, and other materials may enter the core of the helix to form stable complexes with starch. Small amounts of crystalline amylose-lipid V-type complexes are usually found in starches such as corn and wheat, which contain free fatty acids and phospholipids [8–10].

Starch molecules readily depolymerize into glucose monomer units when heated in acidic solutions or when treated with a variety of amylolytic enzymes. They are generally stable under alkaline conditions at moderate temperatures. When heated with amines under alkaline conditions, they undergo complex Maillard reactions to form brown-colored products with caramel-like odors. If the produce is damaged during its processing, storage, or transportation via mechanical injury or is subjected to conditions that can alter its surface morphology and cause starch to be exposed on the surface, the starch will degrade fairly quickly.

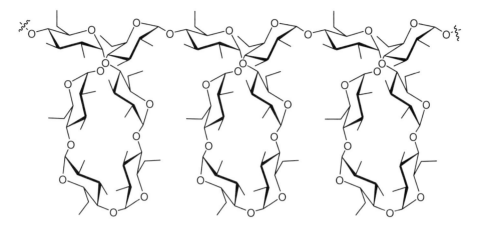

FIGURE 19.6 Helical conformation in amylose and the outer branches of amylopectin. Each turn of the helix comprises about six monomer units.

The presence of many hydroxyl groups on starch permits easy alteration of its properties through chemical derivatization. Acetate esters and carboxymethyl and hydroxypropyl ethers exemplify starch derivatives. Starch polymer aged at constant temperature and moisture level results in starch embrittlement. Differential scanning calorimetric (DSC) studies [11] have shown that this phenomenon is due to structural relaxation of starch chains, leading to decreases in enthalpy and free volume with time. This type of aging is typical of most amorphous polymers [12,13]. The rates of aging seem to vary with polymer structure but the reasons for such differences are not fully understood at present. Gelatinized starch also tends to swell in water, leading to its hydrolytic degradation.

Considerable literature is available indicating that under appropriate conditions the starch polymer is susceptible to biodegradation, photodegradation, or chemical degradation processes. Particularly, biodegradation of starch and starch-based materials has been studied in detail [14–16]. Most efficient and rapid starch degradation usually requires a combination of degradation mechanisms and occurs within living organisms and involves their hydrolytic enzymes. Native starch is unquestionably biodegradable, because microorganisms such as bacteria, fungi, algae, and yeasts readily metabolize it. Additionally, biosynthetic enzymes inherently present and stored within the cells and tissues of fruits and vegetables also participate in the starch degradation process.

Mechanisms for starch biodegradation are numerous and dependent on the type of hydrolytic enzymes present [6]. Enzymes are extremely efficient and specific biocatalysts produced by plants, animals, and microbes and are known for their catalytic efficiency and regio- and stereoselectivity. These biocatalysts are involved in the: (1) biosynthesis of polysaccharides to assemble complex polymeric structures; (2) polymer modifications to render useful functionalities; and (3) biodegradation process (recycling) of natural polymers. A comprehensive monograph on the roles of enzymes in food science has been published [7].

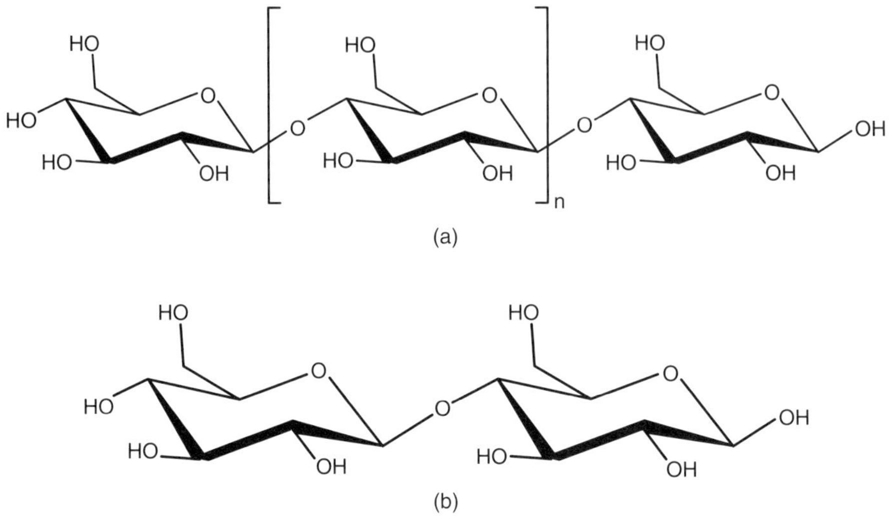

(a)

(b)

FIGURE 19.7 Structure of cellulose (a) and cellobiose (b) indicating β-glucosidic bonds.

19.2.2 CELLULOSE

As opposed to starch, which serves as energy storage molecules, cellulose mainly confers structural integrity to plant cells. The high tensile strength of cellulose allows plants cells to withstand osmotic pressure and resist mechanical stress. For example, woody plants and plants used as a source for textile fibers, such as cotton, exhibit high mechanical strengths. Chemically, cellulose is a linear polymer of glucose subunits (100 to 14,000) linked together via −1,4 bonds forming biological structures that are highly crystalline and water-insoluble (Figure 19.7). Because of their capacity to form inter- and intramolecular hydrogen bonds, adjacent cellulose molecules adhere strongly to one another in parallel arrays of 60 to 70 cellulose chains, all having the same polarity, to form long, rigid, and highly ordered crystalline aggregates called microfibrils (Figure 19.8). Microfibrils range, in lateral dimension, anywhere from 3 to 4 nm in higher plants and up to 20 nm for the microfibrils of the alga *Valonia macrophysa*, which contains up to several hundred cellulose chains [1]. Also, depending on the botanical origin, the degree of crystallinity of cellulose could vary from 0% for totally amorphous cellulose, to 70% for cotton [17] to 100% in *Valonia macrophysa*. Because microfibrils are highly crosslinked they exhibit a high degree of resistance to enzymatic breakdown, often requiring the combined action of enzymes with multiple specificities for complete hydrolysis.

It is generally assumed that the hydrolysis reaction catalyzed by glycosidases, including cellulases and xylanases, proceeds via an acid–base mechanism involving two residues. The first residue acts as a general acid catalyst and protonates the oxygen of the glucosidic bond. The second residue acts as a nucleophile, which either interacts with the oxocarbonium intermediate or promotes the formation of a hydroxide ion from a water molecule. Reactions leading to retention of configuration involve a two-step mechanism, with a double inversion of configuration at the

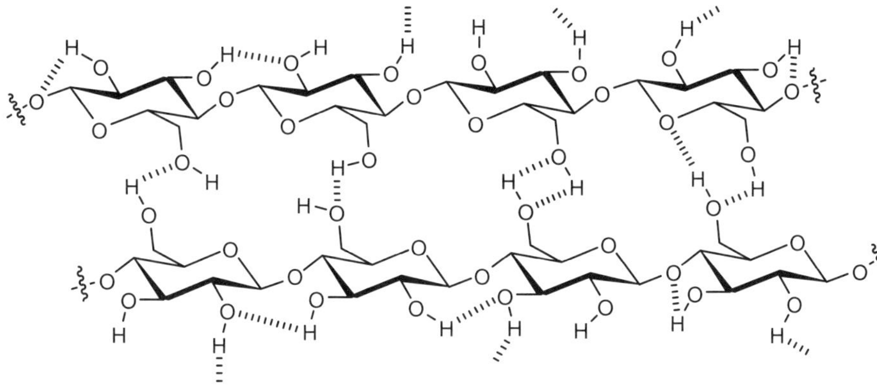

FIGURE 19.8 Inter- and intraspecific hydrogen bonding in two β-linked glucose chains of cellulose.

anomeric carbon, and the formation of an oxocarbonium intermediate [18]. A paradigm of this type of reaction is the mechanism of lysozymes [19]. Reactions leading to inversion of configuration proceed via a single nucleophilic substitution [18].

The complete degradation of crystalline cellulose requires the combined interaction of cellulolytic enzymes of different specificities. Based on the biochemical studies of the cellulase systems derived from many microbial sources [20–32], a cellulose degradation model has been developed to show how cellobiohydrolases, endoglucanases, and β-glucosidases function together (Figure 19.9). According to this model, the endoglucanases would first hydrolyze amorphous regions of the cellulose fibers. The nonreducing ends generated could then be attacked by cellobiohydrolases, which would then proceed with the degradation of the crystalline regions. The β-glucosidases would prevent the accumulation of cellobiose, which inhibits cellobiohydrolases. This model captures the functional role of β-glucosidase enzymes working together without any physical interactions between these enzymes. Studies have shown that these enzymes are capable of behaving as individual proteins without forming stable complexes. However, contrary to these assumptions, it has been shown that a purified stable multienzyme complex (or cellulosome) from *T. reesei,* upon treatment with urea and octylglucoside, dissociated into endoglucanase, a xylanase, and β-glucosidase components [33]. It is not clear whether the physical association between multiple cellulolytic enzymes has any advantage as far as their catalytic efficiency is concerned.

The enzymatic action on native cellulose can be evaluated by a combination of biochemical and spectroscopic techniques. Catalytic activity is usually assayed by measuring the release of soluble reducing sugars, which are expressed as glucose equivalents. The residual insoluble cellulose can also be determined gravimetrically [34], calorimetrically [35], or by turbidimetry [36]. The physical and chemical properties can be determined by crystallinity index, degree of polymerization, alteration of available surface area, or small particle formation. Electron microscopy can reveal surface morphology, showing enzyme adhesion sites [37], patterns of erosion [38], and alterations of fiber or microfibril structure [39–42].

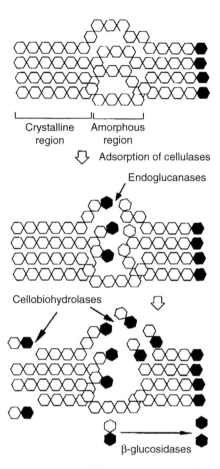

FIGURE 19.9 Cellulose degradation model showing how cellobiohydrolases, endoglucanases, and β-glucosidases function together. Glucose residues are indicated by hexagons; reducing ends are shown in black. (From Béguin, P. and Aubert, J.-P., Cellulases, in *Encyclopedia of Microbiology*, Lederberg, J., Ed., Academic Press, New York, 1992, pp. 467–477. With permission.)

19.2.3 HEMICELLULOSE, PECTIN, AND OTHER INTERACTING CELL WALL POLYMERS

Primary cell walls in plants are complex matrices where fibers are interconnected to other polysaccharide components such as hemicellulose and pectin. The microfibrils are usually embedded in a matrix of hemicellulose and lignin. Hemicellulose is an abundant, heterogeneous group of branched polysaccharides that are tightly bonded to the surface of cellulose microfibrils as well as to each other via hydrogen bonding. The type of hemicelluloses present in the cell wall varies considerably depending on the botanical origin and the stage of the growth cycle. All hemicelluloses have a 14-linked linear backbone of one sugar from which short side

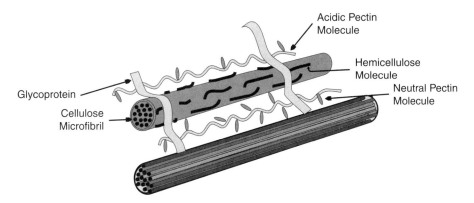

FIGURE 19.10 The primary cell wall in plants represents a complex matrix where fibers are interconnected to other polysaccharide components such as hemicellulose, pectin, and glycoprotein. The microfibrils are usually embedded in a matrix of hemicellulose and lignin.

chains of other sugars protrude, and the variation in these sugar types determines the type of hemicellulose. The backbone sugar molecules of the hemicellulose also provide bonding between hemicellulose and other cell wall matrix components on the cellulose microfibrils. Some of these hemicellulose molecules are cross-linked in turn to acidic pectin molecules (e.g., rhamnogalacturonans) by short neutral pectin molecules (e.g., arabinogalactans). Cell-wall glycoproteins are probably attached to the pectin molecules. The structure of cellulose microfibrils and the associated matrix are depicted in Figure 19.10.

Hemicellulose is composed of complex carbohydrate polymers with xylans and glucomannans as the main components. In most xylans, the xylan backbone carries acetyl, methylglucoronyl, and arabinofuranosyl side chains in varying proportions [43]. The arabinofuranosyl groups can be further esterified by aromatic acids such as ferulic and *p*-coumaric acid [44], which are thought to participate in lignin-hemicellulose crosslinks involving ether linkages [45]. Lignin is a highly branched, random polymer generated by the free-radical condensation of aromatic alcohols [46].

The degradation of hemicellulose is performed by a complex set of enzymes hydrolyzing the xylan or glucomannan backbone and the various bonds of the side chains [47–50]. Lignin is highly resistant to biodegradation and protects cellulose and hemicellulose against enzymatic hydrolysis. Enzymatic attack of lignin involves radical oxidation by peroxidases [46,51].

Xylanases that break down hemicellulosic materials are currently being utilized by industry in many processes, including the hydrolysis of agricultural waste for further bioconversion into alcohol fuels, treatment of animal feed to release pentose sugars, dissolving pulps to obtain cellulose for rayon production, and biobleaching of wood pulps for paper production. Bacteria, yeasts, and fungi are all known to produce xylanase enzymes. Of much interest are xylanases produced by *Bacillus thermotoga* and *Talaromyces byssochlamydoides* because of their thermotolerance (75 to 80°C) [52]. Xylanase breakdown of hemicellulose in wood pulp will produce

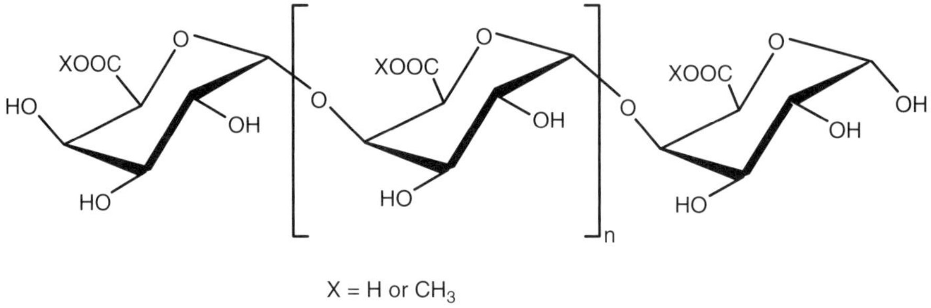

X = H or CH$_3$

FIGURE 19.11 Structure of the pectin polymer.

pulp that is less hydrophilic and less reactive to certain chemicals or crosslinking agents. This might be a useful property to exploit in development of specialty powders or tissues where strong fiber–fiber bonding is undesirable.

19.2.3.1 Pectins

Pectins constitute another major polysaccharide in cell walls of higher plants and are commonly produced during the initial stages of primary cell wall growth in a variety of dicotyledonous and some monocotyledonous plants [53–55]. In plants, pectins function as a hydrating agent and cementing material for the cellulosic network [56]. Pectins are found in relatively large amounts in soft plant tissues under conditions of rapid growth and higher moisture contents. They appear to help with the movement of fluids through the rapidly growing tissues [57]. Pectin is particularly abundant in the middle lamella, the region that serves to cement together the cell walls of adjacent cells. It is this layer that ruptures in some places, forming the intercellular air spaces found in many tissues.

Pectin is a heteropolysaccharide with a main chain consisting of α-1,4 linked poly-D-galactouronic acid units. The carboxylic groups are partially esterified with methanol, and the hydroxyl groups (on C2 and C3) can be substituted with acetyl groups. Pectins are heterogeneous, branched, and highly hydrated polysaccharides that are negatively charged due to the presence of many galacturonic acid residues in the polymer chain (Figure 19.11). Because of their negative charge, pectins form strong ionic bonds in the presence of cations. For example, in the presence of divalent cations such as Ca^{++}, highly crosslinked pectin is produced. There is evidence that similar Ca^{++} crosslinks play a role in holding the cell wall components together, but the exact nature of the bonding that gives the wall its integrity is unknown. In plant cell walls, pectin is linked together with other cell wall polymers such as proteins, hemicellulose, and cellulose. The formation of covalent bonds between pectin and hemicellulose have been reported in the plant cell wall [58–60], but apparently not all pectin is bound in such a manner, as indicated by the presence of unbound pectin recovered in preparations where cell walls have been subjected to extraction with cold water [61–63].

Functionally, pectins are quite similar to collagen in animals, providing firmness and integrity to plant tissues both as a part of the primary cell wall and as the main middle lamella component involved in intercellular adhesion [57,64]. The overall strength of the plant cell wall is ultimately determined by factors such as the orientation, mechanical properties, and links between pectic substances and cellulose fiber, etc. [65]. Some pectin molecules are glycosidically linked to xyloglucan chains that can bind covalently to cellulose [58,60,66]. The firming effect of pectin in fresh tissues involves the formation of free carboxyl groups, which leads to enhanced calcium binding between pectin polymers but decreases the susceptibility of the pectin to depolymerization by β-elimination [67]. It has been reported that in tissues of apples and tomatoes the normal decrease in the degree of methoxylation (DM) (increase in free carboxyl groups) is not accompanied by firming during ripening [68,69]. Softening during the ripening of fleshy fruits is attributed to enzymatic degradation and solubilization of the protopectin [70–79]. It is generally believed that textural changes in some produce occur as cell wall pectins are hydrolyzed by polygalacturonases, as evidenced by the softening in some fruits such as tomatoes that exhibit high levels of polygalacturonase activity [80–82]. However, other investigators found somewhat different results when comparing the firmness of unpeeled kiwifruit and its water-soluble, high methoxyl pectin content [78].

The blanching and degradation of carrot tissues, surprisingly, increased molecular weights of both the water-soluble pectin and the EDTA-soluble pectin obtained from these tissues [83]; however, dehydration without blanching drastically decreased the molecular weight of pectin in both of these fractions. The observed increase in molecular weight in blanched tissues is attributed to the inactivation of pectolytic enzymes. In this regard, many species of plants have shown a preferential loss of either galactose or arabinose during the ripening process [84]. A good portion of these neutral sugar residues can come from pectin side chains, which could then increase the susceptibility of pectin to polygalacturonases and pectin methylesterases by making it more accessible to these enzymes. Loss of side chains would also reduce the entanglement of the pectin molecule, increasing the slippage factor. A variety of glycosidases have been found to remove neutral sugars from pectin side chains [85]. Nonetheless, several other studies suggested that additional mechanisms might be involved in tissue softening [86–91].

Besides hydrolytic cleavage, other nonenzymatic mechanisms have also been proposed for the possible loss of cell wall integrity and pectin degradation [91]. For example, pectin degradation due to alteration in the ionic strength of fluids that solvate the cell well has been suggested. In a recent study, Batisse et al. [92] reported that softening during ripening in cherry fruits does not depend upon pectin depolymerization.

Pectins are among the cell wall components whose collective ability to contain the turgor pressure of the cell wall determines whether growth will take place [93]. As structural components of plant cell walls, native pectins play an important role in many quality aspects of fruit and vegetable products [94]. The size, charge density, charge distribution, and degree of substitution of pectin molecules can be biologically or chemically altered [95]. Pectins are synthesized during the early stages of growth in young, enlarging cell walls [96–97] and are present in various stages of molecular development, growth, and maturity [57]. In the initial stages of synthesis, the carboxyl groups of pectins are highly methylesterified, but the ester groups are later cleaved by pectin methylestarases present in the cell wall [98,99]. Reduction in pectin methylestarase and polygalacturonase activities in tomato fruits results in pectins with a higher degree of esterification and higher molecular weights [100,101]. Reduced pectin methylestarase activity in tomatoes causes an almost complete loss of tissue integrity during fruit senescence but has little effect on fruit firmness during ripening [102]. Reduced pectin methylesterase activity also modifies both accumulation and partitioning of cations between soluble and bound forms of pectin and selectively impairs the accumulation of Mg^{2+} ions over other cations. High-resolution images of the tomato cell wall suggest that it is constructed from at least two independent networks, one based on cellulose/hemicellulose and the other on pectin. Reduction in the cellulose/hemicellulose network does not affect the thickness of the cell wall formed or the spacing of pectin molecules [103]. When tobacco cell walls were grown under unadapted high-salt conditions, pectin molecules were oriented within the wall in a manner similar to cellulose, whereas in an adapted cell wall there was no clear orientation [104]. Esteban et al. [105] reported on the role of pectic substances in the texture maintenance of eggplant fruit. Pectin esterification is also reported to play a role in plant resistance to certain diseases [106].

Although proteins are not discussed in this chapter, it is important to note that in addition to cellulose, hemicellulose, and pectin, the primary cell wall also contains small amounts of proteins or glycoproteins that are enriched in hydroxyproline, an amino acid. Hydroxyproline residues are also found in collagen, the extracellular matrix protein in animals. In the case of plants, however, the hydroxyproline as well as many serine residues are linked to short oligosaccharide side chains, thus forming glycoproteins. Since it is difficult to extract the glycoprotein without destroying the structure of the cell wall, it seems that these molecules, along with cellulose, hemicellulose, and pectin, are tightly integrated in the complex polysaccharide matrix of the cell wall.

For a plant cell to expand or bring about changes in its shape or morphology, the cell wall has to be stretched or deformed. Because cellulose microfibrils exhibit little elasticity, such changes must involve the movement of microfibrils past one another. The possible types of microfibril movements allowed will depend on the orientation of the microfibrils within the primary cell wall as well as on the bonding interactions between matrix macromolecules and the cellulose microfibrils. Cells not only take in nutrients and expel waste products across their plasma membranes, but they also respond to chemical signals in their environment. In the case of plant cells, such signal molecules must penetrate the cell wall. Since the matrix of the wall is a highly hydrated polysaccharide gel (the primary cell wall being 60% water by weight), water, gases, and small, water-soluble molecules penetrate rapidly.

19.3 MOLECULAR MECHANISMS OF CARBOHYDRATE DEGRADATION MEDIATED BY ENZYMES AND MICROBES

Cell and tissue injuries in produce can be triggered or initiated by a single or a combination of factors that include photo, thermal, mechanical and chemical exposures. Once the deterioration is initiated, hydrolytic enzymes from foodborne

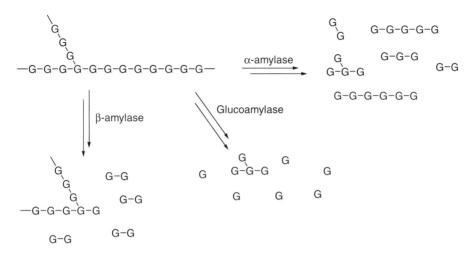

FIGURE 19.12 Possible mode of actions for enzymatic breakdown of an amylopectin polymer mediated via α- and β-amylases and glucoamylase.

microbes such as bacteria and fungi, as well as the biosynthetic enzymes stored within the plant cells, bring about the biodegradation of the plant carbohydrates. The enzyme-mediated biodegradation in turn alters the appearance, color, texture, flavor, and taste of the infected fruits and vegetables [107,108].

19.3.1 STARCH DEPOLYMERASES

The enzyme specificity and the availability of multiple enzyme reactive sites on the starch polymer are two characteristics that offer potential for produce biodegradation. The amylases act on starch polymers to hydrolyze α-1,4 and α-1,6 glucosidic bonds. These bonds are cleaved via an exo or endo mode of hydrolysis. In complex branched polymers such as cellulose and starch, branching creates a large number of nonreducing terminal residues, which require a combination of both exo- and endo-splitting enzymes to achieve a complete polymer hydrolysis. For example, β-amylase (1,4-α-D-glucan maltohydrolase, EC 3.2.1.2), an exo-splitting enzyme, acts only on residues at the nonreducing terminus until it reaches α-1→6 linkage, whereas α–amylase (1,4-α-D-glucan glycanohydrolase, EC 3.2.1.1) is an endo-splitting enzyme that can randomly cleave α-1→4 glycosidic linkages throughout the linear chain. Glucoamylase (1,4-α-D-glucan glucohydrolase, EC 3.2.1.3 or 1,4-α-glucosidase) can act on α-1→4 bonds but can also hydrolyze α-1→6 linkages, albeit at a much slower rate. The enzymatic breakdown of highly branched amylopectin polymer by α- and β-amylases and glucoamylases is depicted in Figure 19.12. Loss of viscosity, which is critical for texture and taste of the juicy fruits, is related to the type of amylase used. In starchy fruits and vegetables, the loss of starch polymer viscosity is much more rapid with α–amylase; however, both –amylase and glucoamylase are slow-acting and would have a lesser impact on the polymer viscosity. It has been observed that the oxidation and etherification of starch significantly alter the extent of starch degradation by commercial amylases without affecting the rate of degradation [109]. The development of genetically modified varieties of fruits and vegetables with starch that is resistant to enzymatic breakdown could play an important part in the diet of diabetic patients. In this regard, preliminary work in our laboratory on the enzymatic modification of granular starch (pH 5.5 from 0 to 8 h) yielded granules that were morphologically coherent and had intact crystalline structures but suffered a 5 to 6% loss of the amorphous region and indicated a slight shift in the glass transition (Tg) value of the starch polymer. However, the most dramatic effect of this treatment was on starch degradation behavior. With time, as the amorphous portion of the starch polymer became more solubilized, crystalline domains in the starch were increasingly resistant to degradation. Such modified starches offer potentially interesting opportunities in the design of food products in which delayed carbohydrate degradation is highly desirable.

19.3.2 CELLULOSE DEPOLYMERASES

The cellulose depolymerization process is somewhat complex due to the physical heterogeneity of the cellulose substrate (degree of crystallinity, available surface area, pore size, etc.), which requires the participation of several regio-selective enzymes. For example, endoglucanases (EC 3.21.4) are inactive against crystalline cellulose but hydrolyze the amorphous region by random hydrolysis of β-glucosidic bonds. This treatment could result in the rapid decrease of viscosity relative to the rate of increase in reducing groups. On the other hand, for cellobiohydrolases (exo-splitting) (EC 3.21.91), which degrade amorphous cellulose by consecutive removal of cellobiose from nonreducing ends, the rate of increase in reducing groups in relation to the decrease in viscosity would be much higher than for the endoglucanases. The exoglucohydrolases (EC 3.2.1.74) consecutively hydrolyze glucose units from the nonreducing end of cellodextrins, which contributes to a rapid decrease in the chain length of the substrate, yielding a cellulosic material with a greatly reduced rate of hydrolysis. Finally, β-glucosidases (EC 3.2.1.21) cleave cellobiose to glucose and remove glucose from the nonreducing end of small cellodextrins. Unlike the exoglucohydrolases, the rate of β-glucosidase increases as the size of substrate decreases, cellobiose being hydrolyzed the fastest. These enzymes work best at pH optimum between 5.5 and 6.0 and are stable at temperatures up to 60°C. These enzymes are sensitive to heavy metals ions, sulfhydryl compounds, and oxidizing and reducing agents, and some exhibit glucose repression. Possible scenarios of the enzymatic breakdown of cellulose polymer are presented in Figure 19.9.

A wide variety of aerobic and anaerobic cellulolytic microorganisms are found in nature, particularly in environments where cellulosic substrate is abundantly available. Fruits and vegetables also produce cellulolytic enzymes, but mostly for the purpose of fruit maturation [110–113]. Only a few microorganisms produce a complete set of enzymes capable of degrading native cellulose efficiently. For efficient cellulose degradation, cellulolytic enzymes usually work in conjunction with the noncellulolytic enzymes to bring about the complete degradation of cellulose, which is ultimately converted into carbon dioxide, water, and the residual biomass under aerobic conditions and into carbon dioxide, methane, and water under anaerobic conditions. In woody plants, cellulose degrades very slowly due to its high

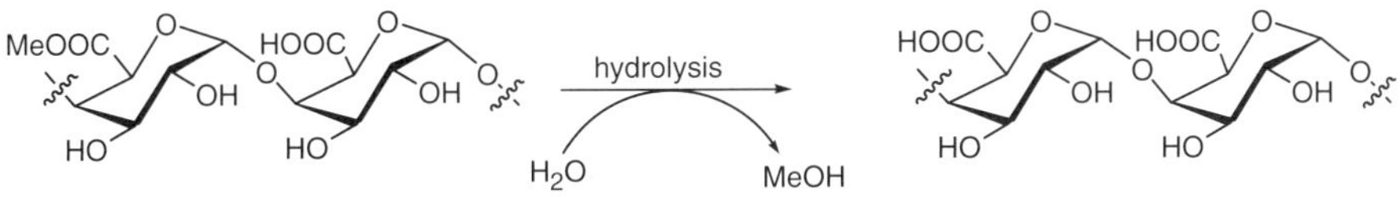

FIGURE 19.13 Enzyme action of pectinesterase (also known as pectin methylesterase, EC 3.1.1.11) on a pectin molecule. The enzyme removes methoxy groups from methylated pectin substrate at pH 7.5, and a proton is released upon hydrolysis of an ester bond.

lignin content. Complete lignin degradation is an oxidative process that is only performed by a few microorganisms, such as the white-rot fungus *Phanerochaete chrysosporium*. However, a larger variety of organisms, in particular actinomycetes [114], are capable of carrying out partial delignification in order to gain access to the cellulosic substrate. *Trichoderma reesei,* an aerobic and highly cellulolytic fungus, is probably the microorganism whose cellulase system has been most thoroughly investigated. Among aerobic, cellulolytic soil bacteria, several species belonging to the genera *Cellulomonas, Pseudomonas, Thermomonospora,* and *Microbispora* have been studied in detail [1–5]. Restricted cellulolytic action may facilitate the access of phytopathogens to plant tissues and may contribute to the leakage of plant cell sap. Due to the inability of cellulose-degrading enzymes to penetrate the cell wall, enzymes are usually secreted or bound to the outside surface of cellulolytic microorganisms. Additionally, enzyme systems generally display a set of typical properties. The systems contain a multiplicity of enzyme components showing a marked synergism against crystalline cellulose. These components often possess a substrate-binding site independent of the catalytic site and are often associated with each other and with the surface of cellulolytic microorganisms.

19.3.3 PECTIN-DEGRADING ENZYMES

Up to 75% of the carboxyl groups in a pectin polymer are generally esterified. Enzymes that degrade pectins are absent in animals but are commonly found in plants and microorganisms. These enzymes are also responsible for softening of many fruits and vegetables, especially citrus fruits and tomatoes. Pectinesterases, polygalacturonases, and pectate lyases are the three main enzymes that catalyze the degradation of pectic substances. Pectinesterases from citrus fruits have been well studied for their role in maturation, ripening, and fruit softening [71–76, 81–84, 95,96,102,105]. Pectinesterase (EC 3.1.1.11), also known as pectin methylesterase, removes methoxy groups from the methylated pectin substrate at pH 7.5, with concomitant release of a proton as the ester bond is hydrolyzed (Figure 19.13). These enzymes typically attack on a position adjacent to a preexisting free carboxyl group in pectin molecules.

The other pectin-degrading enzyme, polygalacturonases (poly-α-1,4-galacturonide glycanohydrolase, EC 3.2.1.15), cleaves glycosidic bonds in hydrated pectic substances (Figure 19.14). These enzymes work best at pHs between 4.5 and 6.0, and their mode of action could either be exo- or an endo-splitting, depending on the substrate type.

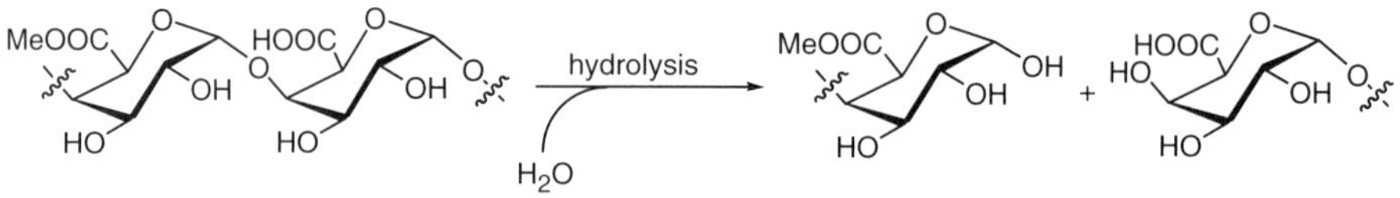

FIGURE 19.14 Action of polygalacturonases (poly-α-1,4-galacturonide glycanohydrolase (EC 3.2.1.15) on the hydrated pectic substances where it cleaves the glycosidic bonds.

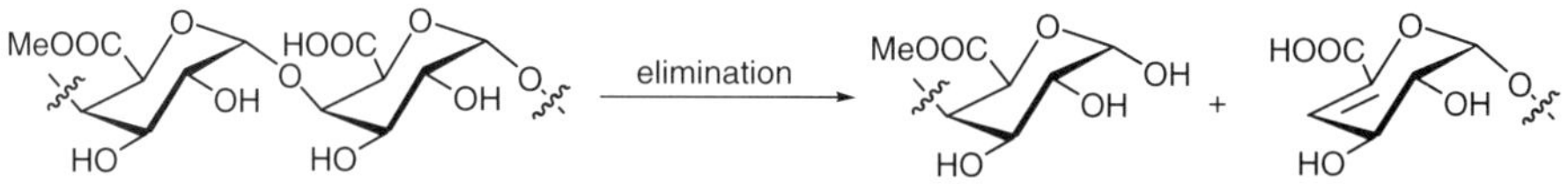

FIGURE 19.15 The pectate lyases (poly(1,4-α-galacturonide), EC 4.2.2.2) work by *trans* elimination of hydrogen from C-4 and C-5 positions in the pectin molecule.

The pectate lyases (poly(1,4-α-galacturonide), EC 4.2.2.2) belong to the lyase group of enzymes, which do not involve water as a substrate. These enzymes work by splitting the glycosidic bond via *trans* elimination of hydrogen from the C-4 and C-5 positions in the sugar molecule (Figure 19.15). This group of enzymes could utilize either pectin or pectic acid as a substrate and could operate in both exo and endo modes. These enzymes have a pH range between 8.5 and 9.5 and appear to require divalent cations such as Ca^{+2} as a cofactor for optimal efficiency. EDTA inhibits the enzyme activity completely.

The knowledge gained from the chemistry of pectin polymers and the role enzymes play in their degradation has transformed the fruit and juice industry worldwide. For example, fungal pectinases from *Aspergillus* spp. are commercially used for partial degradation of pectin in fruit processing, leading to the clarification of fruit juices and wines. More recently, microbial propectinases from *Geotrichum penicillatum* and *Geotrichum candidum* have been discovered [96]. They are very specific endo-polygalacturonases that solubilize all the pectin in a protopectin during extraction from citrus waste. This has replaced a chemical (acid) extraction procedure, which constitutes a significant breakthrough for the citrus industry in terms of food safety and cost saving. The possibility exists to use pectin in biocatalytic reactions exactly as envisioned for starch and cellulosics in health and environmental protection.

19.4 BIOCHEMICAL AND STRUCTURAL CHANGES IN PRODUCE IN RESPONSE TO POSTHARVEST STRESSES AND ITS RELATIONSHIP TO CARBOHYDRATE POLYMERS

Fruits and vegetables are subjected to a wide variety of stresses from the time of harvest through postharvest handling and processing. In response to these stresses, tissues and cells in plants undergo numerous physiological and biochemical changes that affect the quality of the produce. The subject of physiological response of fruits and vegetables to postharvest stresses and its resulting effects has been reviewed [111–112,114]. Postharvest deterioration is particularly acute in tuber plants such as cassava, where wounding and mechanical damage of the tuberous roots cannot be prevented during harvesting. Research indicates that processes such as oxygen stress, carbohydrate metabolism, protein metabolism, and phenolic compound synthesis are all involved in the postharvest deterioration of produce [115].

Two notable changes that fruits and vegetables exhibit during postharvest handling are (1) the accumulation of sugars (monosaccharides, oligisaccharides, and polysaccharides) in cells and tissues and (2) the production of high levels of catalytic enzymes of variable specificities. While some of these activities are part of the plant's normal biochemical processes that are designed to accomplish fruit maturity and ripening, others are a direct response to postharvest stresses, which lead to produce deterioration that affects its quality. For example, in apple fruits, a gradual accumulation of soluble sugars occurs during the first 60 d of harvest, and this increase in sugars can be correlated with an increase in the activity of sucrose-phosphate synthase [116]. Interestingly, other investigators have found that during storage, cell-wall polysaccharides in apples increased, while the total neutral sugar and protein content decreased considerably. However, these activities could be reversed if the fruit was infiltrated with $CaCl_2$ during storage [117]. Liu et al. found that the carbohydrate content in avocado fruits fluctuated during the growth and ripening of fruits [118]. More than half of the fruit's total soluble sugars were composed of the seven-carbon (C7) heptose sugar D-mannoheptulose in both its monomer and its polyol form, perseitol. The balance was accounted for by the more common hexose sugars, glucose and fructose. Results from this research suggest that the C7 sugars play an important role not only in metabolic processes associated with fruit development but also in respiratory processes associated with postharvest physiology and fruit ripening [118]. Higher sugar contents and accumulation of alcohol-insoluble solids containing pectins and other cell wall polysaccharides have also been reported for 'Valencia' oranges and, particularly, for grapefruits [119,120]. Blackberry fruits kept under different environmental conditions and storage periods showed changes in their pH, total titratable acid, and in total soluble solids (polysaccharides) [121].

Scores of biosynthetic enzymes are produced in plants during their growth and development as well during their maturity and ripening [6,7,13]. Numerous enzymes, including amylases, cellulases, pectinases, and glucosidases of variable specificities are also produced in response to the pre- and postharvest stresses in fruits and vegetables. Many of these enzymes play a role in the cell wall polysaccharide deterioration in produce. A number of carbohydrate degradation products such as 3-deoxypent-2-ulose, furan-2-carboxaldehyde and hydroxy-acetaldehyde have been isolated; these serve as precursors of the browning process [122–125]. Asparagus (*Asparagus officinalis* L.) spears, which contain immature tissues, are highly susceptible to deterioration. Postharvest deterioration in the spears is accompanied by changes in respiration rate, soluble carbohydrates, and carbohydrate-degrading enzymatic activities. During initial quality loss, spear tips experience a rapid loss of sucrose and a gradual decrease in glucose concentration, whereas fructose remains unchanged [126]. Irving et al. also reported significantly higher enzyme activities during the ripening of buttercup squash, which led to starch degradation and accumulation of high levels of sucrose in cell walls [127].

Pectin depolymerization or degradation appears to influence both the ripening and sweetening process and helps develop the flesh to a desirable texture in many fruits. The pectinolytic enzymes, which include pectin methylesterases, polygalacturonases, and pectate lyases, are described in a previous section along with their mechanisms of pectin degradation. While maximum hydrolytic activities are present during tissue disruption or wounding, pectin depolymerization, interestingly, differs significantly between different fruits, even when the levels of enzymatic activity in these fruits are similar. Both polygalacturonase and pectin methylesterases cause extensive solubilization and depolymerization of cell wall polysaccharides during the ripening of the avocado fruit. In particular, polygalacturonase is important for polyuronide degradation in the ripening avocado cell walls (partial de-esterification was necessary for the increase in susceptibility of polyuronides to polygalacturonase) [128,129]. Glycosidase and pectin methylesterase activities have been reported during postharvest ripening of apricots [130]. Modifications in cell wall structure during senescence include fragmentation of pectic polymers and hemicelluloses, solubilization of long-chain pectin and loss of pectic sugars. Within hours of harvest, produce cellular responses lead to altered metabolism, which causes reduction of proteins and lipids, accumulation of free amino acids [131,132], and degradation of chlorophyll [133]. Arenas-Campos et al. [134] also noted that textural changes occurred in the preclimacteric stage of ripening of sapote mamey fruit but concluded that the fruit pulp softening was not dependent on pectin methylesterase, polygalacturonase, or β-galactosidase enzymatic activities.

The occurrences of pectic hairy regions in various plant cell wall materials have been noticed. Rhamnogalacturonan oligomers liberated during the degradation of pectin have been isolated and characterized [135]. Many high-molecular-weight pectic polysaccharide fractions isolated from a variety of sources showed considerable resistance to degradation by pectolytic, hemicellulolytic, and cellulolytic enzymes [136]. Enzyme-assisted degradation of the surface membranes of harvested fruits and vegetables improves the water permeability of these surface membranes, which facilitates dehydration and absorption of substances such as sweeteners, stabilizers, preservatives, and flavor enhancers [137–139]. *In vitro* fermentation of two dietary fibers with distinct structural features (pea hull and apple fibers) showed different compositions and physiochemical properties. Fermentation of apple fiber led to a higher production of short-chain fatty acids compared to fermentation of pea fiber. Production of major short-chain fatty acids, acetate, propionate, and butyrate, occurred in both fibers. Uronic acid and arabinose were the most extensively fermented sugars, while xylose and glucose were least fermented [140].

It is interesting to note that nonenzymatic breakdown of pectin has also been proposed. Huber et al. [141] have reported that radical oxygen species, generated either enzymatically or nonenzymatically, might also participate in the scission of pectin and other polysaccharides during ripening and other developmental processes.

Another group of enzymes that are abundantly distributed throughout starchy fruits and vegetables are starch-degrading enzymes. These enzymes have been extensively studied [142–149]. Their mode of action is depicted in Figure 19.12. These enzymes are present in embryonic cells as well as in developing young plants and play a critical role in fruit maturity and ripening, including storage and postharvest deterioration [150]. Kim and Robyt have shown that the enzyme glucoamylase reacts with starch inside the granules (waxy maize, maize, and amylomaize) to produce D-glucose, which is subsequently retained within the granule for other functions [151]. During the postharvest storage of ripened grape flesh, a 52% increase in the α-galactosidase activity was observed over a period of 10 to 15 d [152]. Sugar beet taproots could lose sucrose in storage as a result of mechanical wounding [153]. Interestingly, these enzymes have been shown to adapt well with respect to genetically modified plants. For example, transgenic potatoes exhibiting up to a 25% increase in the degree of branching of amylopectin indicated no apparent change in granular size or morphology, but an increased degree of branching produced 5 to 15% more short chains in the amylopectin [154].

Because cellulose has crystalline and amorphous domains and in plant cell walls it is complexed with other polysaccharides, its degradation requires multiple enzymes of variable specificities. Most of the knowledge pertaining to cellulose degradation has been generated from *in situ* studies on cellulosic substrates and microfibrils utilizing enzymes from a wide variety of sources [155–158]. Several excellent reviews have been written detailing the cellulolytic enzymes and their sources and reaction pathways [1,159,160]. Coughlan has also reviewed aerobic and anaerobic fungal cellulases and their mechanisms of attack on the crystalline cellulose [1,161]. In this regard, it has been shown that while some endoglucanases decreased the average degree of polymerization and improved the alkaline solubility of pure natural cellulose, some cellobiohydrolases exclusively degraded the crystalline portion of the cellulose structure [162,163].

Many soil-borne microbes such as yeasts, fungi, and bacteria are potentially capable of degrading cellulose. During postharvest handling, fruits and vegetables contaminated with cellulose-degrading microbes could cause minor to major damage to the produce's structural integrity. Spoilage of produce by yeasts is directly related to degradation of cellulose, starch, sugars, and protein polymers [164]. During the storage and aging process, exterior surfaces in produce are more susceptible to deterioration. Scanning electron microscopy of the blueberry surface (Figure 19.16) showed substantial changes in surface morphology of a fresh fruit relative to that of old fruits bought at the local grocery store. The fresh surface appears to be quite smooth without any fungal growth (Figure 19.16A), whereas the surface of an aged blueberry had attracted fungal growth (Figure 19.16B) and developed cracks (Figure 19.16C). Further examination at higher magnification revealed that these surfaces also included some areas with heavy damage to surface cuticle/waxes, and perhaps to the cell wall (Figure 19.16D).

Additionally, many biochemical changes occur in fruits and vegetables exposed to soil microbes during both pre- and postharvest [164–166]. For example, *Marasmiius quercophilus*, a white-rot basidiomycete, produces lacases that can effectively bleach the whole leaf [167]. The white-rot basidiomycete *Phanerochaete chrysosporium* also produces extracellular peroxidases that can destroy lignin, which affects the integrity of the lignocellulosic cell walls by making them prone to electrolyte leakage, infection, dehydration, and other postharvest disorders [168]. Johnson et

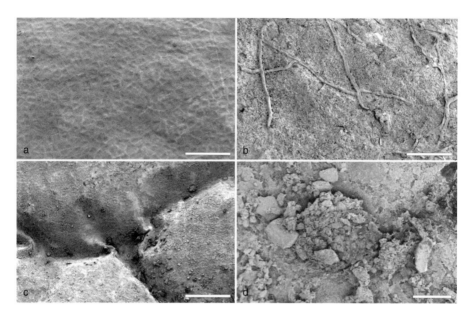

FIGURE 19.16 Scanning electron micrographs of a blueberry surface: fresh surface (a), aged surface indicating microbial growth (b), aged surface with cracks (c), and damaged surface (d). Bars = 50 μm.

al. [169] also observed that the activity of cellulase from *Clostridium thermocellum* was inhibited by the end products of cellulolysis, suggesting sugar repression in these enzymes.

All produce contains carbohydrates, which are a good source of pulp, syrups, gums, juices, gels, fiber, polysaccharides, oligomers, and monomeric sugars as well as adhesives. Each of these substances, either directly or in a chemically modified form, is a source of compounds that have potential use in food or feed and nonfood industrial applications. The use of cell wall-degrading enzymes as a processing aid has become normal practice by the fruit and vegetable processing industry worldwide [170–172]. Cereal beta-amylases with high activities *in situ* are particularly well known to the brewing industry for releasing easily fermentable sugars from cereal grain starches [173]. Two fungal pathogens (*Phomopsis cucurbitae* and *Rhizopus stolonifer*) are also known to produce endo- and exo-polygalacturonases with the ability to macerate netted muskmelon tissue at different stages of fruit development [174].

In response to various postharvest stresses, cell wall-associated polysaccharide polymers also undergo modifications in structure and composition, which changes their functional and material properties, including changes in their rheological, thermal, and hydration properties. For example, fluctuations in ambient temperature could lead to pectin degradation, whereas changes in pH, ionic strength, etc. can affect cell walls during storage and the ripening process [175–179]. It has been shown that in untopped radishes and carrots the loss of postharvest quality, particularly changes in textural properties such as visible wilting and/or yellowing of leaves as well as a decrease in firmness and crispness, was found to be associated with changes in cell wall carbohydrate metabolism. Furthermore, the product quality was closely related to cell wall polysaccharides and textural properties [180].

Various strategies that plants adopt in response to environmental stresses have been studied in detail [181]. Elevated temperatures, for example, generally known to contribute to flesh softening, increase the soluble solids (polysaccharides) content and change the flesh color [182,183]. Because carbohydrates degrade at a faster rate at higher temperatures, postharvest losses are much higher at ambient temperatures [184]. Low levels of chlorophyll accompanied by high chlorophyllase activity were also observed in apples stored at ambient temperatures compared to apples stored at refrigerated temperatures [185]. Interestingly, contrary to the deleterious effects of high temperatures on produce, some investigators found that cucumbers grown in a greenhouse at elevated temperatures during the day could increase their tolerance to postharvest chilling [186].

Preharvest exposure of fruits and vegetables to direct sunlight, with associated high tissue temperatures, can result in differences in internal properties such as sugar content, tissue firmness, oil levels, and mineral content. Fruits with different temperature histories respond differently to postharvest treatments. For example, avocado fruit from exposed sites on a tree have less chilling injury, whereas in citrus and persimmons chilling damage is more pronounced in exposed tissues [187,188]. Cycles of intermittent warming for 1 d at 20°C every 6 d of postharvest storage were found to be useful in maintaining the cellular integrity in peaches during ripening [189].

19.5   STRATEGIES TO MINIMIZE POSTHARVEST DAMAGE TO CARBOHYDRATES

Generally, many efforts have proven to be quite useful in delaying or minimizing the damage caused by postharvest stresses. However, only the few approaches that affect the polysaccharides are mentioned here. Some studies have shown that the surface coating of fruits has been very effective in improving the storage time, delaying carbohydrate degradation, and preventing infections. In this regard, coating several varieties of pears with a carnauba-based wax emulsion suppressed ripening, reduced the incidence of senescent breakdown, greatly improved the finish of the skin [190], and protected fruits against wound pathogens [191]. Chitosan treatment of tomatoes inhibits the growth and the production of pathogenic factors by black-mold rot [192]. Surface coatings have also been effective in preventing the loss of bioactive compounds in highly perishable fruits and vegetables. For example, a starch coating prevented the loss of soluble pectins in radishes [138,193–195].

Exposure to various gases and chemical agents also affects storage and ripening in fruits. For example, ripening in kiwifruits is accompanied by softening of tissues and an increase in soluble solids (polysaccharides). A 2-week exposure to high levels of carbon dioxide substantially lowered the soluble solids production for up to 8 weeks. This effect, however, lasted only for a week when kiwi fruits were stored at 20°C [196]. Postharvest storage of tomatoes under nitrogen for 35 h, under low oxygen, or at low temperature conditions significantly reduced the production of volatile compounds, which occurs as the cell wall is ruptured [197]. In contrast, injury to apples could result from external CO_2, which causes loss of cytoplasmic integrity, coagulation of the protoplast, the loss of organelle structure, and disintegration of the cell wall. Additionally, undigested starch can be found in cells of affected fruits at the hypodermis–cortex boundary [198]. Exposure to exogenous ethylene in watermelon fruit causes activation of cell wall-degrading enzymes and pectin depolymerization. In citrus fruits, continuous exposure to ethylene results in a marked increase in endo-1,4-β-galacturonase (cellulase) and polygalacturonase activities in the calyx abscission zones [199–201].

During storage, 'Ceccona' apricots treated with 1-methylcycloprene for 12 h at 20°C showed delayed softening and as well greatly reduced α-D-galactosidase and beta-D-galactosidase activities [202]. In avocado fruits, 1-methylcyclopropene treatments delayed the ripening and slightly reduced the depolymerization of cell wall matrix polysaccharides hemicellulose andxyloglucans [203]. Many references are found in the literature concerning the effect of gibberellic acid, jasmonate and polyamine treatments of produce and their positive impact on the fruit's storage life, the ability to suppress mold decay, and the reduction of chilling injury [204–207]. Fumigation as well as irradiation has also been suggested as a measure to control postharvest decay and infestation [208–210]. The enzyme catalase has also been reported to function as an antioxidant in the defense response of mandarin fruit to chilling stress. It is believed that heat conditioning may induce catalase activity [211]. Other investigators have reported that hot water treatment (44°C for 15 min) of 'Royal Gala' apples was very effective for controlling insect pests [212].

Under continuous illumination of low-intensity white light, spinach leaves showed increased availability of soluble carbohydrates, especially glucose [213].

Textural changes were more apparent at the preclimacteric stage in the ripening of the sapote mamey fruit, but the fruit pulp softening was not dependent on pectinmethylesterase, polygalacturonase, or β-galactosidase enzyme activity [214].

Produce quality is directly related to biochemical components such as cell wall polysaccharides and their composition, particle size, shape, moisture content, and mechanical properties. Abiotic factors such as soil moisture, temperature, relative humidity, and nutrient availability directly influence texture, flavor, color, and freshness of fruits and vegetables [112]. There are a number of factors that can render serious damage to produce. The producers, processors, and transporters can use common sense and simple approaches to minimize produce damage and the subsequent economic loss and can still deliver the best-quality produce to the consumers. Extended exposure to water can induce hydrolytic activities in produce, which causes quicker ripening and thus much shorter storage time. Since mature fruits contain the enzymatic machinery to metabolize soluble sugars [215], high temperature and high humidity can accelerate the process of ripening and cell wall softening, leading to early loss of texture and flavor. Mechanical damage or bruising during postharvest handling could activate the production of carbohydrolyases, leading to cell wall degradation, which compromises fruit texture and flavor due to the breakdown of sugars. Exposure to high winds can cause dehydration in fruits and vegetables, triggering the production of chlorophyllase, which degrades chlorophyll and results in the loss of green color in plants. Yellow vegetables and dehydrated surfaces on fruits are unappealing to consumers. Soil contamination provides the opportunity for microbial pathogens to thrive. Microbes could tenaciously binds to cell surfaces and produce enzymes to break down cell wall polysaccharides to simple sugars for their own use. If this occurs, poor-quality produce with damaged surfaces will result. Some fungi are well known for producing toxins that make fruit unhealthy and unappealing for consumers.

ACKNOWLEDGEMENT

Authors are thankful to Mr. Danny A. Iman for assistance in preparing this manuscript.

REFERENCES

1. Beguin, P. and Aubert, J.-P, The biological degradation of cellulose, *FEMS Microbiol. Rev.*, 13, 25, 1994.
2. Coughlan, M. P., Enzymatic hydrolysis of cellulose:an overview, *Bioresource Technol.*, 39, 107, 1992.
3. Vandamme, E. J. and Soetaert, W., Biotechnical modification of carbohydrates, *FEMS Microbiol. Rev.*, 16, 163, 1995.
4. Walker, L. P., and Wilson, D. B., Enzymatic hydrolysis of cellulose: an overview, *Bioresource Technol.*, 36, 3, 1991.
5. Aubert, J. P., Bèguin, P., and Millet, J., Eds., *Biochemistry and Genetics of Cellulose Degradation*, Academic Press, New York, 1988, p. 31.
6. Yalpani, M., Developments and prospects in enzyme biopolymer modifications, *Progr. Biotechnol.*, 3, 7, 1987.
7. Whitaker, J. R., Ed., *Principles of Enzymology for Food Science*, 2nd ed., Marcel Dekker, New York, 1994, p. 625.
8. Shogren, R. L., Effect of moisture content on the melting and subsequent physical aging of cornstarch, *Carbohydr. Polym.*, 19, 83, 1992.
9. Chinnaswamy, R., Hanna, M. A., and Zobel, H. F., Microstructural physiochemical and macromolecular changes in extrusion-cooked and retrograded cornstarch, *Cer. Foods World,* 34, 415, 1989.
10. Galliard, T. and Bowler, P., in *Starch: Properties and Potential,* T. Galliard, Ed., John Wiley & Sons, New York, 1987, p. 55.
11. Shogren, R. L., Effect of moisture content on the melting and subsequent physical aging of cornstarch, *Carbohydr. Polym.*, 19, 83, 1992.
12. Hodge, I. M., and Berens, A. R., Effects of annealing and prior history on enthalpy relaxation in glassy polymers: 2. mathmetical modeling, *Macromolecules*, 15, 762, 1982.
13. Hutchinson, J. M., and Kovacs, A., Effects of thermal history on structural recovery of glasses during isobaric heating, *J. Polym. Eng. Sci.,* 24, 1087, 1984.
14. Swanson, C. L. et al., Starch-plastic materials: preparation, physical properties, and biodegradability (a review of recent USDA research), *J. Environ. Polym. Degrad.*, 1, 155, 1993.
15. Imam, S. H., and Snell, W. J., The *Chlamydomonas* cell wall degrading enzyme, lysine, acts on two substrates within the frame-work of the cell wall, *J. Cell Biol.*, 106, 2211, 1998.

16. Imam, S. et al., A study of cornstarch granule digestion by an high molecular weight a-amylase from *Lactobacillus amylovorus*, *Curr. Microbiol.*, 22, 365, 1991.

17. Wood, T. M., Preparation of crystalline, amorphous and dyed cellulose substrates, *Methods Enzymol.*, 160, 19, 1988.

18. Sinnott, M. L., Catalytic mechanisms of enzymatic glycosyl transfers, *Chem. Rev.*, 90, 1171, 1990.

19. Kelley, J. A. et al., X-ray crystallography of the binding of the bacterial cell wall trisaccharide NAM-NAG-NAM to lysozyme, *Nature*, 282, 875, 1979.

20. Eriksson, K.-E., Enzyme mechanisms involved in cellulose hydrolysis by the rot fungus *Sporotrichum pullverulentum, Biotechnol. Bioeng.*, 70, 317, 1978.

21. Eriksson, K.-E. and Pettersson, B., Extracellular enzyme system utilized by the fungus *Sporotrichum pullverulentum (Chrysosporulm lignortum)* for the breakdown of cellulose. I. separation, purification and physicochemical characterization of five endo-I,4-β-glucanases, *Eur. J. Biochem.*, 51, 193, 1975.

22. Wood, T. M. and McCrae, S. I., The cellulase of *Trichoderma koningii:* purification and properties of some endoglucanase components with special reference to their action on cellulose when acting alone or in synergism with the cellobiohydrolases, *Biochem. J.,* 171, 61, 1978.

23. Wood, T. M. and McCrae, S. I., The cellulase of *Penicillium pinophilum, Biochem. J.,* 234, 93, 1986.

24. McHale, A. and Coughlan, M. P., Synergistic hydrolysis of cellulose by components of the extracellular cellulase system of *Talaromyces emersonii, FEBS Lett.*, 17, 319, 1980.

25. Gilkes, N. R. et al., Mode of action and substrate specificities of cellulases from cloned bacterial genes, *J. Biol. Chem.*, 259, 10455, 1984.

26. Bronnenmeier, K. and Staudenbauer, W. L., Cellulose hydrolysis by a highly thermostable endo-I,4-β-glucanase (Avicelase I) from *Clostridium stercorarium, Enzyme Microb. Technol.*, 12, 431, 1990.

27. Bronnenmeier, K., Riicknagel, K. P., and Staudenbauer, W. L., Purification and properties of a novel type of exo-I,4-β-glucanase (Avicelase II) from the cellulolytic thermophile *Clostridium stercorarium, Eur. J. Biochem.*, 200, 379, 1991.

28. Creuzet. N., Berenger, J.-F., and Frixon, C., Characterization of exoglucanase and syncrgistic hydrolysis of cellulose in *Clostridium stercorarium, FEMS Microbiol. Lett.,* 20, 347, 1983.

29. Creuzet, N. and Frixon, C., Purification and characterization of an endoglucanase from a newly isolated thermophilic bacterium, *Biochimie*, 65, 149, 1983.

30. Yablonsky, M. D. et al., Characterization and cloning of the cellulase complex of *Microbispora bispora,* in *Biochemistry and Genetics of Cellulose Degradation*, Aubert, J.-P., Beguin, P., and Millet, J., Eds., Academic Press, New York, 1988, pp. 249–266.

31. Walker, L. P. et al., Fragmentation of cellulose by the major *Thermomonospora fusca* cellulases, *Trichoderma reesei* CBHI, and their mixtures, *Biotechnol. Bioeng.*, 40, 1019, 1992.

32. MacKenzie. C. R., Bilous, D., and Johnson, K. G., Purification and characterization of an exoglucanase from *Streptomyces flavogriseus, Can. J. Microbiol.,* 30, 1171, 1984.

33. Sprey, B., and Lambert, C., Titration curves of cellulases from *Trichoderma reesei:* demonstration of a cellulase-xylanase-β-glucosidase-containing complex, *FEMS Microbiol. Lett.*, 18, 217, 1983.

34. Tailliez, P. et al., Enhanced cellulose fermentation by an asporogenous and ethanol-tolerant mutant of *Clostridium thermocellum, Appl. Environ. Microbiol.*, 55, 207, 1989.
35. Updegraff, D. M., Semimicro determination of cellulose in biological materials, *Anal. Biochem.*, 32, 420, 1969.
36. Johnson, E.A. et al., Saccharification of complex cellulosic substrates by the cellulase system from *Clostridium thermocellum, Appl. Environ. Microbiol.*, 43, 1125, 1982.
37. Chanzy, H., Henrissat, B., and Vuong, R., Colloidal gold labelling of 1,4-β-D-glucan cellobiohydrolase adsorbed on cellulose substrates, *FEBS Lett.*, 172, 193, 1983.
38. Chanzy, H., and Henrissat, B., Unidirectional degradation of *Valonia* cellulose micro-crystals subjected Go cellulase action, *FEBS Lett.*, 184, 285, 1985.
39. Din, N. et al., Non-hydrolytic disruption of cellulose fibres by the binding domain of a bacterial cellulase, *Bio/Technology*, 9, 1096, 1991.
40. Sprey, B., and Bochem, H. P., Electron microscopic observations of cellulose microfibril degradation by endo-cellulase from *Trichoderma reesei, FEMS Microbiol. Lett.*, 78, 183, 1991.
41. Sprey, J. and Bochem, H. P., Effect of endoglucanase and cell cellobiohydrolase from *Trichoderma reesei* on cellulose microfibril structure. *FEMS Microbiol. Lett.*, 97, 113, 1992.
42. Sprey, B., and Bochem, H. P., Formation of cross-fractures in cellulose microfibril structure by an endo-glueanase-cellobiohydrolasc complex from *Trichoderma reesei, FEMS Microbiol. Lett.*, 106, 239, 1993.
43. Joseleau. J. P., Comtat. J., and Ruel, K., Chemical structure of xylans and their interaction in the plant cell walls, in *Progress in Biotechnology*, Vol. 7, *Xylans and Xylanases,* Visser. J. et al., Eds., Elsevier, Amsterdam, 1992, pp. 1–15.
44. Hartley. R. D., and Ford. C. W., Phenolic constituents of plant cell walls and wall biodegradability, in *Plant Cell Wall Polymers: Biogenesis and Biodegradation*, Lewis, N. G. and Paice, M. G., Eds., American Chemical Society, Washington. DC, 1989, pp. 135–145.
45. Scalbert, A. et al., Ether linkage between phenolic acids and lignin fractions from wheat straw, *Phytochemistry*, 24, 1359, 1985.
46. Gold, M., Wariishi, H., and Valli. K., Extracellular peroxidases involved in lignin degradation by the white-rot basidiomycete *Phanerochate chryosporium,* in *Biocatalysis in Agricultural Biotechnology*, Whitaker. J. and Sonnet, P. E., Eds., American Chemical Society, Washington, DC, 1989, pp. 127–140.
47. Biely, P., Microbial xylanolytie systems, *Trends. Biotechnol.*, 3, 286, 1998.
48. Coughlan. M. P. and Hazlewood, J. P., β-1,4-D-xylan-degrading enzyme systems: biochemistry, molecular biology, and applications, *Biotechnol. Appl. Biochem.,* 17, 259, 1993.
49. Wong. K. K. Y., Tan, L. U. I., and Saddler, J. N., Multiplicity of β-1,4-xylanase in microorganisms: functions and applications, *Microbiol. Rev.*, 52, 3115, 1988.
50. Thomson. J. A., Molecular biology of xylan degradation, *FEMS Microbiol. Rev.*, 104, 65, 1993.
51. Kirk. T. K., Biochemistry of lignin degradation by *Phenerochaete chryosporium,* in *Biochemistry and Genetics of Cellulose Degradation*, Aubert, J.-P., Beguin, P., and Millet, J., Eds., Academic Press, New York, 1988, pp. 315–332.
52. Zamost, B. L., Nielsen, H. K., and Starnes, R. L., Thermostable enzymes for industrial applications, *J. Indust. Microbiol.*, 8, 71, 1991.
53. Northcote, D. H., Chemistry of the plant cell wall, *Annu. Rev. Plant Physiol.,* 23, 113, 1972.

54. Jarvis, M. C., Forsyth, W., and Duncan, H. J., A survey of the pectin content of nonlignified monocot cell walls, *Plant Physiol.,* 88, 309, 1988.
55. Hoff, J. E. and Castro, M. D., Chemical composition of potato cell wall, *Agric. Food Chem.,* 17, 1328, 1969.
56. Muralikrishna, G. and Taranathan, R. N., Characterization of pectin polysaccharides from pulse husks, *Food Chem.,* 50, 87, 1994.
57. Glickman, M., *Gum Technology in the Food Industry,* Academic Press, New York, 1969.
58. Keegstra, K. et al., The structure of plant cell walls. III. a model of the walls of suspenson-cultured sycamore cells based on the inter connections of the macromolecular components, *Plant Physiol.,* 51, 188, 1973.
59. Rees, D. A. and Wight, N. J., Molecular cohesion in plant cell walls: methylation analysis of polysaccharides from the cotyledons of white mustard, *Biochem. J.,* 115, 431, 1969.
60. Jarvis, M. C. et al., The polysaccharide structure of potato cell walls: chemical fractionation, *Planta,* 152, 93, 1981.
61. Stevens, B. J. H. and Selvendran, R. P., The isolation and analysis of cell wall material from alcohol insoluble residue of cabbage *(Brassica oleraceae var.15 capitata), J. Sci.Food Agric.,* 31, 1257, 1980.
62. Aspinall, G. O., Craig, J. W. T., and Whyte, J. L., Lemon-peel pectin. I. fractionation and partial hydrolysis of water soluble pectins, *Carhohydr. Res.,* 7, 442, 1968.
63. Knee, M., Metabolism of polygalacturonate in apple fruit cortical tissues during ripening, *Phytochemistry,* 17, 1261, 1978.
64. Wilson, L. G. and Fry, S. C., Extensin: a major cell wall glycoprotein, *Plant Cell Environ.,* 9, 239, 1986.
65. Preston, R. D., Polysaccharide: conformation and cell wall function, *Rev. Plant Physiol.,* 30, 55, 1979.
66. Darvill, A. G. et al., The primary cell wall of flowering plants, in *Biochemistry of Plants,* Tolbert, N. E., Ed., Academic Press, New York, 1980, p. 91.
67. Sajjaanatakul, T., Van Buren, J. P., and Downing, D. L., Effect of methyl ester content on heat degradation of chelator soluble carrot pectin, *J. Food Sci.,* 54, 1272, 1989.
68. Burns, J. K. and Pressy, R., Ca++ in cell wall of ripening tomato and peaches, *J. Am. Soc. Hortic. Sci.,* 112, 783, 1987.
69. O'Beirne, D., van Buren, J. P., and Mattick, L. R., Two distinct fractions from senescent dared apples extracted using non-degradative methods, *J. Food Sci.,* 47, 173, 1982.
70. Labavitch, J. M., Cell wall turnover in plant development, *Annu. Rev. Plant Physiol.,* 32, 385, 1981.
71. Pressey, R., Reevaluation of the changes in polygalacturonases in tomatoes during ripening, *Planta,* 174, 39, 1988.
72. Soda, I. et al., Detection of polygalacturonase in kiwifruit during ripening, *Agric. Biol. Chem.,* 50, 3191, 1986.
73. Dick, A. J. and Labavitch, J. M., Cell wall metabolism in ripening fruit. IV. characterization of the pectic polysaccharides solubilized during softening of 'Bartlett' pear fruit, *Plant Physiol.,* 89, 1394, 1989.
74. Barbier, M. and Thibault, J. F., Polysaccharides of cherry fruit, *Phytochemistry,* 21, 111, 1982.
75. Gross, K. C. and Wallner, S. J., Degradation of cell wall polysaccharide during tomato fruit ripening, *Plant Physiol.,* 63, 117, 1979.
76. Pressey, R., Hinton, D. M., and Avants, J. K., Development of polygalacturonase activity and solubilization of pectin in peaches during ripening, *J. Food Sci.,* 36, 1070, 1971.

77. Huber, D. J., Polyuronide degradation and hemicellulose modifications in ripening in tomato fruits, *Am. Soc. Hortic. Sci.,* 108, 405, 1983.
78. Robertson, G. L. and Swinburne, D., Changes in chorophyll and pectin after storage and canning of kiwifruit, *J. Food Sci.,* 46, 1557, 1981.
79. Mollendroff, L. J. et al., Molecular characteristics of pectin constituents in relation to firmness, extractable juice, and woolliness in nectarines, *Am. Soc. Hortic. Sci.,* 118, 77, 1993.
80. Dellapenna, D. C., Alexander, D. C., and Bennett, A. B., Molecular cloning of tomato fruit polygalacturonase: analysis of polygalacturonase mRNA levels during ripening, *Proc. Natl. Acad. Sci. USA.,* 63, 6420, 1986.
81. Huber, D. J., The role of cell wall hydrolases in fruit softening, *Hortic. Rev.,* 5, 169, 1983.
82. Brady, C. J., Fruit ripening, *Annu. Rev. Plant Physiol.,* 38, 155, 1987.
83. Ben-Shalom, N. et al., Changes in molecular weight of water soluble and EDT-soluble pectin fractions from carrot after heat treatment, *Food Chem.,* 45, 243, 1992.
84. Gross, K. C. and Sams, C. E., Changes in neutral sugar composition during fruit ripening: a species survey, *Phytochemistry,* 23, 2457, 1984.
85. Wallner, S. J. and Walker, J. E., Glycosidases in cell wall degrading extracts of ripening tomato products, *Plant Physiol.,* 55, 94, 1975.
86. Bartley, I. M., Exo-polygalacturonase of apple, *Phytochemistry,* 17, 213, 1978.
87. Huber, D. J., Strawberry fruit softening: the potential roles of polyuronides and hemicelluloses, *J. Food Sci.,* 49, 1310, 1984.
88. Brady, C. J. et al., Interactions between the amount and molecular forms of polygalacturonase, calcium, and firmness in tomato fruits, *Am. Soc. Hortic. Sci.,* 110, 254, 1985.
89. Hall, C. B., Firmness of tomato fruit tissues according to cultivar and ripeness, *J. Am. Soc. Hortic. Sci.,* 112, 663, 1987.
90. Tong, C. B. and Gross, C. K., Ripening characteristics of a tomato mutant, dark green, *J. Am. Soc. Hortic. Sci.,* 114, 635, 1989.
91. Fishman, M. L. et al., Macromolecular components of tomato fruit pectin, *Arch. Biochem. Biophys.,* 274, 179, 1989.
92. Batisse, C., Fils-Lycaon, B., and Buret, M., Pectin changes in ripening cherry fruits, *J. Food Sci.,* 59, 389, 1994.
93. Jarvis, M. C., Structure and properties of pectin gels in plant cell wall, *Plant Cell Environ.,* 7, 153, 1984.
94. Voragen, A. G. J., Schols, H. A., and Pilnik, W., Determination of the degree of methylation and acetylation of pectins by HPLC, *Food Hydrocolloids,* 1, 65, 1986.
95. Kerstez, Z. I., *The Pectic Substances,* Interscience, New York, 1951.
96. Sakai, T. et al., Pectin, pectinase and protopectinase: production, properties and applications, *Adv. Appl. Microbiol.,* 39, 213, 1993.
97. Karr, A. L., Cell wall biogenesis, in *Plant Biochemistry,* Bonner, J. and Varner, J. E., Eds., Academic Press, New York, 1976, p. 405.
98. Kauss, H. and Hassid, W. Z., Enzymatic introduction of methylester groups of pectin, *J. Biol. Chem.,* 242, 3449, 1967.
99. Roberts, K., Structure at the plant cell surface, *Curr. Opin. Cell Biol.,* 2, 920, 1990.
100. Tieman, D. M. et al., An antisense pectin methylesterase gene alters pectin chemistry and soluble solids in tomato fruits, *Plant Cell,* 4, 667, 1992.
101. Dellapenna, D. et al., Polygalacturonase isozymes and pectin depolymerization in transgenic tomato fruits, *Plant Physiol.,* 94, 1882, 1990.
102. Tieman, D. M. and Handa, A. K., Reduction in pectin methylesterase activity modifies tissue integrity and cation levels in ripening tomato *(Lycopersicon esculentum* Mill.) fruits, *Plant Physiol.,* 106, 429, 1994.

103. Wells, B. et al., Structural features of cell walls from tomato cells adapted to grow on the herbicide 2,6-dichlorobenzonitrile, *J. Microsc.*, 173, 155, 1994.
104. McCann, M. C. et al., Changes in pectin structure and localization during the growth of unadapted and NaCI-adapted tobacco cells, *Plant J.*, 5, 773, 1994.
105. Esteban, R. M. et al., Pectin changes during the development and ripening of eggplant fruits, *Food Chem.*, 46, 289, 1993.
106. McMillan, G. P. et al., Potato resistance to soft rot erwinias is related to cell wall pectin esterification, *Physiol. Mol. Plant Pathol.*, 42, 279, 1993.
107. Weichmann, J., Postharvest physiology and secondary metabolism, *J. Appl. Bot. Angewandte Botanik*, 74, 126, 2000.
108. Sams, C. E., Preharvest factors affecting postharvest texture, *Postharv. Biol. Biotechnol.*, 15, 249, 1999.
109. Wolf B. W., Bauer, L. L., and Fahey, G. C. Jr., Effects of chemical modification on in vitro rate and extent of food starch digestion: an attempt to discover a slowly digested starch, *J. Food Agric. Chem.*, 47, 4178, 1999.
110. Coleman, G. S., The metabolism of cellulose, glucose, and starch by the rumen ciliate protozoa *Eudiplodinium maggii*, *J. Gen. Microbiol.*, 107, 159, 1978.
111. Tucker, M. L. et al., Avocado cellulase: nucleotide sequence of a putative full-length cDNA clone and evidence for a small gene family, *Plant Mol. Biol.*, 9, 197, 1987.
112. Jones, T. H. D., de Renobales, M., and Pon, N., Cellulases released during the germination of *Dictyostelium discoideum* spores, *J. Bacteriol.*, 137, 752, 1979.
113. Blume, J. E., and Ennis, H. L., A *Dictyostelium discoideum* cellulase is a member of a spore germination-specific gene family, *J. Biol. Chem.*, 266, 15432, 1991.
114. McCarthy, A. J., Lignocellulose-degrading actinomycetes, *FEMS Microbiol. Rev.*, 46, 145, 1987.
115. Huang, J. et al., Molecular analysis of differentially expressed genes during postharvest deterioration in Casava (Manihot Esculenta Crantz) tuberous roots, *Euphytica*, 120, 85, 2001.
116. Duque, P., Barreiro, M. G., and Arrabaca, J. D., Respiratory metabolism during cold storage of apple fruit. I. sucrose metabolism and glycolysis, *Physiologia Plant.*, 107, 14, 1999.
117. Chardonnet, C. O. et al., Chemical changes in cortical tissue and cell walls of calcium-infiltrated 'golden delicious' apples during storage, *Postharv. Biol. Technol.*, 28, 97, 2003.
118. Liu, X. et al., 'Hass' avocado carbohydrate fluctuations. 2. fruits growth and ripening, *J. Am. Soc. Hortic. Sci.*, 124, 676, 1999.
119. Song, K.J., Echeverria, E., and Lee, H. S., Distribution of sugars and related enzymes in the stem and blossom halves of 'valencia' oranges, *J. Am. Soc. Hortic. Sci.*, 123, 416, 1998.
120. Burns, J. K. and Albrigo, L. G., Time of harvest and method of storage affect on granulation in grapefruit, *Hortscience*, 33, 728, 1998.
121. Antunes, L. E. C., Duarte, J., and De Souza, C. M., Postharvest conservation of blackberry fruits, *Pesquisa Agropecuaria Brasileria*, 38, 413, 2003.
122. Hofmann, T., Characterization of precursors and elucidation of the reaction pathways leading to a novel colored 2h, 7h, 8ah-pyranol[2,3-B]pyran-3-one from pentoses by quantitative studies and application of C-13-labelling experiments, *Carbohydr. Res.*, 313, 215, 1998.
123. Hofmann, T., Quantitative studies on the role of browning precursors in the Maillard reaction of pentoses and hexoses with L-alanine, *Eur. Food Res. Technol.*, 209, 113, 1999.

124. Hofmann, T., Bors, W., and Stettmaier, K., Radical intermediates in the early stage of the nonenzymatic browning reaction of carbohydrates and amino acids, *J. Agric. Food Chem.*, 47, 379, 1999.

125. Yaylayan, V. A. and Keyhani, A., Origin of carbohydrate degradation products in L-alanine D-[C-13]glucose model systems, *J. Agric. Food Chem.*, 48, 2415, 2000.

126. Irving, D. E. and Hurst, P. L., Respiration, soluble carbohydrates and enzymes of carbohydrate metabolism in tip of harvested asparagus spears, *Plant Sci.*, 94, 89, 1993.

127. Irving, D. E., Hurst, P. L., and Ragg, J. S., Changes in carbohydrates and carbohydrate metabolizing enzymes during the development, maturation, and ripening of buttercup squash (*Cucurbita maxima* D. 'Delica'), *Am. Soc. Hortic. Sci. J.*, 122, 3, 1997.

128. Wakabayashi, K., Chun, J. P., Huber, D. J., Extensive solubilization and depolymerization of cell wall polysaccharides during avocado (*Persesa Americana*) ripening involves concerted action of polygalacturonase and pectinmethylesterase, *Physiol. Plant.*, 108, 345, 2000.

129. Wakabayashi, K., Hoson, T., and Huber, D. J., Methyl de-esterification as a major factor regulating the extent of pectin depolymerization during fruit ripening: a comparison of the action of avocado (*Persea Americana*) and tomato (*Lycopersicon esculentum*) polygalacturonases, *J. Plant Physiol.*, 160, 667, 2003.

130. Cardarelli, A. et al., Effects of exogenous propylene on softening, glycosidase, and pectinmethylesterase activity during postharvest ripening of apricots, *J. Agric. Food Chem.*, 50, 1441, 2002.

131. Thakur, B. R., Singh, R. K., and Handa, A. K., Chemistry and uses of pectin: a review, *Crit. Rev. Food Sci. Nutr.*, 37, 47, 1997.

132. King, G. A. and O'Donoghue, E. M., Unravelling senescence: new opportunities for delaying the inevitable in harvested fruits and vegetables, *Trends Food Sci. Technol.*, 6, 385, 1995.

133. Heaton, J. W., and Marangoni, A. G., Chlorophyll degradation in processed foods and senescent plant tissues, *Trends Food Sci. Technol.*, 7, 8, 1996.

134. Arenas-Ocampoo, M. L., Evangelista-Lozano, S. R., Arana-Errasquin, R., Jimenez-Apricio, A., and Davila-Ortiz, G., Softening and biochemical changes of sapote Mamey fruit (*Pourteria sapote*) at different development and ripening stages, *J. Food Biochem.*, 27, 91, 2003.

135. Schols, H. A., Voragen, A. G. J., and Colquhoun, I. J., Isolation and characterization of rhamnogalacturonan oligomers, liberated during degradation of pectic hairy regions by rhamnogalacturonase, *Carbohydr. Res.*, 256, 97, 1994.

136. Schols, H. A. and Voragen, A. G. J., Occurrence of pectic hairy regions in various plant cell wall materials during their degradability by rhamnogalacturonase, *Carbohydr. Res.*, 256, 83, 1994.

137. Poulose, A. J. and Boston, M., Enzyme assisted degradation of surface membranes of harvested fruits and vegetables, U.S. Patent 5 298 256, 1994.

138. Dourtoglou, A. and Tsopelas, P., Postharvest fruit protection using components of natural essential oils in combination with coating waxes, WO patent 93/06735A1, 1993.

139. Poulose, A. J. and Boston, M., Enzyme assisted degradation of surface membranes of harvested fruits and vegetables, U.S. Patent 5 037 662, 1991.

140. Guillon, F. et al., Characterization of residual fibers from fermentation of pea and apple fibers by human fecal bacteria, *J. Sci. Food. Agric.*, 68, 521, 1995.

141. Huber, D. J., Karakurt, Y., and Jeong, J., Pectin degradation in ripening and wounded fruits, *R. Bras. Fisiol. Veg.*, 13, 224, 2001.

142. Planchot, V. et al., Extensive degradation of native starch granules by alpha-amylase from *Aspergillus fumigatus*, *J. Cereal Sci.*, 21, 163, 1995.

143. Offner, A., Bach, A., and Sauvant, D., Quantitative review of in situ starch degradation in the rumen, *Anim. Feed Sci. Technol.*, 106, 93, 2003.

144. Rollings, J., Enzymatic depolymerization of polysaccharides, *Carbohydr. Polym.*, 5, 37, 1985.

145. Gupta, R. et al., Microbial alpha-amylases: a biotechnological perspective, *Process. Biochem.*, 38, 1599, 2003.

146. Pilnik, W. and Rombouts, F., Polysaccharides and food processing, *Carbohydr. Res.*, 142, 93, 1985.

147. Pandy, A. et al., Advances in microbial amylases, *Biotechnol. Appl. Biochem.*, 31, 135, 2000.

148. van der Maarel, M. et al., Properties and applications of starch converting enzymes of alpha amylase family, *J. Biotechnol.*, 94, 137, 2002.

149. Sutherland, I.W., Polysaccharases for microbial exopolysaccharides, *Carbohydr. Polym.*, 38, 319, 1999.

150. Zhu, Z. P. et al., Characterization of starch-debranching enzymes in pea embryos, *Plant Physiol.* 118, 581, 1998.

151. Kim, Y.-K., and Robyt, J., Enzyme modification of starch granules: in situ reaction of glucoamylase to give complete retention of D-glucoside in side the granule, *Carbohydr. Res.*, 318, 129, 1999.

152. Kang, H. C. and Lee, S. H., Characteristics of an alpha-galactosidase associated with grape flesh, *Phytochemistry,* 58, 213, 2001.

153. Rosenkranz, H. et al., In wounded sugar beet (*Beta Vulgaris* L.) tap-root, hexose accumulation correlates with the induction of a vacuolar invertase isoform. *J. Exp. Bot.*, 52, 2381, 2001.

154. Kotstee, A. et al., The influence of an increased degree of branching on the physico-chemical properties of starch from genetically modified potato, *Carbohydr. Polym.*, 37, 173, 1998.

155. Din, N. et al., Non-hydrolytic disruption of cellulose fibers by the binding domain of a bacterial cellulase. *Bio/Technology,* 9, 1096, 1991.

156. Sprey, B. and Bochem, H. P., Electronmicroscopic observations of cellulose microfibril degradation by endocellulase from *Trichoderma reesei, FEMS Microbiol. Lett.,* 78, 183, 1991.

157. Sprey, B., and Bochem, H. P., Effect of endogluconase and cellobiohydrolase *Trichoderma reesei* on cellulose microfibril structure, *FEMS Microbiol. Lett.,* 97, 113, 1992.

158. Sprey, B., and Bochem, H. P., Formation of crossfracture in cellulose microfibril structure by an endogluconase-cellobiohydrolase complex from *Trichoderma reesei, FEMS Microbiol. Lett.,* 106, 239, 1993.

159. Wood, T.M. and Garcia-Campayo, G. V., Biodegradation, in *Physiology of Biodegradative Microorganisms*, Ridge, C., Ed., Kluwer Academic Publishers, Dordrecht, The Netherlands, 1990, pp. 147–161.

160. Walker, L. P., and Wilson, D. B., Enzymatic hydrolysis of cellulose: an overview, *Bioresource Technol.*, 36, 3, 1991.

161. Wood, T. M. et al., Aerobic and anerobic fungal cellulases with special reference to their mode of attack on crystalline cellulose, in *Biochemistry and Genetics of Cellulose Degradation*, Aubert, J.-P., Béguin, P., and Millet, J., Eds., Academic Press, New York, 1988, pp. 31–52.

162. Cao, Yu., and Huimin, T., Effects of cellulase on the modification of cellulose, *Carbohydr. Res.,* 337, 1291, 2002.

163. Tuula, T. T., Crystalline cellulose degradation: new insight into the function of cellobiohydrolases, *Trends Biotechnol.*, 15, 160, 1997.

164. Fleet, G., Spoilage yeasts, *CRC Critical Rev. Biotechnol.*, 12, 1, 1992.

165. Eskin, N. A. M., *Biochemistry of Foods*, Academic Press, New York, 1990, p. 557.

166. Zhang, D. et al., Biochemical changes in iceberg lettuce (*Lactuca sativa*) during postharvest storage at 4°C, *Food Sci. Technol.*, 14, 209, 2000.

167. Tagger, S. C. et al., Phenoloxidases of the white-rot fungus *Marasmiius quercophilus* isolated from an evergreen oak litter (*Quercus ilex* L.), *Enzym. Microbial Technol.*, 23, 372, 1998.

168. Gold, M., Wariishi, I. T., and Valli, K., Extracellular peroxidase involved in lignin degradation by the white-rot *Phanerochaete chrysosporium*, in *Biocatalysis in Agricultural Biotechnology*, Whitaker, J. and Sonnet, P. E., Eds., American Chemical Society, Washington, DC, 1989, pp.127–140.

169. Johnson, E. A., Reese, E. T., and Demain, A. L., Inhibition of *Clostridium thermocellum* cellulase by end products of cellulolysis, *J. Appl. Biochem.*, 4, 64, 1982.

170. Demir, N. et al., The use of commercial pectinase in fruit juice industry. 3. immobilized pectinase for mash treatment, *J. Food Eng.*, 47, 275, 2001.

171. Tijskens, L. M. M. et al., The kinetics of pectinmethylesterase in potatoes and carrots during blanching, *J. Food Eng.*, 34, 371, 1997.

172. Massiot, P., Baron, A., and Drilleau, J. F., Enzymatic hydrolysis of carrot cell-wall polysaccharides, in situ or after isolation as alcohol insoluble residue, *Acta Aliment.*, 21, 293, 1992.

173. Ziegler, P., Cereal beta-amylases, *J. Cereal Sci.*, 29, 195, 1999.

174. Bruton, B. D. et al., Polygalacturonases of a latent and wound postharvest fungal pathogen of muskmelon fruit, *Postharv. Biol. Technol.*, 13, 205, 1998.

175. Kunzek, H., Kabbert, R., and Gloyna, D., Aspects of material science in food processing: changes in plant cell walls of fruits and vegetables, *Eur. Food Res. Technol.*, 208, 233, 1999.

176. Paliyath, G. and Droillard, M. J., The mechanisms of membrane deterioration and disassembly during senescence, *Plant Physiol. Biochem.*, 30, 789, 1992.

177. Almeida, D. P. F. and Huber, D. J., Apoplastic pH and inorganic ion levels in tomato fruit: a potential means for regulation of cell wall metabolism during ripening, *Physiol. Plant.*, 105, 506, 1999.

178. Chun, J. P. and Huber, D. J., Polygalacturonase-mediated soubilization and depolymerization of pectic polymers in tomato fruit cell walls: regulation by pH and ionic conditions, *Plant Physiol.*, 117, 1293, 1998.

179. Moretti, C. L., et al., Chemical composition and physical properties of pericarp, locule, and placental tissues of tomatoes with internal bruising, *J. Am. Soc. Hortic. Sci.*, 123, 656, 1998.

180. Huyskens-Keil, S., Schreiner, M., and Ulrichs, C., Cell wall carbohydrate metabolism of perishable vegetables in pre- and post-harvest, *Acta Hortic.*, 553, 201, 2001.

181. Kozlowski, T. T. and Pallardy, S. G., Acclimation and adaptive responses of woody plants to environmental stresses, *Bot. Rev.*, 68, 270, 2002.

182. Diaz-Perez, J. C., Bautista, S., and Villanueva, R., Quality changes in sapote mamey fruit during ripening and storage, *Postharv. Biol. Technol.*, 18, 67, 2000.

183. Dixon, J., and Hewett, E. W., Temperature effects postharvest color change of apples, *J. Am. Soc. Hortic. Sci.*, 123, 305, 1998.

184. Umar, B., The use of solar cooling to minimize postharvest losses in tropics, *Trop. Sci.*, 38, 74, 1998.

185. Ihl, M., San Martin, A., and Bifani, V., Chlorophyllase and quality of 'Granny Smith' apples during storage, *Gartenbauwissenschaft*, 65, 266, 2000.

186. Hang, H. M., Park, K. W., and Saltveit, M. E., Elevated growing temperatures during the day improve the postharvest chilling tolerance of greenhouse-grown cucumber (*Cucumis sativus*) fruit, *Postharv. Biol. Technol.*, 24, 49, 2002.

187. Woolf, A. B. and Ferguson, I. B., Postharvest response to high fruit temperatures in the field, *Postharv. Biol. Technol.*, 21, 7, 2000.

188. Ferguson, I., Volz, R., and Woolf, A., Preharvest factors affecting physiological disorders of fruits, *Postharv. Biol. Technol.*, 15, 255, 1999.

189. Artes, F., Fernandez-Trujillo, J., and Cano, A., Juice characteristics related to wooliness and ripening during postharvest storage of peaches, *Z. Lebensm.-Unters. –Forsch.*, 208, 282, 1999.

190. Amarante, C. and Banks, N. H., Ripening behavior, postharvest quality, and physiological disorders of coated pears (*Pyrus communis*), *N. Z. J. Crop Hortic.*, 30, 49, 2002.

191. Schirra, M. et al., Epicuticular changes and storage potential of cactus pear fruit following gibberellic acid preharvest sprays and postharvest heat treatment, *Postharv. Biol. Technol.*, 17, 79, 1999.

192. Reddy, M. V. et al., Chitosan effects on blackmold rot and pathogenic factors produced by *Alternaria alternata* in postharvest tomatoes, *J. Am. Soc. Hortic. Sci.*, 125, 742, 2000.

193. Schreiner, M. et al., Effect of packaging and surface coating on primary and secondary plant compounds in fruit and vegetable products, *J. Food Eng.*, 56, 237, 2003.

194. Nisperos, M., Edible coatings for whole and minimally processed fruits and vegetables, in Proc. Joint Meeting of Australia and New Zealand Institutes of Food Science and Technology, 1995, p. 46.

195. Nussinovitch, A., and Lurie, S., Edible coatings for fruits and vegetables, *Postharv. News Info.*, 6, N53, 1995.

196. Irving, D. E., High concentration of carbon dioxide influences kiwifruit ripening, *Postharv. Biol. Technol.*, 2, 109, 1992.

197. Boukobza, A., and Taylor, A. J., Effect of postharvest treatment on flavor volatiles of tomatoes, *Postharv. Biol. Technol.*, 25, 231, 2002.

198. Watkins, C. B., Silsby, K. J., and Goffinet, M. C., Controlled atmosphere and antioxidant effects on external CO_2 injury of 'empire' apples, *Hortscience*, 32, 1242, 1997.

199. Karakurt, Y. and Huber, D. J., Cell wall-degrading enzymes and pectin solubility and depolymerization in immature and ripe watermelon (*Citrullus lanatus*) fruit in response to exogenous ethylene, *Physiol. Plant.*, 116, 398, 2002.

200. Burns, J. K. et al., Endo-1,4-beta-glucanase gene expression and cell wall hydrolase activities during abscission in Valencia orange, *Physiol. Plant.*, 102, 217, 1998.

201. Kazokas, W. C. and Burns, J. K., Cellulase activity and gene expression in citrus fruit abscission zones during and after ethylene treatment, *J. Am. Soc. Hortic. Sci.*, 123, 781, 1998.

202. Botondi, R. et al., Influence of ethylene inhibition by 1-methylcyclopropene on apricot quality, volatile production, and glycosidases activity of low- and high-aroma varieties of apricots, *J. Agric. Food Chem.*, 51, 1189, 2003.

203. Jeong, J., Huber, D. J., and Sargent, S. A., Influence of 1- methycyclopropene (1-mcp) on ripening and cell-wall matrix polysaccharides of avocado (*Persea americana*) fruit, *Postharv. Biol. Technol.*, 25, 241, 2002.
204. Eshel, D. et al., Resistance of gibberellin-treated persimmon fruits to *Alternaria alternata* arises from the reduced ability of the fungus to produce endo-1,4-β-glucanase, *Phytopathology*, 90, 1256, 2000.
205. Brown, G. E., Davis, C., and Chambers, M., Control of citrus green mold with Aspire is impacted by the type of injury, *Postharv. Biol. Technol.*, 18, 57, 2000.
206. Droby, S. et al., Suppressing green mold decay in grapefruit with postharvest jasmonate application, *J. Am. Soc. Hortic. Sci.*, 124, 184, 1999.
207. Gonzalez-Aguilar, G. A., Polyamines induced by hot water treatments reduce chilling injury and decay in pepper fruit, *Postharv. Biol. Technol.*, 18, 19, 2000.
208. Miller, W. R. and Mcdonald, R. E., Amelioration of irradiation injury to Florida grapefruit by pretreatment with vapor heat or fungicides, *Hortscience*, 33, 100, 1998.
209. Miller, W. R. and Mcdonald, R. E., Short-term heat conditioning of grapefruit to alleviate irradiation injury, *Hortscience*, 33, 1224, 1998.
210. Miller, W. R., Mcdonald, R. E., and Chaparro, J., Tolerance of selected orange and mandarin hybrid fruit to low-dose irradiation for quarantine purposes, *Hortscience*, 35, 1288, 2000.
211. Sala, J. M. and Lafuene, M. T., Catalase enzyme activity is related to tolerance of Mandarin fruits to chilling, *Postharv. Biol. Technol.*, 20, 81, 2000.
212. Smith K. J. and Lay-Yee, M., Response of "royal gala" apples to hot water treatment for insect control, *Postharv. Biol. Technol.*, 19, 111, 2000.
213. Toledo, M. E. A. et al., L-ascorbic acid metabolism in spinach (*Spinacia oleracea* L.) during postharvest storage in light and dark, *Postharv. Biol. Technol.*, 28, 47, 2003.
214. Arenas-Ocampoo, M. L. et al., Softening and biochemical changes of Sapote Mamey fruit (*Pourteria sapote*) at different development and ripening stages, *J. Food Biochem.*, 27, 91, 2003.
215. Goren, R. et al., Sugar utilization by citrus juice cells by ^{14}C-sucrose and ^{14}C-fructose feeding analyses, *Plant Physiol. Biochem.*, 38, 507, 2000.

20 Temperature Effects on Produce Degradation

Justin R. Morris and Pamela L. Brady
Institute of Food Science and Engineering, University of Arkansas, Fayetteville, AR

CONTENTS

0-8493-1902-1/05/$0.00+$1.50
© 2005 by CRC Press

20.1   THE NEED FOR TEMPERATURE MANAGEMENT

Good temperature management is the single most important factor in determining the ultimate quality of fresh fruits and vegetables. All fresh produce are living organisms and, for optimum quality, must remain alive and healthy until they are processed or consumed. Plant material can live only as long as the membrane structure and enzymes remain functional. Exposure to undesirable temperatures, either above or below the optimal range for this functioning, results in many physiological disorders. Temperature also influences the rate of ethylene production and utilization as well as the rate of spore germination and growth rate of pathogens. Temperature extremes can cause changes in the functions of the plant tissues, many of which lead to quality loss through product degradation.

In order to maintain the highest possible produce quality, it is necessary to slow respiration and, therefore, product deterioration processes as much as possible.[1] One way to do this is to lower the temperature. As a general rule, each 10°C reduction in temperature lowers respiration rate by a factor of 2 to 4. This can have a significant effect on the keeping quality of produce. For example, an apple or a pear will ripen as much in a day at 21.2°C as it will in a week at 0°C.[1]

Because various fruits and vegetables respond differently to temperature, there is no one ideal temperature for the storage of all produce. However, the many positive effects of lowered temperatures lead to extensive use of cool storage to restrict deterioration without causing abnormal ripening or other undesirable changes (Figure 20.1). In fruits and vegetables not susceptible to cold injury, maximum storage life can be obtained by storage close to the freezing point of the tissue.[2] For chilling-sensitive produce the advantages of reduced respiration and fungal growth must be balanced against potential losses from chilling injury. Table 20.1 and Table 20.2 show the recommended storage temperatures for a variety of fruits and vegetables. The temperature in a storage room should be kept within 1°C of the desired temperature for the produce being stored.[3] Temperatures above the optimal range can shorten storage life; temperatures below the optimal range can cause chilling or freezing injury.

20.2   EFFECTS OF TEMPERATURE

20.2.1   RESPIRATORY ACTIVITY AND STORAGE LIFE

The energy for life processes comes from the food reserves that are accumulated while the commodities are still attached to the plant. The process of converting the food reserves to energy is respiration. Some of the energy produced by respiration goes to maintain life processes and, therefore, to slow degradation. In general, there is an inverse relationship between respiration rate and storage life.[2] Therefore, a product with a low respiration rate usually keeps longer. It is widely accepted that

FIGURE 20.1 The many positive effects of lowered temperatures lead to extensive use of cool storage to restrict deterioration without causing abnormal ripening or other undesirable changes. (Photo courtesy of Western Precooling Systems, Fremont, CA.)

those items that have the shortest storage life respire at high rates (most leafy vegetables), are harvested ripe (berries), or are chilling-sensitive (bananas and cucumbers).

The most important factor in controlling respiration rate is temperature.[4] Sweet corn, fresh summer squash, and lima beans respire about 8 times faster at 21°C than at 0°C; broccoli respires 15 times faster and spinach 11 times faster.[5]

Van't Hoff's rule states that the velocity of a reaction increases two- to threefold for every 10°C rise in temperature. The temperature quotient for a 10°C interval is called the Q_{10}.[6] This value is useful since it allows calculation of the reaction rate at one temperature from a known rate at another temperature. The Q_{10} can be calculated by dividing the reaction rate at a higher temperature by the rate at a 10°C lower temperature, i.e., $Q_{10} = R2/R1$.

The Q_{10} values are useful in predicting quality of fresh fruits and vegetables; however, they have limitations.[4] Because produce respiration rates do not follow ideal behavior, Q_{10} values of fresh fruits and vegetables are usually given within specified temperature ranges. The Q_{10} values of fresh fruits and vegetables are generally lower in higher storage temperature ranges (greater than 10°C) than at lower temperatures, and the Q_{10} concept is not valid at low temperatures for chilling-sensitive products. Values are dependent on time, the physiological state of the material, and the continuing physiological activity of the material.[7]

Typical figures for Q_{10} are shown in Table 20.3. These typical Q_{10} values make it possible to construct a table like Table 20.4 showing the effect of different temperatures on the rates of deterioration and relative shelf life of a typical perishable

TABLE 20.1
Recommended Precooling Methods, Storage Conditions, Approximate Storage Life, and Average Respiration Rates for Selected Fruits

Commodity	Precooling Method[a]	Recommended Storage Conditions Temperature (°C)	Relative Humidity (%)	Approximate Storage Life (at recommended conditions)	Respiration Rate[b] (mg CO_2 kg^{-1} h^{-1})
Apple	R, F, H	−1 to 4	90–95	2–4 months	6 (0°)
Apricot	R, H	−0.5 to 0	90–95	1–2 weeks (3–4 weeks some cultivars)	
Avocado	F, R, H	5–12	85–95	3–6 weeks	35 (5°) 105 (10°)
Banana		13–14	90–95	2–4 days[3]	140 (15°)
Berries					
Blackberries	F to 5° in 4 hours	−0.5 to 0	>90	2–14 days	19 (0°)
Blueberry	F to <10° in 1hr	−0.5 to 0	>90	2 weeks	6 (0°)
Cranberry	F	2	90–95	2–4 months	4 (0°)
Raspberry	F to 1° in 12 hr	−0.5 to 0	>90	2–5 days	17 (2°)
Strawberry	F, R	0	90–95	7 days	16 (0°)
Cherry (Sweet)	R, F, H to <5 in 4hr	−1 to 0	>95	2–4 weeks	8 (0°)
Fig	F	−1 to 0	90–95	1–2 weeks	6 (0°)
Grape					
Vinifera	F	−1 to 0	90–95		3 (0°)
Rotundifolia	F to ≤2° in 12hr	−0.5 to 0	>90	1–4 weeks	10 (2°)
Labrusca	F to <2° in 24hr	−0.5 to 0	85–90		3 (0°)
Grapefruit		12–15	95	6 weeks	<10 (15°)
Kiwi	F after 48–72 hr curing	0	90–95	4–5 months	3 (0°)
Lemon		7–12	85–95	6 months	11 (10°)
Lime		10	95	8 weeks	<10 (10°)

Commodity	Precooling method				
Mango	F, R in 24 hr				
Mature green		10–13	85–90	14–28 days	35 (10°)
Ripe		7–8	90		16 (10°)
Melons					
Casaba, Crenshaw, and Canary	H, F	10	90 – 95	3 weeks	nd
Honeydew		7	95	7–10 days	14 (10°)
Watermelon		10–15	90	2–3 weeks	8 (10°)
Nectarine	F, H	−1 to 0	90–95	14–18 days[3]	5 (0°)
Orange	R, F			12 weeks	4 (0°)
FL and TX		0–1	85–90		
CA and AZ		3–8	90–95		
Passion fruit	R, F to 10°				44 (5°)
Yellow		7–10	90–95	2 weeks	
Purple		3–5		3–5 weeks	
Peach	F, H	−1 to 0	90–95	14–28 days[3]	5 (0°)
Pear		−1	90–94	60–90 days[3]	
Persimmon		−1 to 1	90–95	3 mo	6 (0°)
Pineapple	R, F	7–12	85–95	14–20 days	6 (10°)
Plums and fresh prunes	F, H, R	−1 to 0	90–95	14–28 days[3]	3 (0°)

Note: All values represent averages for the commodities. Values for individual cultivars and differing handling and growing conditions may vary significantly.

[1] For commodities where no precooling method is indicated, precooling is not required. Precooling methods F = forced air; R = refrigerator; H = hydrocooling; V = vacuum; HV = hydrovacuum; I = icing; PI = package icing

[2] Respiration rate provided for temperature indicated in parentheses. n.d. = no data

[3] Data source USDA Handbook 66, 1968 ed.

Developed from data in *The Commercial Storage of Fruits, Vegetables, and Florist and Nursery Crops*, Gross, K.C., Yang, C.Y., and Saltveit, M., Eds., Agricultural Research Service, Beltsville, MD, 2002. Draft version of revised USDA Agriculture Handbook 66 on the USDA website http://www.ba.ars.usda.gov/hb66 /index.html

TABLE 20.2
Recommended Precooling Methods, Storage Conditions, Approximate Storage Life, and Average Respiration Rates for Selected Vegetables

Commodity	Precooling Method[a]	Recommended Storage Conditions		Approximate Storage Life (at recommended conditions)	Respiration Rate[b] (mg CO_2 kg^{-1} h^{-1})
		Temperature (°C)	Relative Humidity (%)		
Asparagus	H	0–2	95–99	14–21 days	60 (0°)
Beans, Snap	H	5–7.5	95–100	8–12 days	34 (5°)
Beets					
With tops	H, F, PI	0	98	10–14 days	5 (0°)
Topped	F	1–2	98		
Broccoli	I (liquid), H, F	0	98–100	2–3 weeks	21 (0°)
Brussel sprouts	V, H, I, F	0	95–100	3–5 weeks	40 (0°)
Cabbage	H, F	0	98–100	3–6 months	5 (0°)
Carrot	H	0	98–100	7–9 months	15 (0°)
Cauliflower	V, H, F	0	95–98	3 weeks	17 (0°)
Celery	H, HV, V	0	> 95	5–7 weeks	15 (0°)
Corn, sweet	V with top icing, H,	0³	95–98³	4–6 days	41 (0°)
Cucumbers	H, F	10–12.5	95	<14 days	26 (15°)
Eggplant	H, F, R	10–12	90–95	14 days	69 (12.5°)
Endive	V, H	0	95–100	2–3 weeks	45 (0°)
Garlic (bulbs)	cured	–1 to 0	60–70	9 months	8 (0°)
Greens for cooking	H, HV, I (liquid, top, or package)	0	95–98	2 weeks	21 (0°) - Spinach
Jicama		12.5–15	80–90	2–4 months	14 (15°)

Kohlrabi					
Topped	H, I, F	0	98–100	2–3 months	10 (0°)
With tops				2–4 weeks	
Leeks	H, I, V	0	95–100	2–3 months	15 (0°)
Lettuce	V, HydroV, H for nonheading types	0	98–100	< 4 weeks	12 (0°)–head / 23 (0°)–leaf
Mushroom	H, F, V to 2–4°	0–1	95	7–9 days	35 (0°)
Okra	H, F	7–10	>90	7–14 days	21 (12.5°)
Onion					
Bulb	A to 0 after drying	0	65–75	1–9 months	3 (0°)
Green	H, F, V to <4° in 4–6h	0	95–98	3–4 weeks	
Peas, green and with edible pods	F, H, V	0	95–98	1–2 weeks	38 (0°)
Peppers	F, H, V	7	90–95	2–3 weeks	12 (10°)
Potatoes	Cure at 20°C, RH 80–100%	7–10 (eating) / 10–15 (frying) / 15–20 (chipping)	95–99 at all temps	2–12 months	16 (10°)
Pumpkin and Winter Squash	R	10–13	50–70	2–3 months	99 (12.5°)
Radish	H at 0 to 4.5°C	0	90–95	3–4 weeks	16 (0°)–topped / 6 (0°)–bunches
Rutabaga	R, F, H, I	0	98–100	4–6 months	5 (0°)
Southern peas					
Unshelled	F	4–5	95	6–8 days	24 (5°)
Shelled	H	4–5		24–48 h	29 (0°)
Spinach	H, I[3]	0[3]	95–100[3]	10–14 days[3]	21 (0°)
Squash, Summer	R, F, H	5–10	95	2 weeks	32 (5°)
Sweet potatoes	Cure at 29±1, 90–97% RH for 4–7 days	13–15	90	1 year	67 (10°) / n.d.

TABLE 20.2 (continued)
Recommended Precooling Methods, Storage Conditions, Approximate Storage Life, and Average Respiration Rates for Selected Vegetables

Commodity	Precooling Method[a]	Recommended Storage Conditions		Approximate Storage Life (at recommended conditions)	Respiration Rate[b] (mg CO$_2$ kg^{-1} h^{-1})
		Temperature (°C)	Relative Humidity (%)		
Tomatillo	F, R	5–10	80–90	≤3 weeks	16 (10°)
Tomatoes	R, F				22 (15°)
Mature green	to 20° for ripening	19–21	90–95		35 (20°)
Ripe	to 12° for storage	13	90–95	7–28 days[3]	
Turnip		0	90–95	4–5 months	8 (0°)

Note: All values represent averages for the commodities. Values for individual cultivars and differing handling and growing conditions may vary significantly.

[1] For commodities where no precooling method is indicated, precooling is not required. Precooling methods F = forced air; R = refrigerator; H = hydrocooling; V = vacuum; HV = hydrovacuum; I = icing; PI = package icing

[2] Respiration rate provided for temperature indicated in parentheses.

[3] Data source USDA Handbook 66, 1968 ed.

Developed from data in *The Commercial Storage of Fruits, Vegetables, and Florist and Nursery Crops*, Gross, K.C., Yang, C.Y., and Saltveit, M., Eds., Agricultural Research Service, Beltsville, MD, 2002. Draft version of revised USDA Agriculture Handbook 66 on the USDA website http://www.ba.ars.usda.gov/hb66 /index.html

---

TABLE 20.3
Typical Figures for Q_{10}

Temperature	Q_{10}
0–10°C	2.5–4.0
10–20°C	2.0–2.5
20–30°C	1.5–2.0
30–40°C	1.0–1.5

Source: Saltveit, M.E., Respiratory metabolism, in *The Commercial Storage of Fruits, Vegetables, and Florist and Nursery Crops,* Gross, K.C., Yang, C.Y. and Saltveit, M., Eds., Agricultural Research Service, Beltsville, MD, 2002, draft version of revised USDA Agriculture Handbook 66, available at http://www.ba.ars.usda.gov/hb66/index.html. (With permission.)

TABLE 20.4
Effect of Temperature on the Rate of Degradation

Temperature (°C)	Assumed Q_{10}	Relative Rate of Degradation	Relative Shelf-life
0	—	1.0	100
10	3.0	3.0	33
20	2.5	7.5	13
30	2.0	15.0	7
40	1.5	22.5	4

Source: Saltveit, M.E., Respiratory metabolism, in *The Commercial Storage of Fruits, Vegetables, and Florist and Nursery Crops,* Gross, K.C., Yang, C.Y. and Saltveit, M., Eds., Agricultural Research Service, Beltsville, MD, 2002, draft version of revised USDA Agriculture Handbook 66, available at http://www.ba.ars.usda.gov/hb66/index.html. (With permission.)

---

commodity. This table can show, for example, that if a commodity has a shelf life of 13 d at 20°C, it can be stored for as long as 100 d at 0°C but will last only about 4 d at 40°C.[6]

The effect of reducing temperature on storage life is not uniform.[6] Only a small improvement in storage life is achieved by small reductions at the upper end of the temperature range. Much larger improvements are obtained by small reductions at the lower temperatures, where even a change of 1°C can have a significant effect.

Excess energy from respiration, i.e., energy not used to maintain life processes, is released in the form of heat, called vital heat. The amount of vital heat varies with the type of product, variety, maturity or stage of ripeness, injuries, temperature, and other stress-related factors. This heat, which is about 673 kcal per mole of sugar utilized, must be considered in any temperature management program since it can have a significant effect on refrigeration requirements.[6] Some commodities have high respiration rates and require considerably more refrigeration to hold them at a specified temperature. Tables such as Tables 20.1 and Table 20.2 that present respiration rates of fruits and vegetables are useful in calculating refrigeration requirements of various products. Vital heat also must be considered in selecting proper methods for cooling, packaging, stacking packages, and refrigerated storage facility design.

20.2.2 RIPENING AND ETHYLENE PRODUCTION AND UTILIZATION

Normal ripening generally occurs within a particular temperature range (commonly 10 to 30°C), although some fruits, for example, some pear varieties, will ripen slowly and satisfactorily at temperatures below 10°C.[2] The best quality of ripe fruit generally develops at about 20°C, the temperature generally considered optimum for ripening most fruits.

Ethylene is a natural product of plant metabolism and is produced by all tissues of higher plants.[8] As a plant hormone, ethylene regulates many aspects of growth, development, and senescence. There is no consistent relationship between the ethylene production capacity of a given commodity and its perishability; however, exposure of most commodities to ethylene accelerates their senescence.

Climateric and nonclimateric fruits differ in their response to applied ethylene and in their ethylene production during ripening. Climateric fruits produce much larger amounts of ethylene during ripening than do nonclimateric fruits. Applying low concentrations of ethylene will hasten ripening of climateric fruits, but the magnitude of the climateric is relatively independent of the concentration of applied ethylene.

Temperature affects the rate of ethylene production.[1] The ethylene synthesizing enzymes, 1-aminocyclopropane carboxylic acid (ACC) oxidase and ACC synthase, are sensitive to low temperatures. Typically, as temperatures are lowered, less ethylene is produced.[3] Temperature also affects the sensitivity of products to ethylene. At lower temperatures longer exposure to a given concentration of ethylene is required to initiate ripening. Figure 20.2 shows the effect of temperature on ethylene-induced floret yellowing of broccoli.

20.2.3 MOISTURE LOSS

As fresh commodities lose water to the surrounding environment, this water loss may lead to shriveling and reduced quality. Three environmental factors affect the rate of produce shrinkage from water loss: (1) temperature of the produce; (2) temperature and relative humidity of the air; and (3) air velocity past the product.[9] The first two factors are related since together they determine the vapor pressure deficit between the fruit and the surrounding air.

Relative humidity (RH) indicates the amount of water vapor in the air as a percentage of the maximum amount of water vapor the air can hold at that temperature. It has a critical effect on the moisture lost from a product. Product moisture loss is rapid at low RH and slower at higher RH.[10] If the RH in a room is 100%, the air in the room is saturated, as is the interior of fruits and vegetables. Thus, there

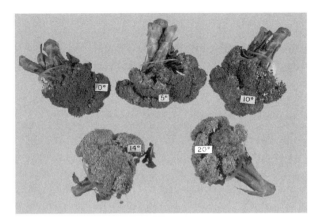

FIGURE 20.2 Ethylene-induced floret yellowing of broccoli is enhanced by increasing temperatures. (Photo courtesy of A.A. Kader, University of California, Davis.)

is a balance in respect to moisture, and virtually no moisture is lost from the produce to the air. If the RH of the room is low, water vapor is readily transferred from the humid interiors of the produce to the relatively dry air. This moisture loss may manifest itself as shriveling and wilting, and many products show visible effects of moisture loss after losing 3 to 5% of their initial weight.[1]

The amount of moisture the air can hold rises as the temperature increases. More water is required to saturate air at 16°C than at 4°C.[10] Therefore, at 16°C and 90% RH the air is drier than at 4°C and 90% RH, and there is more rapid dehydration of produce at the higher temperatures. Because of the relationship of RH to temperature, RH is always expressed with temperature.

Vapor pressure deficit (VPD) is a term used to indicate the combined influence of temperature and relative humidity and is related directly to the rate of water loss from produce.[1] The VPD may be expressed as VPD = VP (100 − RH/100), where VPD = vapor pressure deficit (millimeters or inches of Hg), VP = vapor pressure of water at a given temperature (millimeters or inches Hg), and RH = relative humidity (percentage). It is apparent from the equation that the VPD increases as the VP increases. An increase in VP in the produce would occur with increased temperature since VP is directly related to the temperature of water and most produce is largely water. In addition, the VPD will increase as the RH is lowered. Thus, it would be expected that VPD would be especially high during the hot, dry weather that accompanies harvest in many growing regions.

In addition to a greater VPD at higher temperatures leading to more moisture loss from produce, water evaporates from the produce more quickly at higher temperatures.[11] The dual effect of temperature on RH and increased tendency for evaporation at higher temperatures means that as storage temperature increases the quality of produce tends to decrease more rapidly. Produce in the field at 25°C and 30% RH loses water 36 times faster than it does when cooled and stored at 0°C and 90% RH. Thus, maintaining low product temperature can be a mechanism for reducing water loss and subsequent shriveling and wilting.

Air velocity must be high during cooling to bring about a low VPD as quickly as possible.[11] However, the rate of moisture loss from a product in storage is directly related to air velocity. Therefore, in order to minimize moisture loss, air velocity during storage should be reduced to that needed to maintain desired temperature.

20.2.4 DECAY ORGANISMS

Temperature affects the rate of growth and spread of fungal organisms in and on fruits and vegetables. Postharvest pathogens generally grow best at 20 to 25°C, depending on the fungus species.[12] The maximum temperatures for growth are generally 32 to 38°C, but some species can grow at higher temperatures.

Lower temperatures may be used for the control of fungal disease in one of three ways[12]:

1. They may be used to achieve storage conditions below the minimum temperature for growth of fungus. At –1 to 0°C, the recommended storage temperature for non-chilling-sensitive commodities, only a few fungi grow. *Botrytis cinerea* and *Pencillium expansum* are widely known exceptions since they can grow and cause product rot at 0°C. Some fungi have a minimum temperature for growth of –5 to –2°C. The growth of these fungi cannot be stopped by lowering product temperature since most produce would freeze before it reached a temperature low enough to halt the fungi. However, although these fungi are active at refrigerator storage temperatures, their growth rates are significantly slowed from those observed at higher temperatures.
2. The time between a fungal spore landing on a suitable growth medium and the development of a tiny fungal colony is referred to as the lag phase. On fruit, the lag phase is usually longer than on culture medium since the spore must not only germinate, but also must initiate growth in highly resistant living tissue. Lower temperature may extend the lag phase of the fungal growth so that it has not been completed before the product is consumed. Depending on the fungus species, the lag phase may lengthen from a few hours to several days at optimum temperatures to weeks or months at temperatures near minimum for fungal growth.
3. Lower temperatures may kill the spores of cold-sensitive fungi while they are germinating.

Heat treatments to control pathogens have been the subject of extensive research.[13] The response of fungi to heat treatments varies with such factors as the organism, its metabolic activity, the type of heat used (dry vs. moist), and the duration of the treatment. Heat treatments have been reported to be effective in controlling *Botrytis cinerea* on apples, peppers, tomatoes, and strawberries. Combining fungicides and hot water treatments has been shown to be useful on citrus since the two give better protection than either alone and on tomato and pepper since there are no approved postharvest chemicals for decay control. However, *Penicillium* spp. have been increased in grapefruit as a result of hot water treatment to control Caribbean fruit fly.[13]

20.2.5 VITAMIN LOSSES

Produce is the most important source of many vitamins in the diets of most people. Consumers generally regard any fresh produce as being extremely nutritious and far superior to processed products.[14] This belief may not be true, however, since storage conditions, especially storage temperature, can have a significant effect on vitamin content.

Vitamin C is generally considered the most important vitamin in fruits and vegetables for human nutrition. More than 90% of the vitamin C in human diets is supplied by fruits and vegetables (including potatoes).[15] Vitamin C is most sensitive to destruction when the commodity is subjected to adverse handling and storage conditions. Losses are enhanced by extended storage, higher temperatures, low relative humidity, physical damage, and chilling injury. The review paper by Lee and Kader[15] provides an overview of the effects of preharvest conditions, harvesting, and postharvest handling procedures on vitamin C content of fruits and vegetables.

Lee and Kader[15] reported that delays between harvesting and cooling can result in losses of vitamin C due to water loss and decay. Holding temperature was found to affect the vitamin C content of a variety of produce including tomatoes, kale, spinach, cabbage, snap beans, and broccoli, and all citrus fruits lost vitamin C if stored at high temperatures. The range of temperatures and the extent of vitamin C loss depended on the type of citrus fruit. In general, the extent of loss in ascorbic acid content, the principle biological form of vitamin C, in response to elevated temperatures was greater in vegetables than in acidic fruits, such as citrus, because ascorbic acid is more stable under acidic conditions.

Horticultural crops such as sweet potatoes, bananas, and pineapples that suffer from chilling injury at low temperatures lose significant amounts of ascorbic acid during storage.[15] Destruction of ascorbic acid can occur before any visible symptoms of chilling.

In studies with kale, it has been demonstrated that increased storage temperature and therefore accelerated wilting led to greater losses of vitamin C and carotene.[16,17] Although these data were obtained from studies with kale, it has been accepted that many other horticultural commodities would have similar responses.

20.2.6 SUGAR–STARCH BALANCE

Some produce contains appreciable amounts of carbohydrates that are susceptible to changes in the constituent substances.[7] A prime example of such change is ripening of fruit in which reserve carbohydrates in the form of starch are turned to sugar. At any given temperature starch and sugar are in dynamic equilibrium, and some sugar is degraded to carbon dioxide during respiration[2]:

$$\text{starch} \leftrightarrow \text{sugar} \rightarrow CO_2$$

In some vegetables (e.g., sweet corn and peas) a high sugar content is desirable. These vegetables are harvested immature, when the sugar content is highest, and stored at low temperatures to retard the conversion of sugar to starch. High sugar content is less desirable in other vegetables with high carbohydrate content such as tubers like potatoes, sweet potatoes, and yams and swollen tap roots like carrots, parsnips, horseradish, and turnips.

Low temperature sweetening (LTS) occurs in roots and tubers when some of the breakdown products of starches normally destined for respiration accumulate as excess sugar.[13] Wismer and coworkers[18] provided an in-depth discussion of this phenomenon and proposed mechanisms for its occurrence.

In potatoes, LTS occurs at temperatures lower than 10°C that are often used to minimize respiration and sprouting during storage.[18] The LTS is characterized by a breakdown of starch to reducing sugars. Thus, the sugars that accumulate plus those consumed in respiration equal the amount of starch metabolized. Structural polysaccharides (e.g., cellulose and pectin) are not degraded and do not contribute to the excess sugar. Processed products such as chips made from potatoes that have undergone LTS tend to be dark brown due to the presence of reducing sugars that participate in Maillard browning during frying. Due to the increase in potato tuber sugar content that occurs early in cold storage, and over 2 to 3 months at storage temperatures of 1 to 3°C, tubers can lose as much as 30% of their starch content. Pollock and ap Rees[19] reported that both sucrose and reducing sugars increased within 5 d in tubers stored at 2°C, and after 20 d in storage the sugar content was approximately six times greater than at d 0.

Reconditioning tubers by storing them at 18°C after cold storage may be used to decrease the sugar content and raise the starch levels.[18] However, the response to reconditioning is not consistent or completely restorative and tends to be cultivar-dependent.

20.3 REMOVAL OF FIELD HEAT

20.3.1 WHY PRECOOL?

Temperature control of produce begins in the field. At harvest, the temperature of fruits and vegetables is close to that of ambient air, which, depending on the location and the time of the year, may be as high as 40°C.[2] In order to ensure the lowest possible temperature at harvest, it is generally recommended that harvest of most fruits and vegetables occur in the coolest part of the day, usually early morning. Exceptions to this recommendation is produce such as citrus fruit, which is damaged if handled in the morning when it is turgid, and situations when fruit is harvested late in the afternoon so that it can be transported to a local market during the cooler night hours.[20]

After harvest, produce should be handled to keep it as cool as possible until removed from the field. Exposure of picked fruit to direct sun can result in a significant rise in fruit temperature and an associated loss of quality. Nelson[11] reported that the temperature of grapes held in the sun was as much as 7°C above air temperature while that of shaded fruit remained at least 3°C below air temperature. This temperature range can mean the difference between fruit that is acceptable and fruit that has lost a significant amount of quality.

Time between picking and cooling should be kept to a minimum. Rapid cooling is beneficial because lower temperatures lead to reduced metabolic rates, decreased

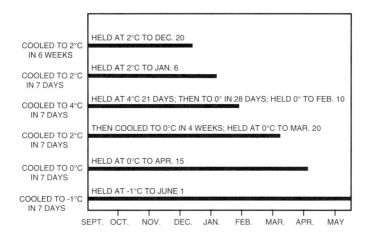

FIGURE 20.3 Normal storage life expectancy for Delicious apples when cooled at different rates and stored at different temperatures. (Adapted from Bramlage, W. J. and Morris, J. R., Storing apples, in *Modern Fruit Science*, 10th ed., Childers, N. F., Morris, J. R., and Sibbett, G. S., Eds., Horticultural Publications, Gainesville, FL, 1985., Chap. 11.)

water loss from the products, and less development of decay caused by pathogens. Harding and Haller[21] demonstrated that harvested peaches held at 23°C lost firmness at a rate of 3 to 4 pounds per day as measured by a Magness-Taylor pressure tester (5/16-inch tip). If the fruit was harvested at 14 pounds firmness, it softened to 2 lbs in just 3 to 4 d. At cooler temperatures, near 0°C, the softening was only a fraction of this rate. As a general rule, more quality is lost in 1 h at 20°C than in 24 h at 0°C.[3]

As calculated for Q_{10}, the generally accepted rule is that the respiration rate of fruit is doubled or tripled for each 10°C rise in temperature. This means that fruit ripens two to three times faster at 10°C than at 0°C and two to three times faster at 20°C than at 10°C.[22] Thus, it is critical that fruit is cooled to storage temperatures as quickly as possible (Figure 20.3). For example, assume that apples are put into storage at 20°C and are cooled to a holding temperature of 0°C in 1 week. The cooling of these apples involves holding the fruit at 20°C for 4 d and then dropping the temperature to 10°C and holding the fruit for 3 d before the final drop to the 0°C holding temperature. The first 4 d at 20°C are comparable to 28 d at 0°C and the 3 d at 10°C are comparable to 6 d at 0°C. In other words, the storage life of apples cooled under these conditions is shortened by 34 days. The conditions would not be quite so extreme in actual practice since the cool-down would actually start immediately, so a 4-d holding period at 20°C would probably not occur. However, apples are often brought into storage at temperatures higher than 20°C.

Precooling is commonly used to refer to any cooling treatment before shipping, storage, or processing.[2] A stricter definition is "cooling methods by which produce is cooled rapidly, generally within 24 hours of harvest." A number of methods are in use for precooling fruits and vegetables; however, the two most commonly used methods are cold air cooling and hydrocooling.

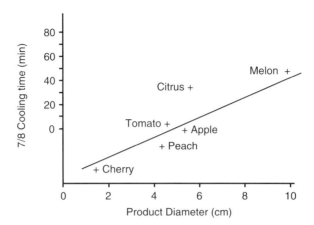

FIGURE 20.4 Effect of product diameter on cooling rate. (Lurie, S., Temperature management, in *Fruit Quality and Its Biological Basis*, Knee, M., Ed., CRC Press, Boca Raton, FL, 2002, Chap. 5.)

20.3.2 PRECOOLING METHODS

Heat transfer occurs by conduction, convection, or radiation.[20] The most common method of heat transfer in precooling is conduction (i.e., heat transfer by direct contact). Since heat moves from a warmer substance to a cooler one, contact between the warm produce and the cooling medium results in the produce being cooled. A product with a large mass and a relatively small surface area will cool more slowly than one with a larger surface area since heat must move from the interior to the surface before it is removed from the product (Figure 20.4). For example, at an airflow rate of 1 L kg^{-1} sec^{-1}, small-diameter grapes can be air cooled in about 2 h, while large-diameter cantaloupes will require more than 5 h.[3,23] With 1.7°C water as the cooling medium, it took more than 30 min to cool a 3-inch-diameter peach from 32 to 4.4°C, but it took about 15 min for the same temperature change with a 2-inch peach.[24]

A critical factor in determining cooling rate is the difference in temperature between the produce and cooling medium.[23] A product is considered "half cool" when its temperature drops so that the difference between product temperature and cooling medium temperature is half of the initial difference. After another half-cooling period the product is "three-quarters cool." Because half-cooling time is a constant value for a given system, the speed of cooling appears to slow as cooling continues (Figure 20.5). Reducing the temperature the last few degrees may take an extended period and is of little practical importance. In comparing cooling times for various cooling methods, the time required to lower the pulp temperature three half-cooling cycles, seven-eighths of the difference between the initial product temperature and cooling medium, is often used.

Mathematically, rate of cooling is often expressed as either the cooling coefficient, C, or the half-cooling time, Z.[2] Theoretically, Z is independent of the initial produce temperature and remains constant throughout the cooling period. Mathematically it is expressed as Z = log$_e$ 0.5/C, where C is a negative value.

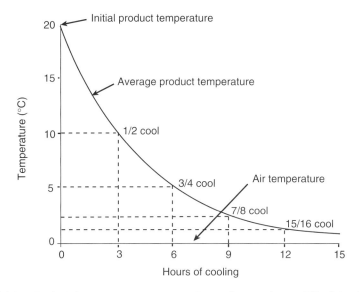

FIGURE 20.5 Typical time–temperature pattern for cooling produce to 7/8 of the difference between its original temperature and the cooling medium in 9 h. (Adapted from Thompson, J.F., Pre-cooling and storage facilities, in *The Commercial Storage of Fruits, Vegetables, and Florist and Nursery Crops,* Gross, K.C., Yang, C.Y., and Saltveit, M., Eds., Agricultural Research Service, Beltsville, MD, 2002, draft version of revised USDA Agriculture Handbook 66, available at http://www.ba.ars.usda.gov/hb66/index.html. With permission.)

The nature of the product is a major determinant in the selection of a cooling method.[25] Different types of produce have different cooling requirements. For example, strawberries and broccoli require near-freezing temperatures, while such low temperatures may damage summer squash and tomatoes. In addition, some products are damaged by exposure to water that is used as the cooling medium in some cooling methods.

20.3.2.1 Cold-Air Cooling

Cold-air cooling, in the forms of room cooling or forced-air cooling, is the most widely adaptable method of precooling and is commonly used for many fruits and vegetables.[24] In this cooling method the cooling medium, refrigerated air, surrounds the produce packed in boxes or pallet bins. Cold air cooling is the least energy-efficient type of cooling but is widely used because it is adaptable to a wide range of products and packaging systems.

Room cooling is most commonly used for products with a relatively long storage life that will be cooled and stored in the same room.[1] Examples of such products include potatoes, sweet potatoes, citrus, and controlled-atmosphere-stored apples. Room cooling is generally sufficient for keeping produce at a low temperature once it has been cooled, but it often does not remove field heat rapidly enough to maintain the quality of highly perishable crops. In the room cooling process, heat is removed slowly from only that produce near the outside of the container.[25] Near the center

FIGURE 20.6 Forced-air cooling facility. The photo on the left shows fans mounted along the wall of the cold room. In the photo on the right, produce is placed in front of the fans so that cold air from the room is pulled across the produce. (Photo courtesy of Western Precooling Systems, Fremont, CA.)

of a container, heat is often generated by natural respiration more rapidly than it can be removed, causing the temperature to rise. Some types of produce, such as strawberries, must be cooled as quickly as possible after harvesting to preserve fresh quality. Even a delay of several hours may be enough to reduce quality considerably. In such cases, room cooling is not fast enough to prevent serious damage.

The rate of cooling is significantly increased if the air is forced through the packages and around each piece of produce, thereby greatly increasing the heat transfer surface area.[2] Forced-air cooling is accomplished by exposing packages of produce in a cooling room to higher air pressure on one side than on the other. This pressure difference forces the cool air through the packages and past the individual units of produce, where it picks up heat, greatly increasing the rate of heat transfer. Depending on the temperature, airflow rate, and type of produce being cooled, forced-air cooling can be from 4 to 10 times faster than room cooling (Figure 20.6).

With both types of cold air methods, cooling time is decreased as airflow rate is increased.[3,23,25] As shown in Figure 20.7, the 7/8 cooling time in still air is more than 7 compared to just over 1 for products cooled with an airflow of 1 cfm/pound of produce.

Causing the air to move faster around the product may shorten cooling time; however, the movement of air over the product during cold air cooling also causes some moisture loss.[23] The loss may be virtually nonexistent for products like citrus fruits that have a low transpiration coefficient, or it may equal several percent of the initial product weight for products with a high transpiration coefficient. If the relative humidity of the cooling air is above 80%, moisture loss during cooling is negligible. An additional advantage of high humidity during cooling is that it will add moisture to the storage boxes and, therefore, decrease the amount of moisture removed from the product during subsequent handling.[3]

Wrapping the product in plastic or packaging it in bags can reduce the loss of moisture to cooling air. However, the use of packaging material can hinder airflow around the product and slow cooling time.[23] The amount of airflow restriction caused by the packaging material should be a consideration in selecting materials.

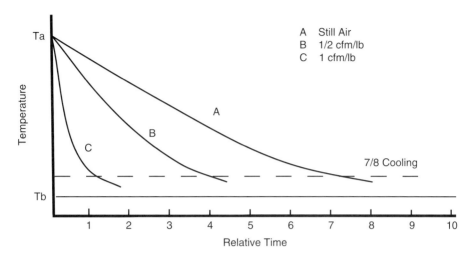

FIGURE 20.7 Response of a typical commodity to airflow rate during cooling. Ta is the beginning temperature of the product; Tb is the air temperature in the cooling room. The time required to lower the produce temperature to 7/8 of the difference between Ta and Tb drops as airflow rate is increased. (From Boyette, M.D., Wilson, L.G., and Estes, E.A. Forced air cooling, Publication AG-414-3 The North Carolina Agricultural Experiment Station, Raleigh, NC, 1989. Use with permission.)

20.3.2.2 Hydrocooling

Hydrocooling is accomplished by moving cold water around the produce with a shower system or by immersing the produce directly in cold water.[23] Because water is a better heat transfer medium than air, hydrocoolers reduce temperature faster than forced-air coolers. For most products, cooling water is kept at 0 to 0.5°C.[3] If the produce is simply immersed in the cold water, the water in contact with the produce will heat up and cooling will be slowed. Therefore, to be effective, a hydrocooler must cause the water to move over and contact all surfaces of the produce. The water must also be free of disease-causing organisms.

There are two types of hydrocoolers.[3] Shower-type coolers rain water down over the produce. They can be built with a slowly moving conveyor for continuous flow operation, or they can be batch-operated. Immersion coolers are best suited for dense products that sink. They usually cool slower than shower coolers because the water flows past the product at a slower rate.

Cooling time in hydrocoolers depends on the diameter of the produce and water flow rate.[3] With the same flow rate, small-diameter products, such as cherries, radishes, and asparagus, cool faster than large items, such as melons. Asparagus spears, being long and narrow, can be hydrocooled in about 2 min, whereas large globular capsicums require 10 min.[20]

Hydrocoolers do not cause moisture loss during cooling and, in fact, can rehydrate slightly wilted produce.[23] Because hydrocooler water can spread spoilage organisms, the water should be obtained from a clean source and treated (usually with hypochlorous acid from sodium hypochlorite or gaseous chlorine) to minimize levels of spoilage organisms.

Since product contact with cooling water is critical, packaging for produce to be hydrocooled must tolerate water contact and allow water to flow through.[23] Plastic or wood containers work well in hydrocoolers.

20.3.2.3 Package Icing

Package icing involves packing the product in crushed or flaked ice. Ice not only removes heat rapidly when first applied to produce but, unlike other cooling methods, it continues to absorb heat as it melts.[23,25] This cooling method has the advantage of maintaining a high humidity around the product, thereby reducing moisture loss. Disadvantages of package icing include its high capital and operating costs, added package weight due to the ice, the necessity that the product package be capable of withstanding constant water contact, and melting ice that may damage or contaminate other products in a shipment of mixed commodities.

20.3.2.4 Vacuum Cooling

Vacuum cooling is achieved by causing water to evaporate from the product, thereby using the latent heat of vaporization as the cooling method.[23] The product is placed in a steel vessel and vacuum pumps reduce pressure in the vessel from an atmospheric pressure of 760 mmHg to 4.6 mmHg. Since water boils at a pressure of 20 to 30 mmHg, depending on its temperature, reduction in the pressure surrounding the product causes moisture evaporation and cooling. At the end of the cooling cycle when the pressure is equal to 4.6 mmHg, water boils at 0°C. If the product is held at this pressure long enough, its temperature will be reduced to 0°C.

Water loss of about 1% causes a 6°C reduction in product temperature.[26] Generally, products lose 2 to 4% of their weight during vacuum cooling, depending on initial temperature. Spraying produce with water, either before starting the cooling process or towards the end of the vacuum operation, minimizes the moisture loss from the product (Figure 20.8).

The rate of cooling by this technique is largely dependent on the surface-to-volume ratio of the produce and the ease with which the product loses water.[2] Leafy vegetables are ideally suited to vacuum cooling. Other vegetables such as asparagus, broccoli, brussels sprouts, mushrooms, and celery also cool well in this system. Fruits and vegetables that have a low surface-to-volume ratio or a surface barrier to water loss, such as a waxy cuticle, lose water slowly and are not good candidates for vacuum cooling.

20.4 RESPONSES TO TEMPERATURE EXTREMES

Produce is frequently exposed to temperature extremes. Some, such as low-temperature storage to extend shelf life and heat treatment to eliminate pests, are intentional. Exposing crops to low temperatures postharvest can be used to kill insect pests.[20] The U.S. Department of Agriculture recommendations for controlling the Mediterranean fruit fly

FIGURE 20.8 Hydrovac coolers minimize moisture loss due to evaporative cooling by spraying produce with water either before the cooling process or towards the end of the vacuum operation. The photo on the left shows produce arriving at the cooling facility. The photo on the right shows produce being loaded into a hydrovac cooler. (Photos courtesy of Western Cooling Systems, Fremont, CA.)

(*Ceratitis capitata*) are to expose fruit to 0°C for 10 d, 1.1°C for 12 d, 1.7°C for 14 d, or 2.2°C for 16 d. However, many fruits suffer chilling injury at these temperatures.

20.4.1 Chilling Injury

20.4.1.1 Occurrence

Low temperatures are the most effective means of extending the shelf life of many commodities. However, some commodities are "chilling-sensitive" in that they suffer injury when cool-stored at temperatures above their freezing point. The term *chilling injury* is used to refer to both the physiological response of the tissue to the low temperature and the resultant symptoms that affect product acceptability.[4] It can occur in the field, during transportation, during storage, during wholesale distribution, in the retail store, and in the home refrigerator. Synonymous terms include *chilling damage*, *chilling disorders*, *low-temperature breakdown*, and *low-temperature disorders*.

Commodities experiencing chilling injury were once thought to be tropical or subtropical crops, since these materials are acclimatized to fairly high temperature and intolerant of low temperatures. However, crops of temperate origin have also been found to develop physiological disorders when exposed to low temperatures.[27] Among the temperate-zone fruits susceptible to chilling injury are apples, cranberries, peaches, nectarines, and plums.

Chilling injury is thought to occur in distinct phases.[28] First, there is the primary response to temperature, usually considered physical in nature, such as membrane alterations. Next, physiological changes occur. Although short-term exposure to chilling temperatures may produce reversible changes, if exposure is continued the changes become harmful and result in the development of symptoms.

Both temperature and duration of exposure are involved in the development of chilling injury.[29,30] In tropical fruits such as bananas, severe injury occurs after exposure to chilling temperatures for only a few hours, and all fruits in a sample suffer injury. In commodities such as apples and grapefruits, injury can take weeks or months of chilling to develop and may be seen only in part of a chilled sample.

Maturity at harvest and degree of ripeness are important factors in determining chilling sensitivity in some fruits. The effects of chilling are cumulative in some commodities, so that low temperatures in transit, or even in the field shortly before harvest, may add to the total effects of chilling that occur in cold storage.

Variations in the effect of chilling injury on tissues are illustrated by the following generalizations[31]:

- Some species are susceptible, whereas others are not.
- Within a susceptible species, some genotypes are more susceptible than others.
- Within a genotype, most, if not all, organs are susceptible, but some organs are more susceptible than others.
- Within an organ, response does not tend to be uniform and is often localized.
- Damage incurred at low temperatures may not become evident until the plant or organ is exposed to nonchilling temperatures.

20.4.1.2 Symptoms

Although chilling of the growing plant may reduce its final yield, damage to produce may appear if chilling occurs after harvest, during transport, or in storage. Some symptoms of chilling injury are only skin-deep (spots, pits, or discoloration), but others involve a more extensive breakdown of internal tissues that may be visible only when the fruit is cut. In addition to the visible symptoms, chilled fruits tend to be more susceptible to decay and may fail to ripen properly.

Some of the more common visual symptoms of chilling injury include the following[28]:

- Surface lesions: pitting, large sunken areas, and discoloration
- Water-soaking of the tissues: disruption of cell structure and accompanying release of substrate favors growth of microorganisms; this commonly occurring symptom in leaves is followed by wilting and desiccation
- Internal discoloration of pulp, vascular strands, and seeds
- Breakdown of tissue
- Failure of fruits to ripen in the expected pattern following removal to ripening conditions
- An accelerated rate of senescence, but with otherwise normal appearance
- Increased susceptibility to decay, especially to organisms not usually found growing on healthy tissue
- A shortened storage life or shelf life due to one or more of the above responses
- Compositional changes
- Loss of growth (sprouting) capacity

Figure 20.9 through Figure 20.14 show chilling injury in a variety of horticultural products.

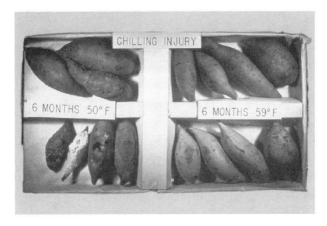

FIGURE 20.9 Chilling injury of two cultivars of sweet potatoes. Roots on left, stored 6 months at 10°C (50°F), show injury symptoms such as shriveling, surface pitting, and fungal decay not seen in roots on right stored for the same period at 15°C (59°F). (Photo courtesy of A.A. Kader, University of California, Davis.)

FIGURE 20.10 Browning under the skin, observed as brown or black streaks in a longitudinal cut, is a dependable indication of chilling injury of bananas. (Photo courtesy of A.A. Kader, University of California, Davis.)

For each chilling-sensitive crop species, there is a unique temperature at or below which chilling injury will occur. This threshold or lowest safe temperature is the basis for the widely circulated "recommended" or "optimum" storage and handling temperatures for various chilling-sensitive crops. The lower the temperature below the threshold temperature and the longer the exposure, the worse the chilling injury will be. The critical threshold temperature below which injury develops ranges from less than 5°C for apples and oranges to 10 to13°C for mangoes and 12 to 13°C for bananas.[30] Cucumbers, eggplants, papayas, and peaches are injured below about 7°C. The critical temperatures and symptoms of chilling injury for a variety of fruit and vegetables are provided in Table 20.5.

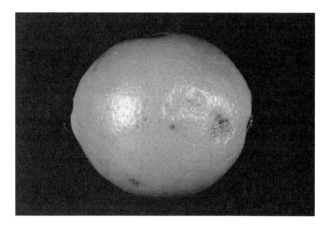

FIGURE 20.11 Symptoms of chilling injury on oranges include pitting and brown discolorations. (Photo courtesy of A.A. Kader, University of California, Davis.)

FIGURE 20.12 Tomatoes stored 3 weeks at 5°C followed by 1 week at 20°C show symptoms of chilling injury including pitting, discoloration, decay, and uneven ripening. (Photo courtesy of C.Y. Wang, Produce Quality and Safety Lab, USDA, Washington, DC.)

Chilling produce is not necessarily all bad.[30] For example, slight chilling can result in improved sweetness and texture of stored sweet potatoes. A cold treatment improves the ripening and sweetness of harvested pears. Night temperatures in the chilling range result in improved color of oranges.

Wade[30] pointed out that attributing tissue damage observed in products stored at lower temperatures to chilling may not always be appropriate. Since chilling extends shelf life, senescent disorders may have time to develop in chilled tissue. Such disorders may never be observed when products are stored at higher temperatures since the development of tissue disintegration or disease may occur before senescence symptoms are observed. Thus, in some cases, damage attributed to chilling

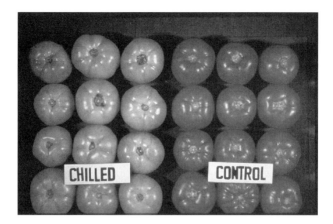

FIGURE 20.13 Failure to ripen and increased susceptibility to decay seen in the tomatoes on the left is the result of chilling injury. (Photo courtesy of A.A. Kader, University of California, Davis.)

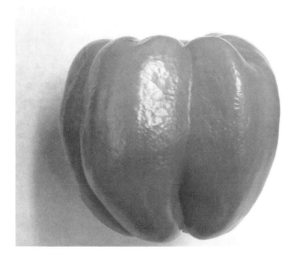

FIGURE 20.14 Sheet pitting is a symptom of chilling injury of pepper stored 1 week at 0°C followed by 3 weeks at 20°C. (Photo courtesy of C.Y. Wang, Produce Quality and Safety Lab, USDA, Washington, DC.)

may actually be normal aging of the tissue. Low temperature may also aggravate an injury resulting from some other cause such as heat or mechanical damage.

20.4.1.3 Responses to Chilling Temperatures

Chilling injury has been the basis of a great deal of research in recent years, yet a review of the literature still reveals no general consensus on the mechanisms leading to this injury. What is known is that chilling leads to a variety of physiological and

TABLE 20.5
Fresh Produce Susceptibility to Chilling Injury When Stored at Low but Nonfreezing Temperatures

Commodity	Lowest Safe Temperature		Symptoms of injury when stored between 0°C and safe temperature[a]
	°C	°F	
Apples, certain cultivars	2–3[b]	36–38	Internal browning, brown core, soggy breakdown, soft scald
Asparagus	0–2	32–36	Dull, gray-green, limp tips
Avocados	4.5–13[b]	40-55	Grayish-brown discoloration of flesh
Bananas	11.5–13[b]	53–56	Dull color when ripened
Bean (lima)	1–4.5	34–40	Rusty brown specks, spots, or areas
Bean (snap)	7	45	Pitting and russeting
Cranberries	2	36	Rubbery texture, red flesh
Cucumbers	7	45	Pitting, water-soaked spots, decay
Eggplants	7	45	Surface scald, alternia rot, blackening of seeds
Ginger	7	45	Softening, tissue breakdown, decay
Guavas	4.5[b]	40	Pulp injury, decay
Grapefruit	10[b]	50	Scald, pitting, watery breakdown
Lemons	11–13[b]	52–55	Pitting, membranous staining, red blotch
Limes	7–9	45–48	Pitting, turning tan with time
Mangoes	10–13	50–55	Grayish scaldlike discoloration of skin, uneven ripening
Melons, cantaloupe	2–5[b]	36–41	Pitting, surface decay
Melons, honeydew	7–10	45–50	Reddish-tan discoloration, pitting, surface decay, failure to ripen
Melons, casaba, Crenshaw and Persian	7–10	45–50	Pitting, surface decay, failure to ripen
Okra	7	45	Discoloration, water-soaked areas, pitting, decay
Olive, fresh	7	45	Internal browning
Oranges	3[b]	38	Pitting, browning stain
Papayas	7	45	Pitting, failure to ripen, off-flavor, decay
Passion fruit	10	50	Dark red discoloration on skin, loss of flavor, decay
Peppers, sweet	7	45	Sheet pitting, alternaria rot on pods and calyxes, darkening of seeds
Pineapples	7–10[b]	45–50	Dull green when ripened, internal browning
Pomegranates	4.5	40	Pitting, external and internal browning
Potatoes	3[2]	38	Mahogany browning, sweetening
Pumpkins and hard-shell squash	10	50	Decay, especially alternaria rot
Sweet potatoes	13	55	Decay, pitting, internal discoloration, hardcore when cooked
Tamarillos	3–4	37–40	Surface pitting, discoloration
Taro	10	50	Internal browning, decay
Tomatoes, ripe	7–10[b]	45–50	Watersoaking and softening, decay
Tomatoes, mature-green	13	55	Poor color when ripe, alternaria rot
Watermelons	4.5	40	Pitting, objectionable flavor

TABLE 20.5
Fresh Produce Susceptibility to Chilling Injury When Stored at Low but Nonfreezing Temperatures (continued)

[a] Symptoms often become apparent only after removal to warm temperatures, as in marketing
[b] Chilling sensitivity may be influenced by factors such as maturity at harvest, degree of ripeness, cultural practices, and handling practices.

Source: Wang, C.Y., Chilling and freezing injury, in *The Commercial Storage of Fruits, Vegetables, and Florist and Nursery Crops,* Gross, K. C., Yang, C. Y., and Saltveit, M., Eds., Agricultural Research Service, Beltsville, MD, 2002, draft version of revised USDA Agriculture Handbook 66, available at http://www.ba.ars.usda.gov/hb66/index.html. (With permission.)

biochemical responses in plant tissue. Reviews such as those by Lyons et al.,[32] Wang,[33] Parkin et al.,[34] and Shewfelt[35] have looked at these responses as a means of identifying possible mechanisms for chilling injury.

Researchers agree that control of chilling injury will not be possible until the mechanism of this injury is fully understood. It also is agreed that, in order to fully define the injury and its causes, a hypothesis for the mechanism of chilling injury must meet several requirements.[34] It must account for the chilling response of all sensitive plant tissues and for the progressively irreversible nature of injury during prolonged periods of chilling stress. It must be consistent with the observation that, in some chill-sensitive plants, chilling injury becomes evident only after the material is transferred from the chilling environment to a warmer one. It must account for the ability of some plants to acclimatize to adverse temperatures and for the varying degrees of susceptibility to chilling injury of members within a genotype. Finally, it must account for the fact that some plant materials are resistant to chilling injury.

Determination of the mechanism of chilling injury may be even further complicated since, even within a single species, more than one mechanism may be involved. For example, tomatoes are most susceptible to chilling injury at the mature green and pink stages.[36] This would suggest that two distinct mechanisms may be involved in the development of these conditions.

20.4.1.3.1 Changes in Cell Membranes

One of the first hypotheses to explain the nature of chilling injury was developed by Lyons and Raison.[37] They proposed that chilling injury was the result of a physical phase transition of membranes from a flexible liquid-crystalline to a solid-gel structure. This theory was based primarily on observations of Arrhenius breaks between 9 and 12°C for respiratory activity in mitochondria isolated from chilling-sensitive tissue (tomato, cucumber, and sweet potato). These breaks were not evident in tissue from chilling-resistant tissues (cauliflower, potato, and beets).

In subsequent work, Raison et al.[38] demonstrated that the activities of membrane-associated enzyme systems were directly correlated to the physical state of the membrane components. Chilling-resistant plants did not show the physical phase transition of membranes or the sudden change in enzyme activity at chilling temperatures.

Although the phase-change hypothesis has been disputed, the involvement of the membrane in the response to chilling has been widely accepted. Lyons et al.[32] and Shewfelt[35] presented discussions of research that investigated membrane responses to low temperature.

Whether the physical phase change of membranes leads to secondary responses or irreversible changes is dependent on such factors as temperature, the length of exposure, and the susceptibility of the plant species to a particular temperature.[33] Exposure of sensitive species to a chilling temperature may result in:

- loss of membrane integrity
- leakage of solutes
- loss of compartmentation
- decrease in the rate of mitochrondrial oxidative activity
- increase in the activation energy of membrane-associated enzymes
- cessation of protoplasmic streaming, reduction in energy supply and uti-lization
- decrease in photosynthetic rate
- disorganization of cellular and subcellular structure
- dysfunction and imbalance of metabolism, accumulation of toxic sub-stances
- manifestation of a variety of chilling injury symptoms

The effects of chilling may take time to develop, especially if the temperature is not far below the critical temperature.[7] These effects may be reversible if the chilling exposure is short but generally become irreversible with prolonged chilling.

Damage to cellular membranes as the primary effect of chilling injury was reaffirmed by studies of ultrastructural changes associated with the development of chilling injury. In tomatoes, plastid mitochondrial compartments degenerated several days after chilling.[39] Some of the early chilling-related ultrastructural changes were reversible with rewarming.

20.4.1.3.2 Stimulation of Ethylene Production

Although ethylene production is stimulated by chilling temperatures in a number of plants, Wang and Adams[40] found that the pathway for ethylene biosynthesis in chilled tissues was the same as that in ripening fruit. The step that was enhanced by the chilling temperature was the synthesis of 1-aminocyclopropane-1-carboxylic acid (ACC), the immediate precursor of ethylene. Increased ethylene production in chilled tissue was the result of increased capacity of the tissue to make ACC. Generally, ACC levels, ACC synthase activity, and ethylene production remained low while the tissue was held at chilling temperatures. However, they increased when the tissue was removed to a warmer temperature. Chilling-induced ethylene production declined after the initial stimulation, even when endogenous ACC levels were still elevated.[40] These data suggest that conversion of ACC to ethylene was the first step damaged by chilling.

Stimulation of ethylene production may not be a primary response to chilling.[40] Inhibition of chilling-induced ethylene production did not result in retardation of chilling injury symptoms. Because ACC synthase, the key enzyme for ethylene production, was stimulated by chilling but is soluble and not membrane-bound, it is unclear whether increased ethylene production is related to the physical phase transition of the membrane.

20.4.1.3.3 Changes in Respiratory Activity

Tissues that are tolerant to low temperatures have a decrease in respiratory activity as temperature decreases.[32] Thus, the respiratory activity remains in balance with glycolysis and other closely associated reactions, and there is little change in the respiratory quotient. However, chill-sensitive products, including citrus fruits, cucumbers, snap beans, and sweet potatoes, exhibit abnormal respiration when their temperature falls below 10 to 12°C.[6] Typically, the respiratory quotient is much higher at these low temperatures for chilling-sensitive crops than for chilling-tolerant ones. Respiration may increase dramatically at the chilling temperature or when the product is returned to nonchilling temperatures. The mechanism of this stimulation is not known but is presumed to be due to uncoupling of oxidative phosphorylation. The enhanced respiration may reflect the cells' efforts to detoxify metabolic intermediates that accumulated during storage and to repair damage to membranes and other subcellular structures. Since enhanced respiration is one symptom of chilling injury, respiratory response may be used as an index of the extent of chilling injury. A sustained increase in respiratory rate after prolonged exposure to chilling temperatures may be indicative of irreversible metabolic disturbance and accumulation of oxidizable intermediates.

In bananas that ripen after harvest, Pantastico et al.[41] found the respiratory response to chilling was much different. Respiration at 20°C was depressed and the climacteric rise delayed following 1 or 2 d at 5°C. After 5 d at 5°C, respiration was completely inhibited, and the climacteric rise was absent. As a result, the fruit lost the capacity to ripen after prolonged chilling.

20.4.1.3.4 Changes in Proteins and Enzyme Activity

Most of the changes in enzyme activity associated with chilling injury occur in enzymes associated with membranes.[33] The Arrhenius plots of the membrane-bound enzyme systems in sensitive plants show a "break" at the same temperature at which the membrane undergoes a phase transition from the liquid-crystalline to the solid-gel state. For chilling-resistant plants, the "break" is either nonexistent or occurs at a lower temperature.

Lyons and Raison[37] and Raison et al.[42] reported that the mitochrondrial succinate oxidase system, succinate dehydrogenase, and cytochrome oxidase of chilling-resistant plant tissues showed constant activation energy over the usual range of biological temperatures. There was an increase in activation energy below 9°C for these same enzymes from chilling-sensitive plant tissues. Raison et al.[42] showed that the temperature-induced change in activation energy observed in chilling-sensitive tissues was due to configurational changes in the enzyme proteins caused by a temperature-dependent phase change in the lipid component of the mitochrondrial membranes.

20.4.1.3.5 Effect on Protoplasmic Streaming

A difference in the protoplasmic streaming response of chilling-sensitive and chilling-insensitive plants was one of the chilling responses first identified by early researchers.[32]

It was observed that protoplasmic streaming ceased in chilling-sensitive plants exposed to chilling temperatures but was basically unaffected by low temperatures in chilling-resistant plant tissues. The reasons for the effects of chilling temperatures on streaming were postulated to be related to one or more of the following[33]: (1) cell lipids and their role in structure and action of the protoplasm; (2) energy supply from respiration to maintain protoplasmic streaming; (3) energy utilization for streaming; (4) protoplasmic viscosity; or (5) difference in sensitivity to chilling temperature of enzyme systems that are responsible for utilizing ATP for streaming.

20.4.1.3.6 Increase in Permeability and Solute Leakage

A number of chilling-sensitive plant tissues have been shown to develop a high rate of electrolyte leakage in response to exposure to chilling temperatures.[33] For example, ion leakage in the mitochondria isolated from chilled sweet potato slices was as much as five times that of unchilled tissue.[43] This difference has led to postulations that increased cell permeability and solute leakage are common responses to chilling temperatures. However, in some chilling-sensitive fruits, such as peaches, electrolyte leakage remains low and does not appear to be altered while the fruit is held at a chilling temperature. It does, however, increase rapidly after the fruit is transferred to ripening temperature.[44] Some chilling-sensitive vegetables, such as bell peppers and eggplants, do not exhibit any increase in electrolyte leakage at critical chilling temperatures.[45] These findings have led to the proposal that high rates of electrolyte leakage are not necessarily a general property of chilling-sensitive plant tissue.

20.4.1.3.7 Accumulation of Metabolites During Cool Storage

Inhibition or disruption of mitochondrial function may lead to an accumulation of pyruvate.[32] This may be metabolized to such compounds as acetaldehyde, ethanol, and acetate. Changes in metabolite concentration may occur as a result of the primary injury of chilling stress or as an outcome of the tissue injury. There is also the possibility that the metabolite accumulation as a result of chilling stress exerts a toxic effect on the tissue.

Hulme et al.[46] observed a positive correlation between the accumulation of oxaloacetate in cool-stored apples and the subsequent development of low-temperature breakdown. Oxaloacetate did not accumulate in varieties of apples that did not suffer from low-temperature breakdown. This correlation was not sustained by other studies that found no differences in the levels of TCA cycle intermediates that could be associated with chill injury.[32]

20.4.1.3.8 Fatty Acid Composition of the Membrane Lipids

It is generally agreed that a physical change in the membranes is the primary response of plants to chilling. However, there is no consensus on whether the composition of membrane lipids plays a deciding role in determining the sensitivity of plant tissue to chilling. Lyons et al.[47] reported that the membrane lipids of chilling-resistant tissues had a higher degree of unsaturated fatty acids than lipids of chilling-sensitive tissues.

However, the relationship of fatty acid composition and chilling sensitivity may not be as clear-cut. A study by Yamaki and Uritani[48] showed that the white potato, a temperate-zone plant, had a lower ratio of unsaturated fatty acids to total fatty acids in the pure mitochondrial fraction than the sweet potato, a tropical plant. As a result of this study and a number of others, it was concluded that fatty acid composition in membrane lipids may not be the only factor that determines phase transition and fluidity of the membrane. Other membrane components, such as sterols and cholesterol or the lipid-protein complex, may also be involved in regulating membrane fluidity.

20.4.1.4 Treatments to Alleviate Chilling Injury

The most obvious way to prevent chilling injury is to store susceptible commodities above the temperature at which the injury will occur. In commercial practice, however, a wide range of commodities is often stored together and at a compromise temperature chosen to minimize chilling injury of susceptible products while extending the shelf life of nonsusceptible products as much as possible.

A number of studies have been conducted to identify treatments to alleviate chilling injury. Some treatments have been applied directly to the commodity, while others involve manipulation of the product's environment.[28] Treatments that have shown promise include ensuring proper humidity levels during cool storage, conditioning the produce at near-chilling temperatures before chilling, intermittent warming treatments during chilling, increased CO_2 in the atmosphere during chilling, pretreatment with calcium or ethylene, and holding under hypobaric conditions during chilling.

Chilling injury may be reduced if the storage humidity is either high or low, depending on the commodity.[30] The mechanism of high humidity prevention of chilling injury may be in preventing drying out of injured tissue. Low humidity may prevent injury by enhancing the loss of volatile toxins from the tissue.

Reduction of the incidence of chilling injury has been observed in some produce when it is held at room temperature for a period before being placed in cold storage.[32] Chilling injury was reduced in sweet pepper fruits stored 5 or 10 d at 10°C prior to subsequent storage at 0°C.[33] A 7-d exposure of grapefruit to 10 or 15°C prevented or significantly reduced chilling injury during storage at 0 or 1°C. Produce showing reduced chilling injury after delayed cold storage were the ones that were ripening at harvest or beginning to ripen just after harvest.[32] Thus, the reduction of chilling injury by delaying storage seems to be related to alteration of the stage of ripening when the fruit enters storage.

The use of high temperatures (38°C or higher) has been shown to be an effective way to increase resistance to low temperatures in some products.[3] In practice, a prestorage high-temperature treatment is often used for insect disinfestation or for fungal control. In some products, however, this elevated temperature also can slow the progression of ripening and, therefore, assist in maintaining quality. A high-temperature prestorage treatment was found to inhibit chilling injury in avocado, citrus fruits, mango, papaya, persimmon, and some apples.

The mechanism by which heat treatment protects against chilling injury is not known. Exposure of some plant tissues to temperature stress for several hours or more has been shown to induce a heat-shock response characterized by reduced total protein synthesis and the production of specialized heat shock proteins (HSP).[13] These HSPs confer thermotolerance so that subsequent exposure to normally lethal high temperature causes little damage. Since subjecting plants to one form of stress can sometimes confer greater tolerance to other stresses, it has been proposed that HSPs may remain in the tissue and provide thermotolerance during cold storage.[3]

Because high temperatures increase membrane leakage but the tissue recovers after the heat stress is removed, it has been postulated that high-temperature treatment also may cause alterations in the cell membrane that induce tolerance to chilling injury.[13] Comparison of the lipid composition of apple plasma membrane and total tissue lipids of tomato showed that samples heated prior to cold storage had more phospholipids and greater fatty acid unsaturation than did unheated fruit. These differences in fatty acid composition would indicate more fluid membranes in the heated fruit. This condition would correspond with reduced indiscriminate leakage in the tissue of heated fruit.

Chilling injury was reduced in some commodities by interrupting cold storage with one or more short periods of warm temperature. This treatment was found effective with apples, citrus, cranberries, cucumbers, nectarines, okra, peaches, plums, potatoes, sweet peppers, and tomatoes.[33] Ben-Arie et al.[49] reported control of woolliness in peaches if the fruit was warmed to room temperature (not specified) for 2 d every 2 weeks during a 6-week storage at 0°C. Lill[50] showed that warming peaches to 12°C was effective in preventing injury for a longer time than was warming to 20°C.

Although intermittent warming has been shown to be effective in reducing chilling injury of some fruits and vegetables, the mechanism of this effect is not well understood. The warming treatment may allow the tissues to metabolize toxic substances that were accumulated during chilling, or warming may allow tissues to restore materials that were depleted during chilling.[30]

In work with apples, Hulme et al.[46] observed that low-temperature breakdown was preceded by an accumulation of oxaloacetic acid. Removing the apples from 0°C storage to 15°C storage for 5 d then returning them to 0°C could reduce both the oxaloacetic acid content and the subsequent intensity of the low-temperature breakdown. They suggested that the low-temperature breakdown was caused by interference in the operation of the Krebs cycle in the tissue.

Reduction of woolliness in peaches and nectarines by intermittent warming seems to be related to the warming treatments leading to more normal ethylene production by the fruit after storage.[3] Chilling injury resulted in reduced ethylene production and altered cell wall hydrolytic enzyme activities. In healthy fruit, polygalacturonase was the major enzyme leading to softening and juiciness. In chill-injured fruit, this enzyme activity was absent, and softening seemed to depend on endoglucanase. At the same time, pectin molecules bound the extracellular juice, causing the woolly condition.

Another method to prevent chilling injury in some fruit is a dual-temperature method. For example, plums are rapidly cooled and stored at 0 to –1°C for 10 to 14 d.[3] The temperature is then raised to 7°C for another 10 to 14 d. This prevents the storage disorders that would develop at 0°C, yet 7°C is still low enough to prevent over-ripening. A reverse procedure in which the fruit is stored first at a warmer temperature and then the temperature is gradually lowered has been shown to prevent injury of some produce. Lemons can be stored at 6 to 8°C if the temperature is gradually lowered over a week to the storage temperature. Avocados can also be stored below 10°C when they are given a step-down temperature treatment. These regimens allow the produce to acclimate and resist low-temperature stress.

Controlled atmospheres, which are depleted in oxygen and/or enriched in carbon dioxide relative to air, are often used in conjunction with cool storage.[30] The effects of these altered storage conditions on chilling injury are variable. With some products atmospheric changes serve to intensify the inhibition of metabolic pathways caused by chilling and lead to increased injury.

The beneficial effects that are sometimes observed when controlled atmospheres are combined with chilling are more difficult to explain.[30] For example, peaches cool-stored in air at 1°C lose the ability to ripen normally when removed from storage. However, peaches stored in 20% volume/volume carbon dioxide retain the ability to ripen. Although controlled experiments have verified that the effect is due to the carbon dioxide, there are many possible sites of carbon dioxide action in the cell and the exact mechanism of this effect is still unclear.

Chemical treatments have been found to reduce the injurious effects of chilling on some produce.[28] Antioxidants such as diphenylamine, ethoxyquin, and butylated hydroxytoluene have been useful in reducing the development of scald in apples. Similar results were seen when oil was applied to the fruit and when treatments of gibberellic acid and phorone were used. The fungicide thiabendazole reduces surface pitting in cool-stored grapefruit. The mode of action for these responses is unknown.

Good correlations have been found between tissue calcium content and the susceptibility of the produce to chilling injury.[30] Produce with relatively high calcium is less likely to suffer from problems due to cool storage than is produce with low calcium. The application of calcium to fruit after harvest reduces the incidence of disorders such as low-temperature breakdown of apples. Variations in tissue calcium may explain why various units of the same commodity do not respond uniformly to chilling.

20.4.2 FREEZING INJURY

The lower limit for normal metabolism is the freezing point of the tissue. Because of the concentration of soluble solids dissolved in the water in the cell sap, freezing of produce occurs at temperatures below 0°C.[20] Soluble solid concentration varies not only with the type of fruit and vegetable but between individuals and even parts of the same fruit or vegetable. It is difficult to cite precise values for freezing points of a particular commodity since the actual freezing point will vary between cultivars, and with such factors as the conditions in which the crop is grown and previous storage history. In general, leafy vegetables, which are fairly low in sugar, freeze at about –0.5°C, while high-sugar fruits generally freeze at –2 to –5°C.

All fruits and vegetables can be classified into three groups based on their sensitivity to freezing[29]:

1. most susceptible: those that are likely to be injured by even one light freezing
2. moderately susceptible: those that will recover from one or two light freezings
3. least susceptible: can be lightly frozen several times without serious injury

TABLE 20.6
Susceptibility of Fresh Fruits and Vegetables to Freezing Injury

Most Susceptible	Moderately Susceptible	Least Susceptible
Apricots	Apples	Beets
Asparagus	Broccoli	Brussels sprouts
Avocados	Carrots	Cabbage, mature and savory
Bananas	Cauliflower	Dates
Beans, snap	Celery	Kale
Berries (except cranberries)	Cranberries	Kohlrabi
Cucumbers	Grapefruit	Parsnips
Eggplant	Grapes	Rutabagas
Lemons and limes	Onions (dry)	Salsify
Lettuce	Oranges	Turnips
Okra	Parsley	
Peaches	Pears	
Peppers, sweet	Peas	
Plums	Radishes	
Potatoes	Spinach	
Squash, summer	Squash, winter	
Sweet potatoes		
Tomatoes		

Source: Wang, C.Y., Chilling and freezing injury, in *The Commercial Storage of Fruits, Vegetables, and Florist and Nursery Crops,* Gross, K.C., Yang, C.Y., and Saltveit, M., Eds., Agricultural Research Service, Beltsville, MD, 2002, draft version of revised USDA Agriculture Handbook 66, available at http://www.ba.ars.usda.gov/hb66/index.html. (With permission.)

Table 20.6 groups vegetables and fruits according to their freezing susceptibility.

Freezing injury is the direct result of unintentional freezing of fresh tissue. Mere exposure of living tissue to temperatures below the freezing point does not necessarily result in injury.[31] Supercooling of some produce to –10°C has been observed without ice formation. However, agitation of a supercooled item can result in ice formation, which is the primary cause of freezing injury. Prevention of freezing injury is best achieved by maintaining storage temperature above the freezing point of the commodity in question. When temperatures have cycled below the freezing temperature, the most prudent action is to bring the tissue temperature above freezing slowly before surveying the possible damage.

Freezing injury is often associated with chilling injury since both occur as a result of exposure to low temperatures and both are characterized by loss of membrane permeability and altered metabolic activity.[35] Some changes in lipid composition, biophysical properties, and altered enzyme activities also are common to both chilling and freezing injury. However, there are differences between the two types of injury.

Freezing injury occurs when the contents of the cell freeze, causing ice crystals to form in the tissues.[29] The most common symptom of freezing injury is a water-soaked appearance (Figure 20.15). Tissues injured by freezing generally lose rigidity and

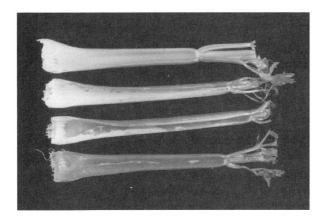

FIGURE 20.15 The water-soaked appearance seen on the lower celery stalks indicates freezing injury. (Photo courtesy of A.A. Kader, University of California, Davis.)

become mushy upon thawing. Once the tissue is frozen, the interchange of metabolites among the various cellular components is seriously hampered.[2]

Ice formation may be extracellular or intracellular.[7] The first ice crystals form in moisture on the surface of the cell wall. The crystals then grow in the intercellular space as water migrates through the cell wall in response to a vapor pressure gradient. As the ice crystals grow, the cells are progressively dehydrated and shrink, or in some cases, the protoplast shrinks away from the cell wall and ice may then form on the inside of the wall. This ice formation and desiccation are not lethal to the material until the ice formation involves up to approximately three-fourths of the water originally present in the cell. This dehydration progressively increases the concentration of the cell solution, lowering its freezing point and providing protection against intracellular freezing.

The process of extracellular freezing is possible only in material with large enough intercellular space to allow ice crystals to form from a considerable proportion of water in the tissue.[7] In tissue with small intercellular spaces, the extent of extracellular freezing is slight before intracellular freezing becomes prominent.

If there is no membrane damage, material with extracellular freezing can thaw without showing freezing injury.[7] However, if membrane damage has occurred as a result of the dehydration associated with extracellular freezing, affected protoplasts eventually disintegrate and die.

If material in which extracellular ice formation has occurred remains at a temperature below the freezing point of the sap, the ice formation spreads into the cytoplasm and vacuolar sap.[7] This freezing disrupts the cytoplasm and nucleus and is always fatal. On thawing, the frozen material collapses and exudes liquid. Intracellular freezing may occur without prior extracellular freezing if ice formation is initiated in material exposed to extreme supercooling.

20.4.3 HIGH TEMPERATURES

Produce exposure to elevated temperatures may occur for a variety of reasons, including exposure to direct sunlight, hot air in the field, or heat treatments used for the eradication of pests. The temperature of the product also may be increased by heat generated during respiration by the product itself. In most cases, increasing the heat of the product, regardless of the source of the heat, may result in produce degradation through a variety of mechanisms. For example, increased temperature can result in more rapid product respiration that can decrease shelf life.

High-temperature injury can result from both chronic exposure for days or weeks to temperatures above typical room temperature (ranging from 30 to 40°C) and acute short-term exposures to higher temperatures.[31] Much less is known about high-temperature injury than about chilling injury since high-temperature injury generally can be prevented with normal handling practices. Short-term injury can be induced during harvesting operations in warm climates and during storage at ambient temperatures in developing nations in the tropics. If the produce is stored at low temperature after exposure to high temperature, the heat damage can be confused with chilling injury, which has similar symptoms.[13]

Elevated product temperatures can have some beneficial effects. These include curing root crops, drying bulb crops, and controlling disease and pests.[20] Many fruits are exposed to high temperatures, often in combination with ethylene or another suitable gas, to initiate or improve fruit ripening or skin color. Lurie[13] presented an in-depth discussion of the mechanisms of postharvest heat treatments for ripening, inducing thermotolerance, reducing chilling injury, and disinfestation of fruits and vegetables.

Three methods are used to apply heat to commodities: hot water, vapor heat, and forced-air treatment.[13] Hot water treatment was originally used for fungal control since fungal spores are either on the surface or in the first few cell layers under the peel of the fruit or vegetable. Hot water dips also have been used for insect disinfestation. Many fruits and vegetables tolerate water temperatures of 50 to 60°C for up to 10 min, but shorter periods at elevated temperatures can control many postharvest plant pathogens.

Vapor heat was developed specifically for insect control and involves heating produce with water vapor at temperatures between 40 and 50°C.[13] Heat transfer is by condensation of hot water vapor on the cooler fruit surface. Hot air treatments can be applied by placing fruits and vegetables in a heated chamber with a ventilating fan or by applying forced hot air. Forced-air heating is slower than hot water or water vapor heating but is faster than a regular heating chamber. It has been used for both insect and fungal disinfestation as well as to study the response of commodities to high temperatures.

The activity of enzymes in fruit and vegetables declines at temperatures above 30°C, but the temperature at which specific enzymes become inactive varies.[2] Many are still active at 35°C, but most are inactivated at 40°C.

When produce is held above 35°C, metabolism becomes abnormal and results in a breakdown of membrane integrity and structure, with disruption of cellular organization and rapid deterioration.[2] The changes are often characterized by a general loss of pigments, and the tissues may develop a watery or translucent appearance.

The short-term effect of elevated temperature on produce respiration is difficult to predict because it is the result of two opposing influences. Enzyme inactivation occurs as a result of protein denaturation occurring at temperatures above 35°C. This results in loss of catalytic activity and a progressive decline in respiratory rate.[7] At the same time, the respiratory rate is increasing due to the direct effect of temperature.

The most extensively studied heat-susceptible fruit is the tomato, which fails to ripen properly at temperatures above 30°C. Increasing storage temperature from 21 to 30°C and to 37°C resulted in abnormal ripening due to inhibition of lycopene development.[51] As storage temperature increased from 21 to 37°C, firmness was more readily retained. Since softening is associated with increased pectolytic enzyme activity, retardation of softening was attributed to inhibition of the activities of these enzymes. Electrolyte leakage, observed on exposure of mature green tomatoes to temperatures above 40°C, indicated that the temperature effect may be mediated through membrane lipids.[52]

Yang et al.[51] added exogenous ethylene to the storage atmosphere of fresh tomatoes to determine whether inhibition of ripening at high temperatures ($\geq 30°C$) was related to reduced ethylene production or reduced ethylene sensitivity. They observed that exogenous ethylene did not overcome inhibition of ethylene formation, ACC synthesis, color development, or softening during storage at 30 or 37°C. Based on these findings, they concluded that ripening inhibition at higher temperatures is not merely the result of decreased production of ACC and ethylene but also of reduced sensitivity to ethylene.

20.5 TEMPERATURE EFFECTS ON SPECIFIC COMMODITIES

Temperature is the most important factor affecting the quality and storage life of all fruits and vegetables. The mechanisms that trigger responses to temperature in produce must be understood relative to the cellular and physiological responses that keep plants alive. However, every species has somewhat unique responses to temperature and its interrelationship with moisture and air flow. The key to managing these responses in handling and storing fresh fruits and vegetables until they are sold or further processed is to understand the conditions needed for each species. The following examples represent fruits and vegetables commonly marketed as fresh produce.

20.5.1 POTATOES, SWEET POTATOES, AND YAMS

Many root crops have a cork layer over the surface called the periderm.[20] This layer serves to protect the interior against microbial invasion and excessive water loss. Curing allows healing of breaks in the periderm that can result from damage due to rough treatment during harvesting and handling. Curing roots to allow wounds to heal before placing them in storage is an extremely important operation to reduce subsequent moisture loss, increase resistance to disease, and help them recover from injury.

Microscopic examination of wounded tissue of sweet potatoes has shown that the first observable change in the wound-healing process is the desiccation of several layers of the outermost parenchyma cells exposed to air after wounding.[53] This is followed by a progressive suberization of the parenchyma cells beneath the desiccated cells. The final stage of the wound healing process is the formation of a wound

TABLE 20.7
Effect of Temperature on Rate of Suberization and
Periderm Formation in Potatoes

Temperature (°C)	Days to Suberization	Days to Periderm Formation after Suberization
≥ 21	1	Data not given
15	2	1
10	3	3
5	5–8	> 10

Note: Average values; actual values vary with cultivar.

Source: derived from data in Thompson, J.F., Postharvest treatments, in *Postharvest Technology of Fruits and Vegetables*, Blackwell Science, Oxford, 1996, Chap. 4.

periderm beneath the suberized parenchyma cells. The wound is healed when the wound periderm is three to seven layers thick.

As can be seen in Table 20.7, both suberization and formation of the periderm are temperature-dependent. Optimal curing conditions for potatoes are around 20°C with RH of 80 to 100%.[54] With an RH of 90%, periderm formation took 6 d at 18°C, 10 d at 15°C, and 12 d at 13°C. Similarly, a RH less than 80% delays curing.[20] Curing at 15°C is commonly recommended to minimize decay. The processes of suberization and periderm formation are similar for sweet potatoes and yams; however, higher temperatures (27 to 29°C for sweet potatoes, 35 to 40°C for yams) are generally recommended.

Frequently, potatoes are harvested in the fall in temperate climates where temperatures are colder than the optimal for curing.[55] However, respiration in storage will heat the tubers and raise the RH. Temperature and RH control during this "sweating" or curing time is managed by using fans to bring outside air into the storage room, depending on the need, during the night or day. Once cured, a process requiring 1 to 2 weeks, the tuber temperature is lowered by 1 to 2°C per day until the desired maintenance temperature and RH are reached. In some cases, two additional steps must be employed: drying of wet potatoes upon entry into storage and warming tubers before removal from storage. Forced movement of air is used to ensure uniform temperature throughout the storage pile. Maintaining 95 to 99% RH is required at all times to minimize shrink and pressure bruising in storage.

Following curing, optimum storage temperatures are determined by the end use of the tubers.[55] The respiration rate of potato tubers is lowest at 2 to 3°C, so this temperature is recommended for long-term storage. Lower temperatures, 0 to 2°C, increase the risk of freezing or chilling injury. Sprouting accelerates at temperatures greater than 4 to 5°C, so seed tubers are commonly stored at these temperatures. Tubers for fresh consumption are stored at 7 to 10°C, to minimize conversion of nonreducing sugars such as starch to reducing sugars such as glucose, which darken during cooking. Tubers for frying are stored at 10 to 15°C, depending on the cultivar and its respective sugar conversion characteristics. Many cultivars for chips accumulate excessive sugar if stored at temperatures less than 15°C. Thus, cultivars for chips often are stored at 15 to 20°C. New cultivars are being developed that will not accumulate sugar at temperatures as low as 5 to 10°C. Tubers can be stored for 2 to 12 months, depending on quality at harvest, quality of storage facilities, variety, and whether or not sprout inhibitors are used. Sprout inhibitors may be applied in the field before senescence begins, on the tubers as they are graded and packaged, or in storage after curing is completed.

Most potatoes are stored in common, air-cooled storage, although refrigeration sometimes is used, mainly as an auxiliary source of cooling.[55] Uniform air and temperature distribution are essential in maintaining good storage conditions. Light should be excluded from potato storage because it causes the production of chlorophyll (which gives the potatoes a green color) and the formation of solanine, a bitter, toxic compound.

Almost all sweet potatoes are susceptible to chilling injury; however, cultivars differ in their level of susceptibility (Figure 20.9).[56] Cured sweet potatoes were less susceptible than uncured ones, possibly due to the beneficial effects of suberization and periderm formation.

20.5.2 GRAPES

Storage temperature has a significant effect on the final quality of grapes.[11] Harvested grapes were found to deteriorate more in 1 h at 32°C than they did in 1 d at 4°C or 1 week at 0°C.

Eastern-type grapes, native American cultivars (*Vitis labruscana* and its hybrids with *Vitis vinifera*), and French-American hybrids cannot be stored successfully for more than 1 to 2 months.[57] However, if storage is desired, they should be stored at –0.6 to 0°C with RH of 80 to 90% immediately after picking. Eastern cultivars are not usually fumigated with sulfur dioxide.

Recommended storage for California-type table grapes (*Vitis vinifera*) is at –1.1 to –0.6°C and 87 to 92% RH.[57] Although some cultivars can be stored satisfactorily at temperatures as low as –2.8°C, cultivars with low sugar content have been damaged by exposure to –1.7°C. Grapes are usually stored in rooms by themselves since the sulfur dioxide fumigation that is needed for mold control during extended storage may be harmful to other products.

Like all fruits, grapes are living tissue carrying on respiration during storage.[11] Compared with most fruits, grapes have a very low respiration rate; however, the heat being evolved by this process can be a significant factor in maintaining fruit quality. For example, if grapes have been cooled to 4°C and then the boxes are stacked so that no heat enters or leaves the space occupied by the fruit, the heat of the fruit will rise 2°C during 8 d of storage. However, if the initial temperature of the fruit is 20°C, then the temperature rise during storage will be about 10°C, a level sufficient to aggravate problems of decay.

Decay is a constant problem during grape handling. Two tools are effective in minimizing this problem: proper temperature management and timely and thorough fumigation with sulfur dioxide.[11] The most common fungal infection of grapes, *Botrytis cinerea*, grows across nearly the entire temperature range in which grapes are handled. Growth of this fungus is slowed as the temperature is lowered until it is almost stopped at the optimum storage temperature. Therefore, it is important that fruit temperature is lowered promptly.

The turgidity of grapes increases as the temperature is lowered.[57] This turgidity can result in splitting of the fruit. With prolonged storage, the fruit loses its turgidity due to moisture loss and begins to soften. Temperature is a determining factor in the amount of water lost from harvested grapes.

At least three symptoms result with water loss from grapes.[11] First to appear are shriveled stems that usually become brittle and break easily when handled. Since the stems of grapes serve as a handle to move the fruit, breaking the stems essentially means the fruit is lost for all practical purposes even though the berries still look and taste good. The rate of stem drying is directly related to temperature and increases logarithmically.

The second symptom of water loss is browning of the stem, which detracts from the appearance of the fruit.[11] The rate of stem browning increases more rapidly with increases in temperature than does the rate of stem drying. From 21 to 38°C there is a threefold increase in stem drying after 8 h, whereas for stem browning the increase is more than fourfold.

The third symptom of water loss is shrinkage and subsequent softening of the berries.[11] At 4 to 5% moisture loss the berries feel softer. This softening is related to temperature before cooling. Grapes held 8 h at 38°C lost 3% of their initial weight and 75% of the berries were rated soft. Berries held at 21°C lost only 0.3% of their weight and only 45% of the berries were rated soft.

Muscadine grapes (*Vitis rotundifolia)* are indigenous to the southern United States and have been under cultivation since the colonial days.[58] Muscadine fruit is produced in tight, compact clusters that do not ripen uniformly. Muscadine grapes for the wine and juice market are usually harvested mechanically, while those for table use are hand-harvested.[57]

Muscadines have a tendency to "shatter" or break away from the stems during harvest. The result is a torn stem cap called a "wet scar" that is an ideal entry point for microorganisms and contributes to shortened storage life. Research has shown that muscadine grapes decayed twice as fast at 20°C as at 10°C and three times as fast at 10°C as at 0°C.[59] Grapes with wet stem scars stored for 1 week at 10°C or 3 weeks at 0°C had 6 to 10 times more decay than grapes with dry stem scars.

A comparison of the effects of storing muscadines at 20, 4.5, and 0°C revealed that decay developed rapidly at the higher temperature, making it impractical to store the grapes for more than 3 d.[57] At the refrigerated temperatures, decay was slowed; however, the fruit softened gradually during a 24-d study. In that study, the percentage of soluble solids, percentage of titratable acidity, individual sugars, and organic acids did not change significantly during storage regardless of storage temperature. Total phenol content increased and pectin decreased during storage, and the higher storage temperature resulted in the greatest changes.

20.5.3 PEACHES AND NECTARINES

Peaches are generally picked before they reach full ripeness and allowed to ripen during shipping and/or storage. They will generally become eating ripe in 2 to 4 d at temperatures of 20 to 27°C.[24] The rate of ripening is about half as fast at 16°C as at 20°C and about half as fast at 10°C as at 16°C. However, when peaches are held at 10°C until they are ripe, flavor does not develop normally.

The problem of woolliness or mealiness in peaches and nectarines has long been recognized as resulting from storage at chilling temperatures.[3] Fruits suffering from these conditions have a good appearance when removed from cold storage and soften normally but have dry, mealy flesh texture when they are ripe. Because woolliness does not affect appearance, it does not become apparent until the fruit is cut.

Storage at 2.2 to 5°C has been shown to result in rapid development of woolliness symptoms, while storage at 0°C delayed symptom development.[60] Von Mollendorff[61] suggested picking maturity is an important factor in determining whether woolliness will develop; however, Luchsinger and Walsh[60] found no relationship between maturity stage and woolliness.

Luchsinger and Walsh[60] observed that storage temperature, not water loss, had the greatest influence on fruit juiciness. Chilling temperatures (0 and 5°C) reduced the fruit's juiciness after as little as 7 d in storage, while storage at 10°C maintained juiciness. Although woolly symptoms developed at 0°C, they took longer to be expressed than in fruit stored at 5°C. The fruit stored at 5°C showed higher ethylene evolution rates and faster softening rates.

One strategy proposed to prevent the development of woolliness is to hold the fruit at ambient temperatures for 36 to 48 h before cooling it to 0°C.[3] The delayed cooling may lead to slightly more softening during storage than is seen with immediate cooling, but the fruit will soften and become juicy following storage. This response may be related to the time at ambient temperature allowing the fruit to maintain its ability to produce ethylene after storage and therefore to initiate normal ripening. Fruit that is chilled immediately loses the ability to produce ethylene after a few weeks in storage.

20.5.4 APPLES

The rate of cooling of apple fruit affects retention of quality, but its importance varies according to variety, harvest maturity, nutritional status of the fruit, and storage history.[62] Apples respire and degrade twice as fast at 4°C as at 0°C.[63] At 15°C they will respire and degrade more than six times faster. Most apple varieties are not sensitive to chilling temperatures and should be stored as close to 0°C as possible. However, varieties that are susceptible to low-temperature disorders should be stored at 2 to 3°C. Temperatures also should be increased for fruit stored in low O_2 controlled atmosphere, since lower temperatures increase risk of low O_2 injury. An additional factor to consider in selecting storage temperatures is the impact of temperature on RH requirements. It is easier to maintain RH greater than 90% at 1°C than at 0°C.

Chilled apples that are suddenly transferred into warm air are likely to "sweat" (i.e., water vapor in the air will condense on them).[63] Sweating also occurs when the doors of a cold storage room are opened, allowing warm, moist air to enter. Sweating itself does not harm the fruit, but it causes wetting, which encourages the growth of fungi and bacteria. Chilled apples should not be allowed to warm and then rechill. To prevent sweating, chilled apples should be allowed to warm gradually inside the storage area, if possible, before they are moved into the open air.

The timing of the climacteric and ripening of apple fruit is advanced by exposure to ethylene.[62] Prevention or slowing of ethylene production, by affecting ethylene synthesis or perception, is a strategy for increasing fruit storability. This delay is achieved primarily by use of low storage temperatures and application of controlled atmosphere storage technologies.

Some varieties of apples are sensitive to low-temperature disorders. Sensitive varieties can develop a number of problems after storage for several months at 0 to 4°C.[63] Brown core, also known as core flush, low-temperature breakdown and internal browning, soft scald, and soggy breakdown are all conditions that may develop in apples. Because each of these conditions is a distinct disorder and each occurs in apples that have been kept for a prolonged time at less than 5°C, it is assumed that each disorder must be derived by a different metabolic route.[10] However, these disorders typically develop only in fruit kept for several months, so risks of low-temperature injury are low for fruit kept in short-term storage (2 to 3 months).

In addition to avoiding long-term storage at low temperatures, other treatments may be used to prevent chilling injury symptoms. For example, low-temperature breakdown is a physiological disease that affects some varieties of apples when stored at temperatures not sufficiently low to freeze the tissue.[46] As a result of this disorder, the fruit develops browning of the cortical tissue, which may eventually affect the whole fruit. It has been observed that low-temperature breakdown is preceded by an accumulation of oxaloacetic acid. Both the oxaloacetic acid accumulation and low-temperature breakdown development can be reduced by interrupting the cold storage with a short period at warmer air temperature (5 d at 15°C after 6 to 8 weeks at 0°C). Hulmes and coworkers[46] suggested this effect occurred because the warming period was sufficient to allow the Krebs cycle to operate "normally," thereby removing keto-acids that accumulated during the slower metabolism of cold storage.

20.5.5 BANANAS

The banana is very susceptible to chilling injury and in some cultivars this injury may occur at temperatures as high as 13°C.[7] Chilling injury of bananas is mainly a peel disorder. Symptoms include subepidermal discoloration visible as brown to black streaks in a longitudinal cut (Figure 20.10), a dull or grayish (smoky) cast on ripe fruit, or failure to ripen. In severe cases the peel turns dark brown or black, and even the flesh can turn brown and develop an off-taste. The pulp of chilled fruit softens unevenly and may be acid, astringent, and lacking in flavor and sweetness. Less severe chilling may have little effect on the green fruit or on the pulp of the ripe fruit, but color development during ripening may be affected, leading to skin that is dull yellow, grayish-yellow, or gray.

20.5.6 CITRUS FRUITS

Many species of citrus fruit are susceptible to chilling injury. Common symptoms include surface pitting (Figure 20.11), browning or darkening of the membranes between the segments, discoloration of the white spongy tissue and, after storage below 2°C, a water-soaked appearance of the flesh.[7] The development and severity of chilling injury in citrus is influenced by both preharvest and postharvest factors.[64] With oranges, the location of production affects the development of chilling injury. Fruit grown in California and Arizona may develop chilling injury when held at temperatures below about 3 to 5°C, while oranges produced in Florida or Texas rarely show chilling injury. Other preharvest factors that affect the development of chilling injury in citrus fruits include cultivar, weather conditions, and even location of fruit on the tree (sun-exposed fruit is more susceptible to chilling injury).

Postharvest, the best means of preventing chilling injury is by storing fruit at nonchilling temperatures.[64] Development of chilling injury symptoms also can be reduced through temperature conditioning before storage, use of high-CO_2 atmospheres (e.g., in controlled atmosphere storage or through the use of wax coatings or plastic film wraps), intermittent warming, and use of benzimidazole fungicides (e.g., thiabendazole and benomyl).

Decay associated with *Pencillium* infection may be reduced in oranges by exposing the fruit to temperatures of 30°C and 90 to 100% RH for several days after harvest.[65] *Pencillium digitatum* infection was inhibited by sealing lemons, pomelos, and grapefruit in plastic bags then exposing them to 34 to 36°C for 3 d. This treatment resulted in lignification and an increase in antifungal materials in the peels.

Insect disinfestation of citrus fruit is usually accomplished by holding the fruit at 0 to 2.2°C for 10 to 16 d before raising the temperature to the normal storage temperature of 6 to 11°C, depending on the cultivar.[66] Since citrus is susceptible to chilling injury, fruits are usually held at 20°C for 3 to 5 d before being placed in the low temperature. This holding period decreases susceptibility of the fruit to chilling injury during the subsequent low-temperature disinfestation treatment.

An alternative method of disinfesting citrus fruit is a high-temperature, forced-air treatment.[66] In this treatment, fruits are heated to a temperature of 44°C over a period of 90 min. They are held at this temperature for an additional 100 min before being cooled to their storage temperature.

20.5.7 TOMATOES

Tomatoes are usually harvested when they are mature green or partly colored. The objective in storage after harvest is to control the rate of ripening. Temperatures below 13°C in the field or in storage may result in damage caused by chilling injury.[67] Chilling injury symptoms include alterations in the rate of color change and the development of such effects as uneven coloring, pitting, breakdown, and poor flavor development (Figure 20.12 and Figure 20.13). The extent of injury depends on the storage temperature, the time the tomatoes are exposed to the temperature, and the stage of fruit ripeness. Green and breaker stage fruits are considered more susceptible to chilling injury than are ripe fruits.[36]

A temperature of 13°C is recommended for slow ripening.[68] At this temperature most cultivars can be kept in good condition for 2 to 6 weeks and change color very slowly. At 15°C the rate of color change increases quite sharply, and above 21°C the rate of maturation and other changes are increased. Tomatoes held at 18°C change color rapidly without excessive softening. Temperatures of 21°C or higher induce rapid ripening and bring about changes in color, softening, and flavor.

When tomatoes are fully ripe, the holding time can be increased by reducing the storage temperature to 10°C.[67] Some experiments have shown that ripe tomatoes can be held satisfactorily at 0 to 4°C. However, some softening may occur at 2°C. Thus, it is generally considered detrimental to hold ripe tomatoes in cold storage for more than a few days.

Maul and coworkers[67] found fruit stored for 2 d at temperatures below 20°C was rated by trained panelists as significantly lower in ripe aroma and tomato flavor than fruit stored at 20°C. Fruit stored at 5°C for 4 d was rated significantly lower in ripe aroma, sweetness, tomato flavor, and significantly higher in sourness than fruit stored at 20°C. The researchers associated the differences in sensory characteristics with concurrent reductions in aroma volatiles measured by a gas chromatograph, changes in chemical composition, and in scores during electronic nose evaluations. Since many of these negative alterations in flavor and aroma parameters occurred prior to the appearance of chilling injury symptoms, Maul et al.[67] proposed that current recommended storage temperatures for tomatoes may be a prime contributor to consumer dissatisfaction with commercially marketed tomatoes.

Holding tomato fruit at elevated temperatures inhibits the ripening processes.[69] If the high temperature is maintained too long, fruit will not ripen normally when returned to temperatures of 20 to 25°C. However, if high temperature is held for a week or less, fruits will ripen after a delay, thus extending shelf life.

The effect of exposing tomato fruit to high temperatures to induce tolerance to chilling temperatures has been the subject of a great deal of research. Findings by Lurie et al.[69,70] indicated that heating tomato fruit to 38°C with 95% RH served not only to reduce fungal rots and control insect pests but also decreased susceptibility of the fruit to chilling injury. However, other researchers[52,71,72] found that this high temperature exposure lowered the quality of the fruit by inducing permanent undesirable physiological changes. Whitaker[71] proposed that partial ripening of the fruit at 20°C prior to cold storage was preferable to heat treatment for providing protection from chilling injury.

20.5.8 FRESH-CUT PRODUCE

Minimal processing has been one of the hottest trends in produce marketing in the last decade. Fresh-cut produce, defined as "any fruit or vegetable or combination thereof that has been physically altered from its original form, but remains in a fresh state," offers the flavor, nutritional value, and freshness of fresh produce along with added convenience.[73]

Cut and packaged produce provides consumers with speed and ease of preparation. However, cut produce can be expected to behave differently from intact product during storage due to its response to wounding and damage to the skin.[74]

Possible results of slicing produce may include increased ethylene production and respiration rates, accelerated senescence, and enzymatic browning.

Storage temperature has a significant effect on the keeping quality of fresh-cut products. Behrsing et al.[75] looked at factors affecting the quality of chopped broccoli florets, broccoli stem pieces, cauliflower florets, julienne and diced carrots, and chopped red coral and butter lettuces. They found that respiratory activity of the chopped vegetables was reduced as storage temperature was lowered from 10 to 4.5 to 1°C. Depending on the product, shelf life was increased between 48 and 154% by storage at 4.5°C rather than at 10°C. Further increases in shelf life of 12.5 to 85% were observed by storage at 1°C rather than 4.5°C. However, chopped butter lettuce held at 1 or 4.5°C showed increased browning over that held at 10°C.

20.6 SUMMARY

Good temperature management is the single most important factor in determining the ultimate quality of fresh fruits and vegetables. Temperature influences the respiration rate of products and, therefore, the speed of product deterioration. It also affects ripening, ethylene production and utilization, product moisture loss, rate of decay, and the balance of starch and sugar contents.

Temperature management of produce begins with harvest in the field and should continue throughout the handling and distribution chain to the consumer's table. For most products, rapid cooling after harvest is important to slow undesirable changes that occur in the produce. A number of methods are in use for this cooling; however, the two used most commonly are cold air cooling and hydrocooling.

Produce is frequently exposed to temperature extremes. Some, such as low temperature storage to extend shelf life and heat treatment to eliminate pests, are intentional. Others, such as high temperatures during harvest, are unintentional. Some produce suffers chilling injury at low temperatures above their freezing point. Some are damaged by high temperatures. Freezing damages all produce. Research efforts continue to discover the optimal temperature and other storage and handling conditions for individual species of fruits and vegetables.

REFERENCES

1. Mitchell, F. G., Cooling horticultural commodities. I. the need for cooling, in *Postharvest Technology of Horticultural Crops,* 2nd ed., Kader, A.A., Ed., University of California Division of Agriculture and Natural Resources Publication 3311, 1991, Chap. 8.

2. Wills, R.B.H. et al., Temperature, in *Postharvest: An Introduction to the Physiology and Handling of Fruit and Vegetables*, AVI Publishing, Westport, CT, 1981, Chap. 4.

3. Lurie, S., Temperature management, in *Fruit Quality and Its Biological Basis*, Knee, M., Ed., CRC Press, Boca Raton, FL, 2002, Chap. 5.

4. Shewfelt, R.L., Postharvest treatment for extending the shelf life of fruits and vegetables*, Food Technol.*, 40(5), 70, 1986.

5. Woodroof, J.G., Harvesting, handling, and storing vegetables for processing, in *Commercial Vegetable Processing*, Luh, B.S. and Woodroof, J.G., Eds., AVI Publishing, Westport, CT, 1975.

6. Saltveit, M.E., Respiratory metabolism, in *The Commercial Storage of Fruits, Vegetables, and Florist and Nursery Crops,* Gross, K.C., Yang, C.Y. and Saltveit, M., Eds., Agricultural Research Service, Beltsville, MD, 2002, draft version of revised USDA Agriculture Handbook 66, available at http://www.ba.ars.usda.gov/hb66/index.html, accessed Mar. 31, 2003.
7. Burton, W.G., *Post-harvest Physiology of Food Crops*, Longman, London, 1982.
8. Kader, A.A., Postharvest biology and technology: an overview, in *Postharvest Technology of Horticultural Crops,* 2nd ed., Kader, A.A., Ed., Publication 3311, University of California Division of Agriculture and Natural Resources, Oakland, CA, 1991, Chap. 3.
9. Nelson, K.E., Baker, G.A., and Gentry, J.P., Relation of decay and bleaching injury of table grapes to storage air velocity and relative humidity and to sulfur dioxide treatment before and during storage, *Am. J. Enol. Vitic.*, 15, 93, 1964.
10. Salunke, D.K., Bolin, H.R., and Reddy, N.R., Postharvest physiological disorder, in *Storage, Processing, and Nutritional Quality of Fruits and Vegetables*, Vol. 1, *Fresh Fruits and Vegetables,* 2nd ed, CRC Press, Boca Raton, FL, 1991, Chap. 10.
11. Nelson, K.E., Handling and harvesting California table grapes for market, Bulletin 1913, Agricultural Experiment Station, University of California, Oakland, CA, 1985.
12. Sommer, N.F., Principles of disease suppression by handling practices, in *Postharvest Technology of Horticultural Crops,* 2nd ed., Kader, A.A., Publication 3311, University of California Division of Agriculture and Natural Resources, Oakland, CA, 1992, Chap. 14.
13. Lurie, S., Postharvest heat treatments of horticultural crops, *Hortic. Rev.,* 22, 91, 1998.
14. Fennama, O., Loss of vitamins in fresh and frozen foods, *Food Technol.*, 31, 32, 1977.
15. Lee, S.K. and Kader, A.A., Preharvest and postharvest factors influencing vitamin C content of horticultural crops, *Postharv. Biol. Technol.*, 20, 207, 2000.
16. Ezell, B.D. and Wilcox, M.S., Loss of vitamin C in fresh vegetables related to wilting and temperature, *J. Agric. Food Chem.*, 7, 507, 1959.
17. Ezell, B.D. and Wilcox, M.S., Loss of carotene in fresh vegetables related to wilting and temperature, *J. Agric. Food Chem.*, 10, 124, 1962.
18. Wismer, W.V., Marangoni, A.G., and Yada, R.Y., Low-temperature sweetening in roots and tubers*, Hortic. Rev.,* 17, 203, 1995.
19. Pollock, C.J. and ap Rees, T., Effect of cold on glucose metabolism by callus and tubers of *Solanum tuberosum, Phytochemistry,* 14, 1903, 1975.
20. Thompson, J.F., Postharvest treatments, in *Postharvest Technology of Fruits and Vegetables*, Blackwell Science, Oxford, 1996, Chap. 4.
21. Harding, P.L. and Haller, M.H., The influence of storage temperature on the dessert and keeping quality of peaches, *Proc. Am. Soc. Hortic. Sci.,* 29, 277, 1932.
22. Bramlage, W.J. and Morris, J.R., Storing apples, in *Modern Fruit Science*, 10th ed., Childers, N.F., Morris, J.R., and Sibbett, G.S., Eds., Horticultural Publications, Gainesville, FL, 1985, Chap. 11.
23. Thompson, J.F., Pre-cooling and storage facilities, in *The Commercial Storage of Fruits, Vegetables, and Florist and Nursery Crops,* Gross, K.C., Yang, C.Y., and Saltveit, M., Eds., Agricultural Research Service, Beltsville, MD, 2002, draft version of revised USDA Agriculture Handbook 66, available at http://www.ba.ars.usda.gov/hb66/index.html, accessed Mar. 31, 2003.
24. Morris, J.R., Mason, C.L., and Jones, B.F., Peach production in Arkansas, Circular 449 (Rev), Arkansas Cooperative Extension Service, Little Rock, AR, 1974.
25. Boyette, M.D., Wilson, L.G., and Estes, E.A., Forced air cooling, Publication AG-414-3, The North Carolina Agricultural Extension Service, Raleigh, NC, 1989.

26. Barger, W.R., Vacuum cooling: a comparison of cooling different vegetables, USDA Market Res. Rep. 600, U.S. Department of Agriculture, Washington, DC, 1963.
27. Bramlage, W.J., Chilling injury of crops of temperate origin, *HortScience* 17, 173, 1982.
28. Morris, L.L., Introduction to symposium: Chilling injury of horticultural crops, *Hort-Science* 17, 161, 1982.
29. Wang, C.Y., Chilling and freezing injury, in *The Commercial Storage of Fruits, Vegetables, and Florist and Nursery Crops,* Gross, K.C., Yang, C.Y., and Saltveit, M., Eds., Agricultural Research Service, Beltsville, MD, 2002, draft version of revised USDA Agriculture Handbook 66, available at http://www.ba.ars.usda.gov/hb66/index. html, accessed Mar. 31, 2003.
30. Wade, N.L., Physiology of cool-storage disorders of fruit and vegetables, in *Low Temperature Stress in Crop Plants,* Lyons, J.M., Graham, D., and Raison, J.K., Eds., Academic Press, New York, 1979, p. 81.
31. Shewfelt, R.L., 1993. Stress physiology: a cellular approach to quality, in *Postharvest Handling: A Systems Approach*, Shewfelt, R.L. and Prussia, S.E., Eds., Academic Press, San Diego, CA, p. 257.
32. Lyons, J.M., Raison, J.K., and Steponkus, P.L., The plant membrane in response to low temperature: an overview, in *Low Temperature Stress in Crop Plants,* Lyons, J.M., Graham, D. and Raison, J.K., Eds., Academic Press, New York, 1979, p. 1.
33. Wang, C.Y., Physiological and biochemical responses of plants to chilling stress, *HortScience*, 17, 173, 1982.
34. Parkin, K.L. et al., Chilling injury: a review of possible mechanisms, *J. Food Qual.,* 11, 253, 1988.
35. Shewfelt, R.L., Response of plant membranes to chilling and freezing, in *Plant Membranes: A Biophysical Approach to Structure, Development, and Senescence,* Lesham, Y.Y., Shewfelt, R.L., Willmer, C.M., and Pantoja, O., Eds., Kluwer Academic Publishers, Dordrecht, Netherlands, 1992, p. 192.
36. Autio, W.R. and Bramlage, W.J., Chilling sensitivity of tomato fruit in relation to ripening and senescence, *J. Am. Soc. Hortic. Sci.,* 111, 201, 1986.
37. Lyons, J.M. and Raison, J.K., Oxidative activity of mitochondria isolated from plant tissues sensitive and resistant to chilling injury, *Plant Physiol.*, 45, 386, 1970.
38. Raison, J.K., Lyons, J.M. and Thomson, W.W., The influence of membranes on the temperature-induced changes in the kinetics of some respiratory enzymes in mito-chondria, *Arch. Biochem. Biophys.*, 142, 83, 1971.
39. Moline, H.E., Ultrastructural changes associated with chilling of tomato fruit, *Phytopathology*, 66, 617, 1976.
40. Wang, C.Y. and Adams, D.O., Ethylene production by chilled cucumbers (*Cucumis sativus* L.), *Plant Physiol.*, 66, 841, 1980.
41. Pantastico, E.B., Grierson, W., and Soule, J., Chilling injury in tropical fruits: I. bananas (*Musa paradisiaca* var. Sapientum cv. Lacatan), *Proc. Am. Soc. Hortic. Sci.*, 11, 83, 1967.
42. Raison, J.K. et al., Temperature-induced phase changes in mitochrondrial membranes detected by spin labeling, *J. Biol. Chem.*, 246, 4036, 1971.
43. Liebermann, M. et al., Biochemical studies of chilling injury in sweet potatoes, *Plant Physiol.*, 33, 307, 1958.
44. Furmanski, R.J. and Buescher, R.W., Influence of chilling on electrolyte leakage and internal conductivity of peach fruits, *HortScience,* 14, 167, 1979.
45. Murata, T. and Tatsumi, Y., Ion leakage in chilled plant tissues, in *Low Temperature Stress In Crop Plants*, Lyons, J.M., Graham, D., and Raison, J.K., Eds., Academic Press, New York, 1979, p. 141.

46. Hulme, A.C., Smith, W.H., and Wooltorton, L.S.C., Biochemical changes associated with the development of low-temperature breakdown in apples, *J. Sci. Food Agric.,* 15, 303, 1964.

47. Lyons, J.M., Wheaton, T.A., and Pratt, H.K., Relationship between physical nature of mitochondrial membranes and chilling sensitivity in plants, *Plant Physiol.,* 39, 262, 1964.

48. Yamaki, S. and Uritani, I., Mechanism of chilling injury in sweet potatoes. V. Biochemical mechanism of chilling injury with special reference to mitochondrial lipid components, *Agric. Biol. Chem.,* 36, 47, 1972.

49. Ben-Arie, R., Lavee, S., and Guelfat-Reich, S., Control of woolly breakdown of Elberta peaches in cold storage by intermittent exposure to room temperature, *J. Am. Soc. Hortic. Sci.,* 95, 801, 1970.

50. Lill, R.E., Alleviation of internal breakdown of nectarines during cold storage by intermittent warming, *Sci. Hortic.,* 25, 241, 1985.

51. Yang, R.-F., Cheng, T.-S., and Shewfelt, R.L., The effect of high temperature and ethylene treatment on the ripening of tomatoes, *J. Plant Physiol.,* 136, 368, 1990.

52. Inaba, M. and Crandall, P.G., Electrolyte leakage as an indicator of high-temperature injury to harvested mature green tomatoes, *J. Am. Soc. Hortic. Sci.,* 113, 96, 1988.

53. Walter, W.M. and Schadel, W.E., Structure and composition of normal skin (periderm) and wound tissue from cured sweet potatoes, *J. Am. Soc. Hortic. Sci.,* 108, 909, 1983.

54. Kim, H.O. and Lee, S.K., Effects of curing and storage conditions on processing quality of potatoes, *Acta Hortic.,* 343, 73, 1993.

55. Voss, R.E., Potatoes, in *The Commercial Storage of Fruits, Vegetables, and Florist and Nursery Crops,* Gross, K.C., Yang, C.Y., and Saltveit, M., Eds., Agricultural Research Service, Beltsville, MD, 2002, draft version of revised USDA Agriculture Handbook 66, available at http://www.ba.ars.usda.gov/hb66/index.html, accessed July 31, 2003.

56. Picha, D.H., Chilling injury, respiration and sugar changes in sweet potatoes stored at low temperature, *J. Am. Soc. Hortic. Sci.,* 112, 497, 1987.

57. Morris, J.R., Grape growing, in *Modern Fruit Science*, 10th ed., Childers, N.F., Morris, J.R., and Sibbett, G.S., Eds., Horticultural Publications, Gainesville, FL, 1985, Chap. 20.

58. Takeda, F., Saunders, M.S., and Saunders, J.A., Physical and chemical changes in muscadine grapes during postharvest storage, *Am. J. Enol. Vitic.,* 34, 180, 1983.

59. Ballinger, W.E. and Nesbitt, W.B., Postharvest decay of muscadine grapes (Carlos) in relation to storage temperature, time, and stem condition, *Am. J. Enol. Vitic.,* 33, 173, 1982.

60. Luchsinger, L.E. and Walsh, C.S., Chilling injury of peach fruit during storage, *Acta Hortic.,* 464, 473, 1998.

61. Von Mollendorff, L.J., Woolliness in peaches and nectarines: a review. 1. Maturity and external factors, *HortScience,* 5, 1, 1987.

62. Watkins, C.B., Kupferman, E., and Rosenberger, D.A., Apple, in *The Commercial Storage of Fruits, Vegetables, and Florist and Nursery Crops,* Gross, K.C., Yang, C.Y., and Saltveit, M., Eds., Agricultural Research Service, Beltsville, MD, 2002, draft version of revised USDA Agriculture Handbook 66, available at at http://www.ba.ars.usda.gov/hb66/index.html, accessed Mar. 31, 2003.

63. Boyette, M.D., Wilson, L.G., and Estes, E.A., Posthavest cooling and handling of apples, Publ. AG-413-1, The North Carolina Agricultural Extension Service, Raleigh, NC, 1990, available at http://www.bae.ncsu.edu/programs/extension/publicat/postharv/ag-413-1/index.html, accessed Aug. 1, 2003.

64. Ritenour, M.A., Orange, in *The Commercial Storage of Fruits, Vegetables, and Florist and Nursery Crops,* Gross, K.C., Yang, C.Y., and Saltveit, M., Eds., Agricultural Research Service, Beltsville, MD, 2002, draft version of revised USDA Agriculture Handbook 66, available at http://www.ba.ars.usda.gov/hb66/index.html, accessed Aug. 25, 2003.
65. Ben-Yehoshua, S., Kim, J.J., and Shapiro, B., Elicitation of resistance to the development of decay in sealed citrus fruit by curing, *Acta Hortic.,* 258, 623, 1989.
66. Lurie, S. and Klein, J.D., Temperature preconditioning, in *The Commercial Storage of Fruits, Vegetables, and Florist and Nursery Crops,* Gross, K.C., Yang, C.Y., and Saltveit, M., Eds., Agricultural Research Service, Beltsville, MD, 2002, draft version of revised USDA Agriculture Handbook 66, available at http://www.ba.ars.usda.gov/hb66/index.html, accessed July 15, 2003.
67. Maul, F. et al., Tomato flavor and aroma quality as affected by storage temperature, *J. Food Sci.,* 65, 1228, 2000.
68. Sargent, S.A. and Moretti, C.L. Tomato, in *The Commercial Storage of Fruits, Vegetables, and Florist and Nursery Crops,* Gross, K.C., Yang, C.Y., and Saltveit, M., Eds., Agricultural Research Service, Beltsville, MD, 2002, draft version of revised USDA Agriculture Handbook 66, available at http://www.ba.ars.usda.gov/hb66/index.html, accessed July 15, 2003.
69. Lurie, S. et al., Prestorage heat treatment of tomatoes prevents chilling injury and reversibly inhibits ripening, *Acta Hortic.,* 343, 283, 1993.
70. Lurie, S. et al. Heat treatment to reduce fungal rots, insect pests and to extend storage, *Acta Hortic.,* 464, 309, 1998.
71. Whitaker, B.D., A reassessment of heat stress as a means of reducing chilling injury in tomato fruit, *Acta Hortic.,* 343, 281, 1993.
72. Fraschina, A.A., Sozzi, G.O., and Cascone, O., Impact of a chronic exposure to a high temperature stress in tomato fruit ripening, *Acta Hortic.,* 464, 515, 1998.
73. Kader, A.A., Quality parameters of fresh-cut fruit and vegetable products, in *Fresh-Cut Fruit and Vegetables: Science, Technology, and Market,* Lamikanra, O., Ed., CRC Press, Boca Raton, FL, 2002, p. 11.
74. Rosen, J.C. and Kader, A.A., Postharvest physiology and quality maintenance of sliced pear and strawberry fruits, *J. Food Sci.,* 54, 656, 1989.
75. Behrsing, J.P. et al., Effect of temperature and size reduction on respiratory activity and shelf life of vegetables, *Acta Hortic.,* 464, 500, 1998.

Index

A

Abdul-Raouf studies, 131, 443
Abe and Watada studies, 307
Abeles studies, 63, 69
Abiosis, 321–326, 464–465
Acaricides, 351–352
Acetic acid fermentation, 253–254
Acidity, 387–389
Ackers studies, 455
Acylated homoserine lactone (AHL)-based
     communication, 400–401
Adams, Wang and, studies, 626
Adams, Winsor and, studies, 243–244
Adams studies, 319, 452
Additive application, 310–314
Additives, 256–257, *see also* Food additives,
     treatments, and preservation technology
Ade-Omowaye studies, 326
Adhesion, bacteria, 443–445
Aesthetic quality, cuticles, 33–34
Afek studies, 65
Afolabi studies, 444
Agar studies, 64, 277, 310
Agent Orange, 371
Aggarwal studies, 34
Agrios studies, 465–467, 470, 476–477, 480
Aguero studies, 6
Aharoni studies, 65
AHL-based communication, *see* Acylated
     homoserine lactone (AHL)-based
     communication
Ahmed, Ahmed and, studies, 301
Ahmed and Shivhare studies, 301
Ahvenainen, Laurila and, studies, 309, 317–318
Ahvenainen and Hurme studies, 319
Ahvenainen studies, 307–308
Air, 245–247, 276–277, 285
Aked studies, 237
Akkaravessapong, 90
Albibi studies, 11
Alfalfa seeds and sprouts, *404,* 492
Alkaline fermentation, 254
Allen, Huang and, studies, 510, 513
Allong studies, 93
Almed studies, 448
Almonds, 231, 529
Aloni studies, 44

Alonso-Salces studies, 209
Altschul studies, 3
Amanatidou studies, 131
Amarante and Banks studies, 284
Amir-Shapira studies, 199
Anabiosis, 327–333
Anaerobic catabolism, 136
Anaerobiosis, 173–174
Analytical methodology progress, 197, 207–208
Anderson and Tong studies, 475
Anderson studies, 228
Andre and Hou studies, 195
Andrews and Hirano studies, 442
Anjaiah studies, 518
Annous studies, 455
Antagonism, 399–400
Anthocyanins, 202–207
Antibiotics, 518–519
Antimicrobial agents, 518–522
Antimicrobials, naturally occurring, 391–392
Antranikian studies, 513
Appearance, 286, *see also* Color and appearance
Appendages, 444
Appendini and Hotchkiss studies, 333
Apples and apple products
    additives, treatments, and preservation, 302,
       309
    bacteria, 422–424, 445, 448–449, 452
    chilling, 247, 619, 621, *624,* 630
    color and appearance, 209
    complex carbohydrates, 581, 586
    cuticles, 28, 33, 41–43
    degradation microstructure, 541, 545–546
    flavor systems, 162
    fluorescent pseudomonads, 484
    freezing, *632*
    herbicides, *364*
    infection outbreaks, *404*
    maturity, ripening, and quality, 59–60, 62,
       65–67, 71
    mechanical injury, 83, 91–93, 99–100
    microbial spoilage, 473–474, 479–480
    nutrient loss, 227, 230–231, 238
    packaging, 122
    pesticides, *348, 358,* 360
    phytonutrients, 231
    postharvest treatment, 247

649